T0295383

Essentials of Chemical Biology

Essentials of Chemical Biology

Structures and Dynamics of Biological Macromolecules *In Vitro* and *In Vivo*

Andrew D. Miller
Department of Chemistry and Biochemistry, Mendel University, Brno, Czech Republic

Julian A. Tanner
School of Biomedical Sciences, University of Hong Kong

Second Edition

Copyright © 2024 by John Wiley & Sons, Inc. All rights reserved.

Published by John Wiley & Sons, Inc., Hoboken, New Jersey.
Published simultaneously in Canada.

No part of this publication may be reproduced, stored in a retrieval system, or transmitted in any form or by any means, electronic, mechanical, pho-tocopying, recording, scanning, or otherwise, except as permitted under Section 107 or 108 of the 1976 United States Copyright Act, without either the prior written permission of the Publisher, or authorization through payment of the appropriate per-copy fee to the Copyright Clearance Center, Inc., 222 Rosewood Drive, Danvers, MA 01923, (978) 750–8400, fax (978) 750–4470, or on the web at www.copyright.com. Requests to the Publisher for permission should be addressed to the Permissions Department, John Wiley & Sons, Inc., 111 River Street, Hoboken, NJ 07030, (201) 748–6011, fax (201) 748–6008, or online at http://www.wiley.com/go/permission.

Trademarks: Wiley and the Wiley logo are trademarks or registered trademarks of John Wiley & Sons, Inc. and/or its affiliates in the United States and other countries and may not be used without written permission. All other trademarks are the property of their respective owners. John Wiley & Sons, Inc. is not associated with any product or vendor mentioned in this book.

Limit of Liability/Disclaimer of Warranty: While the publisher and author have used their best efforts in preparing this book, they make no represen-tations or warranties with respect to the accuracy or completeness of the contents of this book and specifically disclaim any implied warranties of mer-chantability or fitness for a particular purpose. No warranty may be created or extended by sales representatives or written sales materials. The advice and strategies contained herein may not be suitable for your situation. You should consult with a professional where appropriate. Neither the publisher nor author shall be liable for any loss of profit or any other commercial damages, including but not limited to special, incidental, consequential, or other damages. Further, readers should be aware that websites listed in this work may have changed or disappeared between when this work was written and when it is read. Neither the publisher nor authors shall be liable for any loss of profit or any other commercial damages, including but not limited to special, incidental, consequential, or other damages.

For general information on our other products and services or for technical support, please contact our Customer Care Department within the United States at (800) 762–2974, outside the United States at (317) 572–3993 or fax (317) 572–4002.

Wiley also publishes its books in a variety of electronic formats. Some content that appears in print may not be available in electronic formats. For more information about Wiley products, visit our web site at www.wiley.com.

Library of Congress Cataloging-in-Publication Data:
Names: Miller, Andrew, author. | Tanner, Julian, author.
Title: Essentials of chemical biology : structures and dynamics of
   biological macromolecules *In Vitro* and *In Vivo* / Andrew D. Miller,
   Department of Chemistry and Biochemistry, Mendel University, Brno, Czech Republic, Julian A. Tanner,
   School of Biomedical Sciences, University of Hong Kong.
Description: Second edition. | Hoboken, New Jersey : John Wiley & Sons,
   Inc., [2024] | Includes bibliographical references and index.
Identifiers: LCCN 2022053778 | ISBN 9781119437970 (Paperback) | ISBN
   9781119437963 (ePDF) | ISBN 9781119437840 (ePUB)
Subjects: LCSH: Biochemistry. | Molecular biology. | Macromolecules.
Classification: LCC QP514.2 .M55 2023 | DDC 612/.015--dc23/eng/20221215
LC record available at https://lccn.loc.gov/2022053778

Cover Images: © LAGUNA DESIGN/Getty Images; ksenvitaln/Shutterstock; Tefi/Shutterstock
Cover design: Wiley

Set in 10.5pt/12.5pt Minion Pro by Integra Software Services Pvt. Ltd, Pondicherry, India

SKY10063182_122023

To my father, who was the first to show me that chemistry does not have to be about disciplines but can also be about interesting problems waiting to be solved with all that the subject has to offer. In addition, thanks to my mother for her limitless encouragement in all I do or try to do. Thanks also to my children Nadia Nozomi, Tatiana Hikari and Samuel Kiyoshi for having grown up into such lovely people and for being such a primary inspiration for all that I do in my life. Thanks, too, to all my past and present students, co-researchers and collaborators who have made my professional life in chemical biology so enthralling, stimulating and such a great adventure. Finally, thanks to the Czech Ministry of Education, Youth and Sports (MŠMT) for the award of OPVVV Project FIT (Pharmacology, Immunotherapy, nanoToxicology) (CZ.02.1.01/0.0/0.0/15_003/0000495), with financial support from the European Regional Development Fund, which provided me with the necessary time and opportunity to bring into being this second edition.

Andrew D. Miller

To my wife, Ali, for all her love and support, her wonderful positivity and optimism that has encouraged me through all the ups and downs, plus her passion for student learning. She will always be an inspiration for how best to enable students to flourish, excel and enjoy learning. Thanks also to my children Alex, Nicky and Sophia for all their perceptive questions which always remind me of the deep curiosity essential for good science. Thanks, too, to my parents Andrew and Christine, and brother, Alastair, for creating around me an environment where learning and critical thinking were treasured from when I was young. Finally, thanks to all my students and teachers from whom I continue to learn so much.

Julian A. Tanner

Contents

Preface

Mapping the Latest Essentials of Chemical Biology

Since the first edition of this textbook, chemical biology has become established as a major branch of scientific activity devoted to understanding the way biology works at the molecular level. Chemical biology itself remains unashamedly multidisciplinary and chemical biology research is essentially problem driven and not discipline driven. Organic, physical, inorganic and analytical chemistry all contribute towards chemical biology, alongside newer emerging molecular disciplines. Some might say that chemical biology is just another way to rebadge biochemistry. However, such a comment misses the point. Biochemistry may have started as a discipline devoted to the study of individual biological macromolecules, but this discipline has been steadily evolving into increasingly descriptive, empirical studies of larger and larger macromolecular assemblies, structures and interacting molecular networks. The molecular increasingly gives ground to the cellular. In contrast, chemical biology is about chemistry-trained graduates and researchers taking a fundamental interest in the way biology works at the molecular level. Consequently, the focus is on the molecular and the quantitative, where molecular properties are investigated, studied and then gradually linked to macromolecular and cellular behaviour where possible. This is a fundamentally "bottom-up" approach to understanding biology, in keeping with the chemist's natural enthusiasm and appreciation for molecular structure and behaviour first and foremost.

This textbook has been produced with the third/fourth year undergraduate student and young (post) graduate school researcher in mind, namely those who have a solid background in chemical principles and are ready to apply and grow their chemical knowledge to suit a future degree or career interest in chemical biology. In preparing this textbook our objective has not been to try and cover everything currently seen as chemical biology, but instead to ask ourselves what topics and themes should be described as the essentials of chemical biology and how should these be presented in a way most appropriate for those of a chemical rather than a biological orientation. In doing this, we concluded that the true essentials of chemical biology are represented by the structure, characterisation and measureable behaviour of the main biological macromolecules and macromolecular lipid assemblies found in all cells of all organisms. We have also concluded that the activities of small molecules in biology for respiration and primary and secondary metabolism should not be included in the essentials of chemical biology, except where they feature as protein prosthetic groups or otherwise modify macromolecule behaviour. In our view, simple metabolism and metabolite interconversions are the stuff of biochemistry, whilst fascination with secondary metabolism, secondary metabolites and their interconversions has been the traditional preserve of bio-organic chemistry (a subset of organic chemistry).

Hence, in our textbook we begin with structure (Chapter 1) and synthesis (Chapters 2 and 3), then consider how structure is determined (Chapters 4, 5 and 6), followed by a consideration of dynamic behaviour and molecular interactions (Chapters 7, 8 and 9), culminating in molecular evolution and thoughts on the origins of life, quintessentially from the chemistry point of view (Chapter 10). New chapters (Chapters 11, 12, 13 and 14) focus on the ongoing transition of chemical biology research from reductionist studies at the molecular level to studies on molecules in cells *in vitro* then molecules in organisms *in vivo*. This transition is neatly framed in terms of the contribution of chemical biology research to the creation and development of advanced therapeutics and diagnostics for the future management and treatment of chronic diseases. Finally, we conclude (Chapter 15) with the interface between chemical biology and synthetic biology. Armed with such essentials, we hope that readers will be empowered to think about and then tackle any problem of their chosen interest at the chemistry-biology and/or chemistry-medicine interfaces, after a little more detailed and specific reading, of course. Foremost, we remain hopeful that our textbook will provide a valuable tool for chemical biology students and researchers to open the door and step through into the extraordinary world of biology without feeling that they should have to leave their chemical principles behind them.

Glossary of Physical and Chemical Terms

Equation glossary

Chapter 1

Potential energy	V	J	[kg m^2 s^{-2}]
Electrical point charge	q_n	C	
Vacuum permittivity	ε_o	F m^{-1} or C^2 m^{-1} J^{-1}	[C^2 kg^{-1} m^{-3} s^2]
Permittivity of medium	ε	F m^{-1} or C^2 m^{-1} J^{-1}	[C^2 kg^{-1} m^{-3} s^2]
Distance between (charge/nuclear) centres	r	m	
(Electric) dipole moment	μ_n	C m	
Polarisability volume	α_n'	m^3 (Å^3, cm^3)	
Ionisation energies	I_n	J	[kg m^2 s^{-2}]

J is Joule; F is Farad; C is Coulomb

Chapter 4

(Time dependent) induced dipole moment	μ_{ind} or $\mu_{ind}(t)$	C m
(Time dependent) electronic polarisability	$\alpha(\nu_v)$	C m^2 V^{-1}
(Oscillating) electric field (of light)	$E(\nu_v)$	V m^{-1}
Absorbance	A or $A(\lambda)$	arbitrary units
Optical density	$OD(\lambda)$	arbitrary units
Pathlength (optical)	l	cm
Extinction coefficient	ε_{max} or $\varepsilon(\lambda)$	l mol^{-1} cm^{-1}
Biological macromolecular concentration	c_M	mol l^{-1}
Wavelength	λ	nm (m, Å)
Molecular weight (of protein)	M_p	D or kD [g mol^{-1}]
Molecular weight (of nucleotide)	M_{nt}	D or kD [g mol^{-1}]
Concentration (of nucleotide)	c_{nt}	mol l^{-1}
Differential absorbance	$\Delta A(\lambda)$	arbitrary units
Differential molar extinction coefficient	$\Delta\varepsilon(\lambda)$	l mol^{-1} cm^{-1}
Ellipticity	$\theta(\lambda)$	deg
Molar ellipticity	$[\theta(\lambda)]$	deg l mol^{-1} cm^{-1}
Vibrational frequency of light	ν_v	s^{-1}
(Equilibrium) electric field (of light)	E_o	V m^{-1}
Equilibrium polarisability component	$\alpha_o(\nu_v)$	C m^2 V^{-1}
Nuclear oscillation component	$\alpha_R(\nu_R)$	C m^2 V^{-1}

Frequency of vibrational modes (molecular)	ν_R	s^{-1}	
Frequency of emitted light	ν_{em}	s^{-1}	
Planck's constant	h	J s or N m s	$[kg\ m^2\ s^{-1}]$
Reduced Planck's constant	$h/2\pi$ or $\hbar$	J s rad^{-1}	
Speed of light	c	m s^{-1}	
Radiative lifetime (fluorescence)	τ_R	s	
Radiative lifetime (phosphorescence)	$\tau_{R,\ Phor}$	s	
Rate of spontaneous emission (fluorescence)	k_F	s^{-1}	
Rate of internal conversion (fluorescence)	k_{IC}	s^{-1}	
Rate of intersystem crossing (fluorescence)	k_{IS}	s^{-1}	
Rate of quenching (fluorescence) (with Q)	k_q	$M^{-1}\ s^{-1}$	
Fluorescence intensity (no Q)	I_{em} or F_o	arbitrary units	
Fluorescence intensity (in presence of Q)	F	arbitrary units	
Förster length	R_o	m	
Interfluorophore distance	R_F	m	
Fluorescence quantum yield	ϕ_F		
Fluorescence quantum yield (of donor, D)	ϕ_D		
Refractive index	n_R		
X-ray absorption coefficient	μ_{ab}	m^{-1}	
Incident intensity (of X-ray)	I_{xo}	arbitrary units	
Transmitted intensity (of X-rays)	I_x	arbitrary units	

V is Volt (J C^{-1}); D is Daltons; kD is kiloDaltons

Chapter 5

(Nuclear) angular momentum	J	J s rad^{-1}	
Gyromagnetic ratio	γ	rad s^{-1} T^{-1}	
Magnetic moment (z-axis)	μ_z	J T^{-1} or A m^2	
Nuclear magneton	μ_N	J T^{-1}	
Nuclear g-factor	g_I		
Charge (of an electron or proton)	e	C	
Mass (of proton)	m_p	kg	
Externally applied magnetic field strength	B_z	T or N m^{-1} A^{-1}	
Larmor (precession) frequency	ν_L	s^{-1}	
Coupling constant	J	s^{-1} (Hz)	
Boltzmann constant	k	J K^{-1}	$[kg\ m^2\ s^{-2}\ K^{-1}]$
(Absolute) temperature	T	K	
(Scalar) longitudinal relaxation time constant	T_1	s	
Transverse relaxation time constant	T_2	s	
Longitudinal magnetisation: polarisation	$M_z(t)$		
Transverse magnetisation: coherence	$M_y(t)$		
Spectral line width (half peak intensity)	$\Delta\nu_{L,\frac{1}{2}}$	s^{-1} (Hz)	
(Electron) angular momentum	J_e	J s rad^{-1}	
Electron gyromagnetic ratio	γ_e	rad s^{-1} T^{-1}	
Electron magnetic moment	μ_z^e	J T^{-1}	
Bohr magneton	μ_B	J T^{-1}	
g-factor	g_e		
Mass (of an electron)	m_e	kg	

rad is radians (2π); T is Tesla; A is ampere (C s^{-1})

Chapter 6

Distance between lattice planes	d_{hkl}	Å
Scattering length	$b_{x\text{-}ray}$	cm
Distance of resolvable separation		

resolution	d_R	Å	
Charge (of an electron)	e	C	
Electrical potential difference (in field emission gun)	Φ	V or J C^{-1}	[kg m^2 s^{-2} C^{-1}]
Mass (of an electron)	m_e	kg	
Maximum particle size	D	m	
Büttiker-Landauer tunnelling time	τ^{BL}	s	
Variable (z-axis) barrier dimension	s_z	m	
Barrier crossing constant	χ	m^{-1}	
Piezo electric bar changes in length	Δl_p	m	
(Piezo electric biomorph) displacement	Δx_p	m	
(Piezo electric) potential difference	U_p	V or J C^{-1}	[kg m^2 s^{-2} C^{-1}]
(Piezo electric) coefficient	d_{31}	m V^{-1} or C N^{-1}	[C s^2 m^{-1} kg^{-1}]
Tunnelling current	I_T	A	
Van der Waals interactions (tip to surface)	$F_{VDW}(d_z)$	N	[kg m s^{-2}]
Hamaker constant	H	N m or J	[kg m^2 s^{-2}]
Distance (z-axis)	d_z	m	
Radius of tip above surface	R_z	m	
Surface-to-tip interaction forces	F_{ST}	N	[kg m s^{-2}]
Spring constant	c_{ST}	N m^{-1}	[kg s^{-2}]
Young's modulus	E_M	Pa or N m^{-2}	[kg m^{-1} s^{-2}]

Pa is Pascal (N m^{-2})

Chapter 7

Hydrated volume	V_h	cm^3 or m^3	
Macromolecular molecular weight	M_{MM}	D or kD	[g mol^{-1}]
Avogadro's number	N_o	mol^{-1}	
Macromolecular partial specific volume	V_{MM}	cm^3 g^{-1}	
Hydration level	Δ		
Coefficient of translational frictional force	$f_{trans, sph}$	kg s^{-1} or g s^{-1}	
Spherical macromolecular radius	r_{sph}	cm or m	
Viscosity	η	Pa s or N s m^{-2}	[kg m^{-1} s^{-1}, g cm^{-1} s^{-1}]
Coefficient of rotational frictional force	$f_{rot, sph}$	kg m^2 s^{-1} or g cm^2 s^{-1}	
Spherical macromolecular volume	V_{sph}	m^3 or cm^3	
General coefficient of translational frictional force	f_{trans}	kg s^{-1} or g s^{-1}	
General coefficient of rotational frictional force	f_{rot}	kg m^2 s^{-1} or g cm^2 s^{-1}	
Macromolecular flux	J_{MM}	kg m^{-2} s^{-1} or g cm^{-2} s^{-1} mol m^{-2} s^{-1}	
Macromolecular concentration	C_{MM}	kg m^{-3} or g cm^{-3} mol l^{-1}	
Average macromolecular velocity	$\langle \nu_{MM} \rangle$	m s^{-1} or cm s^{-1}	
Macromolecular diffusion coefficient	D_{MM}	m^2 s^{-1} or cm^2 s^{-1}	
Debye length	r_D	m	
Ionic strength	I	mol m^{-3} or mol kg^{-1} M (mol l^{-1})	
Association constant	K_a	M^{-1}	
Dissociation constant	K_d	M	
Moles (of ligand) bound per mole (of receptor)	B	(Mol fraction)	
Total molar quantity (of ligand) bound (to receptor)	m_{RL}	mol	
Total molar quantity (of ligand) added	m_{Lo}	mol	
Total system volume	V_{tot}	m^3, dm^3 (l), cm^3	
Chemical potential of species i	μ_i	J mol^{-1}	[kg m^2 s^{-2} mol^{-1}]

Concentration of species i	c_i	M (mol l^{-1})	
Molar gas constant	R	J K^{-1} mol^{-1}	[kg m^2 s^{-2} K^{-1} mol^{-1}]
Standard free energy change	ΔG°	J mol^{-1}, kJ mol^{-1}	[kg m^2 s^{-2} mol^{-1}]
Standard enthalpy change	ΔH°	J mol^{-1}, kJ mol^{-1}	[kg m^2 s^{-2} mol^{-1}]
Standard entropy change	ΔS°	J mol^{-1} K^{-1}	[kg m^2 s^{-2} mol^{-1} K^{-1}]
(Exchangeable) heat energy	q	J	[kg m^2 s^{-2}]
(Fractional) change in enthalpy	dH	J	
Electric field	E_e	V m^{-1} or J C^{-1} m^{-1}	[kg m s^{-2} C^{-1}]
Electrophoretic velocity	ν_e	m s^{-1}	
Electrophoretic mobility	μ_e	m^2 V^{-1} s^{-1}	[C s kg^{-1}]
Apparent electophoretic mobility	μ_a	m^2 V^{-1} s^{-1}	[C s kg^{-1}]
EOF electophoretic mobility	μ_{EOF}	m^2 V^{-1} s^{-1}	[C s kg^{-1}]
Time to detector	t_e	s	
Effective length (of capillary)	l_e	m	
Total length (of apparatus)	L_e	m	
Applied potential difference	V_e	V or J C^{-1}	[kg m^2 s^{-2} C^{-1}]
Rate of association (complex formation) (*on*-rate)	k_{ass}	M^{-1} s^{-1}	
Rate of dissociation (complex) (*off*-rate)	k_{diss}	s^{-1}	
Resonant angle	Y_t	arc s	
Concentration dependent *on*-rate (complex formation)	k_{on}	s^{-1}	

Chapter 8

Initial rate of (biocatalysis)	ν	M s^{-1}	[mol l^{-1} s^{-1}]
Initial substrate concentration	$[S]$	M	[mol l^{-1}]
Unimolecular rate constant for mechanism step **n**	k_n	s^{-1}	
Bimolecular rate constant for mechanism step **n**	k_n	M^{-1} s^{-1}	[l mol^{-1} s^{-1}]
Michaelis constant	K_m	M	[mol l^{-1}]
Equilibrium dissociation constant for **ES** complex	K_S	M	[mol l^{-1}]
Catalytic constant (when $[S] \gg K_m$)	k_{cat}	s^{-1}	
Maximum initial rate (when $[S] \gg K_m$)	V_{max}	M s^{-1}	[mol l^{-1} s^{-1}]
Specificity constant (when $K_m \gg [S]$)	k_{cat}/K_m	M^{-1} s^{-1}	[l mol^{-1} s^{-1}]
Inhibitor equilibrium dissociation constant	K_I	M	[mol l^{-1}]
Base equilibrium ionisation constant	K_d^B	M	[mol l^{-1}]
Acid equilibrium ionisation constant	K_d^A	M	[mol l^{-1}]
Saddle-point vibration frequency	ν_{TS}	s^{-1}	
Transition state forward decomposition rate constant	$k_C^\ddagger$	s^{-1}	
Quasi-equilibrium association constant	$K_C^\ddagger$	M^{-1}	
Microscopic rate constant	k_P	M^{-1} s^{-1}	
Partition function for molecular population **Z**	q^Z		
Transition state-ground state energy difference	E_o	J	[kg m^2 s^{-2}]
Standard free energy (of activation)	$\Delta G_o^\ddagger$	kJ mol^{-1}	[kg m^2 s^{-2} mol^{-1}]
Free energy (of activation) (from **E** and **S**)	$\Delta G_{ES}^\ddagger$	kJ mol^{-1}	[kg m^2 s^{-2} mol^{-1}]
Free energy (of activation) (from **ES** complex)	$\Delta G_T^\ddagger$	kJ mol^{-1}	[kg m^2 s^{-2} mol^{-1}]
Free energy (of association) of (**E** and **S**)	ΔG_S	kJ mol^{-1}	[kg m^2 s^{-2} mol^{-1}]
Rate constant for electron transfer	k_{ET}	s^{-1}	
Equilibrium association constant (for **D** and **A**)	$K_{a,DA}$	M^{-1}	[l mol^{-1}]
Edge to edge distance (between **D** and **A**)	R_{ET}	m	
Beta value	β_{ET}	m^{-1}	

Chapter 9

Unitary charge of an ion	z		
Accelerating electrostatic potential	V_z	V or J C^{-1}	[kg m^2 s^{-2} C^{-1}]
Velocity of ion travel	ν_z	m s^{-1}	
Ion mass	m	D, kD (or amu)	
Time to detector	t_z	s	
Length (of field-free flight tube)	L_z	m	

Chapter 13

Autocorrelation function decay constant	Γ	s^{-1}	
Nanomolecular diffusion constant	D_{NM}	m^2 s^{-1} or cm^2 s^{-1}	
Autocorrelation function wave vector	q	m^{-1} or cm^{-1}	
Zeta potential	ζ	V or J C^{-1}	[kg m^2 s^{-2} C^{-1}]

Chapter 14

Concentration dependent water relaxivity rate	R_1	mM^{-1} s^{-1}

Chemical glossary

Chapters 13 and 14

AtuFECT01	L-arginyl-2,3-L-diaminopropionic acid-*N*-palmityl-*N*-oleylamide trihydrochloride
CDAN	N^1-cholesteryloxycarbonyl-3,7-diazanonane-1,9-diamine
Chol	Cholesterol
DC-Chol	3β-[*N*-(*N'*,*N'*-dimethylaminoethane)carbamoyl] cholesterol
DLinDMA	1,2-dilinoleyloxy-3-dimethyl-aminopropane
DLin-KC2-DMA	2,2-dilinoleyl-4-(2-dimethyl-aminoethyl)-[1,3]-dioxolane
DLin-MC3-DMA (MC3)	(6Z, 9Z, 28Z, 31Z)-heptatriaconta-6, 9, 28, 31-tetra-en-19-yl 4-(dimethylamino)-butanoate
DODAG	*N'*, *N'*-dioctadecyl-*N*-4,8-diaza-10-aminodecanoyl glycine amide
DODMA	*N*-[1-(2,3-dioleyloxy) propyl]-*N*,*N*-dimethyl ammonium chloride
DOPC	1,2-dioleoyl-*sn*-glycero-3-phosphocholine or dioleoyl L-α-phosphatidylcholine
DOPE	1,2-dioleoyl-*sn*-glycero-3-phosphoethanolamine or dioleoyl L-α-phosphatidylethanolamine
DOPE-Rhoda	1,2-dioleoyl-*sn*-glycero-3-phosphoethanolamine-*N*-(lissamine rhodamine B sulfonyl) or dioleoyl L-α-phosphatidylethanolamine-*N*-(lissamine rhodamine B sulfonyl)
DPhyPE	1,2-diphytanoyl-*sn*-glycero-3-phosphoethanolamine or 1,2-diphytanoyl L-α-phosphatidylethanolamine
DPPC	1,2-dipalmitoyl-*sn*-glycero-3-phosphocholine or dipalmitoyl L-α-phosphatidylcholine
DSPC	1,2-distearoyl-*sn*-glycero-3-phosphocholine or distearoyl L-α-phosphatidylcholine
DSPE	1,2-distearoyl-*sn*-glycero-3-phosphoethanolamine or distearoyl L-α-phosphatidylethanolamine
DS(14-yne)TAP	1,2-(distear-14-ynoyloxy)-3-(trimethylammonium) propane
folate-PEG^{2000}-DSPE	(folate-*N*-ω-polyethylene glycol 2000)-*N*-carboxy-1,2-distearoyl-*sn*-glycero-3-phosphoethanolamine

Gd.DOTA.DSA	Gadolinium (III) 2-(4,7-bis-carboxymethyl-10-[(*N,N*-distearyl-amidomethyl)-*N'*-amidomethyl]-1,4,7,10-tetraazacyclododec-1-yl) acetic acid
GL-67	Lipid 67
PEG^{2000}-C-DMA	3-*N*-(ω-methoxy-polyethylene glycol 2000) carbamoyl-1,2-dimyristyloxypropylamine
PEG^{2000}-C-DMG	3-(ω-methoxy-polyethylene glycol 2000)carbamoyl-1,2-dimyristyl-*sn*-glycerol
PEG^{2000}-DSG	(ω-methoxy-polyethylene glycol 2000)-1,2-distearoyl-*sn*-glycerol
PEG^{2000}-DSPE	(ω-methoxy-polyethylene glycol 2000)-*N*-carboxy-1,2-distearoyl-*sn*-glycero-3-phosphoethanolamine or (ω-methoxy-polyethylene glycol 2000)-*N*-carboxy-distearoyl L-α-phosphatidylethanolamine
PEG^{5000}-DMPE	PEG^{5000} conjugate of dimyristoyl-L-α-phosphatidylethanolamine

About the Companion Website

This book is accompanied by a companion website:

www.wiley.com/go/miller/essentialschembiol2

The website includes:
- List of Figures as PPTs
- List of Tables as PDFs

1

The Structures of Biological Macromolecules and Lipid Assemblies

1.1 General introduction

All living organisms are comprised of cells that may vary considerably in terms of size, shape and appearance. In complex multicellular organisms, many cells are organised into diverse, functional organs to perform a collective function (Figure 1.1). In spite of their wide morphological diversity, all cells of all living organisms, wherever they are located, are comprised of **proteins, carbohydrates, nucleic acids** and **lipid assemblies**. Together, these give a cell form and function. To know and understand the chemistry of these biological macromolecules is to comprehend not only the basic infrastructure of a cell but also of living organisms. In functional terms, macromolecular lipid assemblies provide for compartmentalisation in the form of membrane barriers that not only define the "outer limits" of each cell but also divide up the intracellular environment into different organelles or functional zones (Figure 1.2). Membrane barriers are fluidic and lack rigidity, so proteins provide a supporting and scaffolding function not only in the main fluid bulk of the cell, known as the **cytosol**, but also within organelles. Within the **nucleus**, proteins also provide a nucleic acid packaging function in order to restrain and constrain spectacular quantities of nucleic acids within the nuclear volume. Everywhere in any cell, proteins also perform other individualised functions in outer membranes (e.g. as pores or receptors), in organelle membranes (as selective transporters, redox acceptors or energy transducers), in the cytosol or organelle volumes (as enzyme catalysts, molecular chaperones or "communication and control" centres) and in the nucleus (as regulators and transcribers of the genetic code). The extraordinary variety of protein functions and the "work-horse" like nature of proteins in biology has made them endlessly fascinating to biochemists and now to chemical biologists alike.

Nucleic acids are found in two main classes, namely **deoxyribonucleic acid (DNA)** and **ribonucleic acid (RNA)**. DNA is largely restricted to the nucleus and harbours genetic information that defines the composition and structure of cells and even the multicellular organisation of complex organisms, reaching even beyond that to influence organism behaviour as well. DNA molecules are partly segmented into **genes** that contain coding information for protein structures, but also into many other delineations associated with control over gene use. In fact, the level and sophistication of this control may well be the primary determinant of complexity in multicellular organisms, the more extensive and sophisticated the level of control, the more sophisticated and complex the multicellular organism. By contrast, RNA's most important role is in shuttling information from the nucleus to the cytosol. The primary function of RNA equates to the processing of genetic information from the DNA storage form into actual protein structures. RNA makes possible the central dogma of biology that *genes code for proteins*. Finally, carbohydrates, if not stored in complex forms for primary metabolism, are known to decorate some intracellular proteins and attach to outer membrane proteins, forming a **glycocalyx** covering the surface of many cells,

Essentials of Chemical Biology: Structures and Dynamics of Biological Macromolecules In Vitro *and* In Vivo, Second Edition. Andrew D. Miller and Julian A. Tanner.
© 2024 John Wiley & Sons, Inc. Published 2024 by John Wiley & Sons, Inc.

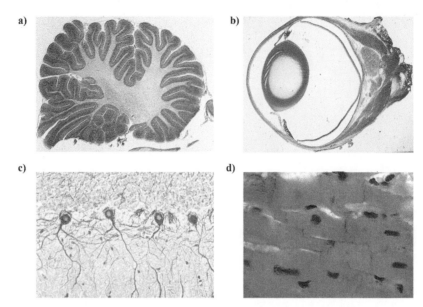

Figure 1.1 Organs and cells. (a) Cross section of mammalian brain showing the complex surface folds. There are an incalculable number of cells that make up the mammalian brain; **(b) Cross section of mammalian eye ball** in which the lens is made of proteins controlled in function by peripheral muscles. There is an enormous morphological and functional diversity between cells required for muscle control, light reception and signal transduction along the optic nerve; **(c) Cross section of mammalian neurological tissue** illustrating the **neuron cell bodies** with complex **axonal/dendritic processes** surrounded by support cells all of a wide range of size, shape, structure and function; **(d) Cross section of mammalian heart tissue** showing clusters of muscle fibres (single cell **myocytes**) that make up the heart wall. Myocytes are multinucleate with a very different shape, composition and function to neurological cells (all illustrations from Philip Harris Ltd).

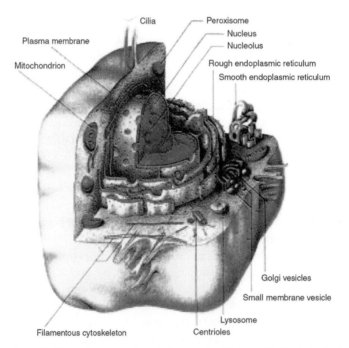

Figure 1.2 General structure of a cell. The main compartments (organelles) into which the interior is partitioned are illustrated. All cells of all organisms are constructed from the main biological macromolecules: **proteins, carbohydrates** and **nucleic acids**, together with macromolecular **lipid** structures that comprise the membranes (illustration from Philip Harris Ltd).

essential for communication between cells. In the plant and insect kingdoms, gigantic carbohydrate assemblies also provide the exo-skeleton framework to which cells are attached, giving form as well as function to plants and insects alike.

In all cases, proteins, carbohydrates and nucleic acids are polymers built from standard basis sets of molecular building blocks. In a similar way, lipid assemblies are built from a standard basis set of lipid building blocks associated through non-covalent bonds. What all biological macromolecules and macromolecular assemblies have in common is that they adopt defined three-dimensional (3D) structures that are the key to their functions (dynamics, binding and reactivity). Remarkably, these 3D structures are not only central to function but they are the result of **weak, non-covalent forces** of association acting together with stereoelectronic properties inherent within each class of polymer or macromolecular assembly. Without structure, function is hard to understand, although structure does not necessarily predict function. Therefore, the chemical biology reader needs to have a feel for the structures of proteins, carbohydrates, nucleic acids and lipid assemblies before embarking on any other part of this fascinating subject. Accordingly, the principles of structure are our main topic in this chapter, concluding with some explanation about critical, weak, non-covalent forces of association that are all so important in shaping and maintaining those structures.

1.2 Protein structures

1.2.1 Primary structure

Proteins are polymers formed primarily from the linear combination of 20 naturally occurring **L-α-amino acids** which are illustrated (see Table 1.1 and Figure 1.3). Almost all known protein structures are constructed from this fundamental set of 20 α-amino acid building blocks, which fall into two main classes, **hydrophobic** and **hydrophilic**, according to the nature of their **side chain** (Table 1.1). Protein architecture is intimately dependent upon having two such opposite sets of α-amino acid building blocks to call upon. Individual α-amino acid building blocks are joined together by a **peptide link** (Figure 1.4). When a small number (2–20) of amino acids are joined together by peptide links to form an unbranched chain then the result is known as an **oligopeptide** (Figure 1.5). However, peptide links can join together

Table 1.1 Structures and simple properties of all naturally occurring L-α-amino acids that are found in all proteins of all organisms. Included are the full name, the **three-letter code** name and the **one-letter code** name. Amino acids are grouped into those with **hydrophobic** side chains (**left panel**) and those with **hydrophilic** side chains (**right panel**). Where appropriate, measured functional group pK_a values are given.

R	Name	Abbrev.	R	Name	Abbrev.	pK_a
	Glycine	Gly (G)		Tyrosine	Tyr (Y)	9.7
	Alanine	Ala (A)		Serine	Ser (S)	15
	Valine	Val (V)				
	Leucine	Leu (L)		Threonine	Thr (T)	15
				Cysteine	Cys (C)	9.1
	Isoleucine	Ile (I)		Aspartic acid	Asp (D)	4.0
	Phenylalanine	Phe (F)		Asparagine	Asn (N)	
	Tryptophan	Trp (W)		Glutamic acid	Glu (E)	4.5
				Glutamine	Gln (Q)	
	Methionine	Met (M)		Lysine	Lys (K)	10.4
				Arginine	Arg (R)	12
	Proline	Pro (P)		Histidine	His (H)	6.0

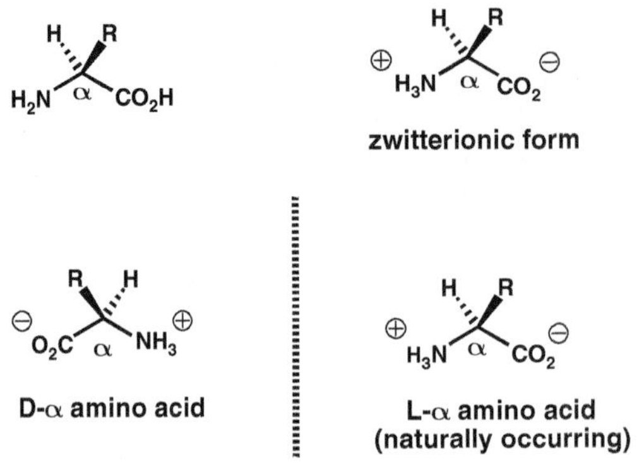

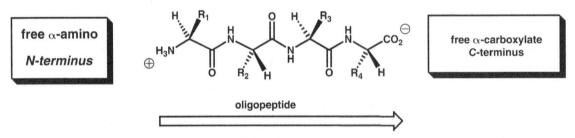

Figure 1.3   Structures of α-amino acids. L-α-amino acids are the preferred monomeric building blocks of proteins.

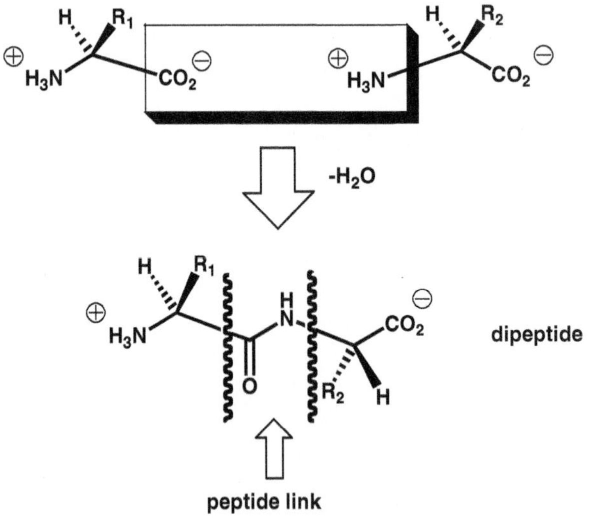

Figure 1.4   Schematic of peptide link formation.

Figure 1.5   General structure of tetrapeptide. By convention the highest priority end is the free *N*-terminus and the lowest priority the free *C*-terminus, giving the backbone a directionality illustrated by the arrow. This convention applies for all peptides, polypeptides and proteins.

anything from 20 to 2000 amino acid residues in length to form substantial unbranched polymeric chains of L-α-amino acids. These are known as **polypeptides**. Within each polypeptide chain, the repeat unit $(\text{-N-C}_\alpha\text{-C(O)-})_n$, neglecting the α-amino acid side chains, is known as the **main chain** or **backbone**, whilst each constituent, linked α-amino acid building block is known as an **amino acid residue**. The order of amino acid residues, going from the free, uncombined α-amino terminal end (*N*-terminus) to the free, uncombined α-carboxyl terminus (*C*-terminus), is known as the **amino acid sequence**.

Quite clearly, each peptide link is in fact a simple secondary amide functional group, but with some unusual properties. In fact, the N, H, C and O atoms of a peptide link, together with each pair of flanking α-carbon atoms, actually

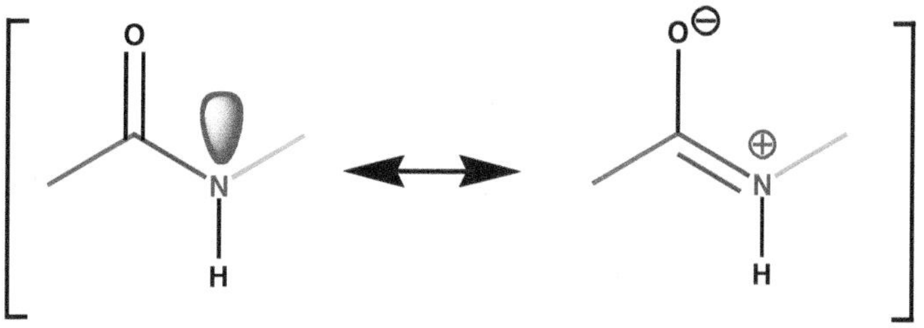

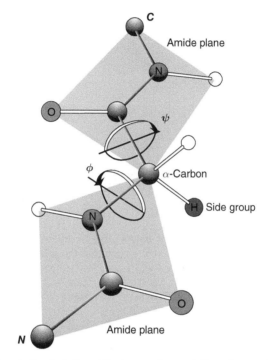

Figure 1.6 **Peptide link resonance structures.** These illustrate the partial double character in the C(0)-N bond (blue) sufficient to prevent free rotation, thereby restricting conformational freedom of peptide or polypeptide backbones.

Figure 1.7 **Peptide link as virtual bond.** The C, O, N and H atoms act as a rigid coplanar unit, equivalent to a single bond (**virtual bond**) so that consecutive peptide links act as rigid coplanar units that pivot around individual C_α-atoms (from Voet *et al.*, 1999, Wiley, Figure 6.4).

form a rigid, coplanar unit that behaves almost like a single bond, owing to restricted rotation about the N-C(O) bond caused by nitrogen atom lone pair resonance and the build up of N-C(O) double bond character (Figure 1.6). For this reason, the peptide link and flanking C_α-atoms together are sometimes referred to as a **virtual bond**. We might say that the C_α-atom of each amino acid residue in a polypeptide chain belongs simultaneously to two such virtual bonds (Figure 1.7). The spatial relationship between each C_α-linked pair of virtual bonds is then defined using the conformational angles ϕ and ψ, which are the main chain dihedral angles subtended about the N(H)-C_α and C_α-C(O) σ-bonds respectively of each amino acid residue (Figures 1.7 and 1.8). Only certain combinations of ϕ and ψ are now allowed owing to steric congestion between the side chains of adjacent amino acid residues (Figure 1.8). Consequently, the overall conformation of a given polypeptide chain is also very restricted, with direct consequences for the 3D structures of proteins. In effect, conformational restrictions imposed by lack of free-rotation in the peptide link and the natures of each peptide-linked amino acid residue, place substantial restrictions upon the conformational freedom of a given polypeptide and hence the range of possible 3D structures that may be formed by any given polypeptide polymer. In fact, the primary structure amino acid sequence of a protein not only influences the 3D structure but also actually determines that structure. In other words, all the necessary "information" for the 3D structure of a protein is "stored" and is available within the primary structure. This is the basis for self-assembly in biology and explains why proteins can be such excellent platforms or "workbenches" for the development of defined functions and the evolution of living organisms.

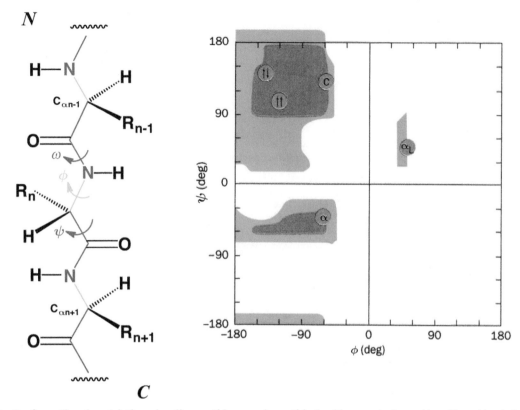

Figure 1.8 **Conformational restrictions in oligopeptide or polypeptide backbones**. Amino acid residue side-chain interactions further restrict free rotation in oligopeptide or polypeptide backbones. Rotational possibilities are defined by allowed values of dihedral angle ϕ subtended about N-C$_\alpha$ bond and ψ subtended about C$_\alpha$-C(O) bond (left).Theoretically allowed angles are shown in **Ramachandran plot** (right), together with positions of actual angles found in real protein secondary structures: α: right-handed α-helix; α_L: **left-handed α-helix**; ↑↑: **parallel β-sheet**; ↑↓: **anti-parallel β-sheet**; **C: collagen** or **P$_{II}$ helix** (see Figures 1.30 and 1.31). (Ramachandran plot from Voet *et al.*, 1999, Wiley, Figure 6.6).

1.2.2 Repetitive secondary structure

If primary structure is amino acid residue sequence, then **secondary structure** represents the first major steps towards a functional 3D structure. Secondary structures are essentially transient 3D structural elements that polypeptides may form in solution and which can interlock or dock together for stability. Polypeptides are capable of forming remarkably beautiful helical structures that are known as the right-handed α-**helix** and the right-handed **3$_{10}$-helix**. The term "right-handed" refers to the way in which the polypeptide main chain traces out the path of a right-handed corkscrew (incidentally, a **left handed α-helix** is possible, but is unknown in natural proteins so far). The α-helix can be a surprisingly sturdy, robust and regular structural feature (Figures 1.9 and 1.10). Typically, α-helices are comprised of up to 35 amino acid residues in length and are very stereo-regular; the ϕ and ψ conformational angles of each amino acid residue in the α-helix are both about -60° in all cases (Figure 1.11). Helices are held together by a regular network of non-covalent hydrogen bonds (see Section 1.6) formed between the peptide bond C = O and N-H groups of neighbouring amino acid residues (Figure 1.12). There are 3.6 amino acid residues per turn, with the result that the hydrogen bonds are formed between the C = O group lone pairs (hydrogen bond acceptors) of *n*-th residues and the N-H groups (hydrogen bond donors) of (*n* + 4)-th residues. The closed loop formed by one of these hydrogen bonds and the intervening stretch of polypeptide main chain contains 13 atoms (Figure 1.12). Hence, the α-helix has also been christened a **3.6$_{13}$-helix**. By contrast, the 3$_{10}$-helix (or α_{II}-**helix**) is effectively a smaller and slightly distorted version of the α-helix but with only three amino acid residues per turn and ten atoms involved in the intervening stretch of polypeptide main chain (Figures 1.13 and 1.14). Hydrogen bonds are therefore formed between the C = O group lone pairs of *n*-th residues and the N-H groups (hydrogen bond donors) of (*n* + 3)-th residues; ϕ and ψ conformational angles are approximately -60° and -30° respectively (Figure 1.15).

Sheet-like structures are the main alternative to helices. The origin of these structures can be found in the behaviour of polypeptide chains when they are fully extended into their β-**strand** conformations. A β-strand has a "pleated" appearance, with the peptide bonds orientated perpendicular to the main chain and with amino acid residue side chains alternating above and below (Figure 1.16). Both ϕ and ψ conformational angles are near 180°, but are typically between -120–150° and +120–150° respectively (Figure 1.17). All β-**strand** conformations are unstable alone, but may be stabilised

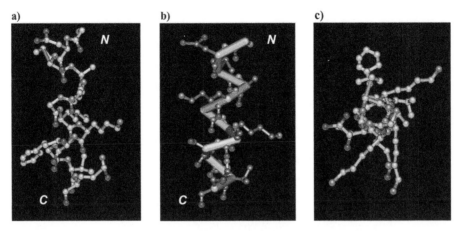

Figure 1.9 **Depictions of an α-helix.** The illustrated α-helix derives from **triose phosphate isomerase** (chicken muscle) (pdb: **1tim**). (**a**) **Ball and stick representation** (side view) of atoms and bonds shown with carbon (**yellow**), nitrogen (**blue**) and oxygen (**red**); (**b**) **CA stick display** of α-carbon backbone, atoms and bonds of amino acid side chains are rendered in **ball and stick representations** with carbon (**grey**), nitrogen (**blue**) and oxygen (**red**); (**c**) **Ball and stick representation** (top view) of atoms and bonds with labels as per (a).

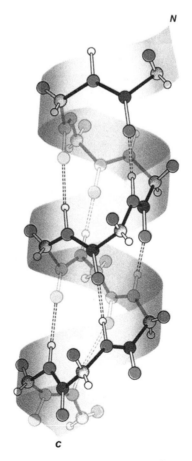

Figure 1.10 **Cartoon rendition of α-helix.** Here the right-hand helix path is illustrated as a **ribbon** over which a **ball and stick representation** of the α-carbon backbone is drawn using the code hydrogen (**white**), carbon (**grey**), nitrogen (**blue**) oxygen (**red**) and side chain atom (**purple**), in order to illustrate general **hydrogen-bonding** patterns in the helix (illustration from Voet *et al.*, 1999, Wiley, Figures 6–8).

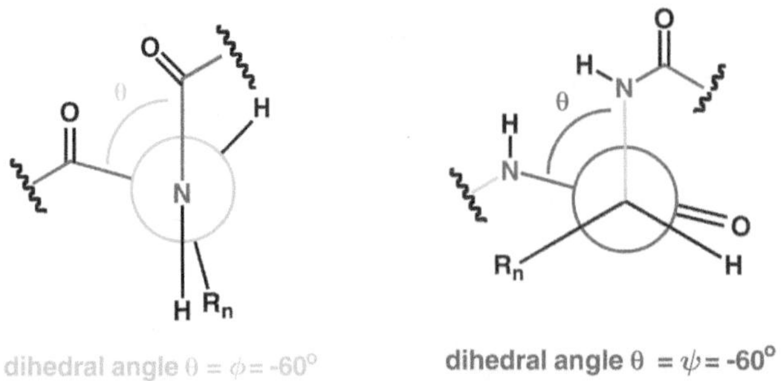

dihedral angle θ = ϕ = -60°   dihedral angle θ = ψ = -60°

Figure 1.11 Newman projections for α-helix. These projections involve N-C$_\alpha$ bonds and C$_\alpha$-C(O) bonds of the α-helix to demonstrate the consequences of highly regular dihedral angles ϕ and ψ respectively. Peptide backbone bonds are colour coded in the same way as Figures 1.6 and 1.8.

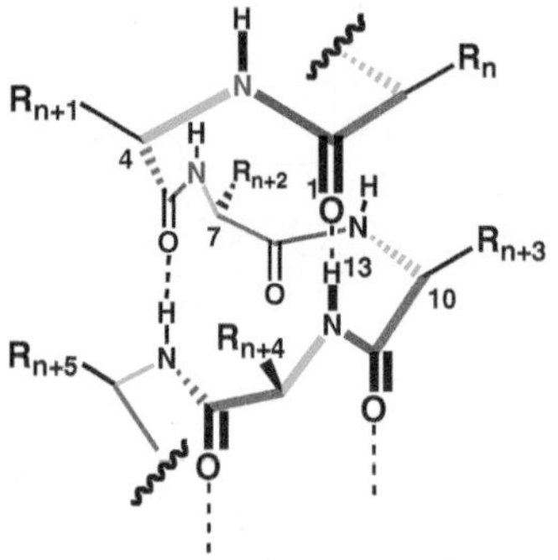

Figure 1.12 Stereo-defined structure of first turn of an α-helix. This demonstrates the atom separation between N-H hydrogen bond donors and C = O hydrogen bond acceptors. The C = O acceptor of each n-th residue forms a hydrogen bond link with the N-H bond donor of the (n + 4)-th residue defining an atom separation of 13 between acceptor O-atom and donor H-atom. Peptide backbone bonds are colour coded in the same way as Figures 1.6 and 1.8.

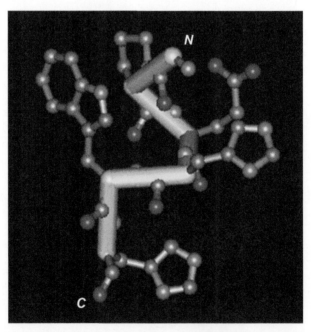

Figure 1.13 Depiction of 3$_{10}$-helix (turn). The illustrated 3$_{10}$-helix derives from triose phosphate isomerase (chicken muscle) (pdb: **1tim**). **CA stick display** of α-carbon backbone, atoms and bonds of amino acid side chains are rendered in **ball and stick representation** with carbon (**grey**), nitrogen (**blue**) and oxygen (**red**).

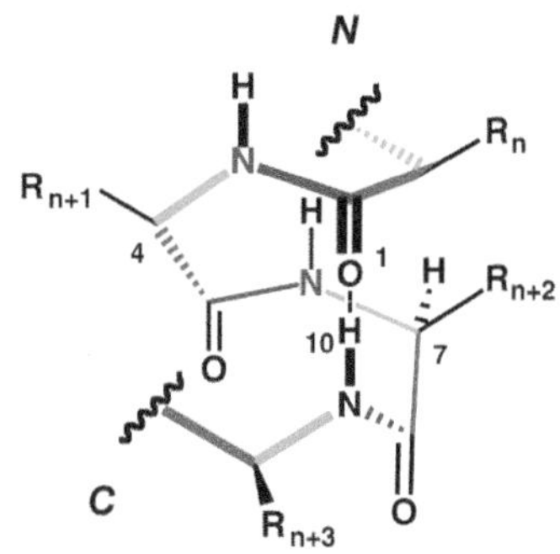

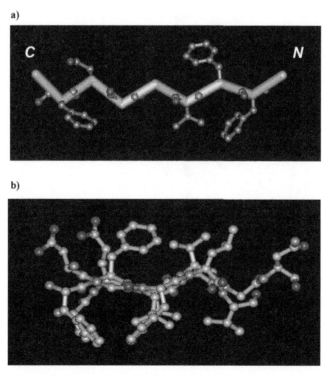

Figure 1.14 **Stereo-defined structure of an a 3$_{10}$-helix.** This structural representation demonstrates. This structural representation demonstrates the atom separation between N-H hydrogen bond donors and C = O hydrogen bond acceptors. The C = O acceptor of each *n*-th residue forms a hydrogen bond link with the N-H bond donor of the (*n* + 3)-th residue defining an atom separation of 10 between acceptor O-atom and donor H-atom. Peptide backbone bonds are colour coded in the same way as Figures 1.6 and 1.8.

dihedral angle $\theta = \phi = -60°$ **dihedral angle $\theta = \psi = -30°$**

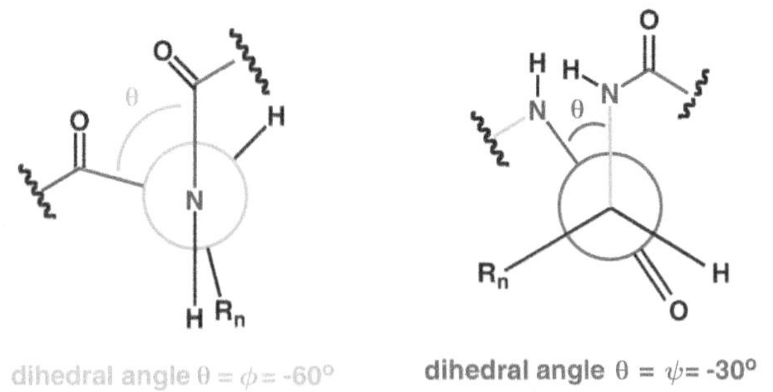

Figure 1.15 **Newman projections for a 3$_{10}$-helix.** These projections involve N-C$_\alpha$ bond and C$_\alpha$-C(O) bonds of the 3$_{10}$-helix to illustrate the result of highly regular dihedral angles ϕ and ψ respectively. Peptide backbone bonds are colour coded in the same way as Figures 1.6 and 1.8. A tighter turn relates to smaller value of ψ.

a)

b)

Figure 1.16 **β-Strand and β-sheet.** The illustrated β-strand and β-sheet derive from. from triose phosphate isomerase (chicken muscle) (pdb: **1tim**). **(a) CA stick display** of α-carbon backbone (side view), atoms and bonds of amino acid side chains are rendered in **ball and stick representation** with carbon (**grey**), nitrogen (**blue**) and oxygen (**red**). β-strand is shown to illustrate "zig-zag" extended conformation; **(b) Ball and stick representation** of β-sheet (side view) is shown with carbon (**grey**), nitrogen (**blue**) and oxygen (**red**) to illustrate "zig-zag" pleating and to show regular arrangement of amino acid residue side chains in close juxtaposition.

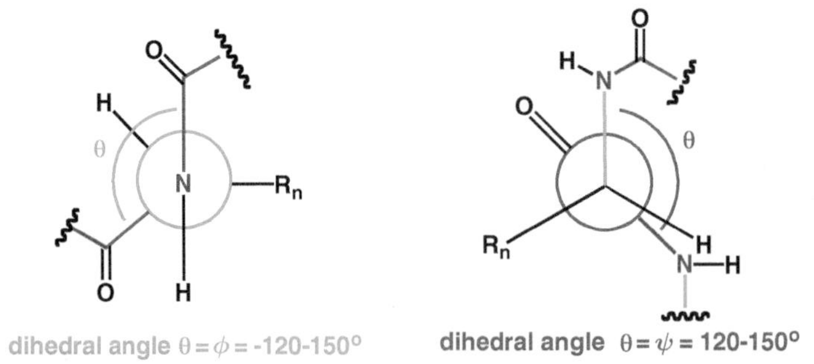

dihedral angle $\theta = \phi = -120\text{-}150°$ dihedral angle $\theta = \psi = 120\text{-}150°$

Figure 1.17   Newman projections for β-strand. These projections involve N-C$_\alpha$ bond and C$_\alpha$-C(O) bonds of a β-strand to demonstrate the consequences of highly regular dihedral angles ϕ and ψ respectively and extending the conformation. Peptide backbone bonds are colour coded in the same way as Figures 1.6 and 1.8.

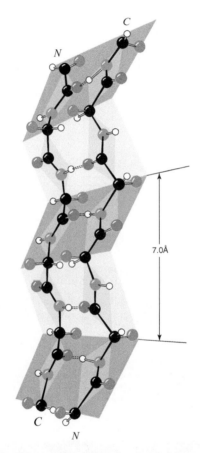

Figure 1.18   Cartoon rendition of a β-sheet. This rendition illustrates the pleating as a sequence of intersecting planes over which a **ball and stick representation** of the α-carbon backbone is drawn using the code hydrogen (**white**), carbon (**black**), nitrogen (**blue**) oxygen (**red**) and side-chain atom (**purple**), in order to illustrate general positioning of side chains above and below the sheet (illustration from Voet *et al.*, 1999, Wiley, Figures 6–10).

by the formation of non-covalent hydrogen bonds between strands, thereby resulting in a β-sheet (Figures 1.16 and 1.18). Such β-sheets may either be **anti-parallel** (β) or **parallel** (β′) depending upon whether the β-strands are orientated in the opposite or same direction with respect to each other (Figures 1.19 and 1.20). The hydrogen bonds that link β-strands together are formed between the same functional groups as in helices. In anti-parallel β-sheets, hydrogen bonds are alternately spaced close together then wide apart; in parallel β-sheets they are evenly spaced throughout (Figure 1.20).

a)

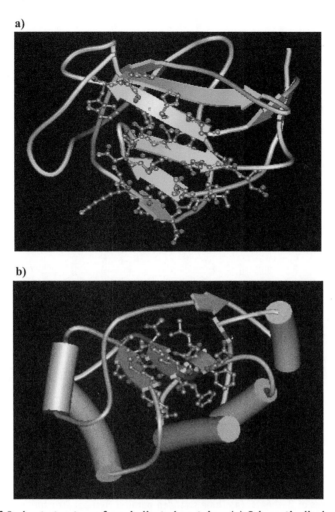

b)

Figure 1.19 **Depiction of β-sheet structures from indicated proteins.** (a) **Schematic display structure** (see Section 1.2.5; each flat arrow is a β-strand with arrow head equal to *C*-terminus of each strand: cylinders are α-helices; remainder represent loops and turns) of **anti-parallel β-sheet segment** of **carbonic anhydrase I** (human erythrocyte) (pdb: **2cab**), atoms and bonds of amino acid side-chains are rendered in **ball and stick representation** with carbon (**grey**), nitrogen (**blue**) and oxygen (**red**); **(b)** **Schematic display structure of parallel β-sheet segment** of triose phosphate isomerase (chicken muscle) (pdb: **1tim**), atoms and bonds of side-chains are rendered as in (a).

1.2.3 Non-repetitive secondary structure

Helices and sheet-like structures are linked and/or held together by turns and loops in a given polypeptide main chain. **Tight turns** in the main chain (also known as β-**bends** or β-**turns**) are very common. These typically involve four amino acid residue units held together by a non-covalent hydrogen bond between C = O group lone pairs of the *n*-th residue and the N-H group of the (*n* + 3)-th residue. Given variations in the possible ϕ and ψ angles of the amino acid residues involved, there are at least six possible variants. However, these are usually divisible into just two main classes **Type I** and **Type II** that differ primarily in the conformation of the peptide link between the second and third residues of the turn (Figures 1.21 and 1.22). **Loops** in the main chain are also very common, but interactions between amino acid residue side chains provide stability rather than peptide link-associated **hydrogen bonding**. Consequently, the path mapped out by the main chain in forming a loop is a good deal less regular than that found in a tight turn (Figure 1.23). Occasionally, **disulfide bridges** in polypeptides replace and/or supplement peptide links. These bridges are formed between the thiol-functional groups of two different cysteine (Cys, C) residues separated by at least two other amino acid residues from each other in the amino acid sequence of a polypeptide. These may be thought of as the polypeptide equivalent of a "tie bar" or some other such reinforcing device. Both right and left-handed spiral forms are known and a series of conformational angles $(\mathcal{X}_1, \mathcal{X}_2, \mathcal{X}_3, \mathcal{X}_2', \mathcal{X}_1')$ define the state of each given disulfide bridge (Figures 1.24 and 1.25).

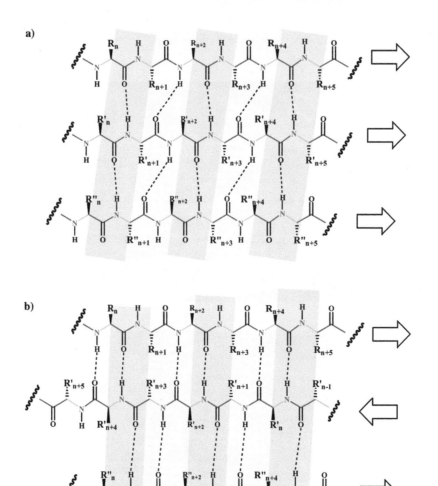

Figure 1.20 Stereo-defined structures of β-sheets. (a) Three stranded **parallel β-sheet** structure showing hydrogen-bonding relationship between parallel β-strands. The N-H donor of each peptide link can form a hydrogen bond with the C = O acceptor of a peptide link in a parallel β-strand. Shading is used to demonstrate pleating and emphasise amino acid residue side-chain orientations with respect to the sheet and with respect to each other. Peptide backbone bonds are colour coded in the same way as Figures 1.6 and 1.8. Arrows define *N* to *C* chain directions; **(b)** Three stranded **anti-parallel β-sheet** structure as for **(a)** except that **hydrogen bonding** occurs between peptide links in neighbouring anti-parallel β-strands.

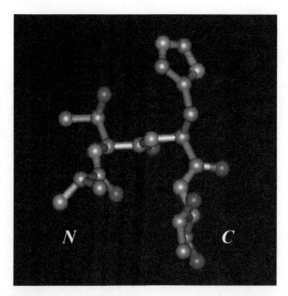

Figure 1.21 Depiction of β-turn (Type I). The illustrated turn derives from carbonic anhydrase I (human erythrocyte) (pdb: **2cab**). **Ball and stick representation** of atoms and bonds with carbon **(grey)**, nitrogen **(blue)** and oxygen **(red)**.

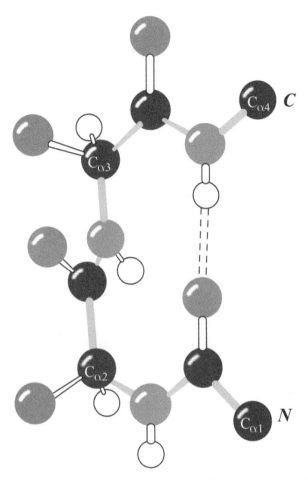

Figure 1.22   Depiction of β-turn (Type II). Ball and stick representation of atoms and bonds with carbon (**black**), nitrogen (**blue**), oxygen (**red**), and side-chain atom (**purple**), to illustrate how the C = O acceptor of the first n-th residue forms a hydrogen bond with the N-H bond donor of the $(n + 3)$-th residue defining an atom separation of 10 between acceptor O-atom and donor H-atom. This is the same of the Type I turn (Figure 1.21). The main difference between Type I and Type II is the orientation of the peptide link joining the $(n + 1)$-th to the $(n + 2)$-th residue (illustration from Voet *et al.*, 1999, Wiley, Figures 6–20b).

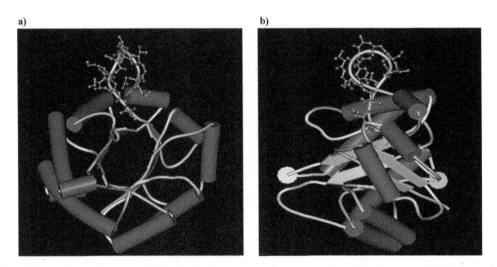

Figure 1.23   Depiction of loop structure. This loop structure derives from triose phosphate isomerase (chicken muscle) (pdb: **1tim**). (**a**) **Schematic display structure** (see Section 1.2.5; each flat arrow is a β-strand with arrow head equal to C-terminus of each strand: cylinders are α-helices: remainder is loops and turns) of triose phosphate isomerase (top view), atoms and bonds of amino acid side chains of key loop are rendered in **ball and stick representation** with carbon (**grey**), nitrogen (**blue**) and oxygen (**red**); (**b**) **Schematic display structure** of triose phosphate isomerase (side view), atoms and bonds of amino acid side chains rendered as in (a).

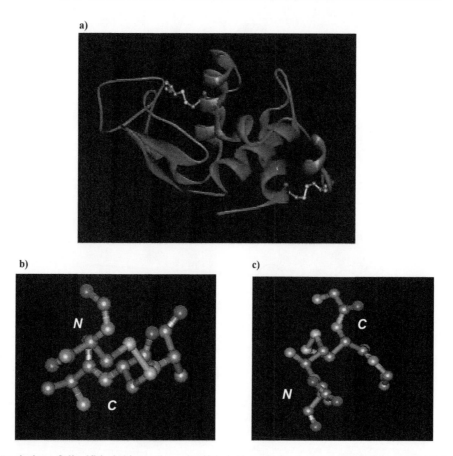

Figure 1.24   Depiction of disulfide bridges. These disulfide bridges derive from **lysozyme** (hen egg-white) (pdb: **6lyz**). (a) **Ribbon display structure** (see Section 1.2.5; each flat helical ribbon is an α-helix; each flat strand is β-strand: thin strands are loops and turns) of lysozyme (side view), atoms and bonds of amino acid side chains of cysteine residues involved in disulfide bridges are rendered in **ball and stick representation** with sulfur (**yellow**) carbon (**grey**), nitrogen (**blue**) and oxygen (**red**); (b) **Ball and stick representation** of cysteine residues forming a **right-handed disulfide bridge**, atoms and bonds labelled as for (a); (c) **Ball and stick representation** of cysteine residues from a **left-handed disulfide bridge**, atoms and bonds labelled as for (a).

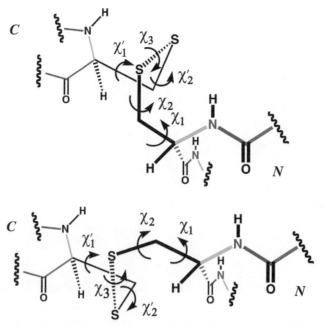

Figure 1.25   Structural description of disulfide bridges. (a) Right-handed disulfide bridge showing all the main conformational angles; (**b**) Left-handed disulfide bridge similarly indicating all the main conformational angles. In both cases, **N** refers to a cysteine residue closest to the *N*-terminus of the polypeptide and **C** refers to the residue closest to the *C*-terminus.

1.2.4 Alternative secondary structures

In certain cases, a polypeptide main chain can map out helical structures that are rather more extended and elongated than the α-helix. These structures are known either as the left-handed P_{II} helix (or collagen helix for reasons that will become apparent). This is unusual in being a left-handed structure when most biological macromolecule structures or substructures are right-handed. The polypeptide main chain is extended so that in appearance it is somewhere in between the topography of an α-helix and a β-strand (Figures 1.26 and 1.27). There are about three amino acid residues per turn of helix, with a rise per residue of about 3 Å between each residue as compared to 1.5 Å in the α-helix. The required ϕ and ψ angles needed to form a P_{II} helix are unusual (Figure 1.8). Accordingly, most polypeptides are unable to adopt an extended P_{II} helical conformation successfully, with the exception of polypeptides comprising a high proportion of glycine and proline residues (Table 1.1).

1.2.5 Tertiary structure

The docking together and mutual stabilisation of 3D secondary structural elements results in an overall 3D fold known as the tertiary structure of a polypeptide. Many proteins are comprised of a single polypeptide chain, hence this fold becomes in effect the tertiary structure of the protein too. The overall shape of this fold is frequently globular and therefore proteins with such a fold are known as globular proteins. Protein folding is known as the process by which proteins acquire their tertiary structure. This field of study is enormously controversial and where globular proteins/polypeptides are concerned, debate still rages about whether secondary structures form first and dock together to form tertiary structure (Framework model) or whether a crude tertiary structure forms first, followed by a process of "side-chain" negotiation to create stabilised secondary structures (Molten-globule model). The latter model appears the more popular today, consistent with the realisation that there are multitudinous "pathways" of protein folding which are themselves highly dependent upon the primary structure of the particular protein/polypeptide of interest. Yet, we still do not know how exactly the primary structure drives the formation of the tertiary structure, so that given a certain primary structure we remain far from sure what the globular tertiary structure will be in many, many cases.

Given the general atomic complexity of protein architecture, several shorthand representations of globular protein/polypeptide 3D structures have been devised, of which the easiest to understand are the ribbon display structures (Figure 1.28). There are other representations too, including surface display, CPK and schematic display structures, all designed to highlight different aspects of protein structure during analysis (Figure 1.29). The illustrated structures

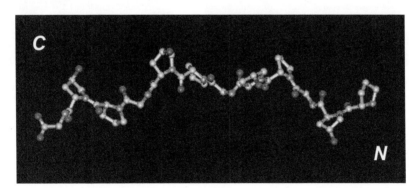

Figure 1.26 Depiction of left-handed P_{II}-helix. This P_{II}-helix derives from **collagen (pdb: 1bkv). Ball and stick representation** of atoms and bonds with carbon (**grey**), nitrogen (**blue**) and oxygen (**red**).

Figure 1.27 Depiction of the extended P_{II}-helix. This depiction emphasises the left-handed character of this helix type (adapted from Voet *et al.*, 1999, Wiley, Figures 6–17).

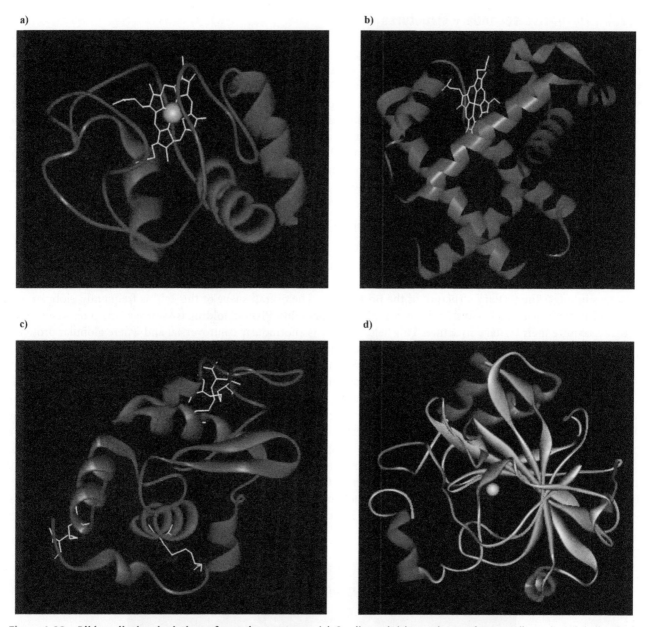

Figure 1.28   Ribbon display depictions of protein structures. (a) *Small metal rich protein* **cytochrome c** (horse heart) (pdb: **1hrc**). Polypeptide backbone is shown as a ribbon (**red**). Stick bonds constituting porphyrin macrocycle (prosthetic group) are highlighted (**yellow**). Van der Waals sphere (**blue**) represents central iron ion; (**b**) *Anti-parallel α-protein* **myoglobin** (sperm whale) (pdb: **1mbi**) with labelling system as for (a); (**c**) *Small SS rich protein* **lysozyme** (hen egg-white) (pdb: **6lyz**). Polypeptide backbone is shown as a ribbon (**red**), stick bonds of cysteine amino acid residues linked by disulfide bridges are also shown (**yellow**); (**d**) *Anti-parallel β-protein* **carbonic anhydrase I** (human erythrocyte) (pdb: **2cab**). Polypeptide backbone is coloured according to secondary structure with α-helix (**red**), anti-parallel β-sheet (**light blue**), loop structures (random coil) (**light grey**). Van der Waals sphere (**yellow**) represents central zinc ion.

show all the main classes of **globular proteins**, namely **small metal-rich proteins, small SS rich proteins, anti-parallel α-proteins, anti-parallel β-proteins** and **parallel α/β-proteins** according to the **Richardson Classification**. This is based on the topographical behaviour of repetitive secondary structure elements in the **tertiary structure** (Figures 1.28 and 1.29). Almost without exception, the interior of a globular protein will contain hydrophobic amino acid residues (e.g. leucine [Leu, L], valine [Val, V], phenylalanine [Phe, F], etc.), whilst the exterior is made up of hydrophobic and hydrophilic amino acid residues, though hydrophilic amino acid residues predominate.

The importance of this will become clear later, but for now let us say that globular proteins have a "waxy" interior and a "soapy" exterior. This "waxy" interior is amazingly crystalline. In fact, the interior packing of hydrophobic amino acid

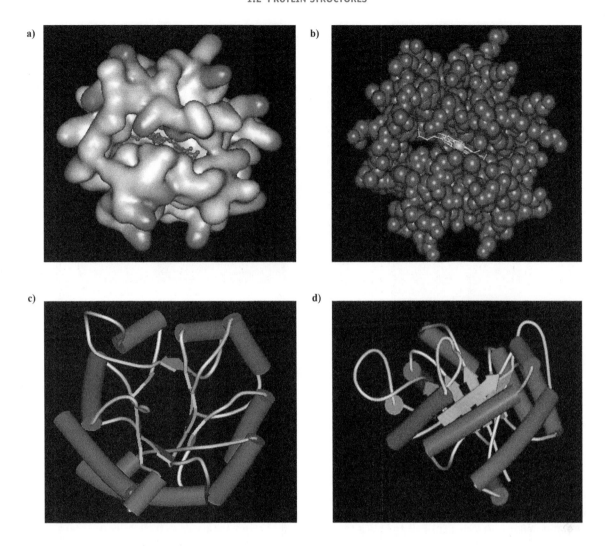

Figure 1.29   Alternative depictions of protein structures. (a) Surface display structure of *small metal rich protein* **cytochrome c** (horse heart) (pdb: **1hrc**) showing **Van der Waals surface** coloured for positive charge (**blue**) and for negative charge (**red**). **Ball and stick representations** of **iron-porphyrin macrocycle (prosthetic group)** are shown (**red**) for each subunit with central iron ion rendered as Van der Waals sphere (**light blue**); **(b) CPK structure** of cytochrome c in which all polypeptide atoms are rendered as Van der Waals spheres (**purple**). Porphyrin and iron ion are shown as in Figure 1.28; **(c) Schematic display structure** (top view) of *parallel α/β -protein* triose phosphate isomerase (chicken muscle) (pdb: **1tim**) with α-helix shown as cylinders (**red**), β-strands as arrowed ribbons (**light blue**), loop structures (random coil) as rods (**light grey**); **(d) Schematic display structure** (side view) of triose phosphate isomerase, otherwise as for (c).

residues is remarkably similar to the crystalline state of organic solids. This amino acid residue distribution indicates that close range non-covalent Van der Waals forces and hydrophobic interactions (see Section 1.6) are critical to the stability of the globular protein/polypeptide chain, whereas the non-covalent hydrogen bond is largely responsible for the formation of repetitive and much non-repetitive secondary structure (except for the disulfide bridge). Of late, there is a realisation that other forces, defined as **weak-polar forces**, may also have a significant role to play in stabilising protein/polypeptide tertiary structure as well (see Section 1.6). Indeed, there may yet be other forces to be discovered. Without doubt globular protein/polypeptide structure remains a rich and fascinating if not controversial area of research.

Globular proteins/polypeptides are frequently employed in cells as functional proteins responsible for respiration, metabolism and communication. However, the vast majority of proteins inside and indeed outside cells have structural, scaffolding functions. In these cases, protein polypeptides generally exhibit an extended rather than globular overall fold and hence proteins are known as **fibrous proteins**. First and foremost amongst these should be considered **collagen**, the main structural component of bone. The collagen molecular fold is created essentially through the association of three

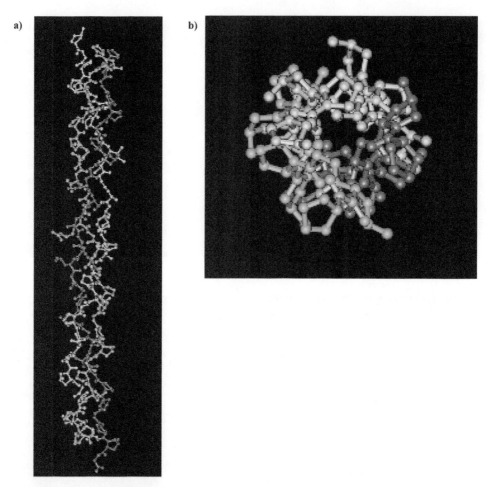

Figure 1.30   Depiction of right-handed collagen triple helix. The illustrated structure derives from collagen (pdb: **1bkv**) **(a) Ball and stick representation** of atoms and bonds of triple helix (side view) with complete, individual polypeptide chain single coloured (**yellow, brown** or **blue**) (side view); **(b) Ball and stick representation** of atoms and bonds of triple helix (top view), polypeptide chains rendered as in (a).

extended left-handed **P$_{II}$ helices** to give a right-handed **P$_{II}$ triple helix** or **collagen triple helix** (Figures 1.30 and 1.31). Each constituent P$_{II}$ helix is linked to neighbours by hydrogen bonds, resulting in high lateral as well as longitudinal stability. Other forces are also thought to be involved, but remain to be fully characterised. Since only polypeptides comprising glycine and proline residues are able to form extended P$_{II}$ helices, owing to the unusual ϕ and ψ angles required, then collagen molecules, too, must contain a disproportionately high level of proline and glycine residues compared with other globular proteins/polypeptides.

Strictly speaking, the term tertiary structure refers to the overall fold of a single polypeptide (as already noted in this section). Yet while a collagen triple helix can be formed from a single polypeptide, the collagen triple helix that makes up the collagen fold is actually constructed from three different collagen polypeptides, linked together initially by disulfide bridges. Therefore, the collagen triple helix that comprises the collagen molecular fold should perhaps be described not as the tertiary structure of collagen but as a form of **quaternary structure** (see Section 1.2.6). Nevertheless, when bone is formed, collagen molecules associate through their collagen triple helices, forming extended bundles that mineralise to form the matrix of bone. Arguably we should characterise these bundles as the collagen quaternary structure instead. In conclusion we would like to point out that there is a small population of single polypeptide proteins whose tertiary structures are actually a combination of globular and fibrous protein folds, including dominant but not exclusive P$_{II}$ triple helical regions. These are regarded as hybrid proteins that are both globular and fibrous proteins simultaneously.

1.2.6 Quaternary structure

Different polypeptide chains may interact to form more complex multi-polypeptide proteins, where each individual polypeptide is known as a **subunit**. Subunit interactions and interrelationships are illustrated for the tetrameric protein haemo-

Figure 1.31 Depiction of the right-handed collagen triple helix. Depiction emphasises how three polypeptides in left-handed P_{II} helical conformations associate to form the triple helical structure wherein each polypeptide adopts a gentle right-handed, rope-like twist to maximise stabilising inter-chain hydrogen bond interactions. Structure is also stabilised by a sheath of ordered water molecules of solvation. (illustration from Voet *et al.*, 1999, Wiley, Figures 6–17).

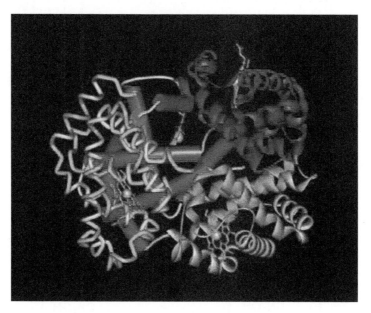

Figure 1.32 Quaternary structure of haemoglobin. The illustrated structure derives from human foetal **haemoglobin** (pdb: **1a3n**). This protein is comprised of four distinct polypeptides (**subunit**s), α_1, α_2, β_1 and β_2 which are rendered two as **ribbon display structures** (**red** and **yellow**; **right-side**), one as a **schematic display structure** (**left-side**, **rear**), and one as **CA stick display** (**left-side**, **front**). **Ball and stick representations** of **iron-porphyrin macrocycle** (**prosthetic group**) are shown (**green**) for each subunit with central iron ion rendered as Van der Waals sphere (**light grey**). Subunits are all associated non-covalently with each other which is usual for polypeptide subunits that comprise a multi-polypeptide protein.

globin (Figure 1.32). In the case of globular proteins, polypeptide chain association allows functions of individual polypeptide elements to be coordinated or indeed supplemented to give the whole molecule the opportunity to perform multiple biological functions. In the case of fibrous proteins, quaternary structure formation enhances overall molecular strength.

1.2.7 Prosthetic groups

Many globular proteins/polypeptides, especially those involved in the catalysis of chemical reactions, also have non-peptidic structures that may be associated covalently or non-covalently with the polypeptide. These are known as **prosthetic groups**. All such prosthetic groups belong with proteins in order to confer particular functionalities that might otherwise not exist. A good example is the iron porphyrin ring in the proteins **haemoglobin** (Figure 1.32), **myoglobin** and **cytochrome c** (Figures 1.28 and 1.29) that enables the first two of these proteins respectively to act in respiration as molecular oxygen carriers/scavengers and the last to act as a redox substrate. Other prosthetic groups will be described in Chapter 4.

1.3 Carbohydrate structures

1.3.1 Primary structure

Carbohydrate polymers are formed from the linear and branched combination of a wide variety of different naturally occurring simple sugars known as **monosaccharides**. These are a reasonably diverse set of molecular building blocks each capable of linking to other monosaccharides in a variety of different ways, with the result that the exquisite hierarchy of structural elements found in protein structures are not so readily duplicated with carbohydrate polymers. Having said this, those carbohydrate homopolymers that are constructed from only one or two monosaccharide building blocks can possess really impressive 3D structures. Therefore, read on.

Monosaccharides may be classified into families according to the number of carbon atoms they contain, usually between three and seven. The **triose** family has the empirical formula $C_3H_6O_3$, the **tetrose** family $C_4H_8O_4$, the **pentose** family $C_5H_{10}O_5$, the **hexose** family ($C_6H_{12}O_6$) and the **heptoses** $C_7H_{14}O_7$. An alternative classification has been to name monosaccharides as either **aldehydo-aldose** or **ketose** sugars, depending upon whether they possess an aldehyde or ketone functional group respectively. By way of illustration, the well-known sugar **glucose** is both a hexose, with six carbon atoms, and an aldehydo-aldose monosaccharide, owing to the aldehyde functional group at carbon atom 1 (C-1) (Figure 1.33). **Fructose** is also a hexose, but is otherwise known as a ketose sugar because of the ketone functional group positioned at carbon atom 2 (C-2) (Figure 1.34). Each exists in two enantiomeric forms (either D or L), as defined by the absolute stereochemistry of the penultimate carbon atom in the chain (in both cases carbon atom 5, C-5) with reference by convention to the C-2 stereochemistry of **D-/L-glyceraldehyde** (Figure 1.35). The D-enantiomers of glucose and fructose tend to predominate in natural carbohydrate polymers. This is also true of most other monosaccharides found in natural carbohydrate polymers as well.

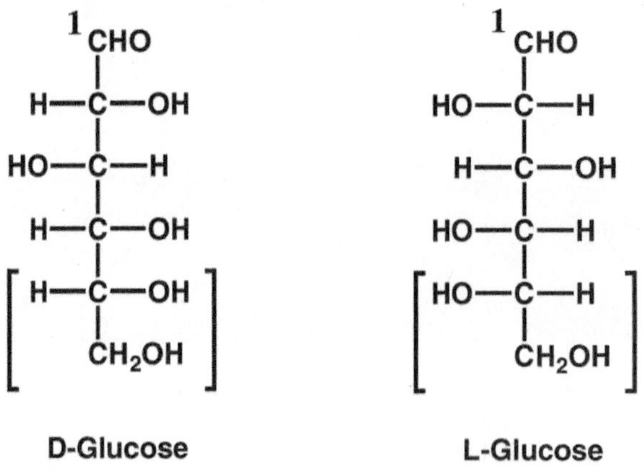

Figure 1.33 D- and L-glucose.

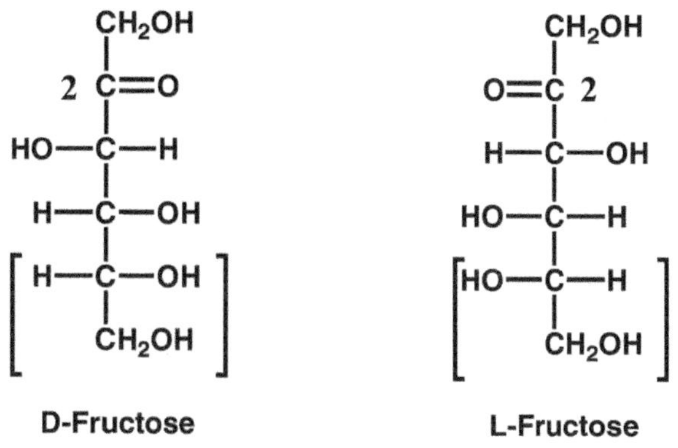

Figure 1.34 D- and L-fructose.

Figure 1.35 **D- and L-glyceraldhyde.** This is depicted to illustrate principles of the **Fischer Projection.** The D and L absolute configuration convention for all sugars refers to the stereochemistry of the bracketed terminal carbons whose configuration is compared with the two enantiomers of glyceraldehyde.

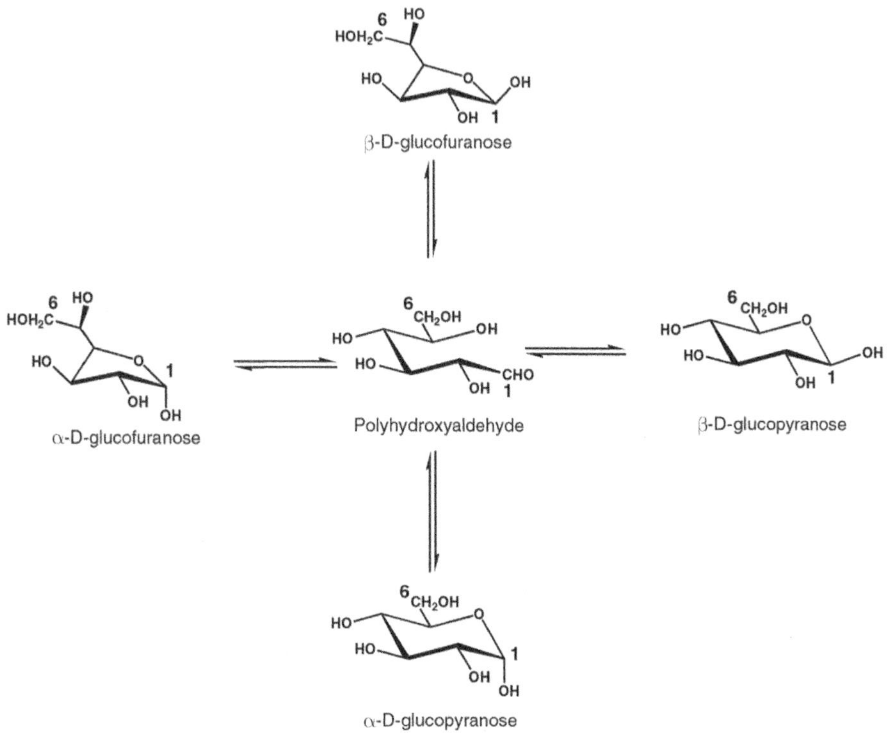

Figure 1.36 **Dynamic equilibria of D-glucose monosaccharide in solution.**

The conformational behaviour of monosaccharides in solution is complicated. In solution, acyclic D-glucose readily converts into cyclic **five-member** (*f*, **furanose**) and/or six-member (*p*, **pyranose**) rings (Figure 1.36). Furanose (*f*) and pyranose (*p*) rings always exist as pairs of **anomers** (α or β). These α and β anomers differ from each other only in their hydroxyl group configuration at the **anomeric carbon**. In D-glucose this is the hemiacetal carbon C-1 (Figure 1.36). Since pyranose rings are largely stable, D-glucose actually prefers to exist in solution as **α-D-glucopyranose** or β-**D-glucopyranose** (Figure 1.36). The relative proportions of either the α-**anomer** or the β-**anomer** respectively in solution depend upon the **anomeric effect**. The anomeric effect may be defined as the thermodynamic preference for one anomer over another, resulting from a combination of internal stereo-electronic effects and solution conditions.

Typically, the opportunity for hyperconjugation involving oxygen lone pairs favours the anomer with axial substituents attached to the anomeric carbon (Figure 1.37), although under aqueous solution conditions, the equatorial β-anomer β-D-glucopyranose is actually favoured over the axial α-anomer α-D-glucopyranose. Conformationally speaking, either anomer may adopt one of two main chair (**C**) conformations. For example, the two main chair conformations of β-**D-glucopyranose** are ^{4}**C**$_1$, where C-4 is above and C-1 is below the plane mapped out by oxygen and carbon atoms 2, 3 and 5, or ^{1}**C**$_4$ where the reverse is true (Figure 1.38). Unsurprisingly perhaps, the all-equatorial ^{4}**C**$_1$ is the more stable of the two and hence is the dominant conformation in solution. By contrast, α-D-glucofuranose or β-D-glucofuranose anomers exist in one of two

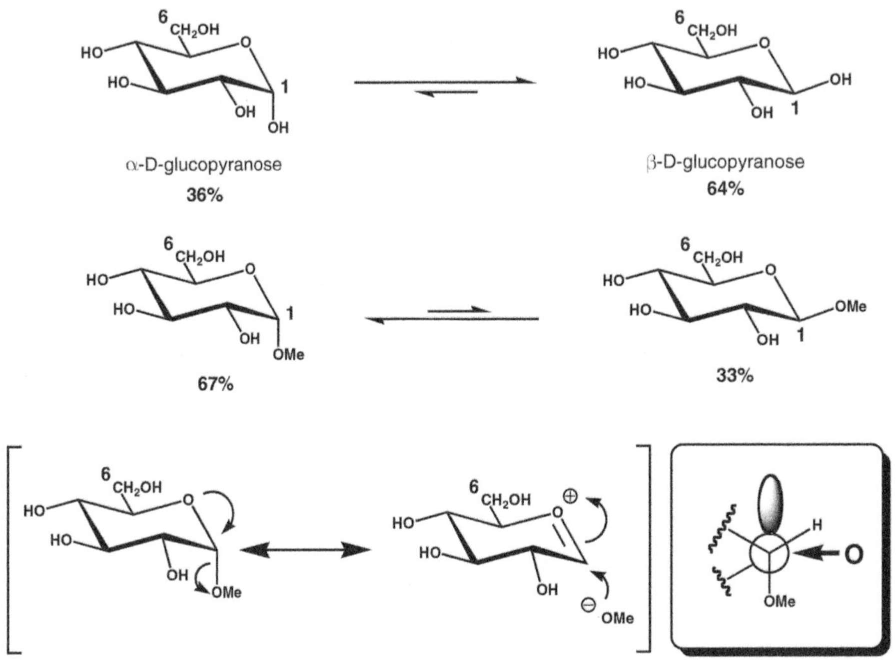

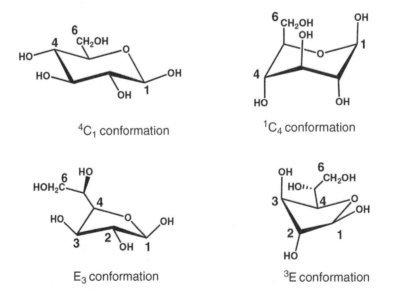

Figure 1.37 Illustration of the anomeric effect in monosaccharides. The anomeric effect is defined as the preference of electronegative functional groups attached to anomeric carbon C-1 to adopt an unexpected axial configuration. In D-glucopyranose, the equatorial β-anomer is favoured but the trend is reversed with alkylation. The origin of the effect is the trans-diequatorial interaction between oxygen lone pair and the σ* orbital of the bond linking C-1 and electronegative function group.

Figure 1.38 Conformational behaviour of D-glucose. Two common cyclic conformations are shown, namely β-**D-glucopyranose** (top) and β-**D-glucofuranose** (bottom). The anomeric carbon C-1 is on right-hand side by convention.

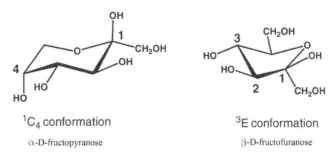

$^1C_4$ conformation

α-D-fructopyranose

^3E conformation

β-D-fructofuranose

Figure 1.39 Conformational behaviour of D-fructose. Two common cyclic conformations are shown, namely β-**D-fructopyranose** (left) and β-**D-fructofuranose** (right). The anomeric carbon C-1 is on right-hand side by convention.

main envelope (**E**) conformations. For example, the two main envelope conformations of β-D-glucofuranose are ^{3}**E**, where C-3 is above the plane mapped out by oxygen and carbon atoms 1, 2 and 4, or **E**$_3$ where the reverse is true (Figure 1.38).

D-fructose shows similar equilibrium behaviour to D-glucose in that cyclic forms are preferred in solution where the anomeric carbon is the hemi-ketal carbon C-2. In the case of D-fructose though, both pyranose and furanose rings are relatively stable and so there is a preference for D-fructose to exist primarily in solution in the cyclic forms α-**D-fructopyranose** or β-**D-fructofuranose** (Figure 1.39). Generally speaking, only the most stable cyclic conformation of either D-glucose of D-fructose will appear in natural carbohydrate polymers. The same is true of the other main monosaccharide building blocks found in natural carbohydrate polymers as well (Figure 1.40). The preferred, stable, cyclic conformations are shown together with the appropriate three-letter monosaccharide code and the short code for preferred conformation in Figure 1.41.

1.3.2 O-glycosidic link

In natural carbohydrate polymers, monosaccharide building blocks are joined together by means of the **O-glycosidic link**. The O-glycosidic link is an ether functional group that originates by convention from the anomeric carbon atom of one monosaccharide residue and terminates at the appropriate carbon atom of a neighbouring residue (Figure 1.42). Each linked monosaccharide building block is known as a **monosaccharide residue**. If an O-glycosidic link is defined as (1→4) then this implies that the link originates from the anomeric carbon C-1 of one monosaccharide residue and terminates at carbon atom C-4 of the next. In common with the peptide bond, the O-glycosidic link has some associated rigidity due to steric congestion between neighbouring monosaccharide residues. The resulting spatial relationships between neighbouring residues can then be defined in terms of two conformational angles ϕ and ψ which are respectively the main dihedral angles subtended about the anomeric carbon to oxygen bond (C_{an}-O) and the following oxygen to carbon bond (O-C) (Figure 1.43). In contrast with proteins, allowed values of ϕ and ψ vary substantially depending upon the identities of the linked monosaccharide residues (Figure 1.44). Therefore carbohydrate polymers comprised of a number of different monosaccharide residues cannot easily form stereo-regular, 3D structures.

By definition, when two monosaccharide building blocks are linked together by an O-glycosidic link, the result is known as a **disaccharide**. Two well-known disaccharides have been shown to illustrate how their structures are defined in terms of the three-letter monosaccharide code and their O-glycosidic links (Figure 1.45). Typically, when 2–20 monosaccharide residues are linked together, in a linear or branched fashion, the resulting carbohydrate polymers are described as **oligosaccharides** (Figure 1.46). When 20–100 monosaccharide residues (usually 80–100) are linked together then resulting polymers are described as **polysaccharides**. Unbranched oligo-/polysaccharides are said to consist of one carbohydrate main chain. Branched oligo-/polysaccharides are said to consist of a main chain to which are attached any number of branch chains. By convention, oligosaccharide and polysaccharide main chains end at the monosaccharide residue that retains a free anomeric carbon not involved in an O-glycosidic link. This terminal residue is known either as a **glycose residue**, if derived from an aldehydo-aldose monosaccharide, or as a **glycoside residue**, if derived from a ketose monosaccharide. The residue at the start of a given oligo-/polysaccharide chain, main or branched, is then known as a **glycosyl group** (Figure 1.46). All monosaccharide residues in between may be called **glycosyl residues**. The complete list of linked monosaccharide residues in travelling from the highest priority glycosyl group(s) to the lowest priority terminal glycose/glycoside residue is known as the **carbohydrate sequence**, or **carbohydrate primary structure**. Where branched polymers are concerned, the longest continuous chain provides the parent sequence, to which are attached branch chains with their own daughter sequences. For each branch chain, daughter sequences begin with glycosyl groups and end where the branch chain meets the main chain.

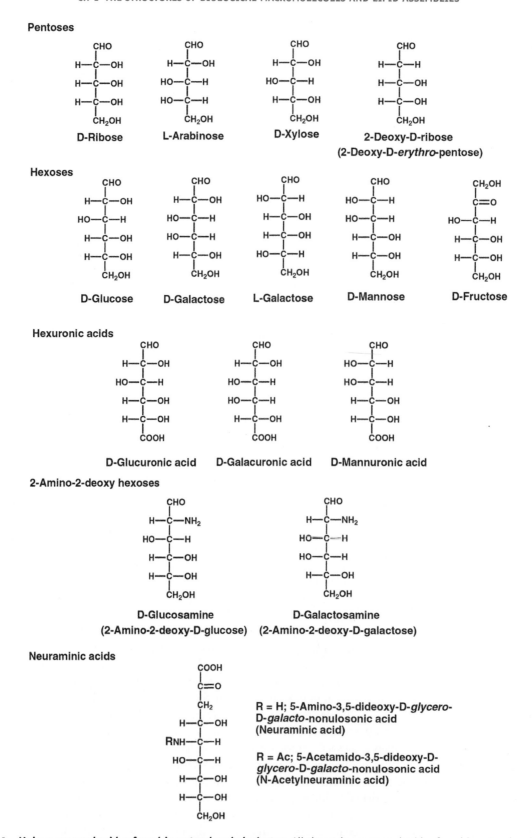

Figure 1.40 Main monosaccharides found in natural carbohydrates. All the main monosaccharides found in natural carbohydrates are illustrated in their Fischer projections to show differences in absolute configurations.

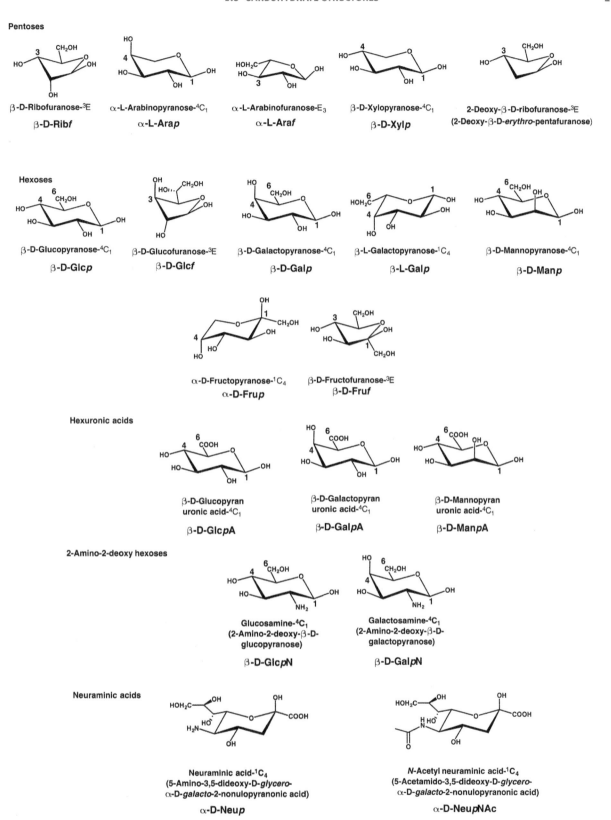

Figure 1.41 Cyclic conformations of main monosaccharides. The preferred cyclic conformations of all the main monosaccharides found in all natural carbohydrates are illustrated.

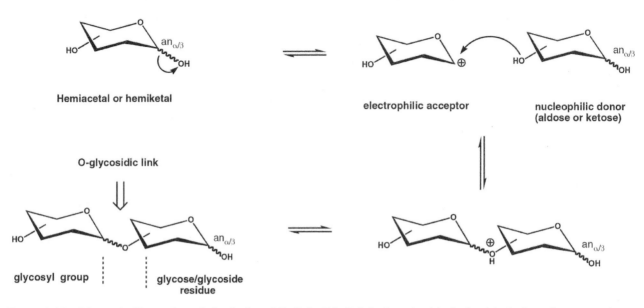

Figure 1.42   Schematic illustration of the O-glycosidic link. This link is formed critically by dehydration of an anomeric carbon centre.

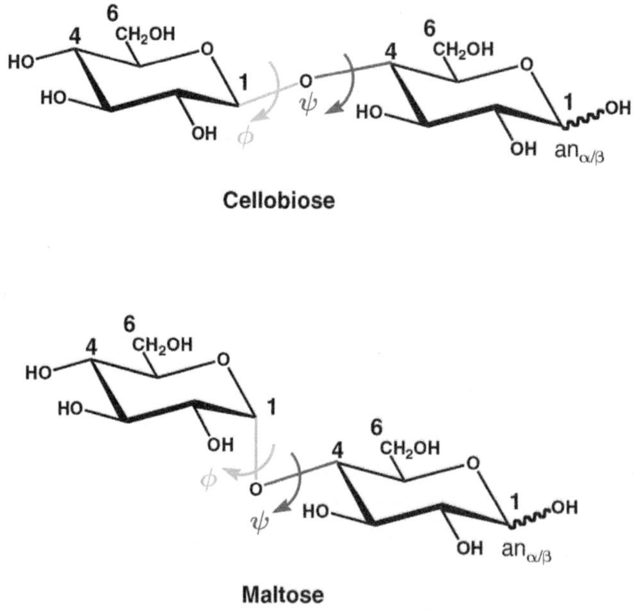

Cellobiose

Maltose

Figure 1.43   Main O-glycosidic link conformational angles. ϕ is dihedral angle subtended about the C_{an}-O bond and ψ is dihedral angle subtended about subsequent O-C bond.

1.3.3 Polysaccharides: secondary, tertiary and quaternary structures

The sequences of many oligosaccharides are often too diverse and too short to encourage the formation of stable 3D structures. Therefore, for the most part, oligosaccharides and shorter polysaccharides exist only as **random coil**. The same is not true of a significant number of the much longer chain polysaccharides. Polysaccharides can also be known as **glycans**, hence polysaccharides containing only one type of monosaccharide residue are known as **homoglycans**, and those with between two and six different types of residue are known as **heteroglycans**. Homoglycans and a few heteroglycan polysaccharides can form extensive periodic secondary structures and these regular structures are usually related geometrically to a helix even if they may not necessarily conform to the conventional idea of a helix. Reg-

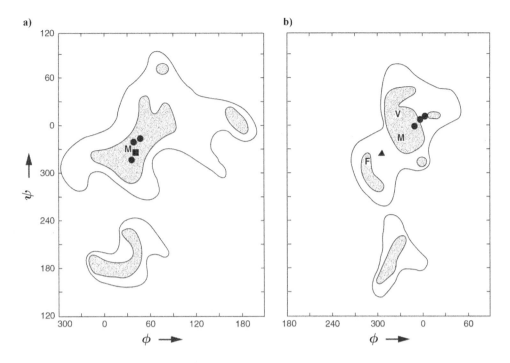

Figure 1.44 **Conformational restrictions in O-glycosidic links.** Allowed values ϕ and ψ angles for O-glycosidic links of **cellobiose** (**a**) and **maltose** (**b**). In (**a**) black squares or circles indicate positions of observed conformational angles in cellobiose or cellobiose units found in larger polysaccharide systems. In (**b**) black triangle or circles indicate position of observed conformational angles in maltose or maltose units found in a larger polysaccharide systems. (adapted from Rees and Smith 1975, Figures 7 and 5).

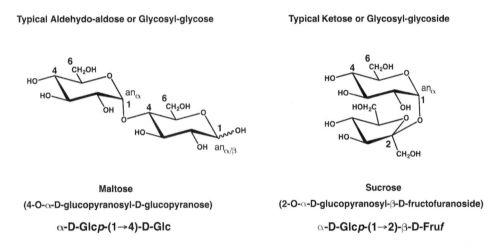

Figure 1.45 **Structures of two main disaccharides.** Maltose is a representative **glycosylglycose** and sucrose a representative **glycosylglycoside,** each with a single O-glycosidic link. Full trivial names are given as well as **three-letter code**-based nomenclature that is used for more complex systems. By convention, the highest priority end is the glycosyl group and the lowest priority the glycose/glycoside residue. This convention applies for all oligosaccharides, polysaccharides and carbohydrates.

ular structures are defined in terms of two parameters that are known as n, the number of glycosyl residues per turn, and h, the projected length of each glycosyl residue on the "helix"-axis (Figure 1.47). Parameter n is either positive or negative, depending upon whether the helix is perceived to be right-handed or left-handed respectively. In the event, almost all homoglycan and heteroglycan polysaccharides that are able to form periodic secondary structures exist only in four main families. These are the **ribbon family** (n = 2 to ± 4; h = approximately 5 Å [approximately length of glycosyl residue]), the **hollow helix family** (n = 2 to ± 10; h = 2.5 to ± 2.5 Å), the **crumpled family** and the **loosely jointed family**. The latter two are rare in biological systems and will not be described further.

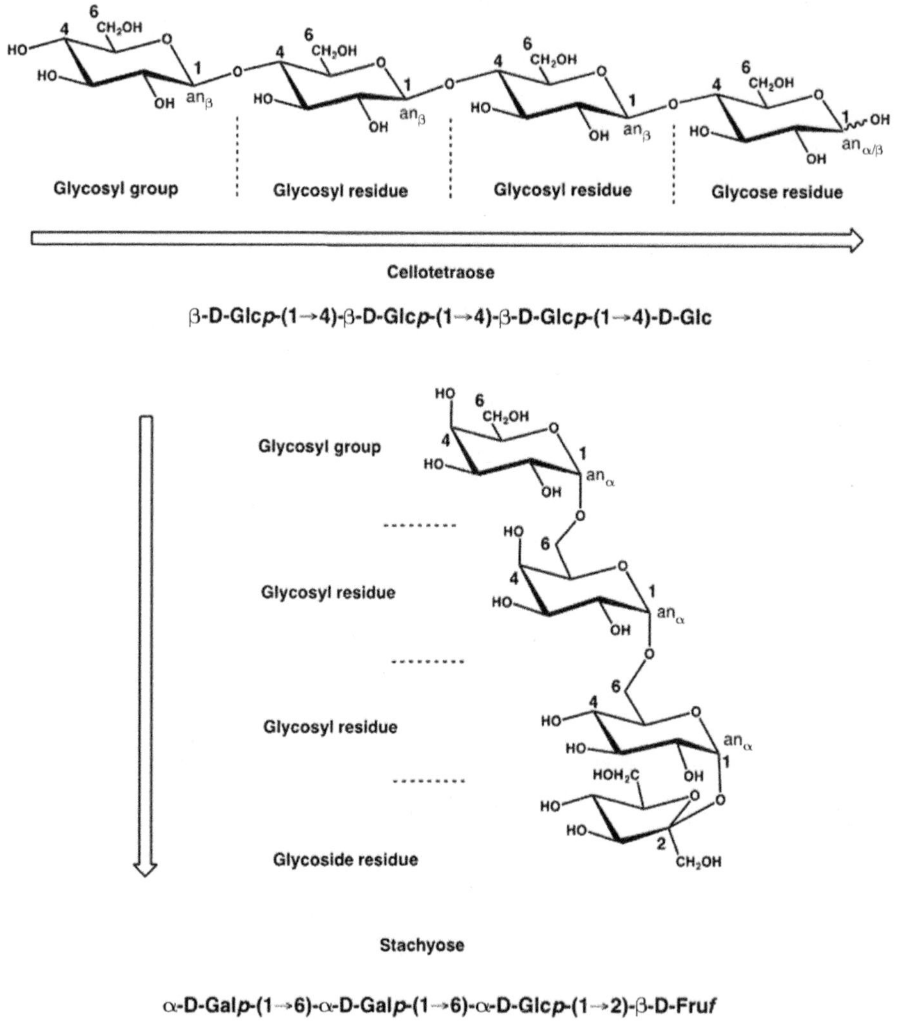

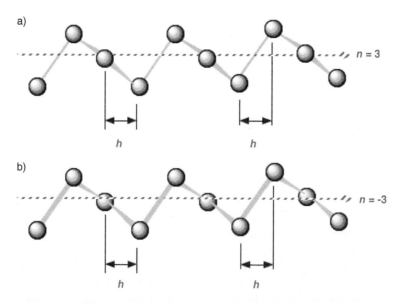

Figure 1.46 **Structures of two main tetrasaccharides. Cellotetraose** is representative of a glycose tetrasaccharide and **stachyose** of glycoside tetrasaccharide, each with three O-glycosidic links. The **three-letter code**-based nomenclature is used. Arrows show chain directions as determined by the convention described in the legend to Figure 1.45.

Figure 1.47 **Depictions of polysaccharide helices.** Simplified descriptions of **right-hand (a)** and **left-hand (b)** polysaccharide **helices** are shown, where **h** is the length of each monosaccharide residue projected on the helix axis, and **n** is the number of residues per turn. A negative value implies left-hand and a positive value implies right-hand. Spheres represent O-glycosidic links between monosaccharide residues.

The most common homoglycans are the plant cell wall polysaccharide components **cellulose** [(1→4)-β-D-glucan] (approximately 5000 residues) and **mannan** [(1→4)-β-D-mannan], the starch component **amylose** [(1→4)-α-D-glucan] (1000–2000 residues), and the insect skeletal polysaccharide **chitin** [(1→4)-β-D-2-*N*-acetylamido-2-deoxyglucan]. In the case of natural cellulose (**cellulose I**), flat ribbons will align parallel with respect to each other and form sheets stabilised through extensive inter-ribbon hydrogen bond networks that are somewhat analogous to the β-sheet structures of proteins. These sheets can further pack in a parallel, staggered fashion giving the appearance of a periodic polysaccharide tertiary/quaternary structure (Figure 1.48). In the main natural form of chitin (β-**chitin**), flat ribbons similarly organise into parallel, hydrogen-bonded sheets that further pack in a parallel, staggered fashion, resulting in a similar periodic tertiary/quaternary structure to that observed with natural cellulose (Figure 1.48). Sheets can also pack antiparallel, giving rise to α-**chitin**, or in a mixed parallel/anti-parallel fashion giving rise to what is known as γ-**chitin**. All three forms of chitin are known in biological systems. Mannan also prefers to form flat ribbons that assemble into

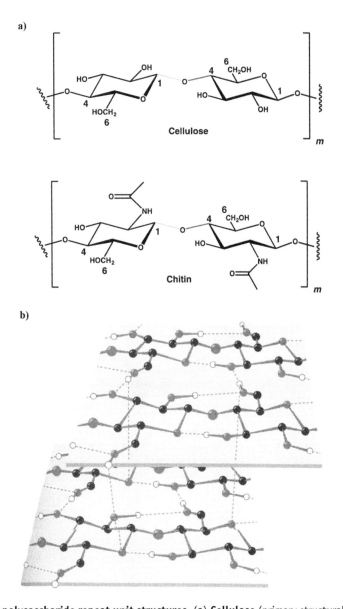

Figure 1.48 **Illustrations of polysaccharide repeat unit structures.** (a) **Cellulose** (primary structural component of plant cell walls) and **chitin** (principal structural component of exoskeletons of invertebrates; crustaceans, insects and spiders: major component of fungal and algal cell walls); (**b**) Cartoon showing how cellulose chains in **ribbon conformations** interact to form hydrogen bonded **sheets**. These sheets (secondary structure) are then packed in a staggered arrangement (**Cellulose I**) to maximise stabilising hydrogen bond contacts between sheets (tertiary structure). Colour code, carbon (**black**), oxygen (**red**) and hydrogen (**white**). In cellulose, sheets are always in parallel alignment. Chitin adopts very similar layered sheet structure, but sheets may be in parallel alignment (similar to **b** above) (β-chitin), or anti-parallel (β-chitin) (illustration from Voet *et al.*, 1999, Wiley, Figures 8–9).

a)

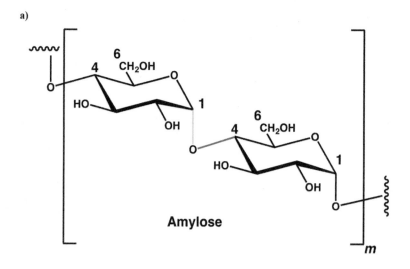

b)

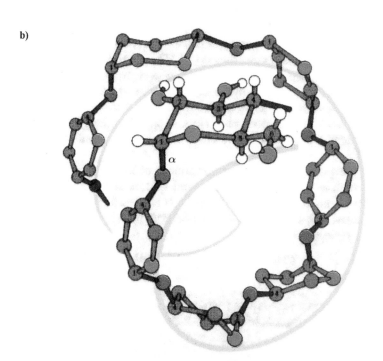

Figure 1.49 **Alternate repeat unit structures.** (a) **Amylose** (primary storage form of glucose in cells); (**b**) Cartoon showing how amylose chains exist in a **hollow helix** (**V-form**) (secondary structure) conformation in the presence of inclusion of a guest molecule such as polyiodide. Such helices become destabilised in the absence of a guest molecule and combine to form double hollow helix structures (**A-form**) as illustrated (see Figure 1.50). Colour code, carbon (**black**), oxygen (**red**) and hydrogen (**white**) (illustration from Voet *et al.*, 1999, Wiley, Figures 8–10).

sheets. By contrast, amylose forms into a variety of hollow helix secondary structures of which the best known is the **V-form** that is a highly compressed form of left-handed hollow helix stabilised by enclosure of small molecules with amphiphilic characteristics such as phenols or polyiodide (Figure 1.49). Otherwise, in the absence of such guest molecules or complexation agents, natural amylose apparently prefers to reside in the form of a left-handed double hollow helix structure known as the **A-form**.

Heteroglycans are usually complex sequences of pentoses, hexoses, 6-deoxyhexoses, hexuronic acids and hexosamines (2-amino-2-deoxy hexoses) that may also be derivatised as sulfates, acetates or methyl ethers at appropriate hydroxyl groups, as well as cyclised. Regular heteroglycans, such as the homoglycans mentioned earlier in this section, can be notably effective at secondary and even tertiary/quaternary structure formation. Two important examples of such heteroglycans are **carrageenan** and **agarose** that both coat the outer surfaces of marine red algae for protec-

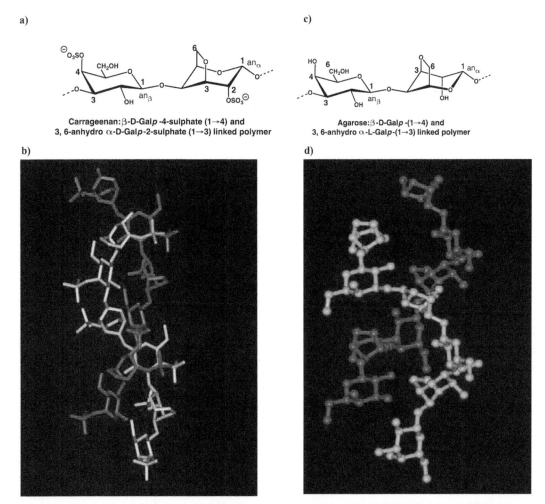

Figure 1.50 **Structures of two important polysaccharides.** Both exhibit ordered secondary structures (**hollow helices, A-form**). **Carrageenan** has the repeat unit structure shown (**a**) and double extended hollow helical secondary structure (**b**) with both separate strands rendered in the stick bond representation (**yellow** and **red**) (pdb: **1car**). **Agarose** has the alternative repeat structure shown (**c**) and double extended hollow helical secondary structure (**d**) with both separate strands rendered in the **ball and stick representation** (**green** and **red**) (pdb: **1aga**).

tion and for the facilitation of metabolite transfer between cells. Both are able to generate A-form helices stabilised by inter-residue hydrogen bonds involving the hydroxyl groups of neighbouring monosaccharide glycosyl residues (Figure 1.50). The ability to form such double helical secondary structures is the reason that carrageenan and agarose polysaccharides are then able to form gels. Left-handed double hollow helix structures generated from different polysaccharide chains link different polysaccharide chains together, creating 3D non-periodic networks of interlinked polysaccharide chains that are resistant to fluid flow and hence have gel behaviour. In effect both carrageenan and agarose exhibit not only secondary structure forming characteristics but also non-periodic tertiary/quaternary structure-forming characteristics as well.

1.4 Nucleic acid structures

As noted in Section 1.1, there are two basic types of nucleic acids found in cells: namely DNA and RNA. DNA represents the ultimate long-term storage form of genetic information because of its greater apparent chemical stability compared with RNA. In the guise of **messenger RNA (mRNA)**, **transfer RNA (tRNA)** and **ribosomal RNA (rRNA)** RNA plays an intermediary role in processing that genetic information locked away in DNA structures into physical reality in the form of the polypeptide/proteins that give individual cells both form and function. The precise interrelationships between DNA, rRNA, tRNA and mRNA will be looked at briefly later on in Section 1.4.5, but for now let us press on with structure.

1.4.1 Primary structures of DNA and RNA

DNA is a mixed polymer made up respectively of **2′-deoxyribonucleotide** (or **deoxynucleotide**) building blocks. Only a single set of four monomeric deoxynucleotide building blocks go to make up all DNA in every cell of every organism. This amazingly small set of building blocks is all that it takes to store the genetic information of all organisms. These deoxynucleotides are composed of **2′-deoxy-β-D-ribofuranose**, linked via an **N-β-D-glycosidic linkage** (originating at anomeric carbon atom C-1′) to a nitrogen heterocyclic **base**, and phosphorylated on carbon atom C-5′ (Figure 1.51). The base may be either a bicyclic purine **adenine** or **guanine**, or a monocyclic pyrimidine **cytosine** or **thymine** (Figure 1.52).

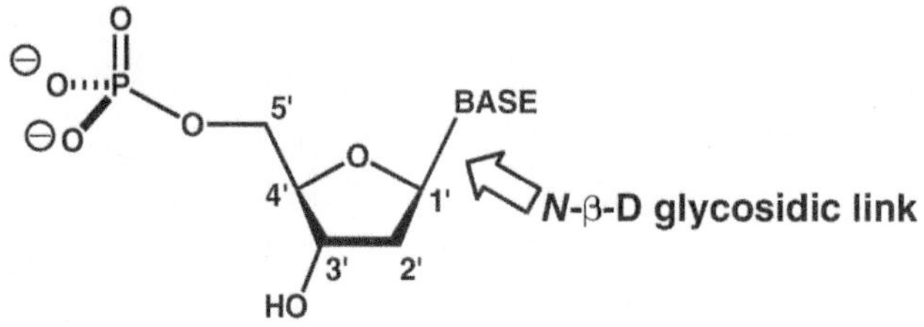

Figure 1.51   2′-Deoxyribonucleotide structure.

Adenine   Guanine   Cytosine   Thymine

Figure 1.52   Bicyclic purine and pyrimidine base structures. Illustration of main bases found in DNA.

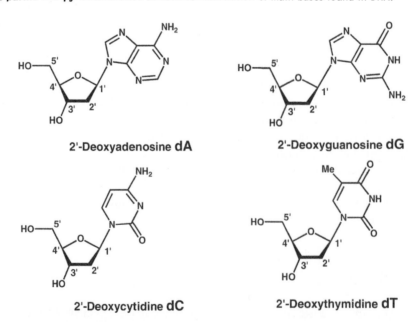

2′-Deoxyadenosine dA   2′-Deoxyguanosine dG

2′-Deoxycytidine dC   2′-Deoxythymidine dT

Figure 1.53   Structures of four main 2′-deoxyribonucleosides of DNA. A 2′-deoxyribonucleoside can also be referred to as a **deoxyribonucleoside** or as a **deoxynucleoside**.

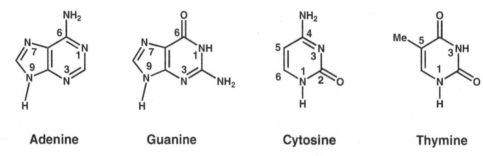

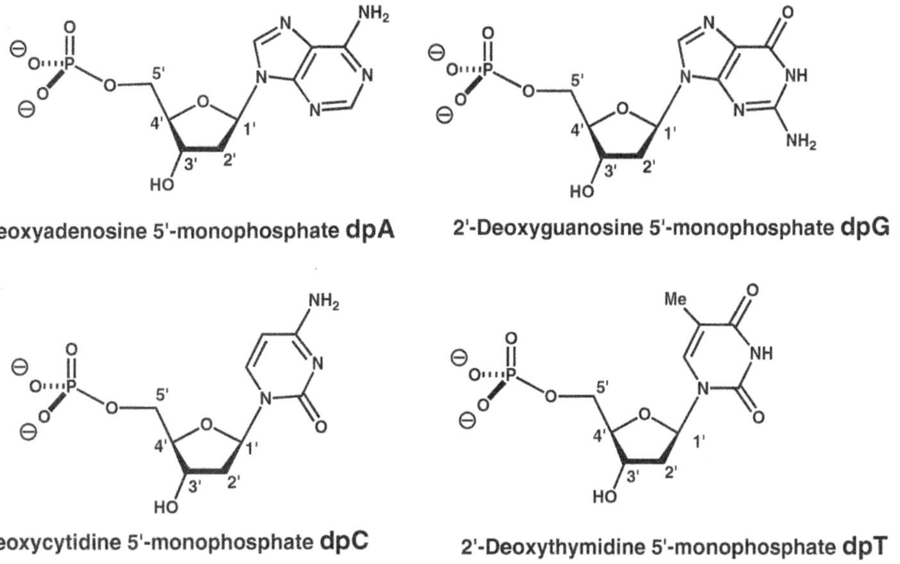

Figure 1.54 Structures of four main 2′-deoxyribonucleoside 5′-monophosphates of DNA. These 2′-deoxyribonucleoside 5′-monophosphates may be referred to as **2′-deoxyribonucleotides, deoxyribonucleotides** or as **deoxynucleotides.**

The combination of 2′-deoxy-β-D-ribofuranose and *N*-linked base alone is known as a **2′-deoxyribonucleoside** (or **deoxynucleoside**) (**dN**); the four DNA deoxynucleosides are known as **2′-deoxyadenosine** (**dA**), **2′-deoxyguanosine** (**dG**), **2′-deoxycytidine** (**dC**) and **2′-deoxythymidine** (**dT**) (Figure 1.53). Hence, the principle deoxynucleotide building blocks from which DNA is constructed are known as **2′-deoxyadenosine 5′-monophosphate** (**dpA**), **2′-deoxyguanosine 5′-monophosphate** (**dpG**), **2′-deoxycytidine 5′-monophosphate** (**dpC**), and **2′-deoxythymidine 5′-monophosphate** (**dpT**) (Figure 1.54).

RNA is similarly a mixed polymer constructed substantially from **ribonucleotide** (or **nucleotide**) building blocks that are very similar to the four monomeric deoxynucleotide building blocks of DNA. The major difference being that **β-D-ribofuranose** is used in place of 2′-deoxy-β-D-ribofuranose (Figure 1.55). In addition, RNA molecules are frequently constructed from an expanded set of nucleotide building blocks. By analogy to DNA, RNA is constructed from four central nucleotide building blocks, namely **adenosine 5′-monophosphate** (**pA**), **guanosine 5′-monophosphate** (**pG**), **cytidine 5′-monophosphate** (**pC**) and **uridine 5′-monophosphate** (**pU**) (Figure 1.56), but this set is also often supplemented by alternative nucleotide building blocks such as **thymidine 5′-monophosphate** (**pT**), **1-methyladenosine 5′-monophosphate** (**pm¹A**), **7-methylguanosine 5′-monophosphate** (**pm⁷G**), **2-N, N-dimethylguanosine 5′-monophosphate** (**pm²₂G**), **5-methylcytidine 5′-monophosphate** (**pm⁵C**) and even **pseudouridine 5′-monophosphate** (**pψ**) (Figure 1.57). The expanded set of nucleotide building blocks used to construct RNA has a significant impact upon the structural diversity of RNA molecules in comparison to DNA.

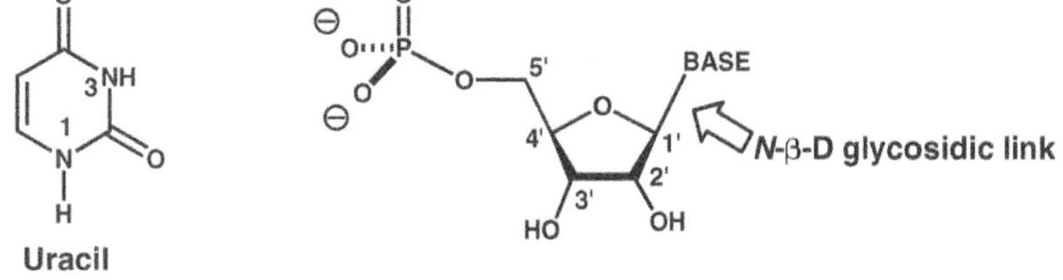

Figure 1.55 Ribonucleotide structure. The structure of the RNA specific pyrimidine base **uracil** is also illustrated.

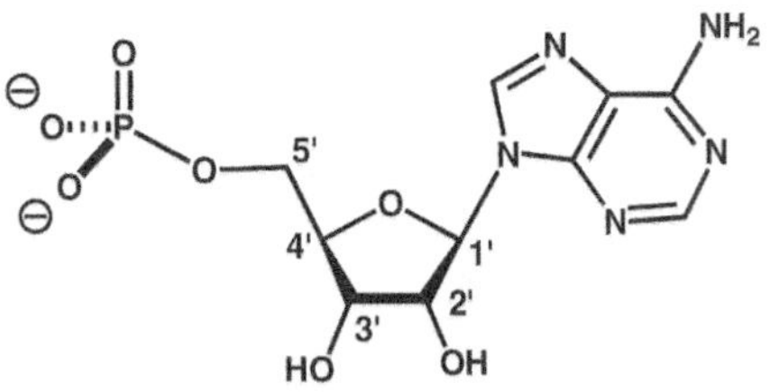

Adenosine 5'-monophosphate pA

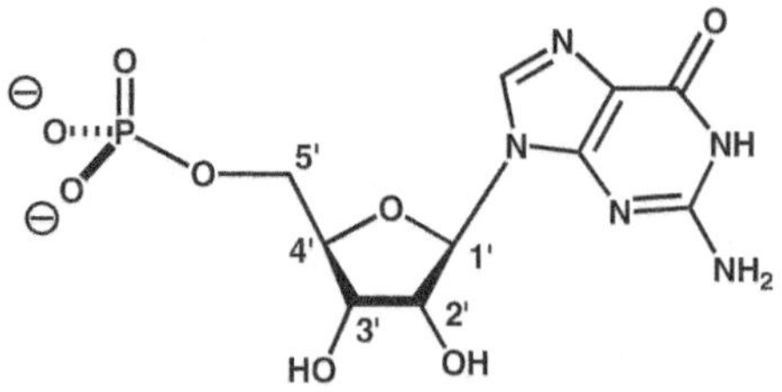

Guanosine 5'-monophosphate pG

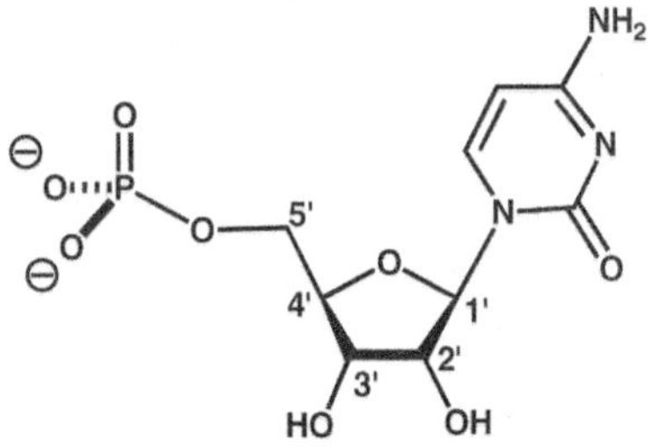

Cytidine 5'-monophosphate pC

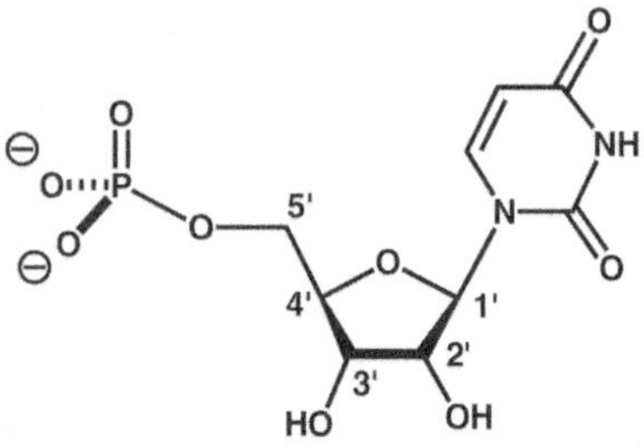

Uridine 5'-monophosphate pU

Figure 1.56   Structures of four main ribonucleoside 5'-monophosphates of RNA. A ribonucleoside 5'-monophosphate can also be referred to as a **ribonucleotide** or **nucleotide**.

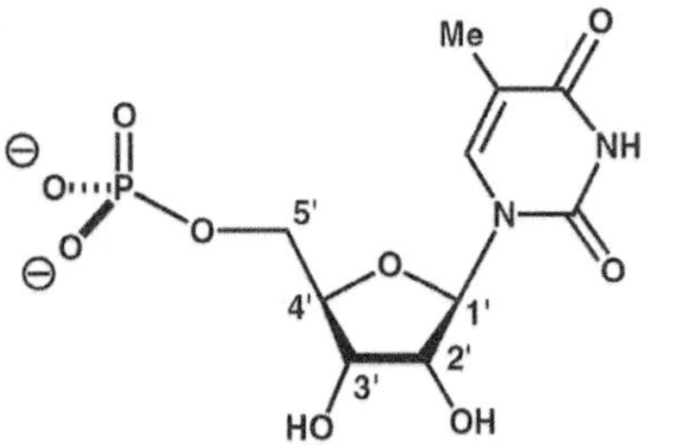

Thymidine 5'-monophosphate pT

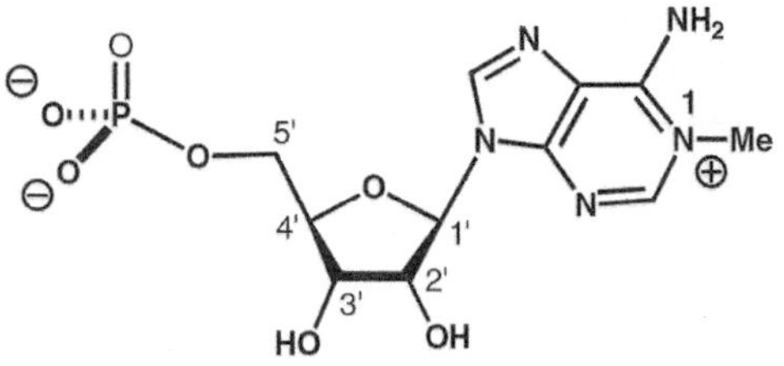

1-Methyladenosine 5'-monophosphate pm^{1}A

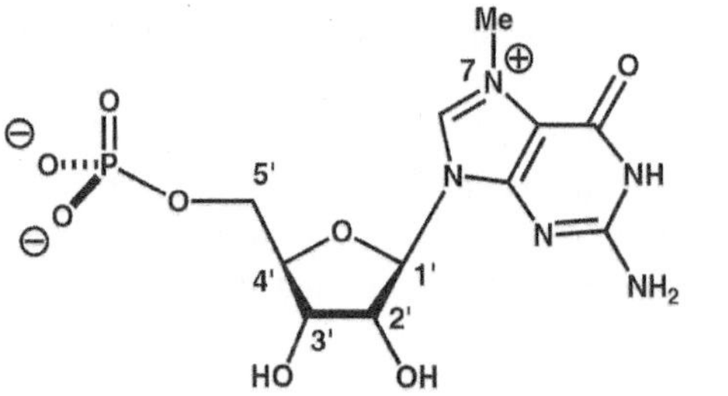

7-Methylguanosine 5'-monophosphate pm^{7}G

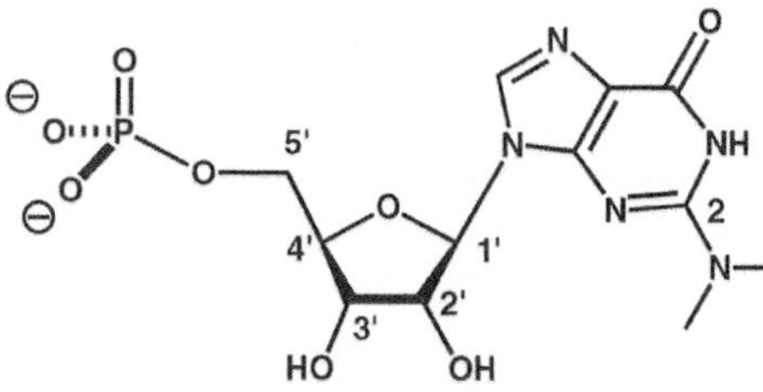

2-N,N-Dimethylguanosine 5'-monophosphate pm$^2_2$G

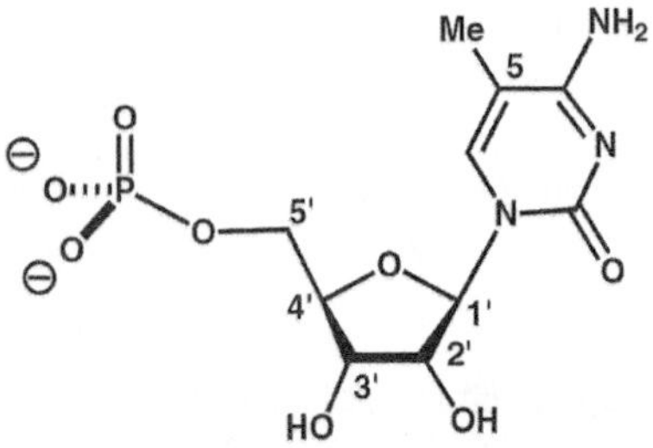

5-Methylcytidine 5'-monophosphate pm^{5}C

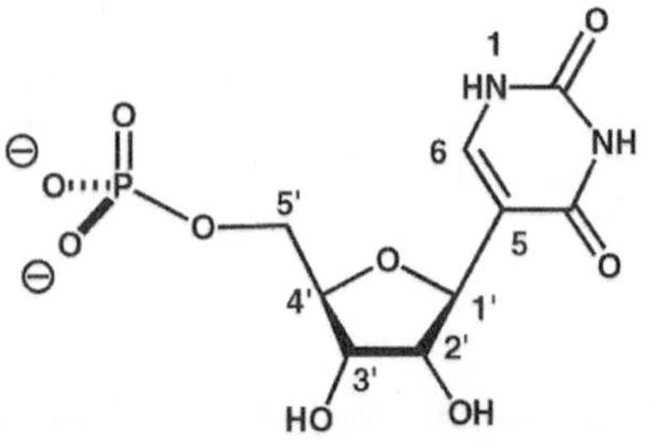

Pseudouridine 5'-monophosphate pψ

Figure 1.57   Structures of less common ribonucleoside 5'-monophosphates of RNA. These less common **ribonucleoside 5'-monophosphates** (also known as ribonucleotides or nucleotides) are equally important for the formation of complex RNA structures.

1.4.2 Phosphodiester link

The deoxynucleotide building blocks of DNA are joined together by **phosphodiester links** (Figure 1.58). Unbranched chains of deoxynucleotide units are formed with ease. A short chain (2–20 units) is known as an **oligodeoxynucleotide**, whilst a long chain (upwards of 20 units) is called a **polydeoxynucleotide**. Each constituent deoxynucleotide unit of DNA is usually called a **deoxynucleotide residue**. The same phosphodiester link in RNA also appears to generate equivalent unbranched chains of nucleotide units, known individually as **nucleotide residues**. By analogy to DNA, a short chain of RNA (2–20 units) is known as an **oligonucleotide,** whilst a long chain of RNA (upwards of 20 units) is called a **polynucleotide**.

In DNA and RNA, the chain of phosphodiester links and sugar rings is known as the **phosphodiester backbone**; the bases may be regarded in both cases almost as "side chains". By convention, DNA or RNA chains begin at the 5′-end (i.e. where carbon atom C-5′ of the terminal residue is not involved in a phosphodiester link) and terminate at the 3′-end (where carbon atom C-3′ is not involved in a phosphodiester link). Each chain is therefore said to run by convention from 5′ to 3′ (5′→3′). Several shorthand conventions are used to describe the sequences of deoxynucleotide or nucleotide residues in DNA and RNA respectively. These include the Fischer, linear alphabetic and condensed alphabetic conventions that draw upon the letter codes for deoxynucleosides or nucleosides as described previously (Figure 1.59).

The phosphodiester link in DNA and RNA has some characteristics in common with the peptide link. Like the peptide link, the phosphodiester link shows considerable conformational rigidity. The link may be thought of as consisting of two atomic segments C$^{4'}$-C$^{3'}$-O$^{3'}$-P (blue) and P-O^{5}-C$^{5'}$-C$^{4'}$ (red) (Figure 1.60). Each segment acts as a rigid, coplanar unit that behaves as a single bond. The phosphodiester link therefore consists of two virtual bonds that pivot about phosphorous. The spatial relation-

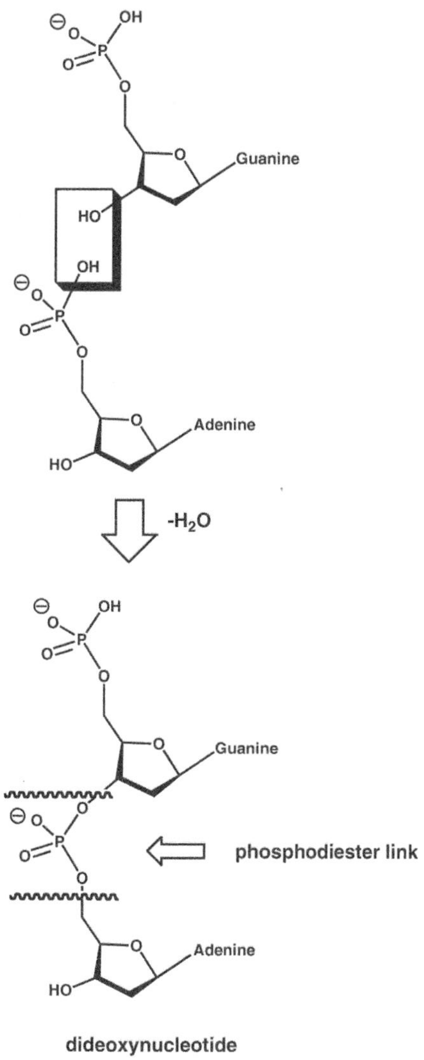

Figure 1.58 Schematic illustration showing formation of a phosphodiester link. These links are formed between deoxynucleotides in order to create DNA. Identical links can be formed between nucleotides in order to create RNA.

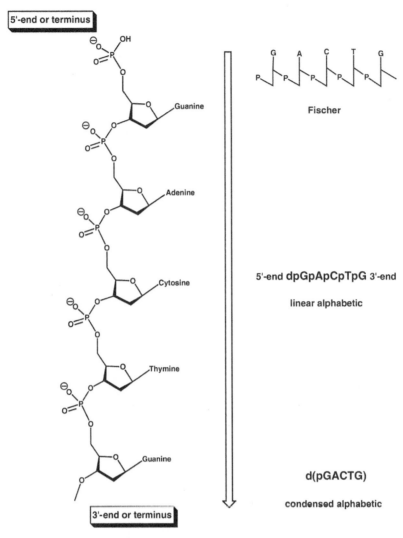

Figure 1.59 Structural illustrations of a polydeoxynucleotide. A full structure is shown (left) and various shorthand structures are shown (right) commonly used to define polydeoxynucleotide structures in more simple ways. Similar shorthand structures exist for polynucleotide structures, except that there is no need "**d**", for "deoxy", to be used.

ship between the virtual bonds which make up each phosphodiester link are defined by conformational angles ζ and α which are the main dihedral angles about the **O3'-P** and **P-O5'** bonds respectively. Both angles ζ and α are approximately +300° (-60°) under most circumstances, causing the phosphodiester link to occupy a **gauche conformation** (Figures 1.61 and 1.62). Favourable anti-periplanar interactions of oxygen atom O-3' lone pairs with the P-O5' bond matched by a similar interaction of oxygen atom O-5' lone pairs with the P-O3' bond are thought to promote this conformation. In addition, the conformational preference for all deoxynucleotide (and nucleotide) residues to adopt the **synclinal (+sc) conformation** in preference to the main **anti-periplanar (ap) conformation** alternative helps to "fix" conformational properties further in DNA and RNA (Figure 1.63).

1.4.3 Secondary structure of DNA

This is where we meet the famous double helix immortalised by many books, articles, television programmes and of course films. The DNA **double helix** is the key element of DNA 3D structure. In the 1950s, when James Watson and Francis Crick first proposed the double helix as the key piece of DNA 3D structure, they generated enormous scientific and popular excitement, since, for the first time, the inheritance of genetic information could be understood explicitly in terms of a real chemical structure. In order to appreciate this structure, there is a requirement to understand more about the heterocyclic bases (see Section 1.4.1) and their unrivalled capacity for specific hydrogen bonding. All these bases are aromatic, but paradoxically prefer keto/amine to enol/imine tautomeric forms (Figure 1.64). The four bases found in DNA show a remarkable pairwise complementarity with respect to each other. Guanine is able to supply two hydrogen

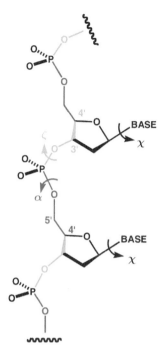

Figure 1.60 **Conformational restrictions in phosphodiester links**. Illustration of polydeoxynucleotide chain highlighting the two atomic segments **C⁴'-C³'-O³'-P** (light blue) and **P-O⁵'-C⁵'-C⁴'** (red). Each segment acts as a rigid, coplanar unit; hence, behaves as a **virtual bond** pivoting at phosphate. Key dihedral angles involved that characterise conformation are ζ, the angle subtended about the **O³'-P** bond and α, the angle subtended about the **P-O⁵'** bond. The same arguments apply in polynucleotide chains too.

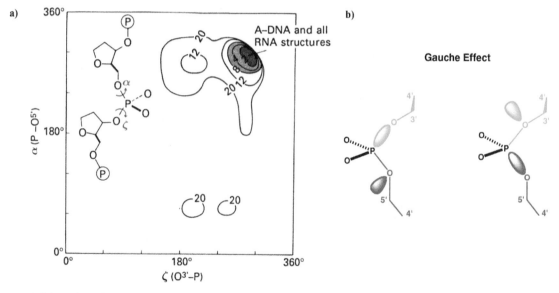

Figure 1.61 **Origins of conformational restrictions in phosphodiester links**. Conformational freedom is heavily restricted in phosphodiester links in particular and nucleic acids in general, owing to lack of free rotation about O³'-P and P-O⁵' bonds. (**a**) Nucleic acid equivalent of the Ramachandran plot illustrating the theoretically allowed angles of ζ and α. (**b**) Free rotation is primarily damped owing to the gauche effect in which lone-pair-σ* orbital overlaps in phosphodiester links generate double bond character in O-P bonds that restricts free rotation. (adapted from Govil 1976, Wiley).

bond donor groups and one acceptor group to complement the two hydrogen bond acceptor groups and one donor group of cytosine, while adenine is able to supply one hydrogen bond donor and one acceptor group to complement the single hydrogen bond acceptor and one donor group of thymine (Figure 1.65). Complementary hydrogen bonding gives rise to the specific **Watson-Crick base pairings, dG.dC** and **dA.dT**, which also show a remarkable **isomorphous geometry** (Figure 1.66). These pairings are not unique and there can be an impressive number of alternatives (Figures 1.67 and 1.68), but they dominate owing to preferable interactions and relative geometries.

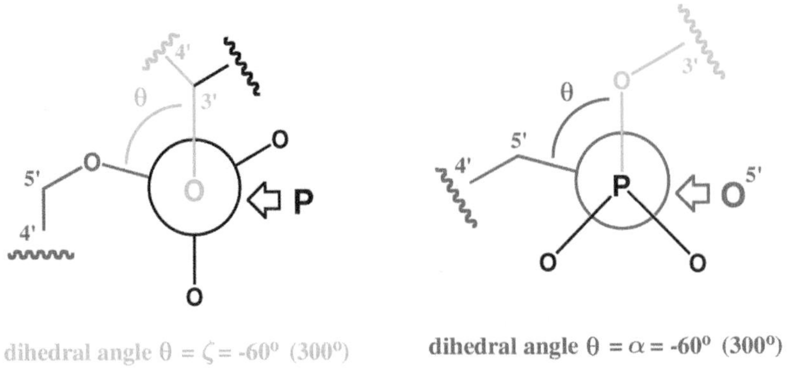

dihedral angle $\theta = \zeta = $ -60° (300°)   dihedral angle $\theta = \alpha = $ -60° (300°)

Figure 1.62   Newman projections for a phosphodiester link. These projections illustrate the dihedral angles subtended about the **O³'-P** and **P-O⁵'** bonds (ζ and α respectively) to demonstrate how such highly regular dihedral angles set up the highly extended phosphodiester backbone conformation. Bonds are colour coded in the same way as Figures 1.60 and 1.61.

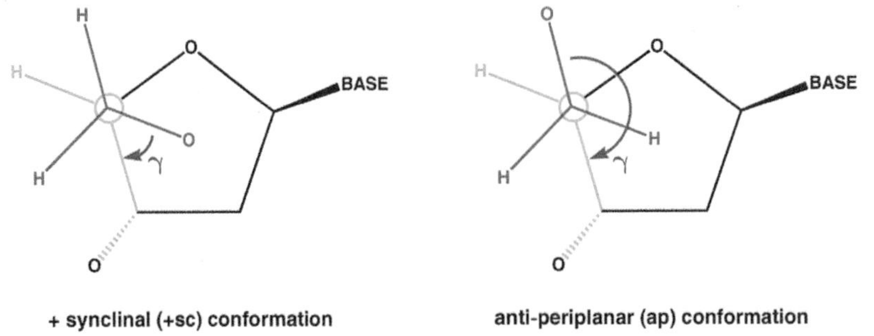

+ synclinal (+sc) conformation   anti-periplanar (ap) conformation

Figure 1.63   Additional conformational preferences in the phosphodiester backbone. Conformational preferences are shown for the conformational angle γ subtended about the **C⁴'-C⁵'** bond. Each deoxynucleotide (or nucleotide) residue adopts either a **+ sc** or **ap** conformation. The former is usually preferred but for exceptional circumstances.

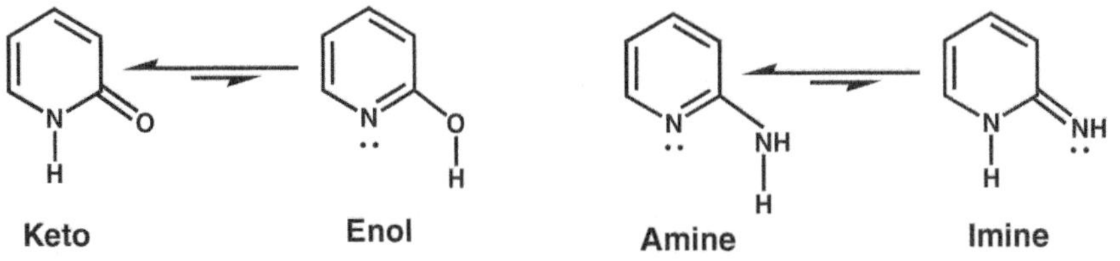

Keto   Enol   Amine   Imine

Figure 1.64   Illustration of the keto-enol and amine-imine tautomeric equilibria.

The chemical biology reader should also be aware that the natural conformational preference of a given polydeoxynucleotide chain (see Section 1.4.2) is to exist in an extended conformation with heterocyclic bases presented in an **anti-conformation** projecting away from the attached 2'-deoxy-β-D-ribofuranose ring (Figure 1.54), in preference to the **syn-conformation**. In the *anti*-conformation specific Watson-Crick base pairings between bases in two independent polydeoxynucleotide chains can be achieved with ease. The optimal arrangement for hydrogen bonding between two independent polydeoxynucleotide chains is for each chain to align anti-parallel with respect to the other (i.e. one chain is orientated in the 5'→3' direction and the other in the 3'→5' direction) and for the two chains to coil around each other in such a way that the backbone of each chain forms a right-handed helix. However, this arrangement is inherently unstable, given the high charge of the phosphodiester backbone, unless every deoxynucleotide residue in one polydeoxynucleotide chain is able to enter into a specific Watson-Crick base pairing arrangement with a deoxynucleotide residue in the other polydeoxynucleotide chain, thereby "cementing" the chains together by a series of specific inter-chain **base pair (bp)** hydrogen bond interactions. In other words, both polydeoxynucleotide chains must be completely complementary to each other in terms of their capacity to form complete Watson-Crick base pairings otherwise the structure will not be stable. The need for such complete base pairing also ensures that both polydeoxynucleotide chains are completely complementary to each other in terms of their deoxynucleotide residue sequence and base composition as well. In these specific Watson-Crick base pairings lies the foundations of the genetic code and the inheritance of genetic information

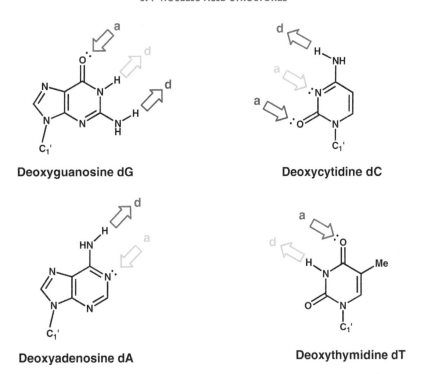

Figure 1.65 Illustration of the matched hydrogen bonding possible between bases. There are clear acceptor (**a**) or donor (**d**) hydrogen-bonding relationships that exist between the deoxynucleoside bases of **deoxyadenosine (dA)** and **deoxythymidine (dT)**, and the deoxynucleoside bases of **deoxyguanosine (dG)** and **deoxycytidine (dC)**.

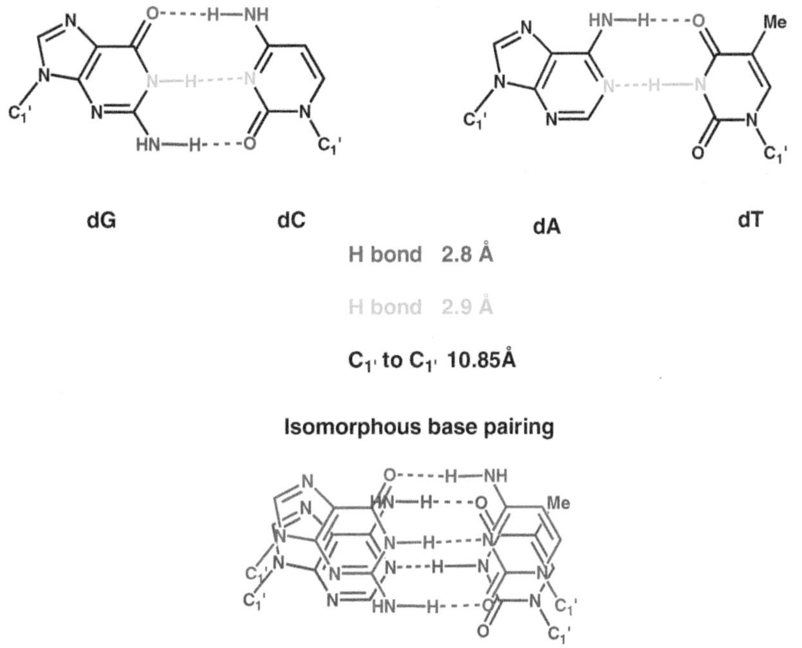

Figure 1.66 Illustrations of specific Watson-Crick base pairings. The specific base-pairings **dG.dC** and **dA.dT** are made possible by highly complementary hydrogen-bonding relationships between complementary deoxynucleoside bases. Overlay structure provides visual demonstration of the **dG.dC/dA.dT isomorphous geometry.**

(see Section 1.4.6). Nevertheless, even given these strict requirements, there is a significant amount of plasticity in the DNA double helix structure that may be essential for a variety of important biological reasons. This means that there can be a number of DNA double helical subtypes. Note that complete Watson-Crick base pairing also creates the possibility for extended π-π **stacking interactions** between sequential base pairs. Such weak covalent interactions doubtless contribute significantly towards DNA double helix stability in addition to Watson-Crick bp hydrogen bonds.

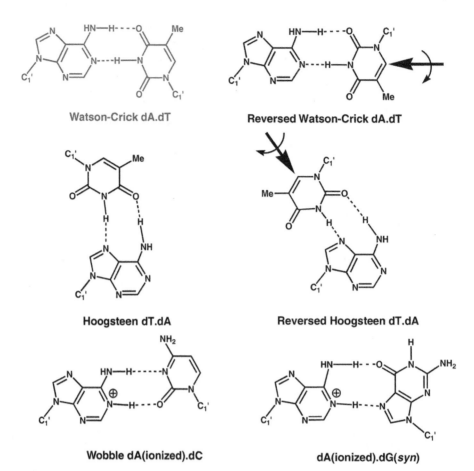

Figure 1.67 Structures of non-Watson-Crick base pairings involving deoxyadenosine. Original **dA.dT** Watson-Crick base pairing is shown (**top left**) for comparison.

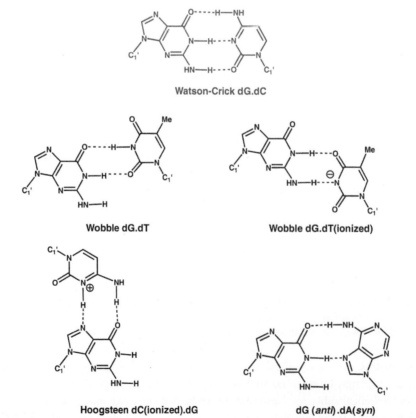

Figure 1.68 Structures of non-Watson-Crick base pairings involving deoxyguanosine. Original **dG.dC** Watson-Crick base pairing is shown (**blue, top**) for comparison.

1.4.3.1 B-form DNA

B-form DNA is the most common form of DNA in solution and is much the most important biologically speaking. The double helical conformation of B-form DNA also conforms most closely with the original model structure for DNA devised by Watson and Crick. The main architectural features of the B-form DNA double helix are illustrated using a **ribbon display**, **rings display** and **ladder display structures** (Figures 1.69 and 1.70). Backbones are shown as ribbons and inter-chain interacting base pairs as either rings or rods perpendicular to the helical axis. There are two structural grooves that corkscrew along the length of the double helix, the wider of which is called the **major groove** and the narrower the **minor groove** (Figure 1.69). The helix and bp parameters of B-form DNA are also summarised (Table 1.2). The helix sense of B-form DNA is right-handed, there are 11.5 bp residues per turn, and the 2′-deoxy-β-D-ribofuranose rings do not adopt the characteristic envelope conformation of furanose rings, but instead exist in a **C2′ endo** twist conformation (Figure 1.71).

In the original model structure for DNA devised by Watson and Crick, base pairs are coplanar and bp planes stack perpendicular to the main helix axis. In reality whilst this is essentially true, there are always distortions away from

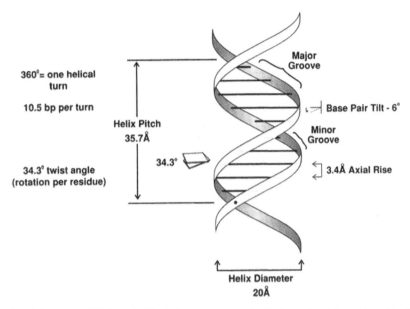

Figure 1.69 **Main structural features of B-form DNA.** A short segment of DNA is depicted in **Ribbon display** format to show ideal dimensions and angles (illustration from Sinden, 1994, Figure 1.12, Elsevier).

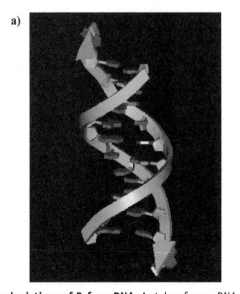 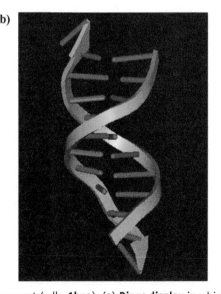

Figure 1.70 **Two depictions of B-form DNA.** As taken from a DNA segment (pdb: **1bna**). (**a**) **Rings display** in which the anti-parallel phosphodiester backbones are shown as arrowed ribbons (**green**); ribose rings (**green**), purine bases (**red**) and pyrimidine (**blue**) bases are shown in structural outline. (**b**) **Ladder display** in which the anti-parallel phosphodiester backbones are shown as arrowed ribbons (**green**); ribose rings are omitted; purine (**red**) and pyrimidine (**blue**) bases are shown as rods.

Table 1.2 **Summary of all the main differences between B-, A- and Z-form DNA.** This in terms of dimensions, angles and conformations.

Parameter	A-DNA	B-DNA	Z-DNA
Helix sense	Right	Right	Left
Residue per turn	11	10 (10.5)	12
Axial rise (Å)	2.55	3.4	3.7
Helix pitch (°)	28	34	45
Base pair tilt (°)	20	-6	7
Rotation per residue (°)	33	36 (34.3)	-30
Diameter of helix (Å)	23	20	18
Glycosidic bond			
dA, dT, dC	*anti*	*anti*	*anti*
dG	*anti*	*anti*	*syn*
Sugar pucker			
dA, dT, dC	C3′ endo	C2′ endo	C2′ endo
dG	C3′ endo	C2′ endo	C3′ endo
Phosphate-phosphate (Å)			
dA, dT, dC	5.9	7.0	7.0
dG	5.9	7.0	5.9

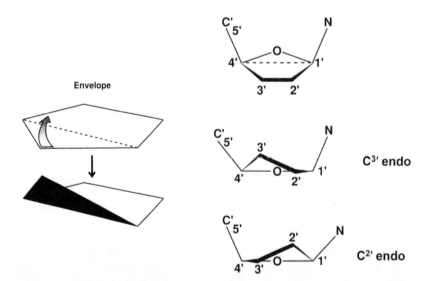

Figure 1.71 Conformational characteristics of 2′-deoxy-β-D-ribofuranose rings. The main conformational preferences in poly-deoxynucleotides of these rings are shown (right). Cartoon illustrations of these main envelope conformations are also shown (left) (illustrations from Sinden, 1994, Figure 1.4, Elsevier).

such ideality. A number of parameters are frequently used as a guide to define these distortions. These parameters include: a description of **helix sense**, whether right-handed or left-handed; **helix diameter** (in Å); **helix pitch** (P), the length of one complete helical turn (in Å); the number of **residues per turn**; **axial** or **average rise** (D_z), the distance between adjacent bp planes (in Å); **bp twist** or **helix rotation** (Ω), the angle of rotation between adjacent bp planes; and **bp tilt** (τ), the angle which a bp plane makes relative to a line drawn perpendicular to the main helix axis (Figure 1.72). In addition, there are other bp parameters that are used to describe more heavily distorted DNA. These are **bp roll** (ρ), the average angle which adjacent base pairs make relative to the main helical axis, and **bp propellor twist** (ω), the angle between the planes of the two paired bases of a bp when the bp is no longer coplanar (Figure 1.72).

1.4.3.2 A-form and Z-form DNA

Other than B-form DNA, there are other forms of DNA that are more distorted from Watson and Crick ideality. These include **A-form**, **C-form**, **D-form** and **T-form DNA**. Of these, **A-form DNA** is the most similar to B-form DNA (Figure 1.73). The only major difference is that the 2′-deoxy-β-D-ribofuranose rings of A-form DNA exist in an

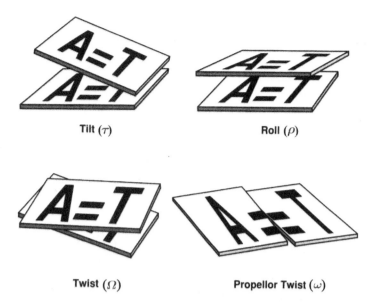

Figure 1.72 Cartoon depictions of bp flexibilities. The main forms of relative bp flexibility are depicted that allow DNA structures to deviate from ideal dimensions and angles (illustrations from Sinden, 1994, Figure 1.13, Elsevier).

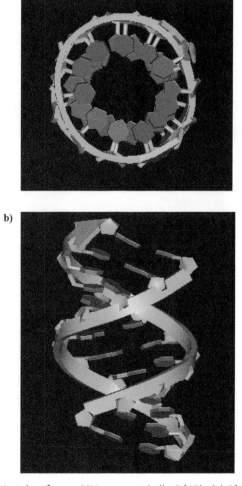

Figure 1.73 Depiction of A-form DNA. As taken from a DNA segment (pdb: **2d47**). (**a**) **Rings display** (top view) in which the anti-parallel phosphodiester backbones are shown as arrowed ribbons (**green**); ribose rings (**green**), purine bases (**red**) and pyrimidine (**blue**); (**b**) **Rings display** (side view).

alternative **C3′ endo** twist conformation (Figures 1.71 and 1.73). Consequently, the helix diameter is wider and there are 11 bp residues per turn (Table 1.2). A-form DNA also possesses shallow major and minor grooves. **Z-form DNA**, by contrast, is radically different from B-form DNA (Table 1.2) (Figure 1.74). The helix sense of Z-form DNA is now left-handed and the backbones of the two polydeoxynucleotide chains map out a zig-zag spiral path. This results from the tendency of deoxyguanosine nucleotide residues to distort so that their 2′-deoxy-β-D-ribofuranose rings adopt a C3′ endo instead of C2′ endo twist conformation (Figure 1.71). But in addition, the guanine bases move to a *syn*-conformation (positioned over the top of their attached furanose rings), instead of the more usual *anti*-conformation (Figure 1.75), and deoxyguanosine nucleotide residues also adjust to adopt an anti-periplanar (**ap**) conformation in preference to the more usual synclinal (**+sc**) conformation (Figure 1.63). A-form and Z-form DNA are not thought to be biologically important except as minor components of otherwise B-form DNA. However, both A-form and Z-form may help assist the formation of DNA supercoiling structures such as those illustrated (Figure 1.76).

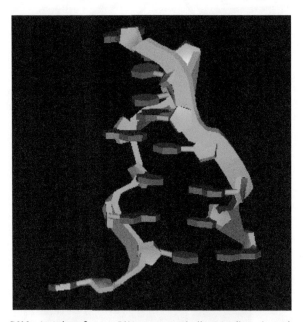

Figure 1.74   Depiction of Z-form DNA. As taken from a DNA segment (pdb: **331d**), using **Rings display** (side view) in which the anti-parallel phosphodiester backbones are shown as arrowed ribbons (**green**); ribose rings (**green**), purine bases (**red**) and pyrimidine (**blue**). This side view demonstrates how the phosphodiester backbone now takes on a "zig-zag" appearance, hence the name of this DNA conformation. The double helix is now left-handed.

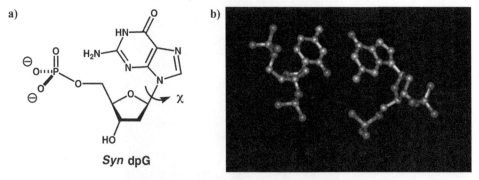

Figure 1.75   Conformational characteristics in Z-form DNA. These result from wholesale deoxynucleotide conformational angle changes in deoxyguanosine nucleotide (**dpG**) residues. The more usual *anti*, +sc, C2′ endo nucleotide conformation gives way to a *syn*, ap, C3′ endo conformation. **(a)** Illustration of *syn* dpG conformation. **(b) Ball and stick** depiction of **pdG**(*syn,* ap, C3′ endo). **pdC** bp as found in Z-form DNA.

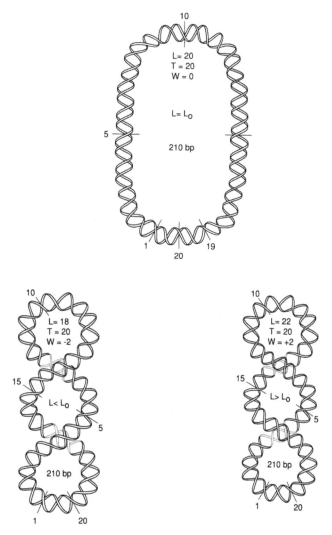

Figure 1.76 Schematic illustrations of covalently closed circular DNA. Such DNA can exist in an open conformation (top) but more usually in a **negative supercoiled** conformation (bottom left) or **positive supercoiled** conformation (bottom right) (illustration adapted from Sinden, 1994, Figure 3.4, Elsevier).

1.4.4 Supercoiling and tertiary structures of DNA

DNA supercoiling provides conformational potential energy for DNA tertiary structure formation such as the development of **DNA cruciform structures** (Figure 1.77). Supercoiling also leads to the creation of **DNA triple helix** (**DNA triplex**) structures that form when an oligodeoxynucleotide chain, with an appropriately complementary deoxynucleotide residue sequence, becomes associated with the major groove of B-form DNA double helix. This is shown using ribbon cartoons (Figure 1.78). Since Watson-Crick base pairings are already involved within the double helix, some alternative base pairings are needed to bind the oligodeoxynucleotide chain. These are known as **tertiary base pairings** or **Hoogsteen structures**. There are four basic stable Hoogsteen structures involved which by convention are described as *dT*.dA.dT, *dC$^+$*.dG.dC, *dA*.dA.dT and *dG*.dG.dC (Figure 1.79). The letter codes in normal type correspond to normal Watson-Crick base pairings, the italic letter code to a tertiary deoxynucleoside in each case. The formation of moderately stable DNA triplex requires that all the bases of the deoxynucleotide chain form correct tertiary base pairings with all the base pairs of the double helix where association is being made. In other words, the oligodeoxynucleotide chain must be complementary with the double helix base pairs with respect to the optimal tertiary base pairings. The existence of these DNA tertiary structures is now well established, but their biological utilities remain mysterious even today. Such tertiary structures may have important functions in the control of gene expression either in positive or negative ways, linking back to comments made in Section 1.1. Evidence remains to be gathered, but this subject is certain to remain of significant interest to chemical biology researchers for a significant time to come.

a)

Inverted Repeat   5' GGAATCGATCTTAAGATCGATTCC 3'
                  3' CCTTAGCTAGAATTCTAGCTAAGG 5'

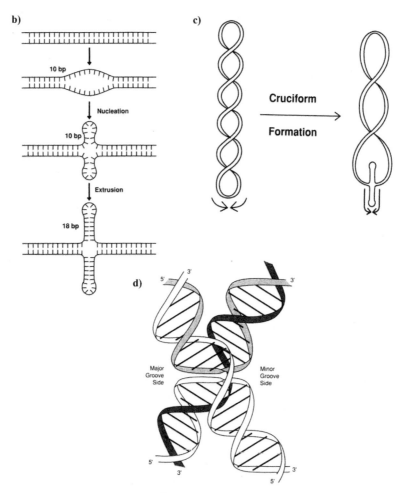

Figure 1.77   **Impact of supercoiling on formation of additional tertiary structure elements in covalently closed circular DNA.** (**a**) Double helical deoxynucleotide **palindrome sequence** (**inverted repeat**) is a prerequisite for **cruciform** formation; (**b**) Schematic diagram to show cruciform formation under conformational pressure of supercoiling as shown in (**c**); (**d**) More detailed ribbon cartoon to illustrate how phosphodiester backbones are "shared" at the cruciform junction (illustrations adapted from Sinden, 1994, Figures 4.1, 4.3, 4.5 and 4.17 respectively, Elsevier).

1.4.5 Secondary and tertiary structures of RNA

Within RNA, the Watson-Crick base pairing equivalents are **G.C** and **A.U** (Figure 1.80) isomorphous with the standard DNA Watson-Crick base pairings (Figure 1.66). Also, the natural conformational preferences of polynucleotide chains are equivalent to polydeoxynucleotide chains. Therefore, RNA may form double helical structures in the same way as DNA. However, owing to the presence of the C2'-hydroxyl functional group in each β-D-ribofuranose ring, polynucleotide chains appear unable to generate sustained double-helical structures and appear able to form only fragmentary A-form or A'-form double helices comprising a few short turns (Table 1.2) (Figure 1.81). The RNA equivalent B-form double helix does not exist, mainly because β-D-ribofuranose rings of oligo and polynucleotides are constrained to adopt a C3' endo conformation that automatically favours A-form type double helices (Table 1.2). As a result, individual mRNA molecules comprised of pG, pC, pA and pU nucleotide residues frequently generate complex, unique tertiary structures consisting of A or A'-form helical regions separated by loops and extensions lacking Watson-Crick base pairing. These mRNA structures could almost be said to rival proteins in their 3D complexity.

The structure of tRNA molecules is a peak of RNA structural variation. All tRNA molecules possess a similar overall tertiary structure known as the **clover-leaf structure** (Figure 1.81). Within the clover leaf, there are secondary structure elements comprised of A or A'-form double helical regions involving RNA equivalent Watson-Crick base pairings (Figure 1.80).

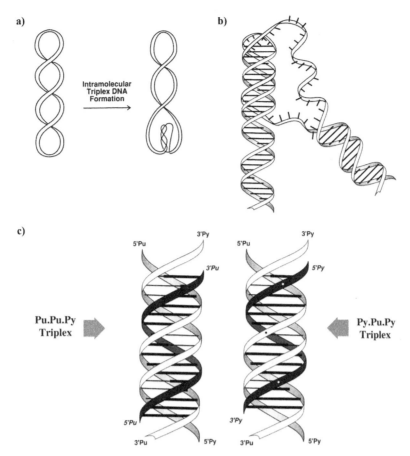

Figure 1.78 **Impact of supercoiling on formation of triplex DNA tertiary structural elements in covalently closed circular DNA.** (**a**) Schematic diagram to show triplex formation under conformational pressure of supercoiling; (**b**) Ribbon cartoon to show how triplex DNA forms after local DNA strand separation occurs; (**c**) More detailed ribbon cartoon to illustrate DNA phosphodiester backbone arrangements in two main types of **DNA triplex.** In both cases triplex is formed when a short, liberated oligodeoxynucleotide chain with the appropriate complementary deoxynucleotide sequence inserts into the major groove of B-form DNA. These triplex types are named after the triplex bps involved in their formation and stabilisation (Figure 1.79) (illustrations adapted from Sinden, 1994, Figures 6.6, 6.4 and 6.1 respectively, Elsevier).

These are separated by loops and bulges comprised substantially of representatives from the expanded set of RNA nucleotide building blocks such as pm^1A, pm^7G, pdm^2G, pm^5C and $p\psi$, which were described previously (Figure 1.57). Furthermore, RNA equivalent Watson-Crick base pairings are also supplemented by base pairings involving the less common nucleotide residues such as pm^1A, pm^7G, pdm^2G, pm^5C, and $p\psi$ (Figure 1.57), together with occasional tertiary base pairings (Figure 1.79). Even RNA equivalent non-Watson-Crick base pairings such as the **G.U Wobble base pairing** may be found (Figure 1.82). Perhaps the main reason that RNA is capable of such structural diversity in comparison with DNA is that numerous stabilising hydrogen-bonding interactions can take place between C2'-hydroxyl groups of the polynucleotide phosphodiester backbone and bases. These interactions presumably act to stabilise tertiary structures at the expense of secondary structural double helical elements.

1.4.6 The genetic code and structure

A general appreciation of the relationship between DNA structure and protein structure, not to mention the functional interplay between the different types of DNA and RNA polymers, is a critical component of biology and essential background knowledge here. In spite of their apparent structural equivalence, the two anti-parallel polydeoxynucleotide chains or **strands** of DNA are not functionally equivalent in biology. By convention, one strand is called the **sense** or **+strand**, the other is called the **antisense**, **complementary** or **-strand**. Therefore, genes must consist of a sense and complementary strand too. Within each gene, the specific sequence of deoxynucleotide residues, identified by their component base nucleoside compositions, in the sense strand (read sequentially in the 5'→3' direction) is known as the **gene sequence** and it is this sequence that codes for a corresponding polypeptide sequence, after decoding with the assistance of the **genetic code**

Table 1.3 **The genetic code.** Two versions are shown, namely the sense strand DNA (**a**) and mRNA versions (**b**).

Sense strand DNA ⟹ mRNA ⟹ Protein

a)

First 2'-deoxynucleotide residue (5'-end)	Second 2'-deoxynucleotide residue				Third 2'-deoxynucleotide residue (3'-end)
	T	C	A	G	
T	TTT Phe	TCT Ser	TAT Tyr	TGT Cys	T
	TTC Phe	TCC Ser	TAC Tyr	TGC Cys	C
	TTA Leu	TCA Ser	TAA Stop	TGA Stop	A
	TTG Leu	TCG Ser	TAG Stop	TGG Trp	G
C	CTT Leu	CCT Pro	CAT His	CGT Agr	T
	CTC Leu	CCC Pro	CAC His	CGC Agr	C
	CTA Leu	CCA Pro	CAA Gin	CGA Agr	A
	CTG Leu	CCG Pro	CAG Gin	CGG Agr	G
A	ATT Ile	ACT Thr	AAT Asn	AGT Ser	T
	ATC Ile	ACC Thr	AAC Asn	AGC Ser	C
	ATA Ile	ACA Thr	AAA Lys	AGA Agr	A
	ATG Met	ACG Thr	AAG Lys	AGG Agr	G
G	GTT Val	GCT Ala	GAT Asp	GGT Gly	T
	GTC Val	GCC Ala	GAC Asp	GGC Gly	C
	GTA Val	GCA Ala	GAA Glu	GGA Gly	A
	GTG Val	GCG Ala	GAG Glu	GGG Gly	G

b)

First nucleotide residue (5'-end)	Second nucleotide residue				Third nucleotide residue (3'-end)
	U	C	A	G	
U	UUU Phe	UCU Ser	UAU Tyr	UGU Cys	U
	UUC Phe	UCC Ser	UAC Tyr	UGC Cys	C
	UUA Leu	UCA Ser	UAA Stop	UGA Stop	A
	UUG Leu	UCG Ser	UAG Stop	UGG Trp	G
C	CUU Leu	CCU Pro	CAU His	CGU Arg	U
	CUC Leu	CCC Pro	CAC His	CGC Arg	C
	CUA Leu	CCA Pro	CAA Gin	CGA Arg	A
	CUG Leu	CCG Pro	CAG Gin	CGG Arg	G
A	AUU Ile	ACU Thr	AAU Asn	AGU Ser	U
	AUC Ile	ACC Thr	AAC Asn	AGC Ser	C
	AUA Ile	ACA Thr	AAA Lys	AGA Arg	A
	AUG Met	ACG Thr	AAG Lys	AGG Arg	G
G	GUU Val	GCU Ala	GAU Asp	GGU Gly	U
	GUC Val	GCC Ala	GAC Asp	GGC Gly	C
	GUA Val	GCA Ala	GAA Glu	GGA Gly	A
	GUG Val	GCG Ala	GAG Glu	GGG Gly	G

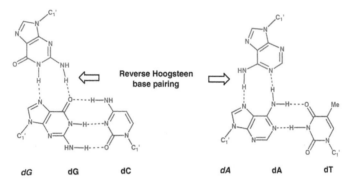

Figure 1.79 **Illustration of triplex base pairs that enable triplex DNA to form. (a)** Pyrimidine:Purine:Pyrimidine (Py:Pu:Py) Hoogsteen/Watson-Crick triplex base pairings can form between the indicated deoxynucleoside bases; **(b)** Purine:Purine:Pyrimidine (Pu:Pu:Py) Reverse Hoogsteen/Watson-Crick triplex base pairings. These can form between the indicated deoxynucleoside bases.

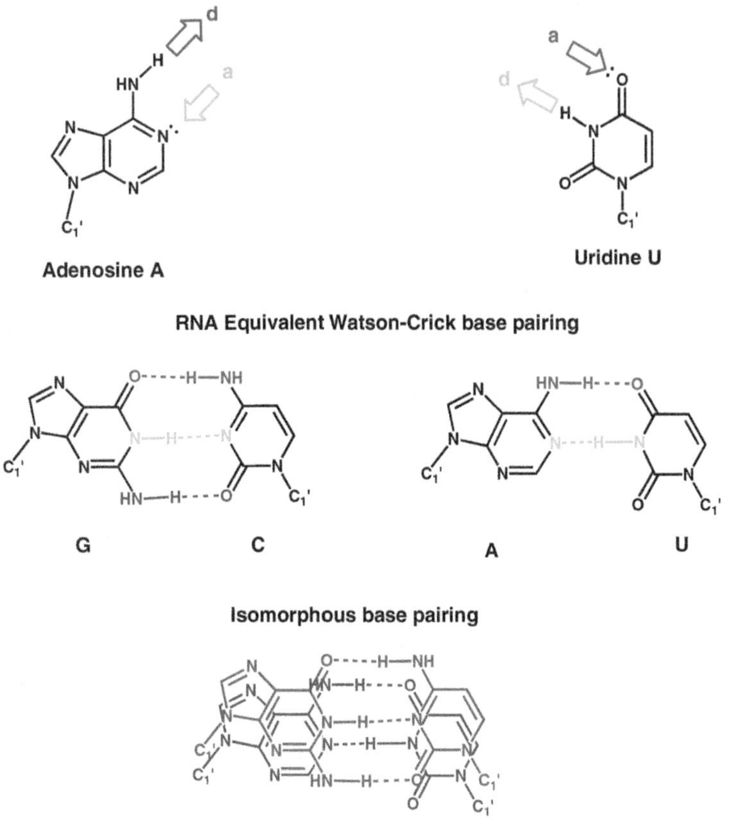

Figure 1.80 **Illustrations of the specific RNA equivalent Watson-Crick base pairings.** These are **G.C** and **A.U** which are made possible by highly complementary hydrogen-bonding relationships between complementary nucleotide residues. Overlay structure provides visual demonstration of the **G.C/A.U isomorphous geometry.**

3′ acceptor stem

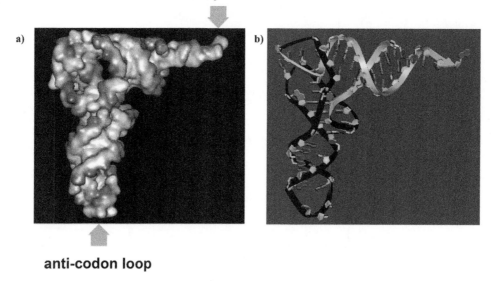

anti-codon loop

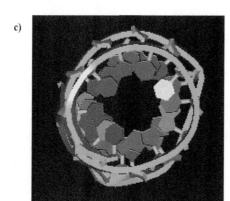

Figure 1.81 **Two depictions of transfer RNA.** A **transfer RNA (tRNA)** (pdb: **1tn2**) is illustrated. (**a**) **Surface display** in which the Van der Waals surface of all atoms is shown and negative charge (red areas). This display clearly demonstrates the bent shape of the molecule; (**b**) **Rings display** in which phosphodiester backbone is shown as an arrowed ribbon (**green/black**); ribose rings (**green**), purine bases (**red**) and pyrimidine (**blue**) bases are shown in structural outline; (**c**) **Rings display** of distorted **A-form RNA helix** that forms part of the acceptor stem of tRNA. Both **anti-codon loop** and **3′-acceptor stem** are illustrated, given their functional importance in **translation** (see Section 1.4.6).

(Table 1.3). According to the principles that underlie the genetic code, all gene sequences are divisible into sequential, non-overlapping groups of three successive deoxynucleotide residues, identified by their component base nucleoside compositions, known as **triplet codons** and each triplet codon codes for one amino acid residue in almost all cases. When the entire set of triplet codons that make up a specific gene sequence are read sequentially (5′→3′ direction) with the assistance of the genetic code from one end of a given gene to the other, then the primary structure (complete amino acid residue sequence) of a corresponding polypeptide/protein is revealed (N→C-terminus direction). In other words, specific linear gene sequences decode to reveal linear amino acid residue sequences. Crucially, mention was made earlier (Section 1.2.5) of the fact that the 3D structure of a polypeptide is determined by the amino acid residue sequence itself. Therefore, any given gene sequence by default specifies not only the primary structure of a polypeptide/protein but also the secondary and tertiary structure as well.

Needless to say, the actual overall process of decoding with the genetic code and subsequent conversion of gene sequences into polypeptide/proteins in cells is enormously complex and dynamic, involving many, many biological macromolecular "actors" such as enzymes (protein catalysts) and various binding proteins. However, the chemical biology reader needs to be aware of the overall process in order to appreciate properly the relationships between DNA and the various forms of RNA described here as mRNA, rRNA and tRNA. The whole process of decoding a sequence of deoxynucleotide residues requires initially that each specific gene sequence of interest should be "copied" from DNA into an mRNA form (Figure 1.83). This copying process is known as **transcription** and takes place in the nucleus, of cells with

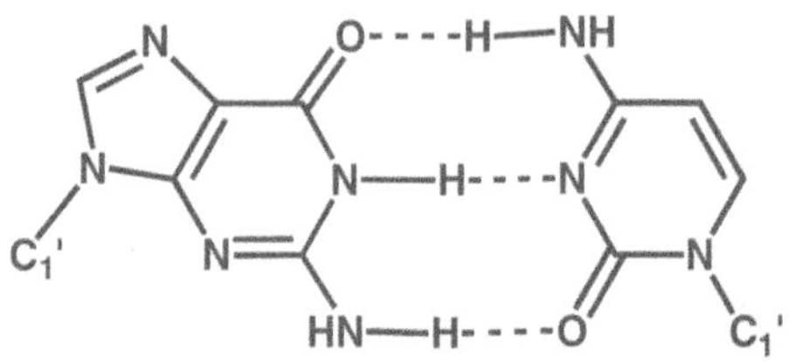

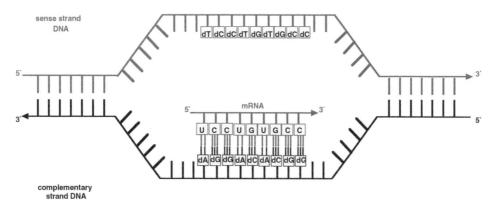

RNA equivalent Watson-Crick G.C

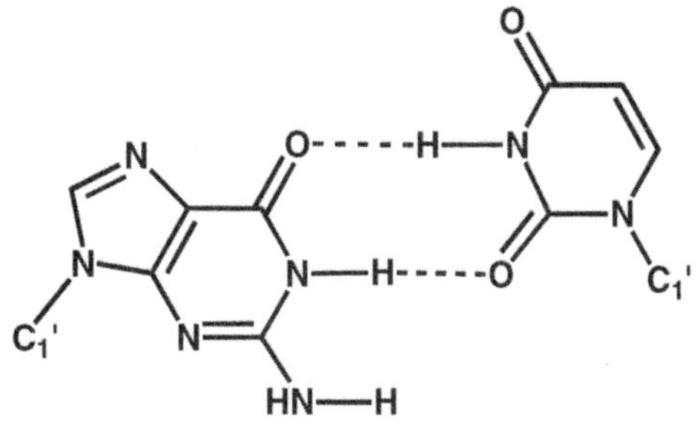

Wobble G.U

Figure 1.82   Structure of G.U, non-Watson-Crick base pairing. This involves the guanine base of guanosine and uracil base of uridine. Original **G.C** RNA equivalent Watson-Crick base pairing is shown (**blue, top**) for comparison.

Figure 1.83   Schematic illustration of a transcription bubble. The **complementary** (**antisense**) **strand** of DNA in the region of a gene acts as a template upon which to synthesise **messenger RNA (mRNA)** with the assistance of Watson-Crick base pairing. Watson-Crick base pairing ensures that the mRNA nucleotide residue sequence is the RNA equivalent of the deoxynucleotide residue sequence found in the sense strand of DNA. Hence coding information in sense strand DNA of genes found in DNA are smoothly transcribed into an mRNA form for translation into polypeptide sequences (see Figure 1.84).

nuclei. Transcription literally involves the faithful rendering of the specific gene sequence in DNA into a portable mRNA copy that migrates to locations in the cytosol for **translation** into a corresponding polypeptide sequence.

Faithful rendering in transcription is made possible by using the complementary gene sequence, located on the complementary strand of DNA, as a molecular template upon which mRNA is assembled according to Watson-Crick base pairing rules (Figure 1.83). Translation takes place in the cytosol at **ribosomes**, gigantic biological macromolecular assemblies that are substantially comprised of rRNA. Ribosomes are workbenches for translation. Amino acids are transported to ribosomes by means of specific tRNA molecules. Each type of specific tRNA molecule covalently binds only one specific amino acid out of the 20, and also possesses an **anti-codon** nucleotide residue sequence that may only bind with an mRNA codon (by Watson-Crick base pairing) that specifically codes for the attached amino acid (as defined by the genetic code).

Such dual functionality allows each type of specific tRNA to simultaneously bind non-covalently by Watson-Crick bp hydrogen bonds to a specific mRNA codon, and covalently with the very amino acid coded for by that mRNA codon. The translation process now begins at the 5'-end of mRNA starting with the first and second triplet codons. Specific tRNA molecules with their amino acids attached bind to mRNA at these codons

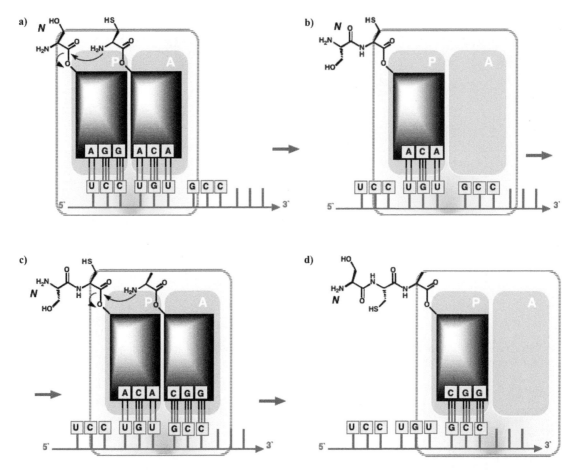

Figure 1.84   Schematic illustration of translation. The mRNA codons are read and converted from nucleotide residue sequences to protein primary structure by means of **cognate aminoacyl-tRNAs**. All mRNA codons are translated at a **ribosome** (prepared from rRNA) that has two main cognate aminoacyl-tRNA binding sites: **P** (**peptidyl**) and **A** (**aminoacyl**). All tRNAs are "adaptors" that can bind a particular mRNA codon through their **anti-codon loop**, using Watson-Crick base pairing, and also associate covalently with the appropriate amino acid residue coded for by the corresponding mRNA codon. When two cognate aminoacyl-tRNA molecules bind mRNA in P and A sites (**a**), then both are close enough for peptide link formation to take place with the emergence of a peptide chain (**b**). As aminoacyl tRNA molecules continue to dock sequentially onto mRNA codons (in the direction 5′→3′) (**c**), and amino acid residues continue to be added (N→C) (**d**), then the peptide chain will elongate progressively from oligo- to polypeptide until a stop signal is reached.

(Figure 1.84). Thereafter, the amino acid associated with the second triplet codon is close enough to attack the α-carboxyl functional group of the amino acid associated with first triplet codon creating a dipeptide. The amino acid-free tRNA attached to the first triplet codon may now dissociate, after which a new specific tRNA molecule with its amino acid attached binds to the third triplet codon in line. This new amino acid associated with the third triplet codon is close enough once again to attack the C-terminal α-carboxyl functional group of the dipeptide associated with the second triplet codon, thereby creating a tripeptide. Now, the amino acid-free tRNA attached to the second triplet codon can dissociate and the process continues in a similar way until specific tRNA molecules with their amino acids attached have bound to all the mRNA codons in turn (as far as the 3′-end of mRNA). The overall effect is to realise the synthesis (N→C-terminus direction) of a polypeptide chain through the sequential association (5′→3′ direction) of specific tRNA molecules to mRNA codons. Clearly the dual functionality of each type of specific tRNA molecule ensures that the sequence of the new polypeptide is precisely that determined by the original DNA gene sequence decoded according to the genetic code. The critical importance of tRNA should now be very clear, indeed. So much so that the question as to how exactly each type of specific tRNA molecule meets with the correct amino acid residue and the correct mRNA codon coding for the attached amino acid, is absolutely central to the fidelity of the link between DNA and protein structure. This remains a subject of enormous current interest in chemical biology circles.

1.5 Macromolecular lipid assemblies

Curiously, there is a perception in some quarters that lipids are just "dull", merely forming retentive membranes that mark the boundaries of cells and the territories of the various organelles or functional zones within them. However, although central to compartmentalisation, lipids and the macromolecular lipid assemblies that they form, including membranes, have a vast and complex dynamic behaviour which underpins cellular metabolism as well as promoting the systems of communication and synergism between living cells in complex, multicellular organisms. There appears to be much yet to learn and study since research into lipids and their macromolecular lipid assemblies has been relatively neglected in comparison to other fields in chemical biology. Nevertheless, the chemical biology reader should appreciate that the influence of macromolecular lipid assemblies on biological function is as equally vast and important as the influence of the biological macromolecules such as proteins, carbohydrates and nucleic acids. Therefore, future chemical biology research into lipids and macromolecular lipid assemblies is certain to be of great significance in developing our understanding of the way biology works at the molecular level.

1.5.1 Monomeric lipid structures

Monomeric lipids may be broadly defined as molecules of intermediate molecular weight (MWt. 100–5000 Da) that contain a substantial portion of aliphatic or aromatic hydrocarbon. Most biologically important lipids belong to a subset of this broad class of molecules known as **complex lipids**. The major members of this subset are the **acylglycerols**, **glycerophospholipids** and **sphingolipids**. Other complex lipids are known, but these have either little structural function or else are of unknown or poorly characterised biological function. Acylglycerols are a combination of the triose glycerol and long-chain fatty acids. Typical natural long chain fatty acids are shown (Figure 1.85). Whilst the

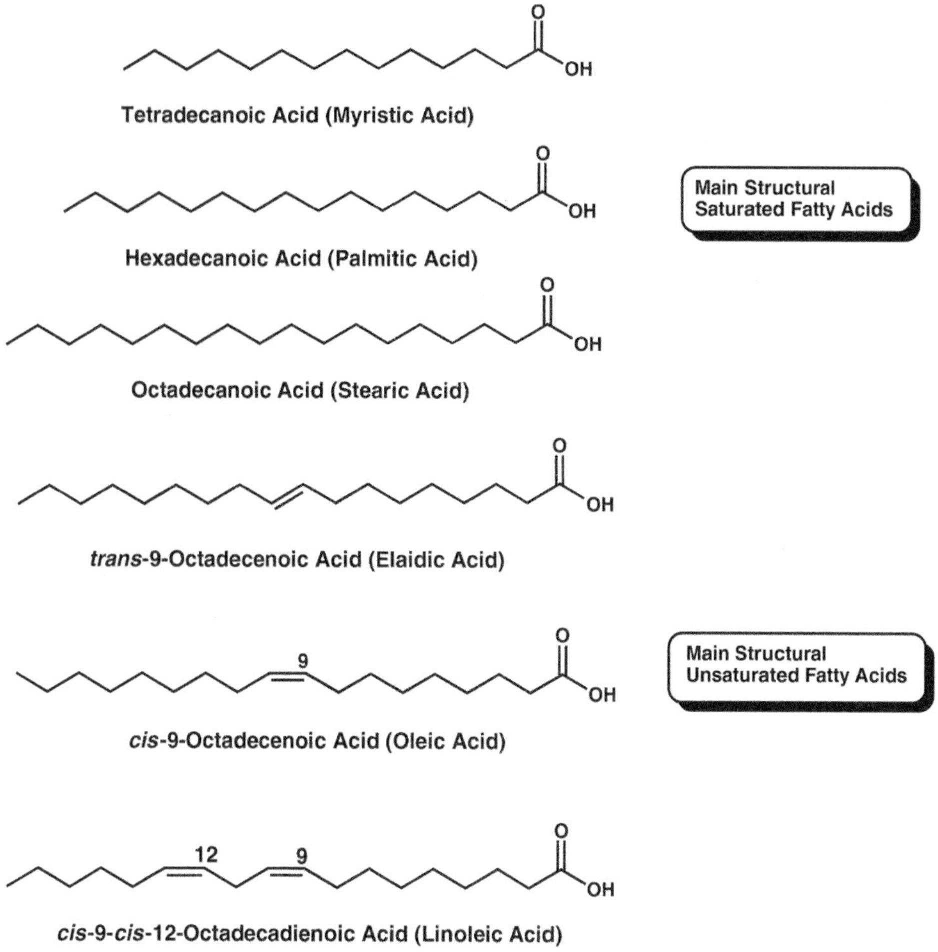

Tetradecanoic Acid (Myristic Acid)

Hexadecanoic Acid (Palmitic Acid)

Main Structural Saturated Fatty Acids

Octadecanoic Acid (Stearic Acid)

trans-9-Octadecenoic Acid (Elaidic Acid)

cis-9-Octadecenoic Acid (Oleic Acid)

Main Structural Unsaturated Fatty Acids

cis-9-*cis*-12-Octadecadienoic Acid (Linoleic Acid)

Figure 1.85 **Structures of main saturated and unsaturated fatty acids in structural lipids that form lipid assemblies.**

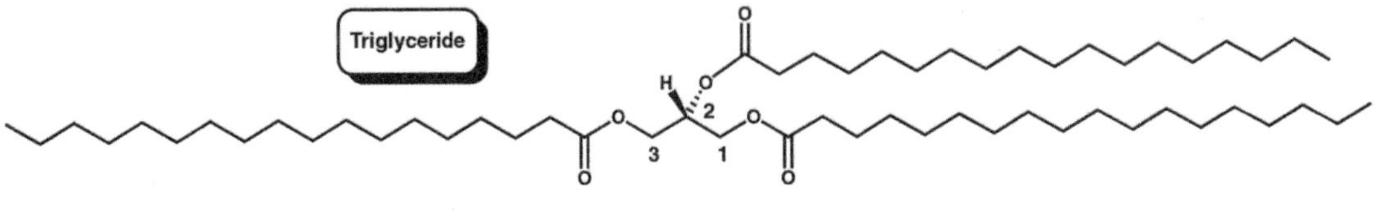

1,2,3-Trioctadecanoyl-*sn*-glycerol

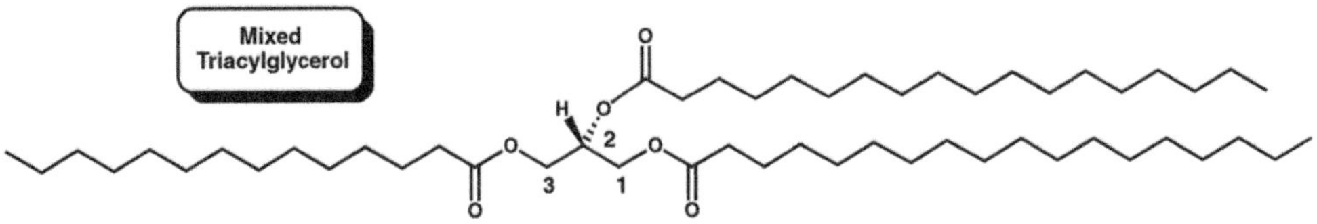

1,2-Dioctadecanoyl-3-tetradecanoyl-*sn*-glycerol

Figure 1.86 **Illustrative structures of main storage lipids.** The main storage lipids are known as **triglycerides** and the mixed **triacylglycerols.**

number of possible acylglycerols is vast, the most biologically significant are the triacylglycerols, in which all three reactive hydroxyl groups of glycerol have been esterified by fatty acids (Figure 1.86). Triacylglycerols are the major storage form of lipids in plants and higher animals. Where triacylglycerols contain identical fatty acid acyl chains they are called **triglycerides** or **simple triacylglycerols**. Where different fatty acid acyl chains are involved, the term **mixed triacyl glycerols** is used (Figure 1.86).

Glycerophospholipids may be thought of as derivatives of triacylglycerols in which the carbon atom C-3 carboxylate ester has been replaced by a phosphate ester. The number of possible glycerophospholipids is also vast, so only the most widely studied and arguably most biologically important are illustrated (Figure 1.87). These glycerophospholipids are important predominantly as constituents of biological membranes. Sphingolipids are a combination of the base sphingosine and long-chain fatty acids. Acylation of the carbon atom C-2 amine group by a fatty acid gives rise to the **ceramides** from which **phosphoceramides** are derived by phosphate ester derivatisation of the C-1 hydroxyl group. Biologically significant members of both types are illustrated (Figure 1.88). Ceramides and phosphoceramides are important constituents of human skin lipids and neural membranes. The phosphoceramides, together with the glycerophospholipids, are collectively known as **phospholipids**.

1.5.2 Lyotropic mesophases of phospholipids

Macromolecular lipid assemblies arise from the amphiphilic character of phospholipids (Figure 1.89). Broadly speaking, all phospholipids contain a compact **polar region**, which is hydrophilic in character, and an extended **chain region**, which is hydrophobic in character. In the presence of water, the tendency of the hydrophobic chain regions to self-associate and simultaneously exclude water leads to macromolecular lipid assemblies that may be described as vast, extended non-covalent structures held together by Van der Waals interactions and the hydrophobic effect. These adopt different phase states depending upon the character of the phospholipid involved and the local conditions. These phospholipid structures in their various phase states are also known as **lyotropic mesophases**. Some of the most well-established lyotropic mesophases have been summarised in Table 1.4 using the **Luzzati nomenclature**. Structures and transitions between mesophases are very relevant to the biological behaviour of lipids and so these will be discussed. However, given the current state of knowledge on lipids and their macromolecular assemblies, there can be no absolute certainty concerning precisely which of the lyotropic mesophases are actually relevant in biology and which are not.

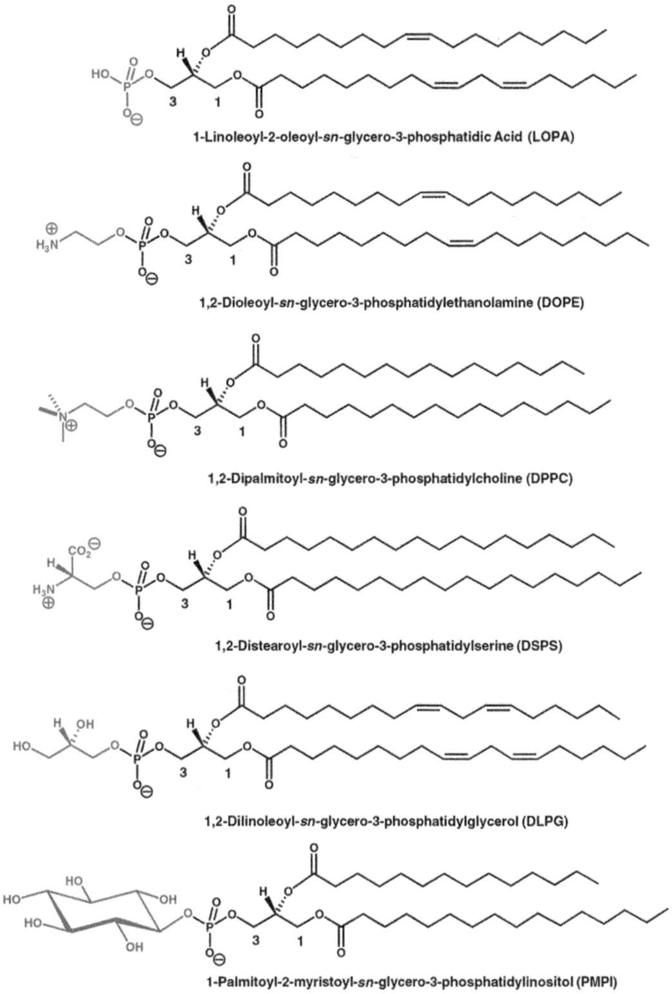

1-Linoleoyl-2-oleoyl-*sn*-glycero-3-phosphatidic Acid (LOPA)

1,2-Dioleoyl-*sn*-glycero-3-phosphatidylethanolamine (DOPE)

1,2-Dipalmitoyl-*sn*-glycero-3-phosphatidylcholine (DPPC)

1,2-Distearoyl-*sn*-glycero-3-phosphatidylserine (DSPS)

1,2-Dilinoleoyl-*sn*-glycero-3-phosphatidylglycerol (DLPG)

1-Palmitoyl-2-myristoyl-*sn*-glycero-3-phosphatidylinositol (PMPI)

Figure 1.87 **Representative first rank structural lipids**. These first rank structural lipids are known as **glycerophospholipids**, which are major and integral components of most lipid assemblies.

Table 1.4 **Summary of the main lyotropic mesophases**. Here are summarised each main phase type with corresponding symmetry name plus the specific physical nature of each mesophase type.

Phase type	Name	Phase structure
Solid-like lamellar		
3D	L_c	3D crystal
2D	L_c^{2D}	2D crystal
	$P_{\beta'}$	Rippled gel
1D	L_β	Untilted gel
	$L_{\beta'}$	Tilted gel
	$L_{\beta I}$	Interdigitated gel
	$L_{\alpha\beta}$	Partial gel
Fluid mesophases		
1D	L_α	Fluid lamellar
2D	H	Hexagonal
	H^C	Complex hexagonal
	R	Rectangular
	M	Oblique
3D	Q	Cubic
	T	Tetragonal
	R	Rhombohedral
	O	Orthorhombic

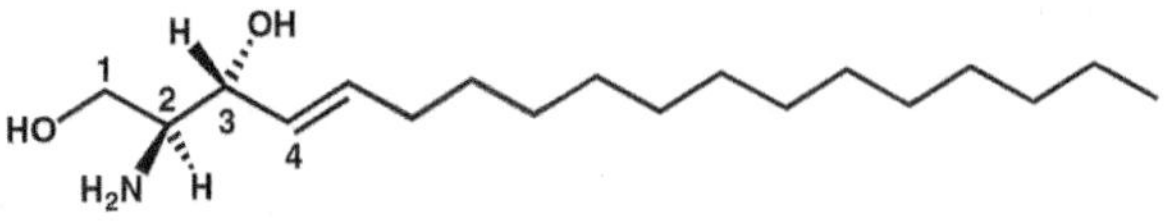

D-erythro-2-amino-4-octadecene-1,3-diol (Sphingosine)

Phosphoceramide

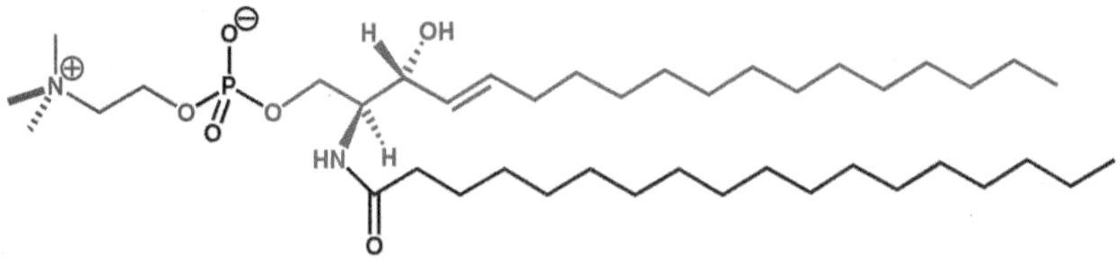

2-Stearoyl-1-phosphatidylcholine sphingosine (Stearoyl sphingomyelin)

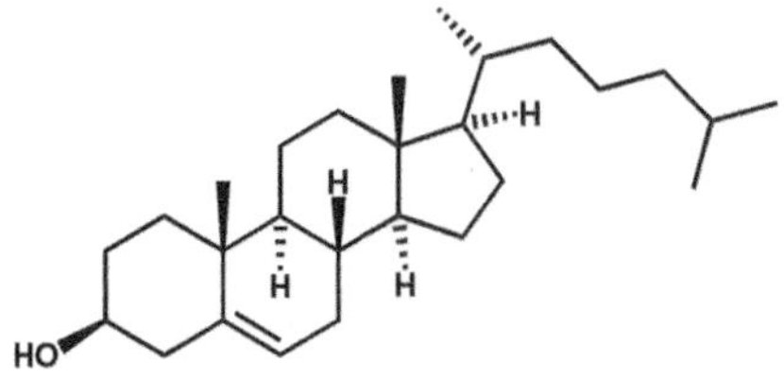

Cholest-5-en-3β-ol (Cholesterol)

Figure 1.88   Representative second rank structural lipids. These are **phosphoceramides** that are formed from the amine diol **sphingosine**. These can be components of some lipid assemblies, but in most cases are not dominant. The structure of **cholesterol**, another critical structural lipid is also shown.

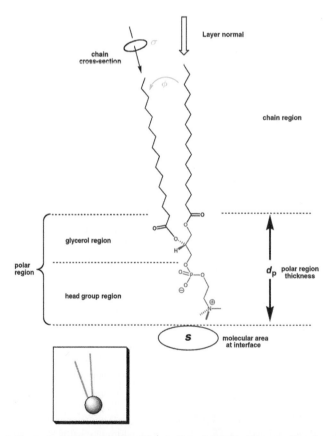

Figure 1.89   Structure of dipalmitoyl L-α-phosphatidylcholine (DPPC). This structure is used to illustrate the main molecular parameters and structural features that dictate the formations of **crystalline lamellar phases** of macromolecular lipid assemblies. A schematic representation of a phospholipid molecule is also shown (**inset**).

1.5.3 Solid-like mesophases

One or more **crystalline lamellar (L$_c$)** phases may be formed by all phospholipids at low temperature and/or low levels of hydration. When long and short-range order is found in three dimensions then the result is a **3D lamellar crystal**, which is a true crystal. The 3D crystalline order results from the close packing of two-dimensional (2D) phospholipid crystalline sheets (Figure 1.90). In all crystalline and ordered states, phospholipid close packing and molecular configuration is defined in terms of a number of parameters. These parameters are σ – the **mean cross-sectional area** of a fatty acid alkyl chain perpendicular to the chain axis, ϕ – the **tilt angle of the chain** with respect to bilayer plane, d_p – the **thickness of the head group region** and S – the **surface area at the bilayer plane** occupied by the individual phospholipid. When the 2D phospholipid crystalline sheets cease to maintain regular stacking arrangements with respect to each other, then 3D crystalline order breaks down, leading to series of 2D crystalline sheets, each irregularly stacked with respect to the next. Such mesophases are known as **2D lamellar crystals** since they still maintain a good deal of crystalline order.

One-dimensional (1D) ordered lamellar phases are known as gel states and occur when phospholipids are arranged into bilayers that then stack into a multilayer with each bilayer separated by water. The fatty acid alkyl side chains are stiff and extended, as in the 3D and 2D lamellar crystal phases, but may undergo hindered rotations about their chain axes. There are a number of types of 1D ordered lamellar phases depending upon the tilt angle ϕ; these are **L$_\beta$** ($\phi = 0$), **L$_{\beta'}$** ($\phi > 0$) and an interdigitated phase **L$_{\beta I}$** ($\phi = 0$) where fatty acid alkyl side chains from different monolayers overlap with each other (Figure 1.91). One other 1D ordered lamellar phase, **L$_\delta$**, is known, in which the fatty acid alkyl side chains cease to be linear, but adopt a helical conformation.

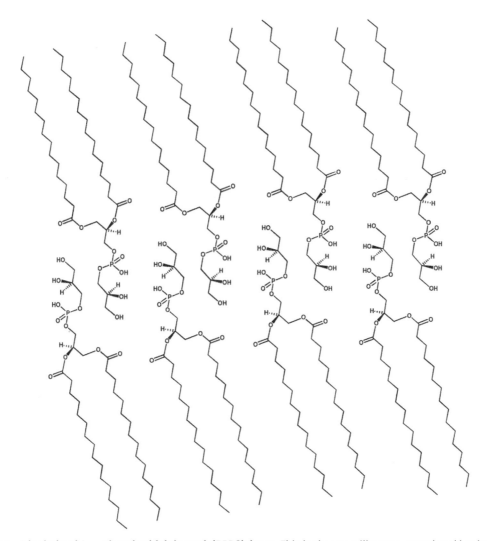

Figure 1.90 Dipalmitoyl L-α-phosphatidylglycerol (DPPG) layer. This is shown to illustrate crystal packing in **2D-lamellar layers**. In lipid assembly terms this can be considered secondary/tertiary structure formation.

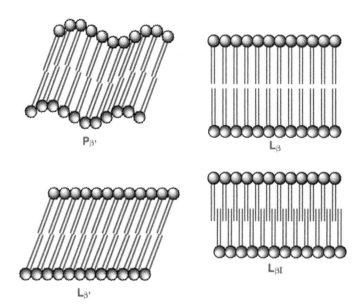

$P_{\beta'}$

L_{β}

$L_{\beta'}$

$L_{\beta I}$

Figure 1.91 **Selection of 2D modulated ordered lamellar phases.** These are partially disordered but exhibit translational ordering in two dimensions. These gel phases exist half way between crystalline L_c states and completely fluid phases such as the $L_{\alpha I}$ and H_{II} phases (Figures 1.93 and 1.94). In lipid assembly terms, these represent the equivalent of secondary/tertiary structure formation.

1.5.4 Fluid mesophases

Under certain conditions of temperature and hydration, solid-like mesophases will undergo a transition into fluid mesophases. The physical conditions under which transitions of this type occur are very important in biological terms. The vast majority of **3D fluid phases** so far identified have **cubic (Q)** symmetry (Table 1.4). Six cubic phases have been characterised so far and these appear to fall into two distinct families: one family based upon periodic minimal surfaces (bicontinuous) and the other on discrete lipid aggregates (micellar). 3D fluid phases may exist as either **Type I** (**normal topology**, oil-in-water) or **Type II** (**inverse topology**, water-in-oil) structures. Frequently, cubic phases of both Type I and Type II exist. For instance, the main bicontinuous phases (**Q230, Q224** and **Q229**) all form Type I and Type II structures consisting of two separate interwoven but unconnected networks of channels or rods which are formed from either fatty acid side chains or water respectively (Figure 1.92). **Type II rhombohedral (R)** or **tetragonal (T)** 3D fluid phases are also known. For example, the R_{II} phase consists of planar 2D hexagonal arrays, formed from aqueous channels, which are then regularly stacked to form a 3D lattice. In a similar way, the T_{II} phase is comprised of planar 2D square arrays that are once again stacked to from the 3D lattice. A debate is now raging amongst some chemical biology researchers concerning the potential existence of cubic phases in cells, indeed there is a proposition that the membranes of cellular organelles may in fact adopt the normal cubic phase, Q_I.

Having said this, membranes in cells are usually considered to adopt **1D fluid phases**, in particular the **fluid lamellar phases, L_α** (Figure 1.93), such as the normal topology Type I structure, $L_{\alpha I}$. This $L_{\alpha I}$ phase is widely considered to represent the default phase state of all biological membranes under normal physiological conditions. Cellular membranes are also thought to be able to adopt certain **2D fluid phases** under certain circumstances, in particular **hexagonal phases, H** (Figure 1.94), such as the inverse topology Type II structure, H_{II}. Both normal and inverse topology hexagonal phases, H_I and H_{II}, have the appearance of stacked cylinders. In the H_I phase, long alkyl acid side chains are contained within the cylinders, whilst polar head groups make up cylinder surfaces; in the H_{II} phase, cylinders of water are bordered by polar head groups and the spaces between cylinders are occupied by long alkyl side chains (Figure 1.94). The interconversion within biological membranes between the $L_{\alpha I}$ fluid phase and inverse hexagonal H_{II} phase is currently considered to be central to the dynamic behaviour of all manner of biological membranes, in particular to allow biological membranes to become temporarily more porous in the H_{II} phase prior to returning to the much less porous $L_{\alpha I}$ phase. In addition, this interconversion appears to be important to facilitate fusion events involving biological membranes as well.

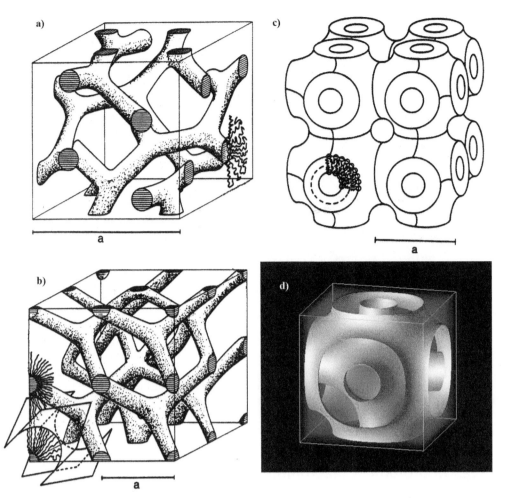

Figure 1.92 Structural representations of the main cubic mesophases formed by lipid assemblies. The mesophases shown are: (a) **Ia3d (Q^{230})**; (b) **Pn3m (Q^{224})**, both of which are **Q$_{II}$** fluid cubic phases; together with (c) **Im3m (Q^{229})**, which is a **Q$_I$** fluid cubic phase. Cartoon (d) provides an alternative representation of the **Im3m (Q^{229})** mesophase. Subscript **I** refers to **normal topology** (lipid "inside") and **II** to **inverse topology** (lipid "outside"). In lipid assembly terms, these represent the equivalent of tertiary structure formation. (illustrations (a)–(c) from Seddon, 1990, Figure 6, Elsevier).

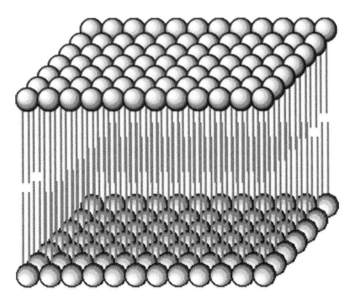

Figure 1.93 Structural representations of the L$_{\alpha I}$ phase. This is the main **fluid lamellar mesophase** formed by lipid assemblies and the primary mesophase adopted by biological membranes. Subscript **I** refers to normal topology (lipid "inside"). In lipid assembly terms, this represents the equivalent of secondary/tertiary structure formation.

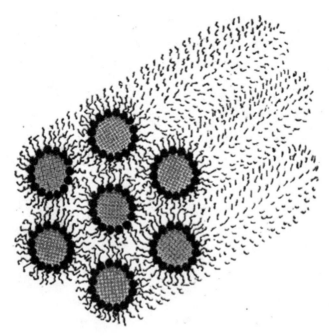

Figure 1.94 Structural representations of the H$_{II}$ phase. This is the main **fluid hexagonal mesophase** formed by lipid assemblies. Subscript **II** refers to inverse topology (lipid "outside"). In lipid assembly terms, these represent the equivalent of secondary/tertiary structure formation (illustration from Seddon, 1990, Figure 3b, Elsevier).

1.6 Structural forces in biological macromolecules

Throughout this chapter, references have been made to the forces that give the structures of biological macromolecules and macromolecular lipid assemblies both form and function. The covalent bond should be well known to all chemical biology readers and need not be discussed further. However, the non-covalent forces are important since these are central to the formation of 3D structure and therefore also to all functions of biological macromolecules and macromolecular lipid assemblies. There are four main types of non-covalent structural forces that matter most in chemical biology, namely **electrostatic**, **Van der Waals** and **dispersion**, hydrogen bonding and **hydrophobic interactions**. These forces will be dealt with in order from first principles.

1.6.1 Electrostatic forces

1.6.1.1 Monopoles

Electrostatic forces are long range, enabling an influence through space between atoms which is longer than any other type of force that will be discussed here. Charges of opposite sign are attracted to each other and charges with the same sign repel each other. The simplest type of electrostatic force exists between two **point charges** (**monopoles**), q_1 and q_2, separated by a **distance r** in a vacuum (Figure 1.95). The strength of interaction between q_1 and q_2 may be described in terms of **potential energy**, V, that is defined by Equation (1.1):

$$V = \frac{q_1 \bullet q_2}{4\pi\varepsilon_0 \bullet r} \qquad (1.1)$$

where ε_0 is a constant known as the **vacuum permittivity**. When q_1 and q_2 are opposite in sign (and are attracted), the value of V is negative and corresponds to the amount of energy required to separate the two point charges to a distance of infinity. If q_1 and q_2 are the same in sign (and repel each other), then the value of V becomes positive and represents the amount of energy required to maintain both charges in position after bringing them together from a distance of infinity. The concept of potential energy is a very useful way of defining the strength of weak forces and provides a

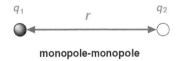

Figure 1.95 Illustration of monopole-monopole interactions. Here q_1 and q_2 are two monopoles, while r is the distance of separation between monopoles.

useful link to thermodynamics as well, as we shall see. By definition, potential energy V is a measure of the ability of a system to "do work" outside the system. Obviously, a system of charges in a state of repulsion is capable of doing work outside the system (hence the positive value of V), whilst a system of charges in a state of attraction is not (hence the negative value of V). The lower the value of V the less able is the system able to do work outside and the more stable is the arrangement of charges.

Biological macromolecules are usually found in a medium (e.g. aqueous buffer) and the nature of the medium typically has a profound effect upon the magnitude of V. Account is taken of this factor by exchanging ε_0 for another constant ε known as the **permittivity of the medium**, as shown in Equation (1.2):

$$V = \frac{q_1 \bullet q_2}{4\pi\varepsilon\bullet r} \tag{1.2}$$

which is a more general version of Equation (1.1). The ratio of ε to ε_0 is known as the **relative permittivity** or **dielectric constant** of the medium.

1.6.1.2 Dipoles

Equation (1.1) needs to be adapted differently if more complicated arrangements of charges are involved. For instance, if a fixed **dipole** and a monopole are allowed to interact in a vacuum separated by a distance r. We usually define a dipole as two opposite point charges, for instance q_1 and $-q_1$, separated by a distance l (Figure 1.96). In this case the electrostatic influence of the dipole is described in terms of a **dipole moment** in Equation (1.3):

$$\mu_1 = q_1 \bullet l \tag{1.3}$$

The potential energy of interaction between dipole moment, μ_1, and point charge, q_2, can be expressed by modifying Equation (1.1) to give Equation (1.4):

$$V = \frac{-\mu_1 \bullet q_2}{4\pi\varepsilon_0} \cdot \frac{1}{r^2} \tag{1.4}$$

Furthermore, the potential energy of interaction between two fixed dipoles (Figure 1.97) can be expressed by a simple expansion of Equation (1.4) to give Equation (1.5).

$$V = \frac{-\mu_1 \bullet \mu_2}{4\pi\varepsilon_0} \cdot \frac{2}{r^3} \tag{1.5}$$

What happens if more complex charge systems are interacting, each involving clusters of four (**quadrupole** or 2^2-pole) or even eight (**octupole** or 2^3-pole) charges? Obviously, the equations must become more complex. However, by comparing the form of Equations (1.1), (1.4) and (1.5), a general proportionality may be deduced. If an **n-pole** charge cluster is able to interact with an adjacent **m-pole** charge cluster, then the potential energy of interaction may be defined by the multipole proportionality in Equation (1.6):

$$V \propto \frac{1}{r^{n+m-1}} \tag{1.6}$$

Thus far, the equations and proportionality discussed in this section were actually derived assuming the dipoles or higher multipoles to be fixed in space. In order to be more realistic, rotation must be allowed for. In that case, if two multipoles were able to rotate freely in the vicinity of each other, then the overall interaction would be characterised by an **average potential energy**, $\langle V \rangle$, of zero. However, mutual potential energy depends on relative orientation. Therefore, the rotation of one dipole is heavily dependent upon the position and rotational behaviour of the other (Figure 1.98). There is no free rotation, even in the gas phase. Consequently, the actual value of $\langle V \rangle$ will be given by Equation (1.7):

$$\langle V \rangle = -\frac{C}{r^6} \tag{1.7}$$

where the constant of proportionality, C, is defined by Equation (1.8):

$$C = \frac{2\mu_1^2 \bullet \mu_2^2}{3\left(4\pi\varepsilon_o\right)^2 \cdot kT} \tag{1.8}$$

in which k is known as the **Boltzmann Constant**. The form of Equations (1.7) and (1.8) give some insights into the effects of **temperature T** on $\langle V \rangle$. The higher is T, the smaller is C and hence the higher is the value of average potential energy $\langle V \rangle$. In other words, increasing T increases system average potential energy $\langle V \rangle$ by disorganising the given system, thereby minimising mutual dipole orientation and attractive interaction effects, that otherwise act to minimise $\langle V \rangle$ and stabilise the system.

Electrostatic interactions are involved in stabilising the structures of biological macromolecules. Globular protein surfaces are covered with charged amino acid residues that interact with each other to form surface stabilising salt-links. Also surface clusters of charges approximate to arrays of monopoles, dipoles and quadrupoles, hence all the electrostatic equations apply. These same surface clusters of monopoles, dipoles and quadrupoles also radiate electrostatic force lines into solution, with consequences for protein-ligand interactions and molecular recognition phenomena (see Chapter 7). In the case of nucleic acids, charge-charge repulsions between the anti-parallel phos-

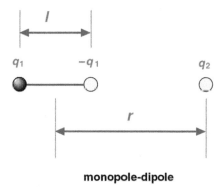

Figure 1.96   Illustration of fixed dipole-monopole interactions. In this case, $q_1/-q_1$ are separated by a distance, l, and represent a fixed dipole, q_2 is a monopole, and r is the distance of separation between dipole midpoint and monopole.

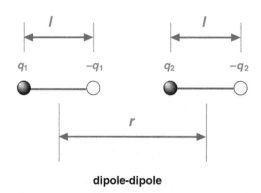

Figure 1.97   Illustration of fixed dipole-fixed dipole interactions. Here $q_1/-q_1$ are separated by a distance, l, and represent one fixed dipole, while $q_2/-q_2$ also separated by a distance, l, represent the other fixed dipole. In this case, r is the distance of separation between dipole midpoints.

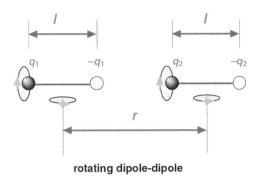

rotating dipole-dipole

Figure 1.98 **Illustration of rotating dipole-rotating dipole interactions.** In this case $q_1/-q_1$ are separated by a distance, l, and represent one rotating dipole, $q_2/-q_2$ also separated by a distance, l, represent the other rotating dipole. Otherwise, r is the distance of separation between dipole midpoints. Both dipoles rotate under mutual influence about dipole midpoints and axes.

phodiester chains could be sufficient to cause chain separation. However, phosphate negative charges are usually counterbalanced by close neighbour counter ions (cations or cationic proteins) that modulate the magnitude of the negative charges. In this case, the charges then provide sufficient charge-charge repulsion to maintain the positions of the anti-parallel phosphodiester chains preventing hydrophobic collapse, without inappropriately perturbing the structure overall.

1.6.2 Van der Waals and dispersion forces

Van der Waals and dispersion forces are typically observed between "closed shell" molecules and are short range in character. Such forces, together with hydrogen bonding (see Section 1.6.3), dominate the landscape of biological macromolecule interactions at short range and play a very dominant role in molecular recognition and binding/catalysis processes involving biological macromolecules and substrates/ligands. These forces are generated by interactions between partial charges in polar functional groups/molecules and induced partial charges in non-polar functional groups/molecules. There are three main contributions to Van der Waals and dispersion forces and these are:

1. *Weak dipole-weak dipole interactions*

2. *Induced dipole-weak dipole interactions*

3. *Induced dipole-induced dipole interactions*

1.6.2.1 Weak dipole-weak dipole interactions

Weak dipoles are associated with any bonds or functional groups involving carbon or hydrogen and an electronegative heteroatom such as peptide, phosphodiester or glycosidic links (as described earlier in this chapter). These weak dipoles interact in a manner described by expressions in the form of Equations (1.7) and (1.8).

1.6.2.2 Induced dipole-weak dipole interactions

When a weak dipole is in the presence of a polarisable functional group/molecule, then the electric field of that dipole will induce a temporary dipole in the polarisable functional group/molecule. The electrostatic influence of the weak dipole may be expressed in terms of a **permanent dipole moment**, μ_1, and that of the induced dipole in terms of an **induced dipole moment**, $\mu_2{}^*$. The potential energy of interaction may then be defined by Equation (1.9):

$$V = -\frac{C'}{r^6} \tag{1.9}$$

where the constant of proportionality, C', is defined with Equation (1.10):

$$C' = \frac{\mu_1^2 \bullet \alpha_2'}{\pi \varepsilon_o} \qquad (1.10)$$

The term α_2' is known as the **polarisability volume** of the functional group/molecule that harbours the induced dipole. Whilst the distance dependency of these induced dipole-weak dipole interactions is similar to dipole-dipole or weak dipole-weak dipole interactions, it is worth noting that there is actually no temperature dependency within the constant C'.

1.6.2.3 Induced dipole-induced dipole interactions

These interactions are driven by **dispersion** or **London forces**, and are the basis of non-polar functional group/molecule interactions. Instantaneous transient dipoles are created in all functional groups or molecules due to electron movement. However, only in truly non-polar functional groups or molecules do such transient dipoles have a major impact such as in the core of a globular protein (see Section 1.2.5). In these cases, an instantaneous dipole in one non-polar functional group or molecule generates an electrostatic field that induces the formation of another instantaneous transient dipole in a non-polar functional group or molecule in the vicinity. The electrostatic influence of the first transient dipole may be defined in terms of **induced dipole moment**, μ_1^*, and that of the second in terms of **induced dipole moment**, μ_2^*. The potential energy between these moments may be expressed in terms of Equation (1.11):

$$V = -\frac{C''}{r^6} \qquad (1.11)$$

Where the constant of proportionality, C'', is given by Equation (1.12):

$$C'' = \frac{2}{3} \cdot \alpha_1' \cdot \alpha_2' \frac{I_1 \cdot I_2}{I_1 + I_2} \qquad (1.12)$$

The terms α_1' and α_2' are polarisability volumes for each of the two functional groups/molecules involved respectively. I_1 and I_2 are the **ionisation energies** of the two functional groups/molecules involved respectively. The form of Equation (1.11) is exactly the same as for 1.9 and 1.7. Therefore, the distance dependencies of dipole-dipole, Van der Waals and dispersion forces are exactly the same. Only the constant of proportionality has a bearing upon the contribution of each type of interaction force to the potential energy of a given system of charges and/or induced charges (see Table 1.5).

Van der Waals and dispersion forces are ubiquitous in stabilising the structures of biological macromolecules and macromolecular lipid assemblies. In the case of globular proteins, Van der Waals and dispersion forces are important stabilising forces within the interior of these proteins owing to the substantial presence of hydrophobic amino acid residues located in the "middle and centre" of globular proteins. These forces may also play a role in stabilising the "interior" bp axial middle of nucleic acids too. However, by far the greatest beneficiary of these forces are macromolecular lipid assemblies, given the vast numbers of interacting alkyl chains within membrane structures.

Table 1.5 **Summary of the main forces.** These are involved in determining the 3D structures of biological macromolecules and also their dynamics, binding behaviour and reactivity. The term r refers to distance between interacting entities.

Interaction type	Distance dependency	V / kJ mol^{-1}	Remarks
Monopole-monopole (ion-ion)	$1/r$	250	Coulombic
Monopole-dipole (ion-dipole)	$1/r^2$	15	
Dipole-dipole	$1/r^3$	2	Static system
	$1/r^6$	0.3	Rotating system
London	$1/r^6$	2	All types of molecules
Hydrogen bond		20	
Repulsive	$1/r^{12}$		Very short range; all types of molecules

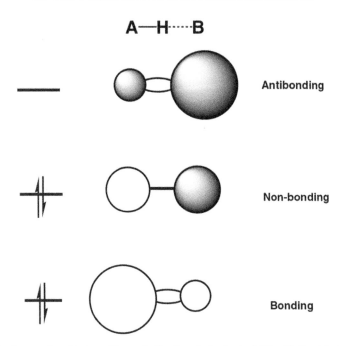

Figure 1.99 **Illustration of hydrogen bond interactions. A, B** = N or O. Contact of **AH** with **B** orbitals leads to the formation of three molecular orbitals according to theory of **linear combination of atomic orbitals (LCAO)**. Relative sizes are as drawn. Orbital filling according to **occupied energy ladder** (left) just favours bonding interactions.

1.6.3 Hydrogen bonding

Hydrogen bond (H-bond) interactions are the shortest range, non-covalent interactions and represent a special type of attractive interaction between closed shell functional groups arising from the atomic arrangement shown in Figure 1.99. **A** and **B** are very electronegative atoms (N, O or F); **B** possesses an available lone pair of electrons in a hybrid lone pair orbital that is then presented co-linear to the **A-H** σ bond axis as shown. An empirical molecular orbital description may then be used to describe the hydrogen bond in the form of a **three-centre, four-electron bond**. According to the linear combination of atomic orbitals, mixing a hydrogen **1s** orbital with a hybrid atomic orbital of **A** and a hybrid lone pair orbital of **B** results in three molecular orbitals as shown in Figure 1.99. Only bonding and non-bonding orbitals are occupied, with the result that a weak bonding interaction is established. This bonding interaction is purely a contact interaction "turned on" when the **A-H** σ bond contacts with the hybrid lone pair orbital of **B** and "turned off" immediately contact is broken. Hydrogen bonding is clearly ubiquitous in stabilising protein secondary structures, nucleic acid double helices (by Watson-Crick base pairing between anti-parallel phosphodiester chains) and in the wide range of homoglycan secondary to quaternary structures.

1.6.4 Hydrophobic interactions

The subject of **hydrophobic interactions** is controversial, but nevertheless appears to be very important for biological macromolecular cohesion in aqueous media, including most especially the cohesion of macromolecular lipid assemblies. The structure of water is critical to the existence of hydrophobic interactions (Figure 1.100). Water molecules consist of a repulsive spherical core centred at oxygen (**d** = 2.4–2.8 Å). Directional hydrogen bonds (V = 20 kJ mol^{-1}) compete with the repulsive core to bring water molecules into close proximity with each other. In ice, hydrogen bonding creates an **ordered lattice structure** (I_h). In liquid water, large-scale order is disrupted, but significant local order is retained. This extends to the solvation of polar and hydrophobic (non-polar) molecules/functional groups in aqueous solution. The solvation of non-polar molecules/functional groups is particularly important for hydrophobic interactions (Figure 1.101). Immediately surrounding any given hydrophobic molecule/functional group, the first solvation shell consists of ice-like hydrogen bonded arrays of water molecules. These "close-packed" water molecules suffer both a loss of hydrogen-bonding potential as well as a loss of entropy as a result of being "locked" in an ordered solvation

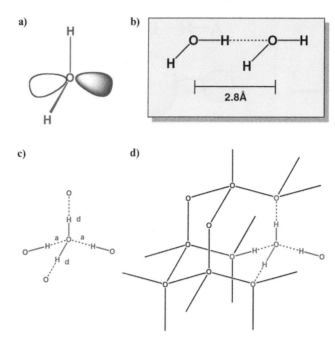

Figure 1.100   Illustrations of water structure. As applicable to solid (ice) and liquid states. (**a**) Water has two **O-H** bonds and two lone pairs; each water molecule may form up to four short hydrogen bonds (**b**) two from **O-H** hydrogen bond donors (d) and two from **O** lone pair hydrogen bond acceptors (a) (**c**); together these water molecules create "adamantane-like" I_h hydrogen bonded structures (**d**).

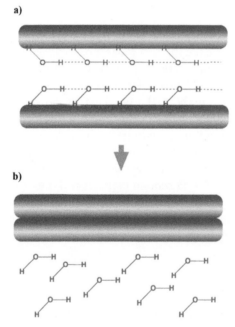

Figure 1.101   Illustration of hydrophobic effect. (**a**) Water forms an imperfect ordered solvent cage around two hydrophobic entities (blue). Each water of solvation is prevented from forming four hydrogen bonds with neighbouring water molecules for steric reasons. Waters of solvation are excluded upon association of two hydrophobic entities (**b**). These increase system entropy by entering bulk solution and release enthalpy through enhanced hydrogen bond formation. Short range Van der Waals interactions between hydrophobic entities may also contribute to system enthalpy.

shell covering the surface of the molecules/functional groups. Therefore, when two hydrophobic molecules/functional groups come into close proximity in aqueous medium, their association is driven by the opportunity to release ordered water molecules from these solvation cages. In effect, water drives the association of hydrophobic entities so as to minimise the hydrophobic exposed surface area that needs to be solvated, maximise entropy gain by releasing water molecules of solvation into the bulk medium and maximise enthalpy gain by allowing additional hydrogen bonds to be created in the bulk. The controversy about hydrophobic interactions is therefore justified. This has less to do with actual bonding interactions between hydrophobic entities and much more to do with the subtleties of water structure and the intermolecular bonding interactions involved. Nevertheless, hydrophobic interactions are obviously crucial for the structural integrity and functions of biological macromolecules (including macromolecular lipid assemblies) in aqueous solution.

The hydrophobic effect is believed to have a significant stabilising effect on globular proteins given the substantial presence of hydrophobic amino acid residues located in the "middle and centre" of globular proteins. Clearly the hydrophobic effect forces may also play a role in stabilising the "interior" bp axial middle of nucleic acids too. But once again, the greatest beneficiaries of the hydrophobic effect are macromolecular lipid assemblies, given the vast numbers of interacting alkyl chains within membrane structures.

1.6.5 Other forces

There are undoubtedly other forces involved in stabilising biological macromolecules and macromolecular lipid assemblies, such as weak-polar forces (see Section 1.2.5) that are yet to be fully described and accounted for. However, although we have said a lot about forces of attraction in Section 1.6, next to nothing has been said about **forces of repulsion**. In fact, when molecules or functional groups are pushed or pulled together, then nuclear and electronic repulsions will eventually begin to dominate over forces of attraction at very short range. Forces of repulsion are very complicated to define and depend heavily upon the nature and electronic structure of the interacting species. Nevertheless, the **Lennard-Jones potential** has been found to be a good overall description of how potential energy varies between species with inter-atomic distance, by taking in to account the behaviour of forces of repulsion. This potential is demonstrated in Equation (1.13):

$$V = \frac{C_n}{r^n} - \frac{C_6}{r^6} \tag{1.13}$$

where C_6 and C_n are constants, and n is a large integer (often given as 12). When n is 12, then Equation (1.13) is said to describe a Lennard-Jones (12,6) potential. Forces of attraction are represented by the negative $1/r^6$ term, since weak forces of attraction typically obey a $1/r^6$ dependence with distance of separation (see Equations (1.7), (1.9) and (1.11)). Forces of repulsion are represented by the positive $1/r^n$ term (where n is usually 12), in line with the fact that such forces can only dominate at very short range. The complete list of the main structural forces in biological macromolecules and macromolecular lipid assemblies is given with distance dependence (Table 1.5), including forces of repulsion.

2

Chemical and Biological Synthesis

2.1 Introduction to synthesis in chemical biology

Chemical synthesis and harnessing biological synthesis are amongst the most powerful tools available to chemists interested in chemical biology. Synthesis begins the process of determining structure and dynamics. Without synthesis, there is insufficient material to study structure. Without synthesis, critical probes are not available to study dynamics. Chemical biology is perhaps unique in requiring both chemical and biological approaches to synthesis; herein rests a key fundamental of this approach.

Traditional chemical synthesis has an incredibly rich heritage and record of success in the directed synthesis of a multiplicity of molecules, from unusual inorganic compounds to complicated and intricate organic molecules, built up from readily available, small molecule starting materials. By contrast, biological synthesis has a much shorter, though no less distinguished, record of success in the directed synthesis of biological macromolecules by means of "factory organisms" such as the bacterium *Escherichia coli* (*E. coli*) that employs complex anabolic intracellular pathways for the directed assembly of biological macromolecules. The directed biological syntheses of many biological macromolecules are heavily dependent on recombinant technologies and gene cloning techniques (Chapter 3).

Although the overall philosophy and objectives of both chemical and biological synthesis are the same (directed synthesis and isolation of a target molecule) the two approaches are obviously radically different. General courses in chemistry are beginning to incorporate certain aspects of biological synthesis, but for the most part biological synthesis is not covered. In Chapter 1, the chemical biology reader was introduced to the structures of the main biological macromolecules and lipid macromolecular assemblies. In this chapter, we aim to provide a balanced introduction to both chemical and biological approaches for the syntheses of biological macromolecules and lipids. We hope that the reader will emerge with not only a proper appreciation about each approach, but also a proper appreciation of their respective limits as well.

2.2 Chemical synthesis of peptides and proteins

2.2.1 Basic principles: peptide synthesis

As a first step to understanding the challenges of peptide synthesis, consider the synthesis of a dipeptide from the constituent amino acids. The amino acid residue sequence must be correct, side reaction products should be prevented from forming, and α-carbon racemisation should be avoided. In order to meet all these requirements, *N*- and *C*-terminal protecting groups are required to ensure chemoselective formation of only the single, desired peptide link

Essentials of Chemical Biology: Structures and Dynamics of Biological Macromolecules In Vitro *and* In Vivo, Second Edition. Andrew D. Miller and Julian A. Tanner.
© 2024 John Wiley & Sons, Inc. Published 2024 by John Wiley & Sons, Inc.
Companion Website: www.wiley.com/go/miller/essentialschembiol2

in coupling. Protection may need to be extended to amino acid residue side chains as well. The synthetic strategy may therefore be considered as a five-stage process (Figure 2.1):

1. Protection of the α-amino group of the *N*-terminal residue

2. Protection of the carboxyl group of the *C*-terminal residue

3. Activation of the carboxyl group of the α-*N*-protected amino acid

4. Coupling (peptide link formation) to give fully-protected dipeptide

5. Deprotection (as appropriate)

For the synthesis of most dipeptides, protection of amino acid residue side-chain functional groups will also be appropriate in order to avoid additional undesirable side reactions.

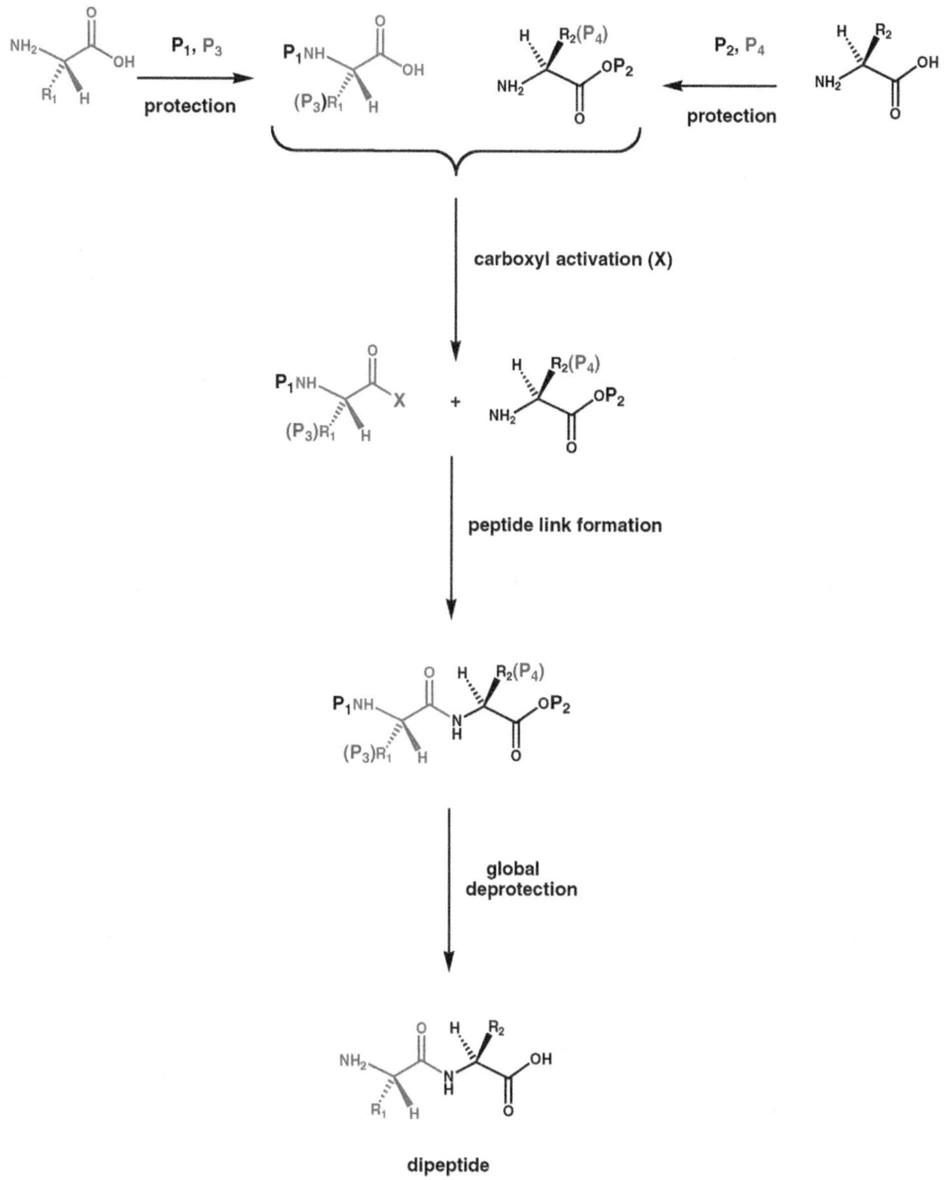

Figure 2.1 Peptide link formation. L-α-amino acids are rich in reactive functional groups. Therefore, chemo-selective peptide link formation is not possible without protecting groups. A future *N*-terminal residue must be α-*N*-protected (**P$_1$**) (with optional **P$_3$** side-chain protection), while future *C*-terminal residue must be α-*C* protected (**P$_2$**) (with optional **P$_4$** side-chain protection). The protected *N*-terminal residue must then be activated to allow peptide link formation to take place. Global deprotection reveals a dipeptide product.

The construction of a tripeptide from a dipeptide requires that step five should only involve selective deprotection of the *N*-terminal α-amino group of a dipeptide. A third amino acid residue may then be added by coupling the free α-amino group of the dipeptide and the free carboxyl group of a selected α-*N*-protected amino acid. Further selective deprotection can then give rise to the coupling of another selected α-*N*-protected amino acid, and so on. The whole process of di-, tri- and tetrapeptide, etc. synthesis relies on **selective deprotection** of the α-*N*-protecting group. Such selective deprotection requires that the conditions of deprotection should be as different as possible (**orthogonal**) to the conditions required for the removal of other protecting groups involved in the synthesis. Therefore, provided that the α-*N*-protecting group can be removed under conditions that do not affect the remaining protecting groups, then this protecting group can be said to be orthogonal with respect to the remaining protecting groups and can be said to be subject to selective deprotection. The use of selective deprotection and orthogonal protecting groups is central to successful solution phase peptide synthesis in order to avoid low yields and horrendous product mixtures.

Furthermore, coupling (peptide link formation) should be performed under very carefully controlled conditions, otherwise α-carbon racemisation becomes a serious problem, especially during coupling. In this instance, each selected *N*-terminal protected amino acid chosen to extend the peptide chain must be activated for peptide link formation (Figure 2.1). Activation renders the α-carbon position more acidic and hence more likely to undergo acid dissociation, leading to racemisation. Clearly, solution phase synthesis using an approach like this becomes a significant challenge for any peptide of more than a few amino acid residues in length. Indeed, the involved and painstaking solution phase synthesis of the nonapeptides **oxytocin** and **vasopressin** by **Vincent du Vigneaud** even resulted in the award of a Nobel Prize for Chemistry in 1955.

2.2.2 Solid phase peptide synthesis

Problems in solution phase peptide synthesis have been largely overcome by the development of **solid phase peptide synthesis (SPPS)**. The chemical principles of peptide link formation and peptide synthesis remain the same, but in SPPS the growing peptide chain is anchored to a solid phase resin, thereby easing the iterative process of peptide bond formation, removing the need for crystallisations and purifications after each step of the synthesis, and in some ways simplifying protection/deprotection problems. SPPS earned **Bruce Merrifield** a Nobel Prize in 1986, and has eased the technical challenges of peptide synthesis to the extent that the process can now be automated.

Solid phase synthesis can be simplified to the following steps (Figure 2.2):

1. The first amino acid (*C*-terminus) is coupled to reactive group X attached to an insoluble matrix (solid support) via its free carboxyl functional group. The amino acid is otherwise α-*N*-protected (and at other functional groups as appropriate).

2. The α-*N*-protecting group is selectively removed from the α-amino position; then a second α-*N*-protected amino acid (also protected at other functional groups as appropriate) is introduced and activated for coupling with the first amino acid bound to the resin.

3. Once the peptide link is formed, the process of α-*N*-deprotection and amino acid coupling continues for as long as required.

4. Finally, the desired length peptide is cleaved from the solid support by chemistry conditions suitable to cleave the linkage between solid support and peptide chain, and remove all the amino acid side-chain protecting groups as well.

2.2.2.1 Solid supports and linkers for SPPS

The solid support should be well solvated to facilitate reactions to take place involving the two phases. The original supports were based on **polystyrene**, but have generally been superseded by **polyamide** resins that have an advantage in that they have a similar polarity to the peptide backbone. More recently, resins based on **polyethylene glycol (PEG)** grafted onto low cross-linked polystyrene, or resins completely based on cross-linked PEG, have been used due to their superior swelling capacity, resulting in a larger solid/liquid interface. Each growing peptide chain is

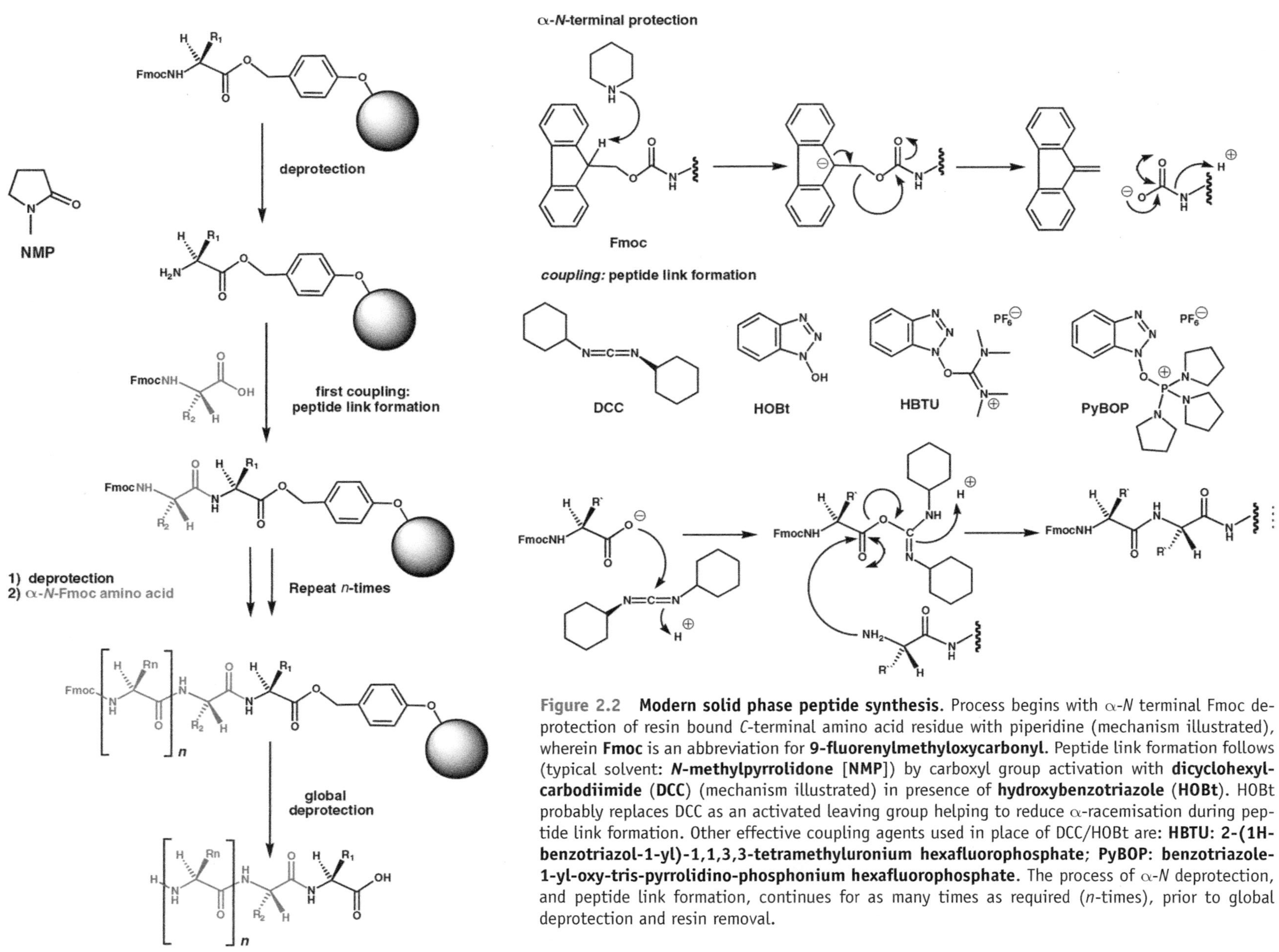

Figure 2.2 **Modern solid phase peptide synthesis.** Process begins with α-*N* terminal Fmoc deprotection of resin bound *C*-terminal amino acid residue with piperidine (mechanism illustrated), wherein **Fmoc** is an abbreviation for **9-fluorenylmethyloxycarbonyl**. Peptide link formation follows (typical solvent: **N-methylpyrrolidone [NMP]**) by carboxyl group activation with **dicyclohexyl-carbodiimide (DCC)** (mechanism illustrated) in presence of **hydroxybenzotriazole (HOBt)**. HOBt probably replaces DCC as an activated leaving group helping to reduce α-racemisation during peptide link formation. Other effective coupling agents used in place of DCC/HOBt are: **HBTU: 2-(1H-benzotriazol-1-yl)-1,1,3,3-tetramethyluronium hexafluorophosphate; PyBOP: benzotriazole-1-yl-oxy-tris-pyrrolidino-phosphonium hexafluorophosphate.** The process of α-*N* deprotection, and peptide link formation, continues for as many times as required (*n*-times), prior to global deprotection and resin removal.

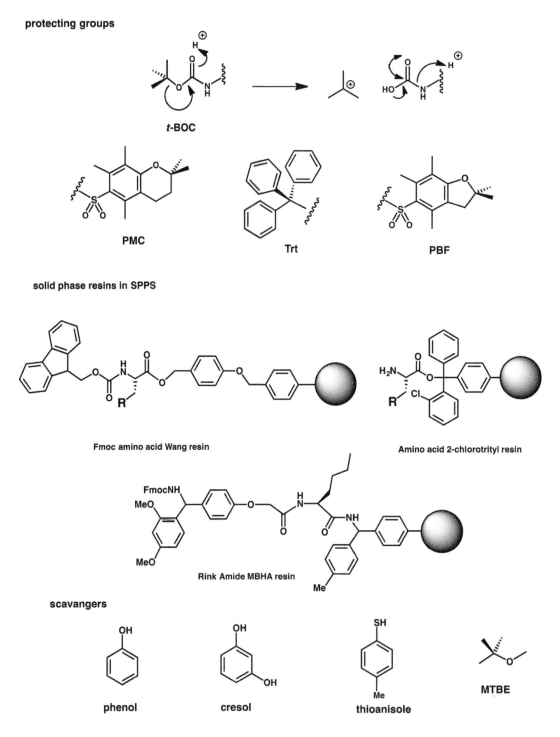

Figure 2.2 Modern solid phase peptide synthesis (Contd.). Most frequently used amino acid side-chain protecting groups are shown: *t*-BOC: *tert*-butyloxycarbonyl; Trt: trityl; PMC: 2,2,5,7,8-pentamethylchroman-6-sulfonyl; PBF: 2,2,4,6,7-pentamethyl-dihydrobenzofuran-5-sulfonyl. These are all acid labile (see illustrated mechanism). Most common solid phase resins are also shown, all of which are labile to acidic release conditions too (the Rink Amide leaves a *C*-terminal amide in place). During global deprotection and resin removal, **scavengers** such as **phenol, cresol** or **thioanisole** are also included to capture reactive cationic species post deprotection. The desired product oligo-/polypeptide is then separated initially by precipitation by means of an agent such as **methyl-*tert*-butyl ether** (**MTBE**) and purified finally by reversed-phase liquid chromatography (see Section 2.6.3).

attached to the solid support by a **resin linker**. The resin linker must be sufficiently robust to be unaffected by all the α-*N*-deprotection and coupling steps involved that comprise a peptide synthesis. Thereafter, the resin linker must be cleaved quantitatively to release the full-length peptide at the end of the synthesis. In the present day, **Wang** (via p-benzyloxybenzyl), **Rink Amide** (an even larger benzyl-based linker) or **super acid-labile** (**2-chlorotrityl**) resins are most commonly used in place of the original polystyrene, since these resins have been adapted to possess fundamentally acid sensitive resin linkers (Figure 2.2).

2.2.2.2 Coupling protected amino acids in SPPS

SPPS relies on very efficient amino acid coupling. Each coupling step involves the linking of an α-*N*-protected amino acid with the α-*N*-deprotected amino terminus of a growing resin-linked peptide chain. Each such coupling step must take place at yields approaching 100% in order that long peptide chains (>20 amino acid residues) may be built up (*C*→*N*) on resin prior to the cleavage of the resin linker. The extent of coupling can be monitored by means of the **ninhydrin test** (Figure 2.3). Ninhydrin reacts with primary α-amines to yield a purple, blue or blue-green colour that may be monitored colorimetrically. Secondary amines such as proline are less reactive with ninhydrin and typically yield a reddish-brown colour. However, the ninhydrin test is a useful diagnostic for incomplete coupling; then a second recoupling step can be performed.

Coupling must also take place without reducing the optical purity of each α-*N*-protected amino acid added per cycle of chain extension. Hence the coupling agents chosen to activate free carboxyl groups during peptide link formation should be capable of suppressing racemisation completely. Carbodiimide **dicyclohexylcarbodiimide** (**DCC**) was the first main coupling agent to be introduced (Figure 2.2). This activates free carboxyl groups through the formation of highly reactive *O*-acylurea intermediates that are subject to facile nucleophilic attack by α-amino functional groups leading to peptide link formation. However, DCC is not sufficient in and of itself to reduce racemisation during coupling, hence **hydroxybenzotriazole** (**HOBt**) was introduced to react with the *O*-acylurea in order to reduce the racemisation risk. The combined use of DCC and HOBt has now largely been superseded by the unitary use of **2-(1H-benzotriazol-1-yl)-1,1,3,3-tetramethyluronium hexafluorophosphate** (**HBTU**) that is today much the most frequently used reagent to activate amino acid coupling in peptide synthesis, along with **benzotriazole-1-yl-oxy-tris-pyrrolidino-phosphonium hexafluorophosphate** (**PyBOP**) (Figure 2.2).

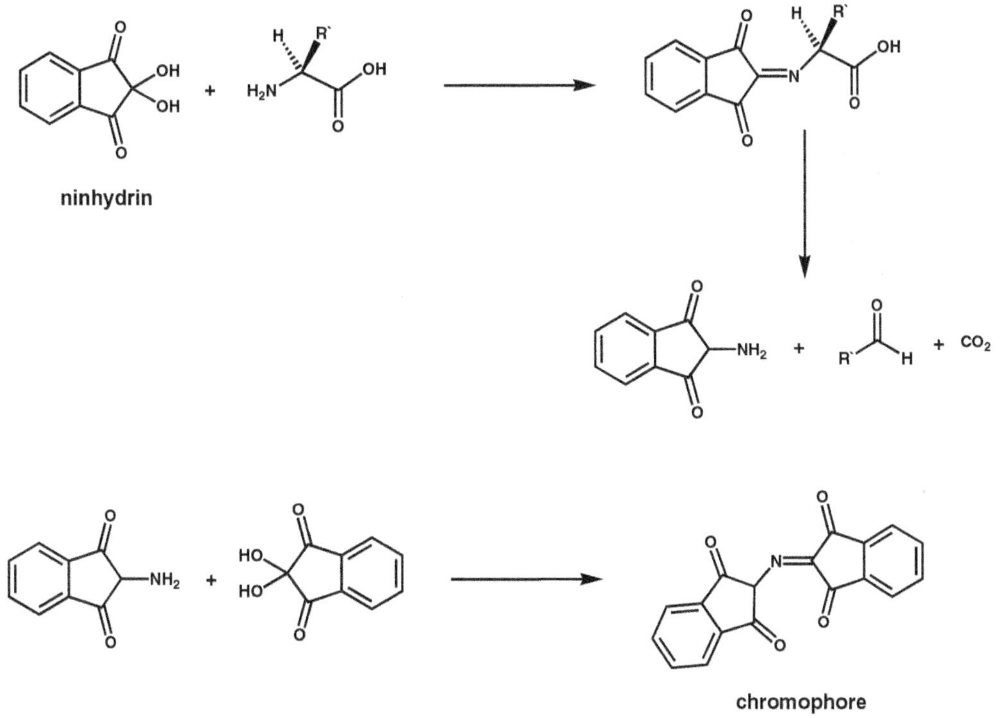

Figure 2.3 Ninhydrin test. Ninhydrin combines with α-*N*-primary amine functional groups to form primary product that combines with a second molecule of ninhydrin to generate a chromogenic chromophore.

2.2.2.3 *Protection/deprotection strategies in SPPS*

Merrifield's original protection strategy was to use the ***tert*-butyloxycarbonyl** (*t*-**BOC**) group for α-*N*-protection on every amino acid added to growing peptide chains, while amino acid side-chain functional groups were given benzyl-protecting groups. The *t*-BOC group is acid labile (using a mixture of **trifluoroacetic** [**TFA**] in dichloromethane) (Figure 2.2). Accordingly, once peptide synthesis had reached full length then separation of the full-length peptide from the resin and simultaneous side-chain deprotection was brought about using liquid hydrogen fluoride, hardly mild resin cleavage conditions. Fortunately, a much milder orthogonal protecting group strategy has since been developed for peptide synthesis with the introduction of **N-9-fluorenylmethyloxycarbonyl** (**Fmoc**) groups for α-*N*-protection. The Fmoc group is sensitive to mild base piperidine (piperidine 20–50% v/v in DMF for 20 mins) but is otherwise stable to decomposition in neutral or acidic conditions (Figure 2.2). Consequently, not only may amino acid side-chain functional groups be protected during peptide synthesis using protecting groups sensitive to only mild acid (or other sets of orthogonal) conditions, but so, too, separation of the full-length peptide from the resin can be adjusted or tuned to take place under similarly mild acid conditions.

Currently, amino acid side-chain functional groups can be protected by a number of different groups such as the **N-benzyloxycarbonyl** (**Z**) amino protecting group that is typically cleaved by simple catalytic hydrogenolysis using a Pd/charcoal catalyst, or by strong acid. In contrast, hydroxy and carboxyl functional groups can be protected by ***tert*-butyl** masking, while ε-amino groups in lysine and the N^1-position of tryptophan can be protected by *t*-BOC groups. In both cases, protecting groups are labile with respect to mild acid (HCl in organic solvents or TFA) treatment. Furthermore, the primary amide side chains of asparagine and glutamine, the thiol of cysteine and the imidazole ring of histidine are commonly masked by **trityl** group association. Trityl protection is also notably labile with respect to mild acid treatment in all the cases listed here. Finally, the unusual guanidine functional group associated with the side chain of arginine is frequently protected by **2,2,5,7,8-pentamethylchroman-6-sulfonyl chloride** (**PMC**) or **2,2,4,6,7-pentamethyl-dihydrobenzofuran-5-sulfonyl chloride** (**PBF**). Both PMC and PBF are mildly acid labile protecting groups as well (Figure 2.2). Clearly the release of resin-bound, full-length peptide can result in the production of highly reactive species such as butyl cations (e.g. *t*Bu-cations and *t*Bu-fluoroacetate). Therefore, **scavengers** such as **phenol**, **cresol** or **thioanisole** are also included in the cleavage cocktail to capture these species and prevent side reactions. The appropriate cleavage mixture is then removed by evaporation and **diethyl ether** or **methyl-*tert*-butyl ether** (**MTBE**) added to precipitate the deprotected peptide product from solution, thereby eliminating many by-products. Thereafter, the crude deprotected peptide needs to be purified to homogeneity by reversed-phase preparative **high-performance liquid chromatography** (**HPLC**) (see Section 2.6.3).

2.2.3 Chemical synthesis of polypeptides

SPPS has allowed relatively facile peptide synthesis for peptides of up to 40 amino acid residues, though it can be used to synthesise polypeptides in excess of 100 amino acid residues in length. Clearly, the synthesis of long polypeptides and proteins should be best accomplished by biological synthesis (see Section 2.7). However, there are instances when chemical synthesis presents a few significant advantages over biological synthesis, particularly where the incorporation of unusual amino acid residues is required, such as **D-amino acids**, **fluorescent-labelled amino acids** (such as Aladan, see Chapter 4), linker moieties and other non-peptidic groups. However, the two approaches need not be exclusive. For instance, it may be envisaged that a combination of chemical and biological synthesis (semi-synthesis) could become increasingly important in future chemical biology applications.

2.2.4 Chemical synthesis of peptide nucleic acids

Peptide nucleic acids (**PNAs**) are synthetic molecules where purine and pyrimidine bases are linked by a polypeptide backbone (Figure 2.4). PNA monomers are typically synthesised using modified protocols for standard peptide synthesis. The *N*-terminus is Fmoc protected, whilst **benzhydryloxycarbonyl** (**BHOC**) is used to protect the exocyclic amino groups on adenine, cytosine and guanine. The backbone consists of repeating *N*-(2-aminoethyl)-glycine units, resulting in a distance between the bases in PNAs similar to natural DNA or RNA containing the usual phosphodi-

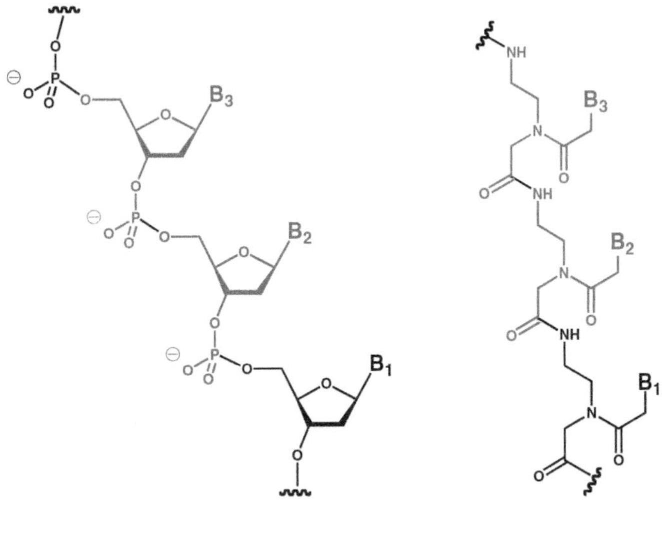

DNA   PNA

Figure 2.4 A comparison of PNA and DNA. PNA contains neither phosphodiester links nor the 2′-deoxy-β-D-fructofuranose rings. Instead, the backbone consists of **N-(2-aminoethyl)-glycine** repeat units conjoined by peptide links. The main purine and pyrimidine bases of DNA are linked to the PNA backbone by methylene carbonyl bridge linkages. Watson-Crick base pair-complementary PNA strands are capable of Watson-Crick base pairing and hence anti-parallel double helix formation.

ester backbone. Bases align perpendicular to the polypeptide backbone, hence PNA may form anti-parallel double helices like DNA and RNA wherein complementary strands are linked by Watson-Crick base pair hydrogen bonding. The uncharged backbone of PNA results in many useful and interesting properties. The lack of charge-charge repulsion in a **PNA/DNA duplex** results in far stronger binding than that observed in a DNA double helix (DNA/DNA duplex). Triple helical complexes can also be formed with DNA that are especially stable when the PNA involved is described as a homopyrimidine. Typically, a **PNA/DNA triplex** has a PNA/DNA ration of 2:1. This structure requires a Watson-Crick PNA/DNA duplex annealed to a second PNA strand by **Py:Pu:Py Hoogsteen/Watson-Crick triplex base pairings** (see Chapter 1). The stability of the PNA/DNA triplex has been crucial to many potential technological applications of PNA. PNA can invade double helical DNA, forming a PNA/DNA triplex with the complementary (antisense) strand of DNA. Although PNAs are most stable as homopyrimidines, low numbers of purines within the sequence can be tolerated. PNAs are also fairly resistant to both nucleases and proteases, and have significant lifetimes *in vitro* and *in vivo*, a useful advantage over standard nucleic acids. Furthermore, PNAs prepared by chemical synthesis can be modified easily by many of the techniques outlined in Chapter 4 with fluorescent groups and other labels. The combination of strong binding to complementary nucleic acid sequences, excellent stability and easy derivatisation make these molecules potentially powerful tools in a variety of applications.

2.3 Chemical synthesis of nucleic acids

Merrifield's SSPS presented a new approach to the chemical synthesis of biological macromolecules and the solid phase approaches were quickly adapted to other fields. **Solid phase oligonucleotide synthesis** (SPONS) is another vital development in chemical synthesis that has had a major impact on the biological sciences. Whereas the main challenge in SPPS was precise formation of peptide links, the challenge for oligonucleotide synthesis was precise formation of corresponding phosphodiester links.

Much of the development in SPONS arose from studies on polyphosphate and nucleotide coenzyme synthesis in the laboratory of Alexander Todd during the 1950s. Not long after, in the early 1960s, Khorana achieved the synthesis of the first sequence-defined oligonucleotides using DCC to activate phosphate-sugar coupling. However, in the late 1960s, Letsinger and co-workers developed the solid phase approach that eventually culminated in the triester chemical approach familiar today.

2.3.1 Chemical synthesis of oligodeoxynucleotides

Oligodeoxynucleotide synthesis today is virtually always carried out on a solid support. Initially, the polymeric solid supports were polystyrene cross-linked with divinylbenzene, but **controlled-pore glass beads** (**CPG beads**) with defined porosities are now used in preference. The 3′-hydroxyl of the first protected deoxynucleotide is precoupled to the solid support and then solid phase, multistage oligodeoxynucleotide synthesis may commence in the 3′→5′ direction, in direct analogy to SPPS (Figure 2.5).

1. The first protected deoxynucleotide attached to the solid support is protected on the 5′-hydroxyl group with **dimethoxytrityl** (**DMT**). Deprotection with **trichloroacetic acid** (**TCA**) results in a free 5′-hydroxyl group and the release of the orange-coloured DMT cation (detected as a measure of the efficiency of deprotection).

2. A 5′-hydroxyl DMT-protected deoxynucleoside with attached 3′-phosphoramidite (Figure 2.6), is then activated for coupling to the 5′-hydroxyl group of the 3′-solid support-bound deoxynucleotide by means of a coupling reagent known as **tetrazole**.

3. Iodine is introduced to oxidise the diester link formed from coupling to a complete phosphodiester link (with P(V) oxidation state). Diester links formed with phosphoramidites are unstable (due to the P(III) oxidation state).

4. Acetic anhydride is then added to block or cap any unreacted 5′-hydroxyl groups so that sequence truncations are not incorporated in later steps (Figure 2.7).

5. With a phosphodiester link formed and 5′-hydroxyl capping performed, the sequential process of 5′-hydroxyl DMT-deprotection, coupling of **5′-DMT-deoxynucleoside-3′-phosphoramidites**, then iodine oxidation and capping again of unreacted 5′-hydroxyl groups, can take place in a repetitive, iterative manner until the desired full-length oligo-/polydeoxynucleotide is prepared.

The entire oligo-/polydeoxynucleotide is then detached from the solid support and deprotected to complete the synthesis. After synthesis of the oligo-/polydeoxynucleotide, purification takes place. At its simplest this just involves desalting, but **polyacrylamide gel electrophoresis** (**PAGE**) (Chapter 3) is the most commonly used approach to purify a synthesised oligodeoxynucleotide with high levels of purity. For higher yields in purification, or for oligodeoxynucleotides with unusual fluorophores or other modifications, then HPLC is the purification approach of choice after synthesis.

2.3.2 Chemical synthesis of oligonucleotides

Oligo- and polynucleotide synthesis is a more significant challenge than oligo- and polydeoxynucleotide synthesis for two major reasons. First, RNA molecules are particularly sensitive to both chemical and enzymatic degradation. Second, the 2′-hydroxyl group in each β-D-ribofuranose ring is potentially reactive and must be protected during chain synthesis. However, the protecting group must also be labile enough to be removed at the end of the synthesis, but stable enough to allow differential protection of the 5′-hydroxyl group during oligo-/polynucleotide synthesis. In other words, 2′- and 5′-hydroxyl protecting groups should be properly orthogonal. This can now be achieved with 5′-hydroxyl DMT-protection and **2′-tert-butyl dimethylsilyl** (**TBDMS**) group protection (Figure 2.8). All 5′-hydroxyl DMT-deprotection steps that take place during oligo-/polynucleotide synthesis require TCA (as with DNA). TBDMS groups are relatively robust during oligo-/polynucleotide synthesis, but are removed easily during final deprotection. At the end of oligo-/polynucleotide synthesis, a three-stage deprotection strategy is used. First, cyanoethyl groups are removed from the phosphodiester internucleotide links using triethylamine. Second, an ammonia/alcohol mixture can be used to cleave the oligo-/polynucleotide from the solid support and also remove exocyclic amino-protecting groups. Third, **tetrabutylammonium fluoride** (**TBAF**) is used to remove the TBDMS groups in a highly chemoselective manner from 2′-hydroxyl positions. Automated oligo- and polynucleotide synthesis is possible using the solid phase method, but this procedure is not so routine as is DNA oligo- and polydeoxynucleotide synthesis.

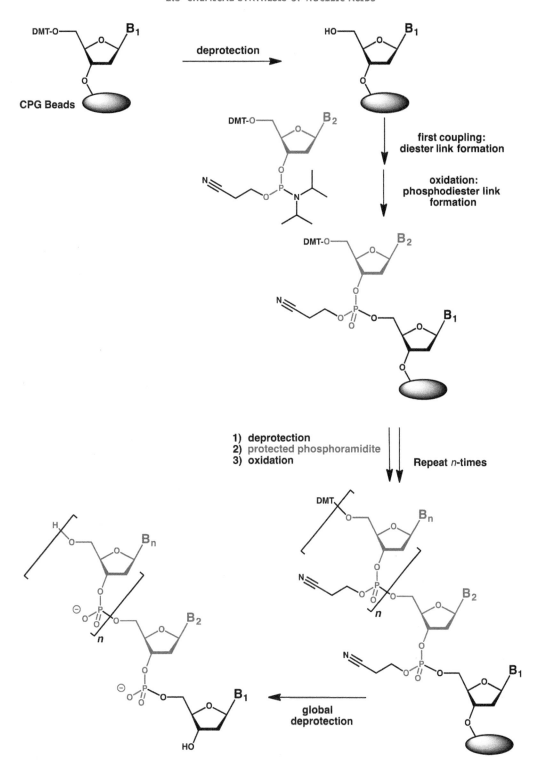

Figure 2.5 Solid phase DNA synthesis. 5′-Hydroxyl **dimethoxytrityl (DMT)**-deprotection of resin bound 3′-terminal deoxynucleoside residue is effected with **trichloroacetic acid (TCA)** (mechanism shown) prior to first coupling reaction enabled by phosphoramidite activation with **tetrazole** (mechanism shown), then oxidation of the newly formed diester linkage to a phosphodiester link. The process of 5′-hydroxyl DMT-deprotection, phosphoramidite coupling and then diester oxidation, continues for as many times as required (*n*-times), prior to global deprotection and resin removal under basic conditions.

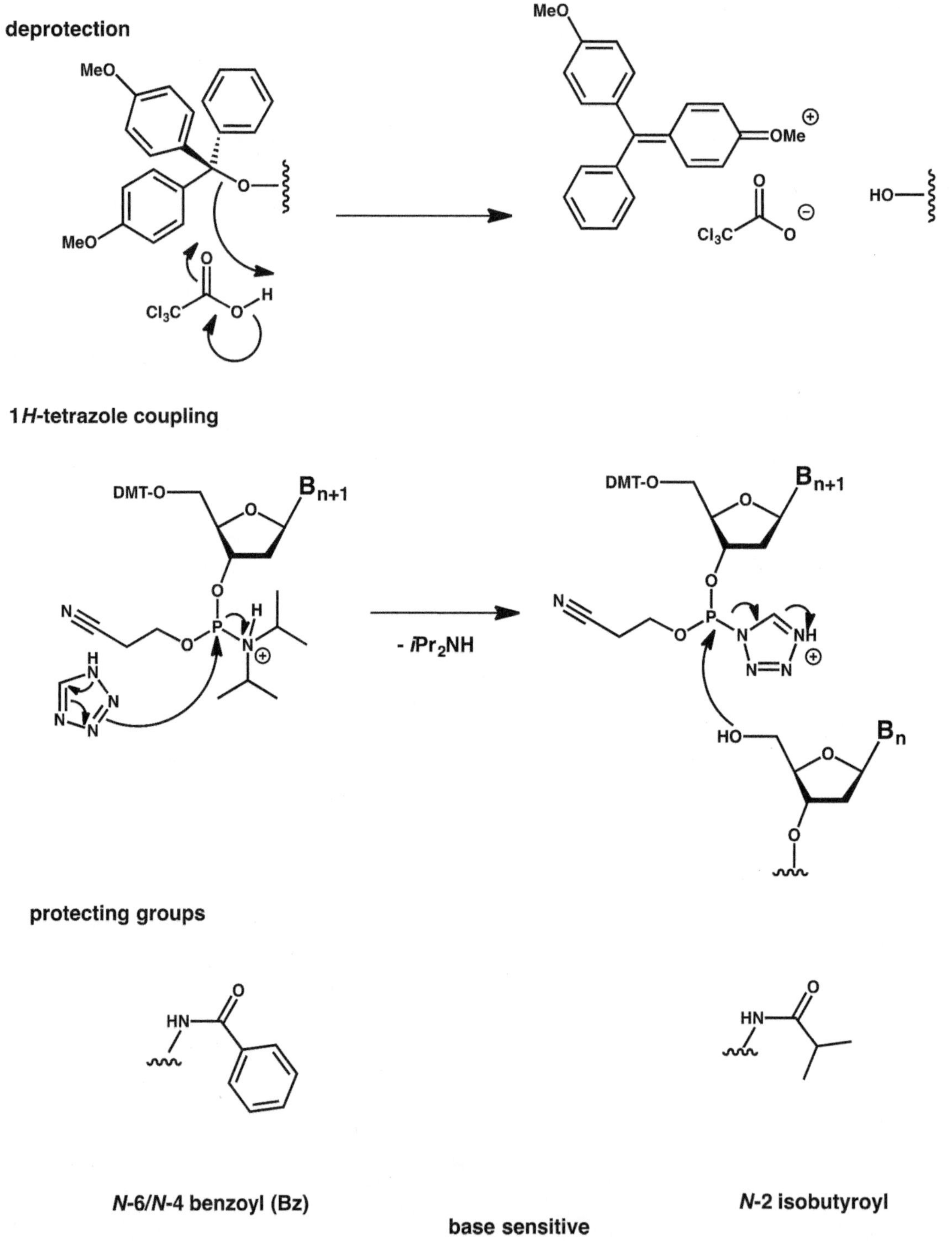

deprotection

1*H*-tetrazole coupling

protecting groups

N-6/N-4 benzoyl (Bz)   **N-2 isobutyroyl**

base sensitive

Figure 2.5   Solid phase DNA synthesis (Contd.). Most frequently used base protecting groups are shown; Bz: *N-6* benzoyl (adenine), *N-4* benzoyl (cytosine); *N-2* isobutyroyl (guanine). All are base sensitive. DNA chain is built up from 3′→5′ on **CPG bead** solid support. Post global deprotection and resin removal, the desired product oligo-/polydeoxynucleotide is then separated initially by precipitation by means of an agent such as **ethanol** and purified finally by **reversed-phase chromatography** or **IEC**, as appropriate (see Sections 2.6.1 and 2.6.3).

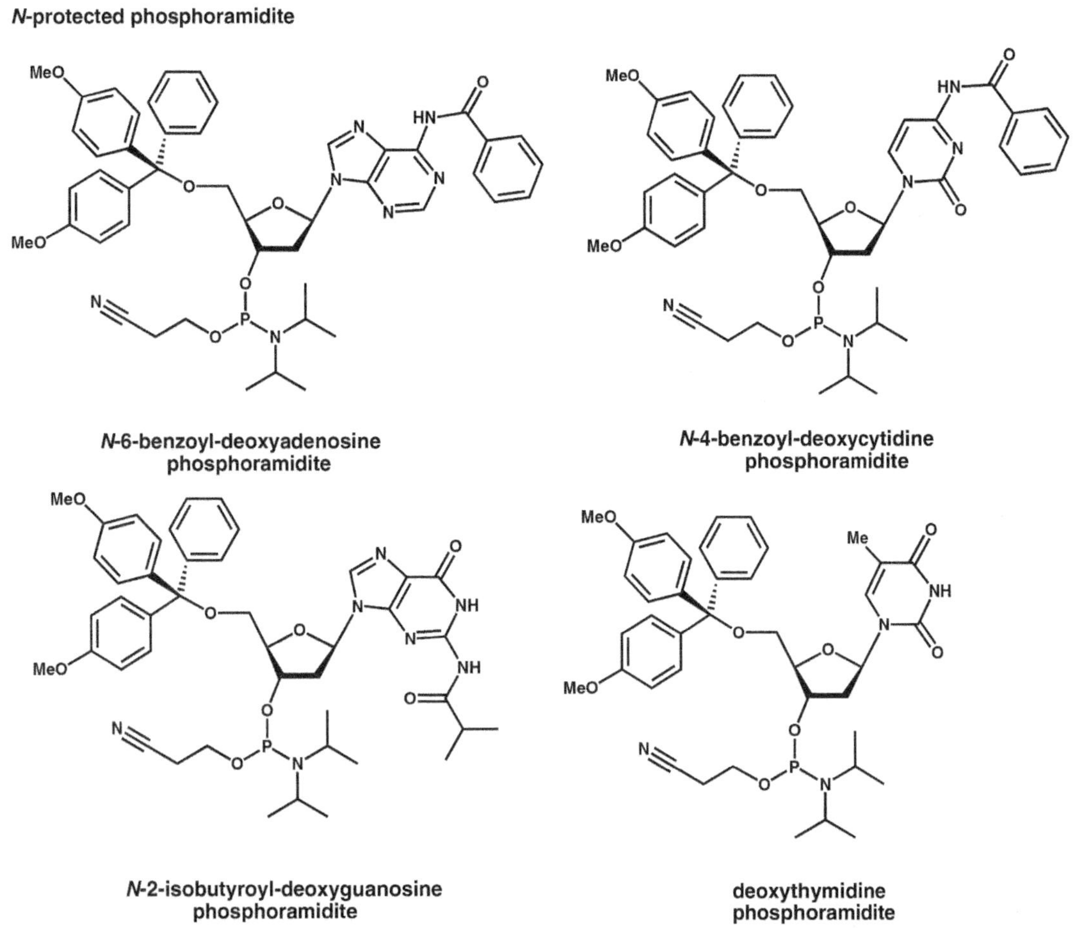

N-protected phosphoramidite

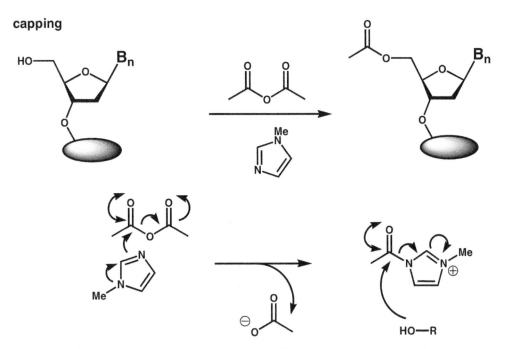

**N-6-benzoyl-deoxyadenosine
phosphoramidite**

**N-4-benzoyl-deoxycytidine
phosphoramidite**

**N-2-isobutyroyl-deoxyguanosine
phosphoramidite**

**deoxythymidine
phosphoramidite**

Figure 2.6   DNA phosphoramidite structures. The four main phosphoramidite "building blocks" used in solid phase DNA synthesis are illustrated, with base protecting groups where appropriate.

capping

Figure 2.7   Capping. Capping is a critical step that can be included after oxidation to cap any 5′-terminal hydroxyl groups that fail to combine with phosphoramidites Capping protects any unreacted 5′-hydroxyl groups and prevents "deletions" being incorporated into DNA during the synthesis. **Acetic anhydride** and **N-methylimidazole** are the capping reagents added for chemoselective combination with 5′-hydroxyl groups.

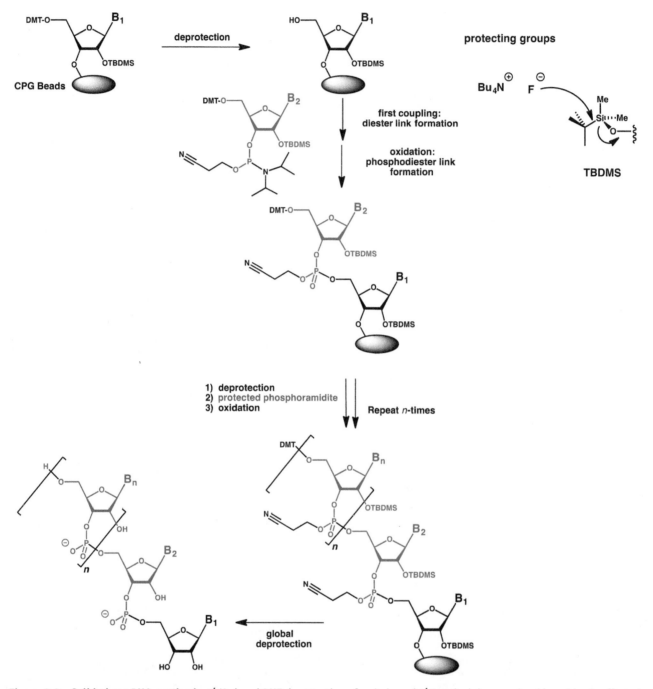

Figure 2.8   Solid phase RNA synthesis. 5′-Hydroxyl DMT-deprotection of resin bound 3′-terminal deoxynucleoside residue is effected with TCA (see Figure 2.5) prior to first coupling reaction enabled by phosphoramidite activation with tetrazole (see Figure 2.5), then oxidation of the newly formed diester linkage to a phosphodiester link. The process of 5′-hydroxyl DMT-deprotection, phosphoramidite coupling and then diester oxidation, continues for as many times as required (*n*-times), prior to global deprotection and resin removal under basic conditions. RNA synthesis requires that 2′-hydroxyl groups are protected during the synthesis by ***tert*-butyl dimethyl silyl (TBDMS)** protecting groups labile only to fluoride treatment from **tetrabutylammonium fluoride (TBAF)** (mechanism shown).

2.3.3 Useful deoxynucleotide/nucleotide modifications

Starting with **backbone modifications**, we will consider first, **phosphorothioate** links that result from the exchange of one non-bridging oxygen atom for a sulfur atom in a given phosphodiester link. The new link becomes a **chiral phosphate** centre where two diastereomers are possible. Phosphorothioate links can be incorporated into either oligo-/polydeoxynucleotides or oligo-/polynucleotides for various reasons. First, the "soft" sulfur can provide insight into metal ion function when compared to the "hard" oxygen atom. Also, phosphorothioates are far more stable to nucleases compared to the normal phosphodiester links, so are useful in cell culture and even for functional studies in animals *in vivo*. Solid-state syntheses can be used to insert phosphorothioate links into oligo-/polydeoxynucleotides, but a sulfurising agent is needed to replace iodine in order to introduce the non-bridging sulfur atom at each link (Figure 2.9). Should sulfur atoms replace both non-bridging oxygen atoms then **achiral phosphorodithioate** links are the result. Such links are very stable to hydrolysis in general, but **phosphorodithioate** links can provoke significant deviations from standard oligo-/polynucleotide structures and so are little used in relative terms. In comparison, **methylphosphonate** linkages that result when a methyl group replaces a non-bridging oxygen atom are undeniably more interesting (Figure 2.10). Unlike the standard phosphodiester links and the thioate links, methylphosphonates are non-ionic and can therefore help to provide insight into the importance of charge to the structure of oligo-/polynucleotides.

Sugar modifications are easily introduced into oligo-/polydeoxynucleotides or oligo-/polynucleotides by solid phase synthesis, and these can have many uses, but particularly in introducing stability to oligo-/polynucleotide structures. Oligo-/polydeoxynucleotides are comprised of 2′-deoxy-β-D-ribofuranose rings that can adopt C2′ endo and C3′ endo conformations (Chapter 1). In contrast, oligo-/polynucleotide β-D-ribofuranose rings only exhibit the C3′ endo conformation. Accordingly, 2′-substituents such as fluoro, amino, methoxy, alloxy, etc., stabilise corresponding oligo-/polynucleotides with respect to conformation and also to hydrolytic attack. Most recently, this concept was extended into the development of **locked nucleic acids** (**LNAs**) that could be profoundly important in RNA therapeutics in future, owing to their conformational and metabolic stability (Figure 2.10). Unsurprisingly, solid phase synthesis also lends itself to the incorporation of **non-natural base modifications** into an oligo-/polydeoxynucleotide or an oligo-/polynucleotide as well. Non-natural bases are useful probes for delineating structure and function. For example, **7-deazapurine nucleotides** that lack the N-7 nitrogen are useful probes for those situations where this nitrogen acts as a hydrogen bond acceptor (e.g. Hoogsteen base pairing, see Chapter 1). Base analogues may instead be fluorescent or dye labelled (see Chapter 4). For instance, **2-aminopurine (AP)** is a useful fluorescent probe that can directly substitute for adenine in dA/A nucleotides and can base pair with thymine in dT nucleotides without affecting the DNA double helical structure (Figure 2.11). Also, **3-methyl-8-isoxanthopterin** is another fluorescent probe that mimics the behaviour of guanine in dG/G nucleosides and is introduced by chemical synthesis. The uses of such fluorescent probes are

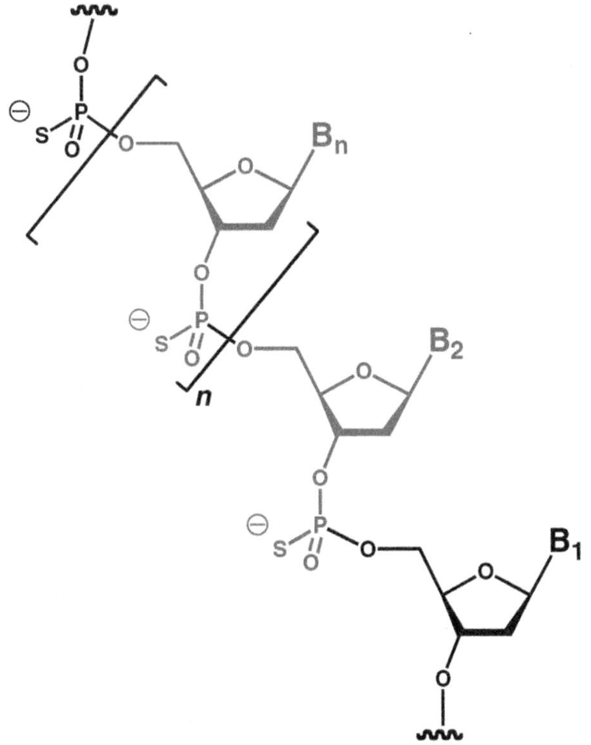

Figure 2.9 Phosphorothioates. Oligo-/polynucleotide or oligo-/polydeoxynucleotide phosphorothioates have extra sources of chirality in that each phosphorothio diester link can exist in either of two diasteromeric forms.

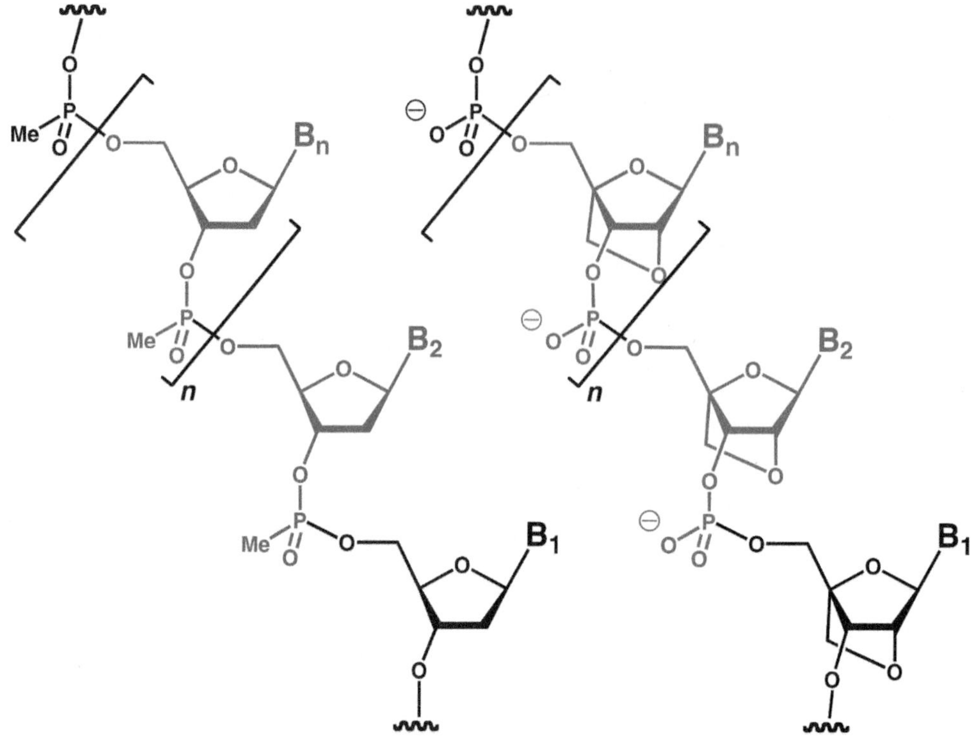

Figure 2.10 **Methylphosphonate and locked nucleic acids.** Oligo-/polynucleotide or oligo-/polydeoxynucleotide **methylphospho-nates** (**left**) have chiral properties in common with phosphorothioate. However, lack of charge on the phosphodiester backbone results in significantly more hydrophobic biophysical characteristics. **Locked nucleic acids** (**LNAs**) are now the subject of great interest since more stable RNA A-helices may be formed between wild-type RNA and base-complementary LNAs.

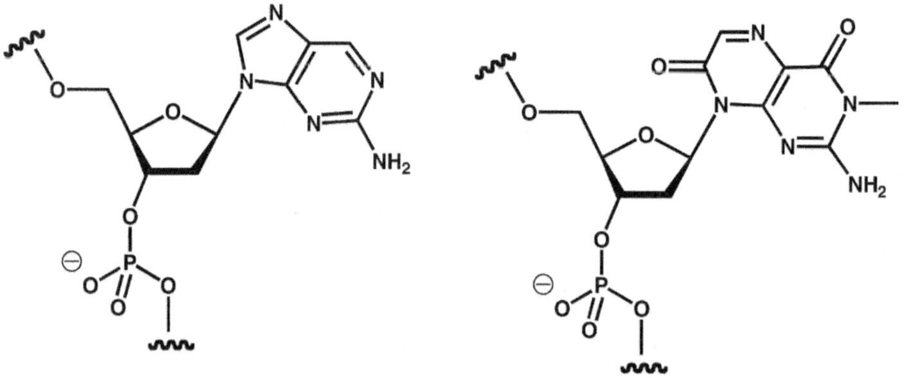

Figure 2.11 **Fluorescent purines.** Two very useful non-invasive DNA fluorophores that can be included by solid phase DNA synthesis are: **2-aminopurine** (**left**) and **3-methyl-8-isoxanthopterin** (**right**). The fluorescence quantum yield depends on the base stacking and both bases are useful probes of conformational change/state in DNA.

accounted for elsewhere (see Chapter 4). Furthermore, **chemical cross-linkers** and **photo-affinity cross-linkers** can also be introduced to bases in oligo-/polydeoxynucleotides or oligo-/polynucleotides during or post-synthesis. These can be used to attach oligo-/polydeoxynucleotides or oligo-/polynucleotides to surfaces or other associated molecules, if required. Alternatively, some cross-linkers may even actually perturb nucleic acid functions by the introduction of significant covalent distortions, such as compounds like the "**nitrogen mustards**" [e.g., **HN2: bis(2-chloroethyl)methylamine; HN3: tris(2-chloroethyl)amine**] or *cis*-**platin** and **carbo-platin** (potent anticancer drugs) (Figure 2.12).

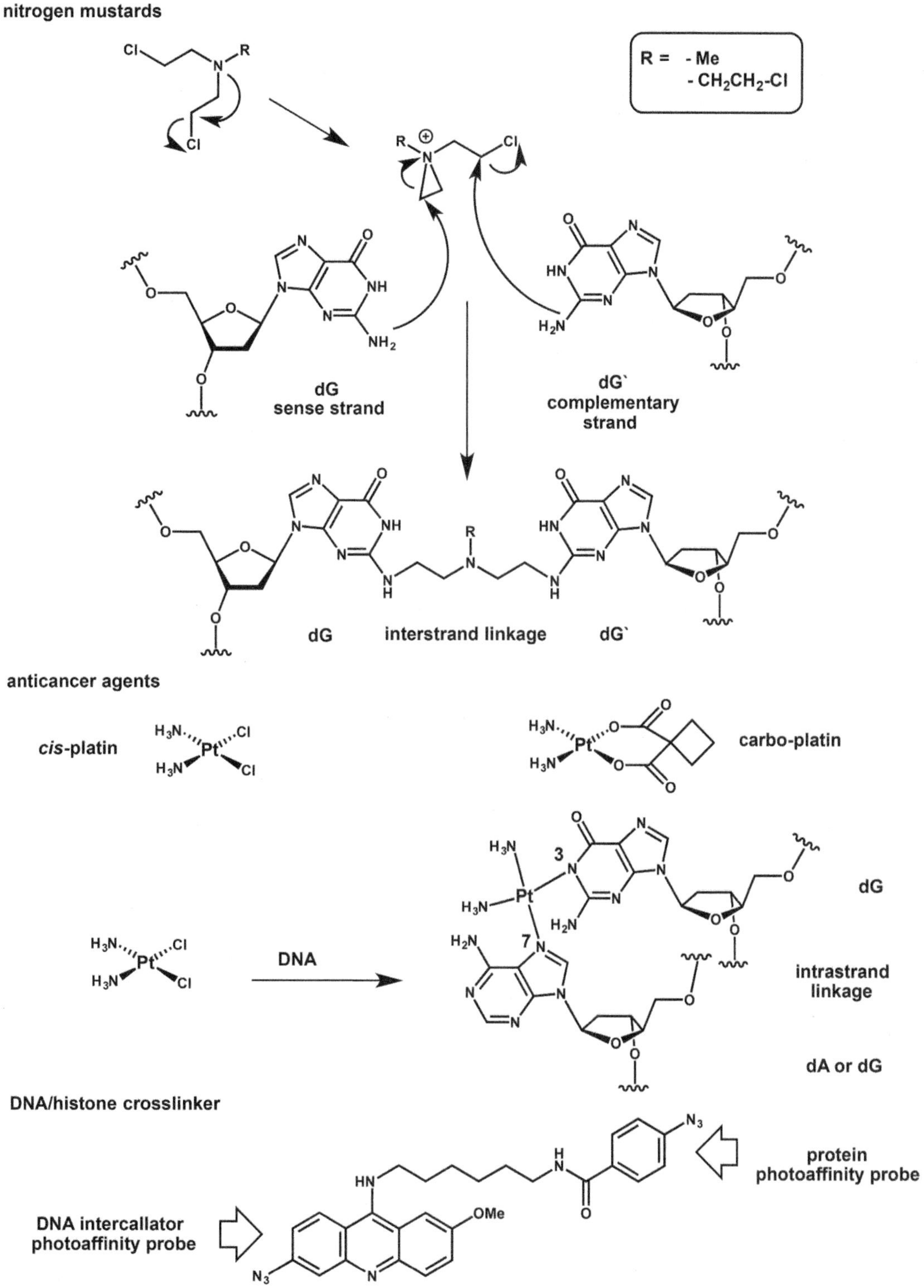

nitrogen mustards

anticancer agents

DNA/histone crosslinker

Figure 2.12 **DNA cross-linking.** The deleterious properties of **nitrogen mustards** are explained through the illustrated interstrand linkage mechanism that makes DNA impossible to duplicate or transcribe. Intrastrand cross-linking is the basis of action for anti-cancer drugs such as **cis-platin** and **carbo-platin**. This is intended to prevent DNA duplication and hence cancer cell division. DNA cross-linking to proteins (such as histones) uses a **non-covalent DNA intercalator** with two azide functional groups. Both azides are activated for covalent coupling under photo-chemical conditions so that DNA subsequently becomes covalently linked to protein.

2.4 Chemical synthesis of oligosaccharides

Whilst solid phase approaches dominate the chemical syntheses of the shorter nucleic acids and polypeptides, the same is not true for oligo-/polysaccharides. The problem is the complexity of oligo-/polysaccharide primary structures. Where homoglycans are concerned, this is not valid, but many oligo-/polysaccharides associated with proteins or lipids to form glycoproteins or glycolipids can be immensely complicated. The complexity comes from the fact that while natural sugars all possess one anomeric carbon that acts as the "electrophilic acceptor" for a glycosidic link, each monosaccharide can present at least two or more different "nucleophilic donor" hydroxyl groups leading to complex branching. Hence, primary structure branching and control of anomeric carbon stereochemistry are two of the great challenges for chemical synthesis of oligo-/polysaccharides. Consequently, oligo-/polysaccharide chemical syntheses demand a range of chemo-selective protecting groups capable of orthogonal, selective deprotections. Special reagents are also required to form glycosidic links with precise stereocontrol, possibly with the anchimeric assistance of neighbouring group protecting groups. Another challenging aspect of oligo-/polysaccharide chemical syntheses is that certain protecting groups and coupling reagents may be useful for the protection and coupling involving only a small range of monosaccharide building blocks. Quite clearly, the chemical synthesis of oligo-/polysaccharides is very much "work in progress" and still awaits the simplifying principles that underlie the chemical syntheses of peptides and nucleic acids. Nevertheless, there are some principles emerging that are worth reviewing with examples.

2.4.1 Protecting groups

Chemical synthesis of oligo-/polysaccharides demands the use of orthogonal protecting groups capable of selective removal. **Benzyl ether** protecting groups have seen widespread use in oligo-/polysaccharide chemistry. The benzyl ether acts as a "permanent" protecting group introduced to protect hydroxyl functional groups not expected to be involved in glycosidic link formation. Hence benzyl ether protected groups are robust enough to remain inert during oligo-/polysaccharide synthesis, only to be removed by mild catalytic hydrogenolysis at the end. Otherwise, oligo-/polysaccharide synthesis makes use of **base-labile protecting groups** such as **acetyl**, **benzoyl** and **pivaloyl** that are frequent monosaccharide C-2 protecting groups. In this capacity, these protecting groups can offer **neighbouring group participation** during glycosidic link formation and so assist in the control of stereochemistry (Figure 2.13). Base labile protecting groups are also intended to provide temporary protection during a synthesis to unmask a key hydroxyl group for reaction at just the right juncture. Other temporary protecting groups that may be used are **acid-labile protecting groups** such as the DMT group. There is also the TBDMS group that may be removed by TBAF. Both DMT and TBDMS have also seen service in oligonucleotide synthesis.

2.4.2 Creating glycosidic links

Coupling of monosaccharides requires activation of the anomeric position in one monosaccharide (**electrophilic acceptor** from **glycosyl donor**) and protection of all functional hydroxyl group positions except for the selected hydroxyl group in the other monosaccharide (**nucleophilic donor** from **glycosyl acceptor**). The very simplest form of this is illustrated in Figure 2.13. Initially, protecting group manipulation is required, followed by coupling that should be as stereocontrolled as possible. Protecting groups are chosen to assist in the current glycosidic link formation, but also with an eye to subsequent glycosidic link formations. Making the right choices of protecting groups and coupling reagents is essentially an empirical exercise based upon a detailed knowledge of oligo-/polysaccharide chemistry. In other words, what will and will not work is still effectively the preserve of an initiated few who possess detailed experience. In recent times, the following synthetic methodologies have emerged as leading ways to effect glycosidic link formation:

1. **glycosyl trichloroacetamidate** coupling

2. **1,2-anhydrosugar-thioglycoside** coupling

3. ***n*-pentenyl glycoside** coupling

The first (1) has been extremely useful in many solution-phase oligo-/polysaccharide syntheses. Glycosyl donors (giving electrophilic acceptors) are activated under very mild conditions by catalytic amounts of **trimethylsilyl triflate (TMSOTf)** or other triflates including **dibutylboron triflate (DBBOTf)**, not to mention **boron trifluoride in diethylether (BF$_3$/Et$_2$O)**. This method of glycosidic link formation is striking for control of the stereochemistry (α or β) of the anomeric centre and hence of the glycosidic link (Figure 2.13). The second method (2) involves 1,2-anhydrosugars that are otherwise known as glycals. Glycals are readily subject to epoxidation with **dimethyldioxirane (DMDO)**, only to react onwards with a thiolate to yield thioglycoside (Figure 2.13). These thioglycosides are robust glycosyl donors that are mobilised for coupling by treatment with thiophilic promoters such as **di-*tert*-butyl peroxide/methyl triflate (DTBP/MeOTf)**. Once again, high α- or β- selectivity can be achieved in glycosylation reactions. Finally, the third method (3) involves ***n*-pentenyl glycoside (NPG)** derivatised glycosyl donors that are activated by strong electrophilic reagents such as ***N*-iodosuccinimide/triethylsilyl triflate (NIS/TESOTf)**. Coupling yields are typically excellent and so is control of stereoselectivity.

coupling: glycosidic link formation

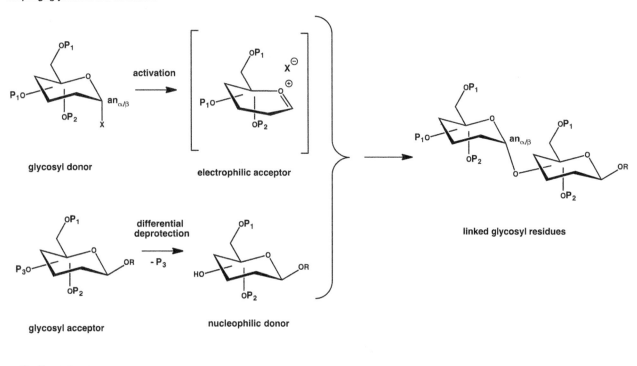

glycosyl donor   *electrophilic acceptor*   *linked glycosyl residues*

glycosyl acceptor   *nucleophilic donor*

protecting groups

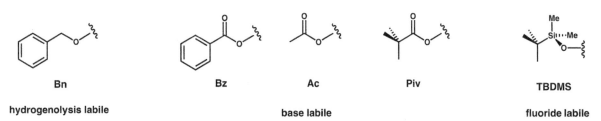

Bn   **Bz**   **Ac**   **Piv**   **TBDMS**

hydrogenolysis labile   base labile   fluoride labile

Figure 2.13 General strategies in oligo-/polysaccharide synthesis. Glycosidic links are created by a two-step process in which a **glycosyl donor** (protected on alcoholic functional groups by protecting groups **P$_1$** and **P$_2$**) becomes activated for coupling by the loss of leaving group **X**, thereby forming an **electrophilic acceptor**. In a separate process, a **glycosyl acceptor** (protected on alcoholic functional groups by protecting groups **P$_1$**, **P$_2$** and **P$_3$**) is differentially deprotected on a selected alcohol functional group, thereby forming a **nucleophilic donor**. In combination, donor and acceptor form a new glycosidic link with anomeric centre stereochemistry determined by conditions, reagents and neighbouring group participation from remaining protecting groups. Standard protecting groups are shown: **Bn:** benzyl (ether); **Bz:** benzoyl (ester); **Ac:** acetyl (ester); **Piv:** pivaloyl (ester); **TBDMS:** *tert*-butyl dimethylsilyl (ether).

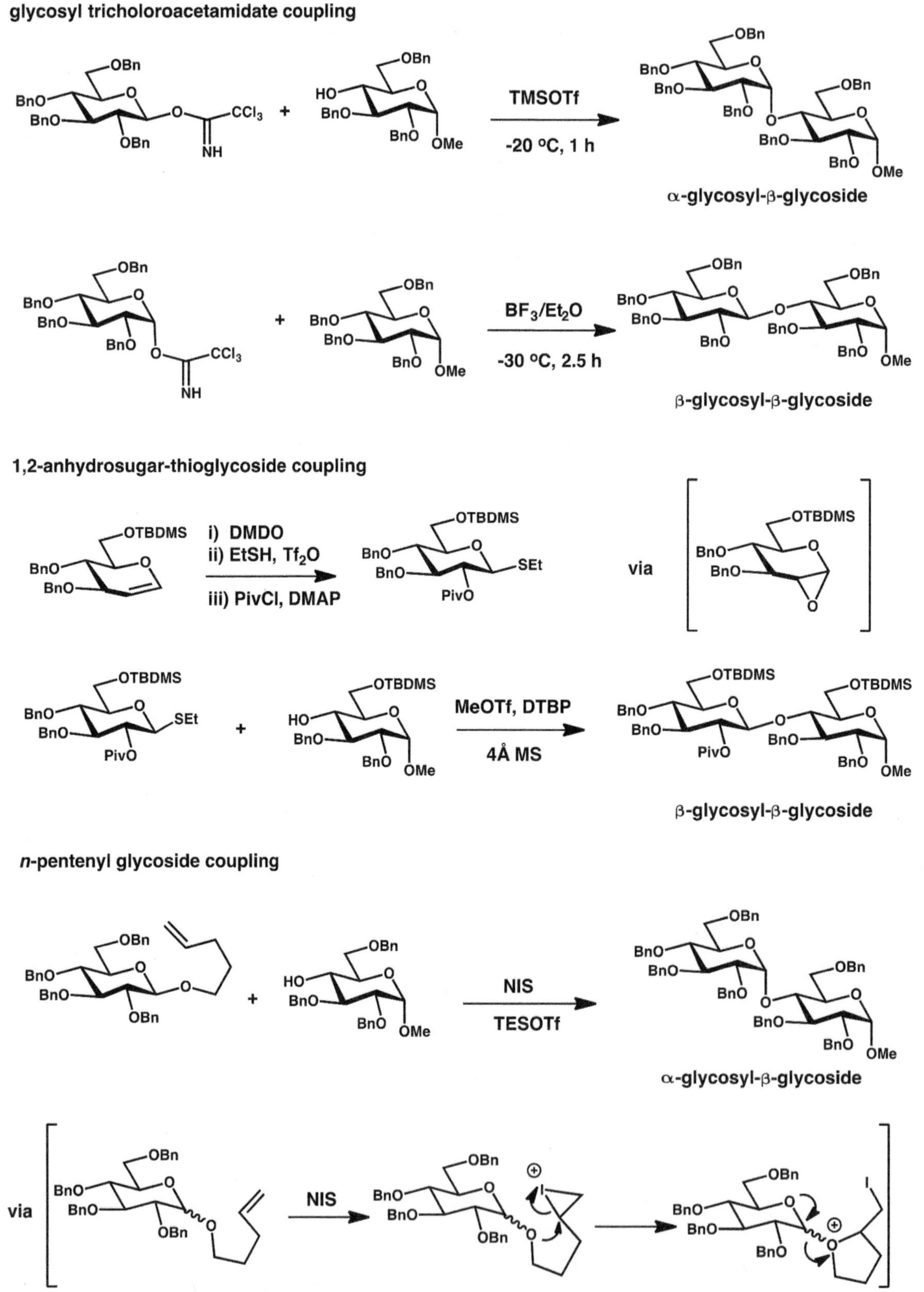

glycosyl tricholoroacetamidate coupling

α-glycosyl-β-glycoside

β-glycosyl-β-glycoside

1,2-anhydrosugar-thioglycoside coupling

β-glycosyl-β-glycoside

***n*-pentenyl glycoside coupling**

α-glycosyl-β-glycoside

via

Figure 2.13   **(Contd.)** Glycosidic links are created by three main chemical strategies. In trichloroacetamidate coupling, the glycosyl donor can be activated by either **trimethylsilyl triflate** (**TMSOTf**) or **boron trifluoride** (**BF₃**) leading to glycosyl-glycoside products of opposite stereochemistry. In anhydrosugar-thioglycoside coupling, the glycosyl donor is activated in several stages, **dimethyldioxirane** (**DMDO**) epoxidation is followed by **thioethanol** (**EtSH**) ring opening aided by **triflic anhydride** (**Tf₂O**), and the alcohol functional group so generated is capped by **pivaloyl chloride** (**PivCl**) activated by **dimethylaminopyridine** (**DMAP**). Coupling then requires further **di-*tert*-butyl peroxide** (**DTBP**) activation of the glycosyl donor assisted by **methyl triflate** (**MeOTf**) and **molecule sieves** (**MS**) (4 Å pore size). In pentenyl glycoside coupling, the glycosyl donor is activated by ***N*-iodosuccinimide** (**NIS**) with **triethylsilyl triflate** (**TESOTf**).

2.4.3 Solid phase oligosaccharide chemistry

Solid phase synthesis is developing surprisingly rapidly given the ease with which much of the solution phase chemistry described in Sections 2.4.1 and 2.4.2 readily adapts for solid phase use. However, imperfect coupling yields and incomplete α/β stereocontrol with glycosidic link formation prevent this approach to oligo-/polysaccharide synthesis becoming routine. The solid support used in synthesis is typically a polymeric variation of the Merrifield resin. Two general approaches are used: either the glycosyl donor or the glycosyl acceptor can be initially immobilised on the solid support (Figure 2.14 gives one example of each). These two approaches can be sufficiently different to demand different coupling reagents, protecting groups and even linker to the solid support. The chemical nature of the linker determines all other protecting groups and coupling manipulations that may be carried out during the entire synthesis, especially since the linker chemistry must be inert (orthogonal) to all other deprotection and coupling steps that take place during the given synthesis. Hence, when a **silyl ether** is used as a linker then obviously other silyl protecting groups cannot be employed elsewhere in the synthesis. Fortunately, silyl deprotection is undoubtedly orthogonal to other oligosaccharide-associated chemistries and these linkers are very resistant to the coupling conditions involving the three main methods of coupling described previously (see Section 2.4.2). In the event, the silyl ether appears to be the most appropriate form of linker when an active glycosyl donor is initially immobilised on the solid support. Otherwise, **thioglycoside** and **ether** linkers appear to be more appropriate when an active glycosyl acceptor is initially bound to solid support instead. Thioglycoside linkers are robust to all but thiophilic reagents and so are cleaved by **N-bromosuccinimide/di-*tert*-butyl pyridine** (**NBS/DTBP**) in MeOH to leave a methylglycoside. Otherwise, the ether linkers can be subject to novel release conditions such as the use of **Grubbs' catalyst** catalysed **metathesis** with **ethene** (Figure 2.15).

Depending upon whether an active glycosyl donor is solid phase attached or an active glycosyl acceptor as appropriate, a regular cycle of synthesis involves the cyclical introduction and coupling of a protected glycosyl acceptor or glycosyl donor respectively, usually preceded by a specific step of selective deprotection to make the coupling possible. In spite of progress in solid phase oligosaccharide synthesis, oligosaccharides of lengths over 4–5 monosaccharide units remain significant challenges for synthesis and there remains no universal protocol, unlike oligopeptide or oligonucleotide synthesis. However, the great impact of solid phase oligo-/polysaccharide synthesis over solution-phase methods is that **libraries** of di- and tri-saccharides can be synthesised. Such libraries can then be used in screening studies to probe a number of problems in chemical biology. The final synthesis (illustrated in Figure 2.15) was also actually performed with full automation as well. Hence the possibility for routine fully automated oligosaccharide synthesis should not be long to arrive.

2.5 Chemical synthesis of lipids

There is, by and large, no need for the chemical synthesis of many of the phospholipids that make up macromolecular lipid assemblies in cells since these are readily available from natural sources and are purified from these by HPLC. In effect, biological syntheses of lipids dominate chemical syntheses. However, there is some value in illustrating the general method of synthesis of acyl glycerides and phospholipids since these synthetic approaches are important in lipid semi-synthesis should a fluorescent tag be required, etc.

Acylglycerols derive from the controlled esterification of hydroxyl functional groups in the alkyltriol known as glycerol. One major problem with the synthesis of pure acylglycerols is that **mono- and diacyl glycerols** easily migrate to adjacent free hydroxyl groups, typically catalysed by acid, base or heat. Therefore, a suitable protecting group strategy is absolutely essential for the chemical synthesis of defined acylglycerols. For instance, glycerol can react with acetone under acidic conditions to give a **1,2-isopropylidene** (**acetonide**), leaving one primary hydroxyl free for subsequent acylation. Alternatively, benzaldehyde can react to give **1,2-** or **1,3-benzaldehyde glyceryl acetals** that can be separated by crystallisation and used for esterification. Typically, acylation can then be performed by **carbonyldiimidazole** (**CDI**) activation of the selected fatty acids (Figure 2.16). Thereafter, acetal deprotection can be followed by further esterification as appropriate. In the case of phospholipid synthesis, this can be reduced to the controlled esterification of L-α-**glycerophosphocholine.CdCl$_2$** (**GPC**) made possible by CDI activation of the selected fatty acids (Figure 2.17). Other phospholipids can then be prepared from this phosphatidylcholine by headgroup **transphosphatidylation** mediated by the enzyme known as **phospholipase D**. Lipid synthesis may appear trivial, but there is nothing trivial about the purification of lipids to homogeneity post synthesis. In this respect reversed-phase HPLC becomes a powerful technique, but using a specially developed **evaporative light scattering** (**ELS**) detector to allow for the detection of eluting lipid analytes that do not possess any strong chromophore (see Section 2.6.3).

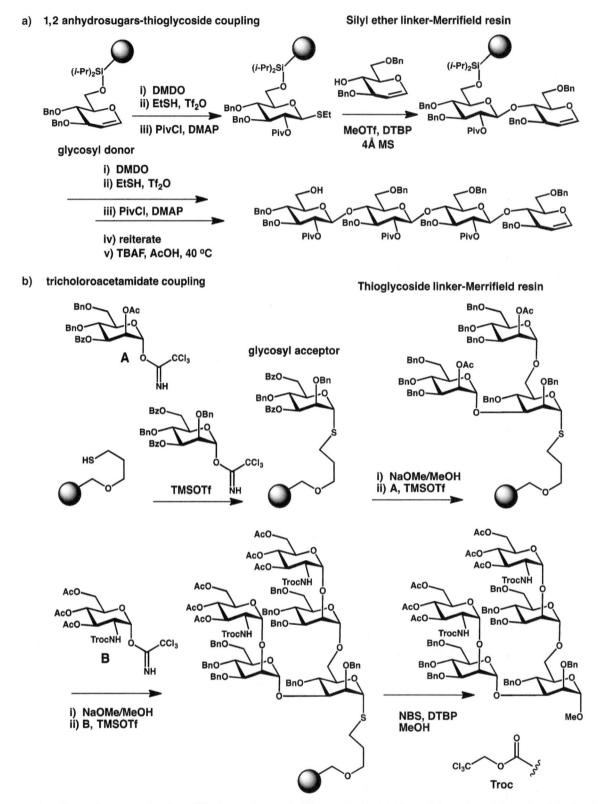

Figure 2.14 Alternative strategies for solid phase oligosaccharide synthesis. (a) Glycosyl donor immobilised by silyl ether linker to resin and activated for coupling by the anhydrosugar-thioglycoside coupling methodology (see Figure 2.13). Final removal of the product is effected with fluoride ions delivered from TBAF; **(b)** Glycosyl acceptor immobilised by thio-ether linker to resin aided by TMSOTf. Repetitive facile deprotection of benzoyl (Bz) protecting groups (sodium methoxide, NaOMe) and further TMSOTf assisted coupling of glycosyl donors to acceptors leads to the final product that is removed from the resin by **di-*tert*-butylpyridine (DTBP)** activation assisted by ***N*-bromosuccinimide (NBS)**. Solid phase oligosaccharide synthesis is an excellent approach for the synthesis of combinatorial libraries of small oligosaccharides for use in screening in biological research. For most indicated protecting groups, see Figure 2.13. Note the additional protecting group mentioned **trichloroethyloxycarbonyl (Troc)**.

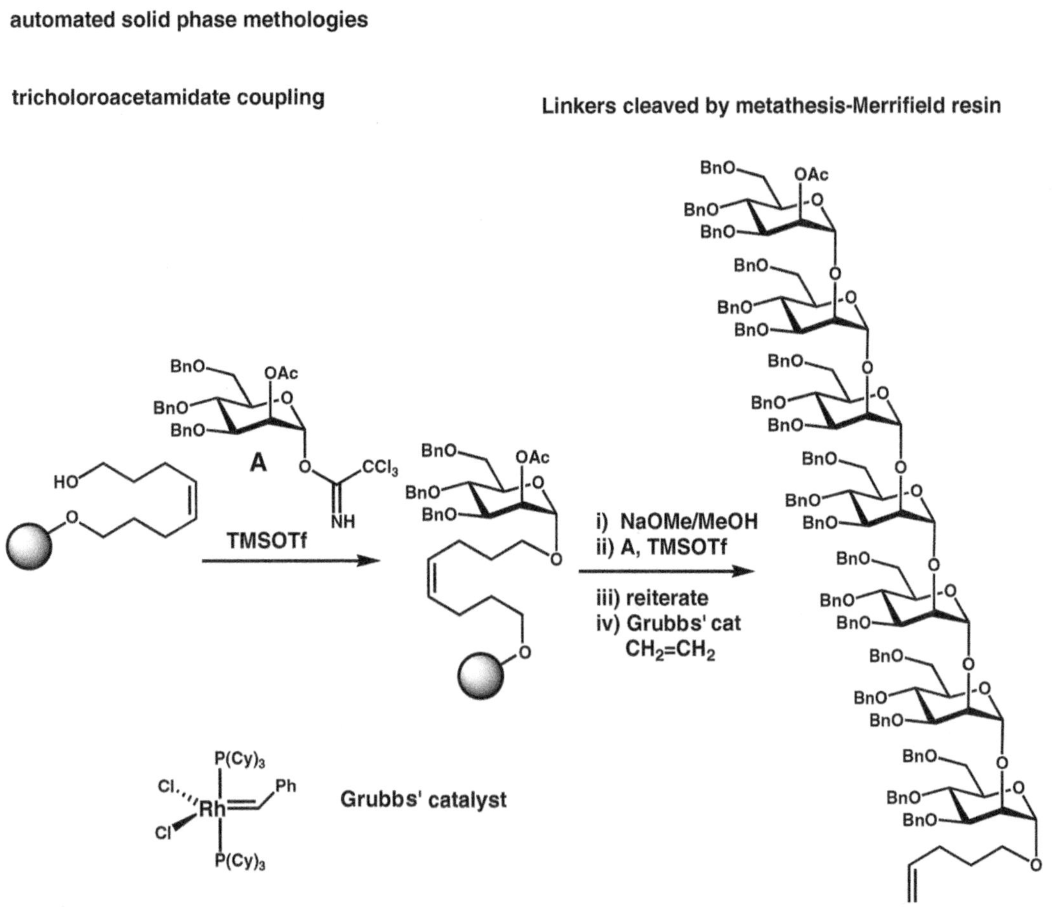

automated solid phase methologies

tricholoroacetamidate coupling

Linkers cleaved by metathesis-Merrifield resin

Figure 2.15 Automated solid phase oligos-polysaccharide synthesis. Glycosyl donor (**A**) is coupled to a solid phase Merrifield-like resin through an alkenyl alcohol linker to set up a glycosyl acceptor attached to solid phase. The double bond allows for clean removal of product at the end of solid phase synthesis by metathesis using **Grubbs' catalyst** in the presence of **ethylene** gas. For protecting groups, see Figure 2.13. TMSOTf is used to activate each added glycosyl donor (**A**) during the synthesis through controlled removal of each trichloroacetamidate.

2.6 Biological synthesis of biological macromolecules

Biological synthesis is used here to describe harnessing the synthetic potential of living organisms for directed synthesis of biological macromolecules of interest. The study of biological macromolecule structure and function is highly dependent on the availability of material to study, hence there must be a huge emphasis on synthesis and purification before anything else is possible. Modern-day biological syntheses, particularly of proteins and also nucleic acids, may frequently require recombinant techniques and the growth of recombinant organisms (in the cases of carbohydrates and lipids, this is not usually the situation) (see Section 2.7 and Chapter 3). Thereafter the main challenge is purification of the biological macromolecule or lipid of interest to homogeneity in the face of all the other cellular components. In effect, since cells (whether recombinant or not) do the actual work of synthesis, then purification becomes our focus. Therefore, the chemical biology reader will be introduced first to those approaches frequently used in the purification of biological macromolecules and lipids from organisms.

2.6.1 Ion exchange chromatography

Ion exchange chromatography (IEC) separates molecules according to differences in their accessible surface charge. The method is very widely used, especially in protein purification, as separation may be applied to any molecule that is charged and soluble in an aqueous system. The mild elution conditions also usually result in high recovery and high retention of biological activity in the eluting biological macromolecule (**analyte**). There are two classes of ion exchange (Figure 2.18). In **anion exchange**, the column carries functional groups bearing a positive charge, and hence separates molecules according to

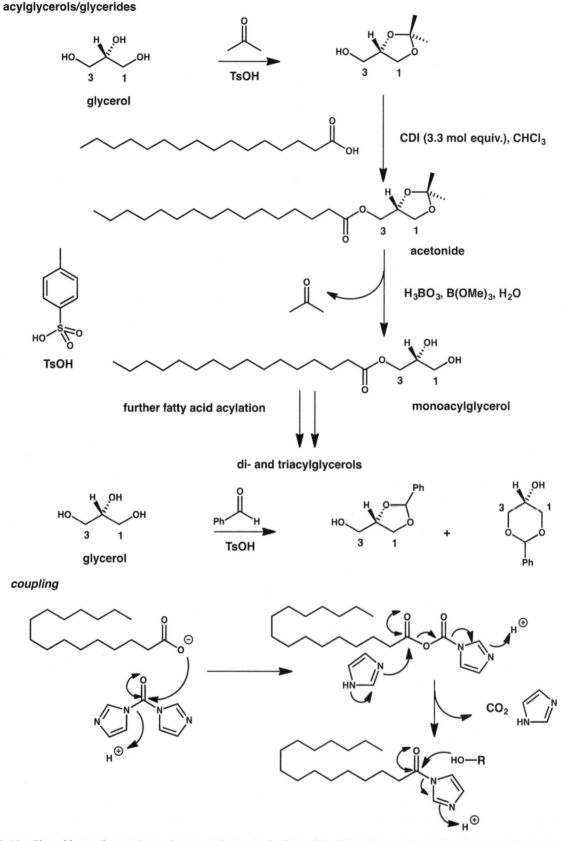

Figure 2.16 Glyceride syntheses. Protecting strategies are typically used in the synthesis of monoacylglycerols. Differential protection is made possible with the assistance of **tosic acid** (**TsOH**) catalysis to give a variety of acetals. Unprotected alcohol functional groups can be esterified with free fatty acids with the assistance of **carbonyldiimidazole** (**CDI**). Deprotection is effected using mild Lewis acids such as **methyl borate/boric acid**. Mixed diacylglycerols can be prepared in a similar way, making use of protecting groups and esterification. Triglycerides can be prepared without the need for protecting groups.

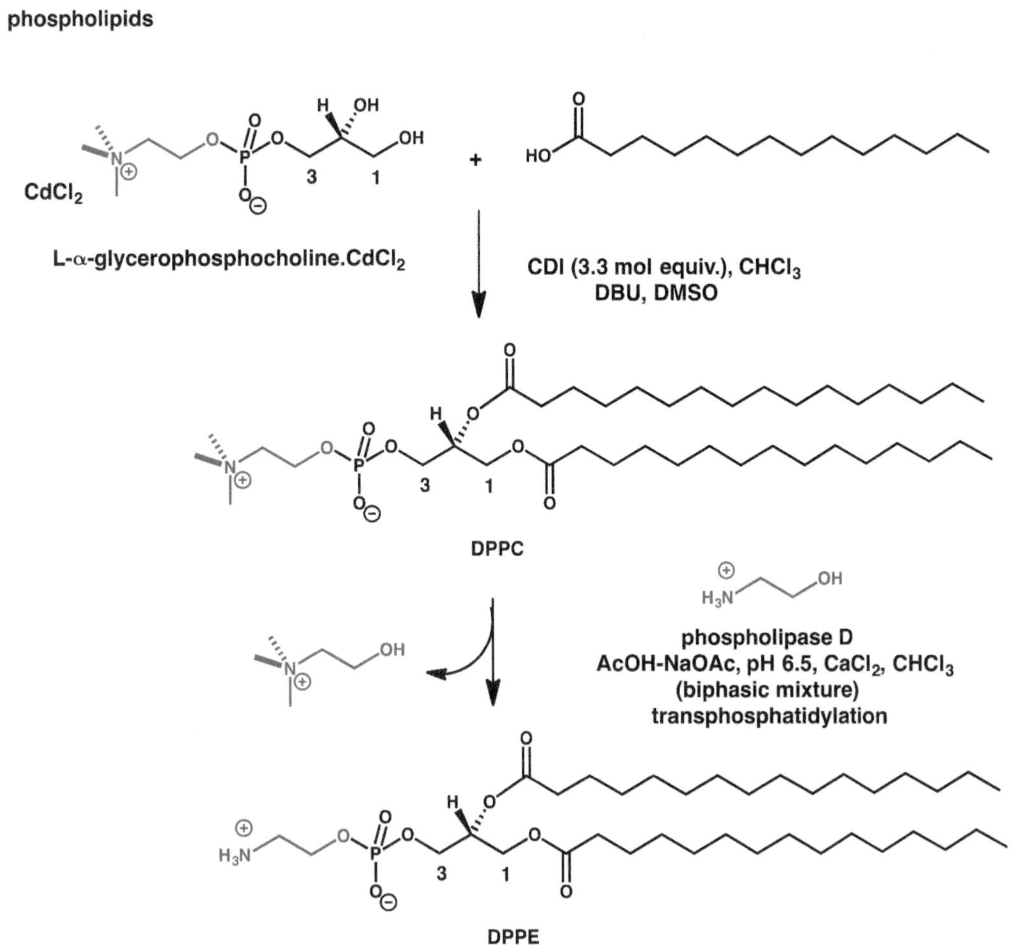

phospholipids

Figure 2.17 Phospholipid syntheses. Starting from **L-α-glycerophosphocholine.CdCL$_2$**, fatty acids are coupled to free alcohol functional groups by means of CDI yielding a **phosphocholine (PC)** product. Head group exchange is made possible with **phospholipase D** enzyme.

the negative (anionic) charges they carry. In **cation exchange**, the column carries functional groups bearing a negative charge, and hence molecules are separated according to the positive (cationic) charges that they carry. Binding is electrostatic in nature between analyte molecules of interest and the column. Since binding only takes place over a relatively short distance, the charges that are accessible on the surface are the most important factors. Both anion and cation exchange resins are classed as being either **weak** or **strong**; this does not refer to the strength of binding but to the effect of pH on the charge of the functional groups. Weak ion exchange resins gain or lose electrical charge with changes in the pH of the **mobile phase** buffer/solution that is used to elute the column. Strong ion exchange resins maintain their charge irrespective of pH changes in the mobile phase.

Functional groups associated with an ion exchange resin require a **counter-ion**; this can either be salt molecules, buffer molecules or the sample analyte itself. Binding to an ion exchange column occurs when the analyte acts directly as the counter-ion to the functional groups on the ion-exchange resin. The usual method of running IEC is to bind the analyte onto the column under low salt conditions, then elute by increasing the salt so that salt molecules displace analyte back into the mobile phase, in so doing picking up a salt counter-ion from the mobile phase (Figure 2.19). Changing pH may also be used together with, or instead of, a salt gradient to elute the analyte, particularly when using weak ion exchange.

The pH of the mobile phase is vitally important in IEC, as most charged amino acid residues on a protein surface are themselves weak ion exchangers. In general, this means that analyte molecules will be more positively charged at lower pH values, and more negatively charged at higher pH values. The **isoelectric point (pI)** is the pH at which the net charge on the molecule is zero; this is a useful starting point when considering which type of ion exchange medium to use for purification (Figure 2.20). An analyte with a high isoelectric point (basic) is positively charged at neutral pH, and therefore using a mobile phase at neutral pH would most likely result in surface positive charge. Hence, the analyte would be best applied to a cation exchange column and *vice versa*. Isoelectric points should only be used as guidelines since the isoelectric point is the pH at which the **net charge** is zero, a value that does not necessarily define **net surface charge**.

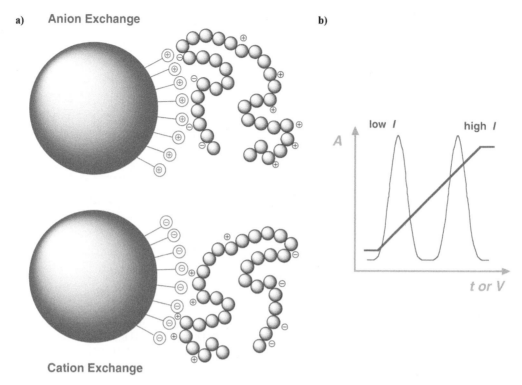

Figure 2.18 Cation and anion exchange chromatography. (a) Anion exchange columns carry a positive charge and bind negatively charged analyte molecules. Cation exchange columns carry a negative charge and bind positively charged analyte molecules; (b) Target analyte ion molecules are eluted (with respect to time t or elution volume V) by increasing the ionic strength of the eluant from low ionic strength (low I) to high ionic strength (high I).

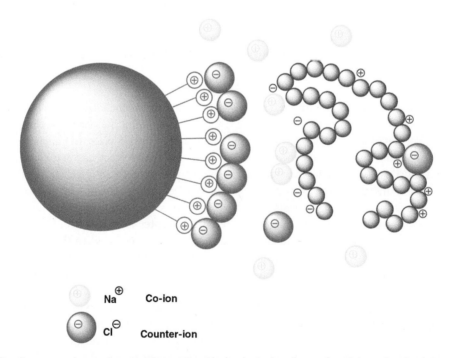

Figure 2.19 Elution from an anion exchange column. Negatively charged surfaces of analyte molecules interact with cationic gel beads at the given pH. As the salt concentration increases, chloride (Cl^-) counter ions help to displace analyte molecules from gel binding in order to encourage elution; sodium (Na^+) co-ions assist elution by associating closely with neutralising the negative charges of newly exposed charged surfaces.

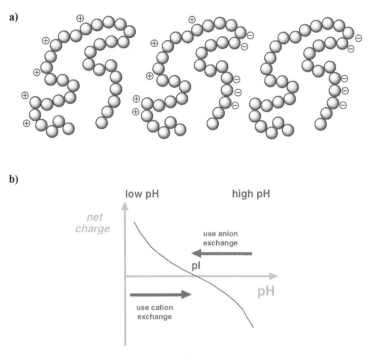

Figure 2.20 Isoelectric point and ion exchange chromatography interactions. (a) pI represents the pH at which the net charge of an analyte molecule in solution is zero (**middle**). At low pH, high protonation confers a positive charge (**left**): at high pH deprotonation confers a negative charge (**right**); (**b**) Plotted variation of net charge as a function of pH. An analyte molecule of interest can be purified by IEC provided that the molecule possesses a net charge. If the pH is such that the analyte has a net positive charge then **cation exchange chromatography** should be used. If the analyte has a net negative charge then **anion exchange chromatography** should apply.

Besides choosing the type of exchange resin, a decision must also be made about whether to use strong or weak ion exchange resins. In most cases either may be used, but in extreme pH conditions (>10 for anion exchange or <4 for cation exchange) weak ion exchange resins lose their charge. Weak ion exchange media tend to take longer to equilibrate, due to the buffering capacity of the column, but are most useful in separations of analytes that bind extremely tightly to strong ion exchange columns and do not elute even with high salt concentrations. This is sometimes the case with large nucleic acids or phospholipids. An alternative elution in these cases is to use weak ion exchange resin followed by elution with acid (for cation exchange) or base (for anion exchange) to neutralise as much charge on the resin as possible. **Chromatofocusing** is essentially a variant of weak anion exchange, often using tertiary or quaternary amines as the stationary phase. Molecules will be separated roughly according to their pI. The analyte is initially bound at a pH above the pI of the target, as for anion exchange. A focusing buffer of a lower pH is then applied to the column; no gradient is required as the focusing buffer will titrate the buffering groups on the ion exchanger. Peak widths as small as 0.05 pH units may be resolved in this way and very high concentrations of proteins may be eluted in a small volume. The method is a powerful analytical probe of surface charge, as well as being an important preparative technique. With regard to method development for ion exchange, it is best to start at a pH suitable for maximal binding, that is, pH 8.5 for anion exchange or pH 4.0 for cation exchange. Initially try a 0–1 M NaCl gradient over 50–100 column volumes. If solubility of the target analyte of interest is a problem in the separation (poor recovery or peak tailing) then addition of 20% water-miscible solvent (methanol, glycerol, isopropanol or acetonitrile) should aid in both reducing hydrophobic interactions with the resin and solubilising the analyte molecules of interest.

2.6.2 Hydrophobic interaction chromatography

Hydrophobic interaction chromatography (HIC) separates biological molecules according to the hydrophobic groups on their surface. It shows many parallels to IEC. HIC is based on hydrophobicity whereas IEC is based on charge; both have mild binding and elution conditions, and are therefore very useful, particularly for protein purification. The HIC resin is similar to the IEC resin, except that the functional group on the surface is a hydrophobic group such as a phenyl.

HIC can be thought of as a high-resolution method with parallels to ammonium sulfate precipitation. Most proteins will not bind to an HIC column using normal buffer conditions since these conditions are not sufficiently hydrophobic. The usual method is therefore to add a **lyotropic salt**, such as ammonium sulfate to the solution, thereby reducing ionic interactions whilst increasing hydrophobic interactions. When hydrophobic surfaces bind to each other, water is released from the surfaces, thereby causing a favourable increase in overall entropy (Figure 2.21, see Chapter 1). Lyotropic salts increase the ordered structure of water, decreasing the entropy, therefore inducing analyte precipitation to compensate for this decrease in entropy. As an entropy driven process, HIC is strongly affected by temperature, generally the higher the temperature the stronger the binding to the column.

The key to HIC is to bind an analyte at an ammonium sulfate concentration as high as possible that does not actually cause precipitation of the analyte; this can be easily determined in some pilot experiments. If the analyte is fairly hydrophobic relative to other contaminants, then HIC should work well at a relatively low ammonium sulfate concentrations. If the target is hydrophilic then it is best to use ammonium sulfate "cuts" initially to remove the most hydrophobic contaminants and then reach a salt concentration high enough to bind the target analyte. Thereafter, a gradient of decreasing ammonium sulfate salt concentration is used to elute the analyte (Figure 2.22). If the analyte is extremely hydrophilic then **negative chromatography** should be considered as a purification method so that the target does not bind, but other molecules do. Some particularly hydrophobic proteins, such as membrane proteins, will bind to the column even without

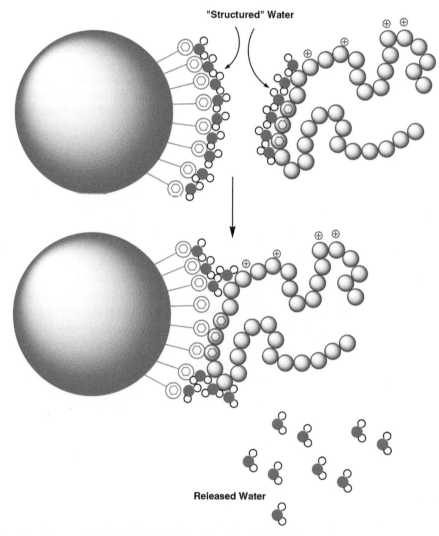

Figure 2.21 Binding of analytes with hydrophobic interaction chromatography media. Structured water covers the hydrophobic functional groups on the gel beads and hydrophobic surfaces of analyte molecules. This water is partially displaced when hydrophobic interactions take place between hydrophobic surfaces (see Hydrophobic interactions, Section **1.6.4**). Interactions are driven in part by entropy gains due to released water.

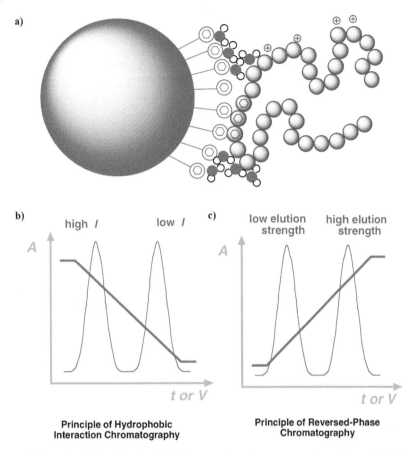

Figure 2.22 **Comparison of the principles of hydrophobic interaction chromatography and reversed-phase chromatography.** (a) In HIC, analyte molecules are bound to the column under high ionic strength (high *I*) conditions (i.e. typically high concentrations of ammonium sulfate). (b) Analyte molecules then are eluted (with respect to time *t* or elution volume *V*) by reducing the ionic strength (low *I*) of the eluant. (**c**) Reversed-phase chromatography works by the reverse principle where the eluant "strength" is increased to encourage elution of more hydrophobic analyte molecules.

ammonium sulfate. In this case, an increasing concentration of a **chaotropic salt** (such as **guanidinium chloride**, **urea** or **isothiocyanate salts**) may be used to elute the analyte. Proteins are the optimal analytes for HIC.

2.6.3 Reversed-phase chromatography

Reversed-phase chromatography separates using an extremely hydrophobic stationary phase that can be varied, together with a polar mobile phase (usually an aqueous solution). In the form of HPLC, this is the most common method for analysis of biological macromolecules and lipids, and for preparative separation of small molecules including lipids, oligo-/polypeptides, and oligo-/polynucleotides or oligo-/polydeoxynucleotides (Figure 2.22). Reversed-phase chromatography is not often used for proteins as both the extremely hydrophobic stationary phase and the organic solvents used for elution tend to trigger irreversible denaturation of all but the most robust proteins. In contrast, HIC does not trigger denaturation because the high salt concentration and only weakly hydrophobic surface usually let the protein retain its correct conformation. Typically, as far as reversed-phase chromatography is concerned, analytes are bound to the stationary phase under a solution of high polarity (e.g. water), after which the polarity of the mobile phase is reduced using a gradient with a water-miscible solvent (often acetonitrile, methanol or isopropanol) in order to enable analyte elution. The pH of the mobile phase is also an important consideration as some analyte functional groups when charged behave as hydrophilic groups, but when uncharged are essentially hydrophobic, and hence changes in pH can significantly affect retention on a reversed-phase column. **Ion-pairing agents** are often used to enhance the interaction of charged groups with the surface. Ion pairing agents [typically **trifluoroacetic acid** (**TFA**) or **formic acid**] comprise both a hydrophobic region (to bind to the stationary phase) and a hydrophilic region (to

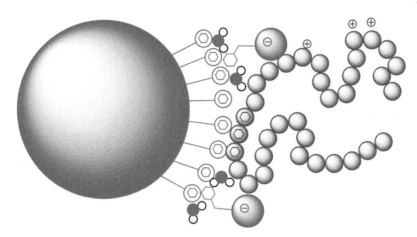

Figure 2.23 **Ion-pairing agents used in reversed-phase chromatography.** Ion-pairing agents, illustrated in pink, enhance the interaction of charged groups with the hydrophobic surface. They also suppress the ionisation of residual silanols on silica-based, reversed-phase media.

ionically interact with any charges on the target analyte) (Figure 2.23). Initially, in attempting reverse phase chromatography for a separation, a relatively steep, broad gradient should be used to find the approximate retention of the target on the column. The start and end points can then be narrowed to the range of interest, and the resolution can often be increased in this way. On the other hand, it is often observed that as the gradient slope is decreased, molecules elute at a lower solvent strength. This is because elution is not strictly binary in nature; at a solvent strength slightly below that expected for full elution from the column, a target analyte can start to slowly move down the column. This effect is more pronounced with shallower gradients, leading to earlier elution.

2.6.4 Gel filtration chromatography

Gel filtration chromatography (GFC) separates molecules according to their size and shape (Figure 2.24). It is also known as **size exclusion chromatography** or **gel permeation chromatography**. The principle behind gel filtration chromatography is different from the other modes of column chromatography in that interactions among analyte, eluant and support should ideally all be equal, that is, all efforts should be made to prevent any interaction between the analyte and the resin. Separation is enabled by particles in the stationary phase, each containing a distribution of pore sizes. Larger molecules are excluded from the pores, and hence run through the column quickly, whereas smaller and smaller molecules have to navigate smaller and smaller pores, resulting in separation according to molecular weight. The range of separation depends on the pore size, certain pore sizes are best for separating in the small biomolecule range (>10 kDa), through to large biosolutes such as proteins (<200 kDa) and even up to the mass of small plasmids, organelles or viruses. This method is the "softest" separation technique since analytes are never actually changing their physical environment, nor actually binding to resin during the chromatography. GFC is often used as a final "polishing step" in a purification procedure. One certain special class of GFC columns are known as **desalting columns** (or buffer exchange columns). A resin is used which completely excludes large analytes, but has no exclusion for salt, solvent or buffer molecules. The column is first equilibrated in the desired buffer that the analyte is to be transferred to, then a sample of analyte is applied, to elute well before the sample salts and buffers that otherwise get caught in the pores of the column. For small volumes, this technique tends to be quicker than **dialysis** and is often useful for unstable molecules where dialysis would take too long. GFC even enables researchers to get a handle on the quaternary structure of proteins, as the columns may be calibrated with known proteins, and an approximate mass directly estimated from the time of elution.

2.6.5 Hydroxyapatite chromatography

Hydroxyapatite (HA) is a crystalline form of **calcium phosphate** $(Ca_5(PO_4)_3OH)_2$ that is sometimes used in analyte purification, particularly of DNA and proteins. Unlike all the other resins, the stationary phase is actually crystalline, and binding

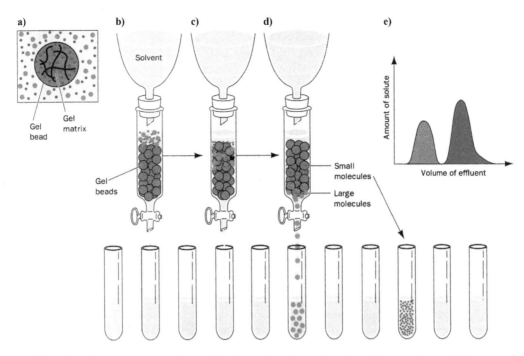

Figure 2.24 Gel filtration chromatography. (a) A **gel bead** consists of a **gel matrix** with pores from which larger molecule analytes (**blue**) are excluded. (**b–d**) Accordingly, larger molecule analytes run faster through the gel than slower molecule analytes (**red**). (**e**) Therefore, larger analyte molecules elute first in advance of smaller analyte molecules (illustration from Voet and Voet, 1995, Wiley, Figures 5–12).

of the analyte to the resin is both partially ionic and partially surface calcium ion specific. The technique is often used to separate proteins that co-purify by many other methods and as a final polishing step. It has advantages in that it can be autoclaved and is therefore useful in strictly sterile work. However, flow rates are slow and the columns do not have long longevity (one problem is that carbon dioxide often binds to HA, causing a crust on the surface of the column). Proteins are usually adsorbed onto the column under low ionic strength phosphate buffer, then an increasing phosphate gradient is used to elute proteins; 0.5 M phosphate is usually sufficient to elute all adsorbed protein. Ceramic hydroxyapatite may replace the original crystalline hydroxyapatite nowadays owing to much improved flow rates and operational column pressures.

2.7 Directed biological synthesis of proteins

The biological synthesis of proteins is central to so much chemical biology research today. Modern-day biological synthesis of proteins requires that all proteins are purified from one organism or another. If particularly large quantities of proteins (mg-g levels) are required then recombinant techniques and the growth of recombinant "factory organisms" is essential (see Chapter 3). In this section, the chemical biology reader will be introduced to the most modern approaches to the biological syntheses of proteins and their purification. Critically, proteins vary widely in structure and physical properties; therefore though the biological syntheses of proteins are broadly similar, they can differ substantially in the details.

2.7.1 Wild-type or recombinant sources

There are two types of source for proteins prior to protein purification: either a natural source, where the protein is found at its natural level, or a recombinant (directed) source. Natural sources were the only option before the advent of molecular biology and recombinant techniques, but remain commonly used when the primary structure of the protein of interest is unknown. Natural sources are also used when the protein is expressed at a naturally high level, or when factors such as correct **post-translational modification** are of particular importance. The alternative approach is to use a recombinant (directed) approach to protein synthesis that requires the creation of a recombinant or genetically engineered "factory organism" (see Chapter 3) to over produce the protein of interest.

Typically, strains of the bacterium *Escherichia coli* (*E. coli*) are harnessed as factory organisms, though there are often problems with protein solubility and correct processing of the protein in *E. coli*. Alternative factory organism systems include **insect cell lines**, **yeast cells** and certain **mammalian cell lines**. Each system has certain advantages and disadvantages.

2.7.2 Expression in *E. coli*; early purification

Once a wild-type or recombinant organism becomes fully grown, then the protein of interest may be purified from the other cell components in a multistep procedure. First, cell walls must be disrupted so as to efficiently release and, if possible, begin to fractionate out the protein of interest. For microbial extracts, **homogenisation** (by bead mill or through high pressure), **sonication**, or the addition of **lysozyme** (protein bio-catalyst or enzyme, see Chapters 1 and 7) are often used to disrupt the cells. Lysozyme catalyses the weakening of the polysaccharide cell wall coat of bacteria. Microbial extracts may then be centrifuged to separate soluble and insoluble fractions. **Non-ionic detergents** such as triton X-100 in buffer are frequently used at this stage to maximise the release of soluble protein of interest. Even at this stage, **protease inhibitors** are essential to prevent digestion of the protein of interest by endogenous biocatalysts (enzymes) suitable for proteolysis. These inhibitors are available as cocktails with broad specificity, or specific inhibitors are available that inhibit serine and acid proteases, metalloproteases and others. A suitable buffer must be chosen to maintain protein stability and activity: often a pH between 7.0 and 8.0, a relatively high salt concentration (0.1–0.5 M) as well as the possible addition of sucrose or glycerol (to around 10%) can aid solubility of the protein. β-**Mercaptoethanol** (β**ME**) or **dithiothreitol** (**DTT**) are used to maintain a reducing environment (under which cystine bridges do not form). More thought needs to be put into the cell-free extract preparation from a eukaryotic system as methods such as differential centrifugation may be used to partially fractionate intracellular organelles prior to release of the protein or interest.

At this stage, **precipitation** is often used as a convenient early step in protein purification to fractionate and concentrate the protein of interest. Three main methods of precipitation are used: **ammonium sulfate**, **organic sulfate** or **polyethyleneglycol** (**PEG**). At high sulfate concentrations, sulfate will bind to water molecules, reducing the amount of water available to shield hydrophobic patches on a protein surface. Sulphate concentrations are defined in terms of percentage "cuts" (0–20% w/v, 20–40% w/v etc.) with the percentage representing the degree of saturation. The wholesale aggregation of the protein of interest will usually be adjusted to occur in one percentage "cut" (i.e. at one approximate level of sulfate saturation). Thereafter, the aggregate is collected by centrifugation and the pellet made ready for further purification. Organic solvents (commonly acetone or ethanol) may also be used to precipitate proteins. Progressively increasing the hydrophobic nature of the solvent by increasing solvent concentration will promote **intermolecular electrostatic interactions**, leading to precipitation. PEG precipitation also works on a similar principle. In some cases, selective heat-induced denaturation is also occasionally used, where slowly increasing temperature may denature impurities, leaving the protein of interest intact in solution.

Methods for the further purification of proteins may be split into two general areas: **classical chromatography** and **affinity chromatography**. Classical chromatography relies on differences in the chemical and physical nature of molecules, whereas affinity chromatography relies on the specific and reversible binding of the target to an affinity ligand immobilised on an insoluble matrix. Although recombinant techniques have significantly increased the relative use of affinity chromatography, classical chromatography may still be used *in toto* or else in combination with affinity chromatography purification steps.

2.7.3 Affinity chromatography

Affinity chromatography uses specific biomolecular recognition between the protein of interest and a molecule bound to the matrix (Figure 2.25). There are a number of different types of affinity chromatography that are useful in purification (see below). Recombinant techniques have made affinity chromatography particularly useful, since **affinity purification tags** can be engineered into recombinant proteins to aid purification.

2.7.3.1 *Immobilised metal affinity chromatography*

Immobilised metal ions may be used to specifically coordinate with proteins so that they may be specifically isolated. The most common use of **immobilised metal affinity chromatography** (**IMAC**) is that the protein is engineered to carry

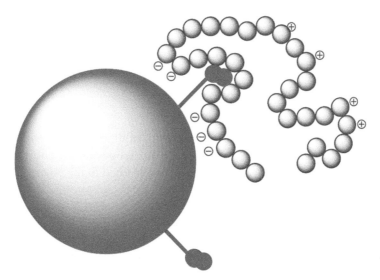

Figure 2.25 Affinity chromatography. Column associated ligands specifically recognise their target analyte molecule of interest, and the target is immobilised on the column by selective binding. The target analyte may be eluted by a variety of methods such as high salt concentrations, pH change or competitive elution with an alternate target for the ligand.

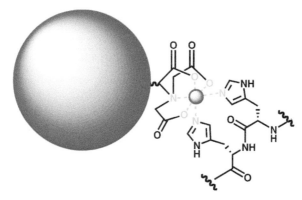

Figure 2.26 Immobilised metal affinity chromatography. Metal ions (for instance, Ni^{2+} ions) are gel-immobilised by chelation to groups covalently attached to the solid support. Histidine (His) residues in proteins (or part of a His-tag) have a high affinity for the immobilised ion. Elution is typically carried out by using a competitively high concentration of imidazole.

a sequence cluster of 4–8 histidine residues, enabling simple purification, usually by interaction with immobilised Ni^{2+} localised on the matrix surface (Figure 2.26). Although Cu^{2+} shows the strongest binding to surface histidines, and Ni^{2+} binds less strongly than Cu^{2+} to polyhistidine sequences, nevertheless Ni^{2+} use allows greater selectivity in purifications. Typically, metal ions (including Cu^{2+}, Zn^{2+}, Ni^{2+}, Co^{2+} or Fe^{2+}) can be immobilised via chelation to nitrilotriacetic acid attached to a solid matrix support and elution is usually made possible with imidazole that competes with the histidine clusters in the protein of interest for co-ordination with the matrix-associated metals.

The engineering of proteins to carry **polyhistidine tails** (*N*-terminal or *C*-terminal **His-tag**) in frame with their primary structure amino acid sequence is achieved by means of recombinant techniques (see Chapter 3) and confers a number of advantages. The His-tag often provides a very simple means of purification of the protein by IMAC. Also, the use of such a small affinity sequence means that **antibodies** may be generated against the protein of interest without the need to remove the tag. Even insoluble proteins may be purified by pre-solubilising the protein in guanidinium chloride or urea and then applying the unfolded protein to metal affinity column purification under the same conditions as above. Refolding may then be attempted post affinity column purification. Clearly, there are some drawbacks in using His-tag purification. For instance, the presence of the His-tag may have a measurable distorting effect on the structure of the protein of interest and even on the function of the protein. Sometimes the protein of interest with His-tag is less soluble than the wild-type protein. Furthermore, the IMAC columns can leak heavy metal ions, together with the protein of interest, which can lead to amino acid side-chain damage from oxidation. This problem can be made worse by the general requirement to avoid using reducing agents such as DTT or high concentrations of βME during IMAC purification. Just occasionally, Co^{2+} has been immobilised in place of Ni^{2+}, via four chelating bonds, in order to increase the specificity of His-tag binding to metal ions and reduce leakage of heavy metal ions into the eluant. This system can even harness natural histidine clusters from chicken lactate dehydrogenase.

2.7.3.2 Glutathione-S-transferase tags

Glutathione S-transferase (GST) was one of the original recombinant affinity tags. The usual role of GST is to catalyse the transfer of **glutathione** tripeptide onto endogenous or xenobiotic substrates that possess electrophilic side chains or groups. GST (usually originating from *Schistosoma japonicum*) may be fused to the protein of interest to enable purification using immobilised reduced glutathione (Figure 2.27). Although the tag is relatively large, GST can confer solubility on some proteins that are insoluble when expressed alone. A further benefit of using GST is that the fusion protein has a simple activity assay. Indeed, active GST will transfer reduced glutathione onto **1-chloro-2, 4-dinitrobenzene (CDNB)**, a reaction easily monitored by UV-visible spectroscopy. GST is naturally dimeric and therefore the fusion protein will probably also be dimeric. Hence a GST tag may upset the natural quaternary structure of some proteins unless removed. Removal of the GST tag is typically carried out by fusing the GST tag to the protein of interest via a short cleavable amino acid residue spacer sequence.

2.7.3.3 Maltose binding protein tags

This tag is based on the natural **maltose binding protein (MBP)** from *E. coli*. The protein of interest may be fused to the maltose binding protein, a tag that enables purification by affinity to immobilised amylose. Maltose may then be used to elute the protein. Unlike GST, MBP is monomeric, but it is one of the largest tags (42 kDa). MBP is one of the best tags for encouraging **solubilisation** of an insoluble or sparingly soluble protein of interest.

2.7.3.4 Biotinylation of proteins

The **biotin-avidin/streptavidin** interaction is one of the strongest non-covalent interactions known and can be used to purify proteins. There are two approaches: the protein can be biotinylated *in vitro* using biotin-ester reagents or a sequence may be used that is naturally biotinylated during synthesis. *In vitro* biotinylation is difficult to control as more than one lysine may be modified and so this is not really an aid to protein purification. **Site-specific biotinylation** is made possible by creating a fusion protein containing a peptide tag that is biotinylated *in vivo* (for example the C-terminal residues of the **biotin carboxyl carrier protein [BCCP]**). The target sequence is biotinylated at lysine 89 in *E. coli* BCCP (with the aid of the **biotin holoenzyme ligase** enzyme). The fusion protein of interest is then recovered

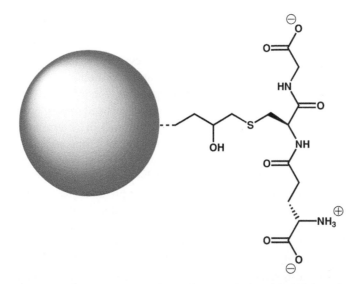

Figure 2.27   Immobilised glutathione. This forms the basis for purification of **glutathione S-transferase (GST)**-tagged proteins. GST binds strongly to the immobilised glutathione, the target protein may be eluted by competitive elution with glutathione.

by avidin-affinity column. Biotin normally exhibits strength of binding to the natural tetrameric avidin/streptavidin that is almost irreversible (K_d appears to be approximately 10^{-13} M). However, avidin resins have now been developed using a resin-bound monomeric form of streptavidin that allows elution of purified proteins using biotin.

2.7.3.5 Intein tags

Inteins are peptide segments of proteins that naturally excise or **self-splice**, severing their covalent links with a parent polypeptide but leaving an intact daughter polypeptide chain behind them by assisting in the formation of a peptide link in place of the original intein segment. This remarkable process has been exploited to design proteins that can in principle self-splice to allow a protein of interest to separate from its attached protein affinity tag. Therefore, a protein of interest may be initially engineered with an intein protein affinity tag, but post affinity column purification the affinity tag may then be encouraged to dissociate, leaving the intact protein of interest behind in a highly purified state. A recent example of an intein protein affinity tag comprises a chitin-binding domain fused to an intein self-splicing element that derives from yeast (*Saccharomyces cerevisiae*). In the presence of thiols such as DTT, βME or cysteine, intein **self-cleavage** can occur without the need for biocatalysis (Figure 2.28). The real beauty of this system is that the self-cleavage reaction can be performed on an affinity column directly after purification on cellulose beads, so that the protein of interest can be eluted directly without any attached protein affinity tags. Naturally, fusion proteins may be engineered with an intein protein affinity tag fused at either the *N*- or *C*-terminus of the protein of interest, as appropriate.

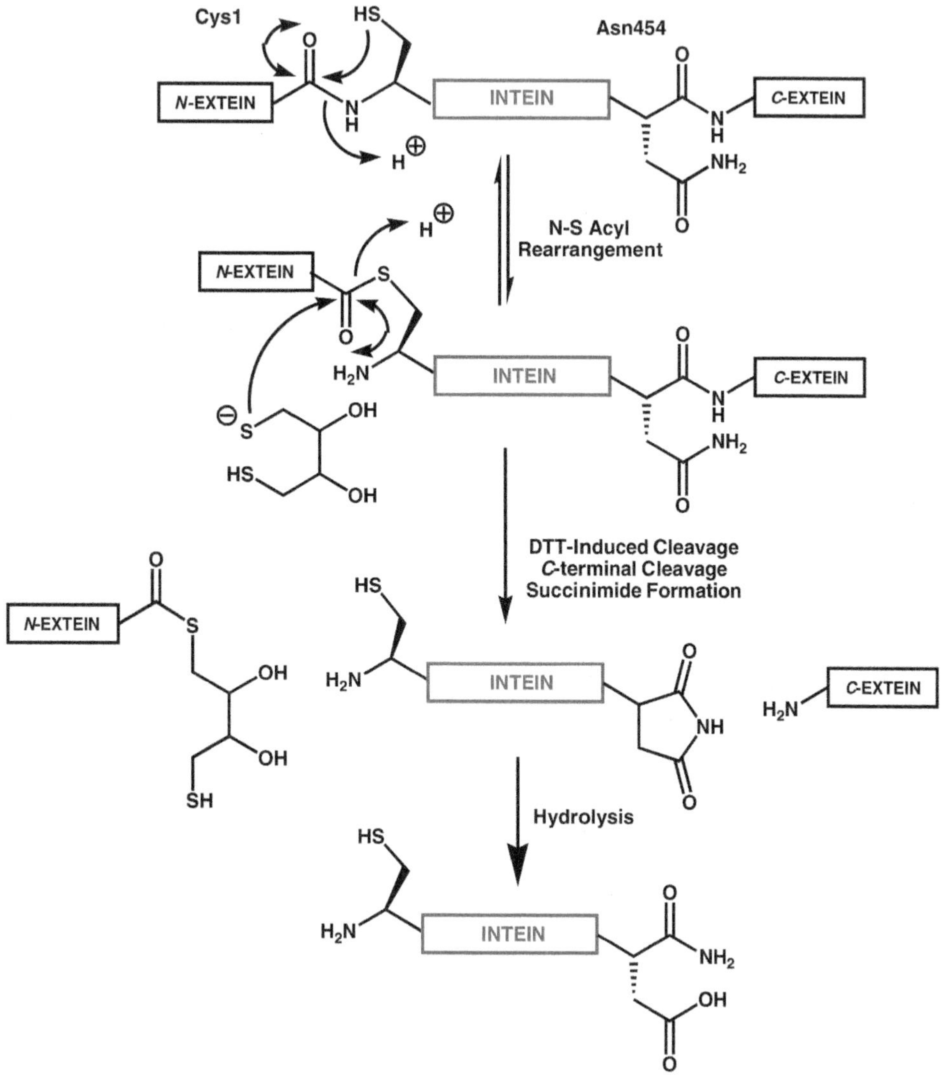

Figure 2.28 Chemical mechanism of intein cleavage. This is intended to take place with di-thiol reducing agent **dithiothreitol (DTT)** present.

2.7.3.6 Other affinity tags and radiolabelling of proteins

There are many other possible affinity tags that one could use in addition to those described in the previous section. Here is a simple survey of the possible diversity. First, there is **Protein A** that binds strongly to **IgG immunoglobulins**. Conversely, mimics of the Z domains of an IgG can be used in fusion proteins in order to bind to protein A columns. Second, there is the **S-peptide tag**, a 15 amino acid residue peptide that binds strongly to a protein derived from pancreatic ribonuclease A. Third, the **Strep-tag** is a 10 amino acid residue peptide that binds strongly to streptavidin. Fourth, **Strep-tag II** is the next generation that is a 9 amino acid residue binding to a streptavidin derivative (**streptactin**). In both cases of strep-tag, fusion proteins are eluted from affinity columns using desthiobiotin, a biotin analogue. Moving on, the **Flag tag** is a small hydrophilic peptide *(N-AspTyrLysAspAspAspAsp-Lys-C)* that may be fused to a protein of interest such that the fusion protein can be purified using an anti-flag affinity resin. Next, the **calmodulin binding peptide** is a relatively small peptide (4 kDa) that binds to immobilised calmodulin protein and elutes from the affinity column under very mild conditions. In fact, binding only requires the presence of Ca^{2+}, while elution is made possible in the same buffer in the presence of EGTA (a particularly effective Ca^{2+} chelator). By contrast, the **T7 tag** consists of an 11 amino acid residue peptide from the T7 *10*-protein and purification is effected using an anti-T7 monoclonal antibody affinity column. However, elution from this column requires very acidic conditions of pH 2.2. Finally, the **cellulose binding domain tag** (**CBD tag**) (~12 kDa) binds strongly to homoglycan cellulose or chitin columns. The fusion protein may then be eluted with ethylene glycol, a very soft method of elution, after which the ethylene glycol may be easily removed by dialysis. A considerable advantage of this tag is that it is particularly cost effective since the resin is very inexpensive, robust and readily reusable.

The CBD tag approach is also well adapted to the labelling of proteins with radioactivity. However, an alternative radiolabelling procedure is to create fusion proteins with tag sequences that are recognised by **kinases** (proteins that act to introduced phosphates to other proteins) such as the *N-ArgArgAlaSerVal-C* tag sequence designed to trigger **cAMP-dependent protein kinase** mediated [^{32}P]-phosphorylation of the tag serine residue. This labelling method requires that the protein of interest, post-purification, should be [^{32}P]-phosphorylated *in vitro*, with the assistance of the appropriate kinase. This method is also very effective, but nowadays there is a considerable move to radiolabel proteins using completely recombinant, technology-driven approaches instead: *in situ* in cells rather than *in vitro*. *In situ* approaches aim to introduce [^{35}S]-labelling that is much milder than [^{32}P]-labelling and more durable owing to a longer radioactive half-life. More recently, some other tags have been developed for real-time detection and quantitation of low abundance proteins and their complexes in cells. For instance, the **HiBiT tag** is an 11 amino acid tag that may bind to a larger subunit LgBiT such that the complex has luciferase activity providing for **bioluminescence** with a furimazine substrate.

2.8 Biological syntheses of nucleic acids, oligosaccharides and lipids

These are also just as important as the biological syntheses of proteins if a holistic view of biological macromolecule and lipid assembly structures and functions is to be obtained. Nucleic acid purification may involve a combination of recombinant techniques and purification, otherwise biological synthetic approaches rely on natural organisms to produce and the chemical biology researcher to isolate, purify and characterise.

2.8.1 Biological synthesis of nucleic acids

If simple isolation of nucleic acids from cells is required then many of the chromatographic methods described in this chapter can be used to isolate and fractionate nucleic acids from natural sources. DNA is typically associated with cationic polyamines or proteins in cells and both can be removed with a phenol solution. Alternatively, proteins can be removed by denudation with **proteases** (proteins that degrade peptide links) or detached with the aid of detergents or chaotropes. Protein-free nucleic acids may then be precipitated with ethanol. Cellular RNA may be isolated distinct from DNA by using pancreatic DNases (proteins that degrade phosphodiester links in DNA) or vice versa, wherein cellular DNA may be isolated from RNA by using RNAses (proteins that degrade phosphodiester links in RNA). Although electrophoresis is the usual approach to separating nucleic acids (see Chapter 3), HPLC is also particularly useful, as is HA chromatography. HA columns have a particular propensity to bind double-stranded DNA molecules very strongly. Therefore HA chromatography is an effective approach for separating double helical DNA from cellular

RNA and proteins. Some modifications of affinity chromatography can also be useful. For example, eukaryotic mRNA typically has poly(A) sequence at its 3'-end. Therefore immobilised-poly(U) on a solid support can represent a powerful tool for the isolation of mRNA from cellular extracts.

Obviously, there are times when there is a requirement to obtain and purify total cellular DNA or RNA (as described in Section 2.8.1). However, there are times when one single DNA or mRNA molecule is required, in which case such a molecule must be identified, amplified and then purified clear of cellular debris. In this second instance, biological synthesis of such DNA or RNA molecules involves intensive use of cloning and other recombinant techniques prior to purification protocols (see Chapter 3).

2.8.2 Biological synthesis of oligosaccharides

Oligo-/polysaccharides can also be purified by many of the methods described in Section 2.7, though they are often found linked to intracellular proteins (not in bacteria), membrane proteins and possibly to some extracellular proteins, too. Some polysaccharides are also closely associated with lipids forming **lipopolysaccharides** (**LPS**) that are extremely prevalent in some bacterial cell walls and can be very immunogenic. Oligo-/polysaccharides are co-precipitated together with nucleic acids and proteins using alcohol and protein linkages cleaved by chemical means (where appropriate). Thereafter, charged proteins and nucleic acids can be removed from the uncharged/weakly charged oligo-/polysaccharides by IEC. Neutral and weakly charged oligo-/polysaccharides pass through columns, whilst the nucleic acids and proteins bind and are removed from the sample. Oligo-/polysaccharides can then be further analysed and purified by some specialised types of anion exchange chromatography and by HPLC. Also, affinity chromatography is possible, for instance using immobilised lectin (typically the jack bean protein **concanavilin A**), a protein that specifically binds glucopyranose and mannopyranose monosaccharide residues.

2.8.3 Biological synthesis of lipids

Typically, the biological synthesis of lipids relies on lipid synthesis in cells and their extraction from macromolecular lipid assemblies with organic solvents. Neutral lipids can be extracted using non-polar solvents such as diethyl ether or chloroform, whereas membrane-associated lipids require more polar solvents such as ethanol or methanol to disrupt intermolecular hydrogen bonding. Phospholipids used commercially in scientific research are typically purified from egg or bovine sources. Cholesterol is typically purified from egg or from wool grease (sheep-derived). Sphingolipids are mostly purified from egg or mammalian tissue sources. Finally, phosphatidylinositols are extracted from soybean or bovine sources. Synthetic lipids are often prepared from a glycerol-3-phosphocholine precursor that is itself typically purified from soybean lecithin. After extraction from a natural source, individual lipids can be purified to homogeneity by HPLC.

3

Molecular Biology as a Toolset for Chemical Biology

3.1 Key concepts in molecular biology

Molecular biology is a powerful tool for the directed synthesis of biological macromolecules, in particular nucleic acids and proteins (see Chapter 2), and in its own right is a highly diverse forward-looking discipline. Too often, however, many chemists have drawn the line at the frontier between protein science and molecular biology. This is, in part, due to the change in language in moving towards a biological discipline. However, chemists have also seriously underestimated the potential of molecular biology to open up new opportunities for chemistry. The serious chemical biology reader cannot afford to make the same mistakes. Molecular biology, with its ever widening and improving range of recombinant techniques, is every bit as essential to the practice of chemical biology research as chemical synthesis in the quest to understand the way biology works at the molecular level. We hope in this chapter to provide a focused introduction to molecular biology, particularly aimed at the chemical biology reader most interested in harnessing molecular biology and recombinant techniques as a tool for biological synthesis.

3.1.1 The central dogma of molecular biology

The well-known **central dogma** of molecular biology, as first discussed by Francis Crick, is that DNA is transcribed into RNA that is translated into protein (see Chapter 1, Figure 3.1). The first of these steps is occasionally reversible, such as in the case of **reverse transcription** of RNA to DNA in the life cycle of some viruses, but there remains no instance of genetic code realised in proteins and polypeptides returning from polypeptide amino acid residue sequences to deoxynucleotide base sequences in DNA. Hence, DNA is central to all biological synthetic strategies since it is from DNA that RNA and polypeptides flow. Therefore, the initial steps should be to identify, isolate and clone those DNA segments of interest in a genome of interest. Typically, DNA segments of interest are **genes**, otherwise known as coding regions, which comprise at least one **open reading frame** (**ORF**), that harbours genetic coding information for a protein of interest, plus other non-coding control elements embedded in the DNA structure outside ORFs. The preliminary steps leading to isolation, cloning and identification of **genomic DNA** are the beginning of any number of activities, including directed expression and purification of the protein coded for by a particular gene, or other more limited activities such as DNA sequencing.

Essentials of Chemical Biology: Structures and Dynamics of Biological Macromolecules In Vitro *and* In Vivo, Second Edition. Andrew D. Miller and Julian A. Tanner.
© 2024 John Wiley & Sons, Inc. Published 2024 by John Wiley & Sons, Inc.
Companion Website: www.wiley.com/go/miller/essentialschembiol2

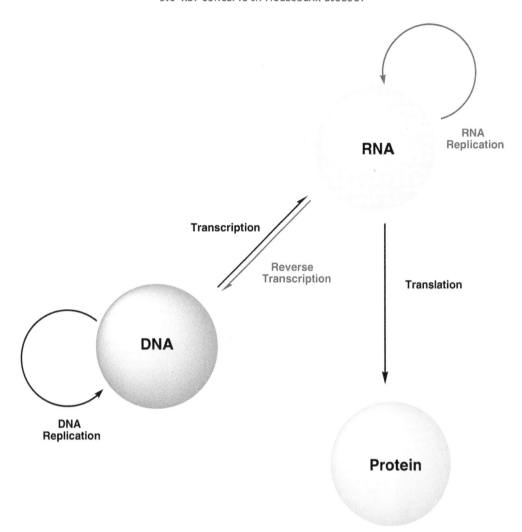

Figure 3.1 The central dogma of molecular biology. Black lines represent information transfers that occur frequently in cells. Blue indicates this exception, such as reverse transcription that occurs with some viruses, and **RNA-directed RNA polymerase** catalysed **RNA replication** that also occurs only in some plants and viruses. Nucleic acids can be both stores and recipients of genetic information, whereas proteins are always only recipients of genetic information.

3.1.2 The difference between prokaryotic and eukaryotic genes

Before proceeding to discuss methods of isolating DNA sequences, it is important to consider some fundamental differences between prokaryotic and eukaryotic DNA. **Prokaryotes** (**bacteria** and **cyanobacteria**) are defined as organisms that lack a distinct nucleus (Figure 3.2). They have a single circular DNA **chromosome**, and may possess smaller circular pieces of DNA called **plasmids** spread throughout the cell that can also code for proteins. Prokaryotes are nearly always unicellular and when grown on agar plates form groups of cells called **colonies**. In contrast, eukaryotic cells have a true nucleus, in which reside multiple DNA chromosomes (DNA bound to and condensed around nuclear proteins), plus numerous other intracellular compartments. These cells are also significantly larger than prokaryotic cells (Figure 3.3). **Eukaryotes** include both some unicellular organisms, such as yeast, and multicellular organisms, including all animals and plants. Whilst prokaryotes and eukaryotes share the same genetic code, there are significant differences between prokaryotes and eukaryotes at the molecular level in that eukaryotes use a vastly more complicated approach to processing mRNA after transcription (**post-transcriptional processing**) for very good reasons.

In prokaryotes, it is usual that the immediate products of mRNA transcription (the primary mRNA transcripts) are translated without any modification. It is even possible that translation can begin at the 5′-end of the mRNA molecule before transcription is even completed at the 3′-end. In some cases, a single mRNA transcript may even harbour the contiguous genetic code for multiple proteins when this transcript has been transcribed from a DNA **operon** (i.e. a

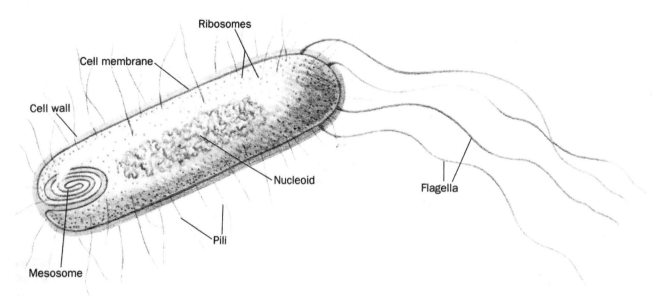

Figure 3.2   Summary illustration of prokaryotic cell such as *Escherichia coli (E. coli)*. Prokaryotic cells do not exhibit discrete compart-mentalisation so transcription and translation can take place concomitantly (Voet and Voet, 1995, with permission from John Wiley & Sons).

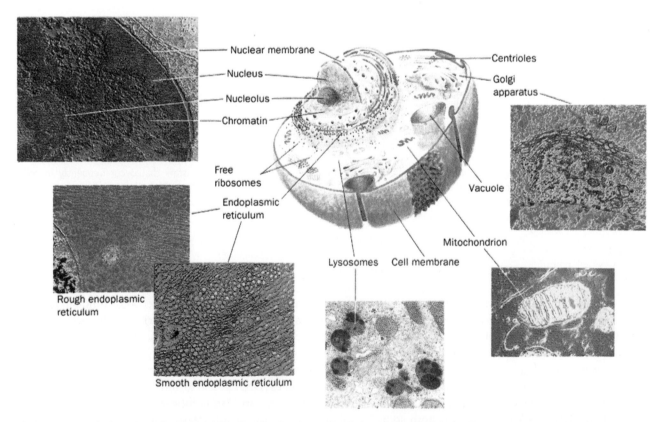

Figure 3.3   Depiction of a eukaryotic (animal) cell. All eukaryotic cells have evolved with significant compartmentalisation, resulting in spatio-temporal separation of transcription and translation. DNA is transcribed to mRNA in the nucleus before being shuttled out of the nucleus for translation in the cytosol where ribosomes are located (Voet and Voet, 1995, with permission from John Wiley & Sons).

contiguous set of genes whose protein products usually have related functions, such as metabolism). Hence, genomic DNA sequences in prokaryotes can be directly translated into protein sequences. By contrast, the same cannot be said in eukaryotes. One of the major reasons for this is that most higher level eukaryotic genes comprise ORFs (also known as **exons**) interspersed with **introns**, which are segments of DNA sequence that do not code for the protein of interest

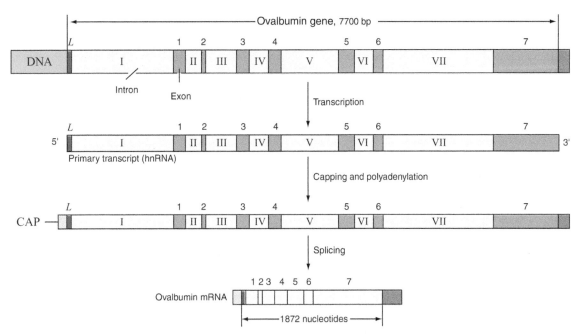

Figure 3.4   Processing of eukaryotic pre-mRNA to mature mRNA. This is demonstrated for the chicken ovalbumin gene. Most higher eukaryotic genes have both **introns** (which do not code for protein) and **exons** (which do code for the protein). There is significant post-transcriptional processing to form mature mRNA prior to translation (illustration from Voet and Voet, 1995, Wiley, Figure 29–35).

and are therefore not usually intended for direct translation with exons (Figure 3.4). In eukaryotes, transcription takes place in the nucleus, and translation takes place in the cytoplasm, so it is impossible to have simultaneous transcription/translation as observed in prokaryotes. Initially, there are two essential post-transcriptional processing events:

1. The primary eukaryotic mRNA transcript is capped with a 5′-5′ triphosphate bridge linked to a 7-methylguanosine nucleotide residue, a process known as **5′ capping**.

2. Post capping, the 3′-end is modified by the addition of a 3′-poly(A) tail, a process known as **3′ polyadenylation**.

Thereafter, the process of intron **splicing** takes place. Introns must be excised and exons "knitted-back together" to create a mature mRNA transcript with a base sequence equivalent to the contiguous base sequences (5′→3′) of the original DNA exons in the eukaryotic gene. In simpler eukaryotes such as yeast, only a few genes contain introns. The situation in a higher eukaryote is quite different and genes contain typically around eight introns whose total sequence length is 4–10 times the coding sequence found in the exons. It remains unclear why there is such extraordinary **redundancy** in higher eukaryotic genomes. To add to the complexity, many higher eukaryotes exhibit **splicing variants** (i.e. where splicing events vary according to the cell type, time of expression, or even other environmental cues).

Hence, if eukaryotic gene structure is so complicated, how can we expect to obtain functionally useful genes for further molecular biology manipulation? The answer to this question lies with mature mRNA transcripts. Such mature transcripts from eukaryotes can be viewed as the functional equivalent of a primary mRNA transcript from a prokaryote. Mature mRNA transcripts contain processed intron-free nucleotide residue sequences that harbour the complete genetic code for a protein of interest. Therefore, if such transcripts can be reverse transcribed into a DNA form, then the resulting DNA should be the functional equivalent of a prokaryotic gene.

3.1.3 The creation of cDNA libraries

Mature mRNA transcripts (sense strand) from eukaryotic cells can be purified and then reverse transcribed, with the assistance of a **reverse transcriptase** enzyme (e.g. from **Moloney Murine Leukaemia Virus [MMLV]**), into **complementary DNAs (cDNAs)** that will anneal with the mRNA transcripts by Watson-Crick base pairing to give anti-parallel **DNA/RNA duplexes** or double helices. The poly(A) tail in each mature mRNA transcript is actually a useful handle for each reverse transcriptase reaction. Thereafter, DNA/RNA duplexes must be broken down with the assistance of RNAase enzymes (specific for the hydrolysis of RNA phosphodiester links) and a sense strand of DNA constructed instead on each cDNA single strand so that equivalent, more stable antiparallel DNA/DNA duplexes are generated instead, with the assistance of a **DNA polymerase** enzyme. In this instance, the **poly(T) tail** in each cDNA molecule turns out to be important for the DNA polymerase reaction.

Overall, provided that this process of reverse transcription and DNA polymerisation is performed starting from a diverse enough population of different mature mRNA transcripts isolated from a given organism, then the resulting collection of DNA/DNA duplexes can represent as complete a source as possible of eukaryotic genomic information for that given organism, all in a functional form ideal for further molecular biology manipulation. This collection of DNA molecules is known as a **cDNA library**. In the final stage of preparation before use, **adapter** DNA sequences are ligated onto both ends of each DNA/DNA duplex with the assistance of a **DNA ligase** enzyme. These adapters contain essential sequence elements for the conversion of every DNA/DNA duplex of the cDNA library into a genomic component that can be integrated with the paraphernalia of recombinant techniques into the complete tool-kit of genomic components associated with molecular biology (see Sections 3.2.2 and 3.2.3).

3.2 Tools and techniques in molecular biology

The key to molecular biology is the isolation, cloning and identification of genomic information in an appropriately useful DNA form. From this all else flows (see Section 3.1). What are the main tools and techniques available?

3.2.1 Plasmid DNA vectors

The **plasmid** is central to almost all molecular biology manipulations of genomic DNA or cDNA. The term plasmid refers to DNA that is able to replicate independently of chromosomal DNA. **Plasmid DNA (pDNA)** systems were first found in prokaryotes and were quickly realised to be ideal potential **vectors** (Figure 3.5). A vector is an autonomous

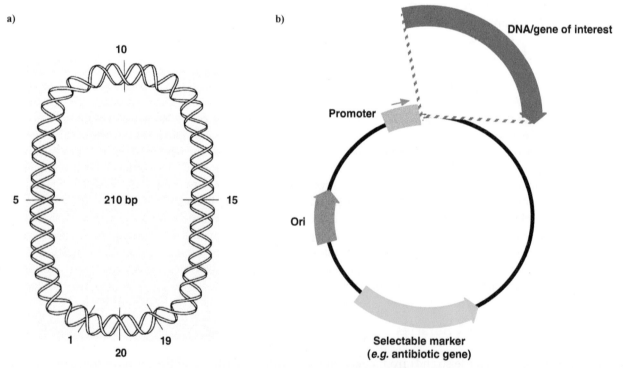

Figure 3.5   Schematic to show cloning vector, the simplest form of pDNA. (a) Diagram of **covalently closed circular DNA (cccD-NA)** (210 bp) to show how **double-stranded DNA (dsDNA)** links up to generate circular **plasmid DNA (pDNA)** molecules (adapted from Sinden, 1994, Figure 3.4); (**b**) Simplified diagram of pDNA cloning vector with **selectable marker**, non-coding **origin of replication (Ori)** and non-coding promoter **element** required for the control of expression of **heterologous (foreign) DNA** (DNA/gene of interest) eventually introduced downstream of the promoter (sense strand 3'-direction).

DNA construct suitable for recombination with **heterologous DNA**, from genomic DNA or cDNA library sources, to form **recombinant pDNA** constructs for the purposes of cloning and/or heterologous gene expression. A vector based on pDNA should have an **origin of replication**, together with one or more **selectable markers**. An origin of replication is a non-coding DNA sequence that defines the position from which pDNA replication commences when appropriate. Selectable markers are genes that code for a selectable trait such as antibiotic resistance. Vectors are divisible into two main types, **cloning vectors** and **expression vectors**, of which the latter can be used to direct expression *in vivo* of a desired protein or non-coding RNA (see Sections 3.3.3.1 and 3.3.3.3). Normal plasmids are only stable with heterologous DNA inserts of up to 10 kb. Other types of plasmids include so-called **phagemids**; these have both a plasmid origin of replication and a single-stranded **phage** origin for the preparation of **single-stranded DNA (ssDNA)**. **Cosmids** are a type of plasmid that combine aspects of plasmid and λ-based vectors; in general cosmids are able to hold much larger sequences of DNA (35–45 kb) and are therefore useful for preparing genomic libraries.

3.2.2 Restriction enzymes

Restriction enzymes (or **restriction endonucleases**) are enzymes that cut **double-stranded DNA (dsDNA)** by hydrolysing two phosphodiester links (one per strand) without altering attached deoxynucleotides. The term restriction originates from the fact that these enzymes were originally discovered as a defense mechanism used by bacteria to restrict infection by certain **bacteriophages**, bacteria-specific viruses. Over 900 restriction enzymes have been discovered to date from various organisms, of which 30–40 are regularly used in a typical molecular biology laboratory. Since they hydrolyse phosphodiester links within a DNA sequence, they are often referred to as restriction endonucleases as opposed to **exonucleases** that cut at the ends of DNA. Restriction enzymes act as "molecular DNA-scissors" for cutting dsDNA at sequence specific **restriction sites**, usually between 4 and 12 bps in length (Figure 3.6). Some restriction endonucleases cleave both DNA strands at the same position across the duplex (resulting in **blunt-ended** DNA fragments), while others perform offset cleavage, thereby leaving 5′ or 3′-base overhangs of defined length at either ends of the cut (causing the formation of **cohesive** or **sticky-ended** DNA fragments). Typically, but not always, restriction endonucleases like to recognise **palindromic** sequences (with a centre of rotational symmetry-see Chapter 1). For example, the famous **EcoRI** restriction enzyme recognises the sequence 5′-d(pGAATTC)-3′ that is rotationally symmetric between **dA** and **dT** in duplex DNA. The cleavage sites are offset between **dG** and **dA** in both sense and complementary DNA strands, resulting in sticky-ended DNA fragments with a 5′-overhang of four deoxynucleotides in length. Another similar such restriction enzyme is known as **XhoI** (Figure 3.6).

3.2.3 DNA ligases

DNA ligases are needed for the covalent **ligation** of heterologous duplex DNA fragments (Figure 3.7). DNA fragments that have been restricted (generated by restriction endonucleases) are rejoined with the aid of ligase enzymes. Both sticky-ended and blunt-ended ligation processes are possible with **T4 DNA ligase**, an **adenosine 5′-triphospate (ATP)**-dependent ligase enzyme, and the *Escherichia coli (E. coli)* **ligase**, which requires an **nicotinamide adenine dinucleotide (NAD)** cofactor rather than ATP. In both cases, the enzymes reintroduce phosphate and regenerate the phosphodiester links removed by restriction. Ligation using sticky ends is a lot easier and simpler in general, so if there is the opportunity for choice then **sticky-end ligation** should be chosen (Figure 3.7). Should this not be possible for any reason, such as lack of sequence complementarity, then the following should apply. Where there are two non-complementary 5′-sticky ends to join then these overhangs should be filled in to form duplex DNA using **Klenow polymerase**, an enzyme derived from **DNA polymerase I** that has had its **5′→3′ exonuclease activity** removed. Where there are two 3′-sticky ends to join then overhangs should be filled in to form duplex DNA using **T4 DNA polymerase**. In either case, blunt-ended duplex DNA fragments are generated that may be conjoined when required by DNA ligase-mediated **blunt-ended ligation**.

3.2.4 Hosts

A host is an organism that carries recombinant pDNA harbouring heterologous DNA. In approaching molecular biology the chemical biology reader should have already noticed that *E. coli* tends to be the most widely used workhorse organism for most experiments in molecular biology experiments, even if the vector is eventually intended for use in other organisms such

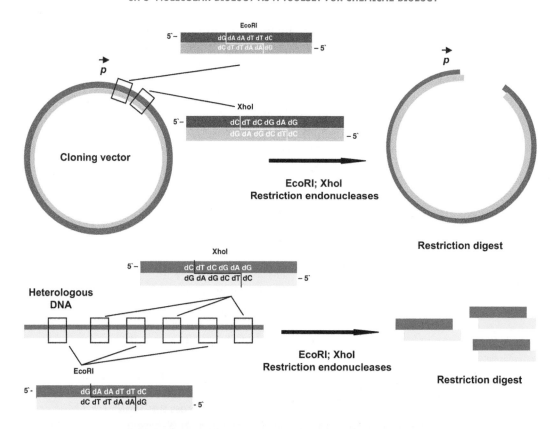

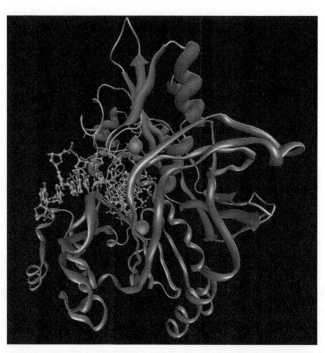

Figure 3.6 Heterologous (foreign) DNA processing. Whether from **genomic DNA** or a **cDNA library** (see Section 3.2), foreign DNA is cut with two different **restriction endonuclease** enzymes (**EcoRI; XhoI**) producing a **restriction digest**. A cloning vector is also cut with the same pair of restriction endonucleases downstream of the promoter element (**p**) (see Figure 3.5). In cloning vector, pDNA sense strand is **dark grey**; complementary strand is in **light grey**. In heterologous DNA, sense strand is in **red**; complementary strand is in **yellow**. EcoRV **restriction endonuclease** rendered in cartoon display with α-carbon backbone traced as a **solid ribbon** (pdb: **1az0**). DNA substrate double helix is shown with all atoms (minus hydrogen) in **ball and stick display**: carbon: green; oxygen: red; nitrogen: blue; phosphorus: purple. Magnesium ions are shown as Van der Waals spheres (light purple). The structure is shown to demonstrate the intimate association between DNA substrate phosphodiester links and restriction endonuclease enzymes that "wrap around" the DNA sequences of importance to cleavage.

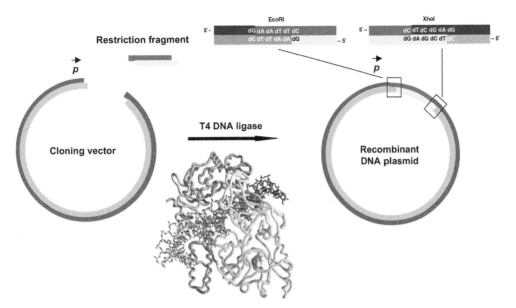

Figure 3.7 **Heterologous DNA restriction digest fragments.** These are created from two restriction endonucleases (EcoRI; XhoI) which are allowed to interact with restriction digest fragments from a cloning vector. Fragments **anneal** by Watson-Crick base pairing. Post annealing, phosphodiester links are regenerated by **T4 DNA ligase** catalysis. The result is a pool of **recombinant plasmid DNA** with original heterologous DNA fragments under expression control from the promoter element in the cloning vector (see Figure 3.5). Insert shows related human DNA ligase rendered in α-carbon backbone display (coloured by domain) and with bound DNA represented in stick display with atoms coloured according to atom type (pdb: **19n**).

as yeast or in cell lines. The reasons for this are partly historical: *E. coli* has been extensively studied, is relatively well understood with regards to the mechanism of gene expression control and was one of the first organisms to be fully sequenced at the DNA level. Furthermore, many of the first plasmids to be used as vectors were initially found in *E. coli*, and large numbers of bacterial strains are readily available that can be used in a variety of circumstances. For molecular biology purposes, *E. coli* has also been attenuated to become a very low-risk microorganism. However, *E. coli* does have a few disadvantages. For instance, *E. coli* has no ability to effect gene splicing and when used for protein biological synthesis many other problems can occur that will be discussed later. Other host or host organisms include yeast, insect cells, mammalian cells and various other cells. However, out of preference most genetic manipulations are still performed in *E. coli*.

3.2.5 Cellular transformation

Exogenous pDNA is introduced into host cells by a process known as **transformation** (Figure 3.8). Cells are rendered **competent** to take up added exogeneous pDNA by pre-treatment with calcium chloride solution on ice. After a short heat shock at 42 °C for two mins, the cellular transformation process is complete and host cells are now recombinant cells. An alternative technique to effect cellular transformation is **electroporation**. In electroporation, ice-cold cells are mixed with pDNA and are then placed in a cuvette, across which a strong potential difference is applied. This promotes a brief increase in cell membrane permeability, so that there is an opportunity for pDNA to traffic into the cytoplasm in a percentage of host cells. Following transformation, bacteria are initially grown in growth medium (broth) for sufficient time so that pDNA gene expression becomes established and genes of the selectable marker are expressed. Bacteria may then be spread on an agar **selection plate** containing antibiotic compatible with the selection marker gene so that only transformed host cells containing the independently expressing pDNA will survive.

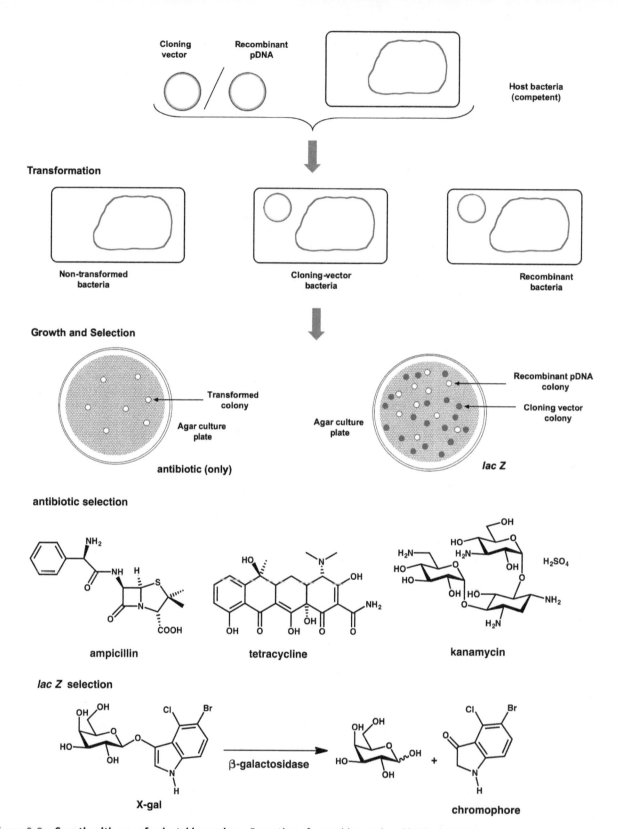

Figure 3.8 **Growth with use of selectable markers.** Formation of recombinant plasmid DNA (pDNA) (see Figures 3.6 and 3.7) is not 100% efficient, such that recombinant and cloning vector pDNA (without heterologous DNA restriction fragment insert) co-exist at the end of the process. Also, host bacterial transformation is not 100% efficient either, so that both transformed and non-transformed bacteria co-exist post transformation. Both of these problems are solved by growth and selection. Bacterial growth is performed in the presence of antibiotic (**antibiotic selection**) on agar culture plates so that only transformed bacteria (1 pDNA/cell) can survive and give rise to observed colonies. Individual colonies must then be selected for further analysis of pDNA sequence to check for the presence of recombinant versus cloning vector pDNA. The **lacZ selection** process allows for the presence of recombinant versus cloning vector pDNA to be determined more directly on the basis that the **lacZ function** is disrupted in recombinant pDNA and not in clonal vector pDNA. Hence if **X-gal** is included in the agar, only bacterial colonies that harbour cloning vector pDNA should develop a blue colour, while those that harbour recombinant pDNA should be white in colour. Individual white colonies should then be selected for further analysis of pDNA sequence. Three main selectable marker antibiotics are shown. X-gal and the effect of β-**galactosidase** (β-**gal**) on X-gal are also illustrated.

3.2.6 Selection

Selectable markers are typically genes that code for resistance to antibiotics such as **ampicillin**, **tetracycline** and **kanamycin** (Figure 3.8). Additional selectable markers may also be used to improve the selection process, such as those related to the inactivation of β-**galactosidase** (β-**gal**) activity (*lacZ* **selection**). In this case, the selection process requires pDNA to harbour an intact *lacZ* gene that codes for the first 146 amino acids of β-gal. The expression of the *lacZ* gene is probed by using the chromogenic substrate **5-bromo-4-chloro-3-indolyl-β-D-galactopyranoside** (**X-gal**), a colourless substance that becomes an intense blue colour due to hydrolysis caused by *lacZ* expressing β-gal enzyme (Figure 3.8). Successful ligation of heterologous DNA (from genomic DNA or cDNA) is designed to take place directly within the *lacZ* gene so that ligation causes X-gal hydrolysis to fail (Figure 3.5). Accordingly, after transformation and plating on agar, X-gal-containing selection plates, those host cell colonies identified as white should possess recombinant pDNA, while those host cells which have turned blue should possess only intact *lacZ* expressing pDNA without heterologous DNA included. *LacZ* selection is not completely reliable for two reasons. First, insertions into the *lacZ* gene do not always totally inactivate the β-gal activity, and occasionally blue colonies do actually harbour recombinant pDNA. Second, occasionally white colonies do not possess recombinant pDNA but are white in colour due to β-gal inactivation caused by excision of a small part of the *lacZ* gene during clonal growth.

3.2.7 pDNA purification

DNA in all forms is generally far more thermally stable than globular proteins and in many ways is easier to work with. The classical approach to pDNA purification of cells has been to bring about initial lysis, typically in the presence of **sodium dodecyl sulfate** (**SDS**) at a high pH. Genomic and pDNA remains in solution, while cell wall material with most of the cellular proteins are precipitated. Post centrifugation, the lysate is then neutralised with potassium acetate. The cleared lysate containing the plasmid DNA is then extracted using **phenol/chloroform/isoamyl alcohol**; **RNAase** is typically added to destroy any contaminating RNA. When particularly pure pDNA is required then **caesium chloride** gradient purification by **ultracentrifugation** can be used. The classical approach is quite cumbersome, time-consuming and phenol is toxic and unpleasant to handle. Therefore, the preparation and purification of pDNA has become much simpler with the widespread use of the chaotropic reagent **guanidinium isothiocyanate** (**GuNCS**) as part of the **miniprep** or **maxiprep** system (Figure 3.9).

The method is performed as follows: initially clonal cells are grown at a small volume (several ml), typically overnight. Thereafter, the bacterial pellet is resuspended in a buffer containing RNAase that will degrade RNA. Cells are then lysed using sodium hydroxide and SDS, causing **open-circle** genomic DNA and proteins to denature, while leaving **supercoiled** pDNA unchanged. Potassium acetate buffer is then used to neutralise the sodium hydroxide, resulting in precipitation of potassium dodecyl sulfate with which genomic DNA coprecipitates. The cleared supernatant is then passed through a **silica-derived glass fibre filter** frit where pDNA is able to bind onto the silica surface in the presence of GuNCS. Finally, the frit is washed and pDNA eluted using a simple aqueous solution in the absence of GuNCS. This miniprep or maxiprep system of pDNA purification and isolation provides ample DNA for other recombinant manipulations. The method may be simply scaled up if large quantities of a particular pDNA are required. Similar guanidinium-silica frit preparation methods may also be used to purify products of a **polymerase chain reaction** (**PCR**) **amplification** reaction, removing the enzymes and excess deoxynucleotides present at the end of the reaction.

3.2.8 Nucleic acid gel electrophoresis

Electrophoresis is the movement of charged molecules through a solid-phase medium under the influence of an electrical potential difference. In molecular biology the solid-phase medium are usually **agarose** or **polyacrylamide gels**. Agarose is a naturally occurring heteroglycan that has been described previously and is used for **agarose gel**

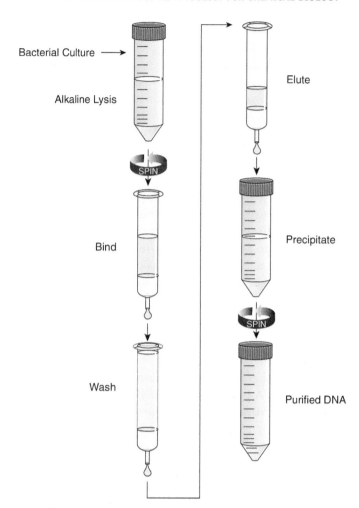

Bacterial Culture →

Alkaline Lysis

Elute

SPIN

Bind

Precipitate

SPIN

Wash

Purified DNA

Figure 3.9 **Recombinant pDNA purification.** Selected transformed clonal bacterial cultures are grown in medium and the pDNA is extracted as shown. Bacterial culture is subject to alkaline lysis. Cellular debris and open-circle DNA is removed and pDNA purified by binding to and eluting from a silica glass fibre filter. Relatively pure pDNA may be obtained following ethanol precipitation and collection of the pellet (Thermo Fisher Scientific, PureLink™ HiPure Plasmid Filter Maxiprep Kit. Last accessed 10 August 2022).

electrophoresis (**AGE**) (Figure 3.10). Polyacrylamide is a synthetic polyamido polymer that provides the essential solid-phase medium for **polyacrylamide gel electrophoresis** (**PAGE**) (see Chapters 2 and 9). AGE is generally used for the separation of polydeoxynucleotides/polynucleotides (>1000 kDa), whereas PAGE is frequently used for the separation of unfolded proteins or oligodeoxynucleotides/oligonucleotides (<100 bp). Therefore, in most molecular biology experiments, agarose is the solid-phase gel medium of choice (Figure 3.10). Agarose gels are very simple to make when compared to polyacrylamide gels. Initially, agarose is mixed with either **tris acetic acid/ethylenediaminetetra-acetic acid** (**TAE**) or **tris boric acid/ethylenediamine-tetraacetic acid** (**TBE**) buffer. Then, the heterogenous mixture is heated (microwaved) to dissolve the agarose and cooled to gelation in a gel-slab mounted with a plastic comb to create sample wells. TAE results in a larger agarose "pore sizes" than TBE. Hence, TAE gels are generally used to resolve DNA >1000 bps, and TBE gels to resolve smaller DNA fragments. However, recovery of DNA is generally poorer from TBE-agarose gels, and therefore TAE often ends up being used for the resolution of smaller DNA fragments too.

In order to run an agarose gel, samples are mixed with a loading buffer, which contains glycerol or Ficoll to increase the density of the sample so that samples sink easily into the wells, plus mobility dyes to aid sample loading and monitoring of the electrophoretic process. After electrophoresis, DNA may be visualised by staining with **ethidium bromide** (Figure 3.11). Ethidium bromide is a base pair **intercalator** that inserts or intercalates between base pairs (maximum 1 ethidium bromide/6.5 base pairs) and in so doing acquires a substantial increase in ϕ_F and hence fluorescence intensity at I_{max} (595 nm) (see Chapter 4). Ethidium bromide is a powerful mutagen. Hence, other staining procedures have been devised. These include silver, methylene blue or acridine orange staining. **Silver staining** has a similar sensitivity to ethidium bromide staining, but the former is able to give a permanent record of the gel, unlike the latter. **Methylene blue** stain-

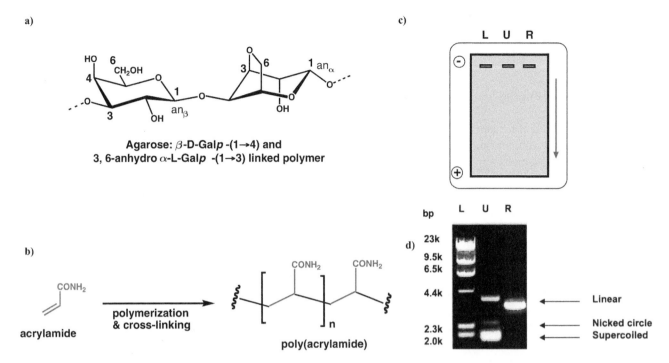

Figure 3.10 **Agarose/poly acrylamide gel electrophoresis.** (a) Agarose is a heteroglycan consisting of a β-**D-galactopyranose** (β-**D-Gal**p) and **3,6-anhydro-α-L-galactopyranose** (3,6-anhydro α-L-Gal*p*) **agarobiose** repeat unit. The total molecular weight is typically 120,000 Da. The heteroglycan readily forms cross-links via intermolecular double helix formation involving different agarose chain intertwined together. (**b**) Polyacrylamide is an artificial polymer prepared and cross-linked from acrylamide monomer. (**c**) Simple schematic of agarose gel electrophoresis set up with wells at the negative end for sample loading: **L: ladder; U: untreated purified pDNA; R: restriction cut pDNA.** Arrow shows direction of movement of DNA under influence of electric field. (**d**) Example of ladder, untreated purified pDNA and restriction cut pDNA run side by side. The identities of the three pDNA species shown in lane U are given.

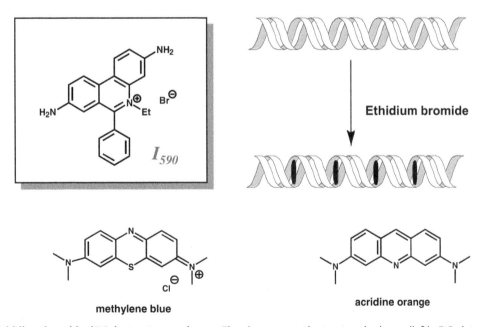

Figure 3.11 **Ethidium bromide (EtBr) structure and uses.** The planar aromatic structure is shown (left). EtBr intercalates between plan parallel Watson-Crick base pairs causing the DNA structure to expand slightly (lengthwise). The maximum possible is 13 EtBr/ nucleotide residue (6.5 EtBr/ base pair). EtBr experiences a significant enhancement in fluorescence properties upon interaction with DNA, but not otherwise. Structures of two other well-known DNA intercalator dyes are shown.

ing is 40 times less sensitive than ethidium bromide, but causes less damage to DNA and is non-carcinogenic. **Acridine orange** is useful for differentiating **single-stranded** (**ss**) and **double-stranded** (**ds**) DNA since dsDNA will fluoresce green under UV light, whilst ssDNA will fluoresce red. The recovery of DNA samples and fragment from electrophoresis gels is relatively simple. There were originally a wide range of methods to recover DNA from agarose gels, but methods have frequently been superseded by the use of the GuNCS miniprep or maxiprep method. The approach in this case here is literally to cut out the DNA band(s) of interest and suspend in guanidinium thiocyanate buffer in order to solubilise the agarose gel before carrying on with the remainder of the purification protocol (see Section 3.2.7).

3.2.9 DNA sequencing

The standard method of sequencing DNA is the **chain-termination sequencing** method (Figure 3.12). In this method, DNA polymerase is used for short time periods to rebuild duplex DNA using the complementary DNA strand as template. The region of sequencing interest is dictated by the choice of sequencing primer that anneals by Watson-Crick base pairing to its complementary sequence in the complementary DNA strand and then acts as an initiation point for DNA sense strand polymerisation and duplex DNA rebuilding.

In early molecular biology research, the standard approach was to set up four simultaneous reactions with equivalent, defined concentrations of complementary DNA strand, primer and four deoxynucleoside triphosphate (dNTP) substrates. Each reaction differed only in having a low mol% of a single radioactively labelled **dideoxy-nucleotide (ddNTP)** substrate (either **ddATP**, **ddGTP**, **ddCTP** or **ddTTP** respectively). These ddNTPs possessed neither 2′- nor 3′-hydroxyl group and therefore prevented further DNA polymerisation whenever they were incorporated into a growing chain. Consequently, once the designated short reaction period had reached completion, each DNA polymerisation reaction was expected to contain a range of incomplete DNA sense strands terminating in a dideoxynucleotide residue. Where ddATP was used, all the incomplete strands were expected to terminate in

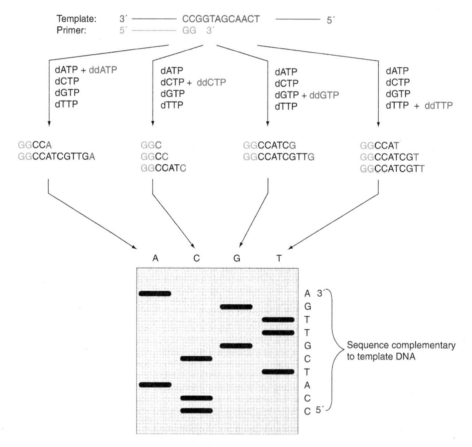

Figure 3.12 DNA sequencing requires template single-stranded DNA and sequencing primers. A selected primer Watson-Crick base pairs to the template and four discrete template-directed DNA polymerisation experiments are begun, each spiked with a different **chain-ter-minating dideoxynucleotide** or **dNTP** (**ddATP, ddCTP, ddGTP** or **ddTTP**). When the polymerisations take place, oligodeoxynucleotide chain synthesis is stopped wherever a ddNTP is incorporated. Hence each different experiment has different chain terminated oligodeoxynucleotide products (as illustrated). If these reactions involve radioactive ddNTPs then the products can be resolved simultaneously by **polyacrylamide gel electrophoresis** (**PAGE**) and imaged by **autoradiography** giving black bands (as above). These black bands together represent a **sequenc-ing ladder** that is read from bottom to top, identifying the DNA sequence (5′→3′) complementary to the **template DNA**, reading directly from the selected primer the identities of successive deoxynucleotide residues according to their base nucleoside composition. (illustration from Voet, Voet and Pratt, 1999, Wiley, Figure 3–23).

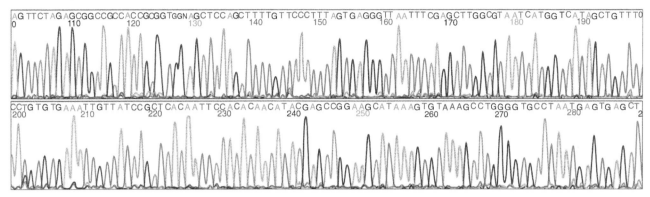

Figure 3.13   Automated DNA sequencing. When ddNTPs are differentially fluorescent-labelled, then only a single template-directed DNA polymerisation reaction is required, spiked with all four ddNTPs. The different length oligodeoxynucleotide products are then resolved individually by **capillary electrophoresis** (**CE**, see Chapter 7) and the identity of the terminating ddNTP can be identified not by horizontal position on a gel (see Figure 3.10), but by colour of extrinsic fluorescence output (**red** for **ddTTP**, **blue** for **ddCTP**, **green** for **ddATP** and **cyan** for **ddGTP**) (see Chapter 4). The collective sequence of fluorescence peaks together represent another version of the **sequencing ladder** that is read from left to right, identifying the DNA sequence (5'→3') (400–800 bases) complementary to the template DNA, reading directly from the selected primer (illustration from Voet, Voet and Pratt, 1999, Wiley, Figure 3–25).

ddA; where ddGTP was used, all were expected to end in ddG; where ddCTP was used, all were supposed to end in ddC and with ddTTP strands all terminated in ddT. The DNA sense strand products of all four reactions were then resolved side by side on polyacrylamide gel and the radioactivity visualised by photographic plate, resulting in the famous **DNA sequencing ladder** wherein each successive ladder line going upwards in order from the bottom of the gel (shorter strands run faster) was used to identify, according to their component base nucleoside compositions, successive deoxynucleotide residues in the DNA sense strand sequence starting at the primer.

Subsequently, sequencing was typically performed with single DNA polymerisation reactions containing all four ddNTPs differentially labelled with four different types of fluorescent label specific for each ddNTP (see Chapter 4) (Figure 3.13). The complete set of individual DNA sense strand products could be resolved by **capillary electrophoresis** (**CE**, see Chapter 7) so that each successive fluorescent label observed and identified from the beginning of the elution run (shorter strands run faster) was able to identify cleanly successive deoxynucleotide residues in the DNA sense strand sequence starting from the primer. Read lengths were typically from 400 to 800 deoxynucleotide residues from the primer.

Nowadays, sequencing can be performed using **high-throughput sequencing** methods. These methods are split into short-read and long-read methods. Short-read methods include **sequencing by synthesis** methods that enable millions of reads per run, with read lengths in the 50–500 bp range. Such approaches are cost-effective and widely used. Long-read methods include **single-molecule real-time** (**SMRT**) **sequencing** and **nanopore sequencing** that read through literally hundreds of thousands of base pairs at a time. Nanopore sequencing, in particular, is performed by passing DNA through a nanopore that measures the ion current as a function of the size, shape and length of DNA sequences. Nanopore sequencing is now available in small portable instruments at low cost, increasing accessibility to long-read methods in a variety of contexts.

3.3   Cloning and identification of genes in DNA

Now that we have briefly introduced some of the most important tools and techniques of molecular biology, we can return to the isolation, cloning and identification of genes of interest from within prokaryotic genomic DNA or eukaryotic cDNA library sequences. There are two main approaches for this:

1. **Direct DNA cloning.** This involves "cutting" DNA from a source (such as genomic DNA), and then "inserting" the desired DNA into plasmid DNA that can be replicated (in effect multiple copied), usually in *E. coli*.

2. **PCR.** This is an enzyme-based method for amplifying (multiple copying) DNA, either from a DNA template, or an RNA template (**reverse transcription polymerase chain reaction** [**RT-PCR**]).

3.3.1 Direct DNA cloning

Direct DNA cloning was originally developed for the isolation and cloning of prokaryotic genes, but can be used for the isolation and cloning of eukaryotic genes as well. The meaning of direct DNA cloning is the isolation and multiple copying of DNA from a source of interest (i.e. genomic DNA primarily purified directly from a prokaryotic organism or DNA established in a eukaryotic cDNA library). With respect to prokaryotic organisms, DNA once isolated must be cut into fragments ("shot-gun" fragments) either by chemical means or by using restriction enzymes (see Section 3.2.2). Restriction fragments of genomic DNA are then ligated into pDNA vectors (see Section 3.2.1) to create recombinant pDNA vectors. With respect to eukaryotic organisms, cDNA library components may be inserted similarly into pDNA to create alternate eukaryotic recombinant pDNA vectors.

Recombinant means that the pDNA is comprised of regions of DNA from at least two different organisms. These recombinant pDNA vectors are used in the transformation of host cells, typically *E. coli*. Multiple copying of the recombinant pDNA vectors then becomes possible with growth of the host, owing to simple **DNA replication**. Since the genetic code is universal, the host is unable to distinguish recombinant DNA from self-(host) DNA and consequently when the host grows and divides, even the recombinant pDNA is copied. If *E. coli* cells are spread on a nutrient plate and allowed to grow, then colonies form such that each cell of a given colony originates from a single parent cell and is a **clone** of the original transformed cell. Each cell of the colony will possess recombinant pDNA (as single plasmid or multiple copy/clonal cell) containing a single, unique restriction fragment not found in any of the other *E. coli* colonies. The result then is known as a **clone library** since each clone harbours a unique restriction fragment from the original genomic DNA of interest.

After making the library, there is a need to identify which of the clones contains the gene of interest. The usual way of doing this has been to use a synthetic ssDNA molecule known as a **probe** that is complementary to a DNA sequence in the target gene of interest. This probe will then be used to bind specifically to the target gene by Watson-Crick base pairing in a process known as **hybridisation** (Figure 3.14). If the DNA sequence is unknown, how is the probe designed? In this instance there are four general approaches:

1. If the amino acid sequence is known for the desired protein, then oligodeoxynucleotide hybridisation probes may be designed with reference to the genetic code (see Chapter 1).

2. If the gene has previously been cloned from a related organism then a previously used probe (known as a **heterologous probe**) may be tried.

3. If a protein is known to be abundant in a particular cell/tissue, then this abundance may also be reflected in its statistical abundance within a clone library.

4. 4 If the protein of interest has been purified before, and an antibody has been raised against the protein, then the clone library will have to be converted into an expression library (see later) and the presence of the desired protein is then screened using this antibody.

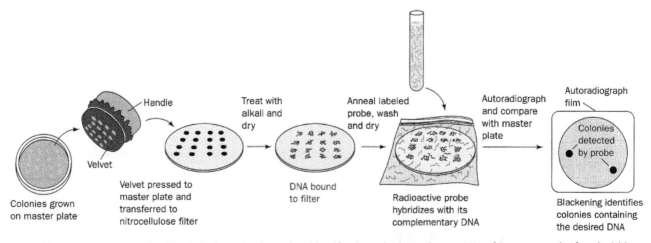

Figure 3.14   Hybridisation in direct cloning. This is used to identify where the heterologous DNA of interest can be found within a **clonal library**. DNA from each clone is isolated by treating each colony with alkali and drying. A labelled probe is then use to hybridise to the target in order to identify in which colony the desired gene/DNA of interest can be found (illustration from Voet, Voet and Pratt, 1999, Wiley, Figure 3–31).

Clearly options (1) and (2) are much preferred since they can give unambiguous identification of a gene of interest within a very small number of clones (possibly one clone) within the clone library.

3.3.2 Polymerase chain reaction

The PCR has revolutionised biological research since its invention in 1986. PCR is a method for amplifying (multiple copying) DNA, from DNA or RNA sources. PCR has many applications, but an important role in the isolation and cloning of eukaryotic genes from cDNA libraries. The mechanism of PCR is shown in Figure 3.15. Duplex **template DNA** is identified (for instance a cDNA library) and initially brought to a very high temperature (94 °C) in order that DNA strands (sense and complementary) may dissociate, a process that is often referred to as **DNA melting** or **denaturation**. The region to be amplified is then defined by the introduction of two short oligodeoxynucleotides, known as **primers**. When the temperature is reduced to 50–60 °C, these primers are able to bind, or **anneal**, anti-parallel to mutually complementary sequences in either DNA strand of the separated duplex template DNA as appropriate, by Watson-Crick base pairing. The temperature is then raised to around 74 °C, optimal operating conditions for a **thermostable DNA polymerase**. This enzyme was originally purified from the hyperthermophile *Thermus aquaticus*, and hence is known as *Taq* **polymerase**.

Taq polymerase reconstructs duplex DNA starting from the primers annealed to either the sense or complementary strands of the original duplex DNA template. The enzyme proceeds in a 5′→3′ fashion taking the deoxynucleoside triphosphates dATP, dCTP, dGTP and dTTP (often referred to as the dNTPs) as substrates for the synthesis of DNA and the reconstruction of DNA duplex. This catalytic process is known as **extension**. From the original duplex DNA template (parent original), two new duplex or dsDNA copies (daughter copies) now exist, wherein one DNA strand from the parent original template is now in each of the two daughter dsDNA copies; PCR is hence a method of **semi-conservative**

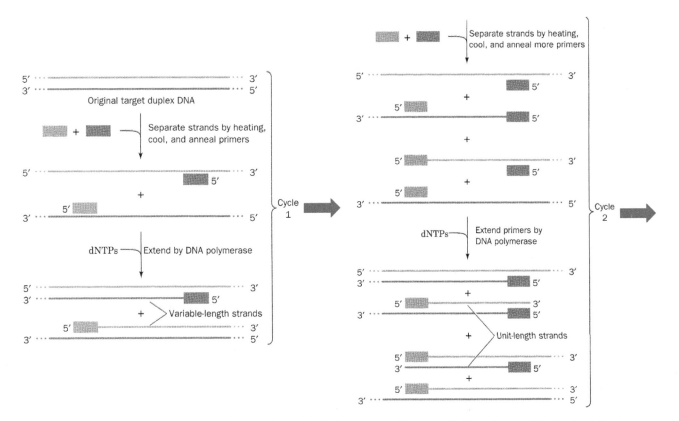

Figure 3.15 Illustration of a PCR reaction. A region of dsDNA for instance (from a cDNA library) is amplified in the following way. Strands are melted by heating and then cooled to anneal two primers. DNA polymerase extends from the primers in the 5′→3′ direction. Thereafter another round of heating, cooling and templated DNA synthesis takes place for as many times as required for the specific amplification of a target DNA sequence of interest (illustration from Voet, Voet and Pratt, 1999, Wiley, Figure 3–32).

replication. The process of denaturing, annealing and extension may then be repeated (typically 20–30 times) and the quantity of dsDNA product expands exponentially, to result in a significant quantity of dsDNA that has been selectively amplified according to the design of the two primers.

Although PCR is powerful, it does have limitations. First, in order to design optimal primers, the base sequences of the regions bordering the target region of interest in the duplex DNA template should be known as well as possible. Second, it is difficult to amplify very large stretches of DNA. As the region of amplification is increased further and further, so is the likelihood of introducing mutations since *Taq* polymerase is not completely faithful (i.e. the **fidelity rate** is not 100%). In practice, PCR amplifications of up to 3000 base pairs (3 kb) are performed easily, and with care and skill this may be extended up to 10 kb. An absolute limit of around 40 kb is the maximum size of region that may be amplified by PCR. However, for most practical purposes, these limits do not pose a major problem. A 6 kb length of DNA coding sequence should code for a protein of approximately 2000 amino acid residues, which corresponds with a protein of 240 kDa assuming an average molecular weight of 120 Da per amino acid residue. Accordingly, one of the main advantages of PCR to the process of isolation and cloning of genes is that direct cloning may be avoided if the gene is quite well known and has been cloned from a related source previously, or else amino acid sequences are known. In this case, heterologous probes may be used to PCR amplify a gene of interest directly from prokaryotic genomic DNA or from a eukaryotic cDNA library **without** the need for any of the previously described direct cloning procedures (see Section 3.3.1). PCR amplification then becomes a potent means to obtain quantities (μg–mg) of a gene of interest irrespective of duplex DNA template source. Obviously, a gene of interest in prokaryotic genomic DNA or eukaryotic cDNA may exceed the limits of PCR amplification, in which case there is no alternative but to effect direct cloning approaches to isolate, clone and identify a gene of interest either from genomic DNA or cDNA libraries as appropriate.

3.3.3 Gene expression and expression vectors

Following the isolation, cloning and identification of genes of interest from within genomic DNA or cDNA libraries by direct cloning or PCR amplification, the next stage is to seek high levels of gene expression and the purification of pure protein in consequence.

3.3.3.1 Expression vectors

Typically, proteins are expressed in *E. coli* using a pDNA expression vector (Figure 3.16) especially adapted for high-level gene expression. These pDNA expression vectors are typically around 5000 base pairs in size and have the following characteristics. As for pDNA cloning vectors, an origin of replication is still required and so is a selectable marker. The origin of replication in this case is a **dA.dT** base pair rich region also containing multiple copies of the palindromic sequence 5′-d(pGATC)-3′, which can be methylated on guanine to control replication. DNA methylation at origins of replication is a well-known way to shut down replication and inactivate a pDNA vector with respect to replication. The naturally occurring origin of replication in the *E. coli* chromosome (there is only a single origin of replication in *E. coli* for the entire chromosome) is **oriC**, a 468 base pair non-coding region that when incorporated into pDNAs supports a single copy per cell. The **pre-replication complex** is a protein complex that binds to oriC allowing replication of the pDNA. When that oriC (often referred to as ori) origin of replication is reduced then there is less control of replication by the pre-replication complex and pDNAs are found in multiple copies per cell. The importance and application of the selectable marker for a pDNA vector have been explained already (see Section 3.2.6). Where the selectable marker confers antibiotic resistance against antibiotics such as **ampicillin**, **kanamycin** and **tetracycline**, then the marker is most commonly known as *Amp*^R, *Kan*^R and *Tet*^R respectively (although other names are used too; see Figure 3.16). *Amp*^R is the most common marker and corresponds to the gene coding for the enzyme β-**lactamase**. One other highly important region of a pDNA expression vector is the **multiple cloning site** (**MCS**). This region is prepared from a synthetic, heterologous DNA fragment engineered to contain as many unique (to the pDNA) restriction sites as possible in order to facilitate ligation of the heterologous gene of interest isolated, cloned and identified from either genomic DNA or a cDNA library by direct cloning or direct PCR amplification (see Sections 3.3.1 and 3.3.2).

There are other important regions in a pDNA expression vector that need to be mentioned as well in order to "complete the tour". Foremost is the **promoter**. Promoters are non-coding stretches of DNA that are located upstream (i.e. to the 5′-side of the sense strand of pDNA) of a heterologous gene of interest. The promoter region controls the transcription of the heterologous gene (5′→3′) into mRNA, so is a very crucial component of any pDNA expression vector. In the early years of molecular biology, a number of natural *E. coli* promoters such as the *lac* and *trp* **promoters** were

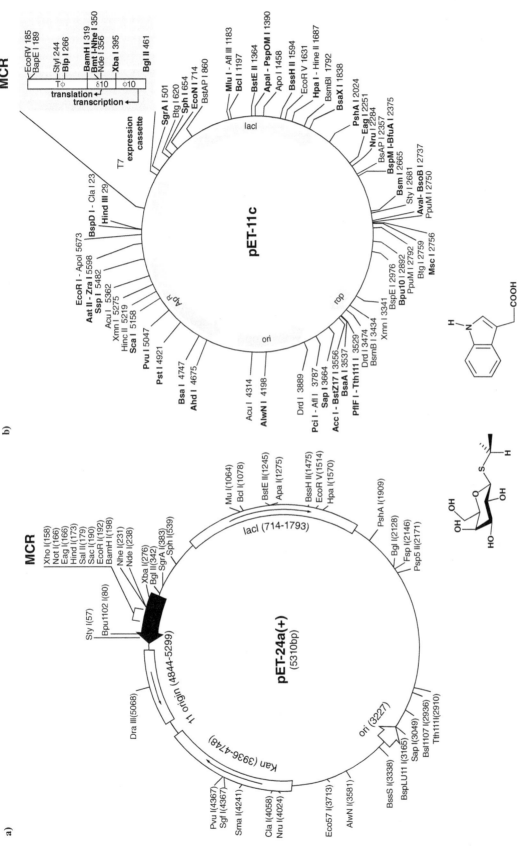

Figure 3.16 **Typical commercial protein expression vectors. (a)** pET-24a (+): the **origin of replication** (*ori*), kanamycin resistance (*Kan*; also known as *Kan*^R) and the **lactose repressor** (*lacI*, also known as *Ap*^R), are shown with the **multiple cloning site (MCR)** (illustration from Merck Biosciences); **(b)** pET-11c: *ori*, *lacI* and *MCR* are clearly visible. So, too, is the **ampicillin resistance** (*Ap*^R, also known as *Amp*, *Amp*^R) region. This protein expression vector also harbours a **T7 promoter** that acts as the promoter to drive the expression of heterologous DNA inserted at the MCR. Both maps have extensive restriction endonuclease mapping (illustration from Merck Biosciences). Small molecule promoters such as **isopropyl-β-D-thiogalactopyranoside (IPTG)** (reverses *lacI* repression) and **indole acetic acid (IAA)** (reverses *trp* promoter repression) are also illustrated.

discovered, as well as phage promoters such as the λp_L **promoter** of **bacteriophage λ**. More recently, artificial promoters such as the ***tac* promoter**, have been created through the fusion of sequence elements derived from original promoters. Promoters may **be inducible**, suggesting that an external agent transcription can trigger transcription. For instance, addition of **isopropyl-β-D-thiogalactopyranoside (IPTG)** to the growth medium induces transcription from the *lac* and *tac* promoters. Similarly, the addition of **indole acetic acid (IAA)** induces the *trp* promoter, while raising the growth temperature (briefly to 42 °C) induces the λp_L **promoter** by inactivating a phage-encoded repressor of transcription. A very powerful promoter element is the **T7 promoter** that needs to be used in association without bacteriophage **T7 RNA polymerase** (see Section 3.3.3.2).

The enzymes responsible for transcription of DNA to mRNA are all **RNA polymerase** enzymes (or more specifically, DNA-dependent RNA polymerases). RNA polymerases bind at promoter sequences in order to initiate transcription. Promoter structure is unique to a given RNA polymerase. Having noted this, promoters also possess common elements such as **TATA boxes** (also known as **Pribnow boxes** in prokaryotes, or **Hogness boxes** in eukaryotes). TATA boxes consist of a 10 deoxynucleotide bp region within a promoter that has a predominance of **dA.dT** base pairs. Accordingly, the duplex DNA structure in this region is primed to separate in response to RNA polymerase "invasion", thereby allowing transcription to begin. **Transcription factors** also play a role in this process, especially in eukaryotes. These factors are defined as proteins that regulate transcription by increasing or decreasing RNA polymerase binding to promoters. Transcription takes place (5′→3′ direction) (see Chapter 1), until a **terminator sequence** is reached that signals RNA polymerase dissociation.

In prokaryotes, many mRNA transcripts are **polycistronic**, this means that the coding regions for several proteins are incorporated sequentially within a single primary mRNA transcript. Proteins coded for in this manner often have related functions. Therefore, the co-transcribed genes are said to belong to an operon. **Operator sites** are found quite frequently near promoters. These are **repressor protein** binding sites to which a repressor protein may bind in order to modulate the transcription of genes, including those of an operon. The mechanism of repression involves the physical halting of the progress of RNA polymerase transcription. Accordingly, small molecular inducers of transcription, such as IPTG and IAA (see above), are now understood to induce or promote transcription by binding to repressor proteins directly, thereby preventing these proteins from binding to operator sites and repressing transcription.

3.3.3.2 *Protein expression strategy*

There are a number of approaches for the expression of proteins in *E. coli*. A modern approach to protein expression is to make use of purpose designed pDNA expression vectors such as the **pET vector** family, freely available from commercial sources (Figure 3.16). This pDNA expression vector family comprises a T7 RNA polymerase promoter for transcription, hence the corresponding *E. coli* host should be a recombinant strain engineered for constitutive expression of the heterologous T7 RNA polymerase. This chromosomal T7 RNA polymerase is under control of the promoter *lacUV5*. The product of the gene *lacI* is the **lactose repressor** that modulates transcription from this *lacUV5* promoter. IPTG is an inducer of transcription from *lac*-family promoters by binding to the lactose repressor protein. Hence, when IPTG is added to recombinant *E. coli* growth medium then transcription (and translation) of the T7 RNA polymerase is promoted. Correspondingly, the T7 RNA polymerase promoter in pET vectors is a hybrid of the wild-type T7 and the *lac* promoters, known as the ***T7lac* promoter**. This, too, is a *lac* family promoter and hence the presence of IPTG also induces transcription (and subsequent translation) for the heterologous gene of interest by T7 RNA polymerase as well. The pET vector family also possesses a **ribosome binding sequence (RBS)** site that provides a means for the primary mRNA transcript to attach to *E. coli* ribosomes post-transcription (see Chapter 1). The RBS site is positioned after the promoter, 7–9 nucleotide residues upstream (i.e. before 5′-end of gene) from the start codon of the heterologous gene of interest. This gene of interest is obviously inserted into a pET family vector by means of the MCS, beyond which is **a transcriptional terminator site** located downstream (i.e. post 3′-end of gene) from the **stop codon** of the heterologous gene of interest. This terminator site obviously indicates the site where the T7 RNA polymerase finishes transcription.

3.3.3.3 *Cloning for RNA synthesis*

Transcription of RNA from DNA is far simpler than translation of RNA into protein, and large RNA molecules can be synthesised using a simple *in vitro* cell free transcription system (see Chapter 2). Typically, the DNA coding for the RNA of interest is inserted into pDNA downstream of a promoter such as the **T3, T7** or **Sp6 promoter**. RNA is then synthesised from the pDNA *in vitro* using the most appropriate **DNA-dependent RNA polymerase** for the promoter (i.e. **T3 RNA polymerase**, T7 RNA polymerase or **Sp6 RNA polymerase**).

3.4 Integrating cloning and expression

The main problem for the chemical biology reader first introduced to molecular biology is to piece together the many tools and techniques involved. The chemical biology reader approaching molecular biology for the first time can be particularly daunted and even put off by the substantial change in tools, techniques and language compared with chemistry. However, once learnt, molecular biology can be very useful and effective as a tool for chemical biology that can be easily implemented with little difficulty. Here is a worked example of cloning and expression of a protein of interest that makes use of concepts and ideas discussed in Sections 3.2 and 3.3.

In our example, large quantities of *E. coli* **iron-superoxide dismutase, SodB**, are required. This is an iron-containing enzyme involved in the removal of superoxide radicals for detoxification (Figure 3.17). The first objective must be to **clone** the gene from the appropriate organism, then to express this gene in fusion with a suitable tag to help purification and also increase the solubility of the protein product. Using web-based *E. coli* **gene databases**, basic information can be obtained about the protein and the gene: for instance, the protein is 192 amino acids (and hence is coded by 576 bases), with a molecular weight of approximately 21 kDa, and a theoretical pI of 5.9. Given the size of the protein, a PCR amplification strategy should be feasible and is preferred for the identification and cloning of the gene. In fact, PCR is probably the only real option

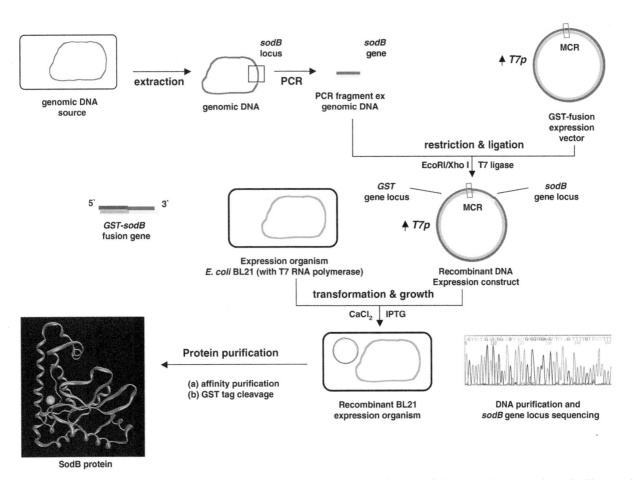

Figure 3.17 Summary of cloning from genomic DNA to glutathione S-transferase – fusion protein expression. The illustrated example concerns cloning of *E. coli* protein **SodB** (pdb: **1isa**). Genomic DNA is isolated by alkaline extraction and the ***sodB* locus** amplified by PCR. The recombinant PCR fragment is subjected to restriction cutting and ligation into the MCR of an appropriate **glutathione S-transferase (GST)**-fusion expression vector. After transformation of *E. coli* **BL21** and growth, DNA sequencing is used to identify complete ***GST-sodB*** recombinant clones, then protein is purified as indicated.

for isolating the gene if no other source is available, such as a eukaryotic cDNA or a prokaryotic clonal library. Accordingly, the first task should be to design primers to amplify the gene from an appropriate genomic source. Within the primers, unique restriction sites must be located in order to help the easy insertion of the PCR product into a pDNA expression vector later on. Furthermore, the strategy to overcome SodB solubility problems during purification requires the introduction of an N-terminal **glutathione S-transferase (GST)**-tag in this case so that the pDNA expression vector is expressing not the ***sodB*** gene but a ***GST-sodB*** fusion gene, giving rise to a more soluble product with translation (see Chapter 2). Hence there is a requirement that the pDNA expression vector should possess a GST-tag sequence in frame with a downstream MCS region (i.e. post 3′-end of GST-tag sequence) into which the PCR product can be inserted by ligation.

3.4.1 Designing forward and reverse primers

The **forward primer** (i.e. initiated before the 5′-end of gene) should be constructed with three or four deoxynucleotide residues prior to a unique EcoRI restriction site, in order to ensure that this restriction site is active, since sites in terminal positions may not be usable restriction sites. Thereafter, the forward primer concludes with the first 20 deoxynucleotide residues (5′→3′) of the known *sodB* gene sequence. The EcoRI restriction site is unique to the pDNA expression vector to be used and is absent from the *sodB* gene sequence as well. The **reverse primer** (i.e. initiated before 5′-end of complementary gene) should also be constructed with three or four deoxynucleotide residues prior to a unique XhoI restriction site, followed by the first 20 deoxynucleotide residues (5′→3′) of the known complementary *sodB* gene sequence. Once again, the XhoI restriction site is unique to the pDNA expression vector to be used and absent from the complementary *sodB* gene sequence.

Hence, the forward primer is present to initiate sense strand DNA synthesis using the complementary (antisense) strand of DNA as a template. By contrast, the reverse primer is present to initiate complementary strand DNA synthesis using the sense strand of DNA as a template (Figure 3.15). The first few times primer design is attempted, it is usually easiest to draw out exactly what is happening to be sure of primer orientation. There is also a need to make sure that the **melting temperatures** of the forward and reverse primers are as similar to each other as possible by extending or shortening those deoxynucleotide stretches post the unique restriction sites, as appropriate. This is important to ensure equal levels of sense and complementary strand DNA synthesis under PCR amplification conditions.

3.4.2 PCR amplification and product isolation

PCR amplification of a gene from prokaryotic DNA is best achieved by prior isolation of genomic DNA from the prokaryote of interest in the following way. In this case the *sodB* gene sequence should be amplified from *E. coli* cells. Therefore, these cells can be washed with PBS buffer, then heated to 95°C for a few mins to allow the genomic DNA to be released without degradation from DNAses. The primers can then be mixed with the released genomic DNA (or even the *E. coli* cells) and all the necessary components for PCR amplification (*Taq* polymerase, 4 dNTPs, Mg^{2+} containing buffer). Conditions and parameters may need some adjusting, most notably the melting temperatures of the primers and the number of cycles used. This melting temperature represents the temperature at which the primer is 50% complexed with the template. This may be calculated from the primer sequence. However, the ideal annealing temperature is 5°C lower than the melting temperature. If the melting temperature is too high then not enough primer will bind to template DNA. If the temperature is too low then non-specific sequences are amplified because primers will bind to similar as well as identical complementary DNA sequences. A typical number of cycles would be 25 amplification cycles. This number is important since if there are too few amplification cycles then there may not be sufficient PCR amplification product to handle or if there are too many amplification cycles then there is serious risk of PCR-based mutations.

Following PCR amplification, a portion of the reaction mix may be run on an agarose gel to check the success of the process. If PCR has been successful, then a clean band should be visible on a stained agarose gel corresponding to the amplified *sodB* gene fragment of approximately 600 bp in length. The simplest way to purify this *sodB* gene fragment would then be by using the guanidinium isothiocyanate-silica method once again.

3.4.3 Ligation and transformation

Extra pDNA expression vector may be obtained if required by mini-prep preparation and then digested with EcoRI and XhoI restriction enzymes. Similarly, the *sodB* gene fragment should be digested with EcoRI and XhoI restriction enzymes. In both cases, digestion should be to completion and products must be purified by agarose gel electrophoresis.

After purification of the two main fragments from each digestion, both fragments may then be ligated together by means of T4 DNA ligase. In the case of our worked example, the pET pDNA expression vector was selected with the idea of ligating the PCR generated *sodB* gene fragment into the vector MCS in frame with the 3′-terminus of a GST-tag (see Section 3.4) thereby generating a new *GST-sodB* fusion gene. After ligation, the newly formed recombinant pDNA expression vector can be transformed into $CaCl_2$ competent *E. coli* cells, or introduced into other appropriate host cells by electroporation. For the transformation of a ligation mixture it is best to use an *E. coli* strain that has been modified so as to be particularly efficient at taking up pDNA, for example the *E. coli* DH10B strain. After transformation the cells should be encouraged to grow without antibiotic for one h at 37 °C, then spread on agar plates for antibiotic selection over a further 16 h incubation at 37 °C.

3.4.4 Validation and sequencing

Successful transformation with recombinant pDNA expression vector should result in the appearance of a number of colonies on the antibiotic selection plate, with none appearing on the control plate. The recombinant pDNA from these colonies needs to be validated and sequenced before proceeding. In order to do this, colony samples should be picked and grown in selective media (5 ml). The pDNA can then be isolated by the guanidine isothiocyanate-silica method (as usual) and then digested with EcoRI and XhoI restriction enzymes in order to see the reappearance of two DNA fragments, one of approximately 600 bp in length. After this, the recombinant 600 bp insert comprising the *sodB* gene should be sequenced as completely as possible by DNA sequencing, using designed sequencing primers, in order to demonstrate that the *sodB* gene has been cloned successfully from genomic DNA sources without PCR errors.

3.4.5 Protein expression

Overexpression of the *GST-sodB* fusion gene should then be made possible as follows. A host *E. coli* strain containing the T7 RNA polymerase is essential for optimal gene expression from the selected pDNA expression vector (pET family). Fortunately, the readily available, attenuated **E. coli BL21** strain has just such a chromosomal copy of this heterologous enzyme and so represents the ideal host for protein expression. Hence, post validation and sequencing of the recombinant pDNA vector, competent *E. coli* BL21 cells can undergo transformation and may then be grown on antibiotic selection plates in order to select for recombinant BL21 colonies. Thereafter, selected colonies may be picked, grown on a small scale (5 ml) at 37 °C until saturation (in the presence of antibiotics), then on a larger scale at 37 °C until the A_{600} **absorbance** is measured to be approx 0.6, after which the inducer IPTG (0.5 mM) should be added to induce substantial *GST-sodB* fusion gene transcription. After a further 2 h of cell growth, samples can then be collected and analysed by **sodium dodecyl sulfate polyacrylamide gel electrophoresis** (**SDS-PAGE**) for a protein overexpression band at the appropriate mass (in this case the mass of a GST-SodB fusion protein should be 48 kDa). Assuming that this band can be seen by SDS-PAGE, then protein purification can begin (see Chapter 2).

3.4.6 Cloning and expressing from eukaryotic genes

For higher eukaryotes, the most common approach to cloning is PCR amplification from cDNA libraries in order to avoid the intron problem. Another challenge when expressing eukaryotic proteins is that eukaryotic proteins frequently do not fold properly in *E. coli* and will not be processed with the correct post-translational modifications; therefore often other eukaryotic expression hosts must be used. Options include using yeast cells, insect cell lines or mammalian cell lines.

3.5 Site-directed mutagenesis

One powerful approach towards analysing protein structure and function is by modifying or mutating the amino acid sequence in a directed manner and then comparing characteristics of the mutant with the wild-type proteins. Before recombinant DNA technology, a number of chemical means were used to modify specific amino acids, or crude methods such as UV radiation were used to introduce random mutations to the DNA coding sequence. However, mutagenesis has never been easier with the applications of recombinant tools and techniques. Hence the practice of **site-directed mutagenesis** can be both PCR and non-PCR based.

3.5.1 PCR-based approaches to mutagenesis

The basic approach towards PCR-directed, site-directed mutagenesis of genomic DNA is the use of primers designed with the desired mutations included. The primer should incorporate the desired **base mismatch**. If the mismatch is to be generated near the 5′-end of the gene of interest then the forward primer can include the mismatch deliberately. Similarly, the mismatch is to be generated near the 3′-end of the gene then the reverse primer can include the mismatch instead. Hence, as the PCR reaction proceeds, then *Taq* polymerase will end up incorporating the mismatch into the gene sequence without interruption (Figure 3.18). Should the desired mismatch involve a central gene location then the approach must be to split the PCR amplification reaction into two. Each PCR reaction uses a central forward or reverse primer containing the desired base mismatch. The result is that the gene of interest is then amplified as two separate, partially overlapping fragments both containing the mutation. These two fragments must then be knitted back together using a further set of primers and a third PCR reaction that will result in a **chimeric product**. Alternatively, there is the **inverse PCR** method in which two primers are introduced, both incorporating the desired base mismatch, but both pointing outwards away from the centre of the gene of interest and instead pointing towards the long way around the pDNA expression vector. However, this has proven to be a very error prone method for site-directed mutagenesis and so should not be used. Hence, where the desired mismatch involves a central gene location, then a non-PCR-based site-directed mutagenesis is desirable.

3.5.2 Non-PCR-based approaches to mutagenesis

Many of these approaches are patented, so it is necessary to use tradenames to describe these useful approaches. Currently popular is the **QuikChange™ system** (Figure 3.19). In this, two primers are designed that anneal respectively to sense and complementary strands of a parent recombinant pDNA expression vector, but are also appropriately

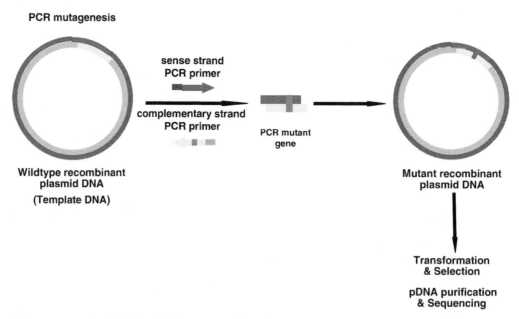

Figure 3.18   PCR mutagenesis. This is the simple way to engineer the primary structure of a protein of interest. Where desired mutations are near the 5′-terminus, the 3′-terminus or the sense strand, then the mismatched primer technique is used. In the illustrated case, the desired mutation is near the 3′-terminus, so a normal sense strand primer is combined with a mismatched complementary strand primer containing a mutation (blue). When the PCR reaction is allowed to proceed with the template DNA, then the mismatch in the complementary strand primer forces a mismatch to appear in both sense and complementary strands of the final **PCR amplification** product, resulting in a **PCR mutant gene**. Restriction cutting and ligation of the mutant PCR product into a cloning vector generates a mutant recombinant pDNA construct ready for transformation and selection, then DNA purification and sequencing of correct mutant recombinant DNA.

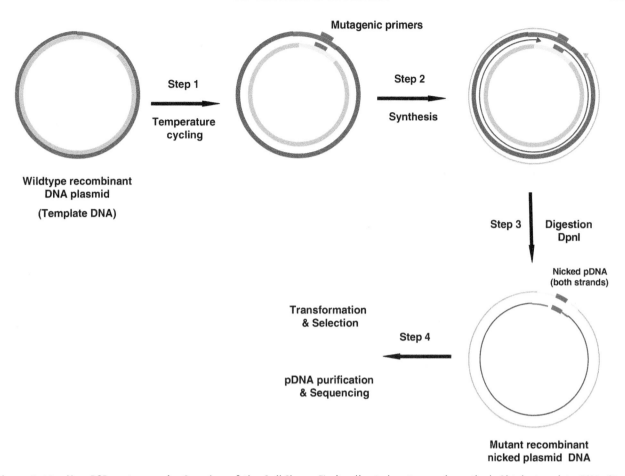

Figure 3.19 **Non-PCR mutagenesis.** Overview of the QuikChange™ site-directed mutagenesis method. Obtain template DNA: Step 1: denature the template DNA and anneal oligodeoxynucleotide mutagenic primers (blue) with desired mutation; Step 2: using the non-strand displacing action of ***PfuTurbo* DNA polymerase**, extend and incorporate the mutagenic primers resulting in nicked circular strands; Step 3: digest the methylated, non-mutated parental template DNA with **DpnI**; Step 4: transform *E. coli* **XL1-Blue** with nicked dsDNA, ready for selection, DNA purification and sequence identification of correct mutant recombinant DNA (illustration was adapted from the manual for the QuikChange™ site-directed mutagenesis method of Stratagene).

modified to introduce a mutation. Necessarily the two primers must point in opposite directions. Hence, in order to effect mutations, a parent vector is then introduced into a mutagenesis mixture containing the two primers and a DNA polymerase is required to assemble a new sense and complementary DNA strand from the two primers using the complementary and sense strands as templates respectively. Both new strands are assembled with high fidelity without displacing the original oligodeoxynucleotide primers. Therefore, the result comprises mutant sense and complementary DNA daughter strands in association with corresponding wild-type complementary and sense DNA parent strands. Daughter strands are clearly distinguishable from the parent strands due to previous **adenine methylation** of the parent vector by *E. coli* (i.e. recombinant pDNA vectors once utilised for *E. coli* transformation become methylated on both strands by their *E. coli* host cells). This methylation allows for the parent strands to be targeted for destruction in the mutagenesis mixture using the restriction enzyme **DpnI** that targets single strand restriction sites, 5′-d(pGm⁶ATC)-3′, containing **6-*N* methylated deoxyadenosine (dm⁶A)** nucleotide residues (see Chapter 1). These restriction sites are frequent enough to allow for complete digestion of the parent strands, leaving mutant daughter strands behind to generate a single mutant recombinant pDNA vector with two staggered nicks (i.e. the phosphodiester link is incomplete), one per strand. Transformation of *E. coli* once more allows these nicks to be repaired and the mutant recombinant pDNA to be replicated. There are various other technologies for site-directed mutagenesis that work by similar approaches.

4
Electronic and Vibrational Spectroscopy

4.1 Electronic and vibrational spectroscopy in chemical biology

Our first chapter on analysing structure is devoted to electronic and vibrational spectroscopy. Compared with **magnetic resonance** (Chapter 5), or **diffraction** and **microscopy** (Chapter 6), **electronic** and **vibrational spectroscopy** are decidedly low resolution techniques, in that atomic-level descriptions of biological macromolecular structures are all but impossible to achieve with these techniques. However, the advantage of electronic spectroscopies in particular is that they are able to provide meaningful data from very small quantities of biological macromolecule (Table 5.3). Therefore, broad brushstroke (Å–nm range) characterisations of biological macromolecular structures can be performed in advance of the more sophisticated, but both more time and material intensive, atomic-level structural characterisations such as **X-ray diffraction** (Chapter 6).

Furthermore, electronic spectroscopies, and to some extent vibration spectroscopy, are also able to demonstrate the occurrence of transconformational changes in the structures of biological macromolecules in response to environmental changes and/or molecular interactions, so providing an additional broad brushstroke view of biological function as well. In the case of **fluorescence spectroscopy**, this view can even extend up to the observations of molecular binding interactions taking place in real time in living cells. Hence, both electronic and vibrational spectroscopies represent "first-pass" structural and even functional characterisation techniques that can contribute to a broad if not necessarily detailed initial appreciation of biological macromolecular structure and function prior to more detailed investigations. Quite often, the absence of adequate quantities of biological macromolecules for studies ensures that information from electronic and vibrational spectroscopies may be, by default, all that is feasible to acquire and therefore currently available. Such realities ensure that electronic and vibrational spectroscopies are certain to play a central linking role from early structural to early functional characterisation for a long while to come.

4.2 Ultraviolet-visible spectroscopy

Ultraviolet (UV)-visible light spectroscopy is one of the oldest and outwardly simplest forms of spectroscopy for the analysis of biological macromolecule structure. However, since UV-visible spectroscopy is based upon electronic excitation then the simplicity of measurement that makes the technique so attractive to use is largely cancelled out by a distinct absence of detailed theory linking measured data through to a detailed structural characterisation. In consequence, UV-visible spectroscopy is a rather blunt instrument for the analysis of biological macromolecular structures. Nonetheless, there is still structural information to be gained. The basis of UV-visible spectroscopy is that when a beam of UV or visible light of a given wavelength passes through a solution of a given solute mounted in a **cuvette** (Figure 4.1), then the intensity of light may become diminished by interactions with dissolved solute or indeed solvent

Essentials of Chemical Biology: Structures and Dynamics of Biological Macromolecules In Vitro *and* In Vivo, Second Edition. Andrew D. Miller and Julian A. Tanner.
© 2024 John Wiley & Sons, Inc. Published 2024 by John Wiley & Sons, Inc.
Companion Website: www.wiley.com/go/miller/essentialschembiol2

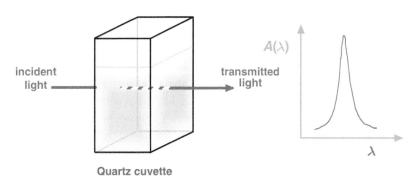

Figure 4.1   Illustration of experimental arrangement for ultraviolet-visible spectroscopy. A sample in a **quartz cuvette** is irradiated with monochromatic incident light at a **wavelength** λ and the amount of light that is absorbed at that wavelength, $A(\lambda)$, is determined by comparison between incident and transmitted light intensities. A plot of $A(\lambda)$, against wavelength λ gives us a typical absorption spectrum.

molecules. These interactions can be considered as inelastic collisions between photons and solute or solvent molecules, followed by **absorption**, then **electronic excitation** (assuming that photon energy coincides with the energy of an acceptable **electronic transition**). The result of these interactions is that the intensity of light transmitted through the solution in the cuvette is diminished.

4.2.1 Transition dipole moments

According to the classical theory of absorption, the interaction of light with a given molecule is thought to result in the induction of dipoles through the interaction of the oscillating electric field of light with polarisable clouds of electrons. This is represented by Equation (4.1):

$$\mu_{\text{ind}} = \alpha(\nu_{\text{v}}) \cdot E(\nu_{\text{v}}) \tag{4.1}$$

where μ_{ind} is the **induced dipole moment**, $E(\nu_{\text{v}})$ the **oscillating electric field of light** and $\alpha(\nu_{\text{v}})$ is the **electronic polarisability** of matter the latter two are a function of the **vibrational frequency of light**, ν_{v}. By analogy, according to quantum theory of absorption, the interaction of light with a given molecule invokes a transition dipole moment, provoking the electronic excitation of a single electron from an initial wavefunction-defined state, ψ_{a}, to a final wavefunction defined state, ψ_{b} (which is by definition an electronic transition). The **transition dipole moment** invoked is given by the following expression (4.2):

$$\left\langle \psi_{\text{a}} \left| \tilde{\mu} \right| \psi_{\text{b}} \right\rangle \tag{4.2}$$

Since all but one of the electrons in the given molecule remain unchanged in state as a result of electronic excitation, then only the wavefunctions involved directly in the electronic transition need be considered in defining the transition dipole moment. In the case of biological macromolecules and macromolecular assemblies, relevant wavefunctions usually correlate to lone pair orbitals, n, and π/π^* molecular orbitals such that only two main types of transition dipole moments need be considered, which are: $\langle n | \tilde{\mu} | \pi^* \rangle$ and $\langle \pi | \tilde{\mu} | \pi^* \rangle$ respectively. The first of these transition dipole moments is in fact zero, consequently corresponding $n{\rightarrow}\pi^*$ electronic transitions are known as weak, **symmetry forbidden transitions**. The second of these transition dipole moments is always non-zero and consequently corresponding $\pi{\rightarrow}\pi^*$ transitions are known as strong, **symmetry allowed transitions**. The symmetry allowed transitions are at least 100 times more intense than symmetry forbidden transitions.

A fall in transmittance resulting from absorption from solution in a cuvette is usually characterised in terms of **absorbance**, $A(\lambda)$, or **optical density**, $OD(\lambda)$, at the given **wavelength**, λ, according to the **Beer-Lambert law** (Equation 4.3):

$$A(\lambda) = c_{\text{M}} \cdot \varepsilon(\lambda) \cdot l \tag{4.3}$$

where *l* is the cuvette **pathlength**, c_M is the **concentration** of biological macromolecule and $\varepsilon(\lambda)$ is the constant of proportionality known as the **extinction coefficient**. Typically, $A(\lambda)$ is far from constant with wavelength and plotting $A(\lambda)$ as a function of λ resulting in a **UV-visible absorption spectrum** (Figure 4.1) that may exhibit characteristics of the solvent, but will certainly exhibit characteristics of the dissolved solute. The primary reason for the variation of $A(\lambda)$ with λ is that neither dissolved solutes nor solvent molecules are able to absorb photon energy uniformly with λ. Instead, there is an extensive pattern of differential absorbance owing to the fact that the absorption characteristics of any given solute or solvent molecule, are substantially dominated by **chromophores**. Chromophores are functional groups or absorbing elements (molecular structures or sub-structures) that absorb strongly at specific values of λ.

In general, the chromophores found typically in solutes, such as proteins, nucleic acids, carbohydrates and lipids (i.e. our biological macromolecules or macromolecular assemblies) absorb at values of $\lambda<300$ nm whilst structures found in water (the standard solvent) tend to absorb at values of $\lambda<170$ nm. In fact, at $\lambda<170$ nm, water has very broad electronic absorption bands, broader than those found in most other solvents. Furthermore, water molecules also interact extensively with dissolved biological macromolecules or assemblies, leading to considerable distortions in the energies of chromophore-associated electronic transitions and hence considerable variations in the values of λ at which electronic excitations take place (see Chapter 1). Consequently, UV-visible absorption spectra of biological macromolecules in water usually take on the appearance of a collection of broad absorption peaks in the range 170–300 nm. As we shall see, UV-visible spectra of globular proteins are the best understood and most widely studied.

4.2.2 UV-visible spectroscopy of proteins

Within globular proteins, the peptide links, amino acid side chains and disulfide bridges are the main characteristic chromophores typical in a globular protein (Table 4.1). In addition, a number of proteins may also be modified by non-peptide prosthetic groups that act as functional **cofactors** to enable protein functional activities. Such prosthetic groups are especially dominant in proteins involved in redox reactions or electron transfer processes. They have an especially rich chromophore behaviour (Figure 4.2) (Table 4.2). Both $n\rightarrow\pi^*$ and $\pi\rightarrow\pi^*$ transitions dominate throughout. Turning to amino acid residues, **tryptophan**, followed by **tyrosine** residues, are much the most significant chromophores and these dominate protein absorption in the UV region. This can be very useful. At a given wavelength, when $A(\lambda)$ is less than 0.5, values of $A(\lambda)$ turn out to be proportional to macromolecular concentration, c_M. Hence, in the case of proteins absorbance measurements made in the UV region at 280 nm, A_{280}, can be used to give a direct measure of protein concentration. Typically, protein concentrations are determined with reference to the A_{280} absorbance of the solution of the protein of interest measured at a fixed concentration of 1 mg ml^{-1} (1 g l^{-1}; 0.1%) (using a 1 cm pathlength cuvette). This $A_{280}^{0.1}$ value may be found from the extinction coefficients of tryptophan and tyrosine amino acid residues (i.e. ε_{max} 5700 and 1300 M^{-1} cm^{-1} respectively), together with the **protein molecular weight** concerned, M_p, according to the following Equation (4.4):

$$A_{280}^{0.1} = (5700 n_{Trp} + 1300 n_{Tyr}) / M_p \tag{4.4}$$

where n_{Trp} and n_{Tyr} are the number of tryptophan and the number of tyrosine amino acid residues per protein molecule. The value of $A_{280}^{0.1}$ for the majority of proteins is approximately 1.0.

Table 4.1 **Summary of the main chromophore/fluorophore residues in proteins and nucleic acids.** The main absorption and fluorescence characteristics are given.

Fluorophore	Conditions	A_{max}/nm	$10^{-3} \times \varepsilon_{max}$	I_{max}/nm	ϕ_F	τ_F/nsec
Tryptophan	aqueous, pH 7	280	5.6	348	0.2	2.6
Tyrosine	aqueous, pH 7	274	1.4	303	0.14	3.6
Phenylalanine	aqueous, pH 7	257	0.2	282	0.04	6.4
Adenine	aqueous, pH 7	260	13.4	321	2.6×10^{-4}	<0.02
Guanine	aqueous, pH 7	275	8.1	329	3.0×10^{-4}	<0.02
Cytosine	aqueous, pH 7	267	6.1	313	0.8×10^{-4}	<0.02
Uracil	aqueous, pH 7	260	9.5	308	0.4×10^{-4}	<0.02

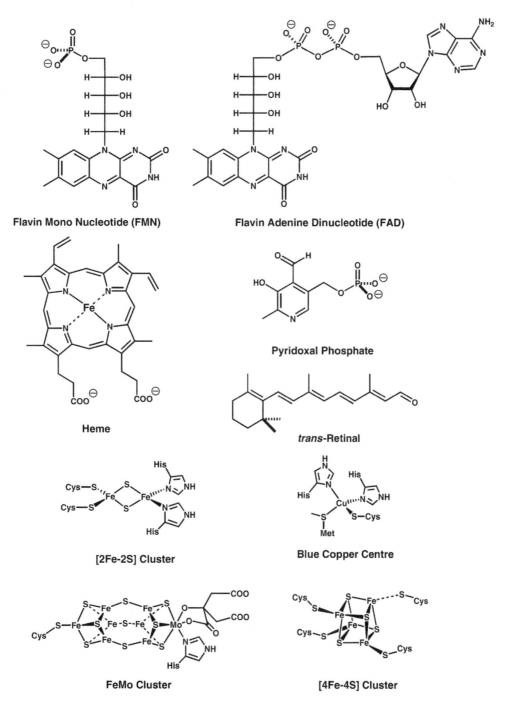

Figure 4.2 **Structures of main prosthetic groups.** These contribute significantly to the UV-visible spectroscopy of proteins. Prosthetic groups are non-amino, acid-based moieties that are covalently attached to the proteins concerned and play an integral part in the structure and function of proteins to which they are covalently attached.

4.2.3 UV-visible spectroscopy of nucleic acids

The UV-visible spectroscopy of nucleic acids is dominated by base absorption. Neither phosphodiester backbone nor β-D-ribofuranose/2′-deoxy-β-D-ribofuranose rings have any significant UV-visible absorbance above 200 nm. In comparison with absorption by the main protein chromophores (not including the prosthetic groups), absorption by each nucleic acid base (Table 4.1) is associated with more than one main $n \rightarrow \pi^*$ and $\pi \rightarrow \pi^*$ transition (in the range 200–300 nm) owing to the fact that each base has low symmetry and several hetero-atom lone pairs (see Chapter 1). Transitions for each individual base in a given nucleic acid tend to overlap and merge into a single broad, strong absorption

Table 4.2 **Summary of the main prosthetic group chromophores found in redox-active or electron transfer proteins/enzymes.** The main absorption and fluorescence characteristics are given. See Figure 4.2 for structures and abbreviations used in the table.

Chromophore	A_{max}/nm	$10^{-4} \times \varepsilon_{max}$	A_{max}/nm	$10^{-4} \times \varepsilon_{max}$
FMN	455	1.3	358	1.1
Blue Cu(II)	781	0.3	625	0.4
Heme Fe(II)	550	2.8		
[2Fe, 2S]	421	1.0	330	1.3
FAD	460	1.3	438	1.5
[4Fe, 4S]	570	0.4	490	0.8
Retinal	498	4.2	350	1.1
Pyridoxal	415	2.6		
FeMo	550	2.2		

band for the nucleic acid polymer as a whole, with an absorbance maximum, A_{max}, at 260 nm (ε_{max} 10,000 M^{-1} cm^{-1} average per nucleotide base). In a similar way for proteins, nucleic acid concentrations in solution may be determined by A_{260} absorbance measurements with reference to the A_{260} value of a 1 mg ml^{-1} (1 g l^{-1}; 0.1%) solution of nucleic acids measured in a 1 cm pathlength cuvette. Unlike proteins, one average value of $A_{260}^{0.1}$ is sufficient to determine the concentration of most nucleic acids of interest, since the bases of each nucleoside or deoxynucleoside building block behave as chromophores in a very similar manner. This average value may be calculated from the **average nucleotide molecular weight**, M_{nt}, according to the following Equation (4.5):

$$A_{260}^{0.1} = 10000 / M_{nt} \tag{4.5}$$

A typical value for M_{nt} is usually 330 Da.

In contrast with proteins and nucleic acids, carbohydrates and lipids possess few substantial inherent chromophores and so neither class of biological macromolecule has particularly rich or useful UV-visible spectroscopic behaviour. Therefore, in Section 4.2.4 we shall focus on structural versus functional information available from the UV-visible spectroscopy of proteins and nucleic acids only.

4.2.4 Structural versus functional information from UV-visible spectroscopy

In the most trivial application, UV-visible spectroscopy provides a means to access the concentration in solution of proteins or nucleic acids. Moreover, calibrated values of A_{260} and A_{280} even provide a means to calculate approximate molecular weights according to Equations (4.4) and (4.5). However, more substantial structural information remains frustratingly difficult to deduce. Nevertheless, UV-visible spectroscopy can be a useful monitor of conformational changes in globular proteins. This is possible owing to the effects that local environment may have on the absorbance characteristics of aromatic amino acid residue chromophores, particularly tryptophan. Aromatic amino acid resides are responsible for clusters of $\pi \rightarrow \pi^*$ transitions absorbing strongly in the region 250–300 nm. The largest contributors to this UV-visible absorbance are tryptophan residues, with more modest contributions coming from tyrosine and phenylalanine residues (Table 4.1). Changes in the local environments of these aromatic residues can often be sufficient to change the UV-visible absorbance behaviour of some or all of these residues. This is particularly true in the event that a globular protein is subject to a range of different physical conditions and/or molecular interactions which can lead to obvious if not substantial trans-conformational changes that can provoke changes in local environments and thereby bring about UV-visible absorbance changes. For instance, many aromatic amino acid residues are often "buried" in the hydrophobic interior of a globular protein and not at the hydrophilic surface. However, situations arise where conditions change and/or molecular interactions take place that may lead to buried aromatic amino acid residues entering more hydrophilic environments and becoming more solvent accessible as a consequence of conformational changes (and vice versa). In such events, energies of both $n \rightarrow \pi^*$ and $\pi \rightarrow \pi^*$ transitions could become altered in response to changes in the local environments of aromatic amino acid residues. Specifically, if the local environment of an aromatic amino acid residue changes from a hydrophobic to a more hydrophilic environment, then a corresponding change in $\varepsilon(\lambda)$ can lead to a reduction in $\pi{\rightarrow}\pi^*$ transition energies associated with an increase in transition wavelength (**red shift**), and an increase in $n{\rightarrow}\pi^*$ transition energies, leading to a decrease in transition wavelength (**blue shift**) (Figure 4.3). (Obviously, the reverse situation is also true if the local environmental changes are from hydrophilic to hydrophobic.) Therefore, in summary, conformational changes in globular proteins may be diagnosed by shifts in the wavelengths of $\pi{\rightarrow}\pi^*$ and $n{\rightarrow}\pi^*$ transitions provided that the aromatic chromophores (especially tryptophan) are sufficiently sensitive to the local environmental changes caused by these conformational changes. Such changes in spectroscopic signature in response to molecular interactions can be very useful to study molecular recognition and binding as well (see Chapter 7).

Protein absorption bands may also be subject to band splitting, or **perturbation**, as a result of electronic interactions between adjacent chromophores (Figure 4.4). To a first approximation, the number of individual absorption bands generated by perturbations will be equivalent to the number of interacting chromophores, although the number of bands may be considerably less if perturbation effects are heavily affected by relative orientations between chromophores. Such reduced perturbation effects created through electronic interactions between peptide links can be used to identify the presence of protein secondary structures under some circumstances. For instance, perturbation effects will identify the presence of α-helical secondary structures in a protein. Owing to the highly regular nature of α-helical secondary structures and the extreme effect of orientation upon peptide link chromophore perturbation effects, α-helical structures are characterised by only two UV-visible $\pi{\rightarrow}\pi^*$ transitions associated with α-helix peptide links that are characterised by:

1. Transition dipole moment parallel to the main helix axis

2. Transition dipole moment perpendicular to the main helix axis

In practical terms, transition (1) is observed as a 190 nm absorption band with a shoulder at 208 nm. Transition (2) is observed only at lower wavelengths with good optical equipment. UV-visible $\pi{\rightarrow}\pi^*$ transitions involving either β-sheets

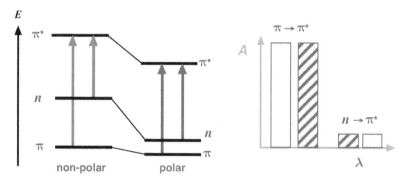

Figure 4.3 Illustration of effect of environmental conditions on $\pi{\rightarrow}\pi^*$ and $n{\rightarrow}\pi^*$ transitions. Displacement of a chromophore from non-polar (white bar) to polar (red hatched bar) conditions results in a **red shift** of the $\pi{\rightarrow}\pi^*$ and a **blue shift** of the $n{\rightarrow}\pi^*$ due to a differential stabilisation of lone pairs, n, compared with π^*-orbitals.

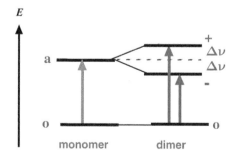

Figure 4.4 Illustration of perturbation effects. These occur when there are electronic interactions between chromophores. Interaction between two monomer chromophores splits the excited states into two sub-states, creating two new displaced electronic transitions. There should be at least as many excited state sub-states as there are electronically interacting chromophores. However, selection rules and orientation effects will diminish the number of transitions observed.

or less regular structures (known broadly as **random coil**) are not subject to extensive chromophore perturbation effects and hence UV-visible spectroscopy is unable to identify the presence of either secondary structural elements directly.

Turning to nucleic acids, the potential of UV-visible spectroscopy for nucleic acid structural analysis has always been considered high in principle. The reason for this is the sheer number of base chromophores available in nucleic acids and their potential for electronic interactions. Given the high degree of structural uniformity in nucleic acids, the total absorbance $A(\lambda)$ of a given sample of nucleic acids may be given by Equation (4.6):

$$A(\lambda) = c_{nt} \cdot l \cdot \sum_i \chi_i \cdot \varepsilon_i(\lambda) \tag{4.6}$$

This is a variation of the Beer-Lambert law (Equation 4.3), where $\varepsilon_i(\lambda)$ is the **molar extinction coefficient of pure nucleotide monomer of type i**, c_{nt} is total **nucleotide concentration** and χ_i is the **mole fraction of nucleotide monomer of type i** in the nucleic acid under investigation. Conformational changes will promote significant deviation from Equation (4.6), primarily by promoting local deviations in $\varepsilon_i(\lambda)$ values away from pure nucleotide monomer values. Local deviations in $\varepsilon_i(\lambda)$ values are primarily the result of local changes in electronic interactions between bases as a result of local geometric distortions, together with some contribution from local environmental changes. In the case of nucleic acids, local environmental changes have much less effect on UV-visible $\pi \to \pi^*$ transitions than changes in electronic interactions between bases. However, the conformational flexibility of nucleic acids turns out to be too high and the resulting geometric distortions too extensive and dynamic to obtain currently meaningful structural information by UV-visible spectroscopy on secondary and tertiary structures in nucleic acids. Nevertheless, the promise remains.

4.3 Circular dichroism spectroscopy

Circular dichroism (**CD**) spectroscopy is in many ways a more sophisticated version of UV-visible spectroscopy and is able to give substantially more structural information than may be obtained by routine UV-visible spectroscopy. Unfortunately, CD spectroscopy gives little meaningful information for carbohydrates or lipids owing to the lack of substantive chromophores in either class of biological macromolecule, once again.

4.3.1 Circularly polarised light

The classical view of light is that of two coupled, mutually perpendicular oscillating fields (electric and magnetic) that propagate together. Typically, the behaviour of either field can be defined in terms of a vector known as the **electric vector** (**E**) and the **magnetic vector** (**B**) respectively (Figure 4.5). The behaviour of the electric vector is critical to electronic spectroscopy and optical activity. Therefore, we will just concentrate on the electric vector for now. In the normal course of events, the electric vector may oscillate in any plane perpendicular to the direction of propagation. **Plane polarised light** is generated by means of a **polariser** that restricts electric vector oscillation to a single plane (Figure 4.5). When two beams of plane polarised light are combined 90° $(\pi/2)$ out of phase with respect to each other, then both electric vectors combine to form a single electric vector that precesses around the direction of propagation, mapping out a pathway that takes on the appearance of either a **left-/right-hand corkscrew**. Provided that both electric vectors are of equal magnitude then the result is either **left-hand** (E_L) or **right-hand** (E_R) **circularly polarised light**. If the electric vectors are of unequal magnitude then the corkscrew becomes asymmetric and the result is either **left-/right-hand elliptically polarised light** respectively (Figure 4.5).

4.3.2 Optical activity and circular dichroism

CD spectroscopy originates from **optical activity**. A given molecule has optical activity if light-molecule interactions are able to alter the physical properties of incident light transmitted through a sample containing that molecule. When left- and right-hand circularly polarised light beams are incident on a sample containing an optically active molecule, then the sample will absorb left- and right-hand circularly polarised light to different extents during transmission (Figure 4.5). The net result of this differential absorption is to transform combined **circularly polarised light** into **elliptically polarised light** (Figure 4.5c and 4.5d). Alternatively, the CD spectrum of a sample containing an optically

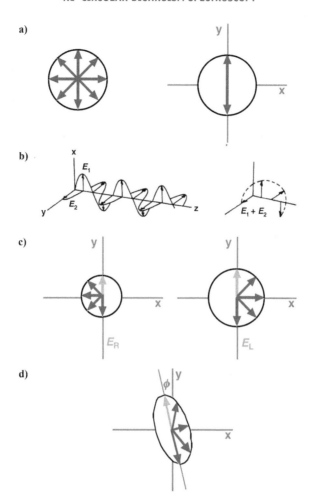

Figure 4.5 Origins of circularly polarised light and generation of ellipticity. (a) If light is viewed approaching an observer along the direction of propagation, then the electric vector oscillates in all possible planes (**left**) until transmission through a polariser (**right**) that restricts oscillations to one plane; (**b**) Two plane-polarised electric vectors $\pi/2$ (90°) out of phase (**left**) but of equal amplitude, interact to generate a combined electric vector ($E_1 + E_2$; **right**) that maps out a helical path, which can be right-handed or left-handed. This is **circularly polarised light**; (**c**) The combined electric vector of right-handed, E_R, and left-handed, E_L, circularly polarised light viewed approaching an observer along the direction of propagation. E_R is shown with smaller amplitude than E_L; (**d**) Constructive interference of E_R and E_L from (c) generates **elliptically polarised light** where the combined electric vector maps out an elliptical path which is left-handed in this case. The angle ϕ is known as the **optical rotation.**

active molecule is typically generated by exposing the sample in a CD spectrometer (Figure 4.6), to alternating beams of $E_L(\lambda)$ and $E_R(\lambda)$ generated by a **Pockels cell**, over a range of different **wavelengths**, λ, and then measuring the **differential absorbance**, $\Delta A(\lambda)$, at each wavelength according to Equation (4.7):

$$\Delta A(\lambda) = A_L(\lambda) - A_R(\lambda) \qquad (4.7)$$

where $A_L(\lambda)$ and $A_R(\lambda)$ are the sample absorbance of $E_L(\lambda)$ and $E_R(\lambda)$ respectively. Values of $\Delta A(\lambda)$ are typically very small and are between 0.03 and 0.003% of total sample absorbance.

4.3.3 The circular dichroism spectrum

The most basic CD spectrum is then generated by plotting $\Delta A(\lambda)$ as a function of λ. CD spectra may be rendered in at least two other ways. The first using the Beer-Lambert law as shown in Equation (4.8):

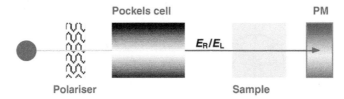

Figure 4.6 Illustration of general experimental arrangement for circular dichroism spectroscopy. The **polariser** and **Pockels cell** generate E_R and E_L circularly polarised light of equal amplitude at any selected wavelength λ. This is passed through a sample and differential absorbance between E_R and E_L is observed as a function of λ. This is a **circular dichroism (CD)** spectrum. Biological macromolecules (proteins and nucleic acids) are usually right-handed and absorb E_R more than E_L. Hence the combination between E_R and E_L after transmission through the sample will generate left-handed elliptically polarised light owing to the smaller amplitude of E_R relative to E_L (Figure 4.5c and 4.5d).

$$\Delta\varepsilon(\lambda) = \varepsilon_L(\lambda) - \varepsilon_R(\lambda) = \frac{\Delta A(\lambda)}{c_M \cdot l} \qquad (4.8)$$

where $\varepsilon_L(\lambda)$ and $\varepsilon_R(\lambda)$ are molar extinction coefficients corresponding to the sample absorbance of $E_L(\lambda)$ and $E_R(\lambda)$ respectively. Hence, a CD spectrum may also be generated by plotting the **differential molar extinction coefficient**, $\Delta\varepsilon(\lambda)$, as a function of λ. Both $\Delta A(\lambda)$ and $\Delta\varepsilon(\lambda)$ plots are related to total sample optical activity, but only the first should be used if combinations of more than one optically active molecule are under investigation in the same sample, since molarity forms part of the dimensions of $\Delta\varepsilon(\lambda)$ which is difficult to define if more than one optically active type of molecule is involved in the sample. The second method relies upon transforming differential absorption data $\Delta A(\lambda)$ into **ellipticity**, $\theta(\lambda)$ and differential molar extinction coefficient $\Delta\varepsilon(\lambda)$ data into **molar ellipticity**, $[\theta(\lambda)]$ by Equations (4.9) and (4.10) respectively:

$$\theta(\lambda) = 2.303 \cdot \frac{180}{2\pi} \cdot \Delta A(\lambda) \qquad (4.9)$$

$$[\theta(\lambda)] = 3300 \cdot \Delta\varepsilon(\lambda) \qquad (4.10)$$

In the same ways as for $\Delta A(\lambda)$ and $\Delta\varepsilon(\lambda)$ plots, only $\theta(\lambda)$ plots should be used if combinations of more than one optically active molecule are under investigation in the same sample, since molarity forms part of the dimensions of $[\theta(\lambda)]$, which once again is difficult to define if more than one optically active type of molecule is involved in the sample.

4.3.4 Structural versus functional information from circular dichroism spectroscopy

One of the most useful ways of using CD spectroscopy is to deduce the relative proportions of secondary structure elements in a globular or even in a fibrous protein. The main structural elements in a globular protein that can be identified separately by CD spectroscopy are defined as α-helix, β-sheets or random coil. Each main secondary structural element is right-handed (see Chapter 1) and consequently absorbs $E_R(\lambda)$ more effectively than $E_L(\lambda)$ in the peptide link region 200–230 nm giving rise to exclusively negative values of $\Delta A(\lambda)$ or $[\theta(\lambda)]$. The CD spectrum of a sample of globular protein of interest can be assumed to be a weighted, linear combination of contributions from α-helix, β-sheets or random coil structure according to Equation (4.11):

$$[\theta(\lambda)] = \chi_\alpha[\theta_\alpha(\lambda)] + \chi_\beta[\theta_\beta(\lambda)] + \chi_r[\theta_r(\lambda)] \qquad (4.11)$$

where χ_α, χ_β, χ_r are **fractional composition terms** and $[\theta_\alpha(\lambda)]$, $[\theta_\beta(\lambda)]$, $[\theta_r(\lambda)]$ are **molar ellipticities measured from polypeptides** in the "pure"-conformation indicated (Figure 4.7). The weakness of this equation is the quality of the molar ellipticity data. Values obtained from "pure"-conformation polypeptides are probably not the same for secondary structure elements in complex tertiary structure environments owing to differences in polypeptide residue lengths and additional tertiary structure interactions. Nevertheless, Equation (4.11) can give meaningful data provided that the basis set of CD molar ellipticities are optimised with reference to the CD spectra of globular proteins with known χ_α, χ_β and χ_r values.

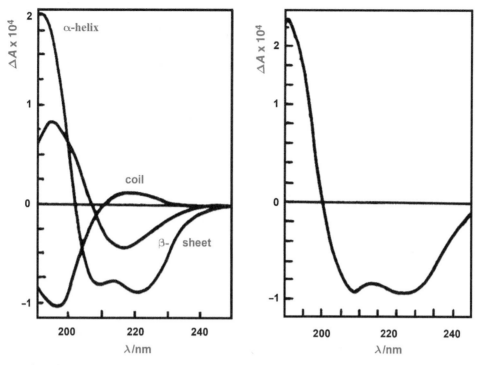

Figure 4.7 Circular dichroism spectra of polypeptides. These are idealised spectra assuming pure conformations as indicated (**left**). These form a typical basis set for the deduction of secondary structure content in a protein sample of interest by mathematical analysis of the recorded CD spectrum (**right**).

In addition to this application to globular proteins, CD spectroscopy is one of the only effective techniques for identifying the presence of left-handed P_{II}-helix structures (ΔA_{max} 225 nm) that make up the anatomy of collagen and collagen-like fibrous proteins (Figure 4.7) (Chapter 1). The high degree of sensitivity of CD spectroscopy to protein secondary structure elements and changes in fractional composition of these elements ensures that CD spectroscopy is also one of the most effective techniques for observing conformational changes of a given protein in solution either as a result of changing conditions or else through interactions with other molecules (Figure 4.8). Once again, such

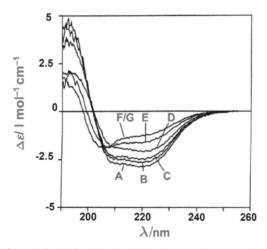

Figure 4.8 pH-Titration of heat shock protein 47 by circular dichroism spectroscopy. The far UV spectral regions of **heat shock protein 47 (Hsp47)** (see Chapter 7) are monitored by CD spectroscopy as a function of pH. Far UV data indicate two **isodichroic points** (at approximately 202 and 208 nm respectively) suggesting a two-stage transition between an **alkali stable state** and an **acid stable state** via a transient intermediate state. CD spectroscopy reveals that Hsp47 undergoes reversible pH-driven trans-conformational changes. **A:** pH 7.2; **B:** pH 6.8; **C:** pH 6.4; **D:** pH 6.3; **E:** pH 6.0; **F:** pH 5.7; **G:** pH 5.0 (from El Thaher *et al.*, 1996, Figure 2a/ Bentham Science).

changes in spectroscopic signature in response to molecular interactions in particular can be very useful to study molecular recognition and binding involving proteins as well; for greater detail see Chapter 7.

The CD spectroscopy of nucleic acids is even more complex than proteins and as such has really not been explored to its fullest potential yet. CD spectra are uniquely dependent on nucleotide composition and sequence in which base-base interaction terms need to be included. For example, the CD spectrum of dinucleotide ApG will be given by an expression in the form of Equation (4.12):

$$2\big[\theta_{ApG}(\lambda)\big] = \big[\theta_A(\lambda)\big] + \big[\theta_G(\lambda)\big] + I_{AG}(\lambda) \tag{4.12}$$

where $[\theta_A(\lambda)]$ and $[\theta_G(\lambda)]$ are **molar ellipticities measured from pure nucleotides** and $I_{AG}(\lambda)$ is a **base-base interaction term**. For more complex nucleic acids, multiple terms are required to compute the CD spectrum of a sample of a nucleic acid of interest. However, given the extensive flexibility inherent in many complex nucleic acid structures, there can be considerable difficulties experienced in making reliable calculations given fluctuations in base-base interaction terms. However, the sensitivity of base-base interaction terms to nucleic acid conformation ensures that CD spectroscopy is also one of the most effective techniques for observing conformational changes of a given nucleic acid in solution, either as a result of changing conditions or else through interactions with other molecules (Figure 4.9). Hence, changes in spectroscopic signature in response to molecular interactions can be turned to good use in studying molecular recognition and binding events involving nucleic acids (see Chapter 7).

4.4 Vibrational spectroscopy

Infrared (IR) spectroscopy functions to probe vibrational transitions (2000–50,000 nm; 5000–200 cm⁻¹, i.e. **wave number**–typical IR spectroscopy units) in the singlet ground electronic state of molecules. The absorption principles of IR spectroscopy are identical to those of UV-visible and CD spectroscopy. Hence the Beer-Lambert law (Equation 4.3) applies. Moreover, absorption band intensities are determined by the transition dipole moment and there are extensive perturbation and coupling effects. Overall though, values of molar extinction coefficients for vibrational transitions, $\varepsilon_V(\lambda)$, are up to 10^2-fold lower in magnitude than UV-visible and CD $\varepsilon(\lambda)$ values, making IR spectroscopy a much less sensitive technique.

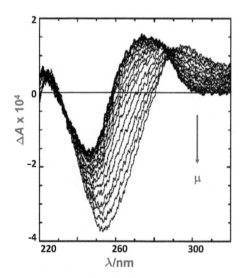

Figure 4.9 **Circular dichroism-titration of pDNA with adenovirus-derived peptide.** CD spectra were recorded of a fixed concentration of plasmid DNA titrated with an increasing concentration of an **adenoviral peptide mu** (μ) known to template with DNA and induce condensation. Data illustrates that the peptide binding induces **base-pair tilting** and subsequent **supercoiling** (Chapters 1 and 7). Arrow shows direction of spectral increase with increasing mu peptide (from Preuss *et al.*, 2003, Figure 4a, Royal Society of Chemistry).

4.4.1 Infrared vibrational modes

In general, there has been some misunderstanding about the nature of the vibrational transitions being observed, given the way IR spectroscopy is frequently referred to in terms of functional group vibrations. In fact, observed vibrational transitions are associated with **normal vibrational modes**. Each normal vibrational mode will in effect include contributions (greater or smaller) from all the atoms in a molecule and not just individual functional groups, owing to the existence of heavy coupling between atoms in a vibrating molecule. Having said this, a normal vibrational mode often involves the substantive vibrations of just two bonded atoms at one location in a molecule, with little motion otherwise elsewhere. Therefore, to a first approximation, a normal vibrational mode may be said to correspond with just the atom motions in a particular type of functional group. As with UV-visible spectroscopy, the extent of IR absorption by a molecule of interest depends entirely upon the **IR transition dipole moment** that should be non-zero for a vibrational transition to be observed. The transition dipole moment is defined by expression (4.13):

$$\langle \psi_o \phi_v | \tilde{\mu} | \psi_o \phi_{v'} \rangle \tag{4.13}$$

where ψ_o is the **singlet ground state electronic wavefunction**, ϕ_v and $\phi_{v'}$ are the **vibrational states** (normal vibrational modes) **pre-** and **post-absorption** respectively.

4.4.2 Structural information from infrared spectroscopy

IR spectroscopy of proteins is often seen as a technique complementary to CD spectroscopy for proteins. The IR spectroscopy of peptide links is particularly rich and useful given the extensive way in which the energies of vibrational transitions associated with the normal vibrational modes centred on atom motions within peptide links, can vary substantially.

The normal vibrational modes of interest for proteins are illustrated in Table 4.3 and Figure 4.10. Variations in **hydrogen bonding** are the primary reason for variations in the energies of vibrational transitions. Hydrogen-bonding arrangements between peptide links differ markedly in pattern and strength, depending upon whether peptide links are located in α-helices or β-sheets. Therefore, the energies of vibrational transitions associated with peptide links in α-helices or β-sheets can vary considerably as well. The situation could become excessively complex given the number of individual peptide links in a given protein. However, α-helices or β-sheets are highly regular and there is substantial coupling between vibrational transitions involving perturbation effects that are extremely orientation dependent. As a result,

Table 4.3 **Summary of the main vibrational transitions associated with the peptide links of protein secondary structural elements.**

Vibration	α-Helix (cm^{-1})	β-Sheet (parallel) (cm^{-1})	β-Sheet (anti-parallel) (cm^{-1})	Random coil (cm^{-1})	H-bond free (cm^{-1})
N-H stretch	3290–3300	3280–3300	3280–3300		3400
Amide I	1650–1660				1670–1700
$\nu(0)$	1650				
$\nu(2\pi/n)$	1652				
$\nu(0,0)$		1645			
$\nu(\pi,0)$		1630			
$\nu(0,\pi)$			1685		
$\nu(\pi,0)$			1632		
ν_0				1656	
Amide II	1540–1550				<1520
$\nu(0)$	1516				
$\nu(2\pi/n)$	1546				
$\nu(0,0)$		1530			
$\nu(\pi,0)$		1550			
$\nu(0,\pi)$			1530		
$\nu(\pi,0)$					
ν_0				1535	

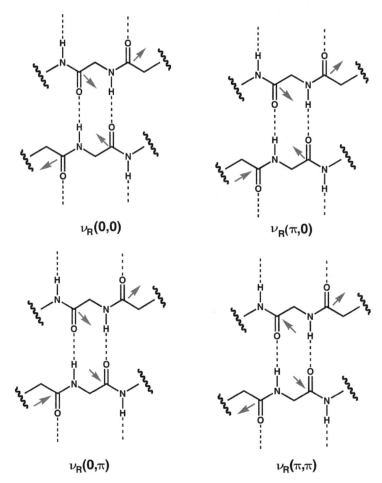

Figure 4.10 **Vibrational modes of β-sheets.** Each mode is differentiated by different in-plane motions (see arrow directions) with out of plane deformations above (red arrow) or below (blue arrow) the plane of the paper.

α-helices, parallel β-sheets and anti-parallel β-sheets are characterised individually by just two main, unique vibrational transitions each (Table 4.3; Figure 4.10), so that the presence of any one secondary structure type can be established with relative ease in a globular protein of interest.

The IR spectroscopy of other biological macromolecules of interest is much less developed. Vibrational transitions in the region 1500–1800 cm^{-1} are associated with normal vibrational modes centred substantially on atom motions in nucleotide bases such as C = O, C = C and C = N stretching vibrations. Energies of these vibrational transitions are very sensitive to base-pair formation owing to hydrogen-bonding effects. However, the IR spectroscopy of nucleic acids has not so far been developed to appreciate the presence of unique vibrational transitions that identify the presence of different types of secondary or tertiary structure in nucleic acids of interest. Otherwise, the IR spectroscopy of carbohydrates and lipids is largely uninformative except to prove the presence of functional groups.

4.4.3 Raman spectroscopy

Raman spectroscopy seeks to analyse vibrational transitions in biological macromolecules in a complementary way to IR spectroscopy. The physical basis of the technique is somewhat different as well. Initially, an intense beam of light of **vibrational frequency**, ν_v, is used to irradiate a sample of molecules of interest in order to classically induce oscillating dipoles of equivalent frequency in the polarisable clouds of electrons. The time dependent magnitude of the **induced dipole**, $\mu_{ind}(t)$, obeys Equation (4.14):

$$\mu_{ind}(t) = \alpha(\nu_v) \cdot E_o \cos 2\pi\nu_v t$$

(4.14)

where $\alpha(\nu_v)$ is the **time dependent electronic polarisability** and E_o the **equilibrium electric field**. This can be divided into an **equilibrium polarisability component**, $\alpha_o(\nu_v)$, and a **nuclear oscillation component**, $\alpha_R(\nu_R)$, which is linked to the **frequency of vibrational modes**, ν_R, in each molecule. These terms are all related according to Equation (4.15):

$$\alpha(\nu_v) = \alpha_o(\nu_v) + \alpha_R(\nu_R)\cos 2\pi\nu_R t \qquad (4.15)$$

By substituting back into Equation (4.14), we obtain Equation (4.16):

$$\mu_{ind}(t) = \left[\alpha_o(\nu_v) \cdot E_o \cos 2\pi\nu_v t\right] + \left[\alpha_R(\nu_R) \cdot E_o \cos 2\pi\nu_R t \cdot \cos 2\pi\nu_v t\right] \qquad (4.16)$$

where the first term in square brackets represents the **induced dipole oscillating at the same frequency**, ν_v, as the incident frequency and the second term is known as the **Raman dipole**. This equation represents the dual reality of the Raman spectroscopy experiment. When molecules of interest in a certain sample are irradiated, the vast majority develop induced dipole oscillations at the same frequency, ν_v, as the incident frequency and then promptly re-emit light at the same frequency in random directions. This effect is simply **classical light scattering**. However, a proportion of molecules in the same sample will funnel energy into or out of a number of normal vibrational modes before re-emission. The result is that a band of light re-emitted from the sample at frequency ν_v due to classical light scattering will be attended by smaller satellite bands at frequencies of $\nu_v + \nu_R$ (**Stokes bands**) and $\nu_v - \nu_R$ (**anti-Stokes bands**). Each pair of satellites can then be identified to assign the **Raman vibrational frequency**, ν_R, of each activated vibrational transition. If the frequency ν_v (electronic oscillation) and ν_R (nuclear oscillation) are very similar, then classical light scattering effects will be minimised in favour of energy funnelling to normal vibrational modes, allowing satellite bands to dominate. This effect is known as **resonance Raman spectroscopy**.

Raman spectroscopy of proteins runs parallel to IR spectroscopy. The same vibrational transitions associated with the same normal vibrational modes centred on atom motions within peptide links, are observed (Table 4.3; Figure 4.10). The same is true for the Raman spectroscopy of nucleic acids as well. Arguably, Raman spectroscopy of a globular protein of interest gives an even more precise characterisation of vibrational transitions than IR spectroscopy, allowing for the clear discrimination and identification of random coil structure as well as α-helix, parallel β-sheet and anti-parallel β-sheet secondary structures.

4.5 Fluorescence spectroscopy

The basic concept of fluorescence spectroscopy is as illustrated (Figure 4.11). Molecular **singlet ground state S_a** is promoted to **singlet excited state S_b** by photon absorption. State S_b may then release a second photon by a process of spontaneous emission known as **fluorescence** to return to singlet ground state S_a. Electronic promotion to singlet

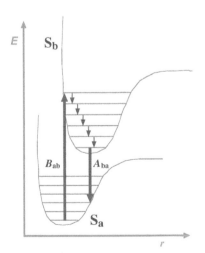

Figure 4.11 General principle of fluorescence. A **singlet ground state**, S_a, absorbs photons at a rate, B_{ab}, and enters **singlet excited state**, S_b, by vertical transition. **Rate of re-emission**, A_{ba}, is slow compared with timescale of vibration and several radiation-less transitions (grey arrows) occur between excited state vibrational states until re-emission (fluorescence) takes place by a second vertical transition.

excited state involves a **vertical transition** that inevitably places the electronically excited molecule in a high-lying vibrational state. A vertical transition arises because electronic transitions occur typically in a time of 10^{-15} s, during which time the molecule is effectively static given that molecules typically vibrate in 10^{-13} s. Furthermore, a typical singlet excited state of any molecule has slightly greater physical dimensions than the singlet ground state since the excited state is created by electronic promotion into antibonding molecular orbitals, causing an overall weakening (i.e. lengthening) of bonds. Subsequent spontaneous emission from this excited state occurs by **second vertical transition**, but only to a high-lying vibrational state of the singlet ground state given the decrease in molecular dimensions with a return to the ground electronic state. Consequently, this second vertical transition involves a smaller energy gap and hence the emission of a photon of lower frequency (higher wavelength). Therefore, fluorescence always involves the emission of a photon of lower energy than that absorbed initially.

4.5.1 Rates of emission and lifetimes

Molecular singlet ground state S_a is promoted to singlet excited state S_b at a **rate of photon absorption**, B_{ab}, related to the transition dipole moment (see expression 4.2) by the following Equation (4.17):

$$B_{ab} = (2\pi / 3\hbar^2)\left|\left\langle \psi_a \left| \tilde{\mu} \right| \psi_b \right\rangle\right|^2 \tag{4.17}$$

where the squared term $\left|\left\langle \psi_a \left| \tilde{\mu} \right| \psi_b \right\rangle\right|^2$ is known as the **ground state dipole strength**, D_{ab}, and $\hbar$ is Planck's constant divided by 2π (also known as the **reduced Planck's constant**). Transition back to the ground state (**fluorescence**) then occurs at a **rate of spontaneous emission**, A_{ba}, that obeys the following Equation (4.18):

$$A_{ba} = (8\pi h \nu_{em}^3 c^{-3}) \cdot B_{ab} \tag{4.18}$$

where the term in brackets, containing **Planck's constant**, h, **frequency of emitted light**, ν_{em}, and **speed of light** c, is known as the **black body radiation constant**. By combining 4.17 and 4.18, we arrive at the important Equation (4.19):

$$A_{ba} = (32\pi^3 \nu_{em}^3 / 3c^3 \hbar) \cdot D_{ab} \tag{4.19}$$

which demonstrates that the rate of spontaneous emission, A_{ba}, is directly proportional to ground state dipole strength, D_{ab}. In other words, the stronger the absorption the faster is the **rate of spontaneous emission** A_{ba} (also known as k_F). This is also related to another important fluorescence property known as the **fluorescence** or **radiative lifetime**, τ_R, of the singlet excited state S_b by Equation (4.20):

$$A_{ba} = \frac{1}{\tau_R} = k_F \tag{4.20}$$

Typical values of τ_R are in the region of nano-seconds.

4.5.2 Effects of non-radiative competition processes

Fluorescence radiation is heavily competitive with a variety of non-radiative processes that reduce the efficiency of spontaneous emission. These processes are known as **internal conversion** (rate k_{IC}), **intersystem crossing** (rate k_{IS}) and **quenching** by an external **quenching agent** typically denoted **Q** (rate $k_q[Q]$) (Figure 4.12). All these non-radiative processes reduce the efficiency of spontaneous emission by depopulating the singlet excited state. A measure of the residual efficiency of spontaneous emission is known as the **fluorescence quantum yield**, ϕ_F, that is related to all the various rate constants according to Equation (4.21):

$$\phi_F = k_F / (k_F + k_{IC} + k_{IS} + k_q[Q]) \tag{4.21}$$

Internal conversion is a pseudo first-order process by which the singlet excited state S_b energy is lost by collisions with solvent molecules or else by transfer between vibrational modes. Inevitably, the rate k_{IC} will increase with increasing

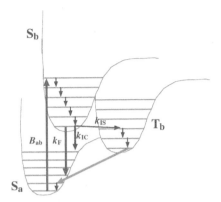

Figure 4.12 Competition between fluorescence and radiation-less processes. These follow the promotion of singlet ground state, S_a, to singlet excited state, S_b, at a rate, B_{ab}. The rate of fluorescence, A_{ba} (k_F), is unaffected by competing radiation-less processes, but the **fluorescence quantum yield**, relating to the intensity of fluorescence, is compromised by high rates of radiation-less **internal conversion (IC)**, k_{IC}, and of **intersystem crossing (IS)**, k_{IS}. IS leads to **triplet excited state**, T_b, which loses energy by **transition forbidden phosphorescent re-emission** to regenerate singlet ground state, S_a.

temperature and vice versa. Quenching is a similar deactivation process in which collision with solute molecules leads to loss in singlet excited state S_b energy. Quenching is a second-order process but can be rendered pseudo first-order if $[Q] \gg [S_b]$. Quenching is only meaningful with solute molecules such as O_2 or I^- that are able to deactivate singlet excited state S_b with almost every collision, in which case the rate constant k_q is at the diffusion limit (see Chapter 7). Finally, intersystem crossing relates to **forbidden spin exchange** that involves the transformation of singlet excited state S_b into a **triplet excited state** T_b that then undergoes slow but spontaneous emission in a process known as **phosphorescence** in order to return to the singlet ground state (Figure 4.12). As mentioned in Section 4.5.1, typical values of τ_R for fluorescence are in the region of nano-seconds. Equation (4.19) suggests that the stronger the photon absorption by S_a becomes, the faster is the rate of spontaneous emission from S_b. Clearly, the theoretical absorption of a photon to bring about a transition from singlet ground state S_a to triple excited state T_b, must be much weaker since the transition is forbidden. Hence, $\tau_{R, Phor}$ must be considerably greater than τ_R for fluorescence. In the event, typical values of $\tau_{R, Phor}$ for phosphorescence are in the region of seconds.

4.5.3 Structural versus functional information from intrinsic fluorescence spectroscopy

The basic structural information available from intrinsic fluorescence spectroscopy of biological macromolecule structures is quite limited compared with information available from even UV-visible or even CD spectroscopy and is largely limited to globular proteins. One of the main reasons for this is that only proteins and nucleic acids possess substantial numbers of chromophores that are potentially capable of absorbing photons leading to fluorescence emission. However, not every chromophore is simultaneously a good **fluorophore** (i.e. chromophore capable of generating fluorescence emission). Indeed, the nucleic acid bases are fundamentally poor fluorophores. In proteins, only tryptophan amino acid residues are substantial fluorophores out of the three main aromatic amino acid chromophores. Moreover, values of ϕ_F are low in water, though values of τ_R are short, hence generally limiting fluorescence utility even further (Table 4.1).

In spite of these limitations, fluorescence spectroscopy of proteins involving **intrinsic tryptophan residue fluorescence** (excitation 295 nm) can be a useful monitor of conformational changes in globular proteins in a way similar to UV-visible spectroscopy of tryptophan residues in globular proteins. If the local environment of a tryptophan residue is changed from hydrophilic (or polar) to hydrophobic, then this leads to an increase in fluorescence emission ($\pi \rightarrow \pi^*$) transition energies associated with a decrease in transition wavelength (**blue shift**) (typically 350 to 330 nm). Since the interior of such a protein may be regarded as hydrophobic, whilst the exterior is hydrophilic, then we can generalise to say that interior "buried" residues are blue shifted compared with "solvent-exposed" residues. Furthermore, interior "buried" residues will not be subject to internal conversion problems caused by random collisions with water molecules, or to collisional deactivation with quenching molecules. Therefore, values of ϕ_F should increase substantially according to Equation (4.21), resulting in significantly enhanced fluorescence emission intensities. Both characteristics ensure that

intrinsic tryptophan fluorescence of proteins becomes a useful tool to probe changes in the conformational state of protein structure in response to changes in environmental conditions and/or the binding of other molecules (Figure 4.13).

Aside from this application, intrinsic tryptophan fluorescence may be used in combination with fluorescence quenching agents to discriminate between those tryptophan residues that are more surface accessible or "solvent-exposed" than others. This relies upon the Stern-Volmer Equation (4.22):

$$\frac{F_o}{F} = 1 + k_q [Q] \cdot \tau_o \tag{4.22}$$

where τ_o is $(k_F + k_{IC} + k_{IS})^{-1}$, that is **excited state lifetime in the absence of quencher molecule Q**, F_o is **fluorescence intensity in the absence of Q** and F is **intensity in the presence of Q**. Data may be presented graphically in the form of a **Stern-Volmer plot** (Figure 4.14) in order to deduce values of k_q. Values of k_q will appear towards the so-called diffusion limit of $10^{10}\,M^{-1}\,s^{-1}$ if the tryptophan is completely surface accessible, but will be much lower if the residue is "buried".

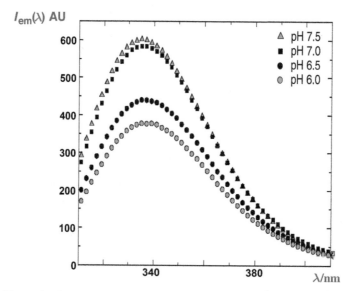

Figure 4.13   pH-Titration of heat shock protein 47 by fluorescence spectroscopy. Intrinsic tryptophan fluorescence spectroscopy (excitation 295 nm) is monitored as a function of pH. Data are indicative of a two-stage transition between an alkali stable state and an acid stable state via a transient intermediate state. Fluorescence spectroscopy, like CD spectroscopy (Figure 4.8), reveals that Hsp47 undergoes reversible pH-driven trans-conformational changes (from Abdul-Wahab et al., 2013, unpublished data).

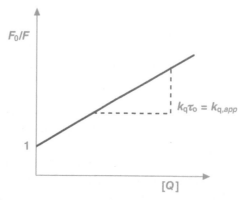

Figure 4.14   Stern-Volmer plot. This is typically used to determine the rate of fluorescence quenching by solute molecules that may function as a **quenching agent, Q.** Quenching also competes with the other factors described in Figure 4.12 to reduce fluorescence quantum yield.

4.5.4 Extrinsic fluorescence and fluorescence resonance energy transfer

Ultimately though, whilst intrinsic fluorescence spectroscopy of biological macromolecules appears to be rather limited, the technique of fluorescence spectroscopy becomes considerably more versatile if **extrinsic fluorophores** are combined with the biological macromolecules of interest. There is now quite an industry in these extrinsic fluorophores (Figure 4.15) as probes of biological function (see Section 4.5.5). The primary reason for this is the fact that whilst typical electronic absorption transitions take place in 10^{-5} s and vibrations in 10^{-13} s (see Section 4.4), values of τ_R vary between 10^{-9} and 10^{-7} s (Table 4.4), which is a slow enough timescale in which to probe a huge range of biologically

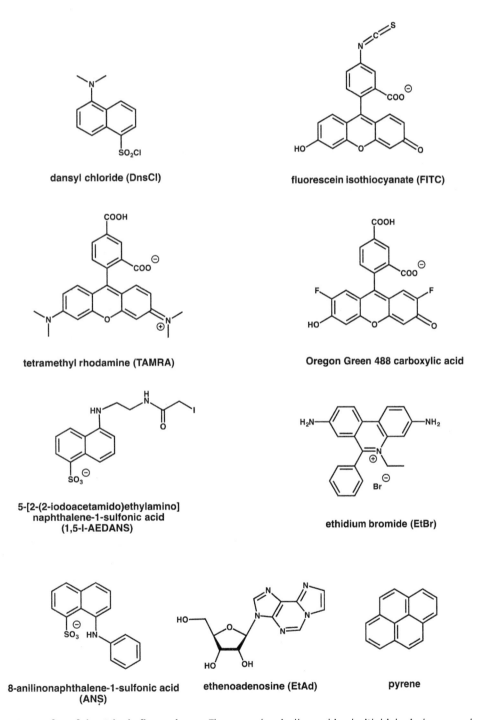

Figure 4.15 Structures of useful extrinsic fluorophores. These are chemically combined with biological macromolecules of interest for structure and/or function investigations.

Table 4.4   Summary of the main extrinsic fluorophores that may be combined with proteins or nucleic acids. Main absorption and fluorescence characteristics are given. See Figure 4.15 for structure abbreviations used in the table.

Fluorophore	A_{max}/nm	$10^{-3} \times \varepsilon_{max}$	I_{max}/nm	ϕ_F	τ_F/nsec
DnsCl	330	3.4	510	0.1	13
1,5-I-AEDANS	360	6.8	480	0.5	15
FITC	495	42	516	0.3	4
ANS	374	6.8	454	0.98	16
Pyrene	342	40	383	0.25	100
EtAd	300	2.6	410	0.40	26
EtBr	315	3.8	600	1	26.5

relevant processes such as protonation, solvent cage relaxation, local conformational changes and other processes coupled to molecular rotation and even translation (see Chapter 7). Therefore, whilst intrinsic fluorescence spectroscopy may be a rather modest technique for structural characterisation of biological macromolecules, **extrinsic fluorescence** spectroscopy is an impressive tool to probe both structure and dynamics of biological molecules, by such techniques as **fluorescence titration binding experiments** for the characterisation of molecular recognition and binding events (see Chapter 7), and **fluorescence resonance energy transfer** (**FRET**) experiments.

FRET experiments involve complementary pairs of fluorophores that are matched to allow coupling to take place between the fluorescence emission transition dipole moment of a given **donor fluorophore** and the absorption transition dipole moment of a corresponding **acceptor fluorophore**, provided that donor and acceptor are sufficiently close in space. This process is typically followed by acceptor fluorescence (Figure 4.16). In other words, provided that a donor and acceptor fluorophore are close enough together in space, then the absorption of a photon by a donor fluorophore can be effectively converted into fluorescence emission from the corresponding acceptor fluorophore by resonance energy transfer between the two fluorophores through space. The **efficiency of the FRET process**, E_{FRET}, is dependent upon the distance between the two fluorophores according to the following Equation (4.23):

$$E_{FRET} = 1 \Big/ \left[1 + \left(R_F / R_o \right)^6 \right] \tag{4.23}$$

where R_F is the **interfluorophore distance** and R_o is the standard **Förster length** for a given donor-acceptor pair (see Chapter 8). The orientation dependence is revealed in the equation for the Förster length (Equation 4.24):

$$R_o = \left[8.8 \times 10^{-28} \cdot \phi_D \cdot \kappa^2 \cdot n_R^{-4} \cdot J(\lambda) \right]^{1/6} \tag{4.24}$$

where ϕ_D is the **fluorescence quantum yield of the donor**, n_R is **refractive index** of the medium, κ is an **orientation parameter** (varying from 0 to 4, but typically set to 0.667 to represent the likelihood of random relative orientations between fluorophores) and $J(\lambda)$ is a **spectral overlap parameter** that governs the overlap between donor fluorescence emission spectrum and the absorption spectrum of the corresponding acceptor. Therefore, successful transition dipole moment coupling leading to resonant energy transfer and acceptor fluorescence relies uniquely upon good spectral overlap, short distances between donor and acceptor fluorophores (10–75 Å), and optimal relative orientations (co-linear is best). Accordingly, FRET effects between complementary pairs of largely extrinsic donor and acceptor fluorophores (Table 4.5) can be employed as a powerful tool for studying the formation of complexes involving different biological macromolecules (see Chapter 7), even in real time within living cells (see Section 4.5.5), but also for interpreting the significance of specific intramolecular motions and normal vibrational modes within single biological macromolecules (see Section 4.5.6).

4.5.5 Probing biological macromolecule functions with extrinsic fluorescence

Extrinsic chemical probes increasingly provide powerful means to study dynamics and function of biological macromolecules and assemblies in biology. Of these, extrinsic fluorophores (or **extrinsic fluorescent probes**) are amongst

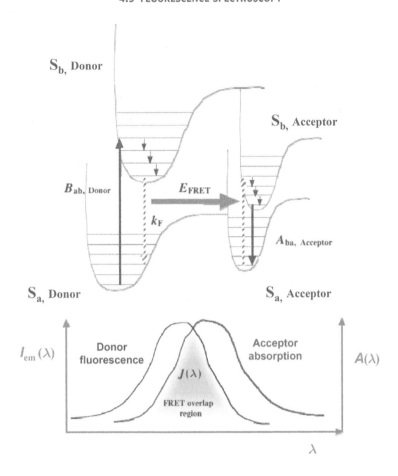

Figure 4.16 **Schematic of principles of fluorescence resonance energy transfer.** Top diagram illustrates the process of **fluorescence resonance energy transfer** (**FRET**) made possible by coupling of the fluorescence emission transition dipole moment of a given **donor fluorophore** with an absorption transition dipole moment of a corresponding **acceptor fluorophore**. This process typically leads to acceptor fluorescence. Bottom diagram illustrates the **spectral overlap**, $J(\lambda)$, between the fluorescence emission spectrum of a donor and the absorption spectrum of a corresponding acceptor. Assuming that both donor and acceptor fluorophores are also close together in space (10–75 Å) and in good relative orientation, then this spectral overlap is essential according to Equation (4.24) in order to allow transition dipole moment coupling and subsequent resonance energy transfer to take place, leading to acceptor fluorescence.

the most potent and useful. Therefore, this section will be devoted to understanding their additional and potential uses. The addition of such probes requires **chemoselective coupling** or **bioconjugation** involving activated fluorescent probe molecules on the one hand and biological macromolecules or macromolecular lipid assemblies of interest on the other hand. A major constraint on all bioconjugation coupling reactions is that they should commonly be possible to perform in an aqueous environment at room temperature, under buffer conditions at around neutral pH.

4.5.5.1 Chemical conjugation of extrinsic fluorescent probes

One of the most common approaches to the bioconjugation of fluorescent probes is the use of **amine-reactive probes** to combine with free primary amine functional groups found in proteins, peptides, oligonucleotides and even lipids. In proteins, target functional groups may be an N-terminal α-amino group or the ε-amino side chain group of lysine amino acid residues (see Chapter 1). The same is true in peptides, while aminosugars are found in oligosaccharides (see Chapter 1). In the case of lipids, the 2-aminoethyl side chains of phosphatidylethanolamines (see Chapter 1) are ideal attachment points for amine-reactive probes. Bioconjugation involving amine functional groups typically takes place best under mildly alkali conditions when these groups are substantially unprotonated (pH 8.5–9.5). The

Table 4.5 Summary of simple donor and acceptor fluorophores competent for FRET studies. See Figure 4.15 for structures and abbreviations used in the table.

Donor fluorophore	Acceptor fluorophore
DnsCl	FITC
FITC	TAMRA
1,5-I-AEDANS	FITC
Tryptophan	1,5-I-AEDANS
Tryptophan	DnsCl

N-terminal α-amino group of an oligo-/polypeptide can bioconjugate under more neutral pH conditions given the fact that N-terminal α-amino groups are more acidic than ε-amino side chain groups and have lower pK_d^A values (see Chapter 8, also known as pK_a values).

Succinimidyl esters are an excellent first choice to activate amine-reactive probes, but their low solubility has led to the alternative use of **sulfonyl chlorides** (Figure 4.17). The resultant sulfonamide link is extremely stable, even more stable than an amide link and will survive even complete protein hydrolysis – a property that can be exploited in protein analysis. The disadvantage of sulfonyl chlorides is that they are unstable in aqueous buffers under mildly alkaline conditions (typically the pH required for reaction with aliphatic amines). Hence, extreme care must be taken to perform bioconjugations with sulfonyl chlorides at low temperatures (approximately 4 °C). Alternatively, amine-reactive probes may be equipped with **isothiocyanate** "traps" from which thiourea links are formed post reaction with amine functional groups, or with aldehydes from which Schiff's base links can be formed with amine functional groups (Figure 4.17).

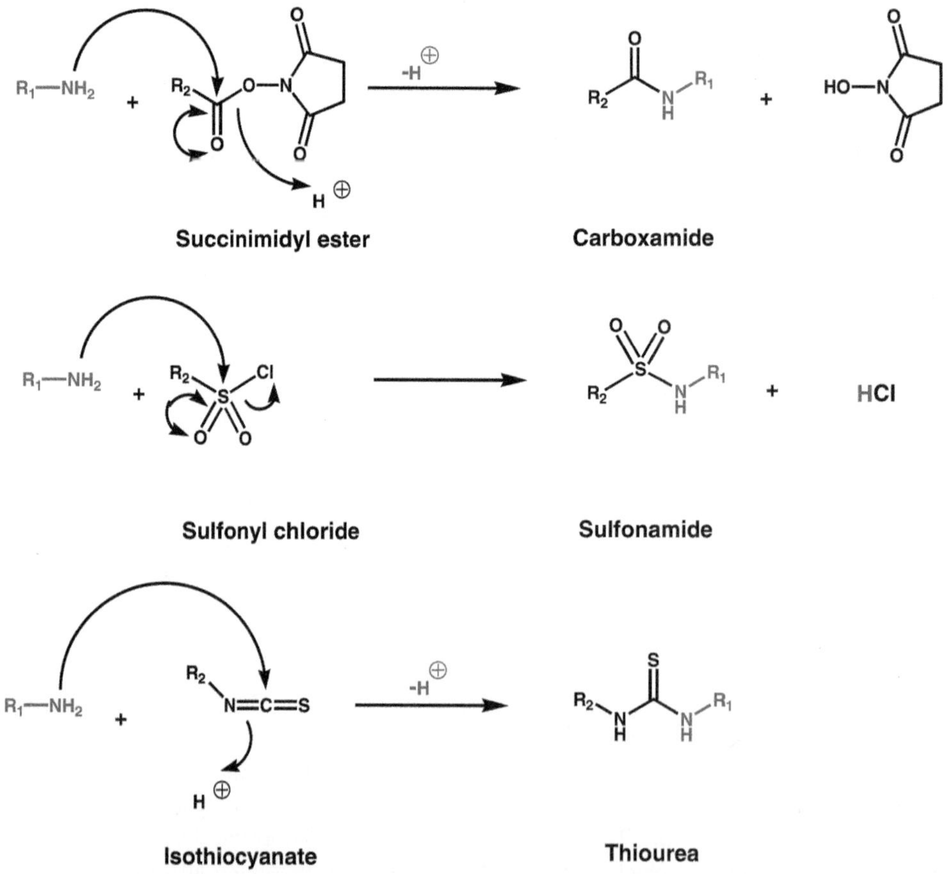

Figure 4.17 Amine reactive probes. Succinimidyl esters, sulfonyl chlorides and **isothiocyanates** can all be used to react with amines in biological macromolecules and lipids.

An alternative approach to bioconjugation of fluorescent probes is the use of **thiol-reactive probes** to combine with free thiol functional groups typically found in proteins and peptides (involving reduced cysteine amino acid residues), although they could be added to oligonucleotides, oligosaccharides and even lipids, if required, on the assumption that thiol functional groups are present. Thiol-reactive probes are typically alkylating reagents such as **symmetrical disulfides, maleimides, alkyl halides, α-halo ketones** or **halo acetamides** (Figure 4.18). Maleimides are probably the first choice as cysteine modifying reagents. Maleimides (usually used as ***N*-ethylmaleimide, NEM**) react quite selectively with cysteine residues in oligo and polypeptides to form **thioethers** (Figure 4.18). Similarly, halo acetamides like **iodoacetamide** react very readily with thiols to form thioethers (Figure 4.18), but are also capable of over-reaction such as with methionine, histidine or even tyrosine residues in oligo- and polypeptides. More control can be exercised again using thiol-exchange reactions making use of symmetrical disulfides (Figure 4.18). However overall, the stability of thiol-reactive probes post bioconjugation is lower than that of amine-reactive probes. Hence, there is a trade-off between stability and the lack of chemo-selectivity comparing thiol-reactive probes to amine-reactive probes.

In addition, bioconjugation of fluorescent probes may be enabled through a variety of functional group activation chemistries as a means to facilitate bioconjugation. For instance, **carboxyl-reactive probes** can combine with carboxyl groups in proteins, peptides, oligonucleotides and lipids, post activation with water-soluble carbodiimides such as **1-ethyl-3-(3-dimethylaminopropyl)carbodiimide (EDAC)** (Figure 4.19). EDAC combines with carboxyl groups resulting in an unstable intermediate – an *O*-acylisourea that rapidly reacts with an amino functional group (either an amine or

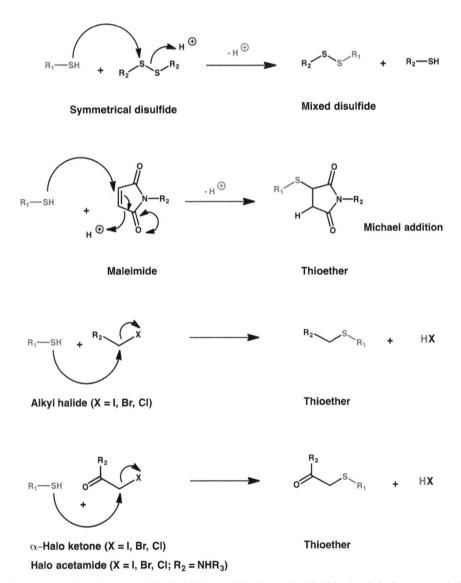

Figure 4.18 Thiol-reactive probes. Symmetrical disulfides, maleimides, alkyl halides, α-halo ketones or **halo acetamides** can all be used to react with thiols in biological macromolecules and lipids.

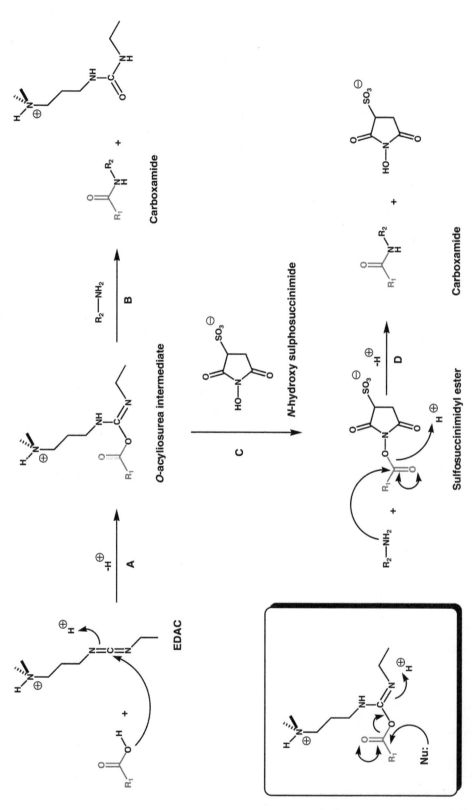

Figure 4.19 **Carboxyl-reactive probes.** (A) Carbodiimide-mediated activation of a carboxyl group, followed by; (B) acylation of a primary amine and formation of *N,N'*-dialkylurea biproduct; (C) Stabilisation of unstable *O*-acylisourea intermediate by *N*-hydroxysulfosuccinimide as part of a more efficient; (D) carbodiimide-mediated activation of a carboxyl group for acylation of a primary amine.

hydrazine) belonging to a carboxyl-reactive probe. The presence of **N-hydroxysulfosuccinimide** improves the coupling efficiency. Alternatively, fluorescent probes with carboxyl groups could be activated with EDAC for the reverse bioconjugation of the probe with amine functional groups in proteins, peptides, oligonucleotides and lipids.

Other highly selective bioconjugations to probes are made possible in oligo-/polypeptides in other ways. For instance, N-terminal serine or threonine residues of an oligo-/polypeptide can be selectively oxidised by oxidative cleavage to give aldehydes (Figure 4.20) that are primed for bioconjugation to fluorescent probes with amines or hydrazines. Furthermore, tyrosine residues can sometimes be selectively modified by *ortho*-nitration using **tetranitromethane** followed by reduction to form an ***o*-aminotyrosine residue** (Figure 4.21). Alternatively, a particularly innovative approach is to use a purified enzyme (**transglutaminase**) to specifically modify glutamine residues with amine-containing probes in the primary amide functional group (Figure 4.22). The presence of an aliphatic spacer between the amine functional group and the fluorescent probe increases reaction efficacy, especially when a cadaverine-spacer is used (as shown).

One extreme view of chemical introduction of an extrinsic fluorescent probe is found in the case of the alanine derivative of the fluorophore **6-dimethylamino-2-acetylnaphthalene (DAN)** (Figure 4.23). This derivative fluorophore, given the trivial name **Aladan**, is incorporated into a polypeptide by solid-phase synthetic chemistry (although a molecular biology technique known as **nonsense suppression** is now available for the introduction of unnatural amino acid

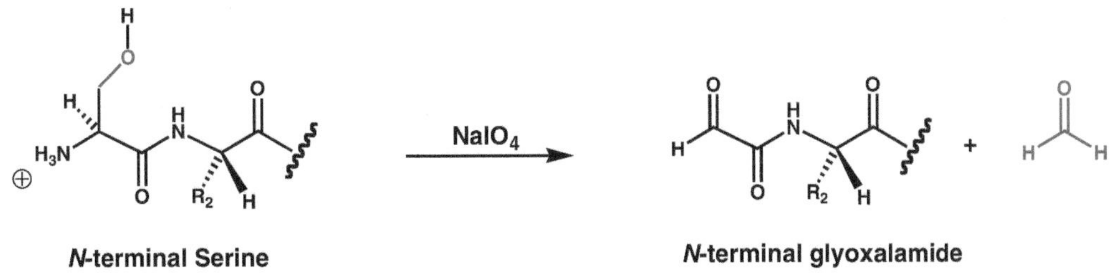

Figure 4.20   Terminal serine-reactive probes. Sodium periodate oxidation of *N*-terminal serine.

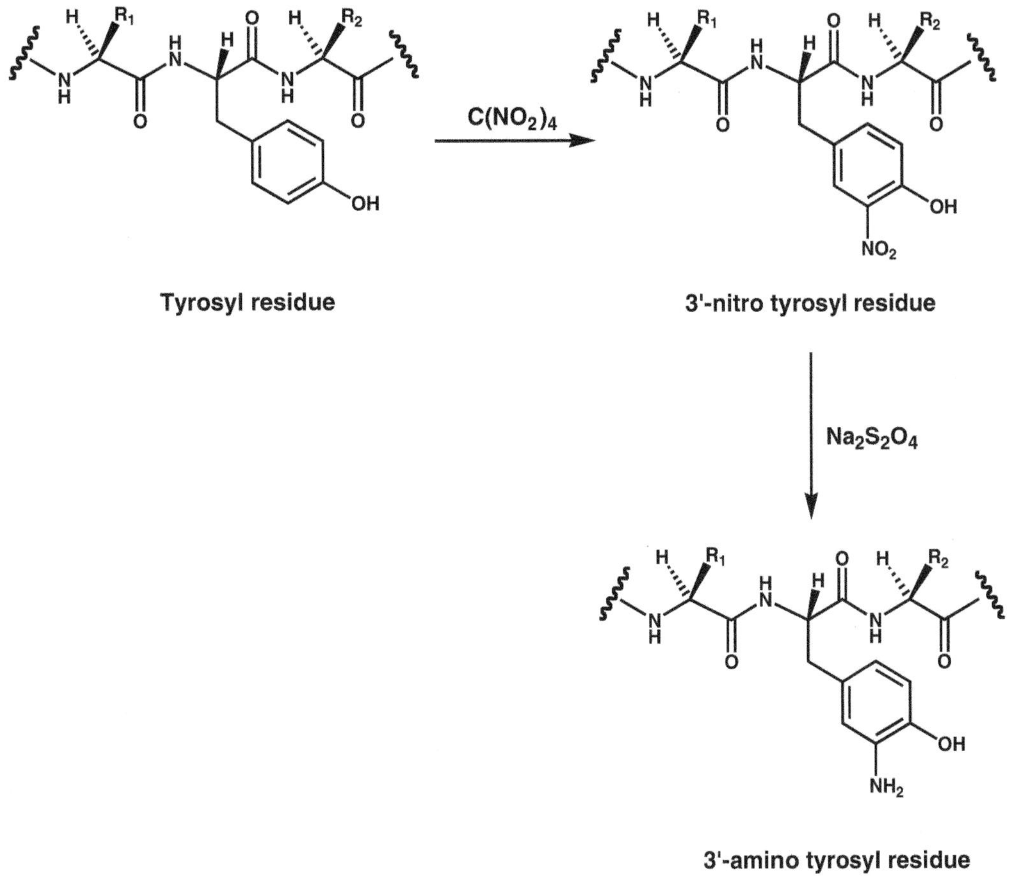

Figure 4.21   Tyrosine reactive probes. Nitration of tyrosine by reaction with **tetranitromethane,** followed by reduction with **sodium dithionite**, to yield an ***o*-aminotyrosine (3'-amino tyrosyl) residue.**

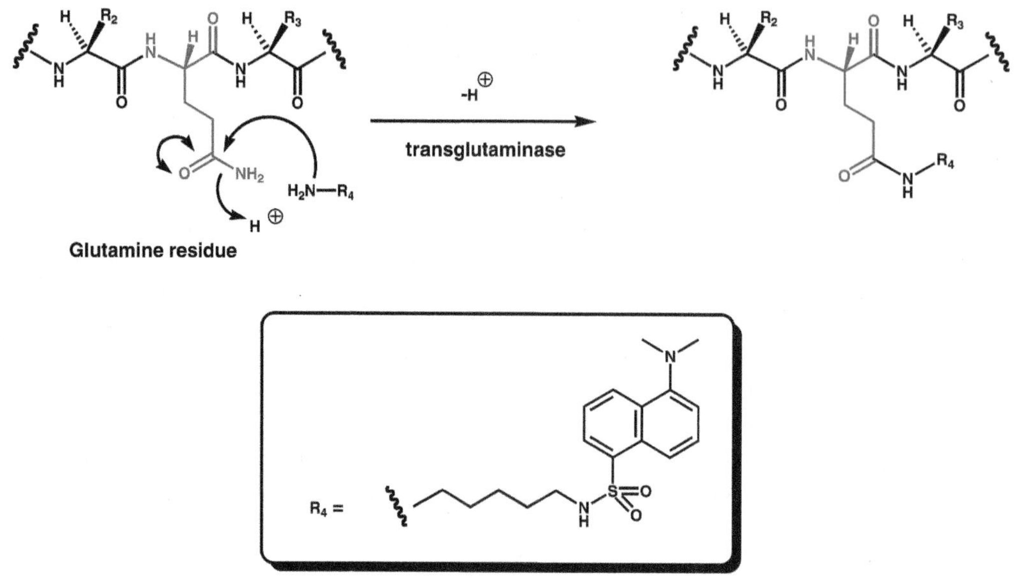

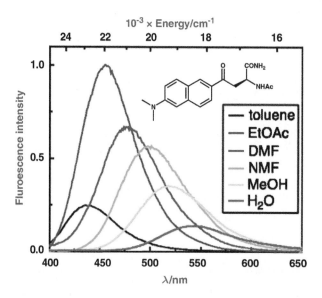

Figure 4.22 **Transamidation of glutamine.** Transglutaminase-mediated labelling of a protein using **dansyl cadaverine**.

Figure 4.23 **Fluorescence of non-natural amino acid Aladan.** Aladan generates a fluorescence signal in a manner strongly dependent on the hydrophobicity/hydrophilicity of the surrounding solvent, showing why it is an excellent probe of environment (from Cohen *et al.*, 2002, Figure 1, American Association for the Advancement of Science).

residues into recombinant proteins) (see Chapter 3 for general background). The fluorescent emission maxima (I_{max}) of Aladan shifts dramatically on different solvent exposures, from 409 nm in heptane to 542 nm in water, yet at the same time remains only mildly changed by variations in pH or salt concentration. This compares to a maximum environment-mediated shift of around 40 nm for intrinsic tryptophan fluorescence. In addition, there is little spectral overlap between extrinsic Aladan fluorescence and intrinsic fluorescence from tryptophan or tyrosine.

4.5.5.2 Biological conjugation of extrinsic fluorescent probes

The chemical approaches to bioconjugation rely upon the chemoselective reactivity of extrinsic fluorescent probes reacting with isolated biological macromolecules or lipids. In the case of proteins, recombinant technologies allow for more biological approaches to fluorescent tagging *in situ*, in cells. This carries great advantages for work in a cellular system since a protein may be specifically labelled and observed within the cellular context. First and foremost of the biological approaches is to express a recombinant protein of interest in a cell as a chimeric or fusion protein in which the protein of interest is expressed conjoined (at *N*-/*C*-terminus as appropriate) to another whole protein with fluorescent properties. The fluorescent protein most used in fusion proteins is **green fluorescent protein (GFP)** from the **jellyfish *Aequorea victoria.*** GFP comprises 238 amino acid residues and is capable of absorbing blue light (A_{max} 395 nm) only to fluoresce green light (I_{max} 509 nm). The protein has no extrinsic prosthetic group or cofactor to act as a fluorophore. Instead, a *p*-hydroxybenzylideneimidazolinone fluorophore is created by direct autocatalytic chemical cyclisation of amino acid residues Ser65, Tyr66 and Gly67 (Figure 4.24). In effect, GFP assembles its own fluorescent cofactor from its own primary structure. This fluorescent cofactor is protected within a cylindrical structure consisting of 11 strands of β-sheet with α-helix at each end, a structure that has been christened the **β-can**. Since the fluorescent cofactor is amino acid residue derived, then the reader should not be surprised to learn that variations in and around the crucial amino acid sequence can lead to other **blue, cyan** and **yellow fluorescent protein** variants (known as **BFP, CFP** and **YFP** respectively). A further alternative to GFP is **DsRed** that is naturally found in the coral *Discosoma*. Wild-type DsRed is an obligate tetramer, but mutant DsRed proteins that are monomeric can now be used in place of GFP in fusion proteins.

Biological and chemical approaches can be combined to yield similar possibilities to the fusion protein approach described above. A tetracysteine motif (**CCXXCC**; where X is a non-cysteine amino acid residue) can be engineered into any protein of interest at any required position, in principle (see Chapter 3). This motif is highly selective for binding to **biarsenical fluorophores** based upon fluorescein. Biarsenical fluorophores are able to permeate cells and remain non-fluorescent until chelated by the tetracysteine motif (Figure 4.25). Biarsenical fluorophores are particularly partial to being across an α-helix with the molecular plane oriented perpendicular to the direction of the helix. Hence, using such fluorophores, an engineered protein of interest can be expressed in cells *in situ*, then tagged *in situ* by any one of a whole range of biarsenical dyes with slightly different structures that are able to fluoresce at different wavelengths.

A final biological approach to extrinsic labelling that is worth noting for the interested chemical biology reader is the use of **cyanine (Cy) dyes** (in particular **Cy3** and **Cy5**). At first sight these look complex, but they can be introduced into DNA by a method known as **nick translation**. According to this method, random nicks (or single strand breaks, i.e. random hydrolyses of phosphodiester links) are introduced into DNA and then repaired using the so-called **Klenow fragment** of DNA polymerase I accessing deoxynucleotides dATP, dGTP and dCTP as repair substrates, together with deoxynucleotide dRTP, where R is uridine substituted at position-5 on the pyrimidine ring with cyanine dye (Figure 4.26). During the repair process, dRTP is used in place of **2'-deoxythymidine 5'-triphosphate (dTTP)** to repair **dA.dT** base pairs. The value of this labelling process is that the amount of nick translation performed can be adjusted to ensure that DNA remains almost completely functional with respect to transcription.

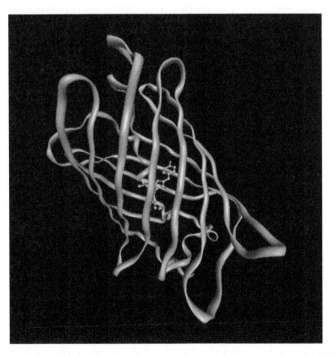

Figure 4.24　Green fluorescent protein. Ribbon display structure of **green fluorescent protein (GFP)** (from *Aequorea victoria*) that possesses 11 β-strands forming a hollow cylinder through which is threaded a helix bearing the chromophore, constructed from amino acid residues Ser65, Tyr66 and Gly67, shown in **ball and stick representation** (pdb: **1emb**).

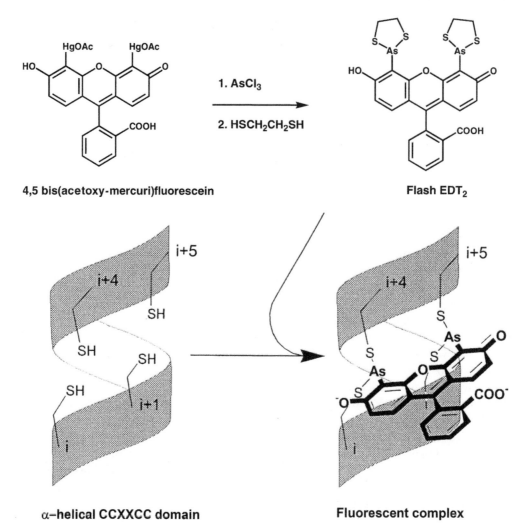

Figure 4.25   Biarsenical fluorophore applications. Synthesis of **Flash EDT₂** and proposed structure of its complex with an α-helical tetracysteine-containing peptide or protein domain. The structure is drawn with the i and i + 4 thiols bridged by one arsenic and the i + 1 and i + 5 thiols bridged by the other (from Griffin *et al.*, 1998, Figure 1, American Association for the Advancement of Science).

4.5.5.3 Selecting extrinsic fluorescent probes

The selection of extrinsic fluorescent probes is driven by the consideration of which biological macromolecule or lipid is to be labelled, the requirement for compatibility between the intended fluorescent probe (in terms of solubility in water, pH sensitivity and so on) and the properties of the molecule to be labelled. Also, choice of the fluorescent probe should be consistent with experimental objectives. For instance, FRET experiments require that extrinsic donor and acceptor fluorophores should be properly matched for their capacity to participate in the FRET effect (see Section 4.5.4).

One of the first extrinsic fluorophores to be developed was **dansyl chloride**, which first emerged in 1951. Dansyl chloride is an amine-reactive probe that is non-fluorescent until bound to amine functional groups with the creation of a sulfonamide link (Figures 4.15 and 4.17). This probe has proven particularly useful in binding interaction experiments, since the values of I_{max} and ϕ_F are so sensitive to the environment (see Chapter 7). There is a substantial blue shift (20–30 nm) and an increase in ϕ_F as a consequence of the transfer of the probe from a more hydrophilic to a more hydrophobic environment, such as the transfer from bulk aqueous conditions to a hydrophobic binding region in a protein (see Figure 4.3, to understand the spectral changes which are equivalent). Subsequently, **fluorescein** was developed that has proved to be a very popular extrinsic fluorophore available in several reactive probe forms that all absorb strongly (494 nm, close to the 488 nm wavelength of the argon-ion laser), fluoresce strongly and are readily water soluble. Direct derivatives of fluorescein including **tetramethyl rhodamine** and **Oregon Green** (Figure 4.15). Fluorescein and these corresponding derivatives are a mainstay of many fluorescent probe experiments, but they do suffer from significant photobleaching, pH-sensitive fluorescence, broad emission spectra and fluorescence quenching post-bioconjugation.

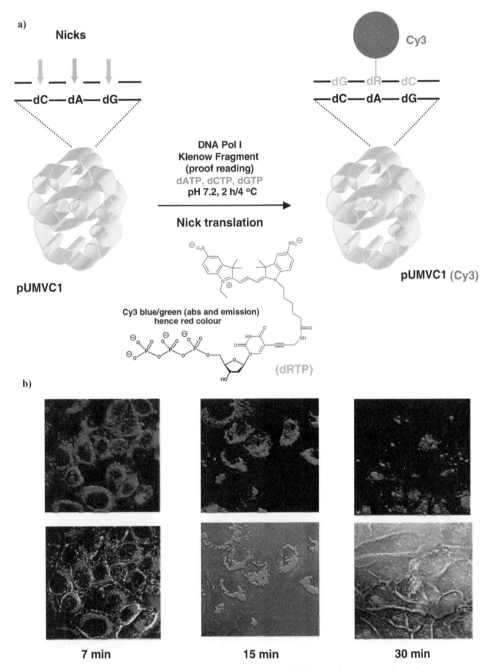

Figure 4.26   DNA nick translation. (a) Plasmid DNA (pDNA; **pUMVC1**) labelling with **Cy3 cyanine dye** was accomplished as shown above by means of an enzymatic procedure. Abbreviations as follows: **dATP: 2′-deoxyadenosine 5′-triphosphate**; **dCTP: 2′-deoxycytidine 5′-triphosphate**; **dGTP: 2′-deoxyguanosine 5′-triphosphate**; (b) The dye-labelled pDNA was then combined into nanometric particles known as liposome:mu:pDNA particles wherein the lipid:mu:pDNA ratio was 12:0.6:1 (m/w/w). These particles (approximately 120 nm in diameter) were added to **human tracheal epithelial (HTE)** cells and incubated for just 2 mins with cells before DNA movements were monitored with time by microscopy. Bottom-line images are same as top-line images except for superposition of cell contrast image to demonstrate sub-cellular localisation. DNA enters the nucleus in approximately 30 mins (adapted from Keller *et al.*, 2003, John Wiley & Sons).

In spite of these problems, fluorescein and derivatives can give excellent cellular tracking and localisation data in cells, particularly where these extrinsic fluorescent probes have been used to label lipids and oligopeptides in conjunction with cyanine dye labelling of DNA (Figure 4.27). Fluorescein and derivatives, with cyanine dyes, have also proven very useful as matching probes in FRET experiments designed to investigate real time bending, folding and conformational dynamics in isolated biological macromolecule systems, particular those involving proteins and nucleic acids (Figure 4.28).

Fluorescein and derivatives are complemented by the **Texas Red** fluorophore, a useful long wavelength extrinsic fluorophore (Figure 4.29). By contrast, coumarin derivatives are useful for shorter UV wavelength excitation, as are **pyrenes** such as **Cascade Blue** (Figure 4.29). One extrinsic fluorophore has proven particularly useful for the probe labelling of oligonucleotides or even nucleotides, namely **N-methylanthranilic acid (MANT)**. This is considered a "small probe" prepared from the alcohol-reactive probe precursor *N*-methylisatoic anhydride that appears to combine with 2′-hydroxyl groups of β-D-ribofuranose rings, introducing the fluorescent MANT moiety without appearing to impair the functional binding behaviour of oligonucleotides or nucleotides (Figures 4.29 and 4.30). Increasingly, the tendency in fluorescent probe design has been to make designs flexible in order that precise wavelengths of absorption and emission can be tailored to requirements by subtle changes in selected structural parameters. For instance, **Alexa Fluor** probes can be tailored to cover a huge range of wavelengths and can be very good choices in many kinds of applications. The most widely used Alexa Fluor dye is Alexa Fluor 488 (the number represents the excitation wavelength) (Figure 4.29). In contrast, **BODIPY** probes have a quite different structure, but once again can be tailored to cover a large range of wavelengths (Figure 4.31). Relatively speaking, the BODIPY probes are less hydrophilic than the Alexa Fluor probes and

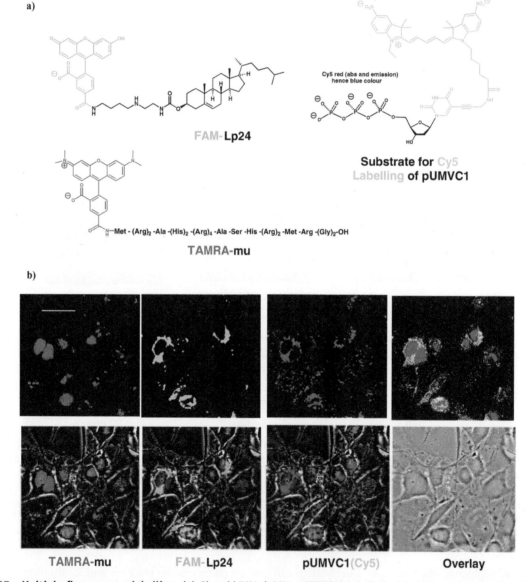

Figure 4.27   Multiple fluorescence labelling. (a) Plasmid DNA (pDNA; **pUMVC1**) labelled with **Cy5 cyanine dye** (see Figure 4.26), mu (μ) peptide labelled with **tetramethyl rhodamine (TAMRA)** by α-*N*-capping (see Figure 4.17); **fluorescein (FAM)** labelling of lipid (see Figure 4.17); **(b)** All three labelled components were then combined together into three-fold labelled liposome:mu:pDNA particles wherein the lipid:mu:pDNA ratio was 12:0.6:1 (m/w/w). These particles (approximately 120 nm in diameter) were added to HTE cells and incubated for 15 mins with cells before three-fold analysis by microscopy. Bottom-line images are same as top-line images except for superposition of cell contrast image to demonstrate sub-cellular localisation. The mu peptide enters the nucleus in 15 mins. The other components remain in the cytosol (adapted from Keller *et al.*, 2003, Figure 3a, John Wiley & Sons).

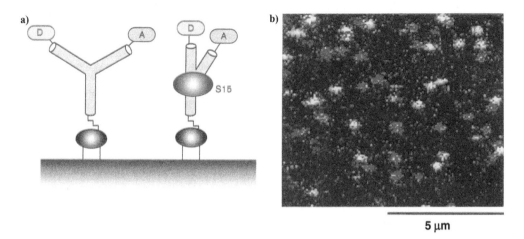

Figure 4.28   Protein induced folding of RNA. (a) Schematic to show RNA junction ("Y" shape cylinders) attached to glass surface at one end and with a **FRET donor-D (fluorescein)** on one arm of the "Y" and a **FRET acceptor-A (Cy3 dye)** on the other arm of the "Y". Binding of ribosomal protein S15 brings arms of the "Y" close enough for a meaningful FRET effect to be observable. **(b)** Glass slide with multiple RNA junction molecules is irradiated at 488 nm. If protein induced folding of RNA takes place, then donor-D fluorescence (green) is quenched through absorption by acceptor-A, leaving red dye colour. Both folding and unfolded junctions appear to be present (adapted from Lilley and Wilson, 2000, Figure 4, Elsevier).

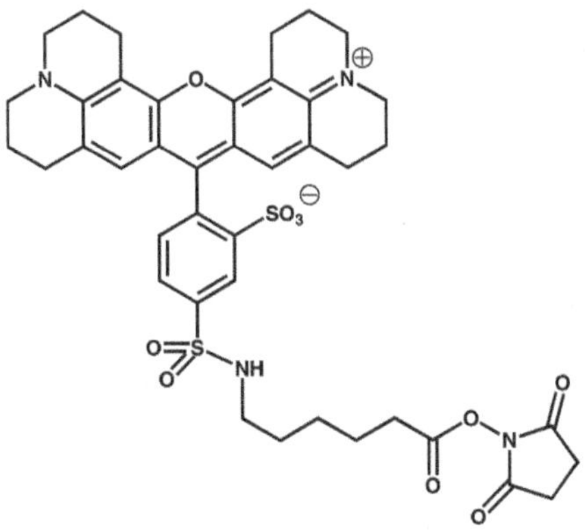

Texas Red-X, succinimidyl ester

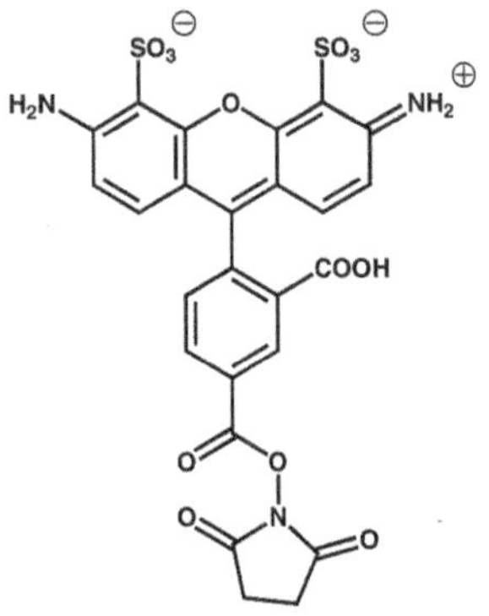

Alexa Fluor 488, succinimidyl ester

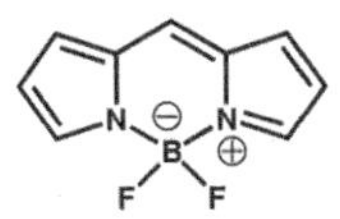

BODIPY, 4,4-difluoro-4-bora-3a, 4a-diaza-s-indacene

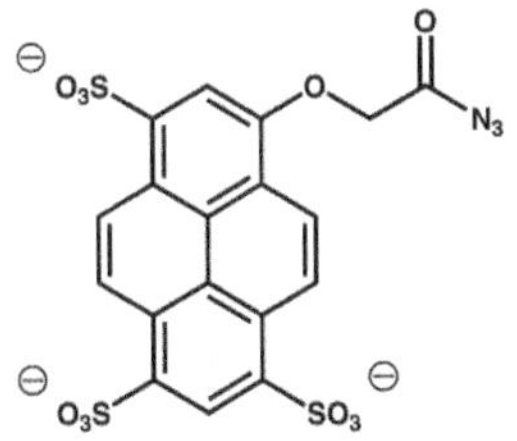

Cascade Blue acetyl azide

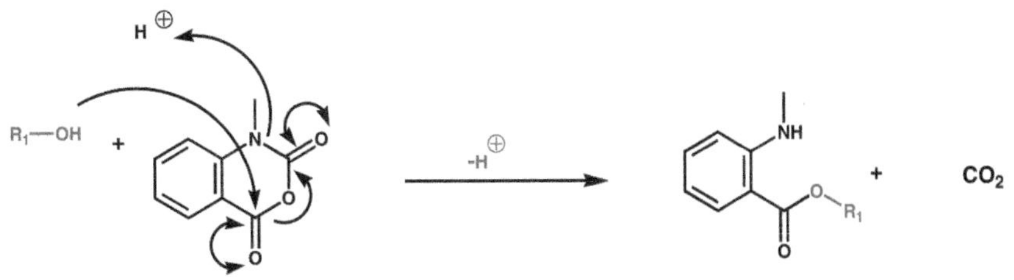

R_1—OH   +

N-methylisatoic anhydride

-H$^{\oplus}$

N-methylanthranilic acid (*N*-MANT)   + CO_2

Figure 4.29   More recent fluorescent probes and tags.

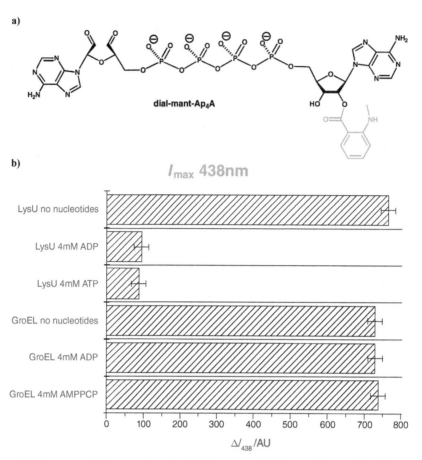

Figure 4.30   Fluorescence binding competition experiments. (a) Structure of **dial-mant-Ap$_4$A** probe. **(b)** Fluorescence binding data when two different proteins are incubated with the probe in the presence and absence of putative site-specific competitive binders. In the case of **LysU** protein (see Chapters 6 and 8), **adenosine 5′-diphosphate (ADP)** and **adenosine 5′-triphosphate (ATP)** are competitive binders and reduce probe binding to basal levels. In the case of **molecule chaperone** protein **GroEL** (see Chapters 5, 6 and 7) the probe is binding allosteric to ADP and **adenosine 5-[β,γ-methylene]-triphosphate (AMPPCP)** (from Wright *et al.*, 2006, Figure 2c, Elsevier).

BODIPY molecules have no net charge. Hence, they are membrane permeable and may enter cells freely, in the same way as extrinsic biarsenical fluorescent probes. Furthermore, BODIPY tagging has also been found to have very little impact upon the electrophoretic mobility of labelled molecules. Consequently, BODIPY probes have become integral to modern-day, automated DNA sequencing (see Chapter 3). The small size of BODIPY probes also makes them potentially useful alternative fluorescent labels for small molecules such as oligonucleotides or nucleotides, in place of MANT labels.

Aladan substitution of internal core amino acid residues provides an approach to characterise the physical characteristics of protein cores. Steady-state fluorescence alone can provide initial insight to the immediate environment of Aladan in the protein core. However, **time-resolved fluorescence spectroscopy** can be used to understand variations in protein core composition and structure as a function of time through the characterisation of Aladan fluorescence intensity and I_{max} changes that are caused by small fluctuations in the relative permittivity, ε, of the protein interior with time (fs–ps timescale). Such spectroscopy is possible since fluorescence lifetimes, τ_R, are typically in the ns-range (see Section 4.5). Also, time-resolved fluorescence spectroscopy can be performed with non-covalently linked extrinsic fluorophores such as **ethidium bromide (EtBr)**. This fluorophore intercalates between the bases of DNA or RNA double helix and in so doing acquires a substantial increase in ϕ_F and hence fluorescence intensity at I_{max} (595 nm). Should there be a disruption or collapse in double helical structure, then intercalation fails and fluorescent intensity drops instantaneously. Accordingly, extrinsic EtBr fluorescence becomes an ideal probe to study the rates of structural collapse and templating of DNA (or RNA) mediated by strongly cationic entities such as certain peptides or proteins (ms–s timescale) (Figure 4.32).

Where proteins are of interest, the use of GFP or GFP-variants (e.g. BFP, CFP, YFP and DsRed) in fusion with proteins of interest is becoming increasingly useful where cellular imaging is concerned, using **fluorescence microscopy**. On the assumption that a GFP or GFP-variant fusion protein retains the biological activities and functions of the wild-type protein of interest (at least to a reasonable extent), then the attached fluorescent protein allows for meaningful *in situ* tracking and localisation of this protein of interest in cells (Figure 4.33), provided that the wavelengths of absorption and emission are clearly distinct from those of the other fluorophores present. This is especially important if multiple individual proteins of

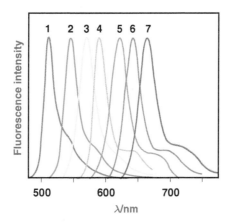

Figure 4.31 BODIPY dyes. Normalised fluorescence emission spectra of (**1**) BODIPY FL; (**2**) BODIPY R6G; (**3**) BODIPY TMR; (**4**) BODIPY 581/591; (**5**) BODIPY TR; (**6**) BODIPY 630/650; (**7**) BODIPY 650/665 fluorophores in methanol (from *Molecular Probes Handbook*, Figure 1.37).

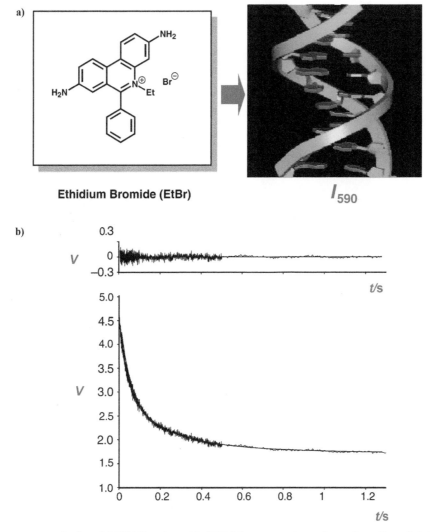

Figure 4.32 Fluorescence exclusion. (a) Ethidium bromide (EtBr) becomes a strong fluorophore through intercalation between DNA bases; (**b**) When DNA is condensed with increasing concentrations of mu peptide (from the adenoviral core) then EtBr is excluded instantaneously, causing fluorescence intensity to drop. At mu:pDNA ratios of 0.6 (w/w) condensation as measured by EtBr exclusion rates are rapid and reproducible. Mu peptide has a template-condensation effect on pDNA and the rates of EtBr exclusion suggest a condensation process similar in type and speed to single-domain protein folding. The top represents the control, the bottom represents the change in EtBr fluorescence signal (as voltage V) with time (from Tecle *et al.*, 2003, Figure 1a, American Chemical Society).

eGFP-*C*-terminal tags

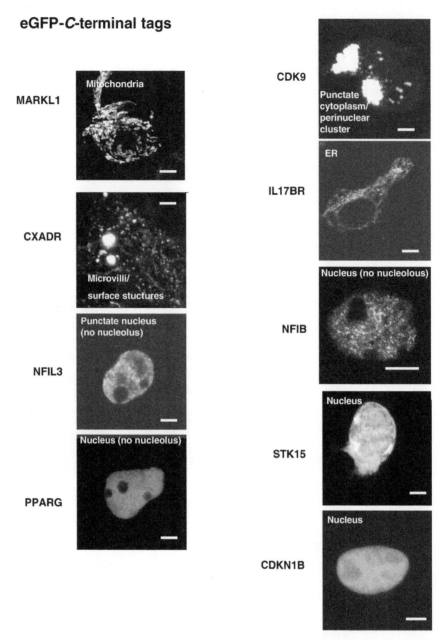

Figure 4.33 Green fluorescent protein-tagging. Complete range of fusion proteins generated by *C*-terminal fusion of GFP (see Figure 4.24) with selection of human proteins expressed by genes from the human genome mapping project. Fusion genes are generated in plasmid DNA(see Chapter 2) and used to transform mammalian cell line by a process of reverse transfection. Subcellular localisation of fusion gene products is then observed by microscopy. All proteins appear in their expected locations given their known functions (Palmer and Freeman, 2004, Figures 2 and 3, adapted from John Wiley & Sons).

interest are under observation simultaneously. Moreover, GFP and GFP variant fusions also offer a powerful way to identify functional binding partners in cells by FRET. In this case, obviously the GFP or GFP variant associated with each potential binding partner protein of interest should be properly matched with the GFP or GFP variant fused with the original protein of interest in order for the FRET effect to be properly observed when binding events take place (see Section 4.5.4).

The corresponding use of biarsenical fluorophore probes is clearly just as diverse but less invasive than the use of GFP or GFP variant fusions. Proteins of interest can be just as easily tracked in cells *in situ* and localised using biarsenical fluorophores. Furthermore, this approach lends itself very easily to differential fluorescent tagging of multiple proteins of interest in one cell, and therefore the identification and visualisation of functional binding partners in cells by FRET experiments, subject to the usual condition that fluorophores are matched to enable the FRET effect to be observed when binding events take place. Moreover, biarsenical fluorophore tags remain fluorescent post protein denaturation during electrophoresis, for instance (provided no reductive dithiols are used in the preparation of the protein) (see Chapter 2). The tags may even be used for affinity chromatography. In the latest manifestation of the technique, known as **chromophore assisted light inactivation (CALI)**, strong illumination of the red biarsenical fluorophore **ReASH**, post chelation to proteins of interest in cells, creates short-lived oxygen radicals that are able to inactivate nearby proteins. Such approaches illustrate how light can be used not only to observe molecules, but to actually control and elucidate molecular function in a real cellular context.

4.5.6 Fluorescence single molecule spectroscopy

All the techniques described elsewhere in this chapter measure the average behaviour of ensembles of molecules. Now, single molecule investigations are becoming possible. Such investigations allow for the dynamic observation of single molecules, thereby revealing behaviour that is often hidden in most studies of biological macromolecular structure and dynamics. Observation of the ensemble (i.e. most techniques used in chemistry) reveals no information about the spread and inhomogeneity in an experimental trajectory. Single molecule experiments can provide a great deal of otherwise untractable, and sometimes surprising, information regarding time trajectories and reaction pathways of biomolecules. Fluctuations and flickering are often observed at the single molecule level that would otherwise be obscured in ensemble experiments.

Single molecule observations are clearly demanding. To observe a single absorbing molecule within trillions of solvent or matrix molecules represents a significant technical problem. In order for this to be possible, two conditions should be fulfilled:

1. Laser beam excitation should be target biological macromolecule specific.

2. Post-excitation, biological macromolecule emission should exceed background.

The simplest approach to meet both conditions is either to immobilise a target biological macromolecule of interest on a surface, or else restrict Brownian motion of the target by making observations in the presence of a gel to substantially attenuate Brownian motion. Alternatively, observations can be made in a flow cell so that photonic bursts of light are generated and observed as biological macromolecules flow through an incident, excitation beam. To meet condition (1), the use of defined wavelength laser beams, also controlled in direction, ensures that an appropriate, matching fluorescent probe can be selected with precision. Meeting condition (2) is more complex. Typically, fluorescence emission is only a tiny fraction of the incident energy used for the excitation, so efficient fluorescence detection is critical for single molecule fluorescence spectroscopy. A number of factors can be considered to reduce background so that single molecule detection becomes possible. For instance, small sample volume, high-efficiency optics and the pre-bleaching of buffers are all helpful. The fluorescent probe should also exhibit a high value of ϕ_F, and D_{ab} (corresponding to an allowed electronic absorption transition) collected over a large absorption cross-section. In all these respects, GFP-fusion protein of interest appears to be an excellent subject for single molecule detection of proteins.

The power of single molecule detection should become obvious with reference to example. The range of observations that can be made using **single molecule spectroscopy (SMS)** is illustrated:

1. The simplest experiment is to observe the position and positional fluctuation of a biological macromolecule of interest. Since the emission wavelength is far greater than the size of the biological macromolecule, each molecule acts as a point light source for fluorescence emission. Observing the position of the biological macromolecule provides a technique for observing protein movements (Figure 4.34a) (e.g. diffusion of lipid molecules within membranes or the diffusion of molecules within gels or sols).

2. Two biological macromolecules of interest can be labelled with "non-interacting" fluorophores. The relative positions of the two molecules can then be monitored to observe protein interactions in real time (Figure 4.34b) (e.g. binding or catalysis can be directly observed in this way).

3. Two biological macromolecules of interest (with "interacting" fluorophores, that come periodically within 2–8 nm of each other) can be probed using single molecule FRET experiments to observe motion in real time (Figures 4.34c, d, g, h) (e.g. intramolecular movements in response to conditions and other molecules, binding interactions and effects of binding upon intermolecular movements, the effects of intramolecular motion on function, such as ion channel opening and closing).

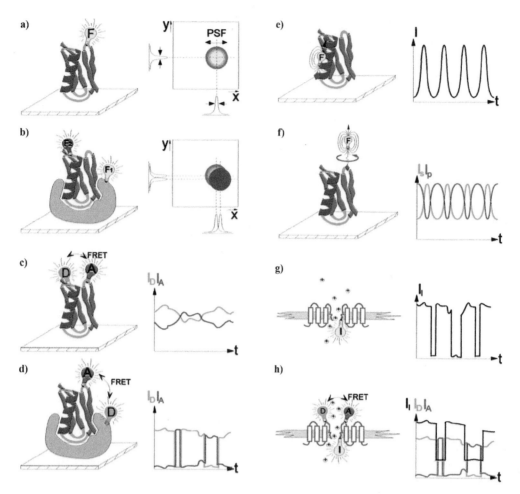

Figure 4.34   Fluorescence spectroscopy of single biomolecules. (a) Localisation of a macromolecule labelled with a single fluorophore, F, with nanometer accuracy. The **point-spread-function** (**PSF**) can be localised within a few tenths of a nanometer; (b) Colocalisation of two macromolecules labelled with two non-interacting fluorophores, F1 and F2. Their distance can be measured by subtracting the centre positions of the two PSFs; (c) Intramolecular detection of conformational changes by spFRET. D and A are donor and acceptor; I_D and I_A are donor and acceptor emission intensities; t is time; (d) Dynamic colocalisation and detection of association or dissociation by intermolecular spFRET. Donor and acceptor intensities are anticorrelated both in (c) and (d); (e) The orientation of a single immobilised dipole can be determined by modulating the excitation polarisation. The fluorescence emission follows the angle modulation; (f) The orientational freedom of motion of a tethered fluorophore can be measured by modulating the excitation polarisation and analysing the emission at orthogonal s and p polarisation detectors. I_s and I_p are emission intensities at s and p detectors; (g) Ion channel labelled with a fluorescence indicator I. Fluctuations in its intensity I_I report on local ion concentration changes; (h) Combination of (c) and (g). D and A report on conformational changes, whereas I reports on ion flux (from Weiss 1999, Figure 1, American Association for the Advancement of Science). Labelling schemes (left) and physical observables (right).

4.   A biological macromolecule with single attached chromophore can be observed using plane-polarised fluorescence light to demonstrate local motion in real time by observing changes in fluorescence emission intensity with time as the orientations of absorption and emission transition dipoles vary from motion (Figures 4.34e, f) (e.g. protein tumbling, discrete intramolecular rotations).

Without doubt, single molecule spectroscopy is a growing field and many of the technical barriers to entry are now removed or understood, so that they can be solved. When combined with many of the other single-molecule techniques available, such as atomic force microscopy (see Chapter 6), then the impact of single molecule detections and analyses can only become more influential, particularly where output from single molecule techniques is sufficiently robust, reproducible and species independent so that single molecule measurements can be related to single cell behaviour and even multi-cellular behaviour. Studies on ion channels and ion channel behaviour certainly have this potential, particular when linked together with electrophysiological investigations of brain matter. Single molecule spectroscopy is sure to become increasingly influential.

4.6 Probing metal centres by absorption spectroscopy

X-ray absorption may also be used to probe the immediate environment of specific metal atom centres in biological macromolecules of interest. X-rays typically have energies ranging from 500 eV to 500 keV, equivalent to wavelengths of 25–0.25 Å. The absorption of X-rays obeys the following Equation (4.25):

$$I_x = I_{xo} \exp(-\mu_{ab} x) \tag{4.25}$$

where I_{xo} is the **incident X-ray intensity**, I_x is the **transmitted intensity**, μ_{ab} an **X-ray absorption coefficient** and x is the **pathlength**. Light at this energy and wavelength is so energetic that absorption of photons can only take place by means of the **photo-electric effect**. According to this effect, absorption occurs when the energy of an incident X-ray photon exceeds the orbital energy of a core electron, allowing the X-ray photon to be absorbed and a **photo-electron** to be ejected. However, only atoms with atomic numbers between 20 and 65 are subject to the photo-electric effect with X-ray photons, hence the X-ray absorption spectrum of molecules, liquids and solids is dominated by metal atom absorption. The X-ray absorption spectrum is known as the **X-ray absorption fine structure** (**XAFS**) spectrum of a biological macromolecule of interest. XAFS spectra are very sensitive to oxidation and coordination state of atoms involved. Moreover, XAFS spectra can yield significant detail about distances (0.02–4 Å) from metal atoms to surrounding atoms, and even identify these surrounding atoms. Unfortunately, spectral information is always averaged over all the metal atoms of a given type in a biological macromolecule of interest. Therefore, the technique of X-ray absorption is most useful only when the biological macromolecule of interest comprises a single metal atom of a given type or at most only a modest cluster of atoms of the same type.

The typical XAFS spectrum is punctuated by the appearance of sharp edges from which two X-ray spectroscopy techniques take their names, namely **X-ray absorption near edge spectroscopy** (**XANES**) and **extended X-ray absorption fine structure spectroscopy** (**EXAFS**), referring to the region of spectral oscillation further from an edge. Each edge corresponds to enhanced X-ray absorption caused by the photo-electric effect and the emission of a photo-electron. Accordingly, the energy of each edge should be metal atom type dependent, and identifiable with X-ray absorption by a given metal atom or a modest metal atom cluster of the same (Figure 4.35). XANES then gives some qualitative insights into the metal atom oxidation state and coordination chemistry, whilst EXAFS analysis gives deeper insights into coordination number, neighbouring atom identities and inter-atomic distances involved. There is a much deeper understanding of EXAFS, so in practice XANES is used only as a qualitative metal atom fingerprint, while EXAFS analysis provides more quantitative information.

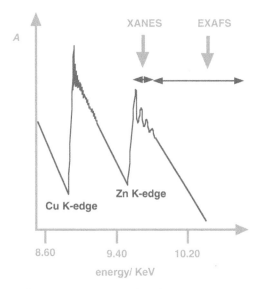

Figure 4.35   X-ray Absorption spectrum of Cu,Zn-metallothionein. The copper and zinc **K-edges** are observed and the **XANES** and **EXAFS** regions are shown for the zinc K-edge.

5

Magnetic Resonance

5.1 Magnetic resonance in chemical biology

Since chemical biology is about understanding the way biology works at the molecular level, then access to detailed 3D structures of biological macromolecules is indispensable. In Chapter 4, the chemical biology reader was introduced to the contribution that electronic spectroscopy can make to an understanding of structure and even function of biological macromolecules. However, electronic spectroscopies, although promising, are unable to give us a truly atomic-level description of any biological macromolecular structure. Therefore, in this chapter and in Chapter 6, the chemical biology reader will be brought face to face with those key, central techniques that have the capacity to generate meaningful atomic-level structures on which an atomic level of functional understanding can be built. All the techniques that will be discussed are surprisingly complementary both in theory and in application. All of these techniques are now totally fundamental to the effective practice of chemical biology.

This chapter is devoted to the contribution that magnetic resonance can make in deriving and understanding the structures of biological macromolecules. In particular, **nuclear magnetic resonance (NMR) spectroscopy** has been one the most beloved and powerful techniques of structural analysis used by chemists. Nowadays the theory and the quality of instrumentation have developed to the point that NMR spectroscopy is also completely relevant and applicable to chemical biology research as well. Although the history of NMR spectroscopy has been dominated by the analysis of small molecule structures (<1000 Da), the structural characterisation of biological macromolecules, in particular of globular proteins, is now fast becoming routine. In writing this chapter, the intention has been to focus on and provide an explanation of the main theories and ideas that underpin the use of NMR spectroscopy in chemical biology research. This is because NMR spectroscopy is fast becoming a generic technique for the determination of 3D structure at atomic-level resolution, so that the 3D structures of many globular proteins, nucleic acids, carbohydrates and macromolecular lipid assemblies, as described in Chapter 1, can be readily accessible using this technique. In this chapter, we shall be making the leap all the way from the simple underlying quantum mechanical theory of NMR spectroscopy to multidimensional NMR spectroscopy that forms the basis of what is also known today as biological NMR spectroscopy.

5.2 Key principles of NMR

This section deals with basic quantum mechanical theory in NMR. For some readers, this may be surplus to requirements. However, in our view, an appreciation of the quantum mechanical theory behind NMR spectroscopy is essential in order to understand the subsequent principles of biological NMR spectroscopy. So. unless the reader is truly familiar with the quantum level NMR theory, please read on.

Essentials of Chemical Biology: Structures and Dynamics of Biological Macromolecules In Vitro *and* In Vivo, Second Edition. Andrew D. Miller and Julian A. Tanner.
© 2024 John Wiley & Sons, Inc. Published 2024 by John Wiley & Sons, Inc.
Companion Website: www.wiley.com/go/miller/essentialschembiol2

5.2.1 Spin angular momentum

In quantum mechanics, each nucleus of an atom possesses the property of **spin**. In other words, each nucleus behaves as if it were a child's spinning top or a gyroscope, spinning freely in one position located on a smooth surface. The top rotates about its main axis, whilst at the same time precessing (Figure 5.1). Each nucleus also functions like a tiny bar magnet radiating magnetic force lines. Hence, we shall define a nucleus as a precessing, rotating, miniature bar magnet (Figure 5.2). Kinetic energy of spin is defined in terms of **angular momentum**, *J*, a quantity that is represented in the form of a vector whose length reflects magnitude and whose orientation reflects direction of spin (i.e. relative to a predefined axis). In the world of quantum mechanics, the **angular momentum vector** is quantised, with allowed magnitude defined by a **spin quantum number**, *I*, as represented by Equation (5.1):

$$J = \left[I(I+1)\right]^{1/2} \hbar \tag{5.1}$$

where $\hbar$ is the **reduced Planck's constant**. The spin quantum number also defines the number of allowed orientations that the spin **angular momentum vector** may adopt relative to a predefined reference axis. This number is equivalent to **2*I* + 1**. For instance, a proton (^1H-nucleus) that has a spin quantum number of ½ will have just two allowed orientations with respect to the predefined reference axis. Each of these distinct orientations is known as a **spin state**.

There is one further consequence of quantum mechanics: according to the **Heisenberg Uncertainty Principle** an atomic or sub-atomic particle cannot simultaneously have both position and momentum defined. Therefore, the angular momentum vector associated with each spin state cannot align with the predefined reference axis but must reside off axis. This condition results in precessional cones, one cone per spin state, as shown for a proton (Figure 5.3). Individually, each precessional cone may be described as a surface representing all the possible orientations that the spin angular momentum vector can adopt in order to satisfy the requirements of quantum mechanics. Alternatively, each cone can be thought of as mapping out the path traversed by the angular momentum vector as it precesses in an even sweep around the predefined reference axis at a fixed frequency.

Figure 5.1 Child's spinning top. This simultaneously rotates around its own primary axis but also precessing around a central axis of gyration.

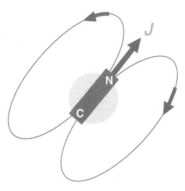

Figure 5.2 Spinning nucleus. This spins with **angular momentum**, **J**, just like the child's spinning top, and also behaves as a miniature bar magnet that projects a magnetic field out into the local environment.

Figure 5.3 Depiction of four spinning nuclei. Spinning nuclei simultaneously spin and precess at a **Larmor frequency**, ν_{L}, therefore angular momentum vectors (red arrows) map out precessional cones. In the absence of any **externally applied magnetic field**, vectors align relative to arbitrary z-axes. According to quantum mechanical rules, magnitude of angular momentum, **J**, is determined by spin quantum number, **I**, and a spinning nucleus can exist in any one of **2I + 1** spin states.

In the absence of an **externally applied magnetic field**, reference axes for each spinning nucleus are arbitrary spin states of a given nucleus that also **degenerate** (i.e. same energy). However, when an external magnetic field is applied then the direction of the external field becomes the reference axis (usually defined as the z-axis) for every spinning nucleus within the influence of that applied field. Interactions between the inherent magnetic field associated with each spin state and an externally applied magnetic field create a situation in which different spin states that may be occupied by given nucleus now possess inherently different energies. This situation is known as **lifting the degeneracy** and is a familiar concept in spectroscopy where externally applied electric or magnetic fields are used routinely to generate energy differences between otherwise degenerate quantum states. Energy differences between spin states are usually small and represented by the energy content of a single radio frequency photon. Such modest energy differences are the entire basis of NMR spectroscopy.

5.2.2 Magnetic moment

The inherent magnetic field strength associated with any given spin state is represented by a **magnetic moment**, μ_z, that is proportional to the **component of spin angular momentum**, J_z, which each spin state projects upon the reference z-axis, defined above as the direction of the externally applied magnetic field. Since spin angular momentum is quantised in terms of magnitude and orientation according to the spin quantum number, **I**, then z-axis components are similarly quantised. Each allowed z-axis component is represented by an individual **magnetic quantum number**, m_I according to Equation (5.2):

$$J_z = m_I \hbar \qquad (5.2)$$

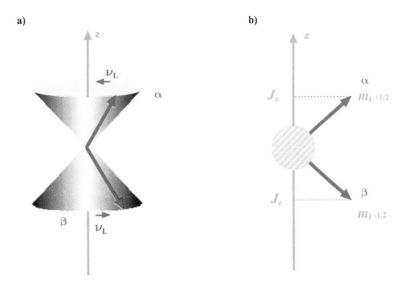

Figure 5.4 Depictions of two different allowed spin states. A spinning nucleus of I 1/2 (e.g. ^1H-nucleus) precesses relative to a reference z-axis supplied by an externally applied magnetic field such that angular momentum vectors (and precessional cones) align with (m_I + 1/2; α-**state**) or against (m_I −1/2; β-**state**) the field direction. Degeneracy between the two allowed (**2I + 1**) spin states is lifted by interaction between intrinsic magnetic fields of these spin states and the externally applied magnetic field.

In order to satisfy the requirement that the number of allowed spin states be equivalent to $2I + 1$, then the allowed values of m_I are equivalent to the series $I, I{-}1,\ldots{-}I$. Hence in the case of a ^1H-nucleus whose spin quantum number is ½, the allowed values of m_I are +½ and − ½. By convention, magnetic quantum numbers denoted with a "+" sign are said to align in generally the same direction as the reference z-axis and those denoted with a "−" sign are said to align in generally the opposite direction. This situation is illustrated for a ^1H-nucleus (Figure 5.4). Both possible states are illustrated on the same diagram, with precessional cones included. The spin state aligned with the applied magnetic field (reference z-axis) (m_I + ½) is known by convention as the α-**state**, the spin state aligned against the applied magnetic field (m_I − ½) is known as the β-**state**.

Magnetic moment relates directly to the allowed z-axis component of angular momentum of any given spin state of a nucleus according to Equation (5.3):

$$\mu_z = \gamma \cdot m_I \hbar \qquad (5.3)$$

where γ is known as the **gyromagnetic ratio**, which in turn is defined by Equation (5.4):

$$\gamma = \frac{g_I \mu_N}{\hbar} \qquad (5.4)$$

in which g_I is known as the **nuclear g-factor** and μ_N the **nuclear magneton**. The nuclear magneton is further defined by Equation (5.5):

$$\mu_N = \frac{e \cdot \hbar}{2m_p} \qquad (5.5)$$

where m_p is the **mass of a proton** and e the **charge**. Substitution of Equation (5.4) into Equation (5.3) gives Equation (5.6):

$$\mu_z = m_I \cdot g_I \mu_N \qquad (5.6)$$

5.2.3 Quantum mechanical description of NMR

The energy differences between different spin states are created by the differential way in which the magnetic moments of given spin states interact with the **externally applied magnetic field of strength B_z**. The **interaction energy attributable to any given spin state, E_{m_I}**, is defined by the product given in Equation (5.7):

$$E_{m_I} = -\mu_z \cdot B_z \tag{5.7}$$

Usually the nuclear g-factor, g_I, is >1 and the ratio γ is positive (Table 5.1). Therefore, taking the ^1H-nucleus as an example once more, the magnetic moment μ_z is positive for the α-state and negative for the β-state according to Equation (5.6). Consequently, the α-state becomes more stable than the β-state by interaction with the applied magnetic field, according to Equation (5.7). This situation is illustrated in the form of an energy level diagram (Figure 5.5) where the energy difference, ΔE, between these two states is then given by a further Equation (5.8):

$$\Delta E = E^{\beta}_{-1/2} - E^{\alpha}_{1/2} = g_I \mu_N B_z \tag{5.8}$$

NMR spectroscopy relies on the fact that transitions from lower to higher energy spin states may be accomplished by interaction with radio frequency photons. When photon energy is equivalent to spin state energy differences, then resonance coupling between nuclear spin and radiation leads to photon absorption and spin-spin transition. The frequency of absorption at resonance is sometimes known as the **Larmor frequency**, ν_L. Hence the resonance condition for NMR spectroscopy can be given in the form of Equation (5.9):

$$\Delta E = h\nu_L = g_I \mu_N B_z \tag{5.9}$$

This resonance condition (Equation 5.9) is such that ν_L is directly proportional to external magnetic field strength, B_z, and to the size of the nuclear g-factor, g_I, while the symbol h is **Planck's constant**. The Larmor frequency, ν_L, has also been linked to the frequency with which spin angular momentum vectors associated with given associated spin states are said to precess evenly around precession cones of the type illustrated in Figure 5.4. Therefore, ν_L may be alternatively called the **Larmor precession frequency**. In such a model, all states that could be occupied by a given nucleus subject to a given externally applied magnetic field will possess the same precession frequency. For a given nucleus, this frequency will be set in proportion to the nuclear g-factor, g_I, but be modulated in direct proportion to the strength of the externally applied magnetic field, B_z. A summary of some nuclear g-factors and spin quantum numbers for a range of different nuclei is given in Table 5.1.

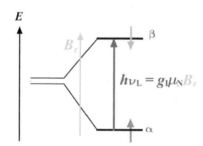

Figure 5.5 Lifting the degeneracy. This process is illustrated with the two allowed ($2I + 1$) spin states of I 1/2 nucleus (e.g. ^1H-nucleus) following introduction of an **externally applied magnetic field** of strength B_z. Two spin states are illustrated by arrows (orange) aligned with (α-state) or against (β-state) the field direction. A vertical spin state transition is shown (red arrow) along with the resonance condition for inter-conversion of α- to β-state.

Table 5.1 Summary of all the main spin active nuclei observed in the biomolecular NMR spectroscopy of biological macromolecules. All these nuclei are possess I values of 1/2. Note that γ is proportional to the nuclear g-factor, g_I.

Isotope	Spin (I)	Natural abundance (%)	$10^7 \times \gamma/\mathrm{rad\ s^{-1}\ T^{-1}}$	Relative sensitivity	NMR frequency (ν_L) at 2.35 T (field strength)
^1H	1/2	99.98	26.75	1.00	100.00
^{13}C	1/2	1.11	6.73	1.59×10^{-2}	25.14
^{15}N	1/2	0.37	−2.71	1.04×10^{-3}	10.13
^{19}F	1/2	100	25.18	0.83	94.08
^{31}P	1/2	100	10.84	6.63×10^{-2}	40.48
^{113}Cd	1/2	12.26	−5.96	1.09×10^{-3}	22.18

5.2.4 Chemical shift and coupling

There are two aspects to NMR spectroscopy that have made it an extremely valuable spectroscopic technique. The first is the phenomenon of **chemical shift** and the second that of **spin-spin coupling**. These topics will be covered very briefly in turn.

5.2.4.1 Chemical shift

Chemical shift may be thought of as the manner in which the resonance frequencies, ν_L, of nuclei that are part of molecular structures vary in a systematic and reproducible way in response to local chemical environment. According to Equation (5.9), changes in externally applied magnetic field strength experienced by a given nucleus must have a direct effect on the energy difference between nuclear spin states, leading to a proportionate change in resonance frequency, ν_L. Chemical shift arises because the strength of the effective magnetic field experienced by any nucleus in a molecular structure appears to vary in response to local movements in neighbouring electron density. In other words, local electronic effects have direct and reproducible effects on ν_L values. Electronic effects are both **shielding** and **deshielding** in character. Shielding effects are generated by the tendency of an externally applied magnetic field to induce electron density to "circulate" in such a way as to create a local magnetic field in opposition to the applied field. The effective magnetic field experienced by any such nucleus is then modulated according to expression (5.10):

$$B_{eff,z} = B_z(1 - \sigma_N) \tag{5.10}$$

where $B_{eff,z}$ is the **effective magnetic field strength** experienced by the nucleus and shielding is characterised by the **shielding parameter**, σ_N, also known as the **chemical shift tensor**. Deshielding arises out of heteroatom σ-bond inductive effects and π-bond ring current effects. The former effect reduces local electron density around a given nucleus, hence increasing the effective field and hence ν_L. The latter effect creates local magnetic fields that cooperate with the applied field to increase the effective magnetic field experienced by nuclei and hence their ν_L values (Figure 5.6). Together, shielding and deshielding effects are primarily responsible for ensuring that ν_L values vary as a direct consequence of local chemical environment and are therefore a direct indication of the nature of that chemical environment. In order to ensure that variation in ν_L values as a function of local chemical environment are standardised between NMR experiments and NMR spectrometers, the δ **chemical shift scale** was introduced. This scale is defined by Equation (5.11) in **parts per million (ppm)**:

$$\delta = \frac{\nu_L - \nu_o}{\nu_o} \cdot 10^6 \, \text{(ppm)} \tag{5.11}$$

where ν_o is the operating frequency (in MHz) of the NMR spectrometer and ν_L the resonance frequency of a nucleus of interest. Typically, chemical shift ranges are small compared to ν_o, therefore ^1H-NMR spectra are recorded over the range 0–12 ppm, ^{13}C-NMR spectra over the range 0–200 ppm and ^{15}N-NMR spectra over the range 1–200 ppm. The nuclei in most biologically relevant molecules will resonate within these ranges with very few exceptions. The δ chemical shift scale is universal and detection of a signal at a given δ value provides a very potent indication of local

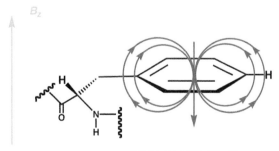

Phenylalanine Residue

Figure 5.6 Ring currents. These are generated by delocalised electron clouds that circulate in such a direction as to create a local magnetic field (**red**) that opposes the effect of the externally applied magnetic field of strength B_z (**light blue**). In aromatic systems, ring current enhances the effective magnetic field experienced by attached ^1H-nuclei, causing enhanced **deshielding**.

chemical environment and even functional group locality as shown (Figure 5.7). The scale is arbitrary and given from a reference value, for instance, in ^1H-NMR spectra the methyl signal of **tetramethylsilane** (**TMS**) is set to 0 ppm.

5.2.4.2 Spin-spin coupling

Spin-spin coupling effects on ν_L values are more subtle than chemical shift effects, but provide a great deal more information about the structure of the local chemical environment than suggested by the gross δ value. In short, electrons like nuclei have the property of spin represented by a spin quantum number, **I**, of ½. Therefore, spin pairing will take place between a given nucleus and the electrons of all associated bonds. Consequently, the spin state of this nucleus will be communicated with and have a direct local effect on the spin state of other nuclei related to the first through-bond. This is known as scalar **spin-spin coupling**. The **electron-coupled spin-spin interaction energy**, $E_\mathcal{J}$ is given by Equation (5.12);

$$E_\mathcal{J} = h\mathcal{J} \cdot I_i I_j \qquad\qquad (5.12)$$

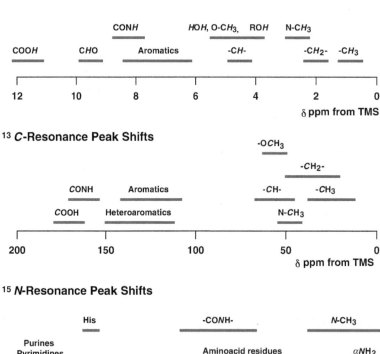

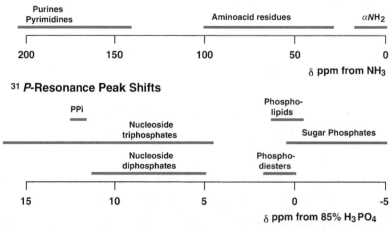

Figure 5.7 **Summary of δ scale chemical shift.** This summary applies for all main functional groups found in biological macromolecules for all main **I** 1/2 nuclei used to probe structure by NMR spectroscopy. Abbreviation: **TMS: tetramethylsilane.**

where I_i and I_j are the **spin quantum numbers of spin-spin coupled nuclei**, and $\mathcal{J}$ is the **scalar coupling constant** (in Hz). Each group of nuclei interrelated by spin-spin coupling is known as a **spin system**. Spin-spin coupling ensures that the spin states of each resonating nucleus in a spin system are evenly split into a number of **spin microstates**, usually closely similar in energy to the parent spin states. This situation is illustrated for two cases where a ^1H-nucleus is coupled with one and then two ^1H-nuclei respectively (Figure 5.8). A practical consequence of spin-spin coupling is that the resonance frequency lines of each nucleus are also evenly split into a number of usually closely related resonance frequency lines, each one of which represents allowed spin-spin transitions between lower energy and higher energy spin microstates (Figure 5.8).

There is a general rule that for each spin state of a resonating nucleus, the number of spin-coupled microstates that are generated by spin-spin coupling with n identical neighbouring nuclei is 2^n. However, the number of allowed spin-spin transitions is only $2nI + 1$, where I is the spin quantum number of neighbouring nuclei. The reason for this is that quantum mechanical selection rules require that there be no changes in the nuclear spin states of coupled nuclei during a given spin-spin transition between two microstates of a given resonating nucleus. So, for instance, if a resonating nucleus is coupled with one neighbouring nucleus ($I = \frac{1}{2}$) then although four spin-coupled microstates are generated in total, only two resonance frequency lines (known as a doublet) will result from allowed spin-spin transitions (Figure 5.8). Similarly, if two identical neighbouring nuclei ($I = \frac{1}{2}$) are involved, then three resonance frequency lines will result (triplet). Furthermore, if two identical neighbouring nuclei ($I = 1$) are involved then five resonance frequency lines will result (quintet) and so on. The spacing between lines is given by $\mathcal{J}$ the scalar-coupling constant (in Hz) and the number and relative intensities follow a distribution denoted by Pascal's triangle, to a first approximation (Figure 5.8). These intensity

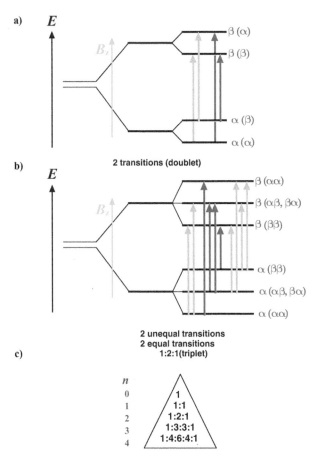

Figure 5.8 **Formation of spin microstates.** This happens in response to spin-spin nuclear coupling. (**a**) Coupling of spinning nucleus of I 1/2 (e.g. ^1H-nucleus) with one neighbouring spinning nucleus of I 1/2 (e.g. ^1H-nucleus). Two new microstates are formed from each main spin state (right). Two spin microstate transitions are allowed (red arrows) and two transitions disallowed (grey arrows); (**b**) Coupling of spinning nucleus of I 1/2 (e.g. ^1H-nucleus) with two equivalent neighbouring nuclei of I 1/2 (e.g. ^1H-nucleus). Four new microstates are formed from each main spin state (right). Four spin microstate transitions (two degenerate) are allowed (red arrows) and six transitions disallowed (grey arrows); (**c**) Pascal's triangle, a useful mnemonic with which to remember the number and intensity of resonance signals that result from spin-spin coupling of n equivalent nuclei of I 1/2 with a main spinning nucleus of I 1/2.

Table 5.2 Summary of main homo- (^1H-^1H) and heteronuclear spin-spin coupling constants. As relevant to the biomolecular NMR spectroscopy of biological macromolecules.

^1H-^1H coupling		
^2J (geminal)	^1H-C-^1H	-12–-15 Hz
^3J (vicinal)	^1H-C-C-^1H	2–14 Hz
	^1H-C=C-^1H	10 Hz (cis); 17 Hz (trans)
	^1H-N-C-^1H	1–10 Hz
		0.5–3 Hz (long range)
^1H-^{13}C coupling		
^1J	^1H-^{13}C(sp^3)	110–130 Hz
^2J	^1H-C-^{13}C	5 Hz
	^1H-C=^{13}C	2 Hz
^1H-^{15}N coupling		
^1J	^1H-^{15}N	89–95 Hz
^2J	^1H-C-^{15}N	15–23 Hz
^1H-^{31}P coupling		
^2J	^1H-O-^{31}P	15–25 Hz
^3J	^1H-C-O-^{31}P	2–20 Hz
^{31}P-^{31}P coupling		
^2J	^{31}P-O-^{31}P	10–30 Hz
^1H-^{19}F coupling		
^2J	^1H-C-^{19}F	40–50 Hz
^3J	^1H-C-C-^{19}F	5–20 Hz

patterns actually correlate directly with the number of allowed spin-spin transitions between microstates that are coincidentally degenerate. For example, when a resonating nucleus is coupled with two identical neighbouring nuclei ($I = \frac{1}{2}$) then inspection of the appropriate spin splitting diagram shows that there are actually four allowed spin-spin transitions between the eight spin-coupled microstates generated, of which two transitions are degenerate, hence the appearance of three resonance frequency lines in the 1:2:1 intensity ratio characteristic of an NMR triplet (Figure 5.8). An important point to be aware of is that these rules only apply when $J \ll \delta$ (in Hz).

Typically, in most organic and biological macromolecules nuclei spin-spin coupling takes place through one, two, three and possibly four consecutive covalent bonds. Such spin-spin coupling behaviour may be called ^1J (**single bond**), ^2J (**geminal**), ^3J (**vicinal**) and ^4J (**long range**) **coupling** respectively. The magnitudes of coupling constants vary substantially (Table 5.2). By definition, **spin systems** are clusters of spin-coupled nuclei that are prevented from, or are otherwise unable to, spin-couple with neighbouring spin systems. There can be enormous varieties of spin systems. For this reason, NMR spectroscopists frequently use a shorthand alphabetic nomenclature to denote the given type of spin system under investigation. Initially, a set of reference nuclei are initially denoted, $\mathbf{A_n}$, where $\mathbf{n}$ is the number of chemically equivalent reference nuclei involved. Spin-coupled nuclei are then denoted as follows. If their resonance frequencies are close to the resonance frequencies of the reference nuclei then the letter codes $\mathbf{B_n}$, usually followed by $\mathbf{C_n}$, are used. Alternatively, if their resonance frequencies are very far from the resonance frequencies of the reference nuclei then the letter codes $\mathbf{X_n}$, usually followed by $\mathbf{Y_n}$, are employed. In the case where the reference nuclei are coupled with two sets of chemically equivalent nuclei, all of which have substantially difference resonance frequencies from each other, then the letter code $\mathbf{M_n}$ may be used. For instance, $CHCl_2$-CH_2Br would be described as an $\mathbf{AB_2}$ spin system; CH_2Br-CF_3 would be described as an $\mathbf{A_2X_3}$ spin system; and $^{13}CHF_3$ as an $\mathbf{AMX_3}$ spin system. We will use this nomenclature later on in this chapter when looking at spin systems found in biological macromolecules, such as the polypeptides of proteins.

5.2.5 Vector description of NMR

On the whole, NMR spectroscopy is a very insensitive technique. The reason is that NMR spectroscopy relies on observing transitions from lower to higher energy spin states. Unfortunately, sensitivity relies on there being a substantial population difference between nuclei in lower energy states and those in higher energy states. However, owing

Table 5.3 **Summary of approximate quantities of biological macromolecules required in order to realise complete structural characterisation using the indicated types of technique.**

Type of structural analysis technique	Approximate quantities required for structural analysis
Electronic (Chapter 4)	μg–mg
Vibrational (Chapter 4)	mg
Magnetic resonance (Chapter 5)	mg–100 mg
X-ray crystallography (Chapter 6)	mg–g
Mass (Chapter 9)	ng

to the very slight energy differences between high and low energy spin states, population differences are frequently small, with the result that NMR spectroscopy requires a large population of nuclei in order to observe any significant photon absorption at all. Compare the amounts of material typically required to observe an NMR spectrum compared with other techniques described previously (Table 5.3). For instance, if a Boltzmann distribution is calculated corresponding to the distribution of ^1H-nuclei between α- and β-spin states using Equation (5.13):

$$N_\beta / N_\alpha = \exp(-h\nu_L / kT) \qquad (5.13)$$

assuming that $h\nu_L/kT$ is approximately 7×10^{-5} for a population of ^1H-nuclei at a temperature of 300 K and an externally applied magnetic field strength of 10 T, the population difference N_β/N_α is only slightly greater than 1. Since ν_L is a function of the nuclear g-factor, then the smaller the g_I of a given nucleus, the smaller still would be the population difference. The only way to reverse this situation is to increase externally applied magnetic field strength, B_z, and in so doing increase ν_L in proportion.

Nevertheless, there is usually a significant enough population between lower and higher energy spin states for NMR spectroscopy to be viable. This population difference averaged over a whole sample population of nuclei subject to an externally applied magnetic field must necessarily result in a net alignment of spin states in the direction of the externally applied magnetic field (reference z-axis). This leads to a net sample magnetisation in the direction of the externally applied magnetic field whose magnitude is determined by the vector sum over all the magnetic moments of the entire population of nuclei under observation. This net magnetisation, M, is frequently referred to as the **bulk magnetisation vector** (Figure 5.9). The existence of M allows for **Fourier transform** (**FT**)-**NMR spectroscopy** to be performed. FT-NMR spectroscopy is the basis of all modern NMR spectroscopy.

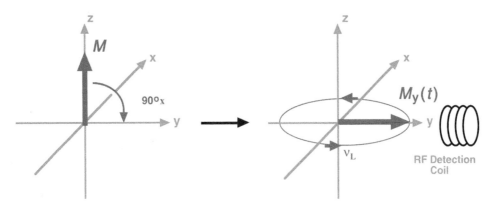

Figure 5.9 **Vector model of NMR spectroscopy. Bulk magnetisation vector M** is the result of complete summation over all z-axis components of magnetic momenta across an entire population of spinning nuclei aligned by an externally applied magnetic field. Illustration is drawn assuming all nuclei to be I 1/2 (e.g. ^1H-nuclei) chemically equivalent, hence precessing with an equivalent Larmor frequency, ν_L, but without spin-spin coupling. A **circular radio frequency pulse** is applied along the x-axis ($90°\text{x}$ **pulse**) (**left**) and **transverse magnetisation**, $M_y(t)$, is realised along the y-axis where radio frequency changes in magnetisation are observed, acquired and stored digitally with time by means of the detection coil.

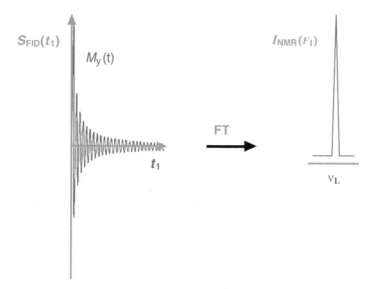

Figure 5.10   Free induction decay. The radio frequency variation of $M_y(t)$ transverse magnetisation, which is observed, acquired and stored digitally with time, is known as a **free induction decay** (**FID**). Stored FID, either singly or averaged, are processed by **Fourier transform** (**FT**) from time domain signal information, $S_{FID}(t_1)$, into frequency domain (spectral) information, $I_{NMR}(F_1)$. Only chemically equivalent nuclei without spin-spin coupling and with an equivalent Larmor frequency, ν_L, are being observed here, hence only a single signal of frequency ν_L will result.

In the simplest possible FT-NMR experiment, the resting sample is comprised of a homogeneous sample population of nuclei precessing at a uniform value of ν_L with a steady state value of M. If this sample is then subject to a **circularly polarised radio frequency pulse** of frequency ν_L along the x-axis, then the NMR resonance condition is satisfied (Equation 5.9) and M is able to precess towards the xy-plane. The pulse duration is sufficient only for M to precess through $90°$ and lie exclusively in the xy-plane. For this reason, such a pulse is referred to as a **$90°$ x-pulse**. The bulk magnetisation vector now lies along the y-axis and begins to disperse in the xy-plane with time. The x component of the bulk magnetisation vector as it evolves with time along the y-axis is usually referred to as **transverse magnetisation** or **coherence**, $M_y(t)$. **Spin-spin relaxation** (see Section 5.2.6) is primarily responsible for the evolution of $M_y(t)$. This evolution is observed by a radio frequency receiver coil mounted coaxial with the y-axis, acquired and then stored as a digitised radio frequency signal that varies as a function of time. This digitised radio frequency signal information is known as a **free induction decay** (**FID**) (Figure 5.10).

Spin-spin relaxation ensures that $M_y(t)$ evolves eventually to zero with time, whilst the original steady state bulk magnetisation vector, M, is regenerated by another relaxation process known as **spin-lattice relaxation** (see Section 5.2.6). After M is fully regenerated, a second $90°$ x-pulse is then applied and a second more or less identical FID may be observed, acquired and stored. This process may be repeated for as many FIDs as are required and then the averaged FID data set is subject to **Fourier transform**. This is an immensely complicated process, but the actual transformation is straightforward to understand. In short, Fourier transform means inverting time domain data (FID data) into frequency domain data (NMR spectrum). In the case of our highly simplified FT-NMR experiment described above, this process will lead to the realisation of an NMR spectrum from the illustrated FID, with a single peak corresponding to a resonance frequency of ν_L (Figure 5.10). In the more usual case, M is a complex vector sum of a wide range of different nuclei in states precessing at different ν_L frequencies owing to a raft of different spin-spin coupling and shielding/deshielding effects. Corresponding FIDs are therefore significantly more complex, but Fourier transform will readily result in a clear NMR spectrum from each FID family provided that there are not too many overlapping ν_L frequencies associated with nuclei in chemically different but accidentally degenerate spin states.

5.2.6 Spin-lattice and spin-spin relaxation

As indicated (Section 5.2.5), successful acquisition of multiple FIDs depends upon the ability of a sample population of nuclei to recover steady state equilibrium between radio frequency pulses. Recovery to equilibrium between pulses relies upon the two main relaxation processes mentioned above, namely spin-lattice and spin-spin relaxation (see

Section 5.2.5). Spin-lattice relaxation is associated with relaying spin energy to the surroundings by such mechanisms as through-space **dipolar coupling** (see Section 5.2.7). This same spin-lattice relaxation is primarily responsible for the restoration of the bulk magnetisation vector, **M**, to its original magnitude and position along the z-axis. The component of the bulk magnetisation vector as it recovers with time along the z-axis is referred to as **longitudinal magnetisation** or **polarisation**, $M_z(t)$. $M_z(t)$ obeys the following proportionality (5.14):

$$M_z(t) \propto 1 - \exp(-t/T_1) \tag{5.14}$$

where t is time, and T_1 the **longitudinal relaxation time constant**. By contrast, spin-spin relaxation is associated with the exchange of spin energies between spin-coupled nuclei over one nuclear transition (**single quantum coherence**) or sometimes over several nuclear transitions (**multiple quantum coherence**) as appropriate. This process is also known as **coherence transfer**. Spin-spin relaxation acts as an indirect mechanism to regenerate the steady state bulk magnetisation vector, **M**. Transverse magnetisation or coherence, $M_y(t)$, obeys the proportionality (5.15):

$$M_y(t) \propto \exp(-t/T_2) \tag{5.15}$$

where t is time, and T_2 the **transverse relaxation time constant**. The transverse relaxation time constant not only has a direct impact on $M_y(t)$ but also has an impact on spectral resolution. Spectral line width at half peak intensity, $\Delta\nu_{L,\frac{1}{2}}$, is given according to expression (5.16):

$$\Delta\nu_{L,\frac{1}{2}} = \frac{1}{\pi \cdot T_2} \tag{5.16}$$

indicating that the smaller the value of T_2, the greater will be the corresponding spectral line width and the worse the spectral resolution. In fact, this expression represents yet another meeting with the Heisenberg uncertainty principle; in other words, as certainty in time increases, uncertainty in frequency also increases. Equation (5.16) is very important for the NMR spectroscopy of biological macromolecules. Many such biological macromolecules are more than large enough and are comprised of more than sufficient spin-coupled nuclei to ensure that T_2 values are very small, leading to substantial line broadening with loss of spectral resolution. This is critical. Without adequate solutions to the "T_2-relaxation problem", NMR spectroscopy of biological macromolecules is technically impossible. Fortunately, there has been ample innovation in NMR spectroscopy over the years to solve this problem, as will be revealed shortly, and in the process bringing biological NMR spectroscopy to life.

5.2.7 Nuclear Overhauser effect

Before moving forward, the chemical biology reader should appreciate that $M_z(t)$ polarisation and recovery towards steady state bulk magnetisation, **M**, by spin-lattice relaxation, are primarily indicative of changes in the levels of spin state occupancy. By contrast, $M_y(t)$ coherence and evolution by spin-spin relaxation are both primarily indicative of transitions between spin state energy levels. In this context, the **nuclear Overhauser effect** (NOE) is a very important through-space effect in NMR spectroscopy that results directly from dipolar coupling-mediated spin-lattice relaxation.

Through-space dipolar coupling originates from the fact that each nucleus is like a precessing, rotating, miniature bar magnet (see Section 5.2.1). Another description for a bar magnet is a dipole. Therefore, nuclei are capable of the equivalent of non-bonded dipole-induced dipole interactions through space. Such dipolar coupling has no effect upon the number and relative energies of spin states, but does influence the relative levels of spin state occupancy, leading to changes in NMR signal intensities. So, **NOE** may be defined as a change in intensity of an NMR signal originating from a given reference nucleus that is dipole-coupled with other nuclei whose spin-spin transitions have been perturbed. Perturbation means saturation in this context. Hence, a change in signal intensity from a reference nucleus of interest should be observed if there is selective irradiation of the dipole-coupled nuclei. By way of illustration, if two nuclei **A** and **X** are close together in space, then irradiation of **X** should provoke an NOE modulation of the signal intensity (S_A) emanating from nucleus **A** according to the proportionality (5.17):

$$\frac{S_A}{S_{Ao}} \propto \frac{T_1}{T_{1,dip-dip}} \tag{5.17}$$

where S_{Ao} represents the **signal intensity prior to irradiation** of X and $T_{1,\text{dip-dip}}$ that **component of spin-lattice relaxation associated with through-space dipolar coupling**. Critically, the term $T_{1,\text{dip-dip}}$ contains an $1/r^6$ dependency characteristic of dipolar-coupling effects and reflects the fact that NOEs must also be subject to the same $1/r^6$ dependency, where r is the inter-nuclear distance in this case. In other words, as two dipole-coupled nuclei get closer then $T_{1,\text{dip-dip}}$ gets shorter, according to a $1/r^6$ dependency, and change in S_A intensity must increase in proportion. Please note that dipole-dipole interactions in general obey a $1/r^6$ dependency (see Section 1.6), and so have little influence beyond 5.0 Å. Accordingly, the magnitude of NOEs can be a powerful tool for showing the proximity of nuclei to each other through space. Typically, NOEs are grouped into three types:

1. **Strong NOEs** correlating with inter-nuclear distances of 1.8–2.7 Å

2. **Medium NOEs** correlating with distances of 1.8–3.3 Å

3. **Weak NOEs** correlating with distances of 1.8–5.0 Å

The reader should now see that in the battle for accurate structural determination of biological macromolecules with their complex 3D structures, the NOE effect is a potentially very powerful tool indeed. In fact, we would go so far as to say that NOE determinations are probably the cornerstone of structure determination experiments by biological NMR spectroscopy once T_2 relaxation problems have been solved.

5.3 Two-dimensional NMR

Armed with a basic knowledge of FT-NMR spectroscopy and NOE theory, we are now in a position to scale the heights of multidimensional NMR spectroscopy. If you have been tempted to skip Section 5.2, please at least be secure about the information in Sections 5.2.6 and 5.2.7 before continuing. Before embarking upon a discussion of multi-dimensional NMR spectroscopy, we shall need to redefine 1D NMR spectroscopy experiments in terms of FT-NMR spectroscopy and the vector description of NMR experiments. In so doing, 1D NMR experiments can be described as consisting of only a single time domain of detection and signal acquisition (t_1) following the application of radio frequency pulse(s). A diagrammatic representation of a general 1D NMR experiment is shown (Figure 5.11). Alternatively, this could be expressed as:

$$\text{preparation} - t_1 \text{ (acquire)}$$

Fourier transform of time domain data, $S_{\text{FID}}(t_1)$, into frequency domain signal intensity data $I_{\text{NMR}}(F_1)$ then yields a characteristic 1D NMR spectrum.

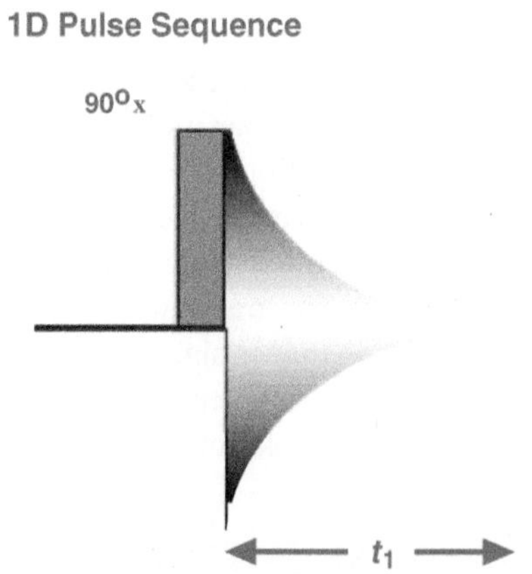

1D Pulse Sequence

Figure 5.11 Alternative diagrammatic representation of general Fourier transform 1D NMR experiment. There is a single 90° pulse, then signal observation and acquisition in time domain, t_1, prior to Fourier transform of time domain signal information, $S_{\text{FID}}(t_1)$, into frequency domain (spectral intensity) information, $I_{\text{NMR}}(F_1)$.

On the basis of this, 2D NMR spectroscopy experiments can be defined as experiments that involve a single time domain for signal evolution (t_1) and an additional time domain for detection and signal acquisition (t_2), together with appropriate radio frequency pulse sequences. The basic experimental form can be expressed as:

$$\text{preparation} - t_1 - M_1 - t_2 \text{ (acquire)}$$

where M_1 is an optional mixing period. Fourier transform of time domain data, $S_{FID}(t_1, t_2)$, into frequency domain signal intensity data $I_{NMR}(F_1, F_2)$ then yields an impressive range of possible 2D NMR spectra. There are three basic types of 2D NMR spectroscopy experiments that will need to be considered here, each of which has a slightly different sequence of pulses, but always involves two domains of time t_1 and t_2. The simplest is **correlated spectroscopy** (**COSY**), followed by **total correlation spectroscopy** (**TOCSY**) and then **nuclear Overhauser effect spectroscopy** (**NOESY**).

5.3.1 Homonuclear 2D COSY and TOCSY experiments

The pulse sequence of a general 2D COSY experiment is shown in Figure 5.12. In time, t_1, transverse magnetisation is generated that evolves with chemical shift. Coherence transfer between spin-coupled nuclei then takes place during the second pulse, after which final coherence is observed and acquired over time t_2. Coherence transfer between spin-spin coupled nuclei is the key element since the purpose of this experiment is to determine those homonuclei of interest (e.g. ^1H-^1H, ^{13}C-^{13}C) that are ^1J, ^2J or ^3J spin-spin coupled with respect to each other in a molecule of interest, although for practical purposes COSY experiments are usually limited to ^3J ^1H-^1H, homonuclear spin-spin coupling.

The appearance of a typical homonuclear 2D COSY experimental spectrum is shown in Figure 5.13. This general appearance shares many common features with other 2D NMR spectra, as we shall see. The spectrum takes the form of a contour map plotted as a function of frequencies F_1 and F_2 (both in ppm). Resonance frequency lines have now given way to peaks, whose intensity and cross sectional area are represented by contour lines in the same way that mountains or valleys are depicted on a geographical survey map. The diagonal cross section of the spectral contour map is equivalent to the 1D NMR spectrum of the given sample under investigation. Critically, cross peaks also exist in the spectral contour map, which are generated by coherence transfer and therefore allow for the identification of nuclei that are spin-coupled by triangulation with the diagonal (Figure 5.13).

The TOCSY experiment bears many features in common with the COSY experiment except for one key essential. A general TOCSY experiment is represented for comparison in Figure 5.12. The key essential is inclusion of a mixing period (denoted τ_m or M_1) involving **spin locking**. Spin locking allows **polarisation transfer** to take place between homonuclei by rendering their spin states temporarily equivalent in order to allow for "isotropic mixing". This is not quite the same as through-space dipolar coupling mediated effects that manifest themselves in NOEs. Instead, polarisation transfer is a form of "warming up" or "increasing the spin temperature" of spin-coupled nuclei (^1J, ^2J, ^3J, ^4J, etc.) by enhancing spin state energy level population imbalances leading to enhanced signal intensities and longer range, spin-spin coupling effects. Hence, in time t_1, coherence is generated that evolves with chemical shift. Polarisation transfer between spin-spin coupled-nuclei then takes place during time τ_m and afterwards final coherence is observed and acquired during t_2.

As a rule of thumb, the longer is τ_m, the longer the distance of polarisation transfer between spin-coupled nuclei, although transfers involving four spin-coupled nuclei usually represent a practical maximum. The main practical difference of polarisation transfer, rather than coherence transfer, is that many more cross peaks are generated in the spectral contour map that allow for the identification of long range as well as short range spin-coupled nuclei by triangulation with the diagonal (Figure 5.13). TOCSY is powerful for the identification of individual spin systems and in a more advanced form is valuable for the identification of homonuclear ^1H-^1H-spin systems in biological macromolecules. For instance, each amino acid residue in a polypeptide represents an individual such homonuclear ^1H-^1H-spin system that can be separately identified and resonance assigned (see Section 5.5.1).

5.3.2 Heteronuclear correlation experiments

Heteronuclear correlation experiments obey the same principles as **homonuclear correlation experiments**, but require more complex execution. The problem with any type of heteronuclear NMR experiment, is that heteronuclei have different nuclear g-factors, g_I, and as a result have substantially different sensitivities given wide potential variations in spin state energy level populations. This situation arises because values of g_I are directly proportional to the Larmor precession frequency, ν_L (according to Equation 5.9) and as values of ν_L decline, so, too, does the spin state

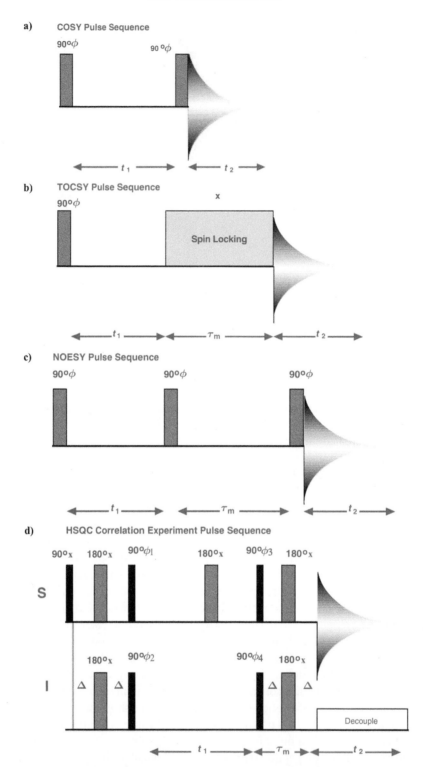

Figure 5.12   Diagrammatic representation of Fourier transform 2D NMR experiments. These experiments involve multiple 90° pulses and signal observation and acquisition in time domain t_2 prior to Fourier transform of time domain information $S_{FID}(t_1, t_2)$ into frequency domain (spectral intensity) information, $I_{NMR}(F_1, F_2)$. Experiments shown are homonuclear (**a**) **COSY**, (**b**) **TOCSY** and (**c**) **NO-ESY** experiments. Typical **heteronuclear correlation experiment** (**d**) is also shown, comprising complex sequences of multiple 90° and 180° pulses involving **source nuclei I** and **destination nuclei S**.

population imbalance between spin states (according to the Boltzmann equation 5.13). In other words, the smaller is g_I, the less is the population difference between spin states and hence the less sensitive is the nucleus concerned. Nevertheless, spin-spin coupling relationships are possible between any number of spin-active nuclei such as ^1H, ^{13}C, ^{15}N, ^{31}P or ^{19}F, making heteronuclear correlation experiments potentially very valuable.

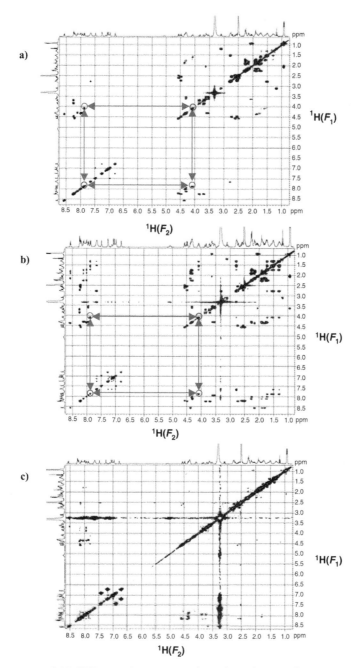

Figure 5.13 Contour map outputs of 2D NMR experiments. Actual $I_{NMR}(F_1, F_2)$ spectral output ^1H-^1H homonuclear (**a**) COSY, (**b**) TOCSY and (**c**) NOESY experiments obtained with a peptide of sequence **VQGEESNDK** (Boraschi loop of **Interleukin-1β**) (see Chapter 7). Output is in the form of contour maps in which peak intensity is indicated by the density of contour lines (in the same way that contour lines represent peaks and valleys in geographical survey maps). In all cases, the diagonal (running from top left to bottom right) reproduces an original 1D spectrum of the molecule under investigation. Triangulation using off-diagonal peaks (**red circles**) identifies those on-diagonal peaks (**blue circles**) that are correlated either through-bond for COSY and TOCSY or through-space for NOESY.

The pulse sequence of a 2D **heteronuclear single quantum coherence** (**HSQC**) correlation experiment is shown to illustrate the additional complexity required in the pulse sequence relative to the pulse sequences of the homonuclear correlation experiments (Figure 5.12). This HSQC correlation experiment requires that there is a preparation period during which nuclei with the highest nuclear *g*-factor (or γ ratio), are subjected to radio frequency pulses in order to effect polarisation transfer to spin-coupled **source nuclei I**, with the lower nuclear *g*-factor (or γ ratio). Thereafter, in time t_1, transverse magnetisation is generated involving nuclei **I** that evolves with chemical shift. Additional pulse sequences during time τ_m then ensure reverse magnetisation transfer from **I** to spin-coupled **destination nuclei S**, and afterwards final coherence is observed involving nuclei **S** and acquired during t_2.

The HSQC experiment is routinely used to identify heteronuclear ^1J spin-spin coupled nuclei such as ^1H-^{13}C or ^1H-^{15}N pairs, and is the basis of many so-called double resonance experiments used for the structural determination of proteins and of other biological macromolecules, as we shall see later. A variation on the HSQC experiment is the **heteronuclear multiple bond correlation** (HMBC) experiment. This is a sensitive technique that may be used to identify heteronuclear ^2J and ^3J spin-spin coupled nuclei.

5.3.3 NOESY experiments

Two-dimensional NOESY experiments are one further variation on the 2D NMR theme. The general 2D NOESY experiment is illustrated (Figure 5.12). The most important feature of the 2D NOESY experiment is the generation of a spectral contour map whose cross peaks are generated in response to NOEs resulting directly from dipolar coupling-mediated, spin-lattice relaxation between nearest neighbour nuclei through space. In this case, a variable length mixing period (denoted τ_m or M_1) is used to allow polarisation transfer to take place under the influence of dipolar coupling. This effect is also known as **cross-relaxation**. Hence in time t_1, transverse magnetisation is generated that evolves with chemical shift. Cross-relaxation then takes place during time τ_m and afterwards final coherence is observed and acquired as a function of time t_2. As indicated, cross peaks are generated by cross-relaxation in the spectral contour map that allows for the identification of dipole-coupled nuclei close together in space (<5 Å) by triangulation with the diagonal (Figure 5.13). The value of τ_m appropriate for a given molecule is directly related to its molecular rotation or tumbling rate, which is related in turn to molecular weight. The larger the molecule, the smaller should be the value of τ_m. Both homonuclear and heteronuclear 2D NOESY experiments are well known. The value of NOESY experiments for determination of quantitative distance information and hence the 3D conformation and structure of biological molecules has already been suggested very strongly. As with correlation experiments, heteronuclear NOEs also exist and are observed by NOESY experiments.

5.4 Multidimensional NMR

So now, at last, we have arrived at the techniques that will underpin the majority of biological NMR spectroscopy and subsequent biological macromolecule 3D structure determination. Key to all techniques are the concepts of polarisation transfer ($M_z(t)$-related longitudinal magnetisation transfers), cross-relaxation (through-space polarisation transfers enabled by through-space dipolar coupling) and coherence transfer ($M_y(t)$-related transverse magnetisation transfers). Clearly, these concepts have all been introduced before, but please try to appreciate the differences between them before advancing further forward into Section 5.4.

5.4.1 Basic principles of 3D experiments

Most 3D experiments are heteronuclear experiments and may be regarded as more sophisticated versions of the 2D HSQC correlation experiment (see Section 5.2.2). A representation of the general 3D correlation experiment concept is shown in Figure 5.14. Alternatively, this can be expressed as:

$$\text{preparation} - t_1 - M_1 - t_2 - M_2 - t_3 \text{ (acquire)}$$

where M_1 and M_2 are mixing periods. Fourier transform of time domain data, $S_{FID}(t_1, t_2, t_3)$ into frequency domain signal intensity data $I_{NMR}(F_1, F_2, F_3)$ results in an extensive array of data from each experiment. The concept may sound simple enough and we hope so. However, it would be inappropriate to deny the fact that the actual complexity involved in a 3D correlation experiment is a sizeable step up even in comparison with the 2D HSQC correlation experiment (Figure 5.12).

5.4.2 Correlation experiments

By way of example, look at the actual pulse sequence for a "routine" **3D HNCA correlation experiment** used to probe polypeptide structure (Figure 5.15) (see Section 5.4.1 for further detail). If the pulse sequence looks complex enough,

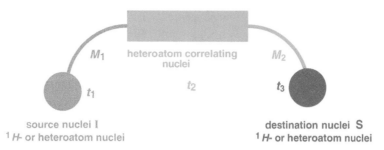

Figure 5.14 Cartoon for general structure of 3D correlation experiments. Pulse sequences are employed to generate transverse magnetisation in source nuclei **I**, that evolves according to chemical shift in time, t_1, prior to magnetisation transfer to **correlating nuclei** in period, M_1. Further pulse sequences promote transverse magnetisation in the correlating nuclei that evolves according to chemical shift in time, t_2, prior to final magnetisation transfer to destination nuclei **S** in period M_2. Final pulse sequence generates transverse magnetisation in the destination nuclei **S** that is observed, acquired and digitised in time t_3. Fourier transform is then used to transform time domain signal information $S_{FID}(t_1, t_2, t_3)$ into frequency domain (spectral intensity) information, $I_{NMR}(F_1, F_2, F_3)$.

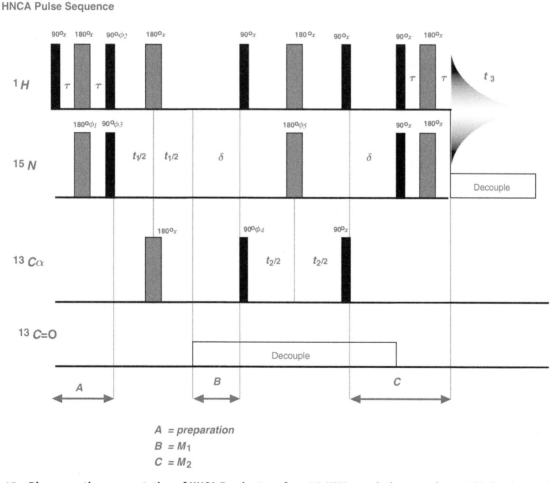

Figure 5.15 Diagrammatic representation of HNCA Fourier transform 3D NMR correlation experiment. This involves multiple 90° and 180° pulses and signal observation and acquisition in time domain t_3 prior to Fourier transform of time domain signal information, $S_{FID}(t_1, t_2, t_3)$ into frequency domain (spectral intensity) information, $I_{NMR}(F_1, F_2, F_3)$.

then have a thought for the data presentation and analysis from such an experiment. Such a 3D correlation experiment generates intensity data $I_{NMR}(F_1, F_2, F_3)$ emanating from the **triple resonance** of three entirely different populations of nuclei, in this case ^1H-, $^{13}C_\alpha$- and ^{15}N-nuclei. In principle, if all the data were to be displayed at once then this would require a 4D means of representation. This is clearly not possible, so data from 3D correlation experiments is normally plotted as a stack or cube of individual 2D correlation experiments. For instance, $I_{NMR}(F_1, F_2, F_3)$ data could be realised as a stack of 2D $I_{NMR}(F_1, F_3)$ contour plots (F_1 and F_3 both in ppm), all corresponding to a different value of F_2 (F_2 also in ppm) (Figure 5.16). Alternatively, $I_{NMR}(F_1, F_2, F_3)$ data could be realised as a stack of 2D $I_{NMR}(F_1, F_2)$ contour plots all corresponding to a different value of F_3 and so on. There can be substantial flexibility in the presentation of data from 3D correlation experiments, depending upon strategy and need. However, the primary need must always be to resolve signals adequately so that any nucleus of interest possesses a unique and unambiguous resonance assignment. In the case of the biomolecular NMR spectroscopy of biological macromolecules, $I_{NMR}(F_1, F_2, F_3)$ data from 3D correlation experiments is normally sufficient to achieve this. For instance, the output of the 3D HNCA correlation experiment should ensure that every ^1HN-nucleus in a protein under investigation has a unique and unambiguous resonance assignment defined in the first instance by its ^1H-chemical shift, but also by the chemical shifts of spin-coupled/correlated $^{13}C_\alpha$- and $H^{15}N$-nuclei. In effect, each ^1HN-nucleus is referenced by a completely unique 3D address, $^1HN(F_{1H}, F_{15N}, F_{13C})$, defined by three resonance frequencies.

5.4.3 Basic principles of 4D experiments

Increasingly, however, as the demands on biomolecular NMR spectroscopy increase, even 3D correlation experiments may not provide sufficient frequency information for any nucleus of interest to possess unique and unambiguous resonance characteristics. In this instance, there is an absolute need for more frequency information; hence a 4D correlation experiment is required. In principle this is merely an extension of the 3D principle to involve another resonating population of nuclei and another frequency domain. A representation of the general 4D correlation experiment concept is shown (Figure 5.17). Alternatively, this can be expressed as:

$$\text{preparation} - t_1 - M_1 - t_2 - M_2 - t_3 - M_3 - t_4 \left(\text{acquire}\right)$$

where M_1, M_2 and M_3 are mixing periods. Fourier transform of time domain data, $S_{FID}(t_1, t_2, t_3, t_4)$ into frequency domain signal intensity data $I_{NMR}(F_1, F_2, F_3, F_4)$ results in an even larger array of data from each experiment than obtained in the typical 3D experiment. In principle, if all the data were to be displayed at once then this would require a 5D means of representation. Since this is absolutely out of the question, data from 4D correlation experiments is normally plotted as a stack or cube of individual 2D correlation experiments in the same way as data from 3D correlation experiments. The only difference is that there is one more frequency to consider, so, for instance, $I_{NMR}(F_1, F_2, F_3, F_4)$ data could be realised as a stack of 2D $I_{NMR}(F_1, F_4)$ contour plots (F_1 and F_4 both in ppm), all corresponding to different

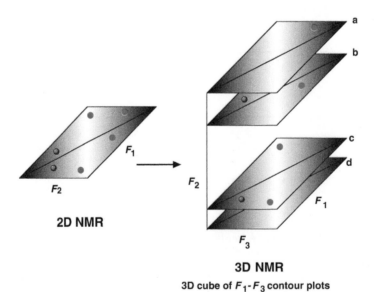

3D cube of F_1-F_3 contour plots

Figure 5.16   Diagrammatic representation of data from Fourier transform 3D NMR correlation experiments. In this representation, frequency domain (spectral) information, $I_{NMR}(F_1, F_2, F_3)$ is plotted as a a stack or cube of 2D NMR $I_{NMR}(F_1, F_3)$ contour plots, each plot resolved at a different value of F_2. Frequency resolution is done to aid resolution of individual resonance signals in order to achieve unique and unambiguous assignment of every resonance signal to resonating nuclei.

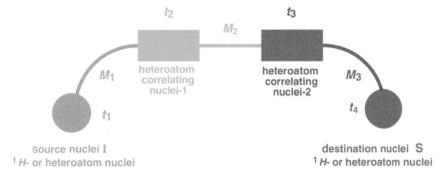

Figure 5.17 Cartoon for general structure of 4D correlation experiments. This is the same as for 3D correlation experiments (Figure 5.14) except that an extra resonant population of heteroatom nuclei are involved in promotion of transverse magnetisation (in time t_3) and magnetisation transfer (during M_3). Final pulse sequence generates transverse magnetisation in the destination nuclei **S** that is observed, acquired and digitised in time t_4. Fourier transform is used to transform time domain signal information, $S_{FID}(t_1, t_2, t_3, t_4)$ into frequency domain (spectral intensity) information, $I_{NMR}(F_1, F_2, F_3, F_4)$.

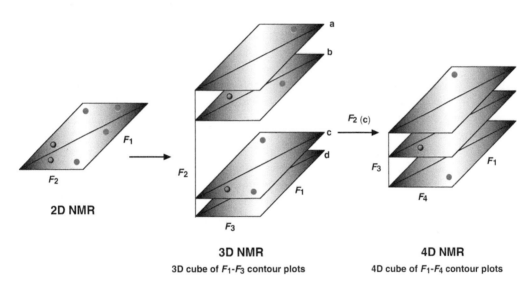

Figure 5.18 Diagrammatic representation of data from Fourier transform 4D NMR correlation experiments. In this representation, frequency domain (spectral) information, $I_{NMR}(F_1, F_2, F_3, F_4)$, is plotted as a stack or cube of 2D NMR $I_{NMR}(F_1, F_4)$ contour plots, each plot resolved at a different value of F_2 and also F_3. Double frequency resolution is carried out when single frequency resolution fails to achieve proper signal resolution and/or unique and unambiguous assignment of every resonance signal to resonating nuclei.

pairs of F_2 and F_3 values (F_2 and F_3 also in ppm) (Figure 5.18). Obviously, there are numerous possibilities for plotting the data in other different ways, again depending upon strategy and need. Nevertheless, after data analysis the main result should be that any nucleus of interest is referenced with a 4D address defined by four resonance frequencies. This should in all but a few instances represent a unique and unambiguous resonance assignment. However, where this does not happen there is always the possibility of advancing to 5D correlation experiments.

5.5 Biological macromolecule structural information

Derivation of biological macromolecular structure by NMR spectroscopy involves the application of a variety of multidimensional NMR spectroscopy experiments, many of which are beyond the scope of this chapter. Nevertheless, there are some basic principles and ideas that the chemical biology reader should be aware of and which we shall attempt to cover. These basic principles and ideas pull very heavily on the discussion in Section 5.3 and 5.4 above. In brief, the structural characterisation of a biological macromolecule by NMR spectroscopy draws upon a similar approach in every single case. The key objectives are:

1. Unambiguous assignment of all possible resonance frequency peaks in biomolecular NMR spectra to all possible resonating nuclei; ^1H-nuclei assignments are especially important, including the identification of spin systems.

2. Identification of all possibly nearest neighbour ^1H-nuclei through space by NOESY experiments; the identification of long distance neighbours with quantitative inter-nuclear distance is most important.

3. Determination of structure by energy minimisation employing through-space constraints from NOESY experiments (long distance neighbours are essential for accurate energy minimisation) as well as angular constraints from coupling constant values.

Each of these key objectives must be achieved successfully and in sequence, otherwise structure determination will itself be unsuccessful.

5.5.1 Analysing protein structures

It is virtually impossible to interpret and assign 1D and 2D NMR spectra of proteins. Typically, polypeptides (>5 kDa) have very short T_2 values, resulting in broad, low resolution spectral lines or peaks, respectively (see Equation 5.16). Furthermore, proteins themselves are biological macromolecular polymers involving amino acid residue building blocks of broadly similar structure and therefore chemical shift characteristics (see Chapter 1). Consequently, 1D and 2D biological NMR spectra of proteins possess an enormous amount of overlapping chemical shift data emanating from almost identical or closely similar amino acid residue spin systems. Having said this, for low molecular weight proteins comprising less than 50 amino acid residues a structure may be determined from 2D NMR spectroscopic data alone. However, for proteins of greater molecular weight, extensive signal overlap and low-resolution can only be solved with 3D and occasionally 4D correlation experiments.

If we are to take full advantage of 3D and 4D experiments though, molecular biology techniques (Chapter 3) must be used to ensure that the protein of interest is effectively universally labelled with spin-active nuclei, in order to avoid the serious sensitivity problem created by the fact that spin-active nuclei of interest are normally found only in low natural abundance (a few per cent). As a general rule of thumb, where a protein is comprised of between 50 and 80 amino acid residues, then 3D experiments involving ^{15}N labelling should suffice. Where more than 80 amino acid residues are involved, combinations of 3D and 4D experiments are essential and require both ^{15}N and ^{13}C labelling. Depending upon the labelling pattern required, universal ^{15}N and/or ^{13}C labelling is accomplished by overexpression of the transgene for the protein of interest in a recombinant organism (usually bacteria) that are grown in minimal medium containing $^{15}NH_4^+$ ions and/or $[^{13}C]$-glucose (see Chapter 3). Labelled proteins of interest are then isolated purely for analysis by standard purification protocols (see Chapter 2). Having carefully prepared such a sample, both 3D and 4D versions of COSY, TOCSY and NOESY correlation experiments (3D and 4D correlation experiments) are then made possible by a viable network of large heteronuclear ^1J spin-spin coupling constants. These enable intense homo- and hetero-nuclear coherence and polarisation transfers to take place, not to mention cross-relaxation (polarisation transfer under the influence of dipolar coupling) (Figure 5.19). Usually, the following sequence of operations is carried out:

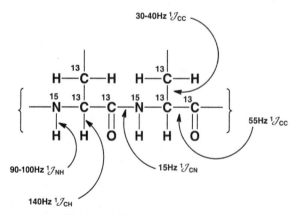

Figure 5.19 Intranuclear coupling. Summary of network of large heteronuclear ^1J coupling constants created by universal labelling of protein of interest with ^{13}C- and ^{15}N-nuclei.

1. 3D/4D correlation experiments are performed to obtain the unique and unambiguous resonance assignment of the ^1H-nuclei of all the amino acid residues in the protein of interest (**spin-system analysis**) Protein amino acid residue spin systems are summarised in Table 5.4. Obviously, with small proteins (<50 amino acid residues), 2D correlation experiments should suffice.

2. 3D/4D correlation experiments such as **HNCA, HNCO, HCACO** and **HCA(CO)N** experiments in particular (see Section 5.5.1.2) are performed to determine neighbouring amino acid residue spin systems and relative connectivities between spin systems. Previous knowledge of the amino acid residue sequence and the judicious use of the most appropriate correlation experiments are essential in order to match spin systems with actual amino acid residues in the sequence and also to generate unambiguous resonance assignments of the ^1H and other heteroatomic nuclei of the peptide links that attach these residues together.

Table 5.4 Spin-system summary for all main L-α-amino acid residues found in proteins. By convention **X** is the $^1HC_\alpha$-nucleus of each residue. For each side chain, **A** denotes the ^1H-nucleus/nuclei furthest from $^1HC_\alpha$. Otherwise, letters close to A in the alphabet imply that subsequent nucleus/nuclei in the chain are of similar chemical shift characteristics to the **A** nucleus/nuclei, letters in the middle of the alphabet imply chemical shift characteristics intermediate between **A** and **X** nuclei, letters close to the end of the alphabet imply chemical shift characteristics close to the **X**-nucleus ($^1HC_\alpha$). The use of the "+" sign indicates separate spin systems.

Common Amino Acid Systems

R	Name	Spin system	R	Name	Spin system
	Gly (G)	AX		Tyr (Y)	AMX + AA'XX'
	Ala (A)	A_3X		Ser (S)	AMX
	Val (V)	A_3B_3MX		Thr (T)	A_3MX
	Leu (L)	A_3B_3MPTX		Cys (C)	AMX
	Ile (I)	$A_3MPT(B_3)X$		Asp (D)	AMX
	Phe (F)	AMX + AMM'XX'		Asn (N)	AMX
	Trp (W)	AMX + A(X)MP + A		Glu (E)	AM(PT)X
				Gln (Q)	AM(PT)X
	Met (M)	AM(PT)X + A3		Lys (K)	$A_2(F_2T_2)MPX$
				Arg (R)	$A_2(T_2)MPX$
	Pro (P)	$A_2(T)_2MPX$		His (H)	AMX + AX

3. 3D/4D NOESY experiments can also be used to identify neighbouring amino acid residue spin systems in a polypeptide chain (**short range distance constraint analysis**). However, they are much more critical for the identification of spin systems that are in close spatial proximity to each other in the polypeptide chain, but separated in terms of sequence (**long range distance constraint analysis**). NOESY experiments rely heavily on extensive unambiguous assignments of as many ^1H and other heteroatomic nuclei in a protein as possible. Therefore (1) and (2) must be completed as much as possible before (3). Again, for small proteins (<50 amino acid residues), 2D NOESY experiments should suffice.

4. Various experiments can be performed to analyse for coupling constants, in particular $^3\boldsymbol{J}$ ^1H-^1H homonuclear and $^3\boldsymbol{J}$ ^1H-^{15}N heteronuclear constants that establish amino acid residue spin system and backbone conformations (**angular constraint analysis**). These experiments provide supplemental information to experiments described under (3) and contribute towards accurate structure determination at the computational stage (see Section 5.5.1.4).

5.5.1.1 *Three-dimensional COSY and TOCSY experiments of proteins*

Multidimensional correlation experiments can be arduous to analyse at the best of times. Therefore, any practitioner of biomolecular NMR spectroscopy will want to do the minimum number of experiments to achieve unique and unambiguous resonance assignment of as many amino acid residue ^1H-nuclei as necessary in order to enable the critical NOESY experiments. Two of the simplest 3D correlation experiments that have been used are **3D ^1H-^{15}N TOCSY-HSQC** and **3D *HCCH*-COSY/TOCSY**. Such 3D correlation experiments are known as **double resonance** experiments in that they generate intensity data $I_{NMR}(\boldsymbol{F}_1, \boldsymbol{F}_2, \boldsymbol{F}_3)$ emanating from the double resonance of two entirely different populations of nuclei, either ^1H- and ^{15}N-nuclei, or ^1H- and $^{13}C_\alpha$-nuclei respectively.

The workings of the 3D ^1H-^{15}N TOCSY-HSQC experiment are illustrated diagrammatically in Figure 5.20. The pulse sequence is designed to develop transverse magnetisation in aliphatic ^1H-nuclei (specifically C_α-^1H atoms) (source nuclei) ($\boldsymbol{F}_1$) that evolves according to chemical shift. Spin-spin coupling then enables coherence transfer involving both intra- and inter-amino acid residue $^1H^{15}N$-nuclei. **Heteronuclear multiple and single quantum coherence (HMQC/HSQC)** is generated, involving ^{15}N-nuclei (**correlating nuclei**) ($\boldsymbol{F}_2$) that is subsequently converted back into transverse magnetisation involving $^1H^{15}N$-nuclei. $^1H^{15}N$-magnetisation becomes modulated by the chemical shift of the directly attached ^{15}N-nuclei and then final coherence is observed, involving $^1H^{15}N$-nuclei (destination nuclei) ($\boldsymbol{F}_3$). Spectral data is usually displayed in the form of individual ^1H-^1H frequency ($\boldsymbol{F}_1$ and $\boldsymbol{F}_3$) contour maps resolved at different ^{15}N frequencies ($\boldsymbol{F}_2$) (all in ppm) (Figure 5.21). Vertically correlated cross peaks allow for the assignment of neighbouring amino acid residue C_α-^1H (*i, i*-1) and $^1H^{15}N$ (*i*) resonance signals, as a function of ^{15}N-chemical shift.

By comparison, 3D *HCCH*-COSY experiments are a little simpler (Figure 5.22). An initial pulse develops transverse magnetisation in aliphatic ^1H-nuclei (specifically $^{13}C_\alpha$-^1H atoms) (source nuclei) ($\boldsymbol{F}_1$) that evolves according to chemical shift before polarisation transfer is enabled to $^1\boldsymbol{J}_{CH}$-coupled ^{13}C-nuclei. Transverse magnetisation is developed involving these directly $^1\boldsymbol{J}_{CH}$-coupled ^{13}C-nuclei and spin-spin $^1\boldsymbol{J}_{CC}$-coupling then enables coherence transfer between adjacent ^{13}C-nuclei for the development of further transverse magnetisation involving ^{13}C-nuclei (all correlating nuclei) ($\boldsymbol{F}_2$). Finally, reverse polarisation transfer is enabled to aliphatic ^1H-nuclei (especially $^1H^{13}C_\beta$ atoms) and final coherence is observed involving $^1H^{13}C$-nuclei (destination nuclei) ($\boldsymbol{F}_3$). The 3D *HCCH*-TOCSY experiment is very similar, but pulse sequences are adapted to allow for more extensive magnetisation transfer between $^1\boldsymbol{J}_{CC}$-coupled ^{13}C-nuclei. In both cases, spectral data is displayed in the form of individual ^1H-^1H frequency ($\boldsymbol{F}_1$ and $\boldsymbol{F}_3$) contour maps resolved at different ^{13}C frequencies ($\boldsymbol{F}_2$) (all in ppm) (Figure 5.23). Horizontally correlated cross peaks allow for the ^1H-assignment of amino acid residue spin systems as a function of ^{13}C-chemical shift.

5.5.1.2 *Three-dimensional HNCA experiment of proteins*

The HNCA experiment has been described in detail already and need not be discussed much further here (see Section 5.3.2), except to note that the HNCA experiment is a **triple resonance** experiment in that it generates intensity data $I_{NMR}(\boldsymbol{F}_1, \boldsymbol{F}_2, \boldsymbol{F}_3)$ emanating from the triple resonance of three entirely different populations of ^1H-, $^{13}C_\alpha$- and ^{15}N-nuclei. The workings of the 3D HNCA experiment are illustrated diagrammatically in Figure 5.24. Frequently, HNCA spectral data is displayed in the form of individual ^{13}C-, ^1H-frequency ($\boldsymbol{F}_2$ and $\boldsymbol{F}_3$) contour maps resolved at different ^{15}N-frequencies ($\boldsymbol{F}_1$) (all in ppm). However, data could equally well be plotted in the form of ^{15}N-, ^1H-frequency ($\boldsymbol{F}_1$ and $\boldsymbol{F}_3$) contour maps if this were more helpful for assignment purposes. Usually the HNCA experiment is combined with other 3D correlation experiments with names such as **HNCO, HCACO** and **HCA(CO)N** experiments. These names

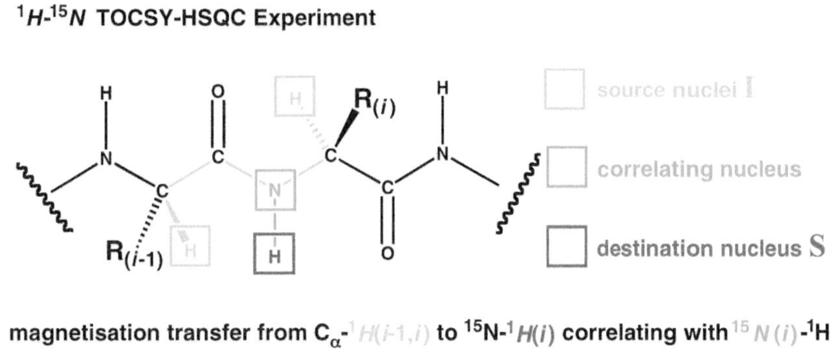

^1H-^{15}N TOCSY-HSQC Experiment

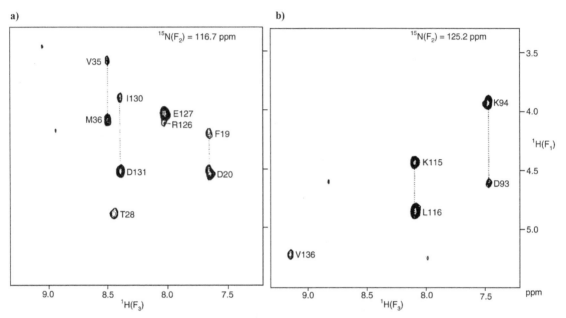

magnetisation transfer from C_α-$^1H(i$-$1,i)$ to ^{15}N-$^1H(i)$ correlating with $^{15}N(i)$-^1H

Figure 5.20   Diagrammatic summary of 3D ^1H-^{15}N TOCSY-HSQC double resonance experiment. Colour notation is identical with Figure 5.14.

Figure 5.21   Contour map outputs of 3D NMR experiments. Two 2D ^1H-^1H contour maps $[I_{NMR}(F_1, F_3)]$ obtained by a ^1H-^{15}N TOCSY-HSQC double resonance experiment involving the small globular protein **calmodulin**, where each plot is resolved at different ^{15}N-chemical shifts (F_2). Peaks indicate intra-residue correlations between ^1H-resonance signals at the given ^{15}N-chemical shift (F_2). Peaks aligned along the same verticals also suggest inter-residue correlations at the givet ^{15}N-chemical shift (F_2). Hence ^1H-^{15}N TOCSY-HSQC experiments can provide unique and unambiguous resonance assignments plus **vertical** correlations between pairs of source C_α ^1H nuclei (i-1 and i-residues) (F_1) and neighbouring destination nuclei $^{15}N^1H$ (i-residue) (F_3) (Kay et al., 1991, Figure 2, Elsevier).

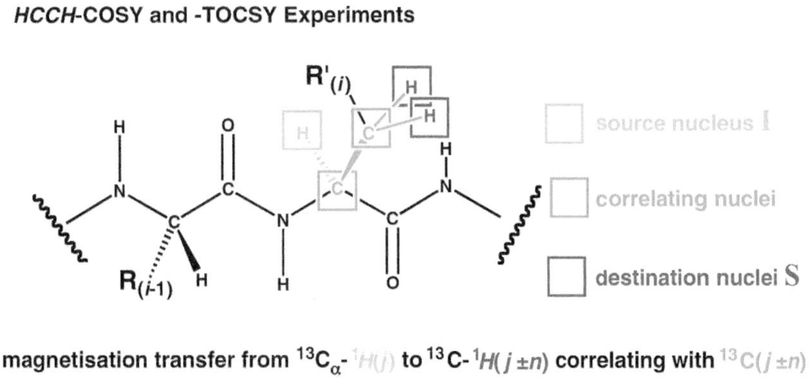

HCCH-COSY and -TOCSY Experiments

magnetisation transfer from $^{13}C_\alpha$-$^1H(j)$ to ^{13}C-$^1H(j\pm n)$ correlating with $^{13}C(j\pm n)$

Figure 5.22   Diagrammatic summary of 3D *HCCH*-COSY and TOCSY double resonance experiments. Colour notation is identical with Figure 5.14.

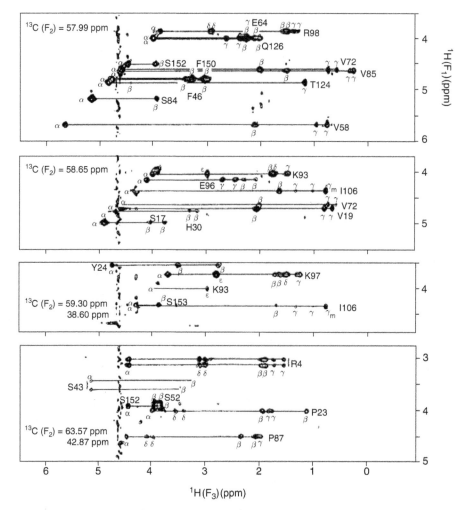

Figure 5.23 Stacked contour map outputs from 3D NMR experiments. Four 2D ^1H-^1H contour maps $[I_{NMR}(F_1, F_3)]$ obtained by a *HCCH*-TOCSY double resonance experiment involving the globular protein interleukin-1β, where each plot is resolved at different ^{13}C-chemical shifts (F_2). Peaks indicate intra-residue correlations between ^1H-resonance signals at the given ^{13}C-chemical shift (F_2). Peaks aligned along the same horizontals also suggest intra-residue side-chain correlations at the given ^{13}C-chemical shift (F_2). Hence *HCCH*-TOCSY experiments provide unique and unambiguous resonance assignments plus **horizontal correlations** between $C_\alpha^{-1}H(j)$ source nuclei (F_1) and neighbouring intra-residue $C_\beta^{-1}H(j + 1)$, $C\gamma^{-1}H(j + 2)$, $C_\delta^{-1}H(j + 3)$ up to $C_\varepsilon^{-1}H(j + 4)$ destination nuclei (F_3) (Clore *et al.*, 1990, Figure 7, American Chemical Society).

are deliberately descriptive and should suggest the correlations that these experiments are intended to establish by analogy with the HNCA experiment described here. All these correlation experiments acting in concert are designed to ensure optimal and mostly complete, unique and unambiguous assignments of ^1H-resonance peaks to ^1H-nuclei in given protein NMR spectra. This level of information is usually sufficient provided that there are enough assignments to allow for adequate NOE experiments to be conducted.

5.5.1.3 Three-dimensional and 4D NOESY experiments of proteins

Both 3D and 4D NOESY experiments are the focal objective of protein NMR spectroscopy and play a key role in knitting ^1H-assignment data into a form suitable for calculating protein 3D structure. Protein structures show a wealth of short range NOEs between nearest neighbour amino acid residues in a polypeptide chain. These short range NOEs can be weak, medium or strong and are highly diagnostic of secondary structure localisation (Figure 5.25). These allow for the determination of the likely secondary structure localisation of each amino acid residue in a protein. This is obviously an essential prerequisite prior to generating a 3D protein structure from NMR spectral data. Peptide link NH groups are critical in generating all these structure-determining short range NOEs. However, the real bases of 3D protein structure determination are long range NOEs. In this respect, 3D and 4D NOESY experiments are critical.

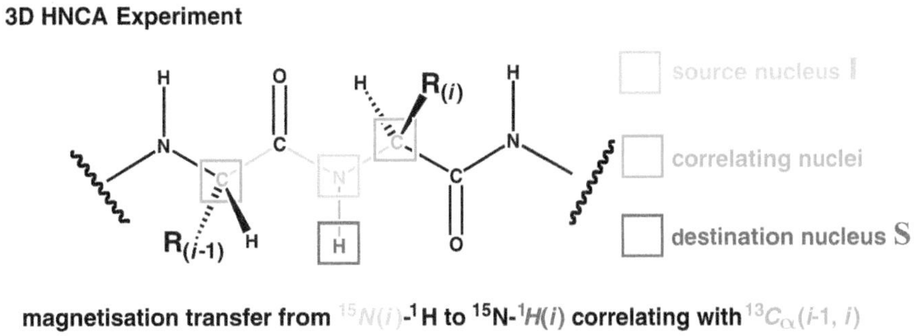

3D HNCA Experiment

magnetisation transfer from $^{15}N(i)$-^{1}H to ^{15}N-$^1H(i)$ correlating with $^{13}C_\alpha (i\text{-}1, i)$

Figure 5.24 **Diagrammatic summary of 3D HNCA triple resonance experiment.** Colour notation is identical with Figure 5.14. This triple resonance experiment is usually used to establish intra-residue correlations between $^{13}C_\alpha$, $^{15}N^1$H and $^{15}N^1H$ resonance signals in conjunction with ^1H-^{15}N TOCSY-HSQC experiments. The experiment generates unique and unambiguous resonance assignments, plus correlations between source $^{15}N^1$H nuclei (*i*-residues) (**F_1**) and neighbouring correlating $^{13}C_\alpha$ nuclei (*i*-1 and *i*-residues) (**F_2**) or destination $^{15}N^1H$ nuclei (*i*-residues) (**F_3**). For instance if **F_2** and **F_3** data are plotted, the output looks like Figure 5.21, with **vertical correlations** between pairs of correlating $^{13}C_\alpha$ nuclei (*i*-1 and *i*-residues) (**F_2**) and neighbouring destination $^{15}N^1H$ nuclei (*i*-residues) (**F_3**).

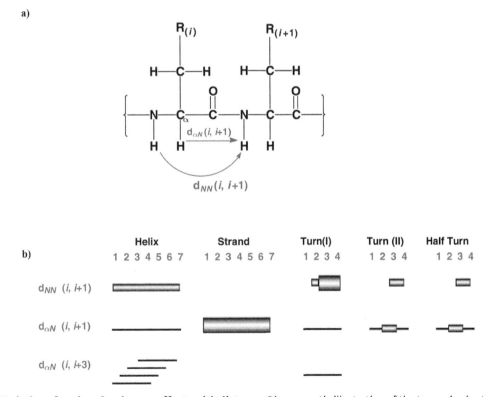

Figure 5.25 **Variation of nuclear Overhauser effects with distance.** Diagrammatic illustration of the two main short range distances, over which NOE distances may be seen in an organised polypeptide chain. (**a**) Short range NOEs involving N*H* and C$_\alpha$*H* protons are classically indicative of secondary structures as illustrated by the different patterns of short range NOEs that are developed by various different secondary structure elements; (**b**) NOEs are classified as strong, medium or weak as reflected by the thickness of each band shown.

The **3D HMQC-NOESY-HMQC (HSQC)** experiment together, with **4D $^{13}C/^{15}N$-edited NOESY** and $^{13}C/^{13}C$-edited **NOESY** experiments, are primarily used to identify long range NOEs between amino acid residues. The HMQC-NOESY-HMQC (HSQC) experiment operates in the following way. Pulses develop initial transverse magnetisation involving ^{15}N-nuclei (source nuclei) (**F_1**) followed by the development of heteronuclear multiple and single quantum coherence (HMQC/HSQC) involving $^1H^{15}N$-nuclei. $^1H^{15}N$-magnetisation becomes modulated by the chemical shift of the directly attached ^{15}N-nuclei. During the subsequent NOESY mixing period, τ_m, $^1H^{15}N$-magnetisation is transferred by cross-relaxation to immediate spatial $^1H^{15}N$-neighbours and becomes converted into further HMQC/HSQC

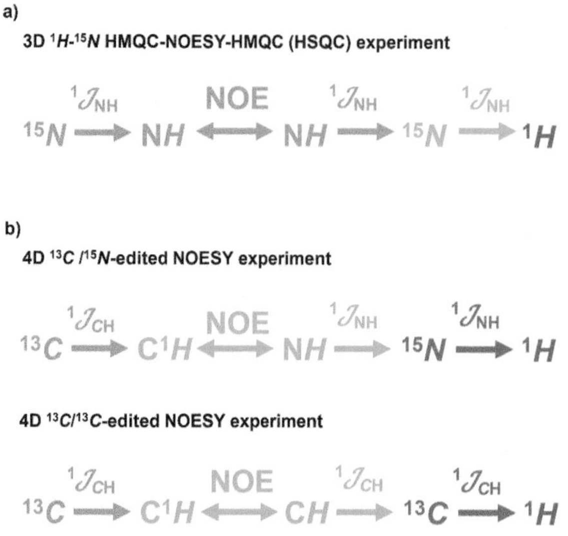

Figure 5.26 **Diagrammatic summary of multidimensional NOESY experiments.** (a) 3D ^1H-^{15}N **HMQC-NOESY-HMQC (HSQC) experiment**; (b) 4D ^{13}C/^{15}N-edited and ^{13}C/^{13}C-edited NOESY experiments. Colour notation is in line with Figure 5.17.

involving $^1H^{15}N$-nuclei (correlating nuclei) (F_2). Final coherence is observed, involving $^1H^{15}N$-nuclei (destination nuclei) (F_3) (Figure 5.26). Spectral data may be presented in a number of ways. For instance, in the form of individual ^{15}N-^{15}N frequency (F_1 and F_2) contour maps resolved according to different ^1H-frequencies (F_3) (Figure 5.27). Each cross peak identifies with impeccable resolution both short or long range NOEs from spatially close peptide link $^1H^{15}N$-nuclei and $^1H^{15}N$-nuclei that have unique and unambiguous resonance assignments as a result of extensive assignment experiments performed in advance (see Sections 5.5.1.1 and 5.5.1.2).

4D ^{13}C/^{15}N-edited NOESY and ^{13}C/^{13}C-edited NOESY experiments are essentially the same, except for one main difference. In the case of the 4D ^{13}C/^{15}N-edited NOESY experiment pulses develop initial transverse magnetisation involving ^{13}C-nuclei (source nuclei) (F_1) followed by the development of HMQC/HSQC involving $^1H^{13}C$-nuclei (correlating nuclei) (F_2). Thereafter there is subsequent NOESY mixing period, τ_m, $^1H^{13}C$-magnetisation is transferred by cross-relaxation to immediate spatial $^1H^{15}N$-neighbours, then develops into further HMQC/HSQC involving $^1H^{15}N$-nuclei (correlating nuclei) (F_3). Final coherence is observed involving $^1H^{15}N$-nuclei (destination nuclei) (F_4) (Figure 5.26). The ^{13}C/^{13}C-edited NOESY experiment is broadly similar to the ^{13}C/^{15}N-edited NOESY experiment except that magnetisation transfer under the influence of cross-relaxation occurs to nearest neighbour $^1H^{13}C$-nuclei instead of $^1H^{15}N$-nuclei, so that ^{13}C-nuclei become second correlating nuclei (F_3) and $^1H^{13}C$-nuclei destination nuclei (F_4) (Figure 5.26).

5.5.1.4 Energy minimisations

The more constraints that are available from NMR data, the more accurate the resulting energy minimised structure will be. Both short and long range NOE data are interpolated by energy minimisation calculations as through-space constraints in order to derive a final protein structure. The other main types of constraints in use are angular constraints, deriving from ^3J ^1H-^1H homonuclear and ^3J ^1H-^{15}N heteronuclear constant data, and distance constraints that derive from chemical shift data which demonstrates the existence of hydrogen bonds and residual dipolar coupling. Energy minimisation is beyond the scope of this book (at this stage at least) so will not be described further, except to say that the process can be imprecise and the number of long range distance constraints heavily influences the accuracy of the structural output.

In consideration of this, NMR energy minimised structures are usually presented *in toto* in the form of an **ensemble** or **family** of different minimised structures that derive from the same basis set of short and long range NOE data. The average of this family usually compares very closely with the protein structure when available, as determined by X-ray crystallography (Figure 5.28). Any major differences between derived structures from the two techniques are generally the result of dynamic regions within a protein structure that are typically "frozen" in place within a crystal environment.

5.5.1.5 Techniques for overcoming the molecular weight limit

From Equation (5.16) it should be obvious that there comes a point when a protein becomes sufficiently large that corresponding T_2 values become sufficiently small for the resulting linewidth to become too broad for any structural

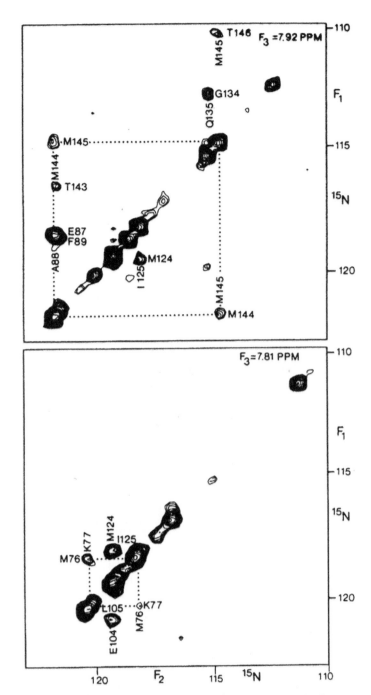

Figure 5.27 **Contour map outputs of multidimensional NOESY experiments.** Two 2D ^{15}N-^{15}N contour maps $[I_{NMR}(F_1, F_2)]$ obtained by a ^1H-^{15}N HMQC-NOESY-HMQC (HSQC) double resonance experiment involving the small globular protein calmodulin, where each plot is resolved at different ^1H-chemical shifts (F_3). Off-diagonal peaks indicate inter-residue through-space correlations between ^{15}N-resonance signals at the given ^1H-chemical shift (F_3). Hence ^1H-^{15}N HMQC-NOESY-HMQC (HSQC) double resonance experiments can provide unique and unambiguous indications of inter-residue proximities by through-space NOE correlations between pairs of $^{15}N^1H$ resonance signals. The appearance of symmetrical peaks (see triangulations above) occurs when NOEs are observed between NH systems of amino acid residues with essentially identical chemical shifts to each other (Ikura *et al.*, 1990, Figure 2, American Chemical Society).

determination to be possible. No matter how high a magnetic field is used, or how many dimensions of experiments are employed, the broad peaks will simply not resolve. Typically, the rule of thumb is that the upper molecular weight limit for protein structure determination is 30 kDa, above which T_2 values are too small for peak-to-peak resolution owing to excessive spectral line broadening. This molecular weight limit is low and for a long time has been seen to be a major blockage to progress in the biomolecular NMR characterisation of protein structure.

a)

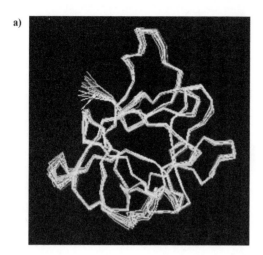

b)

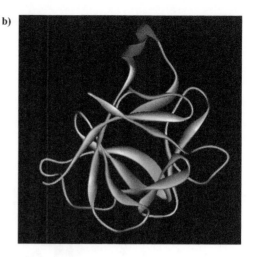

c)

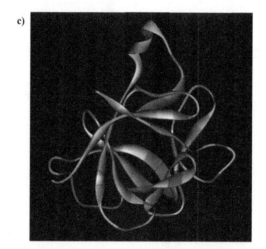

Figure 5.28   Nuclear Overhauser effect distances and protein structure determination. These are used to determine 3D structure of a protein by energy minimisation. (**a**) Gives a representation of an **ensemble** (family) **of structures** for Interleukin 1β produced by energy minimisation with all available NOE distance constraints. The average structure is shown in the form of a **ribbon display structure (b)** (pdb: **7i1b**) which agrees very well with the X-ray crystal structure of the protein. (**c**) The X-ray structure (pdb: **1i1b**) is also illustrated in the form of a **ribbon display structure** but with every different amino acid residue given in a different colour.

However, there is now hope for the future. The partial or full replacement of ^1H- with ^2H-nuclei has been found to result in both longer T_2 values and a reduction in coherence transfer involving $^{13}C_\alpha$ nuclei. Simpler, easier to interpret spectra, are the result. Furthermore, a new class of experiment has been invented, known as **transverse relaxation-optimised spectroscopy** (**TROSY**). TROSY experiments involve the suppression of transverse relaxation by interference in the processes of dipole-dipole coupling and spin-spin coupling through **chemical shift anisotropy** (**CSA**), hence artificially increasing T_2 values, thereby rendering previously undetectable peaks observable. Using these new techniques, structural data for even very large proteins (>100 kDa) is now coming within reach of the biological NMR spectroscopist.

5.5.2 Analysing nucleic acid structures

Nucleic acid NMR spectroscopy is less developed than that of proteins, but many synthetic oligodeoxynucleotides and oligonucleotides have been studied in solution as models of single DNA or RNA strands, hairpins, regular short duplexes, triplexes and even quadruplexes, not to mention structures with irregular bends, bulges or other distorted shapes (see Chapter 1). In all these cases, NMR characterisation is aided by the highly regular nature of nucleic secondary structure, and encouraged by the almost complete absence of X-ray crystal structure information involving such structural features.

Since there are only five main bases (including uracil) in nucleic acids, then the assignments of ^1H-resonance signals to ^1H-nuclei in individual deoxynucleotide or nucleotide residues of a DNA or RNA chain is generally easier to achieve than with proteins. Accordingly, nucleic acid NMR spectroscopy begins with just occasional 2D TOCSY experiments for the unambiguous assignment of ^1H-resonance signals in residues, followed by 2D NOESY experiments in order to detect a small number of key intra- and inter-residue ^1H-^1H distances (<4.5Å) that dominate nucleic acid structure (Figure 5.29). Residue connectivities, relative conformations and even 3D DNA or RNA structures can then be determined directly with ease by energy minimisation using this limited basis set of short range NOE data (see Section 5.5.1.4). If appropriate, 2D NOESY experiments may also be supplemented by ^{31}P-edited 3D experiments where oligodeoxy-/oligonucleotides >25bp are involved, in order to supplement this limited basis set with additional NOE distance constraints. Labelling with ^{15}N-, ^{13}C- and ^2H-nuclei may allow for further multidimensional experiments, which could theoretically allow the size limit to rise to around 120bp.

5.5.3 Analysing carbohydrate structures

Carbohydrate NMR spectroscopy is much less developed even than nucleic acid NMR spectroscopy. However, assignments of aliphatic ^1H-resonance signals in glycosidic residues are possible using routine 2D TOCSY and COSY experiments. Hydroxyl ^1H-nuclei are then assigned by 1D or 2D NOESY experiments. Finally, glycosidic residue connectivities and relative conformations may be identified, making use of 2D NOESY experiments to detect a select number of intra- and inter-residue ^1H-^1H distances emanating from glycosidic linkages which form a selected basis set for structural determination by energy minimisation (Figure 5.30). A primary problem with oligosaccharide NMR spectroscopy is the extensive flexibility and conformational irregularities found in glycosidic linkages. The other problem is that glycosidic residues show only a very narrow chemical shift anisotropy (dispersion) owing to their closely similar chemical structures. Accordingly, carbohydrate NMR spectroscopy requires the use of very high field NMR spectrometers (>600 Mhz) so that there are sufficient numbers of data points (in Hz per p.p.m.) (see Equation. 5.11) to resolve glycosidic residue resonance signals adequately in the critical spectral window (δ_H 2–5 ppm).

5.5.4 Analysing lipid assembly structures

Macromolecular lipid assemblies represent a completely different challenge for NMR spectroscopy. In fact, there is a very strong requirement for either 1D ^2H-NMR or ^{31}P-NMR spectroscopy. However, since phospholipids dominate biological macromolecular assemblies, then ^{31}P-NMR spectroscopy is especially ideal given that the spin quantum number I of ^{31}P-nuclei is ½, and the ^{31}P-natural abundance is essentially 100%. Furthermore, the chemical shift anisotropy (dispersion) is relatively large (see Figure 5.7) and ^1J ^1H-^{31}P coupling constants are large enough to diagnose structural changes within lipid assemblies (Table 5.2). Different macromolecular lipid assemblies exhibit significant differences in the motions of phospholipids that are mirrored by measurable differences in ^{31}P-NMR spectra. This is especially true for the critical $\mathbf{L}_{\alpha I}$ to $\mathbf{H}_{II}$ transition that is characteristic of membrane fusion events.

5.6 Electron paramagnetic resonance spectroscopy: key principles

Electron paramagnetic resonance (EPR) spectroscopy is really in a very primitive state compared with NMR spectroscopy even though the bedrock principles are in fact very similar. Therefore, biological EPR spectroscopy is much less developed and much less important than biological NMR spectroscopy. Nevertheless, EPR spectroscopy is included in order to complete this chapter. In the same way as each nucleus of an atom possesses the property of spin, so, too, does each electron. As a result, each electron behaves as if it were a precessing, rotating, miniature bar magnet with **angular momentum**, J_e, also represented in the form of a vector whose length reflects magnitude and whose orientation reflects direction of spin. Magnitude is quantised with allowed magnitude defined by an **electron spin quantum number, I_e**, as represented by Equation (5.18):

$$J_e = \left[I_e(I_e+1)\right]^{1/2}\hbar \qquad (5.18)$$

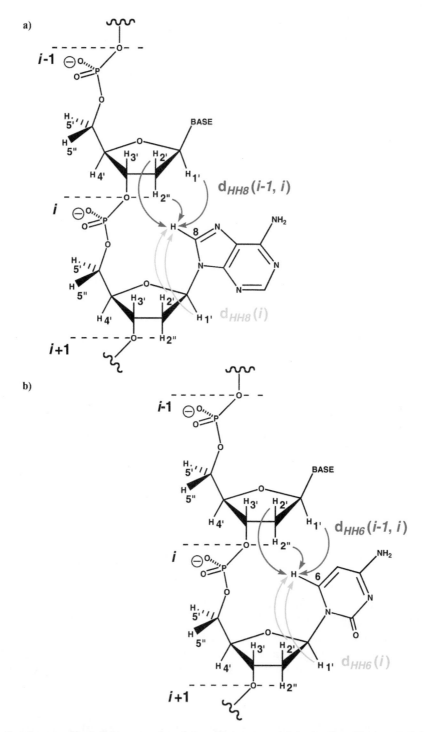

Figure 5.29 Nuclear Overhauser effect distances and nucleic acid structure determination. Diagrammatic illustration of the main short range distances over which NOE distances may be used to determine structure by NMR in an organised oligodeoxynucleotide (or oligonucleotide) single chain or duplex. Note that there are three main types of d_{HH8} (*i-1, i*) (purine) (**a**) or d_{HH6} (*i-1, i*) (pyrimidine) (**b**) distances and two main types of d_{HH8} (*i*) (purine) (**a**) or d_{HH6} (*i*) (pyrimidine) (**b**) distances.

The electron spin quantum number also defines the number of allowed **spin states** that may exist. This number is equivalent to $2I_e + 1$. In the case of the electron, I_e is ½; therefore there are only two allowed spin states that are initially degenerate. Since spin angular momentum is quantised in terms of magnitude and orientation according to the electron spin quantum number, I_e, then z-axis components of angular momentum J_z are similarly quantised. Each allowed z-axis component is represented by an individual **magnetic spin quantum number**, m_s, according to Equation (5.19):

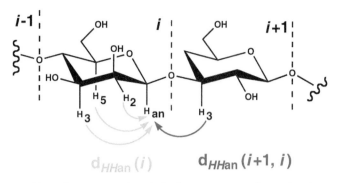

Figure 5.30 Nuclear Overhauser effect distances and carbohydrate structure determination. Diagrammatic illustration of the main short range distances over which NOE distances may be seen in an organised oligosaccharide chain. Note that there are three main types of d_{HHan} (*i*) and one main type of d_{HHan} (*i* + 1, *i*) distance.

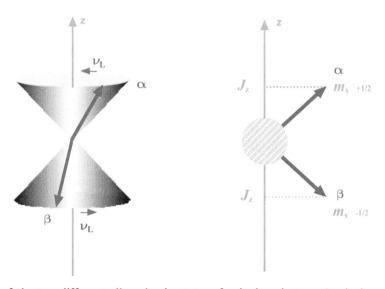

Figure 5.31 Depictions of the two different allowed spin states of spinning electron. A spinning electron (I_e 1/2) precesses relative to a reference z-axis supplied by an externally applied magnetic field such that angular momentum vectors (and precessional cones) align with (m_s **+ 1/2;** α–**state**) or against (m_s **-1/2;** β-**state**) the field direction. Degeneracy between the two allowed spin states is lifted by interaction between intrinsic magnetic fields of these spin states and the external field. The β-state is more stable than the α–state for a spinning electron since γ_e is negative.

$$J_z = m_s \hbar \tag{5.19}$$

In order to satisfy the requirement that the number of allowed spin states be equivalent to $2I_e + 1$, allowed values of m_s can only be +½ and −½. As with nuclei, when an external magnetic field is applied, the two electron spin states all orientate with respect to the field (z-axis) either with the field direction (m_s + ½) or against the field direction (m_s −½) (Figure 5.31). In the process, spin state degeneracy is lifted. By convention, the first state is known as the α-**state** and the second as the β-**state**.

The inherent magnetic field strength associated with any given electron spin state is represented by an **electron magnetic moment**, μ_z^e, which is proportional to J_z according to the Equation (5.20):

$$\mu_z^e = \gamma_e \cdot m_s \hbar \tag{5.20}$$

where γ_e is known as the **electron gyromagnetic ratio**, which in turn is defined by Equation (5.21):

$$\gamma_e = -\frac{g_e \mu_B}{\hbar} \tag{5.21}$$

in which g_e is known as the **g-factor** and μ_B the **Bohr magneton**. The electron gyromagnetic ratio differs from the nuclear equivalent (Equation 5.4) on the basis that nucleus and electron are oppositely charged and hence the electron bar magnet points in the reverse direction to that of a nucleus, thereby reversing the direction of the electron magnetic moment with respect to nuclear magnetic moment (see Figure 5.2). The Bohr magneton is further defined by Equation (5.22):

$$\mu_B = \frac{e \cdot \hbar}{2m_e} \tag{5.22}$$

where m_e is the **mass of an electron** and e is the **electron charge**. Substitution of Equation (5.21) into (5.20) gives Equation (5.23):

$$\mu_z^e = -m_s \cdot g_e \mu_B \tag{5.23}$$

5.6.1 Quantum mechanical description of EPR

The energy differences between different spin states are created by the differential way in which the magnetic moments of given spin states interact with the externally applied magnetic field of strength B_z. The **interaction energy attributable to either spin state, E_{m_s},** is defined by the product given in Equation (5.24):

$$E_{m_s} = -\mu_z^e \cdot B_z \tag{5.24}$$

In comparison with most nuclei, the g-factor is positive but the ratio γ_e is negative. Hence the magnetic moment μ_z^e is positive for the β-state and negative for the α-state according to Equation (5.23). Consequently, the β-state becomes more stable than the α-state by interaction with the externally applied magnetic field, according to Equation (5.24). The energy difference, ΔE, between the α- and β-states is given by Equation (5.25):

$$\Delta E = E_{1/2}^\alpha - E_{-1/2}^\beta = g_e \mu_B B_z \tag{5.25}$$

Clearly this situation (Figure 5.31) is opposite to the situation found with the ^1H-nucleus for instance. Hence the resonance condition for EPR spectroscopy is given by Equation (5.26):

$$\Delta E = h\nu_L = g_e \mu_B B_z \tag{5.26}$$

where the Larmor frequency, ν_L, now lies in the microwave frequency range, typically 10 GHz (wavelength 3 cm) at a magnetic field of approximately 0.3 T.

5.6.2 g-value

In contrast with NMR spectroscopy, EPR spectroscopy is limited to systems in which there is an unpaired electron such as organic radicals or transition metals. Otherwise, the technique is spectroscopically silent. Typically, in an NMR experiment, magnetic field is kept constant and frequency is varied. In EPR spectroscopy, frequency is kept constant and field is varied instead. Consequently, differences in spin state energies and hence resonance condition, due to electron position and environment, cannot be diagnosed by NMR style chemical shift since that is a frequency-based concept. Instead, an alternative concept needs to be originated which is the concept of the **g-value**.

Variations in g-values arise because the strength of the effective magnetic field experienced by any electron in a molecular structure appears to vary in response to local movements in neighbouring electron density. In other words, local electronic effects have direct and reproducible effects on ν_L values. Electronic effects are both shielding and deshielding in character, as with NMR. The effective magnetic field experienced by any such an electron is then modulated according to expression 5.27:

$$B_{eff,z} = B_z(1 - \sigma_e) \tag{5.27}$$

where $B_{eff,z}$ is the effective magnetic field strength experienced by the electron and shielding is characterised by the **electron shielding parameter**, σ_e. However, since B_z is undergoing a continuous field sweep in EPR spectroscopy, when Equation (5.27) is combined with (5.26) to give the resonance condition revised to take account of shielding and deshielding, the result is Equation (5.28):

$$\Delta E = h\nu_L = g_e(1 - \sigma_e)\mu_B B_z \qquad (5.28)$$

where the term $g_e(1 - \sigma_e)$ is known as the g-value. This is very much the EPR equivalent of chemical shift.

5.6.3 Hyperfine splitting

Not only shielding and deshielding but also spin-spin coupling contributes to modulating and perturbing the energies of electron spin states. In the case of EPR, the dominant effect is electron-nucleus spin-spin coupling. This results in **hyperfine splitting** of the EPR signal of interest. There is a general rule that for each spin state of a given resonating electron, the number of spin-coupled microstates that result from spin-spin coupling with n identical neighbouring nuclei is 2^n. However, the number of resonance frequency lines that result from coupling with n identical neighbouring nuclei is only $2nI + 1$, where I is the spin quantum number of the neighbouring nuclei, owing to the selection rules which ensure that there can be no change in coupled nuclear spin state during an electron spin state transition. Hence, for instance, if a resonating electron is coupled with one neighbouring nucleus ($I = \frac{1}{2}$) then although there are four spin-coupled microstates in total, only two resonance frequency lines will result (doublet) (Figure 5.32). Similarly, if two identical neighbouring nuclei ($I = \frac{1}{2}$) are involved then three resonance frequency lines will result (triplet); if two identical neighbouring nuclei ($I = 1$) are involved then five resonance frequency lines will result (quintet); etc. In EPR,

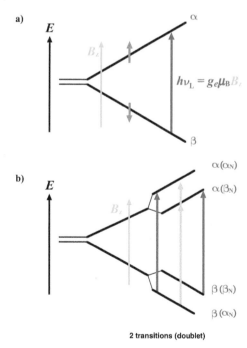

Figure 5.32 Lifting the degeneracy between the two allowed spin states. The degeneracy of the two spin states of the spinning electron (I_e 1/2) is lifted by application of an externally applied magnetic field of strength B_z. (a) Two spin states are illustrated by arrows (orange) aligned against (β-**state**) or with (α-**state**) the field direction. A vertical spin state transition is shown (red arrow) along with the resonance condition for inter-conversion of β- to α-state; (b) Formation of spin microstates in response to spin-spin coupling of electron with one neighbouring spinning nucleus (**N**) of I 1/2 (e.g. ^1H-nucleus). Two new spin microstates are formed from each main spin state. Nuclear spin must be unchanged during electron spin state transitions, hence two spin state transitions are allowed (red arrows) and two transitions disallowed (grey arrows). Note that microstates with anti-parallel spins are the more stable. Both diagrams have been drawn assuming constant frequency irradiation and a variable strength externally applied magnetic field. Note that the opposite is true in the case of NMR spectroscopy (Figure 5.8).

the difference between resonance frequency lines is measured in terms of the **hyperfine coupling constant** (in units of mT). The hyperfine-coupling constant is the EPR equivalent of the NMR coupling constant (in units of Hz). The differences in units reflect the fact that in EPR spectroscopy, frequency is kept constant and field is varied instead, whilst the reverse is true in a typical NMR experiment.

5.6.4 Biological macromolecule structural information by EPR

In spite of the close theoretical relationship between EPR and NMR spectroscopy, EPR has only very narrow applications. The primary reason for this is that the EPR phenomenon is spectroscopically silent unless there are unpaired electrons. Most biological macromolecules are closed shell molecules and contain no unpaired electrons. Therefore, EPR is of little real value for biological macromolecular structure characterisation. The only exception to this rule is that certain prosthetic groups in proteins may contain redox active metal centres/clusters that have transient or even permanent unpaired electrons (see Chapter 4). These metal centres/clusters can be studied by EPR spectroscopy in order to demonstrate the presence of unpaired electrons. Thereafter, EPR data may then be used to derive the relative structural arrangements of metals within centres or clusters and to assign putative distributions of redox states should there be any obvious redox heterogeneity. EPR is also useful to detect transient or even metastable radical formation during biocatalysis (see Chapter 8).

6
Diffraction and Microscopy

6.1 Diffraction and microscopy in chemical biology

This chapter is devoted to the contribution that diffraction and microscopy make in deriving and understanding the structures of biological macromolecules. Without doubt, **X-ray crystallography** is the pre-eminent technique for the structural characterisation of biological macromolecular structure and has been responsible for the derivation of more atomic-level structures of biological macromolecules than any other single technique put together. Hence, X-ray crystallography appears first in this chapter. Thereafter, the attention of the chemical biology reader will be drawn towards **electron microscopy (EM)** and then **scanning-probe microscopy**. Of these, the technique of **cryo-electron microscopy (cryo-EM)** has a realistic potential and capacity to rival X-ray crystallography as the pre-eminent technique for the characterisation of biological macromolecular structures. Scanning probe microscopy is much less developed, but shows an impressive diversity in the size and morphology of 3D biological macromolecular structures that may be studied under a variety of conditions.

This chapter is intended to provide an explanation of the main theories and ideas that underpin the use of diffraction and microscopy in chemical biology research, then to provide a bridge from these main theories and ideas to actual examples of successful 3D structure characterisation. In Chapter 5, we illustrated the concepts of multidimensional NMR spectroscopy with reference to data derived from the structural characterisation of one main protein, **interleukin-1β (IL-1β)**, a key biological messenger of inflammation (see also Chapter 7). In this chapter, illustrations for successful structural characterisations are taken from the fascinating world of heat shock/stress protein research (see also Chapter 7). There is nothing overstated in saying that much of our current understanding of the molecular workings of biology derive from successful structural characterisations made possible by diffraction and/or microscopy. Therefore, read on.

6.2 Key principles of X-ray diffraction

When an intense beam of X-rays is directed at a crystal, the result is a dispersal of the beam known as **X-ray diffraction**. If the crystal is sufficiently ordered, then a regular pattern of dispersal or scattering will be generated, known as a **diffraction pattern**. The task, now solved, of relating this diffraction pattern back to molecular structure within the crystal has been one of enormous complexity. However, nowadays provided biological macromolecules of interest are able to form ordered crystals that diffract X-rays in an ordered and reproducible manner then diffraction patterns of even the most fearsome complexity may be analysed to give accurate predictions of underlying molecular structure in the crystal environment. As a result, **X-ray crystal diffraction** or X-ray crystallography has become arguably the most

Essentials of Chemical Biology: Structures and Dynamics of Biological Macromolecules In Vitro *and* In Vivo, Second Edition. Andrew D. Miller and Julian A. Tanner.
© 2024 John Wiley & Sons, Inc. Published 2024 by John Wiley & Sons, Inc.
Companion Website: www.wiley.com/go/miller/essentialschembiol2

powerful technique available to the chemical biologist for the determination of biological macromolecular structure. Ever more complex structures are being determined, analysed and used to make substantive conclusions about molecular mechanisms in biology. For this reason, it is essential for the chemical biology reader to have a reasonable grasp of the basic principles involved in X-ray crystallography, if for nothing else than to appreciate the advantages and limitations of a technique that has been and remains so pervasive and influential in defining our understanding of biology at a molecular level.

6.2.1 Unit cell

A **crystal** can be described as a 3D periodic arrangement of molecules or a 3D stack of **unit cells** whose edges and vertices form a grid or **direct lattice**. The unit cell is known as the smallest possible unit which maps out the actual crystal structure if translated in any direction, so completely reproducing the 3D environment of electron density present in the crystal structure. The typical unit cell is defined in terms of three dimensions: *a*, *b* and *c*, together with three interdimensional angles: α, β and γ (Figure 6.1). These dimensions and angles are important unit cell characteristics that define the seven main **crystal systems** known as **triclinic, monoclinic, orthorhombic, rhombohedral, tetragonal, hexagonal** and **cubic**. The unit cell should be thought of as the minimum unit that encapsulates all the main features of the crystal. Accordingly, an X-ray diffraction pattern can be thought of as the result of crystal scattering from a single unit cell. In other words, an X-ray diffraction pattern correlates directly with unit cell electron-density. Just how this takes place will be explained.

6.2.2 Bragg's law

The earliest approach to understanding the origins of diffraction patterns was to regard a unit cell as being bisected by stacks of parallel **lattice planes** defined according to the **Miller indices**, integers: *h*, *k* and *l*. Integer *h* represents the number of parallel lattice planes of a given stack that bisect dimension *a* in moving from unit cell origin to the fullest extent of *a*; *k* the number of planes of a given stack that cut dimension *b* in moving from unit cell origin to the fullest extent of *b*; and *l* the number of planes of a given stack that cut dimension *c* in moving from unit cell origin to the fullest extent of *c*. Three different sets of *hkl*-lattice planes are shown in Figure 6.2 to illustrate this point. In classical diffraction theory, each lattice plane of a given stack behaves as a mirror that is capable of "reflecting" X-rays. Assuming that each *hkl*-lattice plane of a given stack is also a fixed distance apart, d_{hkl}, then wave reflection from this stack will be accompanied by **constructive interference** provided that Equation (6.1) is completely obeyed (Figure 6.3):

$$\lambda = 2d_{hkl} \cdot \sin\theta \tag{6.1}$$

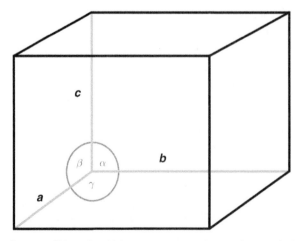

Figure 6.1 The unit cell. The smallest possible unit which maps out actual crystal, reproducing the complete 3D environment of electron density.

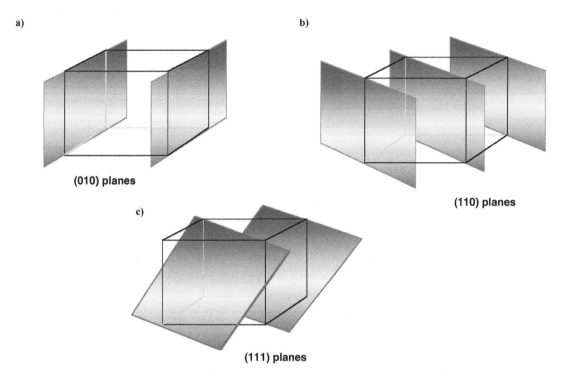

Figure 6.2 **Illustrations of *hkl*-planes**. Three illustrative sets of atomic planes are shown that bisect the unit cell and from which an incident X-ray beam is said to be reflected, giving rise to X-ray scattering or diffraction in 3D space. The **(010) planes** (**a**) are only bisected by movement in the ***b*** axis direction; the **(110) planes** (**b**) are bisected by movement in the ***a*** and ***b*** axis directions; **(111) planes** (**c**) are bisected by movement in all three axis directions.

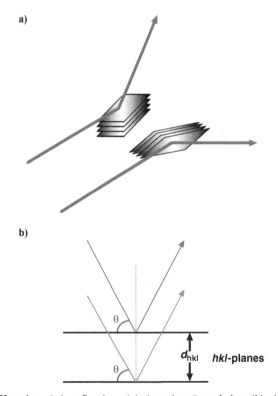

Figure 6.3 **Bragg's law of X-ray diffraction**. Only reflections (**a**) that obey **Bragg's law** (**b**) give rise to observed X-ray scattering or diffraction. Hence X-ray scattering is "quantised" in 3D space according to Bragg's law condition.

where θ is the **glancing angle** and λ the **wavelength of the incident X-ray beam**. Equation 6.1 is known as **Bragg's law**. If Equation 6.1 cannot be obeyed then **destructive interference** would result. Only reflections accompanied by constructive interference would be expected to be visible, giving rise to a regular pattern of X-ray dispersal in 3D space that satisfies Bragg's law. This is one of the great fundamental concepts of diffraction. Clearly, unless d_{hkl} distances are uniform then constructive interference would be impossible and no diffraction pattern would be observed. This underlines the importance of well-ordered crystals and hence unit cells to derive reproducible diffraction patterns of X-ray dispersal or scattering. Curiously, this classical concept has withstood the development of quantum theories and underpins much of our appreciation of X-ray crystal diffraction even today.

6.2.3 Reciprocal lattice

If an X-ray beam could be reflected from every possible stack of *hkl*-lattice planes that bisect the unit cell such that every reflection satisfied Bragg's law, then the total ordered pattern of X-ray scattering would represent the complete diffraction pattern. In reality, this is not possible to achieve without rotating the crystal in the path of the X-ray beam and/or using a range of X-ray wavelengths (see Section 6.3.3). A visual/mathematical depiction of complete Bragg's law reflection from stacks of *hkl*-lattice planes is the **reciprocal lattice**. The reciprocal lattice has imaginary dimensions *a**, *b** and *c** that are inversions of the *a*, *b* and *c* dimensions of the unit cell. Each reciprocal lattice point corresponds to an *hkl*-reflection that satisfies Bragg's law from a given stack of *hkl*-lattice planes that bisect the unit cell. In effect, a given reciprocal lattice represents a complete theoretical diffraction pattern.

The **Ewald sphere** was developed as a tool to construct the directions of X-ray scattering. However, it also serves as a means to determine how to sample the vast majority of reciprocal lattice points, or in other words sample the intensity of X-rays scattered by reflection from every possible stack of *hkl*-lattice planes. In order to construct the Ewald sphere, an incident beam of X-rays is considered as being directed towards the origin, **O**, of the reciprocal lattice. A sphere is then constructed that cuts the origin, **O**, on its surface but whose centre, **M**, marks the position of the crystal. The radius, **MO**, is $1/\lambda$, where λ is the wavelength of the incident X-rays (Figure 6.4). Only reciprocal lattice points that sit on the surface of the sphere correspond with X-ray reflections that may be detected given the orientation of the crystal with respect to the incident X-ray beam and the wavelength of the incident X-rays. If the orientation of the crystal changes, so does the relative orientation of the reciprocal lattice, allowing other reciprocal lattice points to sit on the surface of the sphere. There is a similar consequence if the X-ray wavelength is varied instead. Hence, the Ewald sphere shows how by varying these two

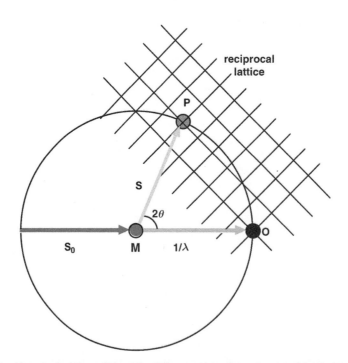

Figure 6.4 The Ewald sphere. The construction of the sphere demonstrates how the complete X-ray scattering pattern (all vertices of **reciprocal lattice**) can be visualised by adjusting: (a) The wavelength (λ) of X-ray beam, S_0, incident upon a crystal mounted at position **M**; (b) The orientation of the crystal relative to beam S_0. Each vertex of the reciprocal lattice corresponds with a different *hkl*-reflection. A given *hkl*-reflection may be visualised only when scattered beam, **S**, cuts the surface of the Ewald sphere at a position, **P**, coincident with a corresponding reciprocal lattice vertex. In principle, λ and crystal orientation at **M** may be adjusted to visualise the vast majority of vertices of a reciprocal lattice and hence the vast majority if not all of the *hkl*-reflections possible from a given mounted crystal (**Laue condition**).

parameters systematically the vast majority of reciprocal lattice points may be sampled and the intensity of the majority of scattered X-ray beams measured.

6.2.4 Structure factors

Another view of X-ray diffraction has grown out of Bragg's law and that is the consideration of the diffraction phenomenon in terms of vectors. This consideration has led to Equation (6.2), which encapsulates the entire relationship between unit cell electron-density and the X-ray diffraction pattern:

$$\rho(xyz) = \frac{1}{V} \sum_h \sum_k \sum_l F(hkl) \cdot \exp\left[-2\pi i (hx + ky + lz)\right] \tag{6.2}$$

where V is the unit cell volume, $\rho(xyz)$ is the **electron density distribution** and $F(hkl)$ is known as the **unit cell structure factor** corresponding to X-ray scattering from a given stack of parallel **hkl**-lattice planes. Each individual unit cell structure factor is a complex number that collectively represents the ability of any given unit cell to generate X-ray scattering from the complete stack of parallel **hkl**-lattice planes. More generally, the complete set of **$F(hkl)$** structure factors represents the ability of the electron density in any given unit cell to interact with and scatter glancing X-ray beams. According to Equation (6.2), the sum over all **$F(hkl)$** structure factors is related directly to $\rho(xyz)$ by Fourier transform. Hence, if sufficient **$F(hkl)$** structure factors can be solved from X-ray diffraction data, then $\rho(xyz)$ can be computed and the crystal structure of the biological macromolecule of interest solved. However, this is not as easy as it sounds.

Any complex number, **F**, can be represented in a complex plane diagram, as illustrated in Figure 6.5, and so takes on the appearance of a vector with the twin properties of magnitude and direction. Magnitude is defined by the length, $|F|$, and direction by the angle, α, subtended with the real axis. Magnitude $|F|$ is a real number and is known generally as the **modulus** of the complex number. Angle α is known as the **phase** of the complex number. All terms are related to each other according to Equation (6.3):

$$F = |F| \cdot \exp\left[i\alpha\right] \tag{6.3}$$

Equation (6.2) may now be expanded to the full form as represented by Equation (6.4):

$$\rho(xyz) = \frac{1}{V} \sum_h \sum_k \sum_l |F(hkl)| \cdot \exp\left[-2\pi i (hx + ky + lz) + i\alpha(hkl)\right] \tag{6.4}$$

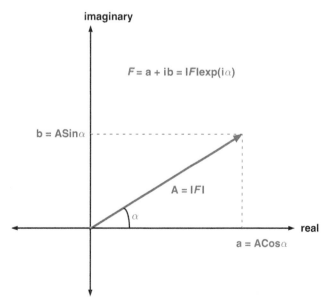

Figure 6.5 Complex plane diagram. An illustration of the two main forms of a complex number **F** and the rendering of a complex number on a complex plane diagram.

where $|F(hkl)|$ is the **modulus of the structure factor** corresponding to X-ray scattering from a given stack of parallel *hkl*-lattice planes and $\alpha(hkl)$ is the **phase of the structure factor** corresponding to X-ray scattering from that same stack of parallel *hkl*-lattice planes. Equation (6.4) can be reduced by complex number manipulation to give "real" expression 6.5:

$$\rho(xyz) = \frac{1}{V} \sum_h \sum_k \sum_l |F(hkl)| \cdot \cos\left[2\pi(hx + ky + lz) - \alpha(hkl)\right] \qquad (6.5)$$

6.2.5 The phase problem

An X-ray diffraction pattern can be defined (from Sections 6.2.2 and 6.2.3) as a regular pattern of X-ray dispersal in 3D space that satisfies Bragg's law (**Laue condition**), wherein each unique direction of scatter corresponds to "reflection" from one unique stack of parallel *hkl*-lattice planes that bisect the unit cell. The discussion in Section 6.2.3 illustrates the possibility of using the Ewald sphere construction to ensure the observation of the maximum number of reflections by varying a few simple parameters. The problem of relating an observed X-ray diffraction pattern to unit cell electron-density now rests with the relationship between actual diffraction data and those X-ray parameters and variables that comprise the right-hand side of expression 6.5. On the one hand, the relationship is surprisingly simple. The intensity of scatter, $I(hkl)$, from a given stack of parallel *hkl*-lattice planes, turns out to be equivalent to $|F(hkl)|^2$, the square of the modulus of the structure factor corresponding to X-ray scattering from that same stack of lattice planes. Fortunately, values of $I(hkl)$ can be measured experimentally with relative ease (see Section 6.3), allowing corresponding values of $|F(hkl)|$ to be determined with relative ease too. Unfortunately, no such easy correlation exists between X-ray scattering data and corresponding values of $\alpha(hkl)$ phases. Indeed, this lack of an easy correlation held up the determination of 3D structures of biological macromolecules by X-ray crystallography for several decades until this **phase problem** could be solved. Amazingly, the phase problem now appears to one of the greatest problems for physics and chemistry that was solved during the last century.

The idea of a vector triangle provided the eventual solution to the phase problem. Three associated vectors with appropriately linked properties should be able to form a triangle in vector space (Figure 6.6). Given the fact that each $F(hkl)$ structure factor, corresponding with a given set of hkl-lattice planes, is also a vector in the complex plane, the idea was advanced that solutions of $F(hkl)$ could be found provided that similar structure-factor-based vector triangle diagrams could be similarly constructed. But how? The most popular approach has been to use the experimental technique of **isomorphous replacement**. The experimental requirements for this will be discussed in Section 6.3, so here we only need define isomorphous replacement as the creation of "heavy atom" derivatives of the biological macromolecule crystal of interest. For the method of isomorphous replacement to work, these heavy atom crystals should be essentially identical with the original biological macromolecule crystal except for the inclusion of heavy atoms (e.g. mercury, Hg; arsenic, As) at reproducibly defined locations within the crystal unit cell.

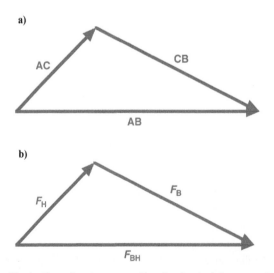

Figure 6.6   Vector triangle diagram. Illustration of vector summation in simple (**a**) or complex (**b**) planes.

The heavy atom crystal can generate an X-ray diffraction pattern by X-ray scattering in the same way as the original biological macromolecule crystal. However, this diffraction pattern is dominated by X-ray scattering from the included heavy atoms; heavy atoms are very intense scattering centres because they are also centres of high electron density. Intensity differences between the heavy atom crystal diffraction pattern and that of the original biological macromolecule crystal will be entirely due to the heavy atom scattering provided that the structure of the biological macromolecule and unit cell parameters are essentially unperturbed by the presence of heavy atoms. Hence, the difference between heavy atom and biological macromolecule crystal diffraction patterns represents a new heavy atom diffraction pattern. Accordingly, we can define three discrete sets of interlinked structure factors for these three interlinked diffraction patterns. If $F_B(hkl)$ represents the structure factor corresponding with each set of *hkl*-lattice planes in the biological macromolecule crystal, then $F_{BH}(hkl)$ represents the same for the heavy atom crystal and $F_H(hkl)$ the same for the heavy atoms alone. The vector triangle relationship between these structure factor terms is illustrated in Figure 6.6. It turns out that provided one of these vectors can be solved completely, then the other two may also be solved, including a complete determination of all the $\alpha_B(hkl)$ **phase angles** for the original biological macromolecule crystal.

In practice, only $F_H(hkl)$ can be solved completely, beginning with a **Patterson function**, *P(uvw),* which is almost identical in form to Equation (6.5), but for the absence of phase angles (angles are all set to zero) and the involvement of an $\left|F_H(hkl)\right|^2$ term, as shown in Equation (6.6):

$$P(uvw) = \frac{1}{V} \sum_h \sum_k \sum_l \left|F_H(hkl)\right|^2 \cdot \cos\left[2\pi(hu+kv+lw)\right] \qquad (6.6)$$

where *u, v* and *w* are **relative coordinates in the unit cell**. To all intents and purposes *u, v* and *w* coordinates should be regarded as being equivalent to the *x, y* and *z* coordinates that reference electron density distribution within the unit cell. The Patterson function transforms heavy atom diffraction pattern intensity data into a **Patterson map** that is used to identify the relative positions of all heavy atoms in the unit cell of a given heavy atom crystal. Relative *u, v* and *w* coordinates are then transposed into *x, y* and *z* coordinates, giving a full positional atomic assignment of heavy metal atoms in the unit cell of a heavy atom crystal. Patterson maps will only give positional assignments for very simple systems involving a limited number of atoms and certainly not for complex biological macromolecules. Therefore, there is almost no alternative to the heavy atom approach. However, armed with heavy atom *x, y* and *z* coordinates a complete solution to $F_H(hkl)$ may then be derived, including the $\alpha_H(hkl)$ phase angles. Exactly how this solution is obtained is beyond the scope of this chapter. However, it is enough to know that a complete solution to $F_H(hkl)$ is usually sufficient to solve $F_B(hkl)$ for the original biological macromolecule crystal using the vector triangle associations described in this section and the following vector construction technique.

6.2.6 Harker construction

After isomorphous replacement, the full complement of X-ray diffraction data should include a complete solution to $F_H(hkl)$ and values of $\left|F_{BH}(hkl)\right|$ and $\left|F_B(hkl)\right|$ from diffraction pattern intensity data. These data may then be analysed with a **Harker construction** (Figure 6.7). For measured diffraction involving each set of *hkl*-lattice planes in the biological macromolecule crystal, a circle of radius $\left|F_{BH}(hkl)\right|$ can be drawn, centred upon the origin of a complex plane diagram and then $F_H(hkl)$ is drawn from this origin. Next, a second circle of radius $\left|F_B(hkl)\right|$ is drawn, centred at the end of the $F_H(hkl)$ vector. The two intersection points of both circles mark the two possible orientations of $F_{BH}(hkl)$ and $F_B(hkl)$ vectors which correctly satisfy the vector triangle relationship that should exist between all three vectors (Figure 6.6). These two possible orientations simultaneously identify two possible but equally probable $\alpha_B(hkl)$ phase angles, both of which correctly allow the correct vector triangle relationship. With a second heavy atom derivative, the two alternatives may be distinguished easily, although poor isomorphism between biological macromolecule and heavy atom crystal may mean that additional heavy atom crystals should be prepared for X-ray diffraction studies in order to remove any chance of ambiguity in phase angle assignments. For this reason, isomorphous replacement may become **multiple isomorphous replacement** (MIR), occasioning the generation and analysis of multiple "heavy atom" derivatives of the biological macromolecule crystal of interest. Once $\alpha_B(hkl)$ phase angles are established by means of Harker constructions, then electron density distribution in the unit cell may be solved directly by means of Equation (6.5), and from this the structure of the biological macromolecule. Obviously, the process is not quite as simple as this and there is plenty of opportunity for error in this entire process. Nevertheless, these basic principles should suffice to allow the interested reader to brave the intricate detail of higher-level X-ray diffraction texts.

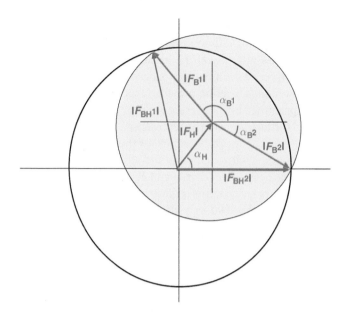

Figure 6.7 **Harker construction.** A solution of the X-ray crystallographic phase problem for each *hkl*-reflection by **Harker construction**. Heavy atom structure factor $F_H(hkl)$ is completely solved by **Patterson function** and plotted on complex plane (white) along with known modulus $|F_{BH}(hkl)|$. The second known modulus $F_B(hkl)$ is then included on a second complex plane (yellow). Intersection points characterise the two possible solutions for $F_B(hkl)$ with two possible solutions for $\alpha_B(hkl)$, one of which is usually eliminated by inspection or with the aid of a second heavy atom derivative.

6.3 Structural information from X-ray diffraction

Section 6.2 was compiled as a theoretical introduction to the more practical problems of obtaining X-ray crystal structures of biological macromolecules. In fact, although X-ray diffraction has proved immensely popular as a means of biological macro-molecule structure determination, the technique is still quite onerous to implement and the possibility for error surprisingly high. These problems will become clear in this section. The first and still the most difficult problem is that of biological mac-romolecule crystallisation. If there are no well-formed crystals then there can be no X-ray crystal diffraction.

6.3.1 Biological macromolecule crystallisation

Crystallisation is empirical and far from rational. A biological macromolecule crystal is slowly precipitated from solu-tion in the expectation that crystals will begin to form. Very pure biological macromolecule samples are required for this process. Typically, the biological macromolecule of interest is dispersed in a buffer solution, sometimes containing an organic solvent such as **2-methyl-2,4-pentanediol** (**MPD**). To this, a **precipitant** such as salt or **polyethyleneglycol** (**PEG**) is added very slowly until high supersaturation is reached, at which point small aggregates will emerge that may act as the nuclei for future crystal growth. Ideally, a low level of supersaturation is then required in order to allow for sustained crystal growth without the complication of excessive numbers of crystal growth nuclei further appearing. This may be achieved by adjustments in pH and temperature.

The simplest technique for biological macromolecule crystallisation is **batch crystallisation**. This is the simple binary combination of biological macromolecule solution and precipitant to create an instantaneous state of high supersatura-tion. This minimal process can give diffraction quality crystals under some circumstances, but many other techniques have needed to be developed. These include the following: **liquid-liquid diffusion**, **vapour diffusion** and **dialysis**. The liquid-liquid diffusion method (Figure 6.8) uses a narrow-bore glass capillary to place a layer of precipitant (approximately 5 μl) in contact with a layer of biological macromolecule solution (approximately 5 μl) over a small surface area. The two layers are intended to gradually diffuse together to bring about crystallisation. There are two alternative vapour diffusion methods, namely: the **hanging drop method** (Figure 6.9) and the **sitting drop method** (Figure 6.10). The former method relies upon surface tension effects to position a combined aliquot of biological macromolecule solution and precipitant (approx-imately 10 μl) on a glass slide, inverted over a well containing precipitant solution (approximately 1 ml), all in a sealed box. The latter method is an alternative method used if surface tension effects are insufficient to sustain a hanging drop during the crystallisation process. In both cases, the separated solutions are expected to diffuse together by slow vapour diffusion

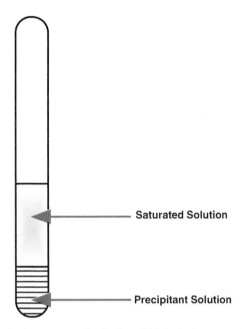

Figure 6.8   Liquid-liquid diffusion method. A saturated solution of biological macromolecule (e.g. protein) is placed in a sealed capillary environment in contact with precipitant solution. Slow mixing and liquid diffusion creates sufficient precipitant gradient in the macromolecule solution to "seed" crystallisation.

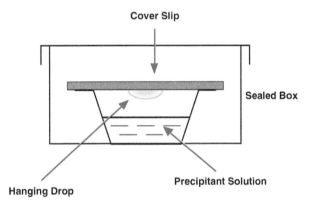

Figure 6.9   Hanging drop-vapour diffusion method. A saturated solution of biological macromolecule (e.g. protein) is placed in a sealed environment suspended above precipitant solution. Very slow vapour diffusion leads to very controlled precipitant gradient in the macromolecule solution that "seeds" crystallisation.

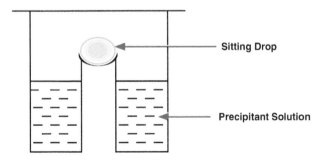

Figure 6.10   Sitting drop-vapour diffusion method. Identical principles to the hanging drop method (see Figure 6.9).

to bring about crystallisation. Finally, the **dialysis method** (Figure 6.11) separates biological macromolecule solution from precipitant by means of a semipermeable membrane through which precipitant will diffuse to initiate crystallisation. This technique is quite versatile in that the precipitant may be altered at will and crystallisations may be performed on a range of different scales.

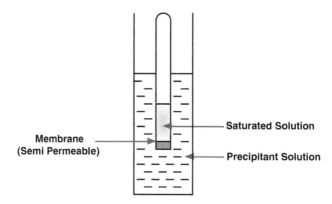

Figure 6.11 Dialysis method. A saturated solution of a biological macromolecule (e.g. protein) is placed in an environment separated from precipitant solution by semi-permeable membrane. Very slow precipitant diffusion across the membrane creates precipitant gradient in the macromolecule solution to "seed" crystallisation.

Once crystals have been obtained, they must be isolated and prepared for X-ray crystal diffraction studies. This, too, is problematic since biological macromolecule crystals are not the dense packed structures familiar to chemists used to working with small molecules. Rather, they are loosely packed with large solvent-filled holes and channels that may occupy up to 50% of the crystal volume. Therefore, biological macromolecule crystals must always be kept in contact with their mother liquor or in contact with a saturated vapour of the same mother liquor in order to ensure crystal integrity, even during exposure to X-ray beams. Hence, crystals are usually introduced into thin-walled capillaries of borosilicate glass or quartz (Figure 6.12) prior to mounting in an X-ray diffraction instrument where they will be subjected to sustained X-ray irradiation.

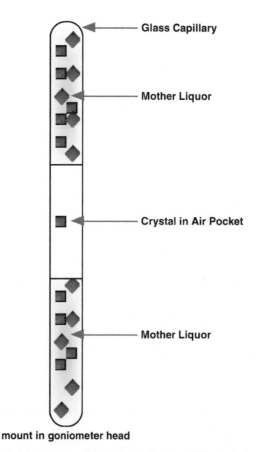

mount in goniometer head

Figure 6.12 Crystal mounting for diffraction. The delicacy of crystals generated from biological macromolecule crystallisation requires that individual crystals for X-ray crystallography be mounted in a **sealed capillary** separated from crystal mother liquors by an air pocket. The capillary is mounted in the **goniometer head** of a **four circle diffractometer** (see Figure 6.15).

6.3.2 X-ray generation

The traditional image of X-ray generation is that of an evacuated cathode-ray tube in which a cathode is at a high negative potential relative with respect to a metal (usually copper) anode. Electrons traverse the evacuated tube from cathode (negative) to anode (positive) under the influence of the potential difference and generate X-rays by high-energy collision with the anode. Collisions displace low-lying electrons that are then replaced by high-lying electrons from within the same atom in a process accompanied by the release of energy in the form of monochromatic X-rays. Copper produces X-rays at three main wavelengths known as $K_{\alpha 1}$, $K_{\alpha 2}$ and K_{β} since they originate from so-called **K-band** electronic transitions (Figure 6.13). The respective wavelengths are 1.54051, 1.54433 and 1.39217 Å. However, modern X-ray crystal diffraction prefers to make use of **synchrotron radiation** in preference to the use of cathode-ray tube generated X-rays. Synchrotrons are particle accelerators that circulate injected particles such as electrons in the form of a particle beam at speeds close to the speed of light. These devices are quite simply enormous and the trajectory of any given particle in a beam may be anything up to a few kilometres in pathlength (Figure 6.14). Whenever the particle beam is forced to change direction in order to complete its trajectory,

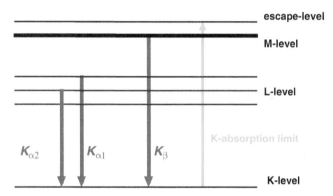

Figure 6.13 K-band transitions. Schematic representation of the orbital energy levels of a Cu-anode mounted in a vacuum-sealed cathode-ray tube, also showing the transitions causing characteristic K-band X-ray wavelengths. Electrons bombard the Cu-anode and in the process the high energy electrons reaching the anode "shoot" electrons out of low-lying orbitals in the anode atoms. Electrons from higher orbitals occupy the empty positions and in the process emit X-ray radiation of defined wavelength.

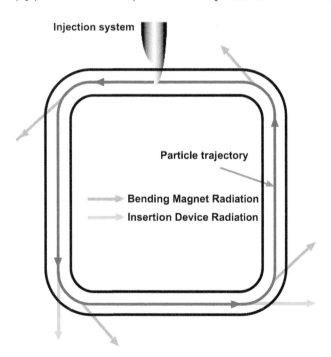

Figure 6.14 Synchrotron X-ray production. Schematic representation of a particle synchrotron storage ring with an injector for the charged particles. Radiation can be obtained from the particles while passing the bending magnets. In the straight sections of the ring, devices can be inserted (e.g. **wiggler** or **undulator**) that produce even higher intensity radiation than the bending magnets.

electromagnetic radiation is released. This radiation covers a wide range of wavelengths owing to the continuous (non-quantised), broad range of particle momenta in the beam, but shows significant intensity even as low as 0.1 Å, a lower limit of X-ray radiation.

Synchrotron X-ray radiation is two orders of magnitude more intense than that from an evacuated cathode-ray tube and tuneable to any desired wavelength in the X-ray spectral range using a monochromator. Hence a given biological macromolecule crystal can be irradiated with high intensity, monochromatic X-rays covering a range of possible X-ray wavelengths (usually around 1 Å), allowing for the sampling of substantial numbers of reciprocal lattice points or, in other words, detection of substantial numbers of X-ray scattering events by reflection from a wide range of *hkl*-lattice planes (see Section 6.2.3). Consequently, the diffraction data necessary to construct unit cell electron density distribution may often be collected from just one single biological macromolecule crystal, provided that radiation damage to the crystal is kept to a minimum. Fortunately, synchrotron derived X-rays with wavelengths close to 1 Å are inherently less damaging than cathode-ray tube derived X-rays close to 1.5 Å in wavelength, owing to the tendency of the shorter wavelength X-rays to be diffracted by a given biological macromolecule crystal rather than to be absorbed causing radiation damage. The general advantages of synchrotron X-ray radiation should now be absolutely clear.

6.3.3 Determination of X-ray diffraction pattern

For a completely accurate crystal structure determination, X-ray scattering must be observed and recorded by reflection from the vast majority of *hkl*-lattice planes associated with a given biological macromolecule crystal. That is to say, that an accurate structural determination is only possible if the vast majority of reciprocal lattice points can be sampled. In order to achieve this, the classical approach has been to use a device like a **four circle diffractometer** (Figure 6.15) in which biological macromolecule crystals are first mounted in a **goniometer head** located at the centre of the diffractometer and then irradiated with an intense beam of X-rays, after which X-ray reflections may be observed sequentially, one at a time in 3D space around the crystal. X-ray detection is effected using a scintillation counter. Throughout the process of detection, the crystal is made to rotate in small oscillation steps in the goniometer head. As the Ewald sphere concept showed in Section 6.2.3, sampling of the reciprocal lattice is maximised with minimal data overlap by a combination of such crystal rotation and the use of a range of possible monochromatic X-ray wavelengths to irradiate the crystal.

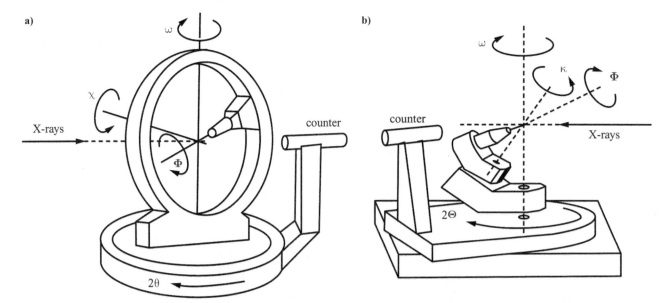

Figure 6.15 Four circle diffractometers. Two examples, (a) and (b), show **single photon counting devices**. The **goniometer head** positions the diffracting crystal in the path of an incident X-ray beam. The devices allow for rotation of the mounted crystal through four complete arcs (designated Φ, ω, 2θ and κ or χ) in order to detect as many *hkl*-reflections (reciprocal lattice points) as possible by diffraction counter prior to crystal destruction by the incident beam (illustration from Drenth, 1994, Figure 2.9, Springer Nature).

Since only one X-ray reflection may be observed at one time, the classical approach to observing an X-ray diffraction pattern is inefficient; typically, several weeks are required to collect sufficient scattering data (i.e. 10^4–10^5 X-ray reflections depending upon the size of the biological macromolecule under investigation). Nevertheless, if time is not limiting and the crystal is stable then the classical approach to X-ray crystal diffraction is still sound and effective. However, the classical approach is now giving way to the use of four circle diffractometer-like devices equipped with **image plates** or **electronic area detectors** capable of sampling more X-ray reflections more rapidly and more accurately.

6.3.4 Heavy atom derivatisation

The introduction of heavy atoms into a biological macromolecule crystal, also known as **heavy atom derivatisation**, is a critical part of X-ray crystal structure determination as indicated in Section 6.2. Critically, derivatisation must as much as possible be isomorphous with respect to the original biological macromolecule crystal, otherwise most of the assumptions underpinning the Harker construction break down and the phase problem remains a problem. Unfortunately, this is a highly empirical process similar to biological macromolecule crystallisation. The biological macromolecule crystal is soaked in a solution containing heavy metal, allowing heavy metal ions or reagents to diffuse through the large solvent-filled holes and channels that make up the crystal. Soaking times will vary from hours to months since the minimum time to reach equilibrium of the reaction is determined by a number of competing factors such as diffusion times through the crystal, ion or reagent accessibility to reactive/combining sites on the biological macromolecule, and the nature of the chemical combining process itself. The process of ensuring true isomorphous replacement can be much enhanced through careful attention to biology or through the use of site-selective modification procedures. For instance, where a protein contains Ca^{2+} or Mg^{2+} ions, heavy atom Sm^{3+} ions are suitable surrogates since they have similar radii. Alternatively, low pH values promote the binding of charged heavy atom reagents such as HgI_4^{2-} or $Au(CN)^{2-}$ to biological macromolecules such as proteins or nucleic acids by minimising charge-charge repulsion. Heavy atoms are frequently thiophilic, consequently genetic engineering of proteins to introduce the amino acid residue cysteine in defined, surface accessible locations in a protein can be a very powerful way of introducing heavy ions such as Hg^{2+} in defined positions, with minimal structural distortion to either protein or crystal.

6.3.5 Fitting an electron density map

After a first complete set of $\alpha_B(hkl)$ phase angles has been determined with data obtained from the MIR method, or equivalent, an electron density map may then be calculated. If sufficient X-ray diffraction data has been acquired then this map will fit the known primary sequence of the biological macromolecule reasonably well, giving a preliminary model of biological macromolecule structure. **Structure refinement** will follow to improve the model. If diffraction data is insufficient to define the electron density map unambiguously, then further improvements in $\alpha_B(hkl)$ phase angle data need to be sought. Structure refinement usually requires techniques such as **solvent flattening**. This relies on knowledge that an electron density map is essentially flat in the solvent regions between biological macromolecules due to the "fluid character" of the solvent molecules in these regions. Ordered solvent molecule structures only occur at interfaces with the biological macromolecule itself or embedded within. Hence, by arbitrarily setting all electron density to a low constant value in the identified solvent regions, a new electron density map will result, from which new $F_B(hkl)$ structure factors and improved $\alpha_B(hkl)_{imp}$ phase angles may be calculated. These improved $\alpha_B(hkl)_{imp}$ phase angles are then combined with experimental $|F_B(hkl)|$ data determined from experimental X-ray scattering intensities in order to recalculate an electron density map. Several further consecutive rounds of solvent flattening may then take place until convergence is achieved.

Alternatively, if the crystal structure contains several biological macromolecules in defined symmetrical arrays (known as regions of **high non-crystallographic symmetry**), then some form of **multi-fold molecular averaging procedure** may be used to refine even quite poor initial $\alpha_B(hkl)_{imp}$ phase angle data and electron density maps. Ultimately, if the refinement process is sufficient, then the electron density map should closely fit the known primary sequence of the biological macromolecule like "hand in glove". Such an example of this close fitting is illustrated in Figure 6.16. In this case the quality of the map is sufficient to allow for detailed structural and functional questions to

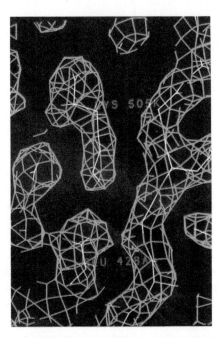

Figure 6.16   Electron density map. An example of an electron density map generated computationally from electron density data that has been derived by the application of the equations and principles described in the main text from X-ray crystallographic scattering data. The electron density map corresponds with part of the active site of an enzyme **LysU** (see Figure 6.19; plus Chapters 7 and 8) from the organism *Escherichia coli* (*E. coli*). This electron density map has been "fitted" with the primary sequence polypeptide chain of LysU (colour code: carbon: **yellow**; oxygen: **red**; nitrogen: **blue**). Once an electron density map has been determined, fitting of the known primary sequence of the biological macromolecule to the electron density map is the final stage that leads to a defined 3D structure (from Onesti *et al.*, 1995, Elsevier).

be asked about the **LysU** protein structure. The general quality of an electron density map is often given by the **common crystallographic R-factor**. This is defined as in Equation (6.7):

$$R = \frac{\sum_h \sum_k \sum_l \left| F_{\mathrm{obs}}\left(hkl\right) \right| - k \cdot \left| F_{\mathrm{calc}}\left(hkl\right) \right|}{\sum_h \sum_k \sum_l \left| F_{\mathrm{obs}}\left(hkl\right) \right|} \tag{6.7}$$

where: $\left| F_{\mathrm{obs}}\left(hkl\right) \right|$ is the **observed modulus of the structure factor** determined from the intensity of X-ray scattering from a given stack of parallel *hkl*-lattice planes, $\left| F_{\mathrm{calc}}\left(hkl\right) \right|$ the **modulus calculated** from a model of the biological macromolecule structure fitted to the refined electron density map and *k* is a **scaling factor**. For model structures with atoms randomly distributed in the unit cell, the R factor should be 0.59. For well-developed, high-resolution model structures (<2 Å resolution), the R-factor should not generally exceed 0.16. This R-factor is an overall number and does not indicate major local errors, but does provide a useful guide to model structure quality.

6.3.6   Biological macromolecule structures by X-ray crystallography

Nowadays, X-ray crystallography has been refined to the point that a sizeable range and diversity of structures have been solved and made available publicly as **protein data bank (pdb) files**. This is a little confusing since pdb files also contain structures of other biological macromolecules in addition to proteins, although the majority are still proteins. These pdb files are basically annotated text files that contain information on the relative coordinates of biological macromolecule atoms in the unit cell, as determined by X-ray diffraction pattern resolutions and electron density fitting. A substantial number of pdb files can be downloaded and visualised using a range of visualisation software programs, also available by downloading from the Internet. Clearly, visualisation requires the use of cartoon depiction methods (see Chapter 1) in order to give clarity, since whole atom representations can be difficult to understand and appreciate. Therefore, relevant software programs specialise in allowing the user to render X-ray crystal structures in a variety of customised cartoon representations suitable for the interest and requirements of the time. The beginning of Chapter 7 documents some truly spectacular X-ray crystal structures; therefore we shall just confine ourselves here to illustrating and discussing structures of three important heat shock/stress proteins from the bacterium *Escherichia coli* (*E. coli*). Typically, many proteins from the bacterium *E. coli* have been studied more extensively than corresponding proteins from other sources, given the predominant use of *E. coli* in genetic engineering, protein overexpression and protein engineering studies (see Chapter 3).

Stress proteins are universal proteins found in all cells of every organism that act to protect proteins in cells from the damaging effects of physiological, environmental and chemical stress. Moreover, stress proteins often play pivotal roles in normal cellular physiology as well. Frequently, stress proteins are some of the most conserved proteins in nature, and close homologues of stress protein families are found in all cells derived from the very oldest bacteria and protozoans up to the most recent eukaryotes. Chemical biology studies on the structure and function of stress proteins are a fundamental part of a wider investigation into the **chemistry of stress**, but the elucidation of their 3D structures by X-ray crystallography have frequently proven challenging. The majority of stress proteins are molecular chaperones that assist the folding/refolding of other proteins without being involved in the final folded state of those other proteins. Although the 3D structure of a given protein is widely considered to be specified by its sequence of amino acid residues, the kinetic process of protein folding frequently needs assistance in living cells; hence the requirement for molecular chaperones. Under stress conditions, molecular chaperones are upregulated in order to protect protein substrates from stressor-induced unfolding and aggregation.

Molecular chaperone GroEL and **co-chaperone GroES** from *E. coli* (also known as **Chaperonin 60 [Cpn60]** and **Chaperonin 10 [Cpn 10]**) are considered to be the archetypal stress protein molecular chaperones. In a wider context, since heat shock is the primary physiological inducer of GroEL and GroES in cells, then GroEL is said to belong to the **heat shock protein 60 (Hsp60)** family of stress proteins, and GroES to the **heat shock protein 10 (Hsp10)** family of stress proteins (60 and 10 refer to approximately molecular weights in kDa of GroEL and GroES subunit polypeptides or monomers respectively). Generally, GroEL and GroES operate together as the **GroEL/ES molecular chaperone machine** that acts to protect a very wide variety of protein substrates from stressor-induced unfolding and aggregation under stress conditions in *E. coli* cells (*in vivo*) and also acts to assist protein folding under normal conditions. Given the complexity of this molecular chaperone machine, the report of the X-ray crystal structure for the GroEL/GroES/(ADP)$_7$ complex in 1997, was regarded as nothing short of a revelation and hailed as a major scientific achievement (Figure 6.17). This structure has since opened the door to an enormous swell of structure/activity studies aimed at understanding everything from molecular recognition (Chapter 7) to the mechanism of GroEL/ES-assisted folding/refolding of model substrate proteins. Starting from this structure, the complete molecule chaperone machine mechanism has now been elucidated in reasonable detail as shown (Figure 6.18).

If there are stress proteins to assist with the problems of stress, there must be stress proteins able to modulate stress responses and assist the return of cells to normal physiology when appropriate. One such stress protein in *E. coli* is **LysU**. Unusually for stress proteins, LysU is a homodimeric **lysyl tRNA synthetase** enzyme that should normally play an essential role in protein biosynthesis in *E. coli* (see Chapters 1 and 3), but is also an excellent synthase for **diadenosine-5′, 5′′′-P^1, P^4-tetraphosphate (Ap$_4$A)** in the presence of Zn^{2+} ions (see Chapters 7 and 8). The X-ray crystal structure of LysU with the amino acid L-lysine bound in both active sites of the dimer was efficiently solved in 1995 (Figure 6.19). Since then, other LysU X-ray crystal structures with other ligands bound have also been solved (Figure 6.19). The determination of the first LysU X-ray crystal structure was remarkable in that the protein was able to form excellent crystals within a few weeks of obtaining sufficient protein for studies, but the subsequent determination of the actual X-ray crystal structure from the diffraction pattern then took years. Usually, the reverse is true. The determination of these LysU X-ray crystal structures has since opened up significant enzymic, mechanistic and even molecular modelling studies designed to understand the way the enzyme operates in molecular recognition, binding and catalysis (see Chapters 7 and 8).

Without doubt, the X-ray crystal structures illustrated here have proven important starting points for the development of an understanding of GroEL/GroES and LysU protein functions. In fact, the same can be said of the X-ray crystal structures of any number of other biological macromolecules. However, X-ray crystal structures should not be used without question. For one thing, the X-ray crystal structure of any one protein is just that, the structure of the protein in a crystallographic state. Although the crystals of biological macromolecules are very open (see Section 6.3.1), there are still realistic crystal-packing forces involved that can actually distort protein structure and/or suppress conformational changes that might be otherwise highly important for biological macromolecular function in solution. For example, the X-ray crystal structure of the GroEL/GroES/(ADP)$_7$ complex failed completely to demonstrate conformational changes in the double doughnut structure of GroEL and the "two-stroke motor" cycle of binding and release of protein substrate (in various states of unfolding) that cooperates together to assist folding/refolding of substrate proteins (Figure 6.18).

In fact, the complete picture was only established later by a combination of X-ray crystallography data and structural data from cryo-electron microscopy, with functional data from other dynamic spectroscopic analyses (see Chapter 7). There is a general rule of thumb here. X-ray crystal structures frequently provide a good foundation for understanding

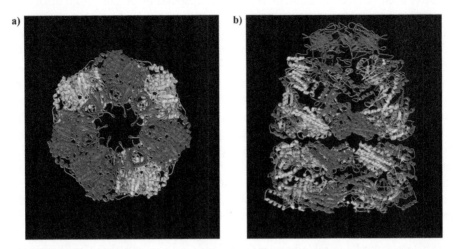

Figure 6.17   X-ray crystal structure of multi-protein complex. The structure shown is of the molecular chaperone GroEL/GroES/ (ADP)$_7$ complex (pdb: **1aon**), key abbreviations: ADP: adenosine 5'-diphosphate; ATP: adenosine 5'-triphosphate in a **Ribbon display structure** depiction to demonstrate the general features of this multi-protein complex. Top view (**a**) and side view (**b**) of complex. The X-ray crystal structure shows how **GroEL** is a homo-oligomer consisting of 14 identical monomers (**red** or **yellow**) (each 57,259 Da) that are assembled into two stacked rings, each consisting of seven monomers with a central cavity for the sequestration of protein sub-strate molecules ("double doughnut" structure). In the case of **GroES**, the X-ray crystal structure shows how GroES is a homo-oligomer consisting of seven identical monomers (**blue**) (each 10,368 kDa) assembled into a single ring (**blue**). In the presence of ADP, GroES forms a tight complex with GroEL and becomes the "cap" of the same GroEL ring to which the seven ADP molecules are bound (space filling representation; **blue**). Each ADP molecule is bound to an individual ADP/ATP binding site in each of the GroEL monomers of that ring. GroEL has a total of 14 individual ADP/ATP binding sites (one ATP/ADP binding site/monomer or seven ATP/ADP binding sites/7mer GroEL ring) (from Jones *et al.*, 2006b, Figures 1a and 1b, Royal Society of Chemistry).

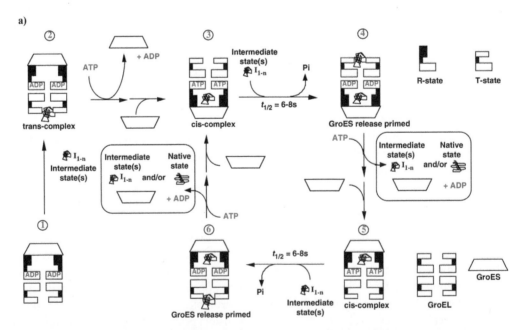

Figure 6.18   Mechanism of GroEL/GroES-assisted refolding of proteins. (a) Complete **GroEL/GroES molecular chaperone machine mechanism** illustrating the regular cycle of binding and release of protein substrate (in various states of unfolding). This regular process of binding and release of protein substrate has been thought at various times to have catalytic effects on protein folding or else provide a means for the folding of substrate proteins at "infinite dilution" in the GroEL central cavities. However, the cycle is much more consistent with a **passive kinetic partitioning mechanism** for molecular chaperone machine activity described below (see **c**). The T-state and R-state nomenclature refer to the conformations of individual GroEL polypeptide subunits in the homo-oligomeric structure. The T-state has a high affinity for substrate protein and the R-state a low affinity (the affinity for ATP is reversed). The term **I**$_{1-n}$ refers to discrete substrate protein folding intermediates, **I**, of between **1** and **n** in number (illustration from Jones *et al.*, 2006b, Figure 1c, Royal Society of Chemistry)

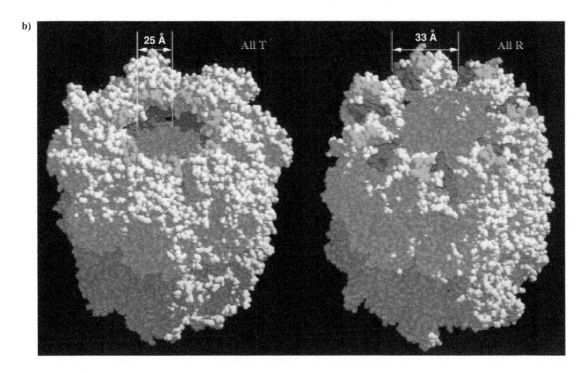

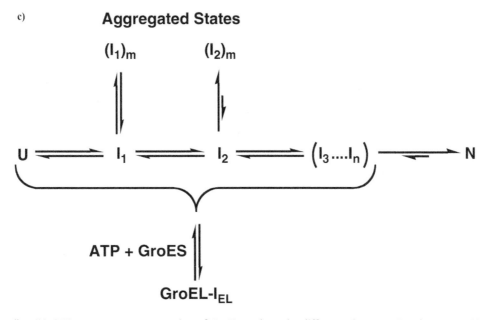

Figure 6.18 (Continued)   (b) **CPK structure** representation of GroEL to show the difference between **T** and **R** states. Hydrophobic binding residues (**green** and **red**) at the mouth of the GroEL cavities drastically change position and move further apart in going from the **T** to **R** states. (c) **Passive kinetic partitioning mechanism** of GroEL/GroES molecular chaperone machine. This assumes that protein folding is initiated at an unfolded state, **U**, which folds through a succession of intermediate states I_1, I_2, $(I_3...I_n)$ before reaching the native state, **N**. States I_1 and I_2 are considered arbitrarily to be unstable to aggregation, forming aggregated states $(I_1)_m$ and $(I_2)_m$ through interaction of their exposed hydrophobic surfaces. GroEL is potentially able to bind to all vulnerable protein folding intermediate states, except **N**, forming a GroEL-bound state **GroEL-I_{EL}**. The nature of this state is a function of the requirement to optimise the free energy of association between GroEL and the given unfolded protein state under the given set of binding conditions. The binding interaction with GroEL is reversed in a controlled manner with the assistance of first ATP and then GroES binding, after which protein substrate is retained by GroEL intra-cavity until ATP hydrolysis is complete ($t_{1/2}$ 6–8 s). Thereafter, protein substrate may be released into free solution ready to rebind again if necessary (see **a**). As a result of this cyclical binding and controlled release into a GroEL cavity and then free solution, steady state concentrations of **U**, I_1, I_2 and $(I_3...I_n)$ are maintained below the critical threshold for aggregation so that these states are free to partition kinetically to **N** (from Smith *et al.*, 1999, Scheme 3, Royal Society of Chemistry).

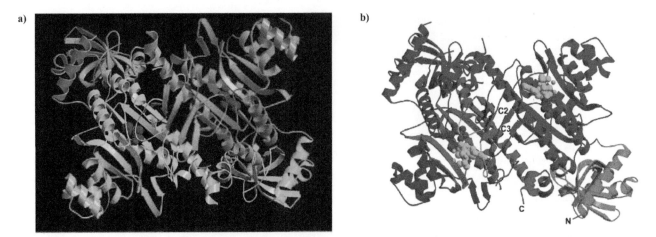

Figure 6.19   X-ray crystallographic images of LysU. (a) **Ribbon display structure** representation of the first X-ray crystal structure of *E. coli* stress protein **lysyl tRNA synthetase (LysU)** (pdb: **1lyl**). LysU is known as a potent **Ap$_4$A synthase** (see Chapters 7 and 8 for additional details). LysU is a homodimer comprised of two identical polypeptide monomers (**green** and **blue**) each with an active site. Successful crystallisation requires the presence of **L-lysine** that binds to both active sites (shown in **CPK structure** representation; **red**). Ap$_4$A may be a signal molecule for stress but is more likely to be involved in stress accommodation, helping cells recover from periods of exposure to physiological, environmental or chemical stressors; (b) **Ribbon display structure** depiction of a later X-ray crystal structure of LysU with substrates L-lysine (**pink**) and β,γ-methylene ATP (**AMPPCP**) (**cyan**) bound in both active sites, together with 3 Mg^{2+} ions (**yellow**), all shown in **CPK structure** representation (pdb: **1e1t**). One complete monomer is shown in **blue**, whereas the other is coloured according to domain structure: *N*-terminal (**orange**), conserved core domain typical to Class II tRNA synthetases (**red**), LysU specific regions (**purple**). In the red domain region, a **motif 2 loop** (**green**) sits over ATP. β-strands **C2** and **C3** in each monomer interact across the homodimer interface. The loop between strands is positioned above and interacts with the motif 2 loop of the opposite monomer (from Hughes *et al.*, 2003, Figure 1, Springer Nature).

in chemical biology, but a proper understanding of dynamic mechanism and function usually requires the application of other techniques, building necessarily upon the knowledge of structure gained from X-ray crystal structures (see Chapters 7 and 8). In this respect, NMR spectroscopy (Chapter 5) has a significant advantage over X-ray crystallography in that the structures of biological macromolecules are determined in solution state. However, it is perhaps gratifying that where X-ray crystal structures are available for comparison with NMR-derived structures, then there is frequently close agreement between them.

6.4 Neutron diffraction

The basic principles of neutron and X-ray diffraction are quite different, although there is some compatibility in terms of data output. X-ray diffraction or scattering may be considered to be a consequence of interactions between the oscillating electromagnetic fields of X-ray photons and the "**clouds of structural electrons**" surrounding atomic nuclei. In the case of **neutron diffraction**, neutrons behave as neutral particles that can pass through the clouds of electrons, but are then diffracted or scattered by "**nuclear force**" interactions. Hence the phenomenon of neutron scattering is essentially electron independent, but atomic number dependent and invariant with angle. Typically, neutron diffraction data may be processed to a resolution of 2–4 Å in common with X-ray diffraction data.

The X-ray scattering power of any given atom can be described in terms of the **X-ray scattering length** of that atom, $b_{\text{x-ray}}$, according to the following simple Equation (6.8):

$$b_{\text{x-ray}} = \text{const} \cdot f_{(0)} \tag{6.8}$$

where $f_{(0)}$ is known as the **atomic scattering factor**, a variable that may be used to quantitate the forward scattering of X-rays, while the constant (**const.**) is equivalent to 2.8×10^{-13} cm. X-ray scattering intensities per atom are related to $b_{\text{x-ray}}^2$. Neutron scattering power may be similarly quantitated, but values of $f_{(0)}$ are substantially different, giving rise to a set of **neutron scattering lengths**, b_{neut}, that are often significantly different to X-ray scattering lengths (Table 6.1). Significantly, the neutron scattering lengths for H and D atoms are much more comparable in magnitude to the scattering lengths of

Table 6.1 **Neutron scattering lengths and atomic scattering lengths of various elements.**

Element	Neutrons $b_{neut} \times 10^{13}$ (cm)	X-rays $b_{x\text{-}ray} \times 10^{13}$ (cm)
H	−3.74	3.8
D	6.67	2.8
C	6.65	16.9
N	9.40	19.7
O	5.80	22.5
P	5.10	42.3
S	2.85	45

other heavier atoms than corresponding X-ray scattering lengths; hence hydrogen atoms are much more clearly visualised by neutron diffraction than by X-ray diffraction. Another significant fact is that b_{neut} values for H and D are negative and positive numbers respectively, indicating that H and D neutron scattering are π-out of phase (180°) even though of similar magnitude. Hence neutron diffraction may even be able to identify isotope locations in different positions. In practice, neutron diffraction of biological macromolecules is rarely performed in the absence of a known X-ray crystal structure and should ideally be performed using the same crystals as those used for X-ray diffraction studies. Furthermore, access to neutron sources usually limits the use of neutron diffraction in structural characterisations, but, when carried out, neutron diffraction studies can reveal a lot of useful information confirming the relative positions of H atoms and the presence of H/D exchange sites, information often poorly revealed by X-ray diffraction data.

6.5 Key principles of electron microscopy

X-ray crystallography is not the only technique that gives detailed information about molecular architecture. The last decade has seen the remarkable rise of electron microscopy, a technique that could soon rival the pre-eminence of X-ray crystallography as the technique of choice for high-resolution studies of biological macromolecule structures. There are three main reasons why this might be possible. First, there is no requirement to obtain biological macromolecule crystals, especially given the advent of cryo-electron microscopy that will be discussed primarily here. Second, electron and X-ray wavelengths can be very similar, giving rise to comparable diffraction effects and hence comparable resolutions. Third, unlike X-rays, charged electrons scattered by interaction with a sample may be refocused using electromagnetic lenses, giving rise to a magnified contrast image that may be detected and viewed. This is not to say that the electron equivalent of a structure factor or a reciprocal lattice ceases to be relevant or useful. Indeed, we might even have some uses for the electron diffraction pattern. However, the capacity to create a contrast image post-diffraction places electron microscopy much less in the realms of a diffraction technique and much more like a sophisticated version of the simple light microscope found in the everyday biology classroom. The basic appearance of a modern-day electron microscope is illustrated (Figure 6.20). An **electron source** is mounted upon an evacuated column in which the **electromagnetic condensing**, **objective** and **projector lenses** are mounted. Two condensing lenses act to focus an intense accelerated beam of electrons upon a sample of interest mounted in a **specimen holder**. Electrons diffracted by interaction with sample molecules are then manipulated by a set of two main objective lenses that are mounted for magnification purposes, followed by two main projector lenses that help to project a magnified contrast image onto a viewing screen or into the waiting aperture of a camera set up to capture the images formed.

6.5.1 Duality of matter

The fact that an accelerated beam of electrons can be used to view magnified contrast images, in the same way as visible light is used to view magnified images in an ordinary light microscope, owes a lot to the famous **duality of matter**. A subatomic particle like an electron possesses both discrete particle-like and electromagnetic wave-like characteristics. As a negatively charged particle, an electron collides readily with gas molecules in air and as such has no capacity to

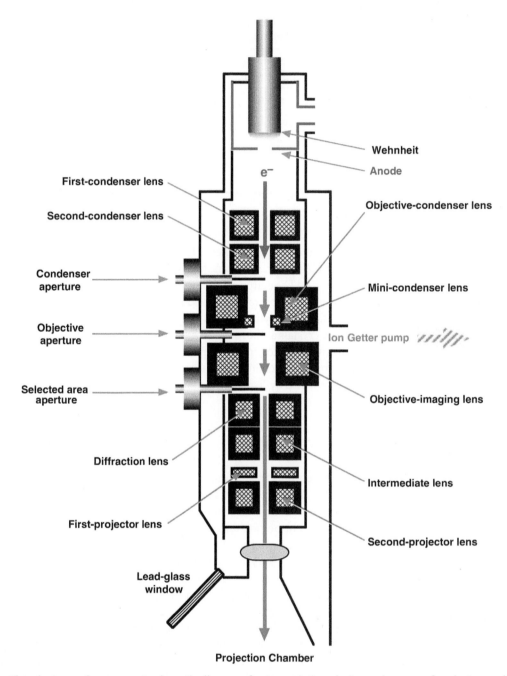

Figure 6.20 **The electron microscope**. A schematic diagram of a transmission electron microscope for **electron microscopy**. The sample is placed in the microscope in the path of an **electron beam** generated by the **field emission gun** (**FEG**) between condenser and objective lenses.

penetrate air to any significant distance. Therefore, free electron motion requires vacuum conditions equivalent to approximately 10^{-3} N m^{-2} (10^{-3} Pa). Such conditions are created in an electron microscope by high vacuum pumps. Furthermore, as a negatively charged particle, an electron can be accelerated by an electrical potential of the sort generated in the **field emission gun** (**FEM**) of an electron microscope (Figure 6.20). Typically, electrons are released from a tungsten filament crystal wire, not unlike the filament of a light bulb, and accelerated by an electrical potential difference of approximately +200 V. Electrons that have been accelerated through a high vacuum under the influence of such a potential form a focused beam, interact with molecules in the mounted sample and then become scattered as if the beam of electrons were a beam of electromagnetic radiation. Beam refocusing leads to the formation of a magnified contrast image (more about this later).

6.5.2 Electron wavelengths

According to the **Abbé relationship** that applies to microscopy, **resolution** or **distance of resolvable separation**, d_R, is related to **numerical aperture**, *NA*, and **radiation wavelength**, λ, according to the following relationship (6.9):

$$d_R = 0.61 \cdot \lambda / NA \qquad (6.9)$$

The value of *NA* can be set by and is a property of the microscope, hence λ is the critical property for ensuring that d_R is as small as possible. Fortunately, the intrinsic wavelength of a beam of electrons generated as described above is sufficient to give a value of d_R in the Å range. This intrinsic wavelength is a property set by the ultimate equation that relates the duality of matter, namely the **de Broglie equation** that relates **subatomic particle-like momentum**, *p*, with concurrent electromagnetic wavelength, λ, where *h* is **Planck's constant**, as shown in Equation (6.10):

$$\lambda = h / p \qquad (6.10)$$

In the case of an accelerated beam of electrons, *p* is related to the energy of each accelerated component electron, $\Phi \cdot e$ (i.e. the product of **electron charge**, *e*, and **electrical potential difference**, Φ), and the intrinsic particle **mass of an electron**, m_e, by Equation (6.11):

$$p = \sqrt{2 m_e \Phi \cdot e} \qquad (6.11)$$

which is a reworked version of the standard, classical dynamics equation for kinetic energy. Numerical substitutions into Equation (6.11), assuming Φ to be +200 V, and then substitution of the resulting value of *p* into Equation (6.10) gives a value for the electromagnetic wavelength λ that translates into a value of d_R as low as 1.4 Å for the most recent electron microscopes. Such a value illustrates that in theory electron microscopy should be able to generate images of biological macromolecules at atomic resolution. In practice this does not always turn out to be quite so, as the following sections on image manipulation will make plain. However, electron microscopy is a rapidly evolving subject area and this theoretical limit is being rapidly approached, suggesting that some day soon, electron microscopy could become the most versatile and significant technique for the characterisation of biological macromolecule structures.

6.5.3 Sample preparation

Nowadays, cryo-electron microscopy, so named for the method of sample preparation and maintenance, is by far the most effective technique of electron microscopy for biological macromolecule characterisation. Therefore, we shall focus on this electron microscopy technique alone from this point onwards. The basis of success in cryo-electron microscopy is very simple, though practically quite demanding. A sample of biological macromolecule of interest in water is flash frozen in such a way that individual molecules become embedded in **vitreous ice**. Vitreous ice comprises solid ice that contains little or no formal crystalline structure. Typically, solid ice possesses substantial tetrahedral structure (see Chapter 1) owing to extensive hydrogen-bonding relationships between water molecules.

However, if liquid water is flash frozen rapidly enough down to a temperature at or below 138 K, then there may be insufficient time for water molecules to orientate and substantially hydrogen bond with each other, leading to a *de facto* solid state without any substantial crystalline structure. In the absence of organised crystalline structural elements, vitreous ice only transmits an incident beam of accelerated electrons without scattering. This creates the ideal situation. When a sample of individual biological macromolecules in vitreous ice is exposed to an incident beam of electrons, only embedded macromolecules are able to scatter electrons and hence produce magnified contrast images after beam refocusing. The vitreous ice should be essentially invisible to the electron beam in theory, and this is substantially true in practice.

Practical realisation of vitreous ice embedded samples of individual biological macromolecules is usually achieved with a **holey carbon-grid** (Figure 6.21), which is a thin carbon-based support film (50–200 nm in dimension) perforated with holes and mounted on a standard electron microscope sample grid. An aqueous sample of a biological macromolecule of interest is applied to the mounted holey carbon grid and flash frozen by virtually instantaneous immersion of the mounted grid in liquid ethane using the illustrated immersion device (Figure 6.21). The result is a thin film of vitreous ice wherein individual biological macromolecules are embedded at semi-regular intervals. This thin film is then mounted in the elec-

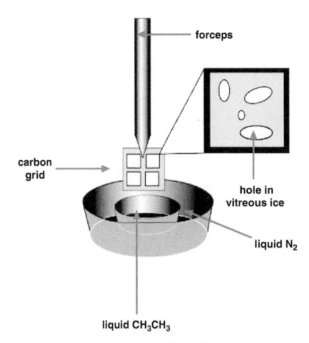

Figure 6.21 Vitreous ice freezing for cryo-electron microscopy. This is effected using a thin carbon support film (50–200 nm) perforated with holes is placed on a standard EM **carbon grid** and then a sample of biological macromolecule is applied and rapidly frozen in **liquid ethane** (<138 K). After this, the vitrified sample is transferred into a **cryoholder** (under liquid nitrogen conditions) and then transferred to the microscope for visualisation. The holes in the thin film allow for the formation of monolayers of biological macromolecules in a range of orientations embedded within a thin (20–60 μm) layer of vitreous ice. The visualisation of such monolayer regions gives the best possible **cryo-EM** images of the embedded biological macromolecules.

tron microscope by means of a specially adapted specimen holder known as a **cryo-holder** which is maintained at 178 K (liquid nitrogen temperature) in order to ensure the continued integrity of the thin film when subjected to a continuous incident beam of accelerated electrons.

6.5.4 Contrast imaging

As described above, electrons diffracted by interaction with individual biological macromolecules embedded in a thin film are manipulated by objective lenses mounted for magnification purposes and afterwards by projector lenses to produce magnified contrast images that are projected onto a viewing screen or into a camera aperture. Such contrast images are generated by direct interactions between accelerated electrons of the incident beam and individual embedded molecules, leading to a variety of **amplitude** and **phase contrast effects** that together create the observed images. Amplitude effects create dark regions within an image that arise from elastic scattering of a proportion of incident beam electrons owing to close atom encounters following thin-film penetration. Phase contrast effects create differential shaded regions within an image that arise alternatively from the partial retardation and subsequent re-phasing of a proportion of incident beam electrons owing to longer range encounters that may take place following thin-film penetration. As a rule, images generated from biological macromolecules embedded in a thin film are formed primarily from phase contrast effects. Incident electrons automatically experience a $\pi/2$ (90°) phase shift as a result of passing through an embedded macromolecule. This effect is then augmented by phase shifts within the focal plane of the objective lenses as well.

6.5.5 Image processing

This is by far the most complex part of the process and is as much error prone potentially as the generation and fitting of electron density maps in X-ray crystallogaphy. Typically, several thousand contrast images are generated from the mounted specimen of a given biological macromolecule, each image corresponding to a 2D projection of an individual 3D molecular object. To a first approximation, all these images can be assumed to be perfect projections of the individual molecule concerned onto an **(x, y) image plane** perpendicular to the incident beam of electrons. Images are **band-pass**

filtered to remove high and low frequency noise features and then normalised to a zero-average density. Both processes are designed to remove the effects of heterogeneity in the thickness and structure of the vitreous ice thin film in which molecules are embedded. The normalisation process ensures that all molecular images may be assumed to have the same z-positioning. Consequently, we could describe image density $P(xy)$ for the image of an individual molecule by using Equation (6.12):

$$P(xy) = \sum_X \sum_Y F(XY) \cdot \exp\left[-2\pi i(Xx + Yy)\right] \qquad (6.12)$$

where $F(XY)$ are **electron diffraction structure factors** corresponding with reflections from all possible XY-structural planes that bisect the given thin-film, embedded molecule whose image has been acquired. This Equation 6.12 is directly analogous with Equation 6.2 that directly correlates unit cell electron density distribution with unit cell $F(hkl)$ structure factors that quantify reflections of X-rays from given hkl-structural planes. However, the chemical biology reader will probably be relieved to hear that any further similarities between electron microscopy and X-ray crystallography need not be considered here. Instead, we shall need to consider how all the thousands of normalised images are put through extensive image processing procedures in order to derive a complete set of 3D images representative of relevant 3D structure(s) of the specimen biological macromolecule of interest.

Image processing is required owing to the simple fact that 3D biological macromolecules flash frozen and immobilised in a thin film of vitreous ice will be immobilised in completely random orientations. Each distinct, individual orientation should be capable of producing a distinct 2D image. The basis of image processing is to compare, contrast and classify all the individual images obtained and use these as the basis for 3D structure generation. The orientation of any 3D object at a fixed point in time may be characterised by six degrees of freedom, namely *(x, y, z)* **Cartesian coordinates** and (α, β, γ) **rotational degrees of freedom** (Figure 6.22). Consequently, any 2D image of a 3D object may be similarly characterised, allowing degrees of freedom parameters to define the relative relationships between a given set of molecular images under investigation. Since all images can be normalised with respect to their z-positioning then only the other five degrees of freedom need be considered in the first instance as image-reference descriptions.

However, the first part of the image processing procedure requires that all images be aligned by in-plane translation and/or rotation. Therefore, the three in-plane (image-plane) degrees of freedom (x, y, α) are eliminated as image-reference descriptions. All that remains to describe relative image relationships are the two remaining rotational degrees of freedom (β, γ) that correspond with rotations out of the image-plane. These two rotational degrees of freedom, subtended about y- and x-axes, are sufficient to classify all 2D images following initial alignment to eliminate the contribution of the in-plane degrees of freedom.

An iterative procedure of **multivariate statistical classifications** and **multi-reference alignment** (MRA) processes is then used to sort aligned images into groups of alike-images characterised by having similar (β, γ)-defined orientations. Each set of alike-images is then segregated and summed as appropriate to form a **classum** (class average), thereby improving signal to noise (signal increases in proportion to number summed per classum; noise in proportion to the square root of number summed per classum). Each classum is uniquely defined and related to another classum by its unique (β, γ)-defined orientation.

The process leading to the generation of a 3D structure now requires that the values of (β, γ) for each classum be identified correctly in order to group classi into related (β, γ)-defined **tilt series**. The values of (β, γ) for each classum are also known as the **Euler angles** of a given classum. Euler angles are usually derived using the **common projection line** (CPL) **theorem**. According to the CPL theorem, a common central line projection may be identified in each given pair of 2D images of the same 3D object. In other words, a single, equivalent line (**sinogram**) may be drawn through a given classum pair that traverses equivalent image densities in equivalent distances. The positional matching of sinograms then allows for the relative (β, γ)-orientations between each classum to be established. Tilt series may then be built up through the positional matching of sinograms corresponding to every possible classum pair. Thereafter, the 3D structure of the biological macromolecule may be computer generated directly from these tilt series. The **resolution of the final structure**, d_R, is directly related to the angular distribution of each tilt series (typically $< 60 - 70°$) and the **numbers of individual classi**, n, that comprise each tilt series. This is summarised by Equation (6.13):

$$d_R = \pi \cdot D / n \qquad (6.13)$$

where D is defined as the **maximum particle size**. Given the substantial requirement for alignment, classification and averaging of 2D images in order to derive the 3D structure(s) of biological macromolecules of interest, inherent molecular symmetry is an enormous advantage in obtaining higher resolution structures. For this reason, cryo-electron

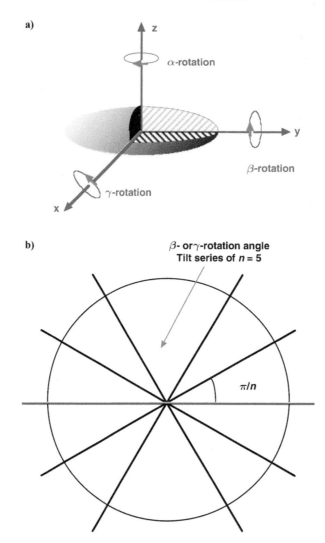

Figure 6.22   Tilt series of cryo-electron microscopy images. Illustration of the basic degrees of freedom that an embedded biological macromolecule may be said to possess (**a**). (**b**) Only variations in β- and γ-rotational degrees of freedom are necessary to classify all images obtained by cryo-EM into (β, γ)-defined **tilt series.** Variations in α-rotation and (x, y, z) Cartesian coordinates are all cancelled out during image processing and manipulation steps.

microscopy derived images of virus particles or highly symmetric protein complexes are amongst the best, highest resolution structures so far obtained by this technique.

6.5.6 Biological macromolecule structures from electron microscopy

Cryo-electron microscopy has progressed significantly in recent years in terms of biological macromolecule structure determination. In fact, many now regard cryo-electron microscopy as a naturally complementary technique to X-ray crystallography. This complementarity with X-ray crystallography has rarely been demonstrated more convincingly than in the characterisation of the *E. coli* GroEL/ES molecular machine (see Section 6.3.6). Cryo-electron microscopy was not only able to demonstrate ATP/ADP-driven conformational changes in the GroEL homo-oligomeric 14-mer (Figure 6.23), not seen by X-ray crystallography, but also in GroEL/GroES/nucleotide complexes, providing structural proof for the changes in molecular shape and topography taking place as part of the "two-stroke motor" mechanism (Figures 6.17 and 6.18) (see Section 6.3.6). The cryo-electron microscopy and X-ray crystallographic images of GroEL/GroES/ATP$_7$ coincide extraordinarily well, so much so that the cryo-electron image could easily pass as the Van der Waals surface description of the X-ray crystallographic structure (Figure 6.24).

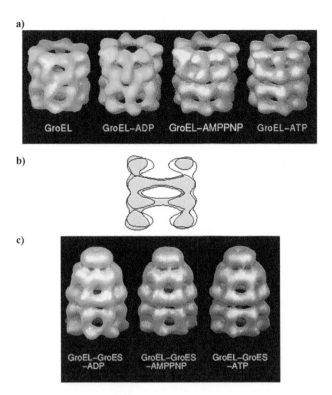

Figure 6.23 Cryo-electron microscopy images of GroEL and GroES. (a) Approximately 30 Å resolution images of GroEL (all **T** state), GroEL/ADP (all **R** state), GroEL/β,γ-**azido-ATP** (**AMPPNP**) (all **R** state), and GroEL/ATP (all **R** state) (see Figure 6.19 for comparison); (**b**) Cross-sectional illustration of GroEL in all **T** state (non-shaded area) and all **R** state (shaded area); (**c**) Approximately 30 Å resolution images of GroEL/GroES/ADP$_7$ (**T** state [bottom ring]; **R** state [top ring]), GroEL/GroES/AMPPNP$_7$ (**T** state [bottom ring]; **R** state [top ring]), and GroEL/GroES/ADP$_7$ (**T** state [bottom ring]; **R** state [top ring]) (see Figures 6.17 and 6.18 for comparison). The variety of conformational variations revealed by cryo-electron microscopy is striking and remarkable (illustrations adapted from Roseman *et al.*, 1996, Elsevier).

Another key success for cryo-electron microscopy has been the visualisation of symmetrical virus structures owing to the high degree of symmetry in virus structures, ideal for characterisation by the sorts of multifold image reconstruction techniques (described in Section 6.5.5) required to construct 3D structures of biological macromolecules from cryo-electron microscopy images. As fine recent examples, please note the illustrated symmetrical protein tegument of **Herpes simplex virus** (**HSV**) and the images of several other virus particles (Figure 6.25). Critically, cryo-electron microscopy has also been able to provide images of proteins embedded in biological membranes such as the **acetylcholine receptor** (**AChR**, see Chapter 7). Critically, in comparison with X-ray crystallography, cryo-electron microscopy avoids the need for biological macromolecular crystallisation and so opens the door to structural analysis of biological macromolecules (especially proteins) that are otherwise not amenable to high-resolution structural analysis owing to the crystallisation problem. However, cryo-electron microscopy could also be said to have two main drawbacks:

1. The need for flash freezing of samples which could be deleterious in some cases.

2. The need for biological macromolecular symmetry for optimal image processing.

Point (2) is the more significant since many macromolecular assemblies do not have the fantastic symmetry of a virus particle or of GroEL or GroES. Therefore, the production of high-resolution images becomes much more difficult where there is lower molecular symmetry, although far from impossible.

In response to these problems, there has been significant progress in the last decade as better and better methods have been developed for electron detection using **complementary metal oxide semiconductor** (**CMOS**) **detector** technologies which have facilitated developments with cryo-EM in particular. Such detectors can read frames continuously at rates in the range of 1–1000 Hz or more. Furthermore, such CMOS detectors are very thin, which helps to reduce backscattering, hence decreasing noise and increasing detector lifetime. This combination of increased frame reading and reduced noise has allowed for higher resolution 3D maps of individual entities to be obtained. Hence, improvements in methods of electron counting and in computational approaches are enabling the transformation of raw 2D images into 3D structures ever more effectively.

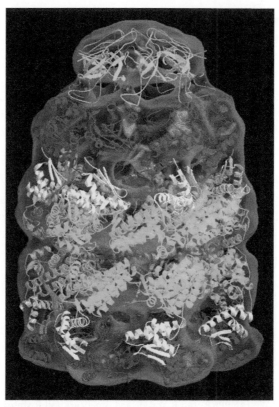

Figure 6.24 Cryo-electron microscopy and X-ray crystallographic images compared. Approximately 30 Å resolution image of GroEL/GroES/(ADP)$_7$ (**T** state [bottom ring]; **R** state [top ring]) within which has been incorporated the approximately 2 Å resolution X-ray crystal structure of GroEL/GroES/(ADP)$_7$ with GroEL and GroES shown in **ribbon display structure** representation (see Figure 6.17 as well). The cryo-electron microscope image has the appearance of a Van der Waals surface representation of the molecular chaperone machine (illustration adapted from Ranson *et al.*, 2001, Elsevier).

Overall, this combination of detectors and software has resulted in a revolution in cryo-EM resolution such that the technique nowadays competes very closely with X-ray crystallography. For example, the structure of an **apoferritin** structure was recently solved to 1.22 Å by cryo-EM. Given that a typical carbon-carbon bond is 1.5 Å then this resolution can be considered to represent true atomic resolution. Apoferritin is often used as a benchmark for cryo-EM due to its 24-fold symmetry which allow for such high-resolution reconstruction. In 2017, the major achievements of cryo-EM, were recognised through the award of the Nobel Prize to Richard Henderson, Jacques Dubochet and Joachim Frank, "for developing cryo-electron microscopy for the high-resolution structure of determination of biomolecules in solution". What is now clear, is that cryo-EM is surpassing X-ray crystallography as the preferred method for the analysis of larger biomolecular complexes. and protein ligand complexes.

6.6 Key principles of scanning probe microscopy

Scanning probe microscopy will increasingly become another class of imaging technique able to provide equivalent and even complementary structural information to cryo-electron microscopy or X-ray crystallography. There are numerous variations upon the scanning probe theme, but two techniques in particular have shown some promise in the imaging of biological macromolecules, namely **scanning tunnelling microscopy (STM)** and **atomic force microscopy (AFM)**.

6.6.1 Scanning tunnelling microscopy concept

The underlying principles of STM are essentially very simple indeed. If a sharp metal tip is placed within a **distance**, d_T, of a few Å from a conducting sample surface and a **bias voltage**, U_z, is applied between tip and surface, a **tunnelling current**, I_T, will be established between tip and surface due to a **quantum mechanical tunnelling** effect that is generated under conditions of "pre-mechanical contact" between tip and sample (Figure 6.26). The current I_T may be used

a)

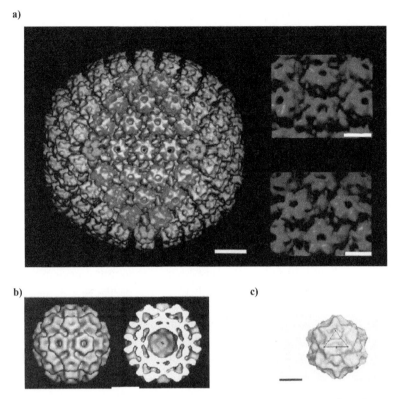

Figure 6.25 **Cryo-electron microscopy images of virus particles.** (a) Contoured surface at 24 Å resolution of the **herpes simplex virus-1 (HSV-1)** capsid tegument (**left**) rendered visually interesting with **depth cueing**, **deblobbing** (noise removal) and the addition of **false colour**. This is achieved with **XIMDISP**, a visualisation tool to aid structure determination. The tegument surface comprises 162 **capsomeres** divided into 12 pentons (**orange**) and 150 hexons (**red**). The hexon homo-oligomers of 6 VP5 (150 kDa) monomers and the pentons are homo-oligomers of 5 VP5 monomers. Additional nodular triplexes (**blue**) reside at sites of three-fold symmetry. The capsid encloses **double-stranded DNA (dsDNA)**. Bar is 20 nm. Close-ups of hexons and pentons are also illustrated with bar now 30 nm (**top and bottom right**) (illustrations from Conway *et al.*, 1996, Figure 3, Elsevier); (**b**) Contoured surface at 23 Å resolution of the **cucumber mosaic virus (CMV)** capsid (**left**) and particle in cross section (**right**) revealing 11 nm inner cavity. The capsid encloses **RNA**. Bar is 10 nm (illustration from Wikoff *et al.*, 1997, Figure 1a, Elsevier); (**c**) Contoured surface at 23 Å resolution of the **canine parvovirus (CPV)** capsid. The capsid is comprised of 60 VP1, VP2 and VP3 monomers and encloses **single-stranded DNA (ssDNA)**. Bar is 20 nm (Agbandje *et al.*, 1995, Elsevier).

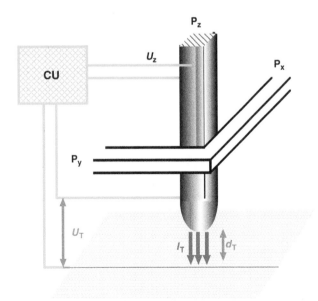

Figure 6.26 **Set up for scanning tunnelling microscopy.** The scanning of a surface takes place in the **near field regime** in which **piezo electric drives** P_x and P_y raster the **STM tip** across the surface of a sample. Piezo electric drive, P_z, ensures that the tip remains a short distance, d_T (a few Å), from surface, encouraging a **tunnelling current**, I_T, to be established under the influence of a tip to surface potential U_T

either to probe physical properties locally at the sample surface or may be used to control the distance of separation d_T between tip and sample surface. Typically, the tip is scanned across a sample surface. Exquisite control of lateral and vertical tip motion is performed with subatomic accuracy by **piezo electric drives** (Figure 6.26). Piezo drives P_x and P_y scan the tip laterally across the surface, whilst voltage supplied by CU to piezo drive P_z seeks to adjust vertical tip position as appropriate. The high resolution of STM scanning derives from characteristics of the quantum mechanical tunnelling effect. Since tunnelling only takes place when d_T is a few Å, then STM may be described as a **near-field imaging** technique whose resolution is neither diffraction nor wavelength limited in contrast to X-ray crystallography and cryo-electron microscopy. The potential for STM-mediated biological macromolecule structural characterisation should now be clear, especially since STM measurements can be made equally well in air or liquid conditions, in addition to more standard *in vacuo* conditions. This is only possible because free electrons are not part of the imaging process.

6.6.2 Electron tunnelling

In a classical sense, if an electron with **total energy**, E, meets a **potential barrier**, V_o, of higher energy then the electron is unable to pass through. However, in a quantum mechanical sense, the electron has a finite probability of location beyond the barrier. This finite relocation probability is a consequence of quantum mechanical tunnelling. Tunnelling is both **elastic** or **inelastic**, depending upon whether energy is preserved or lost as a result of the tunnelling process respectively. The kinetics of tunnelling in relationship to biological electron transfer processes will be discussed in a Chapter 8, so only rudimentary concepts relevant to STM will be mentioned here. In STM, the tunnelling barrier is in fact a 3D potential barrier of arbitrary shape. However, the basic principles do not differ substantially from tunnelling through a uniform uni-dimensional barrier (Figure 6.27). The Schrödinger equation for electron tunnelling through such a simple barrier is given by Equation (6.14):

$$-\frac{h^2}{2m_e} \cdot \frac{d^2\psi_3}{dz^2} = E \cdot \psi_3 \tag{6.14}$$

where **wavefunction** ψ_3 has the solution (6.15):

$$\psi_3 = D_T \cdot \exp(ikz) \tag{6.15}$$

with **integer quantum number**, k, and **barrier tunnelling constant**, D_T. This simple quantum mechanical description allows for the introduction of a number of other important quantum mechanical tunnelling parameters. The first of these is the **barrier transmission coefficient**, T_T, that may be described in terms of **incident current density**, j_i, and **transmitted current density**, j_t, where j_t is equivalent to $|D_T|^2$, as shown in Equation (6.16):

$$T_T = \frac{j_t}{j_i} = \frac{|D_T|^2}{j_i} \tag{6.16}$$

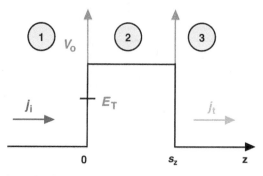

Figure 6.27   1D Tunnelling barrier. Schematic illustration of a tunnelling barrier of length s_z. **Incident current density**, j_i, is transformed into **transmitted current density**, j_t, by tunnelling through the barrier which takes place even though electron energy does not exceed E_T, a value less than the **potential barrier**, V_o

The squared modulus of D_T is the equivalent of a squared-wavefunction, otherwise familiar as the probability density function characteristic of atomic and molecular orbitals. Another important parameter is the **Büttiker-Landauer tunnelling time**, τ^{BL}, that is expressed in terms of Equation (6.17):

$$\tau^{BL} = \frac{m_e \cdot s_z}{\chi \cdot \hbar} \tag{6.17}$$

where s_z is the **variable z-axis barrier dimension** (Figure 6.27), m_e is electron mass and χ represents a **barrier crossing constant** dependent upon total electron energy E and potential barrier, V_o, according to Equation (6.18):

$$\chi = \left[2m_e (V_o - E) \right]^{1/2} / \hbar \tag{6.18}$$

Hence if the actual energetic barrier height (V_o - E) is 4 eV and the z-axis barrier dimension, s_z, is 5 Å, then time, τ^{BL}, corresponds with 4×10^{-16} s. In other words, quantum mechanical tunnelling is an almost instantaneous process. Such a property is almost ideal as the basis for a high-resolution imaging technique.

6.6.3 Piezo electric drives

The heart of control in an STM apparatus is the exquisite sensitivity and fine control of distances exercised by the P_x, P_y and P_z **piezo electric drives** that control tip motion. These drives are either in the form of **bars**, **tubes** or **bimorphs** all comprised of piezo electric ceramics in contact with Macor. A piezo electric bar **changes in length**, Δl_p, as a function of the **potential difference**, U_p, applied between two opposing bar electrodes according to Equation (6.19):

$$\Delta l_p = d_{31} \cdot \frac{l_p}{h_t} \cdot U_p \tag{6.19}$$

where h_t is **bar thickness**, l_p is **bar length** and d_{31} is a **piezo electric coefficient**. Typically, these three bars are combined together to form a **piezo electric tripod** (as illustrated in Figure 6.26), that together render simultaneous control of lateral and vertical tip motion. Bars may be replaced by tubes, in which case Equation (6.19) still applies, but h_t now corresponds to tube wall thickness and U_p to the potential difference applied between two opposing electrodes mounted inside and outside of the tube respectively (Figure 6.28). The tube is more sensitive than the bar owing to much lower tube wall thickness compared with bar thickness. The piezo electric bimorph also appears to be a particularly sensitive physical form of piezo electric drive. To all intents the bimorph is the equivalent of a bimetallic strip. Clamped at both ends, the bimorph comprises two different materials and expands non-uniformly in response to an applied potential difference, U_p, giving a **displacement**, Δx_p, expressed in terms of Equation (6.20):

$$\Delta x_p = \frac{3}{8} d_{31} \cdot \left(\frac{l_p}{h_t} \right)^2 \cdot U_p \tag{6.20}$$

where h_t is **bimorph thickness** and l_p is now **bimorph length** (Figure 6.28). An ultimate extension of the use of piezo electric tubes is the realisation of a **single-tube scanner** (Figure 6.28). The piezo electric tube is fixed at one end and a tip is attached to the other end. Opposing, external electrode pairs (X_{ac}, X_{dc}) and (Y_{ac}, Y_{dc}) are excited by alternating current and direct current sources as indicated, so as to induce tube bending motions that bring about lateral tip motion in x- or y-directions respectively. Tube expansions or contractions along the main tube axis generate vertical tip motion in the z-direction. Such single-tube scanners now appear to be some of the most potent STM devices for high-speed scanning and low thermal drift.

6.6.4 Scanning tunnelling microscopy scanning modes

STM imaging modes are either **constant current imaging** (CCI) or else **constant height imaging** (CHI). The CCI mode is the most commonly used method of imaging. The STM tip is scanned laterally (x- and y-directions) across a sample surface by variations in the respective U_x and U_y potential differences applied to drives P_x and P_y. Vertical (z-direction) tip position is simultaneously adjusted by appropriate changes in the U_z potential applied to drive P_z, keeping the tunnelling current I_T constant between tip and surface. Recorded (U_x, U_y) U_z values are then translated into a $(x, y)z$

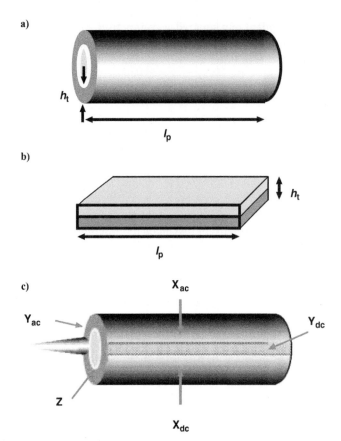

Figure 6.28 **Scanning tunnelling microscopy tubes and biomorphs.** Illustrations of a **piezo electric tube: (a)**, **biomorph (b)** and a complete **single tube scanner (c)**. The tube and biomorph are explained in the text. In the case of the scanner, **external electrodes** (Y_{ac}, Y_{dc}, X_{ac}, X_{dc}) are applied parallel to the tube axis. As voltage is applied to an external electrode, the tube bends away from that electrode. Otherwise, voltage applied to an inside **Z** electrode causes uniform tube elongation.

topographic image. To a first approximation, I_T may be kept constant by maintaining a constant distance between tip and sample surface. Therefore, the topographic image is usually assumed to supply an accurate rendition of real surface atomic arrangements and is considered to represent a real surface contour map. However, in reality, the CCI mode contour map is in fact a constant current surface map that does not necessarily have to correspond with the real surface contours. In the CCI mode, tunnelling current I_T is proportional to the following exponential proportionality (6.21):

$$I_T \propto \exp(-2\chi \cdot s_z) \tag{6.21}$$

The upper limit of CCI mode resolution is determined by the characteristics of the STM tip, assuming that a surface contour map is indeed produced. Where the tip is a semiconductor material, tip-to-surface interactions are presumed mediated by a single p_z orbital. Where the tip is a d-band metal then interactions are presumed mediated by a single frontier d_z^2 orbital. In either case, single orbital-mediated interactions should lead to at least atomic level resolution of surface features.

The orthogonal CHI mode is a variable I_T mode of scanning well suited to real time data collection. Whilst the more common CCI mode is a typically high surface area technique designed to present a surface contour map after data processing, the CHI mode is better seen as a local flat area technique that monitors local dynamic processes as a function of I_T with time such as a receptor-ligand binding or a chemical transformation event. The vertical (z-direction) tip position is not directly available from CCI mode data unless extracted from a detailed knowledge of absolute I_T plotted as a function of lateral x- and y-directions.

6.6.5 Origins of atomic force microscopy

The origins of atomic force microscopy (AFM), also known as **scanning force microscopy (SFM)**, are best appreciated by considering the intermolecular forces involved when a tip is scanned across a surface at **distance, d_z,** of a few Å between tip and surface. If d_z is greater than 10 Å, Van der Waals forces dominate. If d_z is less than 10 Å, then meaningful orbital overlaps begin to take place and quantum mechanical exchange correlation forces apply. Total combined Van der Waals interactions between tip and surface primarily obey a distance relationship (6.22):

$$F_{VDW}(d_z) = -H \cdot (\frac{R_z}{6d_z^2})$$

(6.22)

where R_z is **radius of tip** above a surface and H is a material-dependent constant known as a **Hamaker constant**. AFM becomes a practical reality when F_{VDW} can be monitored as a function of lateral x- and y-directions while an AFM-specific tip is scanned over a surface of interest. One of the primary advantages of AFM that becomes particularly valuable for the imaging of biological macromolecules is that intermolecular forces can be observed irrespective of the nature of tip or sample surface. There is no requirement for tips and sample surfaces to be of uniformly conducting materials in order for forces to be detected. Hence AFM will apply equally well to insulating and conducting samples. Biological macromolecules are typically insulating materials. Obviously, there are other potential surface-to-tip forces that could act upon an AFM-specific tip but F_{VDW} are much the most relevant for the imaging of biological macromolecules.

6.6.6 Atomic force microscopy cantilever

The heart of force detection in an AFM apparatus is a scanning-cantilever microfabricated with a tip for force sensing (Figure 6.29). **Forces generated by surface-to-tip interactions, F_{ST}, lead to vertical tip displacements, Δz,** that are related to each other in a quasi-static way and to a first approximation by a form of the **Hooke's law,** Equation (6.23):

$$F_{ST} = c_{ST} \cdot \Delta z$$

(6.23)

where the so-called **spring constant,** c_{ST} (typically $10-10^5 \, Nm^{-1}$), is further described according to the following Equation (6.24):

$$c_{ST} = \frac{E_M}{4} \cdot w_c \frac{h_c^3}{l_c^3}$$

(6.24)

The term E_M is known as **Young's modulus:** w_c, h_c and l_c are **width, thickness** and **length** of the cantilever respectively. The ideal for an AFM apparatus is a cantilever that possesses the smallest possible spring constant so that any given surface-to-tip force, F_{ST}, results in the largest possible vertical tip displacement, Δz. The best way to accomplish this is to microfabricate a cantilever from Si, SiO_2 or $(Si)_3(N_3)_4$ with substantial length l_c (1–4 mm), modest width w_c (100 μm) and minimal thickness h_c (approximately 1–10 μm). In addition, the sensing tip of the cantilever should be well defined and essentially monatomic. The result would be a cantilever with a spring constant of

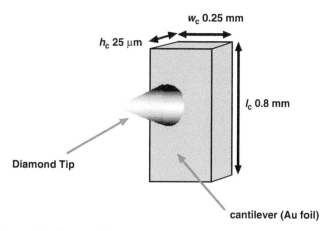

Figure 6.29 Cantilever and diamond tip in atomic force microscopy. Schematic illustration to illustrate potential salient dimensions of an optimal cantilever.

$0.1–1\,\mathrm{N\,m^{-1}}$ whose sensing tip may be displaced by several Å under the influence of surface-to-tip forces as low as 10^{-10} N. Such tip displacements under the influence of surface-to-tip forces may be detected in a number of different ways, of which optical **laser beam deflection** is becoming quite popular and informative (Figure 6.30). Laser light is reflected from a mirror mounted on the back of the cantilever. When a vertical tip displacement takes place, causing simultaneous mirror deflection, a **position sensitive detector** (**PSD**) is able to analyse resulting laser beam deflection and hence compute the original vertical tip displacement, Δz. Such a detection system can be sensitive to Δz displacements of <1 Å.

6.6.7 Atomic force microscopy scanning modes

AFM force imaging modes are closely related to the STM modes described previously. The modes are either **constant force imaging** (**CFI**) or **variable deflection imaging** (**VDI**) that are directly analogous to the CCI and CHI modes respectively, used routinely in STM. CFI is the primary scanning mode for AFM analyses and will be discussed first. In STM, a mobile STM-tip scans a fixed sample surface of interest. More often than not in AFM the reverse is true and a moving sample surface is scanned against a fixed-position AFM-tip. In order to do this, piezo electric drives scan the sample surface laterally (x- and y-directions) against the fixed AFM-tip position by variations in the respective U_x and U_y potential differences applied to drives $\mathbf{P}_x$ and $\mathbf{P}_y$. In the CFI mode, vertical (z-direction) adjustments of the sample surface are then made by appropriate changes in U_z potential applied to drive $\mathbf{P}_z$ so as to maintain the surface-to-tip force, F_{ST}, constant and the vertical tip displacement, Δz, constant as well. Recorded $(U_x, U_y)U_z$ values may then be translated into a $(x, y)z$ topographic image. Once again, to a first approximation a constant surface-to-tip force, F_{ST}, can be considered to result from a constant surface-to-tip distance. Hence this topographic image could be thought of as an accurate rendition of real surface atomic arrangements and should therefore represent a real surface contour map. However, in reality, the CFI mode contour map is in fact a constant force surface that does not necessarily correspond with the real surface. Nevertheless, CFI mode contour maps frequently do provide meaningful and useful images of biological macromolecules (see Section 6.6.8) in complete contrast with the alternate VDI mode that appears to provide little information of direct utility for the characterisation of biological macromolecule structure or function. This is in spite of the fact that the VDI mode is directly analogous to the CHI mode used routinely in STM as a local flat area technique for the monitoring of local dynamic processes such as receptor-ligand binding or chemical transformations.

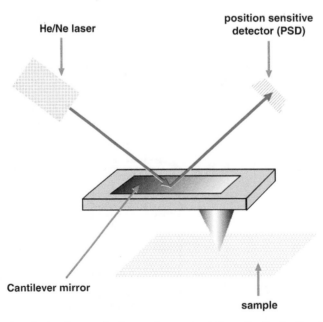

Figure 6.30 **Laser beam detection device in atomic force microscopy.** Schematic illustration to show how force evolution between sample and tip may be detected by displacements in the position of a laser beam reflected from a mirror fixed to the rear of the cantilever. Movements in laser position are a very sensitive measure of surface-to-tip distance changes, Δz, as a result of changes in surface-to-tip forces, F_{ST}.

6.6.8 Biological structural information from scanning tunnelling microscopy and atomic force microscopy

STM and AFM are the most direct techniques for structural analysis because image generation does not involve substantial amounts of mathematical processing in order to obtain an image to view. Of these two techniques, AFM now holds the most utility for the visualisation of biological macromolecules in ambient temperatures and even in buffer conditions. Compared with X-ray crystallography and cryo-electron microscopy, AFM is a technique still in development rather than fully mature. Nevertheless, we believe that it is necessary to include this technique here because of the growing improvements in image generation, the inherent potential for nanometric to atomic level characterisation of biological macromolecular samples with this technique and also because of the potential impact that this technique can have on biological macromolecular structure characterisation. The AFM images of GroEL and GroES should suffice to illustrate how close this technique can come in visualisation and resolution power to cryo-electron microscopy (Figure 6.31). Compare these images with the 3D rendered images of GroEL derived from cryo-electron microscopy previously (Figures 6.23 and 6.24). Moreover, the unique power of AFM to image biological macromolecular samples from nanometric to atomic level characterisation should also become clear from the two AFM images of collagen and cellulose fibres (see Chapter 1) (Figure 6.31), in comparison to these same high-resolution images of GroEL and GroES. Finally, the AFM images of DNA (Figure 6.31), in association with enzymes or a cationic surface, illustrate the power of AFM to render images of biological macromolecules that have no large-scale molecular symmetry, with minimal effort. This is due to the fact that no complex mathematical deconvolutions of data sets as in X-ray crystallography, or intensive image processing as in cryo-electron microscopy, are required to produce AFM images. Literally, meaningful images can be generated in the time it takes to scan an AFM cantilever over a fixed sample surface of interest to which the biological macromolecule is attached.

In recent years, **high speed AFM** (**HS-AFM**) has been developed. HS-AFM allows for the molecular imaging of actual protein molecule dynamics with high spatiotemporal resolution without any marker. For HS-AFM to be effective, the cantilever oscillation amplitude needs to be held constant during lateral (x- and y-direction) raster scanning by moving the sample stage in the z-direction under feedback control. For HS-AFM the feedback loop response speed is increased through short cantilevers with high resonant frequencies that act to minimise the impact of the AFM tip on the sample. HS-AFM can now be used to study proteins, nucleic acids, biopolymers and can even be used potentially for mechanical measurements of entire cells and for whole cell live imaging.

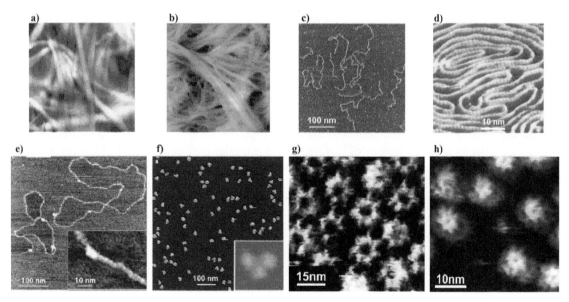

Figure 6.31 Atomic force microscope images at different levels of resolution. All images were generated with false colour: (a) **Collagen fibrils** (Chapter 1); (b) **Cellulose fibres** (Chapter 1); (c) **plasmid DNA** (Chapter 3); (d) DNA attached to cationic surface (Chapter 1); (e) DNA adsorbed onto mica and observed in "**humid air**" with **RNA polymerase** enzyme attached (**bright spot**) attached: (f) **IgG antibody** proteins observed using the fluid "**tapping mode**" in buffer pH 4.0 (see Chapter 7); (g) **GroEL** adsorbed onto mica and observed in buffer pH 7.5 (see Figures 6.18 and 6.23; also Chapter 7); (h) **GroES** adsorbed onto mica and observed in buffer pH 7.5 (see Figures 6.17 and 6.23; also Chapter 7). Considering the simplicity of the technique, images are impressive, not the least when they are recorded in aqueous buffer. Illustrations (a)–(b) Bruker Nano GmbH,www.jpk.com/index.html, last accessed 4 August 2022; (c) A.A. Baker, University of Bristol; (e) from Guckenberger *et al.*, 1994, American Association for the Advancement of Science (AAAS); (f) Digital Instruments,www.di.com, last accessed 4 August 2022; (g)–(h) Muller *et al.*, 2002, Elsevier.

7

Molecular Recognition and Binding

7.1 Molecular recognition and binding in chemical biology

Wherever one cares to look in biology, function and activity is founded upon **molecular recognition and binding** events. These events usually involve interactions between a peptide, protein, nucleic acid, carbohydrate or lipid molecule (**ligand**) and complementary binding sites found in corresponding cognate acceptor molecules (**receptors**), typically proteins, located in lipid membranes or at other key interfaces. Such receptor-ligand interactions are then followed either by chemical catalysis if the receptor is a bio-catalyst (see Chapter 8), or else provoke trans-conformational changes in the receptor that then elicit a range of alternate biological responses. Amazingly the same **non-covalent forces** that create and maintain the structure of biological macromolecules and assemblies (**electrostatic forces, Van der Waals forces, hydrogen bonding** and **hydrophobic interactions**) are the very same that are involved in molecular recognition and binding events (see Chapter 1). However, the way in which these different forces cooperate together to produce the diversity of molecular recognition and binding events found in biology is breathtaking. Hence the chemical biology reader is required to develop a sound understanding of the principles of molecular recognition and binding events in order to begin the journey towards an understanding of the way biology works at the molecular level. Therefore, the objective is this chapter is to map out essential concepts in molecular recognition and binding events, with reference to a few useful biological examples, so that the reader may then have the necessary background to go forward and study other examples of molecular recognition and binding in biology.

7.1.1 Roles of molecular recognition and binding

Molecular recognition and binding events found in biology are ubiquitous, diverse and pivotal. But there are common themes and principles. In order to impart a flavour of this, we will take a brief look at a number of interlocking but diverse examples of biological molecular recognition and binding events. These examples come from fields as diverse as **neurotransmission, bio-catalysis** (see Chapter 8), **immunity** (antibody recognition), **autoimmunity, inflammation, chromatin condensation,** all the way through to the **control of gene expression.** Truly ubiquitous, diverse and pivotal.

7.1.1.1 Acetylcholine, receptor and esterase

Brain tissue is comprised substantially of **neuronal cells** and **glial cells**. The former are the active component of brain tissue responsible for the transmission of sensory, emotional and motor information between brain regions and into the periphery (body) in the form of electrical pulses. The latter are support cells that ensure the metabolic stability

Essentials of Chemical Biology: Structures and Dynamics of Biological Macromolecules In Vitro *and* In Vivo, Second Edition. Andrew D. Miller and Julian A. Tanner.
© 2024 John Wiley & Sons, Inc. Published 2024 by John Wiley & Sons, Inc.
Companion Website: www.wiley.com/go/miller/essentialschembiol2

and structural integrity of neuronal cells. Neuronal cells communicate with each other through **synaptic junctions** that comprise a narrow intercellular gap (**synaptic cleft**) between membrane regions of two adjacent neuronal cells known as the **presynaptic membrane** and the **postsynaptic membrane** (Figure 7.1). Small molecule messengers known as **neurotransmitters** provide communication across the gap between pre- and postsynaptic membranes. In short, electrical pulses travelling down the presynaptic neuron reach the presynaptic membrane where they trigger Ca^{2+}-influx across the membrane, leading to the migration of **synaptic vesicles** containing neuronal transmitters towards the presynaptic membrane. When they reach the membrane, transport vesicles fuse with the presynaptic membrane resulting in the release of substantial neurotransmitter into the synaptic cleft. Neurotransmitter must diffuse across the synaptic junction in order to be recognised by and become bound to binding sites found in **neurotransmitter receptor proteins**. The molecular recognition and binding of the neurotransmitter often facilitates the conversion of the receptor into an ion-channel protein able to re-initiate the propagation of electrical pulses, but this time through the postsynaptic neuron. Such a process is critical to the function of all neuronal cells.

One of the main neurotransmitters in the brain is known as **acetylcholine** (**ACh**) that is recognised and bound by the postsynaptic **acetylcholine receptor** (**AChR**) (Figure 7.2). The molecular recognition and binding of ACh by AChR is a seminal event in neurochemistry and as such is archetypical of all binding events between the many known neurotransmitters and their cognate receptor proteins. Neurotransmitters and their interactions with their cognate receptors are central to the chemistry of the brain and deficiencies (or excesses) of neurotransmitter substances are frequently associated with neurological disease. Indeed, the famous neurodegenerative **Alzheimer disease** (**AD**) has long been associated with deficiencies in neurological levels of ACh, as have various types of clinical depression and memory loss. The exact details and consequences of the molecular recognition and binding of ACh by AChR have yet to be properly elucidated (in spite of the clear importance). However, the molecular recognition and binding of ACh by enzyme biocatalyst **acetylcholine esterase** (**AChE**) is much better known owing to excellent high-resolution X-ray crystallographic data

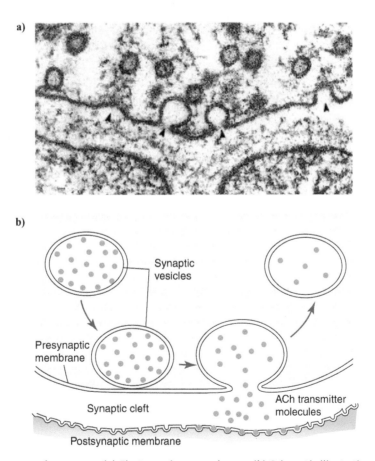

Figure 7.1 Acetylcholine neuronal synapses. (**a**) Electron microscope image; (**b**) Schematic illustration of **synaptic junction** drawn from electron microscope image illustrating the deposition of **acetylcholine** (**ACh**) **transmitter** in the **synaptic cleft** from **synaptic vesicles**. (Voet and Voet, 1995, reproduced from John Wiley & Sons).

a)

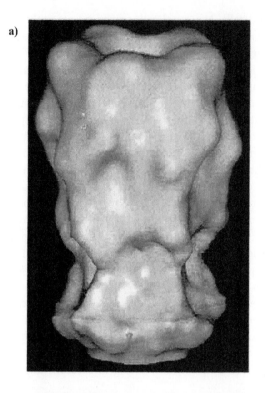

b)

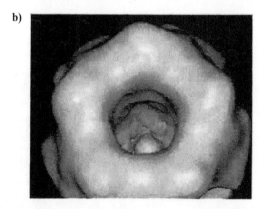

Figure 7.2 Electron microscopy images of acetylcholine receptor. Images of **acetylcholine receptor (AChR)** are as follows: (**a**) Side view, bottom band is the membrane region; (**b**) Top view from the synapse side. (Voet and Voet, 1995, reproduced from John Wiley & Sons).

(Figure 7.3), and in many ways is just as significant. AChE will be discussed again (Chapter 8); therefore all that we need to note here is that AChE has a crucial housekeeping function to metabolise excess ACh remaining in the synaptic cleft post-AChR activation, thereby minimising the possibility of neurotransmitter-mediated excitotoxic damage to neighbouring neuronal cells.

7.1.1.2 Adaptive immunity, antibodies and myasthenia gravis

The immune system can seem impossibly complex and indeed **the study of the immune system (immunology)** is without doubt one of the most complex arenas of biology, comprising a veritable feast of molecular recognition and binding events. In spite of this, there are some useful simplifications once again in that key components of the immune system, such as **antibodies**, are amazingly similar in structure. The schematic shows the main architectural features of all antibodies (Figure 7.4). Four polypeptides, two **heavy chains** and two **light chains**, are linked together covalently by disulfide bonds. Critically, both heavy and both light chains have **constant regions** that differ

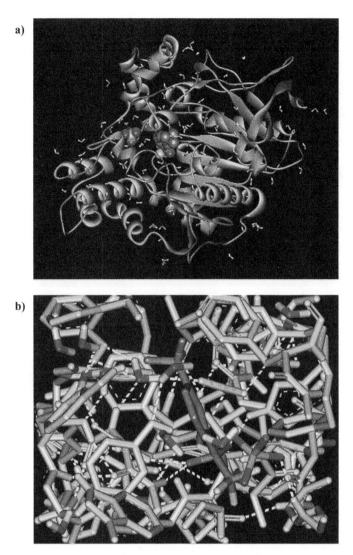

Figure 7.3 X-ray crystallographic structure of acetylcholine esterase. The depictions all derive from one structure of **acetylcholine esterase (AChE)** (pdb: **1amn**). **(a) Ribbon display structure** depiction with X-ray crystallographic water molecules illustrated and bound substrate depicted in the **CPK structure** mode. Colour coding is as follows: hydrogen: **white**; oxygen: **red**; nitrogen: **blue**; carbon: **grey**. In this structure, sulfate (sulfur: yellow) is also included; **(b)** Close-up of active site to show dense hydrogen bonding network (bright yellow). Bound substrates and water molecules are shown in **stick representation.** The protein polypeptide is also shown in **stick representation.**

little between antibodies, irrespective of subclass or type. In addition, both heavy and light chains also have **variable regions** that do vary substantially in amino acid residue sequence between different types of antibody. Each variable region of a given type of antibody can act as variable affinity binding site (**antigen binding site**) for the molecular recognition and binding of a complementary protein or peptide known as an **antigen**. Antigens are **immune system stimulatory** (**immunogenic**) compounds that mobilise the adaptive immune system in response to the formation of antibody-antigen complexes. Immune system diversity is so high in effect that every possible antigen can expect to be matched by a corresponding type of antibody. This antibody is able to recognise and bind to that specific antigen to the exclusion of all others, resulting in a completely specific immune response to that antigen. Such selectivity requires that the antibody type diversity of the immune system must exceed 10^{12}. The chemical biology reader can only marvel that so much diversity in molecular recognition and binding behaviour is possible as a consequence of multiple subtle variations in the amino acid residue compositions of different antibody variable regions.

Remarkably, in the case of the normal functioning of the immune system, self-proteins and peptides do not normally become immune stimulatory antigens. But in autoimmune diseases they do. One such example is **myasthenia gravis (MG)** in which antibodies are generated that target peptide regions (**epitopes**) on the AChR protein, causing destruction of the protein and loss of neuronal cell function (Figure 7.5). If only the process of molecular recognition and binding

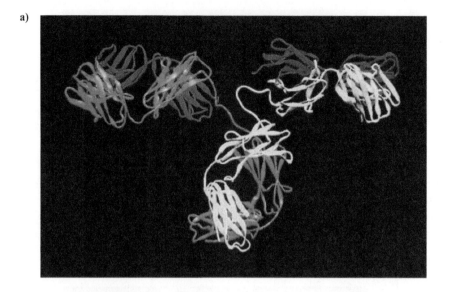

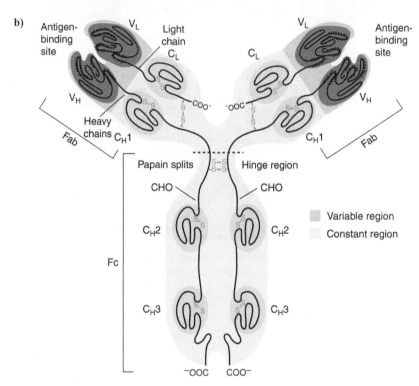

Figure 7.4   Depictions of antibody structures. (a) X-ray crystal structure; **(b) Schematic diagram** drawn from X-ray crystal structure illustrating the conserved structural features of all antibodies. (illustrations (a) and (b) from Voet, Voet and Pratt, 1999, Wiley, Figures 7–33 and 7–34 respectively).

could be adequately understood in this case, then the ravaging consequences of this neurological disorder could be avoided by the design of appropriate inhibitors. An alternative approach might be to study the multicentre interaction of cytokine **interleukin 1β (IL-1β)** (see Chapter 5) with its **cognate receptor IL-1R**. The molecular recognition and binding of IL-1β to IL-1R is an essential signal for inflammation by stimulating the proliferation of immune system cells as well as other agents of inflammation. As with the immune system, the inflammatory system is equally diverse and complex, but underpinned by sequences of molecular recognition and binding events. By studying the interaction between IL-1β and IL-1R (Figure 7.6), there is the promise of being able to identify inhibitors to a molecular interaction that play a very significant role in generating potentially harmful inflammatory effects during the progress of neurodegenerative diseases such as AD and MG.

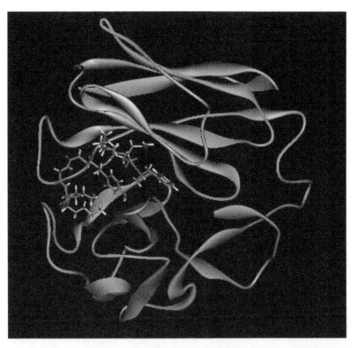

Figure 7.5   X-ray crystallographic structure of antibody fragments. Ribbon display structure illustration of an **F$_v$** (**V$_H$** + **V$_L$**; see Figure 7.4) fragment that binds an AChR-derived peptide (pdb: **1f3r**), with peptide depicted in **stick representation** with carbon (**grey**), nitrogen (**blue**) and oxygen (**red**). This molecular recognition and binding event initiates a cascade of events that leads to **autoimmunity** and **self-destruction** of the **AChR**.

a)   b)

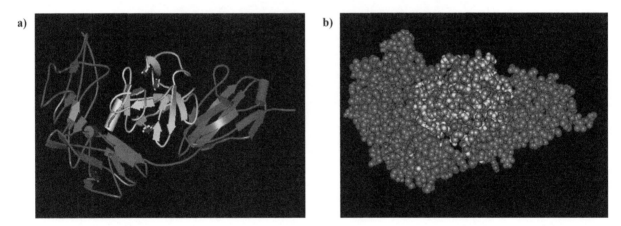

Figure 7.6   X-ray crystallographic structure of cytokine interleukin-1β. Here **interleukin-1β** (**IL-1β**) is shown binding to its cognate **IL-1 receptor** (**IL-1R**) (pdb; **1itb**). (**a**) Side view of IL-1β (**schematic display structure; yellow**) interacting with IL-1R (**schematic display structure; red**); (**b**) Side view of the same but both IL-1β and IL-1R are rendered in **CPK structure** mode to show density of molecular packing.

7.1.1.3 DNA packaging and expression control

One of the great triumphs of molecular recognition and binding in biology is the controlled packaging for storage and unpackaging for function of DNA in eukaryotic cell nuclei. The secret to this is the phenomenal ordering of DNA into **chromosomes** by means of highly cationic proteins. Chromosomes comprise **chromatin filaments** that themselves comprise myriad **nucleosome core particles** (Figure 7.7). Each nucleosome particle contains four **histone proteins** (**H2A, H2B, H3** and **H4**) that are tightly packed and able to distort DNA double helix to wrap around each set of four proteins twice. Each core particle is then clipped together by a rod of **H1 histone protein**. Four nucleosome core particles in a row form a zig-zag arrangement that closes up to form a **solenoid** with six core particles per turn. This

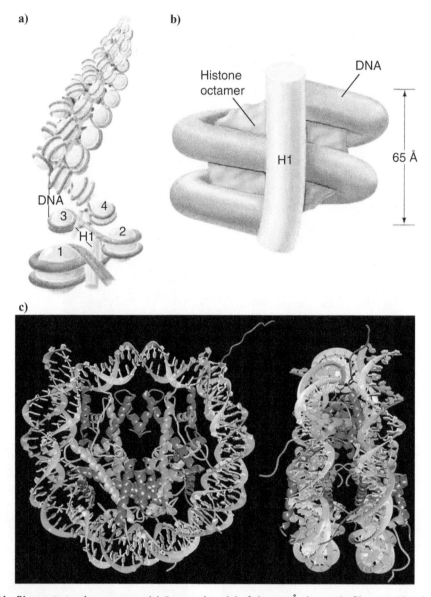

Figure 7.7 Chromatin filaments to chromosome. (a) Proposed model of the 300 Å chromatin filament. The zigzag pattern of **nucleosomes** (1,2,3,4) closes to form a **solenoid** with ~6 nucleosomes per turn; (b) Model of **histone H1** binding to the DNA of the nucleosome; (c) X-ray crystal structure of the **nucleosome core particle.** Two views are illustrated (top left and side right) showing **histone octamer** in **ribbon display structure** depictions (**H2A: yellow; H2B: red; H3: blue; H4: green**). (illustrations (a), (b) and (c) from Voet, Voet and Pratt, 1999, Wiley, Figures 23–48, 23–45 and 23–44 respectively).

solenoid is the basic repeat structure of each chromatin filament. The phenomenal charge-charge recognition by H2A, H2B, H3 and H4 is largely responsible for the extraordinary capacity of these proteins to distort the DNA helical axis from linear to bent with a high degree of curvature. However, little is known about the molecular processes of assembly and disassembly of chromatin filaments before and after gene expression events.

In DNA terms, the other great triumph of molecular recognition and binding is the control of gene expression through DNA-sequence selective protein binding. Proteins that bind DNA and control expression are known as **transcription factors**. One such is the **GCN4 protein** of yeast that was perhaps the first to be studied in molecular detail (Figure 7.8). The extraordinary fact is that DNA has impressively few genes in the total length, especially where eukaryotes are concerned. In total there are 33,000 human genes as against approximately 30,000 nematode genes and a few thousand in most bacteria. Given the fact that there is relatively little difference between the **human genome** and the **nematode genome** in terms of overall gene number then the old adage must apply, that is, it is not how many you have but how you use them. In other words, the high level of expression control (both temporal and cellular) during human development

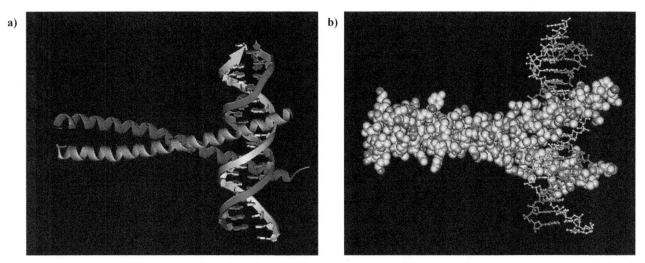

Figure 7.8 **X-ray crystallographic structure of a eukaryotic transcription factor.** The structure of **transcription factor GCN4** includes interactions with **double-stranded DNA** (pdb: **1ysa**). **(a)** Two helices of protein are shown (**red** and **yellow**) in **ribbon display structure**: two strands of DNA are shown in **rings display**; **(b)** Alternative depiction in which protein is shown in **CPK structure** mode (carbon: **grey**; oxygen: **red**; nitrogen: **blue**). Two strands of DNA (**blue** and **yellow**) are shown in **ball and stick representation**.

as compared to the situation in nematode development must be the presiding reason that humans are many-fold more complex as organisms than nematodes. Transcription factors are central to that process of expression control, although not the complete reason. The molecular recognition and binding of DNA by GCN4 rests with the remarkable **scissors grip** where cationic amino acid residues in the *N*-terminal regions of each extended α-helix charge complement anionic phosphates of the phosphodiester backbone and these residues, together with other amino acid residues, make intimate contact with base-pair specific deoxynucleotide sequences within DNA, involving interactions along the DNA major groove. These sequences are non-coding sequences that do not constitute part of any gene but represent control sequences involved in regulating and modulating gene expression through specific interaction with proteins such as transcription factors.

7.1.2 Theoretical framework for molecular recognition and binding events

From a pharmaceutical point of view, any molecular recognition and binding event, such as those mentioned above, could represent a critical drug target to alleviate some disease symptoms or even cure the disease, with obvious benefits for patients. Hence a proper molecular level understanding of recognition and binding events is not just basic knowledge, but may also be a critical prerequisite for successful disease treatment. Therefore, the chemical biology reader can surely have no doubts concerning the importance of research in this arena. Unfortunately, we cannot usually study molecular recognition and binding events without a considerable effort in terms of the synthesis/overexpression and purification of the key biological macromolecular components concerned (see Chapters 2 and 3), followed by structural characterisation where possible and desirable (see Chapters 4, 5 and 6). Only once the appropriate components are available and structurally understood, can studies on molecular recognition and binding events truly begin to make use of a theoretical framework. This theoretical framework will be covered in the next part of the chapter. Before we begin, however, we make one important point. In any discussion (theoretical or practical) about molecular recognition and binding events, there is often a regrettable tendency to ignore the entire preceding molecular recognition process and concentrate only upon binding. Consequently, most discussions about molecular recognition and binding events barely cover the key issues of molecular motions (**translations**, **rotations**, **conformational changes**) prior to binding and how these are perturbed during long-range and then short-range molecular recognition in order to guide a given ligand into making a productive binding encounter with a given cognate receptor. Accordingly, we are presenting a theoretical framework that begins with a short introduction to long-range (early)- and short-range (late)- molecular recognition behaviour before continuing on with binding events.

7.1.2.1 *Motion in solution*

All biological macromolecules in water medium translate by random or **Brownian motion** in the absence of any applied force. Motion becomes organised the moment an external force is applied by means of an external electric or magnetic field, or as a consequence of gravitational or centrifugal effects. Brownian motion represents the origin and basis of biological **macromolecular diffusion**, a term that represents the combined processes of bulk biological macromolecular movement in a given aqueous buffer medium. The translation and rotation of biological macromolecules also has an inverse effect upon overall water medium properties as well. The primary reason for this is that individual biological macromolecules interact with and bind to substantial numbers of water molecules in aqueous buffer medium. Furthermore, in aqueous buffer medium, counter ions and biological macromolecules also associate together such that their motions are coupled. As a result, the movement properties of biological macromolecules are an average of small molecule and large macromolecular properties.

Water hydration and counter ion solvation of biological macromolecules considerably alter the translational and rotational dynamics of biological macromolecules. The effect can be correlated with a number of different solution properties of a given biological macromolecule, such as the **hydrated volume** V_h. The term V_h is given by Equation (7.1):

$$V_h = (M/N_o) \cdot (\bar{V}_{MM} - \delta\bar{V}_{H_2O})$$

(7.1)

where M is the **macromolecular molecular weight**, N_o is **Avogadro's number**, $\bar{V}_{MM}$ is the **macromolecular partial specific volume** (typically about 0.73 cm³g⁻¹ for a globular protein and 0.5 for nucleic acids with an Na⁺ counter ion per phosphodiester link), and $\delta\bar{V}_{H_2O}$ is the **hydration level** (typically 0.3–0.4 for a globular protein). Hydration layers are also typically 2.8 Å thick at the interfaces between biological macromolecules and the aqueous buffer medium in which they reside. Accordingly, hydration effectively corresponds to a **surface-bound monolayer** of water around the entire accessible surface area of a given biological macromolecule, with other water molecules being bound only transiently beyond the confines of this monolayer. In reality, most biological macromolecules do not have a "smooth", surface so a monolayer distribution is not observed, save as a surface-average phenomenon. Instead, macromolecules have a certain surface "roughness" with the result that water molecules of hydration cluster around hydrophilic regions and charges to a greater extent and around hydrophobic regions to a lesser extent (Figure 7.9), for instance six or seven water molecules will cluster around a charged amino acid residue in a protein, but only one water molecule engages tightly with a hydrophobic amino acid residue.

Given the highly interactive nature of water molecules and counter ions with biological macromolecules in aqueous buffer medium, translational or rotational movements of biological macromolecules through buffer medium are significantly impeded. In other words, there are **frictional forces** operating between macromolecules and aqueous buffer medium due to intermolecular interactions that act to retard the rate of macromolecular movement through the aqueous buffer medium and macromolecular rotation. These intermolecular interactions ensure that as each biological macromolecule moves forward then surface-bound and even remote water molecules become displaced. The extent of displacement diminishes with distance from the macromolecule (Figure 7.10). Frictional forces are quantified through the simple definition that a frictional force equals the force required to maintain velocity or angular velocity constant. From this simple definition, equations may be derived.

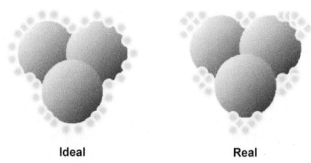

Ideal   Real

Figure 7.9 Smooth surface hydration versus differential surface hydration. A generic biological macromolecule is represented as summation of grey spheres, water as light blue spheres.

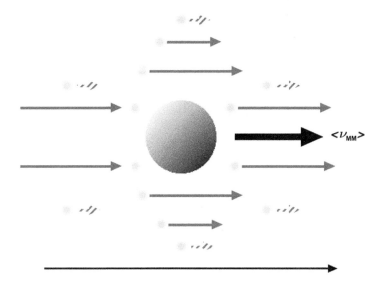

Figure 7.10 Illustration of stick boundary conditions. Solute biological macromolecule, shown as single grey sphere, moves with an **average macromolecular velocity** $<\nu_{MM}>$ through water (shown as light blue spheres). Water molecules in immediate hydration layer move at the same average velocity due to tight hydration interactions. Under **slip boundary conditions**, water molecules do not possess hydration interactions and therefore do not move with the biological macromolecule at all.

In the case of the translational frictional forces opposing translational motion, there are two extremes, known as the **stick boundary condition (strong intermolecular interactions)** and the **slip boundary condition (negligible intermolecular interactions)** that are characterised by the following two basic Equations (7.2) and (7.3) respectively:

$$f_{\text{trans,sph}} = 6\pi\eta \cdot r_{\text{sph}} \tag{7.2}$$

$$f_{\text{trans,sph}} = 4\pi\eta \cdot r_{\text{sph}} \tag{7.3}$$

where $f_{\text{trans,sph}}$ is the **coefficient of translational frictional force** acting on a spherical macromolecule, r_{sph} is the **spherical macromolecular radius** and η is the **viscosity** of the aqueous buffer medium. Equation (7.2) is known as **Stokes law**. Under stick boundary conditions, the water molecules that comprise the immediate hydration layer interact so well with a given biological macromolecule that they move at the same pace as that macromolecule. Under slip boundary conditions, the water molecules that comprise the immediate hydration layer interact to a negligible extent and so have no forward velocity imparted from any given macromolecule (Figure 7.10). In the case of the rotational friction forces that oppose rotational motion, the **coefficient of rotational friction force**, $f_{\text{rot,sph}}$, acting on a spherical molecule tends to zero under slip boundary conditions. Therefore, we only need consider the situation under stick boundary conditions. In this instance, Equation (7.4) applies:

$$f_{\text{rot,sph}} = 6\eta \cdot V_{\text{sph}} \tag{7.4}$$

where V_{sph} is the **spherical macromolecular volume**.

However, it should be obvious that a smooth, perfect sphere makes for a poor approximation of biological macromolecular shape. As a result, the above Equations (7.2), (7.3) and (7.4) do not account very accurately for the frictional forces acting upon a given biological macromolecule in motion. Fortunately, this situation may be resolved to a first approximation for globular proteins by introducing a monolayer of hydration to the sphere and then applying the stick boundary conditions without reservation. In this instance, translational and rotational frictional forces acting on a given biological macromolecule in motion in aqueous buffer medium are then defined by the revised Equations (7.5) and (7.6):

$$f_{\text{trans}} = 6\pi\eta \cdot (3V_{\text{h}} / 4\pi)^{1/3} \tag{7.5}$$

$$f_{\text{rot}} = 6\eta \cdot V_{\text{h}} \tag{7.6}$$

Unfortunately, a uniformly hydrated sphere provides a rather poor approximation of many other biological macromolecules in solution. Instead, an **oblate ellipsoid** or a **prolate ellipsoid** give a much closer approximation to the shape of many biological macromolecules in solution that are not spherical but are compact, globular or irregular rigid bodies (Figure 7.11). In this event, there is a further need to modify the equations that define translational and rotational friction forces. Hence Equation (7.5) becomes Equation (7.7):

$$f_{trans} = 6\pi\eta \cdot F_{trans} \cdot (3V_h / 4\pi)^{1/3} \qquad (7.7)$$

where F_{trans} is known as the translational **Shape factor** or **Perrin factor**. In a similar way Equation (7.6) becomes either:

$$f_{rot, a} = 6\eta \cdot F_{rot, a} \cdot V_h \qquad (7.8)$$

$$f_{rot, b} = 6\eta \cdot F_{rot, b} \cdot V_h \qquad (7.9)$$

respectively, depending upon whether there is primary rotation about the **a**-axis or the **b**-axis of the ellipsoid in solution. In all cases, shape factors are >0. In the event that the biological macromolecule takes on an extended linear form in aqueous buffer solution, then the translational friction forces can be defined on the assumption that the linear form is comprised of linked spherical but non-interacting segments by a simple variation of Equation (7.5):

$$f_{trans} = N \cdot \zeta = N \cdot 6\pi\eta \cdot (3V_{seg,h} / 4\pi)^{1/3} \qquad (7.10)$$

where N is the **segment number**, $V_{seg,h}$ is the **hydrated segment volume**.

Having now discussed hydrated volume, molecular shape and the frictional forces that oppose rotational and translational motion, we are now ready to discuss **diffusion**: the complete set of processes (including Brownian motion) that together bring about the bulk movement of biological macromolecules from one place to another in aqueous buffer solution. The processes that comprise diffusion are quantified by means of the concept of **flux**. Flux is defined as the **rate of mass transport across a unit surface area** (Figure 7.12). Hence, for a two-component system of biological macromolecule in water, the **flux of macromolecular solute**, J_{MM}, in water traversing a unit surface area (1 m² or 1 cm² as appropriate) can be defined by Equation (7.11):

$$J_{MM} = C_{MM} \cdot \langle \nu_{MM} \rangle \qquad (7.11)$$

where C_{MM} is biological **macromolecular concentration** and $\langle \nu_{MM} \rangle$ is **average macromolecular velocity**. Whilst this view of flux gives the correct dimensions and value for J_{MM}, there is a regrettable lack of directionality in this expression. Therefore, flux is often redefined in terms of time-dependent changes in the macromolecular concentration gradient along a defined axis (by convention the x-axis) (Figure 7.12). Since we have selected one from a possible three Cartesian axes, then the concentration gradient must be described in the form of a partial differential proportionality expression:

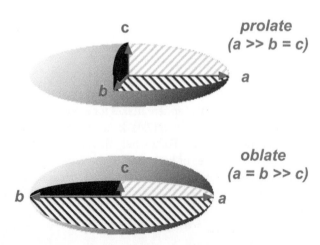

Figure 7.11 **Two non-spheroidal representations of biological macromolecules in solution.**

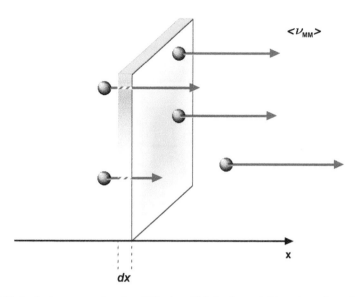

Figure 7.12 **Illustration of biological macromolecular diffusion.** This is shown taking place through an area of **depth** *dx*.

$$J_{MM} \propto -\left(\frac{\delta C_{MM}}{\delta x}\right)_t \tag{7.12}$$

Converting this into an equation, the result is:

$$J_{MM} = -D_{MM}\left(\frac{\delta C_{MM}}{\delta x}\right)_t \tag{7.13}$$

where D_{MM} is known as a **macromolecular diffusion coefficient**. Equation (7.13) is known as **Fick's first law of diffusion**. For biological macromolecules, D_{MM} is usually 10^{-7} to 10^{-6} cm^2 s^{-1} for macromolecules moving according to Brownian motion. This is a lower limit for macromolecular diffusion that is undeniably slow. However, macromolecular diffusion is much enhanced when an external force is applied and also when a molecular recognition and binding event takes place, in which case molecular motion leading to binding is accelerated, as described in the following section. Before such perturbations to Brownian motion are discussed, it is worth noting that there is an attractively simple relationship between those coefficients of translational friction force f_{trans}, expressed according to molecular shape by Equations (7.5), (7.7) or (7.10), and the corresponding **diffusion coefficient at infinite macromolecular dilution**, $D_{o\,MM}$. This relationship is given by Equation (7.14) as follows:

$$D_{o\,MM} = \frac{kT}{f_{trans}} \tag{7.14}$$

where k is the **Boltzmann constant**. Equation (7.14) is known as the **Einstein-Sutherland equation** and shows that $D_{o\,MM}$ is dependent both on temperature T and viscosity η. Therefore, values of D_{oMM} are usually standardised to measurements made with biological macromolecules dispersed in pure water at 20 °C and not aqueous buffer medium.

7.1.2.2 Long-range molecular recognition

Biological macromolecules in translation by Brownian motion alone are inevitably very slow and infrequent. The expression that governs the **encounter rate**, k_a, between two types of biological macromolecule **A** and **B** in aqueous buffer medium is given as follows:

$$k_a = 4\pi N_o \cdot r_{AB} \cdot (D_A + D_B) \cdot 10^3 \, (M^{-1}s^{-1}) \tag{7.15}$$

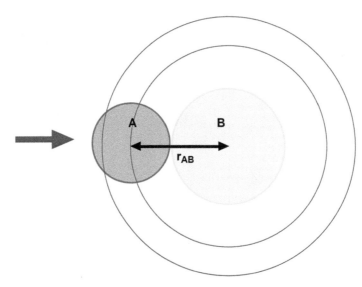

Figure 7.13   Molecular encounters in solution. Ligand **A** collides with the **biological macromolecular receptor B** by traversing through a spherical surface of area $4\pi r_{AB}^2$ that is concentric about **B**. The total flux of **A** through this spherical surface, considering that **B** is also moving, leads to a derivation of the encounter rate k_a (Equation 7.15).

where N_o is **Avogadro's number**, r_{AB} the **sum of both macromolecular radii**, while D_A and D_B are **diffusion constants** for **A** and **B** respectively (Figure 7.13). Note that both r_{AB} and the diffusion constants are expressed in cm units, therefore the 10^3 term is included in Equation (7.15) in order that the encounter rate can be expressed in units of M⁻¹ s⁻¹ (i.e. l mol⁻¹ s⁻¹; where $1 l = 10^3 cm^3$). Typically, values of k_a emerge at 10^8–10^9 M⁻¹ s⁻¹. While this number of molecular encounters per unit concentration per unit time may seem large, there is no certainty that these will be productive binding encounters without additional interventions from long-range molecular recognition processes.

In the case of biological molecular recognition and binding events, recognition is often promoted significantly by the complementary electrostatic surface properties of biological ligands and their cognate receptors. The surfaces of biological macromolecules can be treated as systems of point charges and electrostatic dipoles that interact with each other. Point charges, in particular, provide for especially long-range attractive electrostatic interactions ($1/r$ – dependency; where r is the intermolecular distance) (see Chapter 1 for the expression for the potential energy of monopole-monopole interactions) and are likely to play key roles in mediating the long-range molecular recognition between biological ligands and their cognate receptors. Attractive electrostatic effects not only enhance encounter rates by increasing effective values of diffusion constants, but may also "steer" ligands and cognate receptors to make productive binding encounters with each other owing to the presence of uneven, but complementary, distributions of surface point charges and electrostatic dipoles (Figure 7.14).

These attractive electrostatic interactions are attenuated in aqueous buffer medium due to ionic screening effects from buffer salts. Any biological macromolecule associates with counter ions and also attracts a cloud of more weakly associated ions and counter ions at greater distance from the macromolecular surface. The presence of these weakly associated ions and counter ions alters the **permittivity**, ε, of the aqueous buffer medium and in so doing reduces the magnitude of electrostatic attractions (see Equation 1.2). The attenuating effect of ions and counter ions on electrostatic interactions in aqueous buffer medium can be viewed another way through the **Debye-Hückel theory** that gives us Equation (7.16):

$$r_D = \left(\frac{\varepsilon \cdot kT}{2\rho \cdot N_o \cdot e^2 \cdot I} \right)^{1/2} \tag{7.16}$$

where r_D is known as the **Debye length**, ρ is the aqueous buffer medium **density** and I is the **ionic strength** of the buffer medium in question. The Debye length represents the upper distance limit of separation between two point charges beyond which electrostatic interactions become negligible owing to the decline in electrostatic potential V that takes place as charges separate to greater and greater distances from each other (see Equation 1.2). Equation (7.16) includes several terms relating to the nature of the aqueous buffer medium in which the point charges may be located,

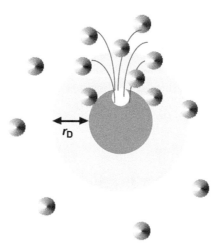

Figure 7.14 Molecular recognition to binding. Electrostatic field lines radiate from systems of point charges and dipoles that are located over the entire surface of the biological macromolecular receptor (red). Influence of charge-dipole systems diminishes to insignificance beyond the **Debye length** r_D from the macromolecular surface. Charge-dipole systems are extensive in the vicinity of the binding site (white region) and radiate electrostatic lines of force (blue) that assist the close-range navigation of ligand molecules (grey) into binding interactions within the cognate receptor binding site by means of close-range and contact forces.

of which ionic strength is especially influential upon the value of r_D. Critically, r_D varies at 5–10 Å in aqueous buffer media as values of I vary at 0.4–0.1 mol kg^{-1}. Such values of I are typical of many physiological aqueous buffer media and therefore 5–10 Å represents an approximate upper distance limit for long-range molecular recognition between cognate **biological macromolecular receptors** and their ligands in aqueous buffer medium (Figure 7.14).

7.1.2.3 Short-range molecular recognition and binding

Once ligands and their cognate receptors come into close proximity, there are a number of factors that come into play in order to maximise the opportunities for productive binding encounters. First and foremost, complementary attractive long-range electrostatic interactions are replaced by complementary attractive short-range electrostatic and van der Waals interactions that also act to "steer" ligands and cognate receptors to make productive binding encounters. This is made all the more possible because macromolecular collisions in aqueous buffer medium usually result in the formation of long-lived, non-covalent encounter complexes (in contrast with small molecules) owing to the extensive disruption of water of hydration at the time of collision, followed by the reinstatement of water of hydration in such a way as to make a new solvent cage around the entire encounter complex, locking ligands and cognate receptors into extended proximity (Figure 7.14). The two macromolecules may then explore one another through numerous short-range collisions and surface diffusion until attractive intermolecular interactions are optimised through docking of ligands into appropriate binding sites in their cognate receptors. Binding strength (see below) is then further optimised during docking by the formation of strong contact interactions such as hydrogen bonds and by careful adaptation between ligands and receptor binding sites. This maximises both attractive short-range interactions and also contributes to binding strength from the hydrophobic effect through the wholesale exclusion of water molecules of hydration from the binding interface (Figure 7.14) (see Chapter 1).

7.2 Theoretical models of binding

Having considered something of the theories of biological macromolecular motion, collisions and encounters in an aqueous buffer medium, we now require a framework with which to be able to study and measure the effectiveness of final binding events themselves. Fundamentally, "everything can be said to bind to everything else" to a greater or lesser extent. Molecules in a biological milieu are continuously associating with and dissociating from neighbour-ing molecules. The vast majority of these binding events are too weak and transient to measure effectively. These are known as **non-specific binding events** (or **non-specific interactions**). Studies of molecular recognition are primar-ily concerned with the tiny minority of strong, stable interactions otherwise known as **specific binding events** (or **specific interactions**). These stand "head and shoulders" above the non-specific background, a bit like spectral peaks projecting above background noise. Specific binding events are the cornerstone of chemical biology.

7.2.1 Single site, single affinity binding

In order to study binding, theoretical models, analytical equations and appropriate constants are needed to character-ise the strength of a given binding event. **Single site, single affinity binding** is the easiest form of binding to model. In that case the binding equilibrium is represented by the following binding scheme (Scheme 7.1), where **R** is receptor (enzyme, protein, etc.) and **L** is ligand. For this simple binding equilibrium, we can define two **equilibrium binding constants** that give a measure of the equilibrium position, either to the right-hand side of the equation or to the left-hand side. These two constants are the **association constant**, K_a, given by the Equation (7.17):

$$K_a = \frac{[RL]}{[R] \cdot [L]} (M^{-1})$$

(7.17)

or the **dissociation constant**, K_d, given by the Equation (7.18):

$$K_d = \frac{[R] \cdot [L]}{[RL]} (M)$$

(7.18)

Either constant will do but they are reciprocals of each other. The principle units of K_a can be defined as per molar (M^{-1}) and the units of K_d as molar (M) (NB: M = mol l^{-1} or mol dm^{-3}). These constants define the strength and integrity of molecular recognition at the single binding site under the equilibrium conditions being investigated. Generally speaking, K_a values of 10^5–10^6 M^{-1} or higher correspond with tight, specific binding equilibria. In other words, K_d values of 10^{-5}–10^{-6} M (or 10–1 μM where μM = μmol l^{-3}). Much of our understanding and apprecia-tion of molecular recognition events derives from our being able to measure and interpret the magnitude of these constants. Therefore, a good deal of this chapter will be concerned with equilibrium binding constants and how we measure them.

In order to measure any constant, we need an equation that relates that constant to variables that may be deter-mined by experiment. For a single site, single affinity binding model, there is a relatively simple way of deriving such an equation. We define **total concentration of receptor** $[R]_o$ according to Equation (7.19):

$$[R]_o = [RL] + [R]$$

(7.19)

so that total receptor concentration is a sum of the concentrations of ligand-bound and ligand-free states. If we now substitute for $[R]$ from Equation (7.17) then we achieve the following:

$$[R]_o = [RL] + \frac{[RL]}{K_a [L]}$$

(7.20)

that rearranges to:

$$[RL] = [R]_o \left(\frac{K_a [L]}{1 + K_a [L]} \right)$$

(7.21)

$$R + L \rightleftharpoons RL$$

Scheme 7.1

Finally, let us introduce **B** that represents the number of moles of ligand **L** bound per mole of receptor **R** (**B** may also be defined as the combined mole fraction of ligand **L** bound to receptor). This is given mathematically by the simple ratio represented by Equation (7.22):

$$B = \frac{[RL]}{[R]_o} \qquad (7.22)$$

Substituting for [**RL**] from Equation (7.21) then gives us the following expression that happily relates K_a to experimental variables:

$$B = \frac{K_a \cdot [L]}{1 + K_a \cdot [L]} \qquad (7.23)$$

Equation (7.23) is a classical rectangular hyperbola as shown by a typical plot of **B** versus [**L**] (Figure 7.15). This plot is sometimes known as a **binding isotherm**. A similar rectangular hyperbola is also seen in biocatalyst studies, reflecting the fact that if sufficient ligand is present, then all the active sites in a receptor should be effectively occupied or saturated with ligand on a continuous basis with time. The binding isotherm and related functions is absolutely fundamental to biological molecular recognition and binding. Very commonly, Equation (7.23) is transposed into a linear form:

$$B = 1 - \frac{1}{K_a} \cdot \frac{B}{[L]} \qquad (7.24)$$

which is known as a **Scatchard equation**. This linear form is also illustrated (Figure 7.16). It is impressive how often this very simple equation surfaces in the analysis of equilibrium binding constants, as we shall see. This equation also has the necessary virtue that the two variables **B** and [**L**] may be measured with relative ease by a wide range of possible techniques as described later on in this chapter.

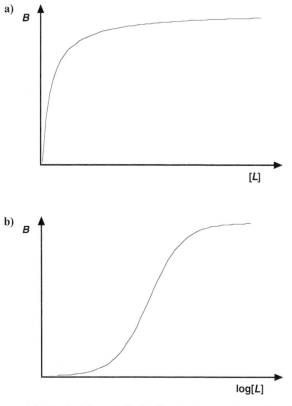

Figure 7.15 Ideal binding isotherms. (a) Classical hyperbolic binding isotherm obtained by plotting values of **B** against [**L**]; **(b)** Classical semi-log plot obtained by plotting values of **B** against **log[L]**. Appearance of sigmoidal shape implies that binding interactions between ligand and receptor are >70% saturated and are therefore appropriate to derive accurate **association constant** K_a or **dissociation constant** K_d values.

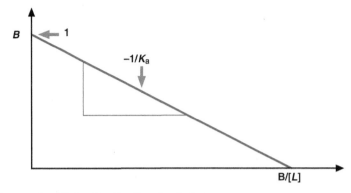

Figure 7.16   Ideal single site, single affinity binding Scatchard plot.

7.2.2 Independent multiple site, equal affinity binding

The Scatchard equation can be very simply adapted for **independent multiple site, single affinity binding** as well. In this case, the binding equilibrium equation can be represented by the following binding scheme (Scheme 7.2), where n corresponds to the total number of independent binding sites located on the receptor **R**. Equal site affinity is critical. If the affinity of each binding site was different for each ligand **L**, then we would be forced to use a **multiple site, variable affinity binding model,** which is a great deal more complicated (see Section 7.2.3). However, where independent binding sites are involved of approximately equal affinity then the combined mole fraction for binding can be simply expressed as:

$$B = \sum_{i=n}^{i=1} [RL]_i / [R]_o \tag{7.25}$$

By analogy to the derivation of Equation (7.23), we can then substitute for the sum of $[RL]_i$ terms with a sum of rectangular hyperbolic expressions as follows:

$$B = \sum_{n}^{1} \frac{K_a \cdot [L]}{1 + K_a \cdot [L]} \tag{7.26}$$

where K_a is now the **site association constant** for *each* individual i-th binding site. K_a is identical for each individual binding site, hence the simple sum expression can be converted into the simple product expression below:

$$B = n \cdot \frac{K_a \cdot [L]}{1 + K_a \cdot [L]} \tag{7.27}$$

Equation (7.27) can then be transposed into the linear form:

$$B = n - \frac{1}{K_a} \cdot \frac{B}{[L]} \tag{7.28}$$

This linear form Equation (7.28) is also illustrated in Figure 7.17. Note that B becomes equal to n at complete saturation binding when $[L]$ is in large excess and all receptor-binding sites are occupied.

7.2.3 Independent multiple site, variable affinity binding

Independent multiple site, variable affinity binding is necessarily much more complicated and demands some quite fearsome mathematics to model correctly. For this reason, we would try and apply simpler binding models wherever possible. In the case of independent multiple site, variable affinity binding, binding behaviour cannot be reduced to a single equilibrium binding event, but is comprised of a series of binding equilibria that each fit the general scheme (Scheme 7.3), where the total number of independent binding sites of variable affinity located on receptor **R** is n and

$$\mathbf{R} \quad + \quad n\mathbf{L} \quad \rightleftharpoons \quad \mathbf{RL}_n$$

Scheme 7.2

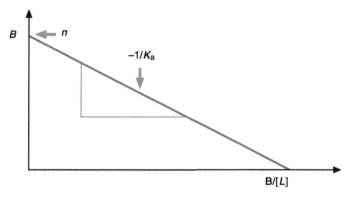

Figure 7.17 **Ideal independent multiple site, single affinity binding Scatchard plot.**

the final saturating binding equilibrium corresponds with the situation when i equals n. Each binding equilibrium for a given site has independent **stoichiometric equilibrium binding constants**. By analogy to Equation (7.17), the equation for each **stoichiometric association constant, $K_{a,i}$,** is as follows:

$$K_{a,i} = \frac{\left[RL_i\right]}{\left[RL_{i-1}\right]\cdot\left[L\right]} \tag{7.29}$$

In comparison with the previous section, the relationship between B and the complete set of $K_{a,i}$ constants does not involve a simple sum converting through into a simple product expression, but is instead a complex polynomial expression of degree n:

$$B = \frac{K_{a,1}\cdot\left[L\right] + 2K_{a,1}K_{a,2}\cdot\left[L\right]^2 + \ldots i(K_{a,1}K_{a,2}\ldots K_{a,i})\cdot\left[L\right]^i + \ldots n(K_{a,1}K_{a,2}\ldots K_{a,n})\cdot\left[L\right]^n}{1 + K_{a,1}\cdot[L] + K_{a,1}K_{a,2}\cdot[L]^2 + \ldots(K_{a,1}K_{a,2}\ldots K_{a,i})\cdot\left[L\right]^i + (K_{a,1}K_{a,2}\ldots K_{a,n})\cdot\left[L\right]^n} \tag{7.30}$$

undeniably a complex expression. There are n terms in this equation corresponding with the n roots of the polynomial representing the number of independent binding sites in the receptor **R**. This is known as a **real sites equation**, with good reason as we shall see. In order to solve such an equation, the polynomial needs to be equated with another algebraic expression. This new expression is known as a **ghost sites equation**, which has all the appearance of Equation (7.26), except comprised of **stoichiometric ghost site association constants** that are all different from each other:

$$B = \frac{K_\alpha \cdot\left[L\right]}{1 + K_\alpha \cdot\left[L\right]} + \frac{K_\beta \cdot\left[L\right]}{1 + K_\beta \cdot\left[L\right]} + \ldots\frac{K_\nu \cdot\left[L\right]}{1 + K_\nu \cdot\left[L\right]} = \sum_\nu^\alpha \frac{K_\omega \cdot\left[L\right]}{1 + K_\omega \cdot\left[L\right]} \tag{7.31}$$

Do not confuse K_α through to K_ω with the **stoichiometric real site association constants** in Equation (7.30). The ghost site constants are not real numbers, but are in fact complex numbers, pure mathematical conveniences that allow us to write Equation (7.31) in a form that is more amenable to solution and then equate this to the polynomial expression (7.30).

A simple example should serve to illustrate how equations of the form (7.30) and (7.31) can be used in combination to solve for stoichiometric real site association constants. Imagine we have a receptor **R** with two independent binding sites of two different affinities. In a real scenario, four equilibria and two stoichiometric association constants account for the interaction between two separate ligands **L** and this receptor (Figure 7.18). Two independent equilibria then make up a parallel imaginary scenario (Figure 7.18). A real sites equation (polynomial of degree 2) may be written to account for the real scenario and then directly equated with a ghost sites equation containing two summed terms harbouring two different complex numbers K_α and K_β as shown (see Equation 7.32):

$$\frac{K_{a,1}\cdot\left[L\right] + 2K_{a,1}K_{a,2}\cdot\left[L\right]^2}{1 + K_{a,1}\cdot\left[L\right] + K_{a,1}K_{a,2}\cdot\left[L\right]^2} = B = \frac{K_\alpha \cdot\left[L\right]}{1 + K_\alpha \cdot\left[L\right]} + \frac{K_\beta \cdot\left[L\right]}{1 + K_\beta \cdot\left[L\right]} \tag{7.32}$$

$$RL_{i-1} \quad + \quad L \quad \rightleftharpoons \quad RL_i$$

Scheme 7.3

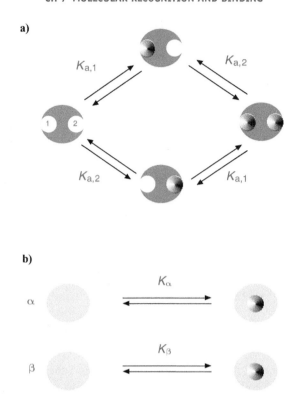

Figure 7.18   Independent multiple site, variable affinity binding equilibria. (a) Diagram illustrates the emergence of four different equilibrium constants to describe the binding interactions between ligands (grey) and two independent binding sites of significantly different affinities found in a biological macromolecular receptor (red); **(b)** Imaginary "parallel" single site, single affinity ghost sites equilibria used to help derive a unique solution for equilibrium constants applicable to situation shown in (a).

Experimental values of B and $[L]$ may then be used to solve for K_α and K_β in the right-hand equality, which are in turn used to extract values of $K_{a,1}$ and $K_{a,2}$ from the left-hand equality. It should be clear how this approach can be expanded in general to include receptors with more than two independent binding sites of different affinities using additional roots of the polynomial on the left and additional imaginary terms on the right.

7.2.4 Dependent multiple site, cooperative binding and Hill equation

The moment that we lift the constraint of independent binding sites, the number of possible interlinked equilibria, each with its own stoichiometric equilibrium binding constants, might be expected to jump alarmingly, together with the complexity of the analysis. However, perhaps surprisingly, when there is a tendency for ligand binding at any one receptor-binding site to enhance ligand binding persistently at neighbouring binding sites (**positive cooperativity**) then binding behaviour can be accounted for in terms of a considerably reduced form of Equation (7.30) as shown (Equation 7.33):

$$B = \frac{n\left(K_{a,1}K_{a,2}\ldots K_{a,n}\right)\cdot[L]^n}{1+\left(K_{a,1}K_{a,2}\ldots K_{a,n}\right)\cdot[L]^n} \tag{7.33}$$

Such a simplification is possible because positive cooperativity ensures that only end point saturation binding equilibria can dominate ligand-binding behaviour and hence only the highest power terms are relevant in an expression for B. Equation (7.33) can be simplified even further to give Equation (7.34):

$$\frac{B}{n} = \frac{[L]^n}{K'+[L]^n} \tag{7.34}$$

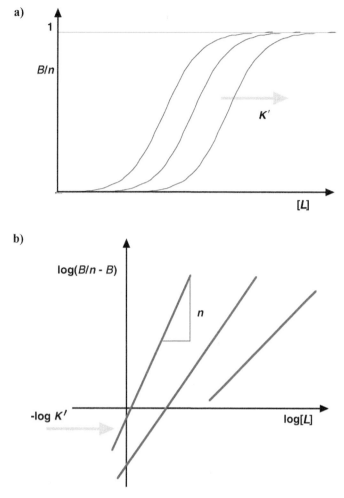

Figure 7.19 Dependent multiple site, variable affinity cooperative binding equilibria. (a) Classical sigmoidal binding isotherms indicative of ligand-receptor interactions that involve strong **positive cooperativity.** Curves move to right as composite equilibrium constant K' increases; **(b)** Linear **Hill plots** derived from sigmoidal data illustrated in (a). Gradients and intercepts define values of n and K' respectively.

that is in the form of a sigmoid (Figure 7.19). The linearised form of this equation is:

$$\log\left(\frac{B}{n-B}\right) = n \cdot \log[L] - \log[K'] \qquad (7.35)$$

This is the **Hill equation** for cooperative binding of ligands by a receptor (see Chapter 8 for the Hill equation equivalent for cooperative biocatalysis). The term n represents the maximum value of B at saturation as well as the effective number of binding sites involved in cooperative binding activity (Figure 7.19). For the sake of completeness, when there is a tendency for ligand binding at any one receptor-binding site to reduce ligand binding at neighbouring binding sites (**negative cooperativity**) persistently, then binding behaviour can also be accounted for in terms of a considerably reduced form of Equation (7.30), that is equivalent to Equation (7.23).

7.3 Analysing molecular recognition and binding

Now that we have covered the main theoretical models of binding and molecular recognition events, including their analytical equations and equilibrium binding constants, we must show how equilibrium binding constants can be derived from the experimental equivalents of B and $[L]$.

7.3.1 Equilibrium dialysis

This is the most ideal method for determining accurate experiment evaluation of B and $[L]$ and hence equilibrium dissociation and association constants. In brief, receptor R and ligand L are combined on one side of a semi-permeable membrane in a dialysis device. Ligand L is able to traverse the membrane but not the receptor R. Therefore, when the system is allowed to reach equilibrium, then the receptor/ligand binding equilibrium on the one side of the membrane is matched by an equilibrated pool of free ligand L on the other side of the membrane (Figure 7.20). The concentration of free ligand $[L]$ on either side of the semi-permeable membrane is equivalent. Given this, very accurate values of B may be determined as follows. First, **free ligand concentration $[L]$** is measured, after which the **total molar quantity of ligand bound to receptor**, m_{RL}, is measured, knowing the **total volume V_o**, of the system and the **total molar quantity of ligand added initially**, m_{L_o}, as follows:

$$m_{L_o} - V_o [L] = m_{RL} \tag{7.36}$$

From this, the corresponding value of B may be determined according to the following relationship:

$$B = \frac{[RL]}{[R]_0} = \frac{m_{RL}}{m_{R_o}} \tag{7.37}$$

provided that the **total molar quantity of receptor m_{Ro}** present is known accurately as well. By combining expressions 7.36 and 7.37 we obtain the following:

$$B = \frac{\left(m_{L_o} - V_o [L] \right)}{m_{R_o}} \tag{7.38}$$

Provided that values of B are then determined under identical experimental conditions, but with different values of m_{L_o}, then an accurate binding isotherm will be generated that obeys the usual hyperbolic binding curve given by Equations (7.23) or (7.27), or alternatively the sigmoidal curve associated with positive cooperativity (given by Equation 7.34). However, whilst equilibrium dialysis is optimal for the determination of equilibrium binding constants, this technique is frequently impracticable for studying the interaction of ligands with most biological macromolecules since these are usually too unstable and usually unavailable in the quantities required to effect a complete set of equilibrium dialysis experiments.

7.3.2 Titration methodologies

There are many ways in which to achieve a **titration binding experiment**, but the underlying principles are quite similar to each other. Frequently, a fixed concentration of receptor R is titrated with ligand L until approximately all the receptor binding sites are occupied; a state known as **saturation** (sometimes it may be more appropriate to do the reverse, titrating a fixed concentration of ligand L with receptor R until saturation is achieved). Progress towards saturation is monitored using any one of a number of physical or spectroscopic techniques. In each case, we are looking for a progressive change, Δx, in a selected physical property or spectroscopic signature. At each stage of the titration,

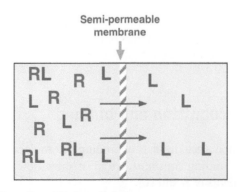

Figure 7.20 **Diagrammatic representation of equilibrium dialysis.**

Δx is identifiable with B provided that the change is a direct consequence of ligand binding alone and equilibrium has been properly established. In this case, Δx at each stage in the titration is a sum of the physical property/spectroscopic signature of **bound ligand**, x_b, and that of **free ligand**, x_f, in the presence of receptor. This is illustrated by Equation (7.39):

$$\Delta x = \alpha x_b + (1 - \alpha)x_f - x_{back} \tag{7.39}$$

where α corresponds to the **fraction of ligand bound** at a given stage in the titration and x_{back} to background contribution from free ligand (in the absence of receptor) and free receptor (in the absence of ligand). When Δx is plotted as a function of **total ligand concentration** $[L]_o$ an experimental binding isotherm will result (Figure 7.21), or else a sigmoidal isotherm, depending upon whether or not positive cooperativity is involved in binding (Figure 7.21). At this point, a decision needs to be taken about which theoretical binding model is most appropriate for the receptor **R** and ligand **L** under investigation, after which data is fitted with the most appropriate analytical equation to extract equilibrium binding constants.

7.3.2.1 Titration data estimations

A number of factors will affect values of estimated equilibrium binding constants. Obviously, the theoretical binding model should be as close to reality as possible. We would normally start with the simplest model first and make things

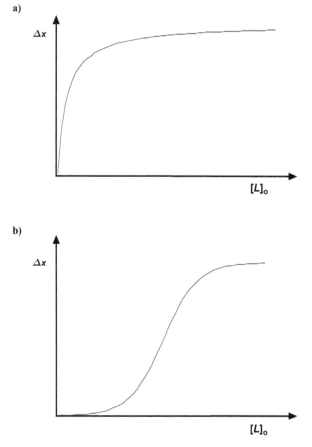

Figure 7.21 Experimental titration binding isotherms. (a) Hyperbolic titration binding isotherm obtained by titrating fixed **[R]** with increasing values of **[L]**$_o$, and then plotting observed Δx against **[L]**$_o$. Hyperbolic binding isotherm is observed characteristically when **single/independent multiple site, single affinity** binding interactions occur; **(b)** Sigmoidal titration binding isotherm is characteristic of **dependent multiple site, variable affinity** binding interactions taking place between ligand and receptor with strong **positive cooperativity.**

more complicated if necessary. However, please note that most experimental binding data is plotted using $[L]_o$ whereas the actual analytical equations involve only free ligand concentration $[L]$. Most experimental methods for observing binding events do not explicitly involve a precise determination of $[L]$, but instead assume that $[L]$ and $[L]_o$ are approximately the same. Obviously, they are not, especially at concentrations of $[L]_o$ less than any K_d values. Fortunately, we can avoid problems with using this approximation provided that the experimental data set converges to saturation as closely as possible. The best way to verify this is to plot Δx versus $\log[L]_o$ (**semi-log plot**). If the experimental data plots out in the form of a sigmoid approaching an asymptotic plateau at high values of $\log[L]_o$ (see Figure 7.15; where Δx is equivalent to B), then we can be sure that receptor **R** binding sites are approaching saturation and that the experimental data set is adequate for the determination of binding constants. Obviously, this semi-log plot diagnostic approach only applies when strong positive cooperativity is **not** involved in binding. Alternatively, performing titrations at concentrations of receptor equivalent to approximate K_d values increases the chance for saturation too.

7.3.2.2 Physical properties vs spectroscopic signatures

Typical physical properties that can change with receptor-ligand binding interactions are, for example: (1) **enthalpy (heat)**; (2) **overall charge**; and (3) **solution refractive index changes**. These changes in physical property can be observed from low to saturating concentrations of ligand by means of techniques such as **isothermal titration calorimetry** (ITC), **capillary electrophoresis**, and **resonant mirror biosensing** respectively (see Sections 7.3.3, 7.3.4 and 7.3.5 respectively). Spectroscopic signatures that change with receptor-ligand binding interactions are numerous, but most frequently used are **fluorescence, nuclear magnetic resonance, circular dichroism** and **UV-visible** spectroscopic signatures in this context. The corresponding spectroscopic techniques have been described previously (Chapters 4 and 5) so will not be covered significantly here, except where they feature in worked examples of studies on molecular recognition and binding events. Where physical properties are used to study receptor-ligand binding interactions then observed changes are typically in the physical state of the whole system of interacting molecules. Where spectroscopic signatures are used to study receptor-ligand binding interactions then observed changes are typically caused only by changes in the spectroscopic behaviour of selected spectroscopically active functional group(s) as a consequence of local environment changes experienced upon binding. Having said this, there are otherwise no fundamental differences in approach between studies on molecular recognition and binding events conducted using changes in physical properties or changes in spectroscopic signatures to report on receptor-ligand binding interactions.

7.3.3 Isothermal titration calorimetry and binding thermodynamics

ITC is almost the ultimate titration methodology in that this technique is based entirely upon titration of heat energy and then deconvolution of this information into equilibrium binding constant information. However, the real beauty of this technique is that it engages directly with the thermodynamics of receptor-ligand binding interactions.

7.3.3.1 Equilibrium thermodynamics of molecular recognition and binding

Every species, i, in aqueous solution is credited with a **chemical potential**, $\mu_i(\text{aq})$, at a given **concentration**, c_i, that is defined according to Equation (7.40):

$$\mu_i(\text{aq}) = \mu_i^o(\text{aq}) + RT\ln(\gamma_i \cdot c_i / c_r) \tag{7.40}$$

where $\mu_i^o(\text{aq})$ is the **standard chemical potential** at a **standard concentration**, c_r, of 1 M ($1\,\text{mol}\,\text{l}^{-1}$) and γ_i is an **interaction parameter**. An **ideal solution** is defined as one in which there are no solute-solute interactions and for which the interaction parameter is therefore equivalent to 1. Clearly solute-solute interactions must exist in reality for molecular recognition and binding events to take place so there is no such thing as a completely ideal solution involving biological macromolecules. Essentially, $\mu_i^o(\text{aq})$ is a function of solute-solvent interactions, while γ_i is a function of solute-solute interactions that tend to a value of 1 as interactions decline to zero.

Consider the original binding equilibrium (Scheme 7.1). By definition, equilibrium is reached when there is no chemical potential difference between both sides of the equilibrium. That is:

$$\mu_R^{eq} + \mu_L^{eq} = \mu_{RL}^{eq} \tag{7.41}$$

By substitution into Equation (7.41) from Equation (7.40) and the reorganisation of terms, we arrive at the following, containing a left-hand side expression that is a difference of standard chemical potentials:

$$\mu_{RL}^{o} - \mu_{R}^{o} - \mu_{L}^{o} = RT \ln\left(\frac{[R]_{eq}[L]_{eq}}{[RL]_{eq} \cdot c_r}\right) \tag{7.42}$$

By definition, the left-hand side of the equation is equivalent to the **standard free energy change of binding** $\Delta G^{o}(T)_{bind}$, therefore Equation (7.42) can be further adapted to give:

$$\Delta G^{o}(T)_{bind} = -RT \ln\left(\frac{[RL]_{eq} \cdot c_r}{[R]_{eq}[L]_{eq}}\right) \tag{7.43}$$

In the brackets is the **absolute association constant** K_a^{o} that is related directly to the experimental association constant introduced in Equation (7.17) according to the following simple equation:

$$K_a = \frac{K_a^{o}}{c_r} \tag{7.44}$$

Hence the key thermodynamics equation for binding is:

$$\Delta G^{o}(T)_{bind} = -RT \ln(K_a \cdot c_r) \tag{7.45}$$

By means of this equation, measured association constant values can be converted directly into a key thermodynamic parameter. Other binding thermodynamic parameters may be obtained through thermodynamic equations or preferentially measured by ITC. The key parameters are **standard enthalpy change of binding**: ΔH_{bind}^{o} and: **standard entropy change of binding** ΔS_{bind}^{o}. These quantities relate directly to: $\Delta G^{o}(T)_{bind}$ according to the well-known and central thermodynamic relationship:

$$\Delta G^{o}(T)_{bind} = \Delta H_{bind}^{o} - T\Delta S_{bind}^{o} \tag{7.46}$$

7.3.3.2 Enthalpy of binding and isothermal titration calorimetry

By definition, a **change in enthalpy** of a given system dH equates to **total heat energy**, q, added (or subtracted) provided that this heat energy is transferred under conditions of constant temperature and pressure and that the system is **thermodynamically closed** (meaning that no chemical matter is added or subtracted). The fundamental equation that defines this situation is:

$$dH = q + Vdp \tag{7.47}$$

where V is **constant closed system volume** and dp is **change in pressure**, which should be zero for most situations involving biological macromolecular interactions. In other words, enthalpy change corresponds directly to heat energy added to or subtracted from a closed system of interest, such as the mixture of a given ligand and its cognate biological macromolecular receptor interacting with each other in solution.

An ITC titration experiment seeks to measure the total heat energy, q, exchanged (i.e. either taken up or released) each time an aliquot of ligand $L[dn^{o}(L)]$ is injected into a stirred and equilibrated solution of cognate receptor, **R**, in a calorimeter vessel (Figure 7.22). The total heat energy exchanged per injection is then given by:

$$q/dn^{o}(L) = \Delta H_{bind}^{o} \cdot \frac{\left[1/2 + \left\{[1-(1/2)\cdot(1+K_d)]-[RL]/2\right\}\right]}{\left\{[RL]^2 - 2[RL](1-K_d) + (1+K_d)^2\right\}^{1/2}} \tag{7.48}$$

The derivation of this equation is well beyond the scope of this book. However, this equation does show that the primary experimental output from an ITC heat titration experiment is: ΔH_{bind}^{o} and the experimental dissociation constant is K_d. The total heat energy exchanged as a function of ligand-receptor interactions is at its greatest at the beginning of a given titration since most of the molecules in each injected aliquot of ligand bind to receptor and either

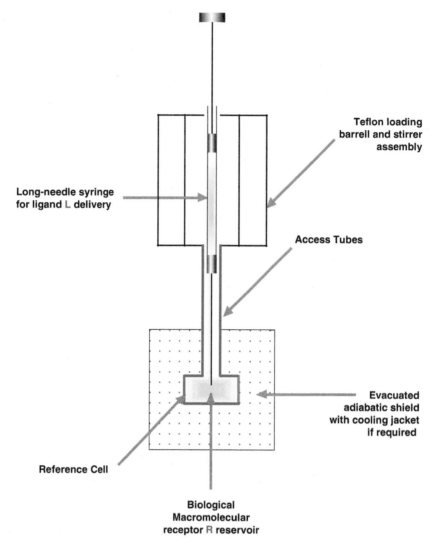

Figure 7.22   Schematic diagram of isothermal titration calorimeter. A solution of receptor **R** is maintained in an **isothermal titration calorimeter** (**ITC**) cell within an adiabatic shield and ligand **L** is injected in with stirring from above.

take up or release heat energy, as appropriate (Figure 7.23). As the titration proceeds, the total heat energy exchanged must decline since more of the ligand binding sites become occupied and are prevented from participating in heat exchanges. The total heat energy exchanged will tend to zero as the titration reaches completion, provided that there is no significant contribution to total heat energy, q, exchanged from other concomitant effects with each injection, such as dilution effects (dilution effects can be very significant depending upon the biological macromolecular ligands and/ or receptors that are being diluted during the titration process).

7.3.3.3 Van't Hoff relationships

A key thermodynamic equation for the characterisation of binding events is the **Van't Hoff relationship** that is a partial derivative equation as shown:

$$\left(\frac{\delta \ln K_a}{\delta T}\right)_{p,V} = \frac{\Delta H_{bind}}{RT^2} \tag{7.49}$$

Equation (7.49) integrates in a number of ways, of which the most commonly used is:

$$\ln K_a = -\frac{\Delta H_{bind}}{RT} + \frac{\Delta S_{bind}}{R} \tag{7.50}$$

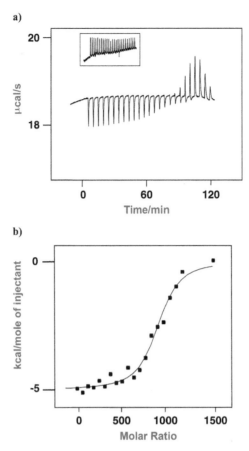

Figure 7.23 **Data output from isothermal titration calorimetry.** The ITC output data from interaction of **adenoviral mu, μ, peptide** (ligand **L**) with **plasmid DNA** (receptor **R**). (a) Heat exchange data obtained in real time. Negative peak implies that combination of **L** with **R** causes heat evolution from cell (exothermic); positive peak implies that the combination of **L** with **R** causes heat absorption by the cell (endothermic); (b) ITC software analysis output. Data fits with Equation (7.48). (adapted from Keller *et al.*, 2002, Figure 4, American Chemical Society).

Equation (7.50) relates equilibrium association constant with **enthalpy of binding** ΔH_{bind} and **entropy of binding** ΔS_{bind}. Provided that both are temperature independent properties then a linear relationship will be observed when experimental association constant data are plotted against absolute temperature using Equation (7.50) (Figure 7.24). The gradient and intercept of this linear Van't Hoff plot then gives us the equivalent of $\Delta H^{\text{o}}_{\text{bind}}$ and $\Delta S^{\text{o}}_{\text{bind}}$ values. Such a linear Van't Hoff plot is an essential tool for the derivation of thermodynamic parameters from equilibrium association constant data obtained by experimental methods other than ITC, such as those described in Sections 7.3.4 and 7.3.5 respectively. Obviously, if we calculate $\Delta H^{\text{o}}_{\text{bind}}$ and $\Delta S^{\text{o}}_{\text{bind}}$ values first, using Equation (7.48), then $\Delta G^{\text{o}}(T)_{\text{bind}}$ may be calculated in turn using Equation (7.46).

Quite frequently, however, the Van't Hoff plot is not linear but has a curvature. In this case, data must be fitted by means of the series integration shown:

$$\ln K_{\text{a}} = a + b(1/T) + c \ln T \tag{7.51}$$

where a, b and c are **variable series constants**. The constant c then relates onwards to the **heat capacity change on binding** $\Delta C_{\text{p bind}}$ according to Equation (7.52):

$$\Delta C_{\text{p bind}} = Rc \tag{7.52}$$

When the Van't Hoff plot is *linear*, then c becomes zero and $\Delta C_{\text{p bind}}$ is also zero. In other words when ΔH_{bind} is independent of absolute temperature, then there is no change in heat capacity of the combined ligand and receptor system after ligands bind to receptors. Critically, when the Van't Hoff plot has a negative curvature, then c becomes

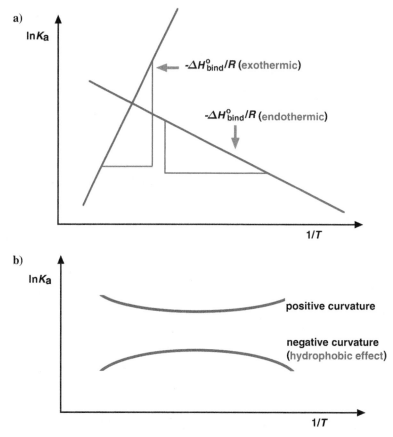

Figure 7.24 Van't Hoff Isochore plots. (a) Classical linear plots obtained if ΔH_{bind} is temperature independent over the temperature range studied; **(b)** Curved plots that characterise the situation when ΔH_{bind} is temperature dependent. Negative curvature is characteristic of the involvement of hydrophobic effect in receptor-ligand binding interactions.

negative and also $\Delta C_{p\,bind}$ becomes negative (Figure 7.24). In other words, there is a loss in system heat capacity after ligands bind to receptors. Such a thermodynamic signature usually suggests that the hydrophobic effect is making a significant contribution towards ligand-receptor binding interactions. In fact, this thermodynamic signature is perhaps the only major way to prove that the hydrophobic effect is involved in mediating binding interactions. The reason for this loss of system heat capacity upon binding is the main requirement of the hydrophobic effect that highly ordered solvation cages of water that surround hydrophobic functional groups should be displaced into free solution in order for these hydrophobic functional groups to make intimate contact with each other (see Chapter 1).

7.3.4 Capillary electrophoresis

Capillary electrophoresis (CE) is one of the most developed examples of emerging "lab-on-a-chip" technologies that have direct applications to chemical biology owing to their capacity to give meaningful information generated from only the smallest quantities of biological macromolecular material (μg levels). The basis of the technique is the very high resolution of charged species of different overall charge owing to the unique flow properties of buffer solutions in **microbore capillaries (micro-capillaries)** fashioned from silicon oxide/hydroxide and charge-charge interactions that take place between charged solutes and the inner bore of a given micro-capillary. A micro-capillary is shown diagrammatically (Figure 7.25). When an electrical potential difference is applied across such a micro-capillary, then there is a bulk movement of buffer solution from anode (positive) to cathode (negative) affected by **electro-osmotic flow (EOF)**. This bulk movement of liquid in the micro-capillary is driven by the negative surface charge on the interior of the capillary bore due to ionisation of silicon hydroxide functional groups (SiO⁻/SiOH), an effect that increases with increasing pH. This surface charge is compensated for by a complementary cationic **Stern layer** that acts to repel other cationic species towards the cathode and retard the motion of anionic counter ion species in the

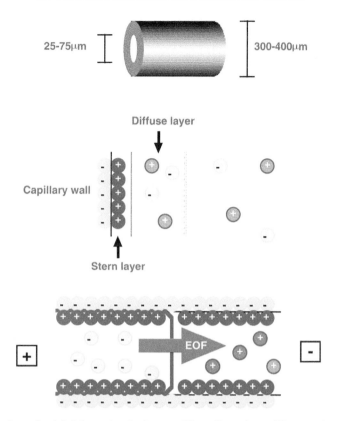

Figure 7.25 Capillary electrophoresis. (a) Schematic of **micro-capillary** dimensions; **(b)** Ion-solute structure at the capillary wall; **(c)** Schematic illustration of **electro-osmotic flow (EOF)** created in a micro-capillary because of capillary bore surface charges.

same direction owing to partial attraction. The repulsion of cationic species towards the cathode dominates and they "drag" the buffer solution in a concerted manner toward the cathode as a result of interactions between cation water-solvation cages and water molecules in the bulk solution. EOF has a characteristic flat flow profile compared with typical non-linear laminar flow (Figure 7.25).

Overlaid on top of EOF is the effect of **electrophoretic mobility**. Electrophoresis can be defined as the differential movement of charged species (ions) by attraction or repulsion in an **electric field** of **strength, E_e**, according to Equation (7.53):

$$\nu_e = \mu_e \cdot E_e \tag{7.53}$$

where ν_e is **electrophoretic velocity** and μ_e is **electrophoretic mobility**. Hence, any ionic species moving in a microbore capillary under the influence of an applied electrical potential difference will in fact move with an **apparent electrophoretic mobility, μ_a**, given by Equation (7.54):

$$\mu_a = \mu_e + \mu_{EOF} \tag{7.54}$$

where μ_{EOF} is **EOF electrophoretic mobility**. Typically, EOF mobility is greater (approximately ten-fold) than electrophoretic mobility. In the case of cationic species that migrate faster than anionic species in a microbore capillary, EOF mobility and electrophoretic mobility may coincide more closely. The **time-to-detector, t_e**, that it takes for a given charged (anionic or cationic) species to migrate from the anode (positive) end of a microbore capillary to a detector positioned at the cathode (negative) end of the capillary is given by Equation (7.55):

$$t_e = \frac{1}{\mu_a} \cdot l_e \cdot \frac{L_e}{V_e} \tag{7.55}$$

where l_e is the **effective length** of the capillary from anode to detector, L_e is **total length** and V_e is the **applied potential difference** from anode to cathode along the length of the capillary.

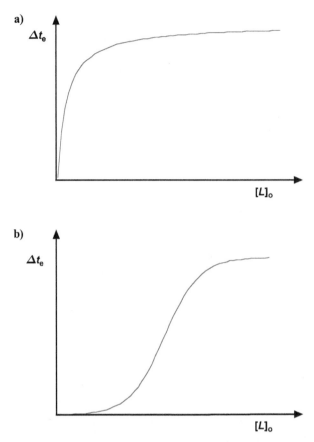

Figure 7.26 Experimental titration binding isotherms. (a) Hyperbolic titration binding isotherm obtained by titrating fixed **[R]** with increasing values of **[L]**$_o$, and then plotting observed Δt_e against **[L]**$_o$. Hyperbolic binding isotherm occurs with single/ independent multiple site, single affinity binding interactions; **(b)** Sigmoidal titration binding isotherm is characteristic of dependent multiple site, variable affinity binding interactions with strong positive cooperativity.

The basis for using CE to determine association constants for receptor-ligand interactions is a change in overall charge following the binding of ligand **L** to receptor **R** due to the formation of complementary ionic associations and hydrogen bonds. Such a change in overall charge should be reflected by a change in apparent electrophoretic mobility, μ_a, of the receptor-ligand **RL** complex compared with the mobility of the free receptor **R** in the absence of ligand **L**, and an inverse change in the time-to-detector, t_e, according to Equation (7.55). Therefore, when a fixed concentration of receptor **R** is titrated with ligand **L** until approximately all the receptor binding sites are occupied, the progress towards saturation may be observed by progressive change in the magnitude of the physical property time-to-detector, Δt_e (selected physical property), as a function of total ligand concentration, **[L]**$_o$. Hence when Δt_e is plotted as a function of total ligand concentration **[L]**$_o$ at fixed receptor concentration **[R]**, then a hyperbolic or sigmoidal binding isotherm will result that can be analysed in the most appropriate way (Figure 7.26). Obviously, if required, the titration experiment may be performed in reverse using a fixed concentration of ligand **L** titrated with an increasing concentration of receptor **R**. In that case, the change in apparent electrophoretic mobility, μ_a, of the receptor-ligand **RL** complex compared with the mobility of the free ligand **L** in the absence of receptor **R** is what matters.

7.3.5 Resonant mirror biosensing (surface plasmon resonance)

Resonant mirror biosensing is a superb technique for the analysis of receptor-ligand interactions in real time. Therefore, association constants may be determined directly from kinetic constants. This may sound a little strange, but let us consider the following binding equilibrium (Scheme 7.4). By definition, equilibrium is reached when the **rate of association**, k_{ass} (M^{-1} s^{-1}), of receptor with ligand is equalled by the **rate of dissociation**, k_{diss} (s^{-1}), of receptor ligand complex. Hence:

$$k_{ass} \cdot [R][L] = k_{diss} \cdot [RL] \tag{7.56}$$

$$R \quad + \quad L \quad \underset{k_{\text{diss}}}{\overset{k_{\text{ass}}}{\rightleftharpoons}} \quad RL$$

Scheme 7.4

This simple relationship rearranges to give:

$$\frac{k_{\text{ass}}}{k_{\text{diss}}} = \frac{[RL]}{[R][L]} = K_a \tag{7.57}$$

Therefore, equilibrium binding constants such as K_a or indeed K_d can be determined from a simple ratio of rate constants. Equations such as 7.57 are a form of **Haldane relationship**, although the term Haldane relationship is more usually applied to expressions in reaction kinetics that link rate constants with an equilibrium constant (see Chapter 8). If the receptor **R** is immobilised (imm) for any reason by covalent attachment to a solid phase, then the relationship in Equation (7.57) becomes slightly modified:

$$\frac{k_{\text{ass}}}{k_{\text{diss}}} = \frac{[RL]_{\text{imm}}}{[R]_{\text{imm}}[L]} = K_a \tag{7.58}$$

Obviously, an equivalent expression could be written for the reverse situation when the ligand **L** is immobilised (imm) instead. In either case, to a first approximation the association constant written in Equation (7.57) is said to be essentially identical to the constant defined in Equation (7.58). This is important since immobilisation of receptor **R** (or ligand **L**) to a solid phase resin is central to resonant mirror biosensing.

The principle of resonant mirror biosensing is as follows. This is a two-phase technique that involves laser light refracted through a prism block over a range of angles (Figure 7.27). For the most part, incident light is refracted through a prism block and exits without loss of intensity due to **total internal reflection**. However, at one unique angle (**resonant angle**), an **evanescent wave** is propagated (quantum mechanical tunnelling effect) into a **resonant layer** attached to the prism block, such that the exit light intensity is significantly reduced. The exact resonant angle is a function of the refractive index of the resonant layer. In fact, the resonant angle is extremely sensitive to the refractive index so that even only modest changes to the refractive index (10^{-3}) of the resonant layer will result in measurable changes in the resonant angle. The resonant layer comprises a **hydrogel** that is functionalised for the covalent coupling of a biological macro-molecule, such as a receptor **R** molecule of interest. After **R** has been immobilised at a certain concentration, an aliquot of ligand **L** may be added that diffuses into the hydrogel and consequently causes some immediate perturbation to the refractive index of the resonant layer. However, subsequent molecular recognition and binding events often result in an even more substantial perturbation to the refractive index of the resonant layer, an effect that can be monitored in **real time** (s) by observing corresponding changes in resonant angle over a period of seconds–minutes (Figure 7.28). Such a plot of resonant angle change as a function of time represents a receptor-ligand association phase that obeys the following exponential:

$$Y_t = (Y_\infty - Y_o) \cdot [1 - \exp(-k_{\text{on}}t)] + Y_o \tag{7.59}$$

where Y_t is the **resonant angle** at **time**, t, Y_∞ is the **final angle** and Y_o is the **initial angle**. The term k_{on} is a **first order**, total ligand concentration $[L]_o$ **dependent, rate constant** for receptor-ligand association.

Once a given receptor-ligand association phase is complete, the hydrogel is washed with buffer to promote dissociation of ligand **L**, causing relaxation of the refractive index of the resonant layer back towards its initial receptor immobilised value. Relaxation of the refractive index is monitored in real time by observing the corresponding relaxation of the resonant angle over a similar period of seconds–minutes (Figure 7.28). In this case, a plot of resonance angle change as a function of time represents a receptor-ligand dissociation phase that obeys the following exponential:

$$Y_t = Y_\infty \cdot \exp(-k_{\text{diss}}t) + Y_o \tag{7.60}$$

For a given total ligand concentration, $[L]_o$, the complete plot of resonant angle change data against time is known as a **sensogram**. Sensograms are usually obtained using a wide range of different total ligand concentrations $[L]_o$ in order to derive accurate values of k_{ass} and k_{on} and hence K_a, as follows. A definitive value of k_{ass} may be determined from the

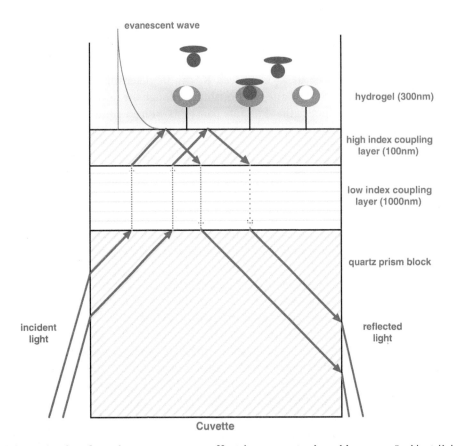

Figure 7.27 Schematic of surface plasmon resonance effect in resonant mirror biosensor. Incident light is normally subject to total internal reflection in the prism block except for losses due to **evanescent wave** penetration of the **hydrogel layer** at the **resonant angle.** Changes in resonant angle due to receptor-ligand interactions are the basis for the real time observation of molecular recognition and association/dissociation events.

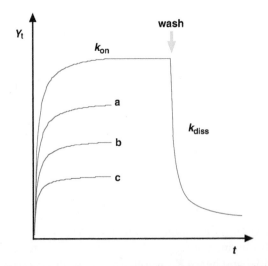

Figure 7.28 Schematic illustration of sensograms from resonant mirror biosensor. Real time observations of change in **resonant angle** Y_t with receptor-ligand association (**a, b, c,** etc.) are used to determine values of unimolecular rate constant k_{on} as a function of **total ligand concentration** $[L]_o$. A wash step is then introduced to promote real time dissociation of ligand from receptor. Subsequent real time changes in Y_t with dissociation are used to determine values of unimolecular rate constant k_{diss}.

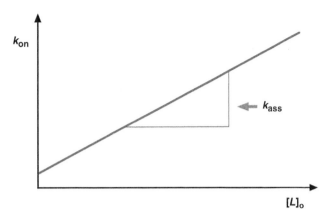

Figure 7.29 **Determination of kinetic constants from resonant mirror biosensor sensograms.** Illustration of plot of k_{on} as a function of total ligand concentration $[L]_o$. The gradient gives a value for the bimolecular rate constant k_{ass}.

association phases of as many sensograms as possible, making use of Equation (7.59) to derive k_{on} values and then these data are plotted versus $[L]_o$ according to Equation (7.61):

$$k_{on} = k_{diss} + k_{ass} \cdot [L]_o \tag{7.61}$$

The gradient of this plot represents a definitive value for k_{ass}, assuming that the plot is linear (Figure 7.29). By contrast, a definitive value of k_{diss} may be obtained either from the intercept of the k_{on} versus $[L]_o$ plot (Figure 7.29), or more accurately as an average of individual values obtained from the dissociation phases of as many sensograms as possible, making use of Equation (7.60). Finally, definitive values of kinetic constants can then be processed using Equation (7.58) to give meaningful values of K_a generated once again from only the smallest quantities of biological macromolecular material (μg levels). The ability of resonant mirror biosensing to give both kinetic and thermodynamic information about biological molecular recognition and binding events is currently almost unique and so is correspondingly very important.

7.4 Biological molecular recognition studies

In this concluding section, we shall look at a small number of molecular recognition events in order to exemplify the theoretical discussions above. Nowadays, there are myriad studies that could be given, but we have chosen to focus on a few examples, interesting for their contrasting demonstrations of molecular recognition and binding events.

7.4.1 LysU enzyme substrate recognition

Many enzymes (protein bio-catalysts) are well characterised as biocatalysts, but quite often key molecular recognition and binding events that precede biocatalysis are overlooked. This cannot be said to be true of **LysU**. LysU is a homodimeric **lysyl tRNA synthetase** enzyme and one of the primary enzymes involved in **diadenosine-5′, 5′′′-P^1,P^4-tetraphosphate** (Ap$_4$A) **synthesis** in the bacterium *Escherichia coli* (*E. coli*) (see Chapter 6, Figure 6.19). The enzymic surface mechanism of **LysU-catalysed Ap$_4$A synthesis** is well known and is illustrated in Figure 7.30. The first step (**Step 1**) is very specific and only adenosine 5′-triphosphate (ATP) or deoxy-ATP analogues are able to act as first nucleotide substrates to form a **lysyl-adenylate intermediate**. The second step (**Step 2**) is unusual in requiring the use of the Zn^{2+} ion (see Chapter 8). The X-ray crystal structures of LysU demonstrate a host of hydrogen-bonding interactions responsible for determining the strength and specificity of binding between substrates and enzyme active site in **Step 1** (Figure 7.31). Also, cationic amino acid residues are present in the vicinity of the polyphosphate chain for electrostatic (long- and short-range) interactions; similarly, anionic amino acid residues are present in the vicinity of L-lysine for equivalent electrostatic interactions. Complementarities between enzyme active sites and substrates are usually extensive at the level of charge and hydrogen bonding in order to ensure optimal specificity of molecular recognition and binding.

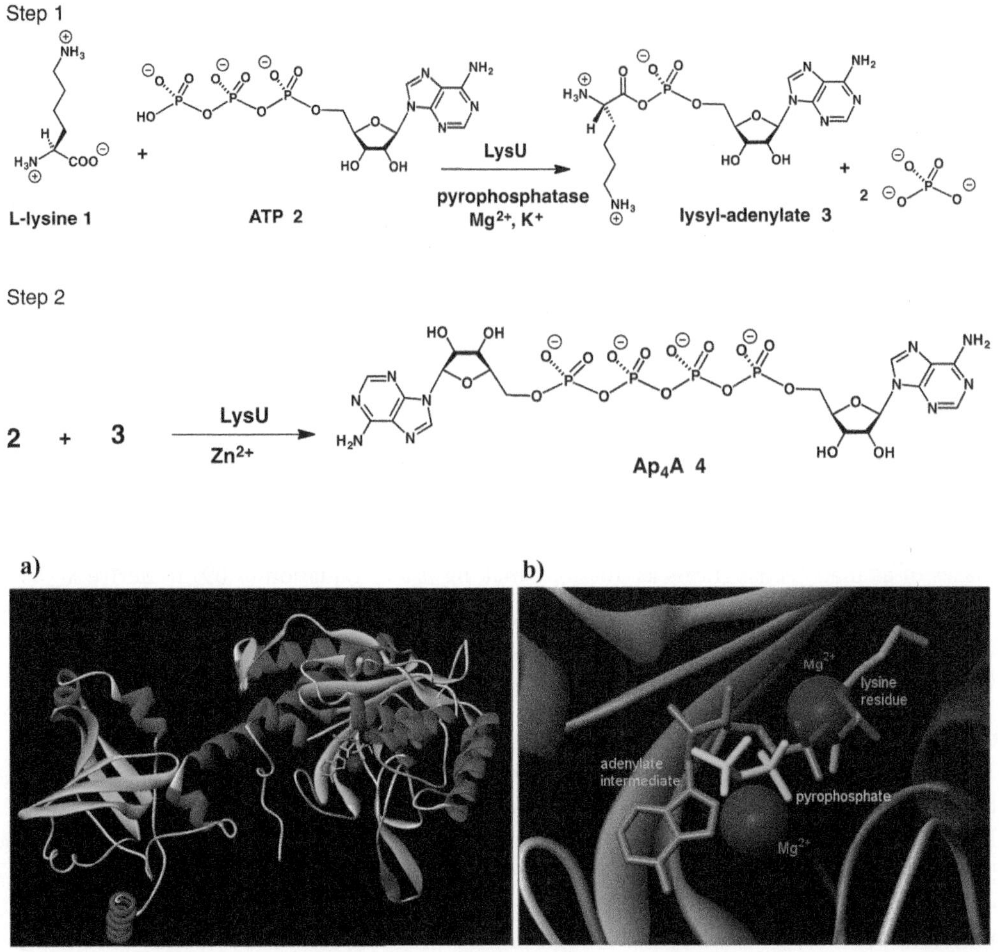

Step 1

L-lysine 1   ATP 2   LysU pyrophosphatase Mg²⁺, K⁺   lysyl-adenylate 3

Step 2

2 + 3   LysU Zn²⁺   Ap₄A 4

a)   b)

Figure 7.30 **Enzymatic mechanism of LysU protein.** Illustration of **Step 1** and **Step 2** of enzymic surface mechanism of **LysU-catalysed diadenosine-5′, 5‴-P^1,P^4-tetraphosphate (Ap$_4$A, 4) synthesis** (see Chapters 6 and 8). **(a) Ribbon display structure** illustration of LysU monomer with bound **lysyl-adenylate** (intermediate, **3**) and **pyrophosphate** leaving group (that is later hydrolysed to two phosphates [Step 1] by pyrophosphatase) rendered in **stick representation** (**orange**: adenosine 5′-monophosphate (AMP) moiety; **green**: lysyl moiety; **light blue**: pyrophosphate). Mg²⁺ ions are rendered as Van der Waals spheres (**purple**); **(b)** Active site close-up of LysU monomer structure shown in **(a)** (pdb: **1e1t**).

The molecular recognition and binding behaviour of the first step has now been studied in some detail and is proving very interesting. Catalysis requires a precise order of substrate binding that was demonstrated by a combination of **fluorescence titration binding experiments** and **ITC titration binding experiments**. Initially, the presence of Mg²⁺ ions was found to be essential for L-lysine binding to either active site of the LysU dimer. Thereafter, the binding of L-lysine at both sites was found to be responsible for a substantial structural rearrangement of the C-terminal active site domain of each monomer, as observed by CD spectroscopy (Figure 7.32). The change in LysU conformation elicited by L-lysine binding was also sufficient to induce a significant change in intrinsic tryptophan fluorescence upon L-lysine binding. Therefore, intrinsic tryptophan fluorescence titration binding experiments were used successfully to determine the site dissociation constant, K_d, for L-lysine and the binding stoichiometry (K_d, 7.96 ± 1.40 μM; 0.95 ± 0.1 L-lysine/monomer) (Figure 7.32). ITC binding experiments revealed a similar value for K_d and binding stoichiometry (K_d 4.95 ± 2.54 μM; 0.87 ± 0.14 L-lysine/monomer) (Figure 7.33). Critically, ATP was unable to bind to LysU until after the association of L-lysine and Mg²⁺ ions. Note how binding stoichiometry is determined either by curve fitting with the appropriate binding equation or else directly from plots (Figure 7.32).

ATP binding was studied by ITC binding experiment in the absence of any other spectroscopic signature that could be titrated to saturation. ITC studies were performed using hydrolysis resistant β,γ-**methylene-ATP (AMPCP)**, in place of ATP, in order to avoid complications in ITC measurements from ATP hydrolysis (Figure 7.33). In this case the value for K_d was found to be 1.22 ± 0.15 μM and the stoichiometry 0.51 ± 0.01 AMPPCP/monomer. Only one ATP molecule was found to bind per LysU dimer, suggesting that ATP binding was accompanied by strong negative cooperativity,

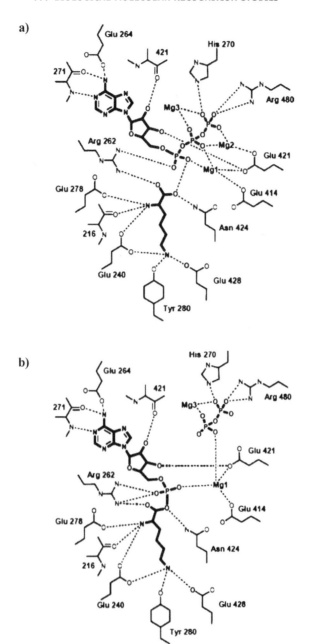

Figure 7.31 **Two-dimensional schematic representation of the active site of LysU.** H-bond interactions are shown for the ternary complex between **L-lysine**, **ATP** (**1** and **2**, Figure 7.30) and LysU (**a**), and for the complex between the lysyl-adenylate intermediate (**3**, Figure 7.30) and LysU (**b**) (see Figure 7.30). The **motif 2** residue **Arg 262** is critical for the molecular recognition and binding of both L-lysine and ATP α-phosphate to LysU. Several conserved residues in the **motif 2 loop** (residues 264–271) assume an ordered conformation only upon ATP binding. Positions of the Mg²⁺ binding sites are also shown. There are three Mg²⁺ ions **Mg1**, **Mg2** and **Mg3** in (**a**) and two Mg²⁺ ions **Mg1** and **Mg3** in (**b**). (Illustrations from Desogus *et al.*, 2000, Figure 4, American Chemical Society).

leading to **half-of-sites binding**. Consequently, the ordered process of LysU molecular recognition and binding of substrates requires that **half-of-sites catalysis** takes place subsequently. In order to understand the basis of half-of-sites binding, theoretical molecular modelling studies were performed using the X-ray crystal structures of LysU as a starting point. Molecular modelling is largely beyond our scope here, but it is worth noting that **molecular mechanics Poisson-Boltzmann Surface Area** (**MM-PBSA**) modelling experiments were able to demonstrate that both LysU active sites are not structurally and functionally equivalent to each other even before the above experiments are performed. In the event, ATP was calculated to bind to one site (**site 2**), ~25 kcal mol⁻¹ more tightly than to the other (**site 1**). This difference is more than adequate to account for the negative cooperativity and half-of-sites binding observed experimentally. Moreover, molecular dynamics simulations (see Chapter 5) were able to indicate that binding of ATP to one active site (**site 2**) was

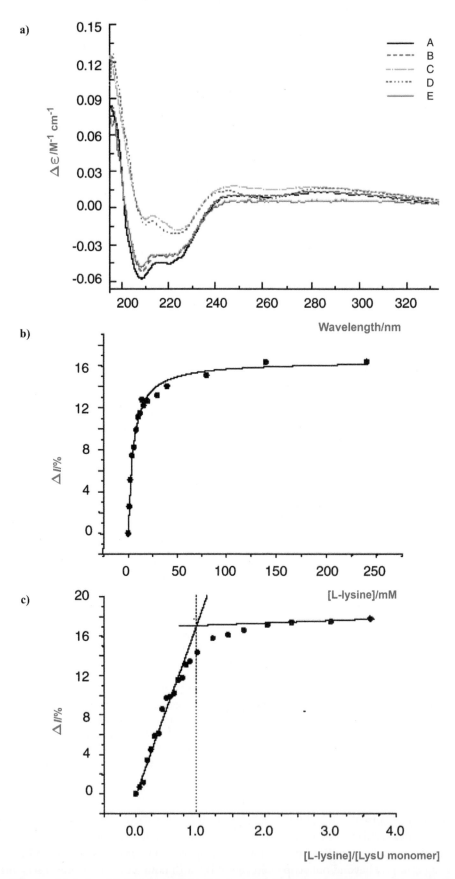

Figure 7.32 Effects of L-lysine binding to LysU in the presence of ions. (a) CD spectra of LysU ($2\,\mu M$) in neutral buffer in the presence of no substrates (**A**, black); 10 mM $MgCl_2$, (**B**, red); 1 mM L-lysine, 10 mM $MgCl_2$ (**C**, green); 1 mM L-lysine, 10 mM $MgCl_2$, $250\,\mu M$ β,γ-**methylene ATP** (**AMPPCP**) (**D**, blue); 10 mM $MgCl_2$, $250\,\mu M$ AMPPCP (**E**, yellow-green); (**b**) **Fluorescence titration binding** of LysU ($0.5\,\mu M$)) with L-lysine in buffer pH 8.0 in the presence of 10 mM $MgCl_2$; (**c**) High concentration fluorescence titration binding of LysU ($17\,\mu M$) with L-lysine in buffer pH 8.0 in the presence of 10 mM $MgCl_2$. Linear regions of data are extrapolated to intercept (**Klotz transition**) that defines saturation binding stoichiometry. (Hughes *et al.*, 2003, Figure 2, Springer Nature).

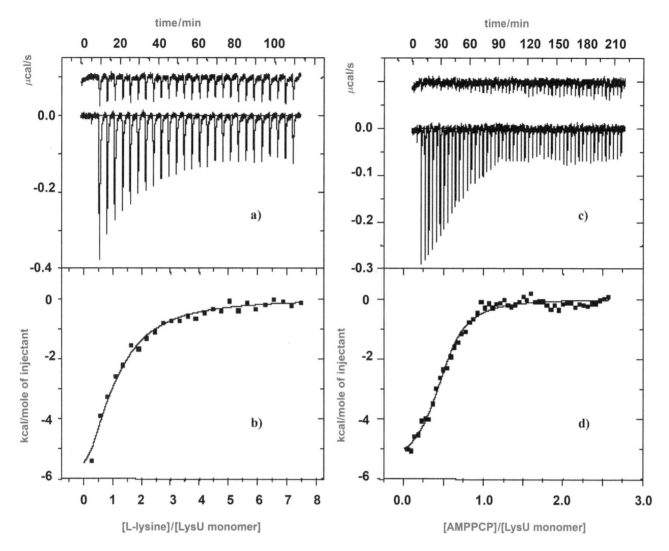

Figure 7.33 Ligand binding experiments using isothermal titration calorimetry. The direct effects of Mg^{2+} and K^+ ions on the binding of L-lysine and AMPPCP by LysU. (**a**) Calorimetric titration profile deriving from the titration of LysU (6 μM monomer) with injected aliquots of L-lysine in buffer pH 8.0 at 20 °C in the presence of 10 mM $MgCl_2$ (upper trace; control titration in absence of LysU); (**b**) Heat absorbed per mole of L-lysine titrant versus [L-lysine]/[LysU monomer]; (**c**) Calorimetric titration profile deriving from the titration of LysU (20 μM monomer) with injected aliquots of AMPPCP in buffer pH 8.0 at 20 °C in the presence of 1 mM L-lysine, 10 mM $MgCl_2$; (**d**) Heat absorbed per mole of L-lysine titrant versus [AMPPCP]/[LysU monomer]. (Hughes *et al.*, 2003, Figure 3, Springer Nature).

capable of inducing **arginine 269 (Arg 269)** closure of the same site, thereby causing an opening up of the other active site (**site 1**) via a trans-conformational relay system operating through the dimer-interface by means of physical contacts between the **C2-C3 loops** of each monomer and corresponding **motif 2 loops** in the corresponding monomer, proximal to each active site (Figure 7.34). Such a conformational opening of **site 1** would be sufficient explanation for impaired ATP binding affinity and **half-of-sites reactivity**.

Clearly, the results of modelling studies should always be correlated, where possible, with experimental observations. However, the combination of experimental techniques with molecular modelling represents an important chemical biology case study for the investigation of the molecular recognition and binding events that preceed catalysis in an enzyme biocatalyst mechanism. Not only has a clear **substrate order-of-addition** been established for LysU but also a half-of-sites binding mechanism. Obviously, the precise combination of experimental techniques and modelling studies will vary according to biocatalyst under investigation, but the outcomes should be illustrative of the need to combine as many techniques as possible in order to understand them. Not only does conformational opening of LysU site 1 now appear to be an important reason that site 1 has a lower affinity for ATP than site 2 and does not experimentally bind ATP,

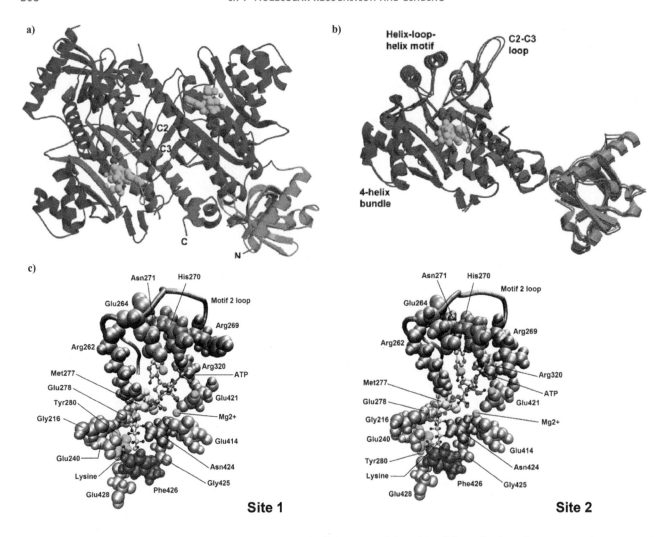

Figure 7.34 LysU half-of-sites binding of ATP and half-of-sites reactivity. (a) **Ribbon display** of X-ray crystal structure of LysU with substrates L-lysine (**pink**) and AMPPCP (**cyan**) bound in both active sites together with three Mg^{2+} ions (**yellow**), all shown in **CPK structure representations. Monomer 1** is shown in **blue, monomer 2** is coloured according to domain structure; *N*-terminal (**orange**), conserved core domain typical to Class II tRNA synthetases (**red**), LysU specific regions (**purple**). In the central domain region, a motif 2 loop (**green**) sits over ATP. β-Strands **C2** and **C3** in each monomer interact across the homodimer interface. The intervening **C2-C3 loop** is positioned above and interacts with the motif 2 loop of the opposite monomer; (**b**) Superposition of **monomer 1** and **monomer 2** post molecular mechanics simulation (1 ns). Colour code is as in (a) monomer 2. Superposition involves fitting the conserved core domain. There is a relative movement of several loops and domains, resulting in asymmetry; (**c**) **Asymmetry observed in active sites.** In **site 1** (monomer 1), the active site is "open", in **site 2** (monomer 2), the active site is "closed" (**Arg 269** of motif 2 loop closes the binding pocket) and binds nucleotide more tightly than the "open" site. Active site closure in monomer 2 appears to result in the displacement of attached motif 2 loop, leading to coupled displacements in the C2-C3 loop, the **C2** β-strand and hence the motif 2 loop of monomer 1. As a result, site 1 (monomer 1) is opened. Such structural and binding asymmetry seen in molecular simulation accounts for **experimental half-of-sites binding** of nucleotide (Figure 7.33) and suggests a mechanism for **half-of-sites reactivity**, where only one active site at a time is involved in the binding of ATP and catalysis of Step 1 (Figure 7.30). (Illustrations (a) and (b) from Hughes *et al.,* 2003, Figure 1, Springer Nature; (c) and (d) related to Chen *et al.,* 2013, Figure 1).

but also lysyl-adenylate formation and pyrophosphate release in site 2 appears likely to trigger a subsequent reverse set of trans-conformational changes across the dimer interface, allowing site 1 to bind ATP, close up and initiate Step 1 catalysis of lysyl-adenylate formation in turn. We appear to be making significant progress in understanding the molecular recognition and binding behaviour of LysU.

7.4.2 Stress protein molecular chaperones

Stress protein molecular chaperones have already been introduced (see Chapter 6). Such proteins have a rich and varied molecular recognition and binding behaviour. Important examples include **GroEL** and the procollagen/collagen molecular chaperone protein known as **heat shock protein 47** (**Hsp47**).

7.4.2.1 GroEL

GroEL is a paradox in molecular recognition and binding. A wide range of different unfolded protein substrates bind to GroEL, with K_d values in the µM-nM range. Such binding strength should clearly be characterised as specific and yet GroEL has a promiscuous capacity with **GroES** to assist the folding/refolding of a diverse range of protein substrates. How is this possible? How is this possible to investigate? In order to simplify the investigation, but at the same time retain valuable information, extrinsic fluorescence titration binding experiments were performed using a series of peptide surrogates for unfolded protein substrates. Peptide binding interactions with GroEL were reported in each case by means of a **dansyl** (**Dns**) group (see Chapter 4) covalently attached to each peptide. Dns groups show a marked shift in I_{max} and an increase in fluorescence emission intensity with increase in quantum yield ϕ_F upon transfer from a free, aqueous hydrophilic environment to a more hydrophobic ligand-bound environment.

Initially, GroEL was found to prefer interactions with peptides comprising sequences or arrays of amino acid residues that are amphiphilic, comprising hydrophobic and polar neutral/cationic residues. Sequences or arrays exclusively polar in character or else harbouring an excess of anionic residues were not recognised or bound at all (Figure 7.35). Following this, the role of secondary structure in molecular recognition and binding was investigated by the systematic preparation of a series of six peptides divisible into two groups known as the **AMPH series**, designed with a graded ability to form a cationic **amphiphilic α-helical structure** in solution, and the **NON-AMPH series**, designed with the graded ability to form a non-amphiphilic α-helical structure in solution (Figure 7.35). Extrinsic fluorescence titration binding experiments were performed using all six peptides. Owing to the molecular weight size difference between GroEL and peptides, and the potential number of peptide binding sites, fluorescence titration experiments were actually performed in reverse, keeping fixed peptide concentrations and titrating with increasing GroEL until fluorescence saturation was achieved (Figure 7.36).

From such experiments, values of K_d were determined, with the discovery that the most α-helical AMPH series peptide (**AMPH⁺**) was much the most effective GroEL "substrate" with a binding interaction affinity (K_d 5 nM), as strong if not stronger than known binding interactions between GroEL and other protein "substrates". Extrinsic fluorescence titration binding experiments under a range of different conditions were also able to show the importance of the hydrophobic effect in peptide-GroEL association (Figure 7.37). Similar titration experiments were used to show that peptides and GroES compete for the same binding sites on GroEL in accordance with the known GroEL/GroES mechanism (Figure 7.37) (see Figures 6.17 and 6.18). Finally, the sum total of all these experiments was the main conclusion that GroEL has a preference to bind regions or subdomains in an unfolded protein substrate that have a propensity to form polar neutral and/or cationic, amphiphilic secondary structural elements in solution. This robust conclusion appears to provide the required explanation for the general-specific molecular recognition and binding behaviour of GroEL.

GroEL has an almost predatory enthusiasm for the binding and hydrolysis of ATP as part of the known GroEL/GroES mechanism (Figures 6.17 and 6.18). Indeed, ATP binding sites reside in each subunit polypeptide of the GroEL 14-mer and binding of ATP at a given site invokes considerable positive cooperativity in the binding of ATP by the remaining six subunits within the corresponding ring of seven. Cooperativity is boosted by a **T** to **R** state transition in the conformation of each subunit in response to binding of increasing amounts of ATP where the **R** state has a higher affinity for ATP binding than the **T** state (Figure 6.18). This positive cooperativity in binding ATP is highly analogous to another famous example of positive cooperativity, namely the binding of oxygen to haem-iron in the hetero-tetramer **haemoglobin**, the oxygen carrying protein found in red blood cells (see Chapter 1). In both cases, cooperativity in binding could be demonstrated by binding isotherms and Hill equations (see Section 7.2.4).

7.4.2.2 Hsp47

Hsp47 is a molecular chaperone specific to procollagen biogenesis and so is almost the exact opposite of GroEL. Furthermore, Hsp47 performs molecular chaperone functions either as a monomeric protein (47 kDa) or as a trimer (141 kDa)

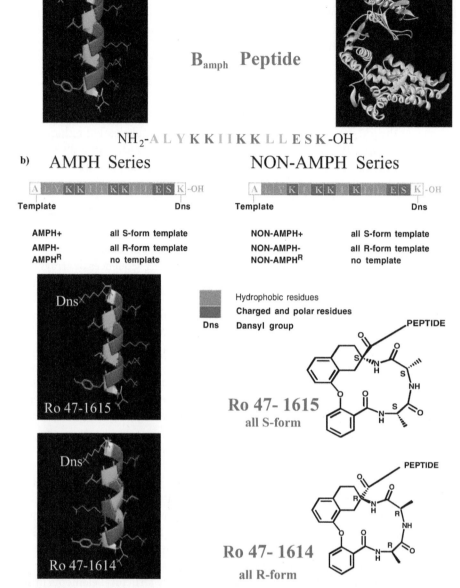

Figure 7.35 **Substrate peptide binding to molecular chaperone GroEL.** (a) Basic amphiphilic (B$_{amph}$) peptide, the optimal first round substrate peptide. Hydrophilic residues (**red**) and hydrophobic residues (**green**) are disposed for B$_{amph}$ to have the ability to form an **amphiphilic α-helix** in solution (**left**), considered to be potentially optimal to interact with exposed hydrophobic residue "patches" in GroEL (coloured regions in GroEL monomer in **ribbon display structure** representation, **right**); (**b**) Summary of **AMPH** and **NON-AMPH** series peptides with helix forming template (Ro 47–1615) and control template (Ro 47–1614) (**right**) and proposed amphiphilic (**AMPH**) (**upper left**) and non-amphiphilic (**NON-AMPH**) helical structures (**lower left**) illustrated. (Peptide structure illustrations in (a) and (b) are from Preuss *et al.*, 1999, Figure 3, American Chemical Society).

(Figure 7.38). However, Hsp47 is also a paradox in molecular recognition and binding. The protein is actually a member of the **serine protease inhibitor** (**serpin**) superfamily, one of the most widely studied and structurally characterised protein families. The main function of the serpin superfamily is clearly completely different to that of a procollagen/collagen molecular chaperone. How is this possible? How do we investigate? Almost from the start, molecular recognition and binding studies were seen to be a key way to reach an understanding about the function and role of Hsp47 in procollagen biogenesis. Such studies were enabled by intrinsic fluorescence and **CD titration binding experiments** to probe the interactions between Hsp47 and **procollagen-like model peptide (PPG)$_{10}$** to Hsp47 (Figure 7.38). From such experiments, values of K_d were determined that were surprisingly strong (approximately 800 nM). Moreover, CD titration

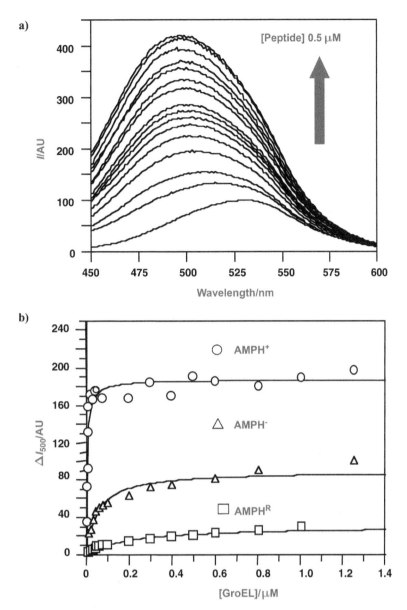

Figure 7.36 **Substrate peptide association with GroEL.** Binding interactions are studied by fluorescence titration binding assays. (**a**) Fluorescence emission spectra from fluorescence titration binding experiment with fixed concentration of AMPH⁺ peptide (0.5 μM) titrated with GroEL until saturation in buffer pH 7.5 at 20 °C. Note I_{max} blue shift and increase in emission intensity as titration progresses; (**b**) Three binding isotherms from three fluorescence titration binding experiments involving the indicated peptides under conditions described in (a). (From Preuss *et al.*, 1999, Figure 6, American Chemical Society).

binding experiments also demonstrated that peptide binding resulted in an increase in P_{II} helix character, actually helping **procollagen triple helix** (molecular rope, see Chapter 1) to form (Figure 7.39). CD titration binding experiments were also of value in demonstrating the effect of pH upon peptide-Hsp47 interactions. Measured binding affinities were found to drop by over 10^3 in moving from neutral-mildly alkali to mildly acidic conditions (pH ≤ 6.3) (Figure 7.39).

Careful **CD** and **fluorescence pH-titration experiments** were found to be useful in understanding this behaviour. By means of these experiments, Hsp47 was observed to undergo reversible pH-driven trans-conformational changes from one main conformational state (**alkali state**), at neutral-mildly alkali pH, to an alternative main conformational state (**acid state**), at pH 6 and below, in a process mediated via a putative transitional state (**intermediate state**) (see Figures 4.8 and 4.13) (see Chapter 4). The close correlation between binding affinity and conformational data suggested the important conclusion that pH-driven trans-conformational changes are the reason for the changes in binding affinity with pH (**conformational pH switch mechanism**), implying that only the alkali state of Hsp47 is competent to bind

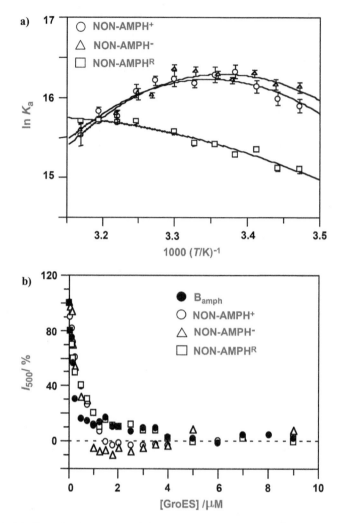

Figure 7.37   Forces involved in binding of substrate peptides to GroEL. (a) Standard **Van't Hoff plots** prepared from data obtained through fluorescence titration binding experiments. Negative curvature suggests that the hydrophobic effect plays an important part in NON-AMPH series peptide binding to GroEL, and by implication AMPH series peptides, too; **(b) Competition binding experiments** in which interaction of fixed concentration of peptides with GroEL (0.5 μM) is studied in the presence of increasing GroES. Fixed concentrations of ATP, $MgCl_2$ and KCl are also present. GroES competes very effectively for binding to GroEL, consistent with shared GroEL binding sites (see Figure 6.18). Competition experiments give important validation that peptides are realistic alternate substrates for proteins in molecular recognition and binding studies involving GroEL. (Preuss *et al.*, 1999, Figures 8 and 9c, American Chemical Society).

procollagen-like model peptide $(PPG)_{10}$ and by implication procollagen/collagen. Importantly, the results of all these titration experiments and the conformational pH switch mechanism actually appear to make sense biologically. Hsp47 is known to bind to and assist procollagen assembly in the endoplasmic reticulum (neutral-mildly alkali pH) and then remain bound to assist protein trafficking to the *cis*-golgi (approximately pH 6.4), where procollagen molecules then dissociate from Hsp47 to begin the process of fibrillisation prior to export from cells. Hsp47 is then apparently returned to the endoplasmic reticulum in order to repeat the process.

Since the conformational pH-switch mechanism is central to the biological activity of Hsp47, further understanding is essential. Detailed site directed mutagenesis (protein engineering) studies are a powerful way forward (see Chapter 3). In the case of Hsp47, histidine amino acid residues found in **histidine clusters** (**breach**, **shutter** and **gate** region) were systematically mutated in turn to alanine residues. Results demonstrated that **breach** and **gate** histidines appeared to control and modulate the conformational pH-switch, with **H191** acting as a **trigger residue** to initiate the correct pH-driven conformational change process (Figure 7.39). Selective mutations of **tryptophan residues** to phenylalanine residues were then used to determine those sub-domains of Hsp47 most susceptible to conformational change with pH. In the event, the A β-sheet region around **W100**, centred close to **H191**, was found to be the most influenced by conformational change as a function of pH (Figure 7.39). Importantly, this region underpins the anticipated binding site region of Hsp47 for collagen/procollagen

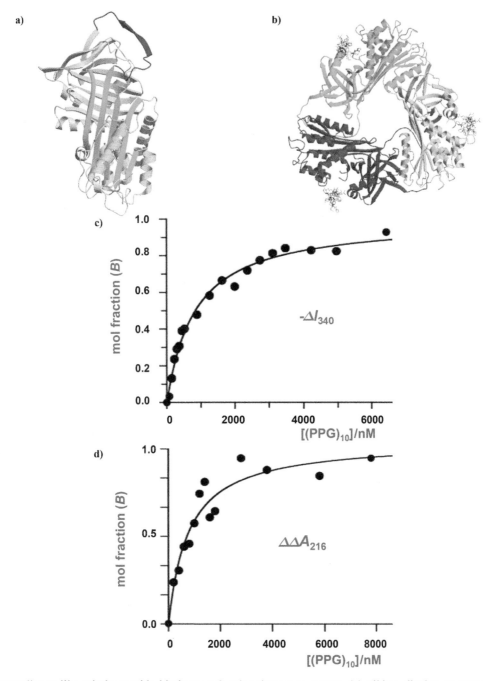

Figure 7.38 **Procollagen-like mimic peptide binds to molecular chaperone Hsp47.** (a) **Ribbon display structure** model of monomeric heat shock protein 47 (Hsp47) with classical serpin fold. View is from bottom illustrating the **reactive centre loop (RCL) (red)** and the 5-strand A β-sheet shown (**green**); (**b**) **Ribbon display structure** model of Hsp47 homo-oligomeric trimer (monomers in **red, yellow** and **green**) formed by the insertion of the RCL of one Hsp47 monomer into the 5-strand A β-sheet of another Hsp47 monomer (head-to-tail interactions); (**c**) **Intrinsic tryptophan fluorescence binding isotherm** from fluorescence titration binding experiment using fixed concentration of Hsp47 titrated with mimic peptide (**PPG**)$_{10}$ until saturation in buffer pH 7.5 at 20°C; (**d**) **CD binding isotherm** from **CD titration binding experiment** using fixed concentration of Hsp47 titrated with (PPG)$_{10}$ until saturation in buffer pH 7.5 at 20°C (illustrations (a) and (b) are from Dafforn *et al.*, 2001, Figures 1a and 6b, Elsevier; (c) and (d) are adapted from Dafforn *et al.*, 2001, Figure 4a).

and procollagen-like model peptide (PPG)$_{10}$. Therefore, the results all seem to interlock to give a complete understanding. So why should a serpin superfamily member become a molecular chaperone involved in procollagen biogenesis? In fact, all serpin superfamily members studied to date have a remarkable conformational plasticity. According to the data described in this section, this plasticity also seems to be ideally suited to molecular chaperone duties as well.

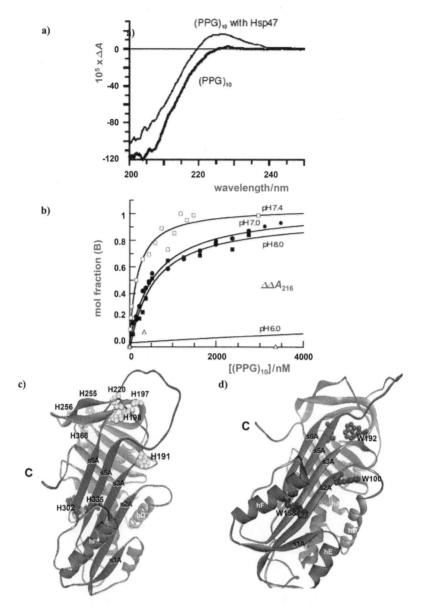

Figure 7.39 **Implications of procollagen-like mimic peptide interactions with Hsp47.** (a) Endpoint of CD titration binding experiment (see Figure 7.38d) illustrating the change in $(PPG)_{10}$ CD spectrum in the presence and absence of Hsp47; (b) Repeat of CD titration binding experiment (see Figure 7.38d) in buffer at different pH; (c) **Ribbon display structure** of Hsp47 to show **histidine residue clusters (ball and stick representation)** for protein engineering. These are breach (**yellow**), gate (**green**) and shutter (**pink**) clusters; (d) **Ribbon display structure** of Hsp47 to show **tryptophan residues (ball and stick representation, purple)** for protein engineering. (Illustrations (a) and (b) are from Dafforn *et al.*, 2001, Figures 4b and 4d, Elsevier; (c) and (d) are adapted from Abdul-Wahab *et al.*, 2013, Figure 1).

7.4.3 Complementary (antisense) peptides

By definition, a sense peptide is one whose sequence is coded for by a nucleotide residue sequence (read 5′→3′) of sense **messenger RNA (mRNA)** whose sequence contains the same coding information as the sense strand of DNA. Conversely, a **complementary (antisense) peptide** is coded for by the nucleotide residue sequence (read 5′→3′) of **complementary (antisense) mRNA** with the same sequence information as the complementary (antisense) strand of DNA. Frequently, sense and complementary (antisense) peptides are capable of specific interactions in a process that may involve an amino acid interaction code embedded within the genetic code and its complement (Figure 7.40). One application of sense-complementary (antisense) peptide interactions has been in the design of **complementary (antisense) peptide mini-receptor inhibitors** of proteins binding to cognate receptors. At its simplest level, a complementary (antisense) peptide inhibitor is designed first by looking at the mRNA (or DNA sense strand) sequence coding for a particular target region of interest in a given target protein. From this, the complementary (antisense) sequence can then be deduced by Watson-Crick base pairing rules and the corresponding peptide sequence determined from the complementary (antisense) mRNA (or DNA complementary [antisense] strand) sequence according to the genetic code (Chapter 1). Successful complementary (antisense) peptide mini-receptor inhibitors have been designed against a variety of proteins, including the cytokine interleukin-1β (IL-1β) involved in mediating inflammation (see Chapters 5 and 7).

In the case of IL-1β, a first complementary (antisense) peptide inhibitor was designed against an exposed, functional surface loop (β-**bulge**) known as the **Boraschi loop** (Figure 7.41). Although biological activity of the complementary (antisense) peptide and close relatives could be shown with some ease, evidence of specific binding of peptide to IL-1β was essential. How was this achieved? Given difficulties in obtaining adequate quantities of IL-1β, an analysis technique was required able to yield data with minimal sample quantity, hence resonant-mirror biosensing was used. The inter-

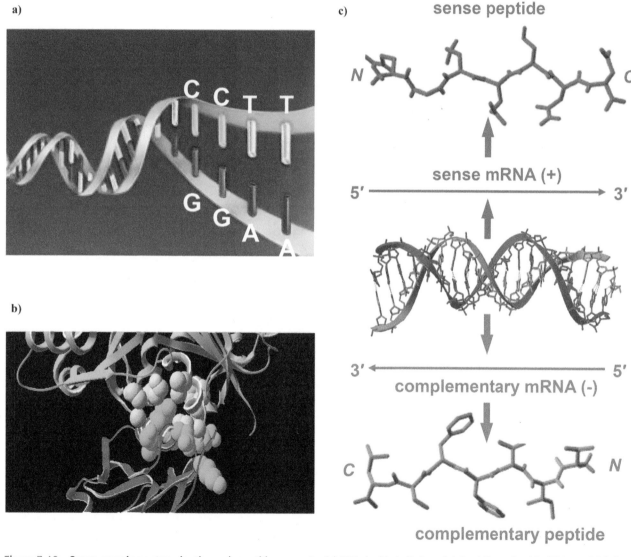

Figure 7.40 Sense-complementary (antisense) peptide concepts. (a) DNA double helix is maintained throughout by **Watson-Crick base pairings; (b)** Protein-protein interactions may be driven in part by a peptide/protein equivalent of Watson-Crick base pairings; **(c) Specific sense-complementary (antisense) peptide interactions** may provide the basis for understanding. A sense peptide is coded for by a nucleotide residue sequence in sense mRNA that has the same sequence information as the sense strand of DNA. A corresponding **complementary (antisense) peptide** is coded for by the nucleotide residue sequence in complementary (antisense) mRNA that has the same sequence information as the complementary (antisense) strand of DNA. Sense and corresponding complementary (antisense) peptides are capable of specific interactions, suggesting that there may be a peptide/protein binding interaction code associated with the genetic code and its complement.

a)

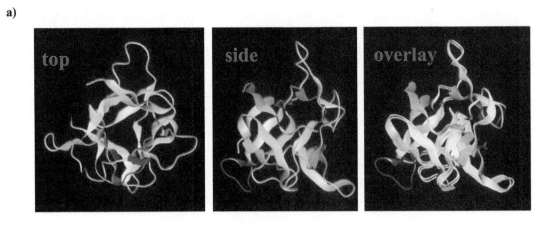

b)

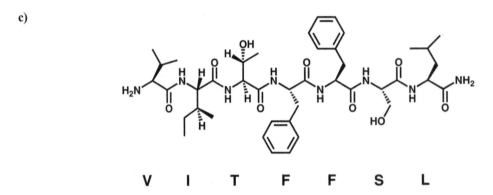

Figure 7.41 Derivation of complementary (antisense) peptides. (a) Three **ribbon display structure** views of interleukin-1β (IL-1β) X-ray structure (pdb: **1i1b**), showing (**top and side**) key receptor binding residue regions (**yellow**) and the **Boraschi loop** (**red**). **Overlay** structure involves superposition of **interleukin-1 receptor antagonist** (**IL-1ra**) X-ray structure (pdb: **1ilt**) upon IL-1β (side view) to demonstrate the general structural similarity between these protein family member proteins, but also the absence of Boraschi loop from IL-1ra. IL-1ra is the only known natural inhibitor of IL-1β; **(b)** mRNA sequence of Boraschi loop and decoded amino acid residue sequence (**red**) set alongside deduced mRNA sequence of complementary (antisense) peptide and decoded amino acid residue sequence (**blue**); **(c)** Structure of complementary (antisense) peptide corresponding with the Boraschi loop; a potential **complementary (antisense) peptide mini-receptor inhibitor** of IL-1β.

actions of immobilised IL-1β with increasing quantities of complementary (antisense) peptide inhibitor were analysed in real time (Figure 7.42), and values of K_d were determined that were reasonably strong (approximately 5 μM). A re-ordered version of the complementary (antisense) peptide (same amino acid composition; different residue order) was shown not to bind to immobilised IL-1β. Furthermore, the complementary (antisense) peptide itself was shown not to bind to other proteins, including **interleukin-1 receptor antagonist (IL-1ra)** that has the same fold as IL-1β but without a corresponding Boraschi loop (Figure 7.41). Finally, a peptide with the sequence of the Boraschi loop was shown to compete with immobilised IL-1β for binding to the complementary (antisense) peptide, providing proof of specific complementary (antisense) peptide binding to the Boraschi surface loop (Figure 7.42). Complementary (antisense) peptides still attract significant controversy given the current absence of an unambiguous mechanism of interaction between a sense and corresponding complementary (antisense) peptide. However, resonant mirror biosensing has enabled huge progress to be made in understanding. A wider role for sense-complementary (antisense) peptide interactions in biology remains to be identified.

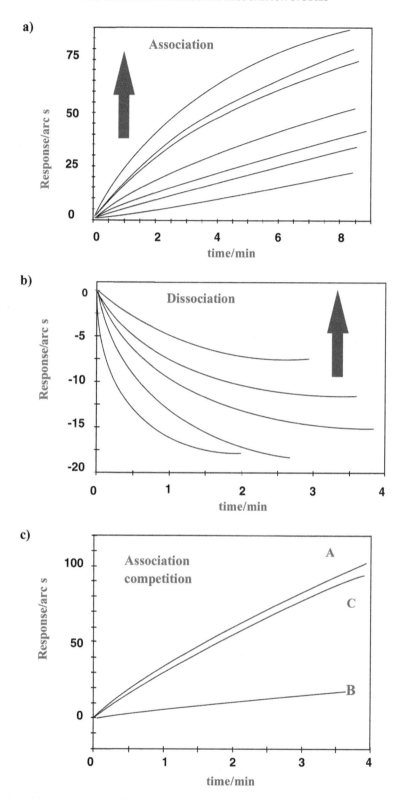

Figure 7.42 **Resonant mirror biosensor assays.** (a) Set of **association curves** for the association of complementary (antisense) peptide **VITFFSL** with immobilised IL-1β (0.39 μM); (b) Set of **dissociation curves** for the dissociation of complementary (antisense) peptide **VITFFSL** from immobilised IL-1β (0.39 μM) at neutral pH and 37 °C; (c) Competition binding experiment showing association curves for complementary (antisense) peptide with immobilised IL-1β (0.39 μM) at neutral pH and 37 °C either alone (**A**), in the presence of a peptide with the Boraschi loop sequence (**B**) or with a peptide with a reordered Boraschi loop sequence (**C**). Experiment demonstrates specific interactions between complementary (antisense) peptide and Boraschi loop competed with by a peptide with the Boraschi loop sequence. Complementary (antisense) peptide will also not associate with a variety of control proteins, including IL-1ra, further demonstrating peptide binding specificity for immobilised IL-1β. (Heal *et al.*, 2002a, Figure 4, John Wiley & Sons).

8

Kinetics and Catalysis

8.1 Catalysis in chemical biology

According to some theories about the origins of life, the key to the creation of organisms is molecular complexity that is sufficient to give self-organisation (see Chapter 10). However, self-organisation alone is insufficient to give life. Instead, self-organisation needs to be partnered with the capacity to accelerate or catalyse chemical inter-conversions as well. This capacity to catalyse chemical inter-conversions is known as **catalysis**. Catalysis is frequently performed by a **catalyst** that is usually defined as an entity that enhances the rate of a given chemical reaction in both forward and reverse directions without being itself permanently changed in the process. Therefore, a **biocatalyst** is a biologically relevant catalyst. Typically, biocatalysts accelerate biological chemical reactions with relative rate enhancements of between 10^5 and 10^{10} relative to the non-catalysed reactions. Catalysis is universal to all cells of all organisms and the range and diversity of known biocatalysts is simply staggering. Biocatalysts are clearly an absolute fundamental for both the origin of life and the promulgation of life.

Biocatalysts are overwhelmingly proteins (**enzymes**) and sometimes RNA-related nucleic acids (**ribozymes**). They catalyse an amazing diversity of reactions for myriads of different important reasons. Enzymes are centre stage in **energy metabolism**, which is the process by which chemical potential is generated and stored in coupling the synthesis of **adenosine 5′-triphosphate (ATP)** (the preferred "form" of stored chemical energy in all cells) with the step-wise degradation and/or reorganisation of covalent bonds of **primary nutrients** such as **glucose** (Chapter 1). For instance, **triose phosphate isomerase (TIM)** (see Chapter 1) catalyses the seemingly innocuous inter-conversion between **dihydroxyacetone phosphate** and **glyceraldehyde-3-phosphate** in the catabolic pathway known as **glycolysis** (Figure 8.1, Table 8.1). Yet surprisingly, TIM is now known to be a "perfect enzyme" (see Section 8.4.8) that makes the inter-conversion possible at a rate which is literally as fast as substrate reaches the enzyme active site. Indeed, without TIM, the glycolysis pathway would be unable to deliver on a net gain of two ATP molecules for each glucose molecule consumed (see Figure 8.1, Table 8.1). In a similar vein, the dimeric enzyme **malate dehydrogenase (MDH)** catalyses the mere reduction of a carbonyl functional group in **oxalic acid** to give **malic acid**, making use of the cofactor **nicotinamide adenine dinucleotide, reduced form (NADH)** (Figure 8.2, Table 8.1), yet this inter-conversion establishes closure of the amphibolic **tricarboxylic acid (TCA) cycle** that takes metabolites from glycolysis and delivers on a net gain of reducing cofactor molecules for each complete rotation through the TCA cycle. There are other types of catabolic enzymes such as **chymotrypsin** that digest polypeptides into oligopeptides within the gut (Figure 8.3, Table 8.1), for absorption across the gut wall into the bloodstream. Alternatively, **ribonuclease A (RNAse A)** does for RNA polynucleotides what chymotrypsin does for polypeptides, albeit by a very different mechanism (Figure 8.4, Table 8.1). Enzymes are not only involved in **catabolism** (i.e. breaking down), but also play a role in **anabolism** (i.e. building up) of monomeric building blocks required for biological macromolecular assembly.

Essentials of Chemical Biology: Structures and Dynamics of Biological Macromolecules In Vitro *and* In Vivo, Second Edition. Andrew D. Miller and Julian A. Tanner.
© 2024 John Wiley & Sons, Inc. Published 2024 by John Wiley & Sons, Inc.
Companion Website: www.wiley.com/go/miller/essentialschembiol2

In this respect, examples include **alanine α-racemase (AlaR)**, **glutamate α-decarboxylase (GadB)** and **aspartate transaminase (AspAT)** enzymes that all make use of the cofactor **pyridoxal phosphate (PLP)** (Figure 8.5, Table 8.1).

Many enzymes have protective functions, frequently outside cells. For instance, **lysozyme** (see Chapter 1), which is produced externally in tears, functions to catalyse the hydrolysis of O-glycosidic links in complex polysaccharides of bacterial cell walls, in order to weaken and destroy those cell walls, kill the bacteria and protect the surface of the eye from infection (Figure 8.6, Table 8.1). **Chloramphenicol acetyl transferase (CAT)** is an example of a bacterial enzyme evolved to inactivate the antibiotic **chloramphenicol** by controlled acetylation (Figure 8.7, Table 8.1). Then there is **superoxide dismutase (SOD)** that is an astonishing di-metal enzyme family that catalyses the disproportionation of **superoxide radicals** into **oxygen** and **hydrogen peroxide** with extraordinary efficiency (Figure 8.8, Table 8.1). Superoxide radicals are generated when the reduction of oxygen to water during **respiration** is incomplete, which is surprisingly frequent. These radicals are lethal to all the biological macromolecules and macromolecular lipid assemblies in a cell unless they are dealt with promptly by SOD. In fact, certain cells of the immune system actually use superoxide radicals deliberately in a controlled fashion to destroy invading microorganisms. Now let us consider **carbonic anhydrase (CA)** (see Chapter 1) that has been described as one of the most efficient biocatalysts of all. This enzyme catalyses the simple inter-conversion of **carbonic acid** into water and **carbon dioxide** (Figure 8.9, Table 8.1). Why is this enzyme protective? Quite simply, CA acts to preserve the

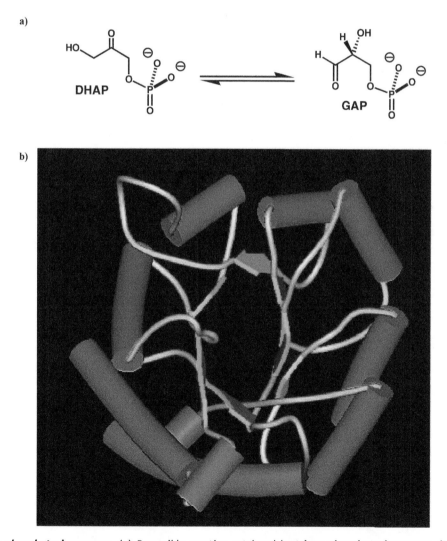

Figure 8.1 **Triose phosphate isomerase.** (a) Reversible reaction catalysed by **triose phosphate isomerase (TIM)** where **DHAP** is **dihydroxyacetone phosphate**, and **GAP** is **glyceraldehyde-3-phosphate**; (b) **Schematic display structure** of chicken muscle TIM (top view) (pdb: **1tim**); (c) Chemical illustration of **glycolysis** — a primary catabolic pathway able to yield two molecules of ATP/glucose consumed, made possible only through the catalytic activity of TIM. Other abbreviations are as follows: **HK: hexokinase; PGI: phosphoglucose isomerase; PFK: phosphofructokinase; GAPDH: glyceraldehyde-3-phosphate dehydrogenase; PGK: phosphoglycerate kinase; PGM: phosphoglycerate mutase; PK: pyruvate kinase; ADP: adenosine 5′-diphosphate.**

c)

Figure 8.1 (Continued)

Table 8.1 Summary of kinetic constants for the enzymes featured in Chapter 8. All the data were from the enzyme database: BRENDA (http://www.brenda-enzymes.info) except where indicated: (a) Value of k_{cat}/K_m estimated assuming all substrate added in buffer is available to enzyme (i.e. has not been sequestered as acetal or ketal derivatives, as appropriate); (b) k_{cat} data from Fields et al., 2006; (c) k_{cat} data from rat source; (d) k_{cat} data from Masaki et al., 2001; (e) k_{cat} data from Day and Shaw, 1992; (f) Kinetic data from bovine source; (g) Kinetic data from human source; (h) All kinetic data from Wright et al., 2006. Abbreviations: **GAP**: glyceraldehyde-3-phosphate; **DHAP**: dihydroxyacetone phosphate; **OAA**: oxaloacetate; **NADH**: nicotinamide adenine dinucleotide, reduced form; **poly(pC)**: polycytidylic acid; **2′,3′-cCMP**: cyclic cytidine 2′,3′-monophosphate; **acetyl-CoA**: acetyl coenzyme A; **chloramp**: chloramphenicol; **thio-ACh**: thio-acetylcholine; **ATP**: adenosine 5′-triphosphate; **ADP**: adenosine 5′-diphosphate; **AMP**: adenosine 5′-monophosphate; **cAMP**: cyclic adenosine 3′,5′-monophosphate; **GTP**: guanosine 5′-triphosphate; **ITP**: inosine 5′-triphosphate. Roman numerals denote compounds illustrated in the set of structures accompanying this table (see following two pages).

Enzyme	Substrate	k_{cat} (s⁻¹)	K_m (M)	k_{cat}/K_m (M⁻¹ s⁻¹)	Inhibitor	K_i (M)
I. TIM (chicken) E.C. 5.3.1.1 1tim (pdb)	GAP	4300 (pH 7.6)	0.47×10^{-3} (pH 7.6)	9.0×10^{6}	AsO₄³⁻	11×10^{-3}
	DHAP	430 (pH 7.6)	0.97×10^{-3} (pH 7.6)	4.4×10^{5}	Ia	—
				1×10^{8} (a)	Ib	—
II. mMDH (porcine) E.C. 1.1.1.37 4mdh (pdb)	OAA	—	0.03×10^{-3} (pH 7.5)	2×10^{7}	ADP	1.34×10^{-3}
	NADH	400 (pH 7.2) (b)	0.02×10^{-3} (pH 7.5)		AMP	0.95×10^{-3}
					cAMP	0.56×10^{-3}
III. α-chymotrypsin (bovine pancreas) E.C. 3.4.21.1 4cha (pdb)	IIIa	110 (pH 8.3)	7.4×10^{-6} (pH 8.3)	1.5×10^{7}	IIIb	2.5×10^{-8}
					IIIc	irrev.
IV. RNAse A (bovine pancreas) E.C. 3.1.27.5 7rsa (pdb)	poly(pC)	—	0.46×10^{-3}		IVa	—
	2′,3′-cCMP	2.5 (c)	0.46×10^{-3} (pH 5.5)	5.4×10^{3}	IVb	—
V. AlaR (B. Stearotherm ophilus) E.C. 5.1.1.1 2sfp (pdb)	L-Ala	1.1	2.7×10^{-3}	4×10^{2}	Va	6×10^{-3}
					Vb	—
VI. GadB (E. coli) E.C. 4.1.1.15 1pmo and 1pmm (pdb)	L-Glu	9.5	0.5×10^{-3} (pH 4.6)	1.9×10^{4}	NO?	—
					O₂?	—
VII. AspAT (E. coli) E.C. 2.6.1.1 2aat (pdb)	L-Asp	259 (pH 8.4)	1.3×10^{-3} (pH 8.0)	2.0×10^{5}	VIIa?	—
					VIIb?	—
VIII. lysozyme (hen egg) E.C. 3.2.1.17 6lyz (pdb)	VIIIa	0.05 (d)	2.4×10^{-5} (d)	2×10^{3}	VIIIb	8×10^{-8}
IX. CAT III (E. coli) E.C. 2.3.1.28 1cla (pdb)	acetyl-CoA	600 (e)	0.093×10^{-3}	6.4×10^{6}	IXa	irrev.
	chloramp		0.012×10^{-3}	—		
X. MnSOD (human) E.C. 1.15.1.1 1n0j (pdb)	O₂⁻	100000 (f)	0.355×10^{-3} (f)	2.8×10^{8}	ClO₄⁻	—
					Xa	
XI. CA1 (human) E.C. 4.2.1.1 2cab (pdb)	CO₂	800 (pH 7.2)	3.6×10^{-3} (pH 7.2)	2.2×10^{5}	XIa	0.40×10^{-6}
					XIb	0.01×10^{-6}
					XIc	0.09×10^{-6}
					XId	0.003×10^{-6}
XII. AChE (T. californica) E.C. 3.1.1.7 1amn (pdb)	thio-ACh	6500 (g)	0.046×10^{-3} (g)	1.4×10^{8}	XIIa	0.175×10^{-3}
					XIIb	0.003×10^{-6}
					XIIc	irrev.
XIII. LysU (h) (E. coli) E.C. 6.1.1.6 1lyl (pdb)	ATP	2.7	7×10^{-3}	3.8×10^{2}	GTP	—
	L-Lys	1.8	23×10^{-6}	7.8×10^{4}	ITP	—

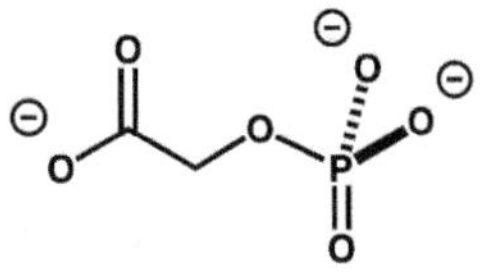

Phosphoglycolate **Ia**

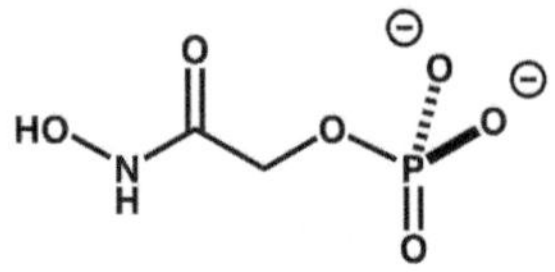

Phosphoglycolhydroxamate **Ib**

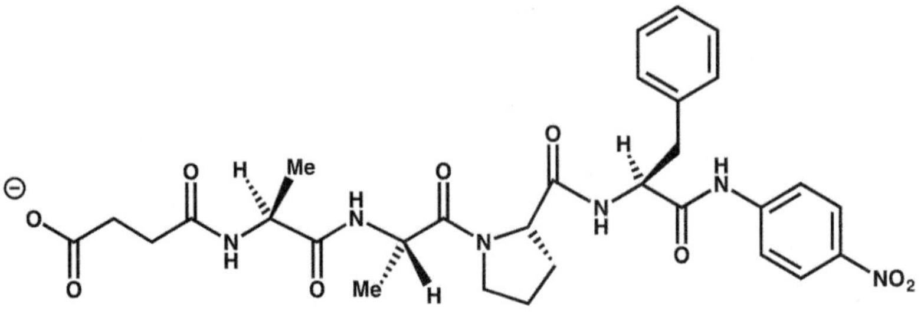

N-succinyl AAPF-*p*-nitroanilide **IIIa**

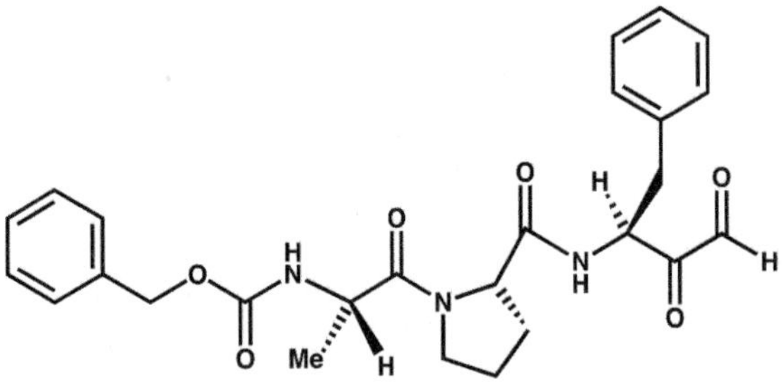

Z-APF-glyoxal **IIIb**

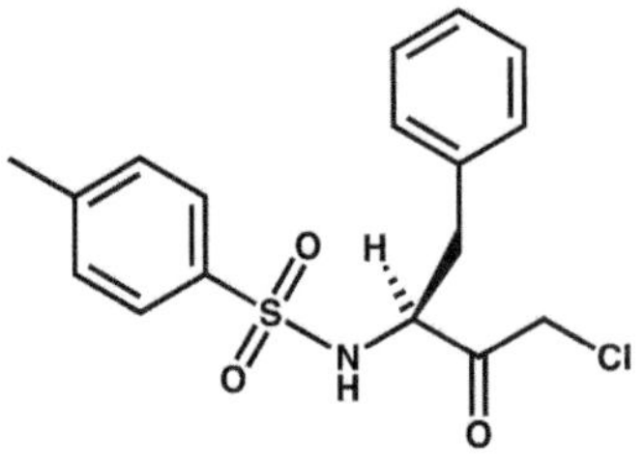

N-Tos-F-CH₂Cl **IIIc**

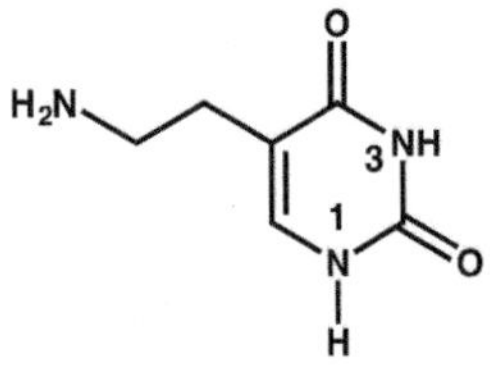

5 amino ethyl uracil **IVa**

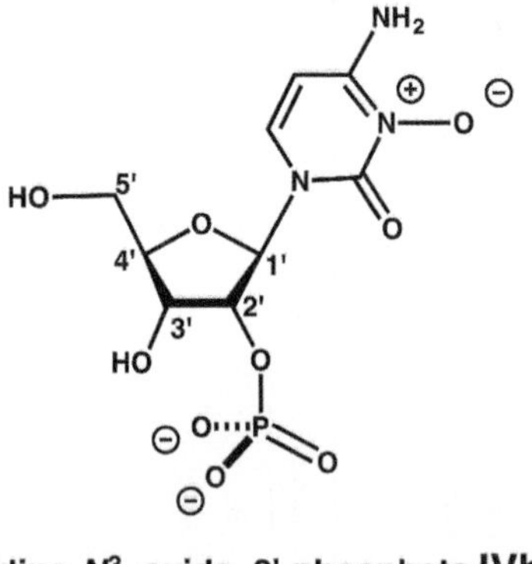

cytidine-N^3-oxide 2'-phosphate **IVb**

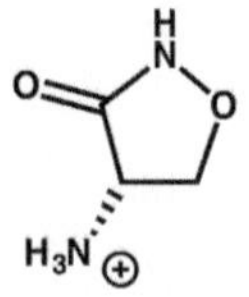

L-cycloserine **Va**

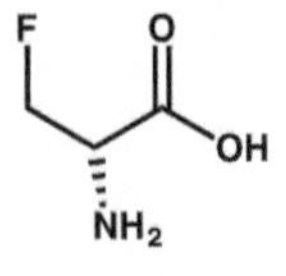

3-Fluoro D-alanine **Vb**

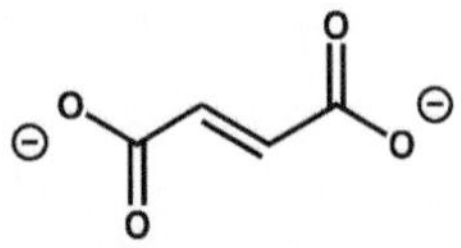

Fumarate **VIIa**

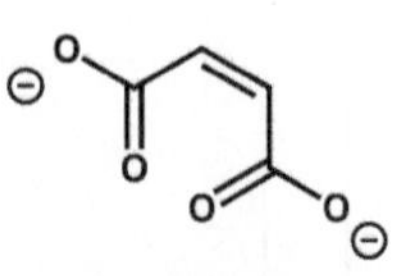

Maleate **VIIb**

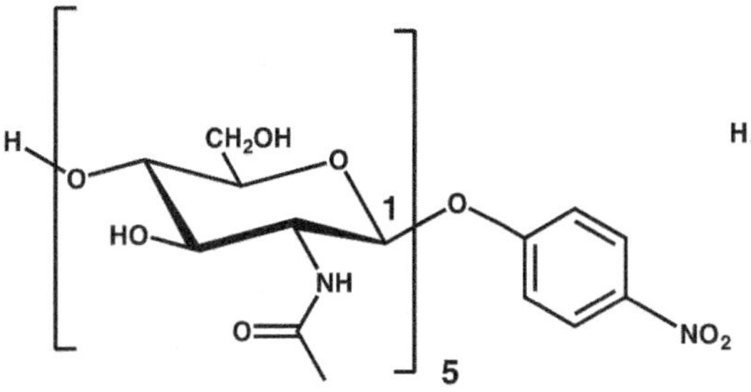

p-nitrophenyl penta N-acetyl-β-D-
glucosaminide NAG₅-PNP **VIIIa**

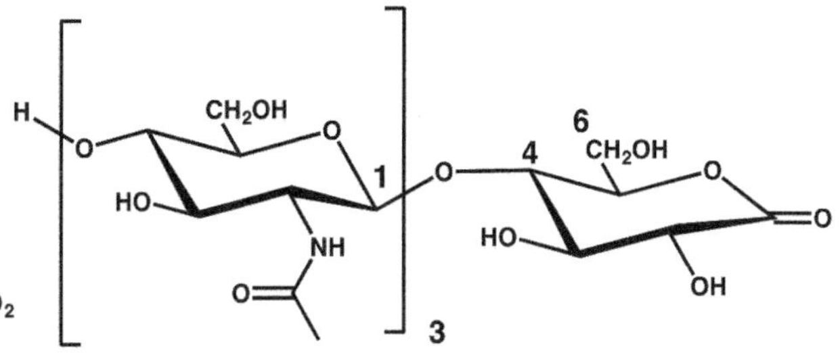

NAG₄ lactone **VIIIb**

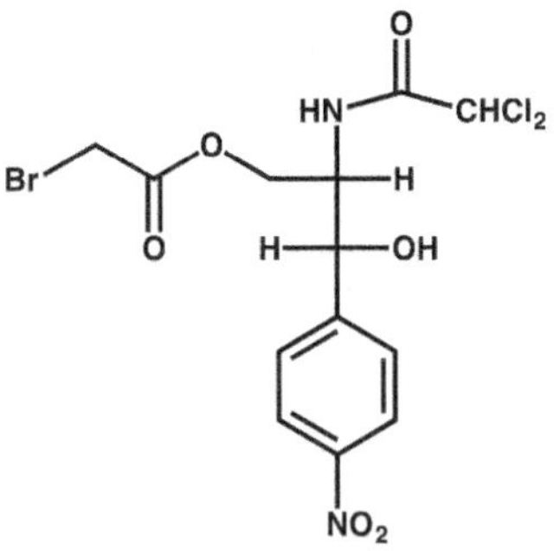

3-(bromoacetyl)chloramphenicol **IXa**

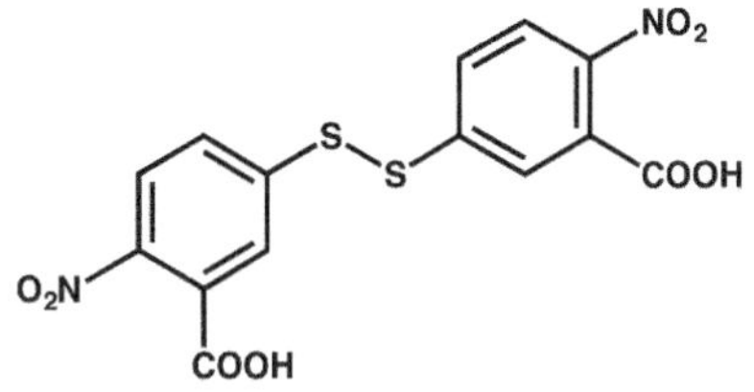

5,5'-dithio bis (2-nitrobenzoic acid) (DTNB) **Xa**

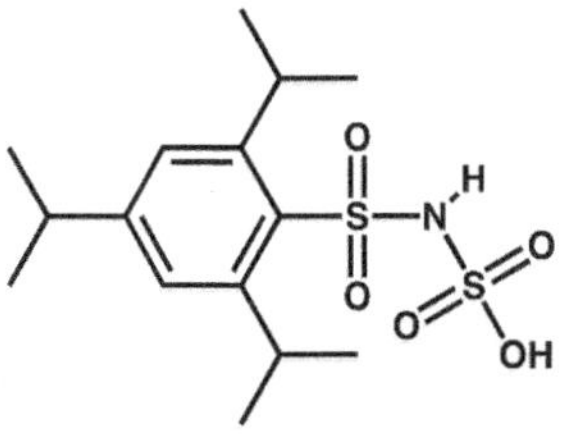

2,4,6-triisopropylphenyl sulphonyl sulfamic acid **XIa**

2,5-dichlorophenyl sulphonyl sulfamic acid **XIb**

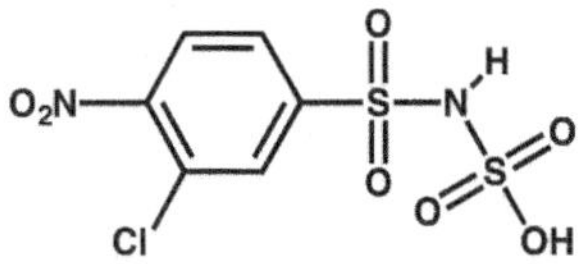

3-chloro-4-nitrophenyl sulphonyl sulfamic acid **XIc**

pentafluorophenyl sulphonyl sulfamic acid **XId**

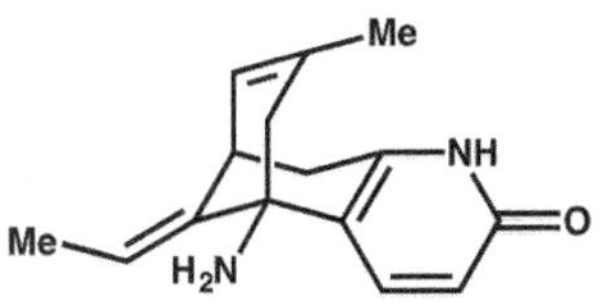

(-) huperzine A **XIIa**

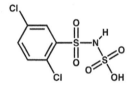

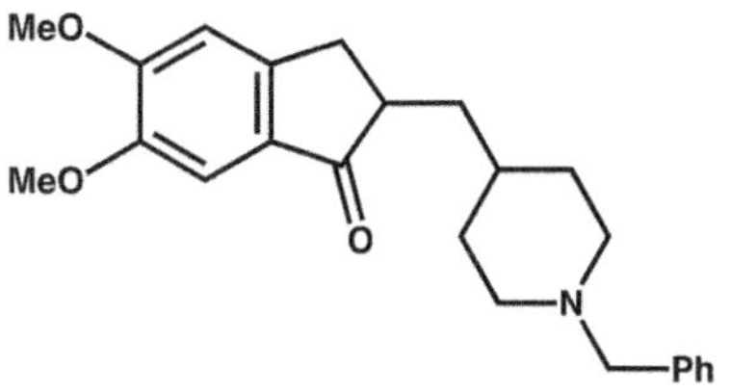

(R,S)-1-benzyl-4-[(5,6-dimethoxy-1-indanon)
-2-yl]methylpiperidine HCl **XIIb**

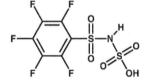

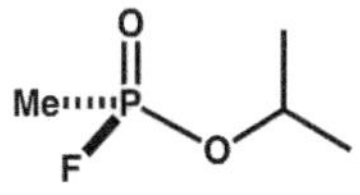

Sarin (nerve gas) **XIIc**

pH of blood at around pH 7, avoiding the problem of acidification that would otherwise take place as dissolved carbon dioxide levels rise in blood following glucose metabolism and TCA cycle activity in cells. Blood acidification or acidosis is spectacularly lethal. Finally, in Chapter 7, we have already met **acetylcholine esterase (AChE)**, without which excess of the neurotransmitter **acetylcholine** would not be removed by hydrolysis from relevant synaptic clefts (Figure 8.10, Table 8.1), leading to uncontrolled neurotransmission associated with such neurological conditions as seizures and epilepsy.

The chemical biology reader should also be aware that there are numbers of enzymes that can have multiple functions and that there are also biocatalysts which are not enzymes. In the first class, the enzyme **LysU** makes for an interesting case in point (see Chapter 7). This is a **lysyl tRNA synthetase** enzyme in the first instance that should have the expected capacity to couple the naturally available amino acid, **L-lysine**, to appropriate cognate tRNAs bearing anti-codon sequences complementary to lysine codons (see Section 1.6.1). However, in the presence of zinc ions, Zn^{2+}, the function of this enzyme becomes altered to catalyse the biosynthesis of **diadenosine-5′, 5′′′-P^1, P^4-tetraphosphate (Ap_4A)**, followed by **diadenosine-5′, 5′′′-P^1, P^3-triphosphate (Ap_3A)** in a sequential fashion (Figure 8.11, Table 8.1). Finally, the discovery that RNA itself can act as a biocatalyst (**ribozyme**) was a huge revelation several years ago, leading to many suggestions that ribozymes may be the first biocatalysts formed in advance of enzymes (Figure 8.12). Accordingly, there has been much discussion about an RNA world that precedes that of the protein world in a molecular theory of evolution (see Chapter 10).

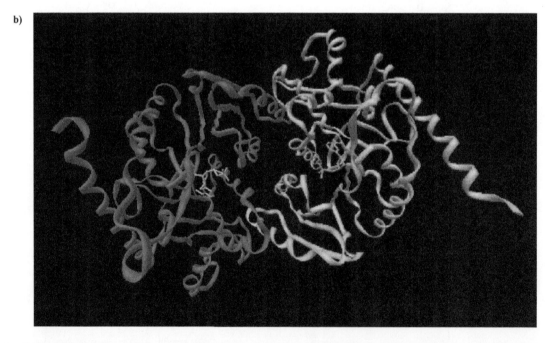

a)

oxalate

NADH

L-malate

b)

Figure 8.2 Malate dehydrogenase. (a) Reversible reaction catalysed by **malate dehydrogenase (MDH)** where **NADH** is **nicotinamide adenine dinucleotide, reduced form; (b) Ribbon display structure** of porcine heart **mitochondrial MDH (mMDH)** (side view) (pdb: **4mdh**). The homodimeric protein consists of two polypeptide chains (**yellow** and **red**), with **ball and stick representation** (**blue**) showing **nicotinamide adenine dinucleotide, oxidised form (NAD⁺)**, in both independent catalytic sites, illustrated in a ball and stick (blue) representation; **(c)** Chemical illustration of the **tricarboxylic acid (TCA) cycle** to demonstrate the importance of MDH catalysis in cycle closure. Other abbreviations are: **PDH: pyruvate dehyrogenase; CS: citrate synthase; CoASH or HSCoA: coenzyme A.**

c)

Figure 8.2 (*Continued*)

a)

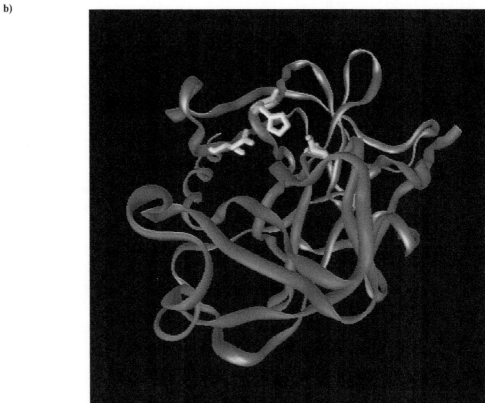

b)

Figure 8.3 α-**Chymotrypsin.** (**a**) Hydrolytic reaction catalysed by α-**chymotrypsin**, where amino acid residue side chain R$_2$ is hydrophobic or aromatic in character; (**b**) **Ribbon display structure** of α-chymotrypsin (bovine pancreatic) (pdb: **4cha**) in which key active site residues **D102**, **H57** and **S195** (left to right) involved in biocatalysis are shown (**yellow**) rendered in a **stick representation** (see Figure 8.50(b)).

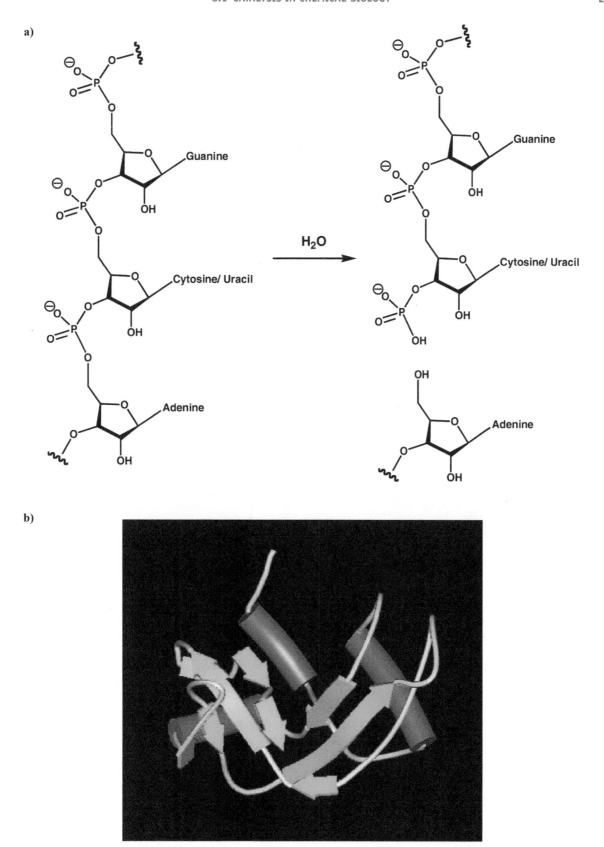

Figure 8.4   Ribonuclease A. (**a**) Hydrolytic reaction catalysed by **ribonuclease A** (**RNAse A**), note that hydrolytic cleavage occurs at the 3'-side of a pyrimidine nucleoside; (**b**) **Schematic display structure** of bovine pancreatic RNAse A (side view) (pdb: **7rsa**).

a)

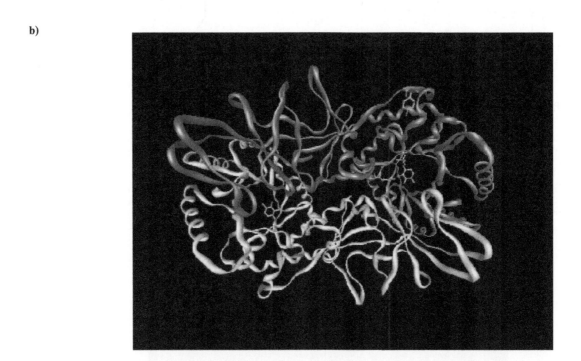

b)

Figure 8.5 Three anabolic enzymes. Alanine racemase. (a) Racemisation reaction catalysed by **alanine racemase (AlaR)**, note this requires the cofactor **pyridoxal phosphate (PLP)** in order to take place; **(b) Ribbon display structure** of homodimeric AlaR (*Bacillus Stearothermophilus*) (pdb: **2sfp**) showing both polypeptide chains (**yellow** and **red**). PLP appears bound in both independent catalytic sites rendered by **ball and stick representation (blue)**; in one catalytic site, key catalytic residues **Y265** and **K39** (top to bottom) are shown, also rendered by **ball and stick representation (green)**. **Glutamate α-decarboxylase. (c)** α-Decarboxylation reaction catalysed by **glutamate α-decarboxylase (GadB)** that requires the cofactor PLP in order to generate product **γ-amino butyric acid (GABA)**; **(d) Ribbon display structure** of **T-state** (neutral pH) homohexameric *Escherichia coli* (*E. coli*) GadB showing all polypeptide chains (two in **green, yellow** and **blue**) (pdb: **1pmo**); **(e) Ribbon display structure** of **R-state** (acidic pH) homohexameric GadB showing all polypeptide chains (two in **green, yellow** and **blue**) and six *N*-terminal α-helices (1–14, **red**) that are key sites for protonation and chloride ion binding induced **positive cooperativity** (pdb: **1pmm**). **Aspartate transaminase. (f)** Transamination reaction catalysed by **aspartate transaminase (AspAT)** that requires the PLP in order to take place; **(g) Ribbon display structure** of homodimeric *E. coli* AspAT (pdb: **2aat**) showing one of two polypeptide chains. PLP appears bound in independent catalytic site rendered by **ball and stick representation (yellow)**; key substrate binding residues **R398** and **R332** (right to left) are shown (**red**), also rendered by **ball and stick representation**.

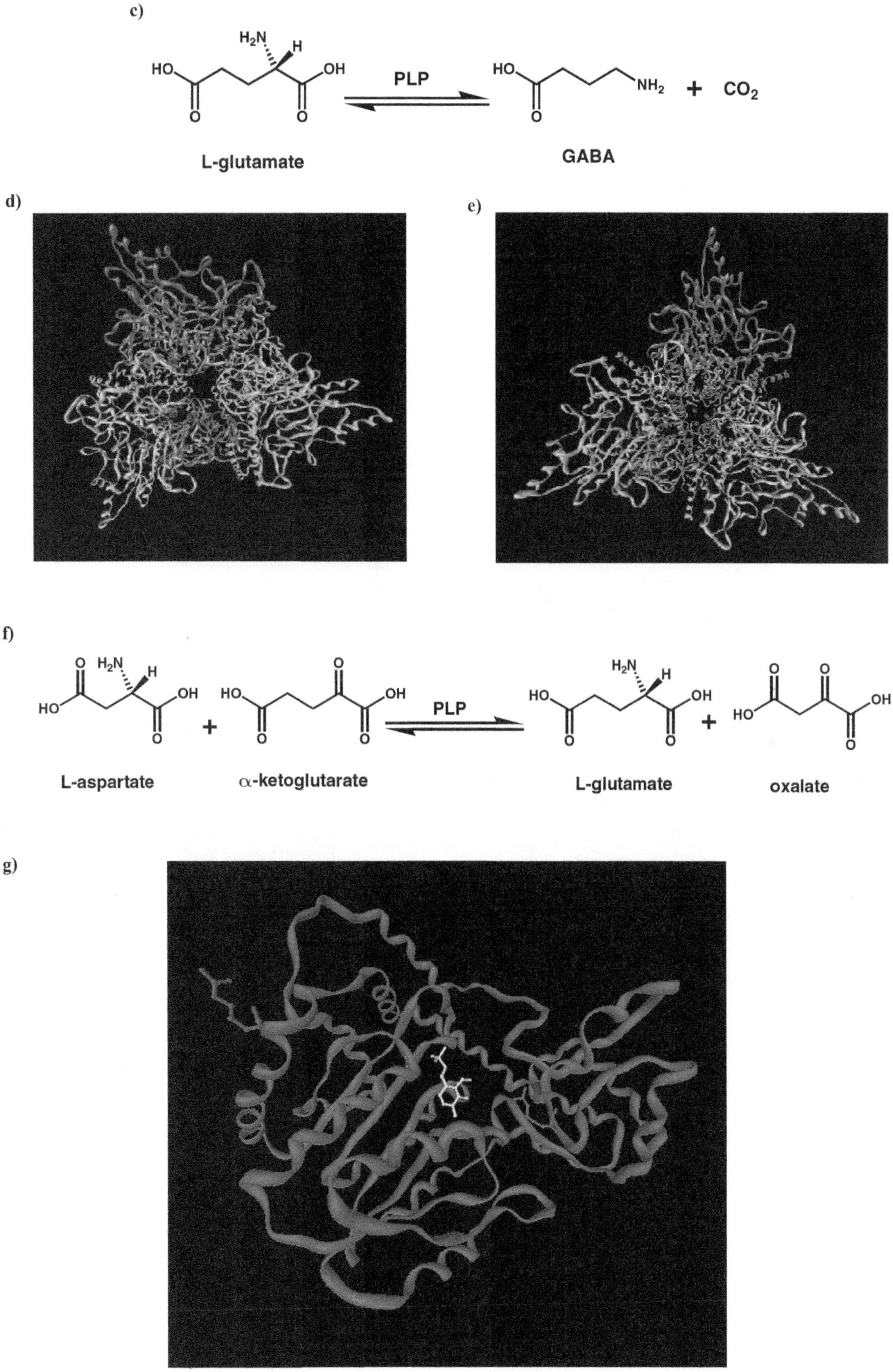

c)

L-glutamate   **GABA**   +   **CO₂**

d)

e)

f)

L-aspartate   **α-ketoglutarate**   **L-glutamate**   **oxalate**

g)

Figure 8.5   (Continued)

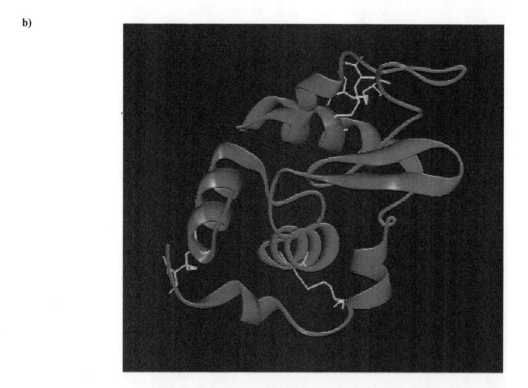

tetra-NAG

Figure 8.6 **Lysozyme.** (a) Hydrolysis reaction catalysed by **lysozyme**; (b) **Ribbon display structure** of hen egg-white lysozyme (pdb: **6lyz**) wherein the four **disulfide bonds** and their corresponding **cysteine residues** are rendered in a **stick representation** (**yellow**). The catalytic site is a broad cleft through the polypeptide.

a)

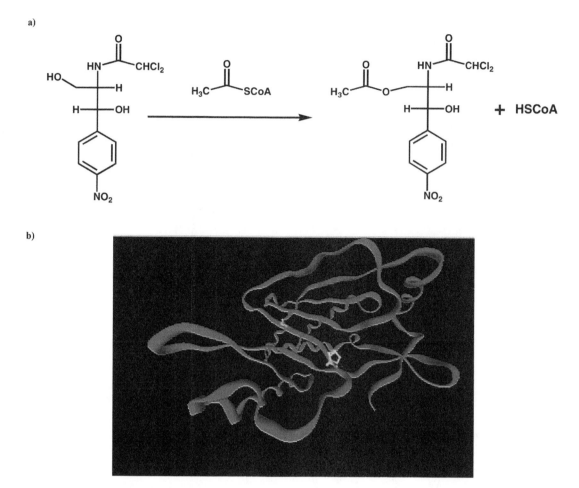

Figure 8.7 **Chloramphenicol acetyl transferase.** (a) Acylation reaction catalysed by **chloramphenicol acetyl transferase (CAT)**; (b) **Ribbon display structure** of Type III, *E. coli* CAT (pdb: **1cla**) wherein two key catalytic residues **H195** and **S148** (right to left) are displayed (**yellow**), in **ball and stick representation**.

a)

$$2O_2^{\cdot-} \quad + \quad 2H^{\oplus} \quad \longrightarrow \quad H_2O_2 \quad + \quad O_2$$

b)

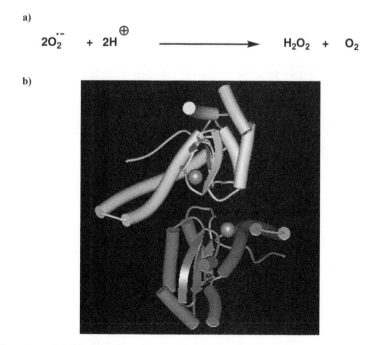

Figure 8.8 **Super oxide dismutase.** (a) Disproportionation reaction catalysed by **super oxide dismutase (SOD)**; (b) **Schematic display structure** of homotetrameric human mitochondrial **manganese SOD (MnSOD)** (pdb:**1n0j**), illustrating two out of the four identical polypeptide chains (**orange, yellow**) and manganese ions ($Mn^{2+/3+}$) (**purple**) rendered as Van der Waals spheres for complete clarity.

a)

$$CO_2 \; + \; H_2O \;\; \rightleftharpoons \;\; HCO_3^{\ominus} \; + \; H^{\oplus}$$

b)

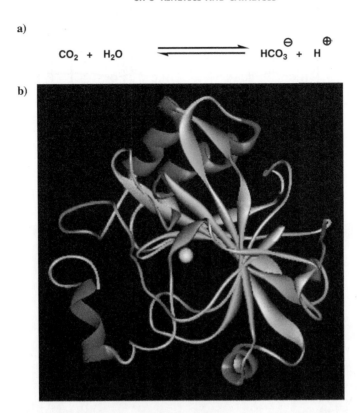

Figure 8.9   Carbonic anhydrase. (a) Reversible hydration reaction catalysed by **carbonic anhydrase (CA)**; (b) **Ribbon display structure** of monomeric CA (**CA1**, human erythrocyte) (pdb: **2cab**) illustrating α-helices (**red**) and β-sheet (**blue**) plus a central zinc ion (Zn²⁺) (**yellow**) rendered as Van der Waals sphere for complete clarity.

a)

AChE

b)

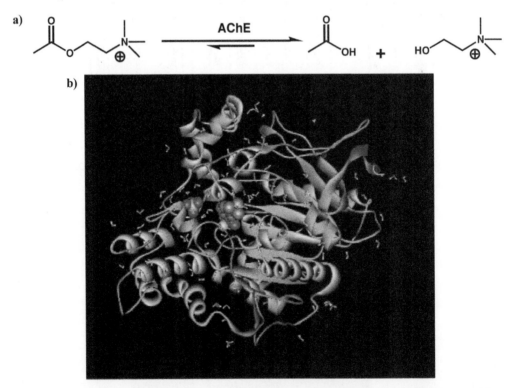

Figure 8.10   Acetylcholine esterase. (a) Hydrolysis reaction catalysed by **acetylcholine esterase (AChE)**; (b) **Ribbon display structure** of *Torpedo californica* AChE (pdb: **1amn**) illustrating crystallographic water molecules (oxygen: **red**; hydrogen: **white**) plus sulfate ion and substrate analogue bound in catalytic site. Both these are rendered as **CPK structures** (red: oxygen; **yellow**: sulfur; **grey**: carbon; **blue**: nitrogen).

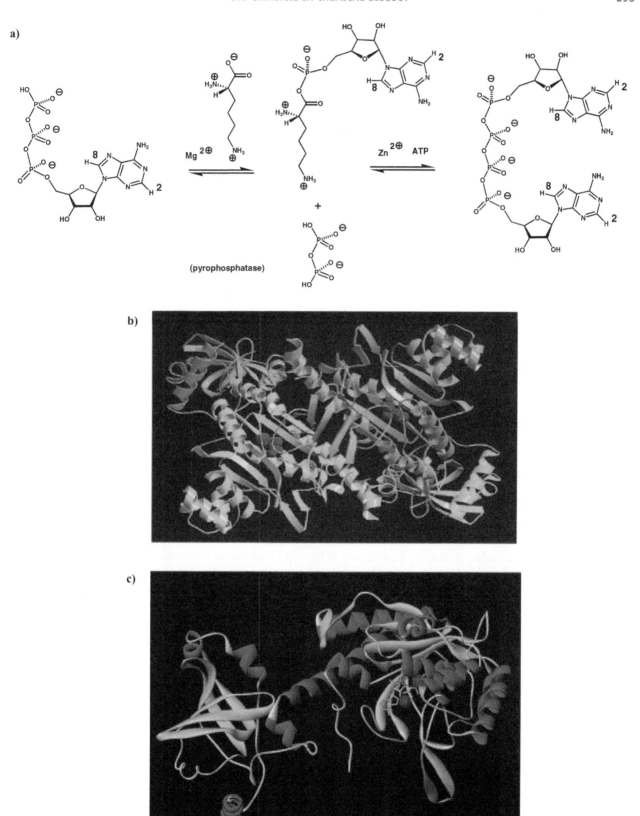

Figure 8.11   Lysyl tRNA synthetase. (a) Diadenosine-5′, 5‴-*P*¹, *P*⁴-tetraphosphate (Ap₄A) synthesis reaction catalysed by **lysyl tRNA synthetase (LysU)** isozyme. Note that pyrophosphate leaving group once formed is later hydrolysed to two phosphates by pyrophospha-tase; **(b) Ribbon display structure** of homodimeric *E. coli* LysU (pdb: **1lyl**) showing both polypeptides (**blue** and **green**) and catalytic site bound **L-lysine (red)** rendered as **CPK structures; (c) Ribbon display structure** of LysU monomer illustrating α-helices (**red**) and β-sheets (**blue**) with AMP (**orange**) and L-lysine (**green**) moieties of **lysyl-adenylate** intermediate bound in the catalytic site with two Mg²⁺ ions (**purple**). The Mg²⁺ ions are rendered as Van der Waals spheres, while intermediate is rendered in **stick representation** (pdb: **1e1t**).

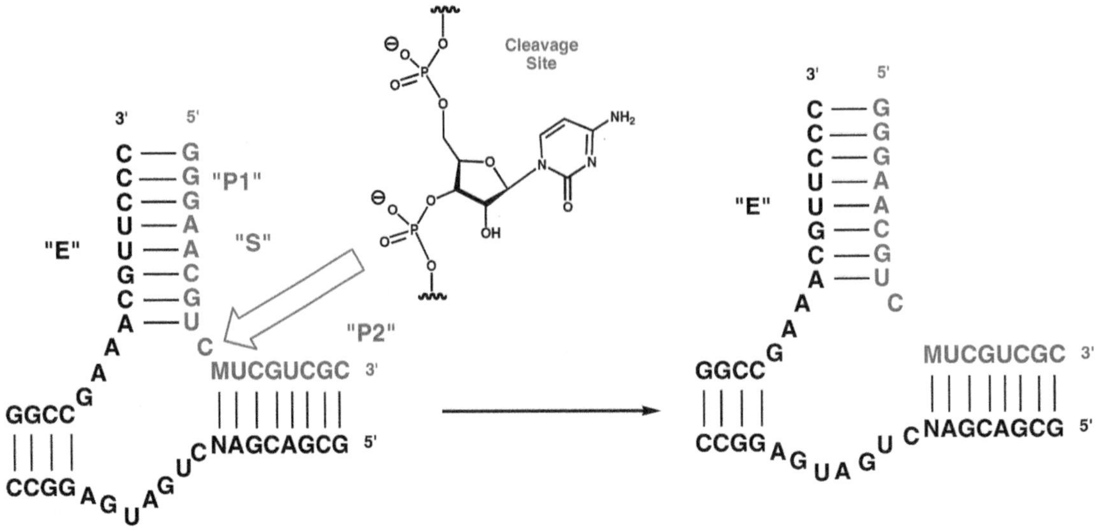

Figure 8.12 Ribozyme. Hydrolysis reaction catalysed by **ribozyme** that consists of an **enzyme "E" strand** and a **substrate "S" strand**. Hydrolysis takes place below a **cytidine nucleotide residue**, converting "S" into two **product strands** "P1" and "P2" that dissociate from the "E" strand.

8.1.1 Simple principles in biocatalysis

The **active** or **catalytic site** of any biocatalyst represents the region of structure devoted to biocatalysis in any given biocatalyst. Catalytic sites are usually surprisingly small and frequently occupy a very small area/volume of the overall structure. During the first stage of biocatalysis, a **substrate** molecule must be recognised and bound to a catalytic site in a process of **molecular recognition and binding** that is completely equivalent to those processes described and discussed in detail in Chapter 7. Therefore, Chapter 7 is sufficient to appreciate the first stage of biocatalysis. During the second stage of biocatalysis, a bound substrate must be transformed into a bound **product** by biocatalysis and then released into the bulk solution. The ways and means of biocatalysis will form the principle subject matter of this chapter.

But first, in order to study biocatalysis, there needs to be a ready supply of a biocatalyst of interest made available through techniques such as those described in Chapter 3. Structure is always very helpful to interpret function (Chapters 4, 5 and 6). After this, there need to be techniques of analysis and a sound theoretical framework with which to interpret biocatalysis data and elaborate those key mechanisms of biocatalyts that make biocatalysis possible. For this reason, we will begin this chapter with a detailed discussion about ways to acquire and analyse biocatalytic data using various models of biocatalysis. Following this, we will take a look at those theories of biocatalysis that help explain how biocatalysts are able to be such effective catalysts of those chemical reactions required for life. Throughout this chapter, we will refer to the examples of biocatalysts introduced to the chemical biology reader already, as an aid to understanding. In reality, the route to a real understanding of biocatalysis is long and involved for each and every biocatalyst under investigation. Little wonder that there remains much to do before we can eventually arrive at a complete understanding of known biocatalysts, let alone all those yet unknown.

8.1.2 Steady state kinetics in biocatalysis

The study of reaction rates and catalysis is known as **kinetics**. Typically, the catalytic effect of any one biocatalyst is evaluated under **steady state** and/or **pre-steady-state** conditions (see Section 8.3). Steady state kinetic analyses are used to understand the surface mechanistic behaviour of a given biocatalyst that is operating at **stasis** and capable of performing multiple catalytic turnovers. Stasis implies that rate measurements (i.e. data describing rates of conversion of a given substrate into product) are restricted to short time intervals over which the concentrations of substrates and reactants do not change significantly. Therefore, stasis also implies that kinetic analysis takes place while there is metastable equilibrium between substrate(s), product(s) and/or various biocatalyst species. Steady state analysis is very representative of biocatalyst behaviour under conditions of normal metabolism, although a rather blunt tool for the investigation of detailed mechanistic steps associated with a given biocatalytic process taking place in the catalytic site of a given biocatalyst. Nonetheless, steady state kinetic analyses usually require relatively simple apparatus and simple procedures to implement and so are performed routinely.

8.1.3 Steady state bioassays

Steady state kinetic analyses can be as diverse as the range of biocatalysts themselves. Rate enhancements brought about by any one biocatalyst are usually assessed by means of a **bioassay**. There is at least one unique bioassay for each biocatalyst. A bioassay is used to generate rate data, namely the variation in **initial rate**, ν, of substrate to product conversion as a function of **initial substrate concentration** [S], while keeping the biocatalyst concentration constant throughout (Figure 8.13). Rate data may itself be determined as a function of temperature, pH, ionic strength and even as a function of other fixed concentrations of second, third or even fourth substrates. This rate data is then processed by means of any number of appropriate **steady state kinetics equations** to give standard biocatalytic parameters (that will be described below) and an indication of the surface mechanism by which the biocatalyst brings about biocatalysis. These standard parameters allow for simple comparisons between biocatalyst performances.

The majority of steady state bioassays tend to be **photometric assays**, but many other **steady state bioassays** including ^1H-NMR **assays**, **high-performance liquid chromatography (HPLC) assays**, **radiometric assays**, **oxygen electrode assays**, and **gel assays** may be used. In all cases, the main principle is to determine how values of ν (directly or indirectly) vary as a function of [S], while keeping the biocatalyst concentration constant throughout. Photometric steady state bioassays make use of some form of **chromogenic substrate** or **reagent** that becomes chemically altered into a chromophore during biocatalysis and as a result absorbs significantly at a particular wavelength of absorption. The initial rate of formation of product is usually said to correlate with the initial rate of formation of the chromophore as determined by **UV-visible spectroscopy**. Steady state bioassays of AChE and CA both make use of chromogenic reagents (Figures 8.14 and 8.15). A steady state bioassay of lysozyme instead makes use of a chromogenic substrate that may also be detected by fluorescence as well. Hence, the initial rate of product formation then correlates with the initial rate of formation of the fluorophore as determined by **fluorescence spectroscopy** (Figure 8.16). In the case of TIM, the actual reaction catalysed is impossible to observe with a chromogenic substrate or reagent directly, but a **coupled assay system** can be used in which one product is irreversibly transformed into a follow-on product through biocatalytic reduction using the cofactor NADH, a **reverse-chromogenic reagent** (Figure 8.17). The initial loss of NADH by oxidation to NAD$^+$ is said to correlate inversely with the initial rate of TIM-catalysed dihydroxyacetone phosphate (DHAP) product formation. Where the enzyme is concerned, NADH is an intrinsic part of the direct catalytic mechanism and so initial loss of NADH with time correlates inversely, but directly with the initial rate of MDH-catalysed oxalic acid product formation (Figure 8.18).

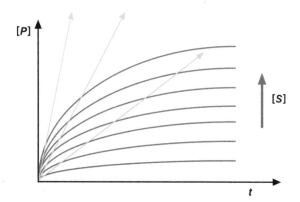

Figure 8.13 **Schematic diagram to illustrate the concept of initial rates.** In the presence of a fixed concentration of biocatalyst, variable concentrations of substrate, **S**, are converted into product, **P**, over a **time, t.** Steady state analyses require initial concentrations of substrate, **[S]**, and initial rates of product formation, ν, that derive from the slopes of tangents (blue arrows) to initial slopes as illustrated.

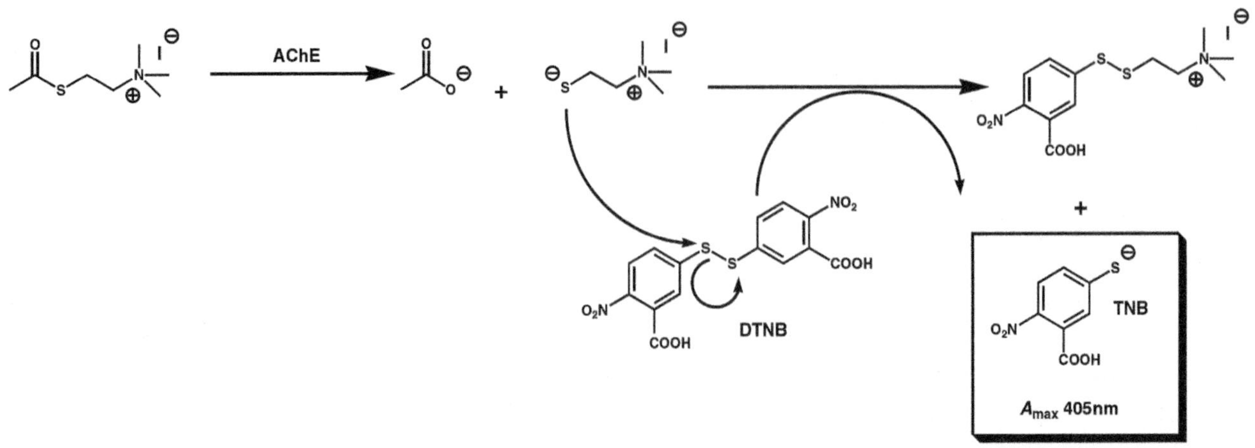

Figure 8.14   Acetylcholine esterase colourimetric assay. This assay system involves **thio-acetylcholine** that is hydrolysed to thio-choline. This in turn combines with colourless reagent **5,5′-dithio-bis(nitrobenzoic acid)** (**DTNB**) to form yellow-coloured **5-thio-2-nitro-benzoic acid** (**TNB**).

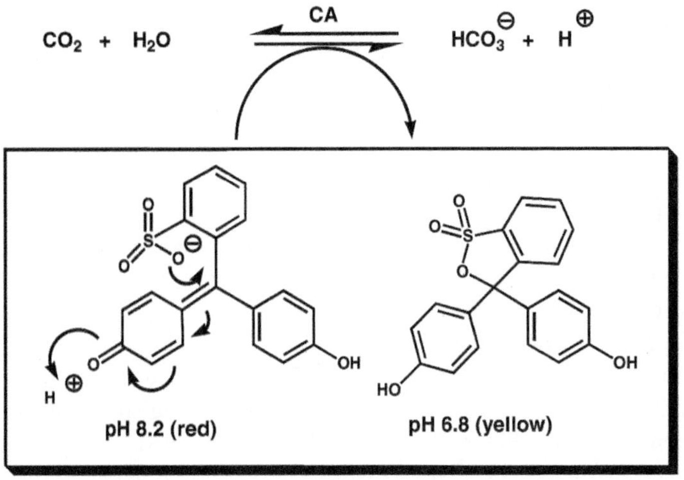

Figure 8.15   Carbonic anhydrase colourimetric assay. This assay involves changes in pH with enzyme catalysis followed with great precision and accuracy by a **pH-sensitive dye** that changes absorbance profile with pH-change.

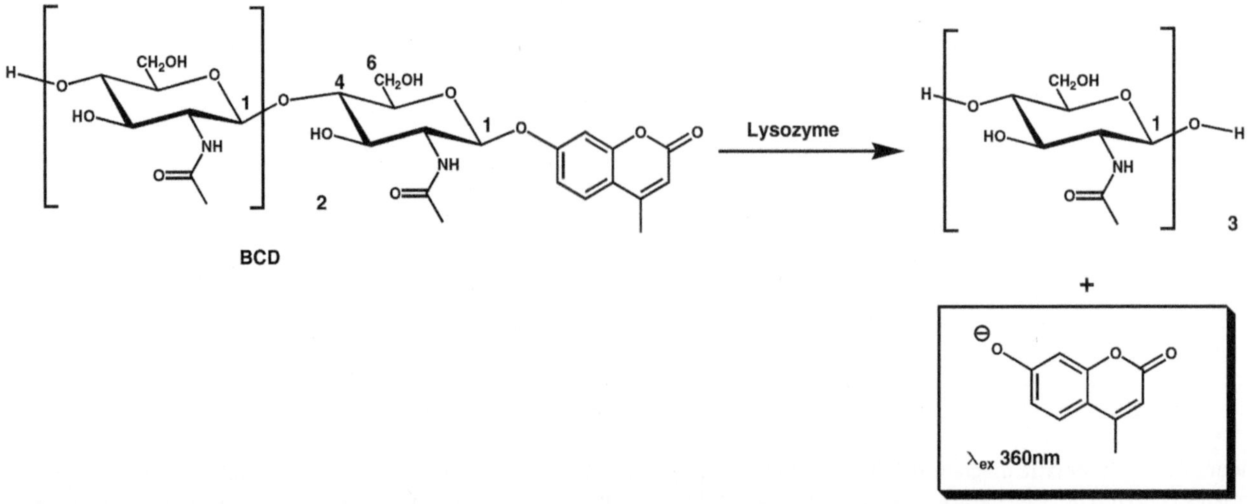

Figure 8.16   Lysozyme fluorimetric assay. This assay is performed under mildly acidic conditions but the appearance of **charged fluorophore** requires pH-quenching to pH 9.

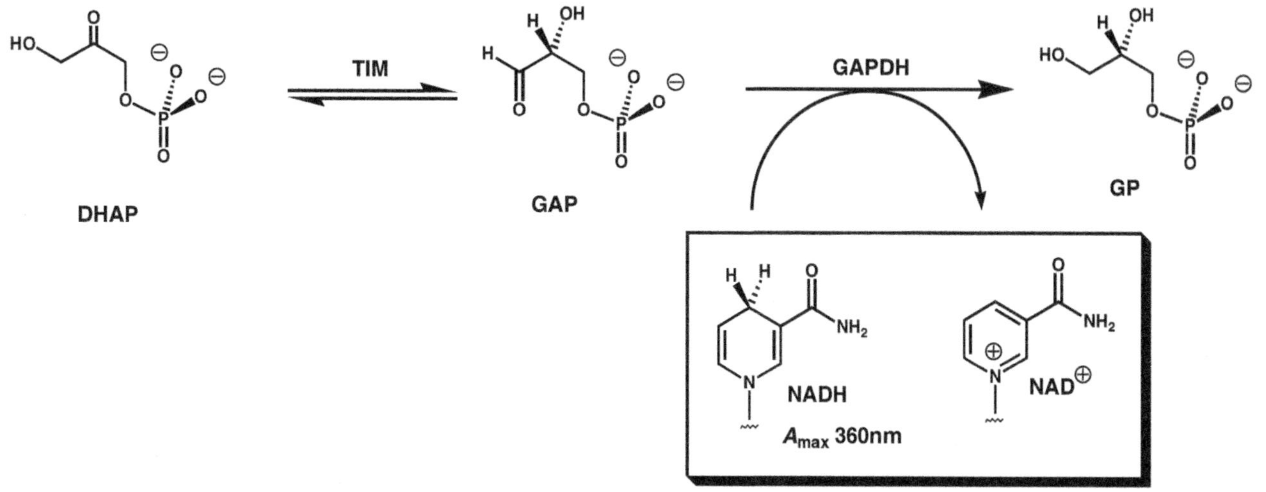

Figure 8.17 **Triose phosphate isomerase enzyme coupled assay.** This is a **GAPDH colourimetric coupled assay** for detection purposes. In this assay, DHAP is converted enzymically to GAP that is onward converted to **glycerol phosphate (GP)** by means of the coupled enzyme GAPDH that uses the **reverse-colourimetric reductant** NADH.

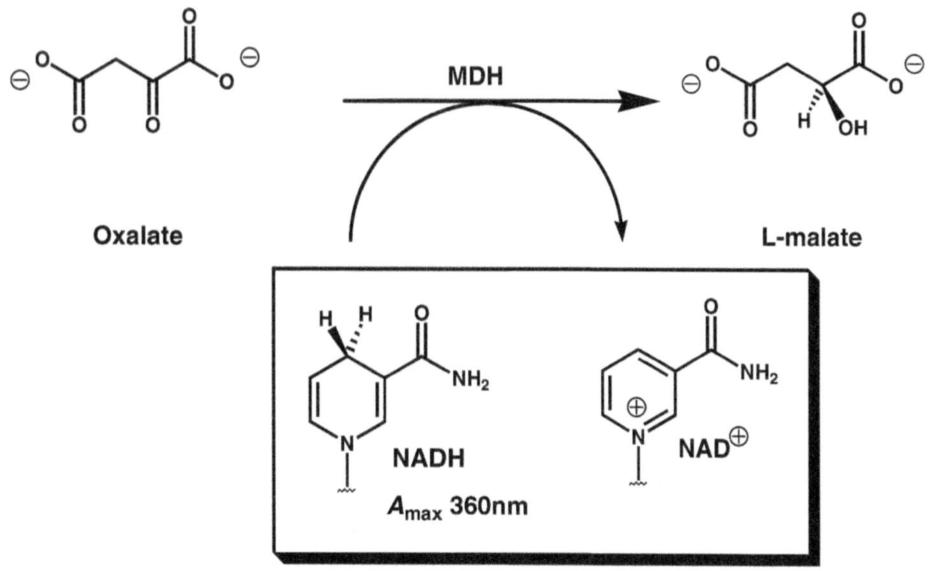

Figure 8.18 **Malate dehydrogensase colourimetric assay.** In this assay, **oxalate** is onward converted to **L-malate** by means of enzyme MDH that uses the reverse colourimetric reductant NADH to effect catalytic reduction.

HPLC steady state bioassays make use of a change in physical elution time when substrate is converted to product. For instance, a steady state bioassay for the enzyme LysU (see Chapters 6 and 7) makes use of the fact that substrate ATP and product Ap_4A elute at different times from an HPLC column. Hence, the initial rate of formation of Ap_4A correlates either to the initial rate of Ap_4A peak formation or inversely to the initial rate of ATP peak disappearance (Figure 8.19). The LysU catalysed conversion of Ap_4A to Ap_3A can also be observed by means of the same HPLC bioassay. A similar principle applies to the *^1H-NMR* **steady state bioassay** in which the initial rate of formation of Ap_4A correlates either to the initial rate of appearance of Ap_4A-purine signals observed by 1D *^1H-NMR* spectroscopy or inversely to the initial rate of disappearance of ATP-purine signals (Figure 8.20). An **oxygen electrode steady state bioassay** is usually required where biocatalysts that evolve or consume oxygen are involved. Where the enzyme SOD is concerned, the appropriate steady state bioassay requires the use of a calibrated **oxygen electrode** to measure the initial rate of oxygen produced as

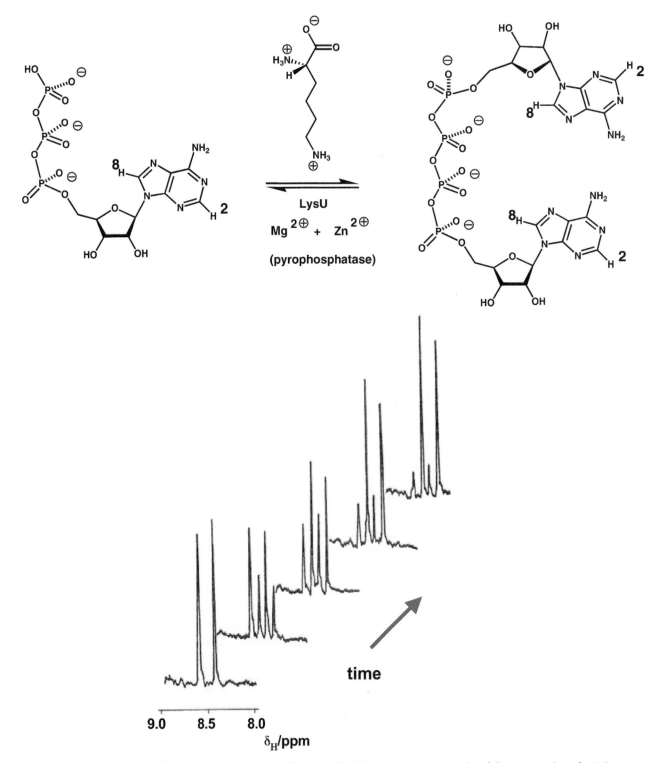

Figure 8.19 Lysyl tRNA synthetase ¹H-NMR assay. In this assay ¹H-NMR spectroscopy is used to follow conversion of ATP into Ap₄A when LysU enzymatic reactions are run in a 5 mm NMR tube. The ¹H-resonance signals of adenine ring are shifted **upfield** with conversion of ATP (δ_H 8.55 [*H*-2] and 8.29 [*H*-8] ppm) to Ap₄A (δ_H 8.40 [*H*-2] and 8.18 [*H*-8] ppm). Rates of conversion are followed by changes in appropriate ¹H-NMR signal peak areas as a function of time. (Theoclitou *et al.*, 1996, Figure 2, Royal Society of Chemistry).

a function of initial superoxide concentration (Figure 8.21). **Radiometric steady state bioassays** normally involve the transformation of a radioactive substrate into product and the clean isolation of the latter from the former. In the case of the bioassay for CAT, the product is much more hydrophobic than the substrate and may be extracted cleanly into organic solvents (Figure 8.22). Hence, the initial accumulation of [¹⁴C] in the organic extract is said to correlate with

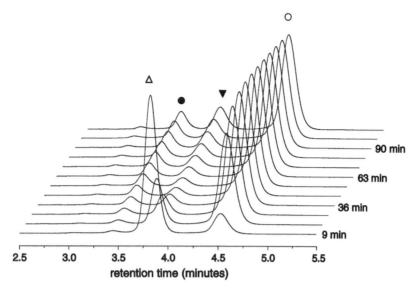

Figure 8.20 Lysyl tRNA synthetase HPLC assay. In this assay, LysU converts ATP ($\triangle$) into Ap$_4$A (O) (see Figure 8.8). Rates of conversion are followed by changes in appropriate HPLC peak area as a function of time. LysU has recently been shown to further catalyse the onwards formation of Ap$_3$A ($\blacktriangledown$) from Ap$_4$A (O) via an ADP ($\bullet$) intermediate. This second stage reaction is clearly visible by HPLC assay too.

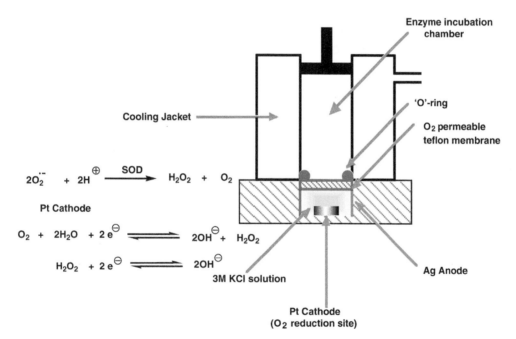

Figure 8.21 Superoxide dismutase oxygen electrode assay. In this assay, SOD generates **hydrogen peroxide (H$_2$O$_2$)** and **oxygen** from superoxide. A calibrated **oxygen electrode** is then used to measure the initial rate of oxygen produced as a function of initial SOD concentration. Calibration correlates current generated from oxygen reduction at the Pt cathode to oxygen produced by SOD.

the initial formation of product. Finally, **gel steady state bioassays** form the basis of ribozyme analyses owing to the difference in molecular weight between substrate S and product P1 polynucleotides. In this case the initial formation of product is determined by the initial increase in product band intensities or inversely to the initial decrease in substrate band intensities (Figure 8.23).

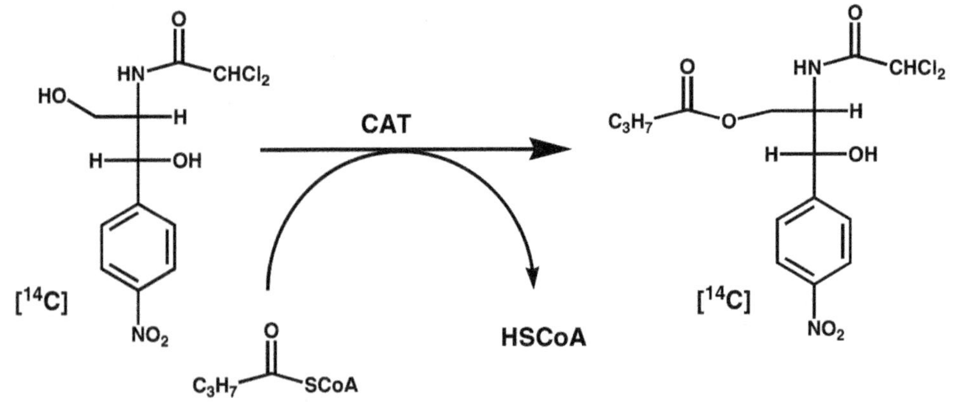

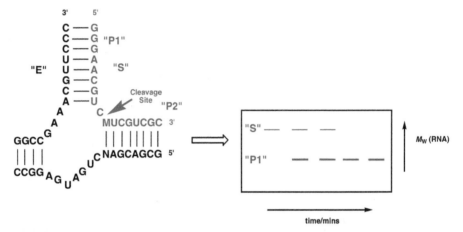

Figure 8.22 Chloramphenicol acetyl transferase radiographic assay. In this assay, butyrylation of [^{14}C]-chloramphenicol by CAT transformation results in a more hydrophobic product that is separated from substrate by extraction into organic solvent.

Figure 8.23 Ribozyme gel assay. According to this assay, one strand of the ribozyme acts as the biocatalyst, **E**, and the other strand acts as the substrate, **S**. **S** is specifically cleaved at the indicated site, generating two product oligonucleotides **P1** and **P2**. The conversion of **S** into lower molecular weight product **P1** can be followed by **gel electrophoresis** with time.

8.2 Steady state kinetic schemes

All steady state kinetics equations are derived by careful analyses from different steady state kinetic schemes that seek to define the likely surface mechanism by which any one biocatalyst may bring about biocatalysis. In all cases, final steady state kinetics equations seek to define a graphical relationship between ν and [S] that varies depending upon differences in the surface mechanism of biocatalysis. Therefore, a steady state equation that most closely accounts for the observed relationship between ν and [S] data not only allows for the most accurate determination of biocatalytic parameters, but also informs us of the most probable surface mechanism. There can be as many steady state kinetics equations as there are kinetic schemes, but there comes a point when greater complexity brings diminishing returns in understanding. For this reason, we shall only present the chemical biology reader with those schemes and equations that are still prevalent and which we consider the most important in general. Wherever possible, reference will be made to the most appropriate kinetic scheme and steady state equations for the analysis of the steady state kinetic behaviour of those biocatalyst examples introduced earlier in the chapter (Section 8.1).

8.2.1 Simple steady state kinetics and Michaelis-Menten equation

The simplest possible kinetic scheme is the **Uni Uni kinetic scheme,** where Uni indicates one substrate and the second Uni indicates the evolution of only a single product (Scheme 8.1). The main feature of this highly simplified kinetic

$$E \;+\; S \;\underset{k_{-1}}{\overset{k_1}{\rightleftharpoons}}\; ES \;\overset{k_2}{\longrightarrow}\; E \;+\; P$$

$$K_S$$

Scheme 8.1

scheme is the presence of an intermediate chemical species designated **ES** that is a complex (essentially non-covalent) between biocatalyst **E** and substrate molecule **S**. This is sometimes referred to as the **Michaelis complex** for reasons that will become apparent shortly. Given such a simple mechanistic scheme, ν can be expressed as in Equation (8.1):

$$\nu = k_2 \left[ES \right] \tag{8.1}$$

For the purposes of the following analysis, biocatalysis is assumed to be irreversible and each biocatalyst **E** possesses only a single catalytic site. Hence, if the **total concentration of biocatalyst** is $[E]_o$, then this must be the sum of **free enzyme**, $[E]$, and **Michaelis complex**, $[ES]$, **concentrations**, as indicated in Equation (8.2):

$$\left[E \right]_o = \left[E \right] + \left[ES \right] \tag{8.2}$$

If Equation (8.1) is then divided through by Equation (8.2) we arrive at Equation (8.3):

$$\frac{\nu}{\left[E \right]_o} = \frac{k_2 \left[ES \right]}{\left[E \right] + \left[ES \right]} \tag{8.3}$$

In order to solve Equation (8.3), expressions for $[E]$ and $[ES]$ are required that are obtained by applying **Briggs-Haldane steady state** principles. According to these principles, biocatalysis rapidly attains a condition of stasis under which all biocatalyst species are at a constant equilibrium concentration. In other words $[E]$ and $[ES]$ are constant with time. Stasis is reflected by Equation (8.4):

$$k_1 \left[E \right] \left[S \right] = k_2 \left[ES \right] + k_{-1} \left[ES \right] \tag{8.4}$$

that rearranges to give the steady state expression 8.5:

$$\left[ES \right] = \frac{k_1}{k_2 + k_{-1}} \left[E \right] \left[S \right] \tag{8.5}$$

Hence, substituting for $[ES]$ in Equation (8.3) followed by rationalisation of terms, we arrive at Equation (8.6):

$$\nu = \frac{k_2 \left[E \right]_o \cdot \left[S \right]}{\left(\dfrac{k_2 + k_{-1}}{k_1} \right) + \left[S \right]} \tag{8.6}$$

This equation may be further simplified by substituting for the bracketed aggregate of rate constants by the constant K_m where:

$$K_m = \left(\frac{k_2 + k_{-1}}{k_1} \right) = k_2 + K_S \tag{8.7}$$

The result is Equation (8.8):

$$\nu = \frac{k_{cat} \left[E \right]_o \cdot \left[S \right]}{K_m + \left[S \right]} \tag{8.8}$$

where K_m is known as the **Michaelis constant**. Compared with Equation (8.6), there is one further change in that rate constant k_2 has been relabelled as k_{cat}, which is known as the **catalytic constant** or **turnover number**. Equation (8.8) is known as the **Michaelis-Menten** or **Henri-Michaelis-Menten** equation after the pioneers of biocatalysis who originally derived this equation.

The Michaelis-Menten equation classically describes a rectangular hyperbola and has the form shown (Figure 8.24). In this graphical form K_m can be described in simple terms as the value of $[S]$ corresponding to an initial rate of catalysis equivalent to $V_{max}/2$, where V_{max} is the maximum initial rate of catalysis that is possible. A similar rectangular hyperbola

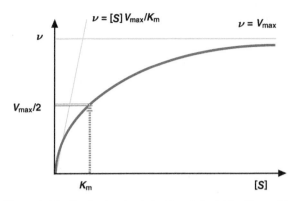

Figure 8.24   **Basic steady state biocatalyst kinetic profile.**

is also seen in molecular recognition studies, reflecting the fact that if sufficient substrate is present, then all the active sites in a biocatalyst should be effectively occupied or saturated with reacting substrate on a continuous basis with time. The Michaelis-Menten equation and associated rectangular hyperbolic graphical depiction are absolutely fundamental to steady state biocatalysis. Although, this equation has been derived by a steady state analysis of the simplest possible kinetic model for biocatalyst action, the form of this equation is frequently retained irrespective of the complexity of the kinetic model, as we shall see. The reason for this is simple, biocatalysis always involves saturation at high values of [S], hence some form of saturating function is essential to describe the catalytic behaviour of a given biocatalyst, and the Michaelis-Menten equation is almost the simplest possible description of a saturating function.

There are two important limits to Equation (8.8). In the first instance, when [S] » K_m then Equation (8.8) reduces to:

$$\nu = k_{cat}\left[E\right]_o = V_{max} \tag{8.9}$$

Alternatively, when K_m » [S] then Equation (8.8) reduces to:

$$\nu = \left(\frac{k_{cat}}{K_m}\right)\left[E\right]_o \cdot \left[S\right] \tag{8.10}$$

However, at low [S] almost all the biological catalyst is free of bound substrate. Therefore, $[E]_o$ may be substituted for by [E], the concentration of free catalyst in solution. This results in the following Equation (8.11):

$$\nu = \left(\frac{k_{cat}}{K_m}\right)\left[E\right] \cdot \left[S\right] \tag{8.11}$$

where (k_{cat}/K_m) represents the **specificity constant** for a given substrate of a given biological catalyst. In practice, Equation (8.11) applies over a wider range of [S] values than those obeying the inequality that K_m » [S] which is extremely useful for reaction kinetic analyses.

8.2.2 Interpretation of k_{cat} and K_m

In the Section 8.2.1, the term K_m has been defined in terms of the value of [S] corresponding to an initial rate of catalysis equivalent to V_{max} /2, or in terms of an aggregate of rate constants. However, a more subtle, all embracing definition of K_m can be given with reference to Equation (8.7). At the limit where $k_{-1} \gg k_2$, the K_m reduces to K_S, an **ES complex equilibrium dissociation constant**. Therefore, in general terms K_m can be considered as either a real or apparent equilibrium dissociation constant (a K_d equivalent, see Chapter 7) that provides a direct measure of the amount of biocatalyst "locked up" in binding to a substrate, reaction intermediate and even the product during a catalytic process.

As to k_{cat}, this has been called the catalytic constant or turnover number and as Equation (8.9) makes clear, k_{cat} is also a first order rate constant that defines maximum catalytic rate at saturating concentrations of substrate per catalytic binding site in a biocatalyst per unit time. However, with reference to Equation (8.1), k_{cat} is more precisely identified as the first order rate constant for the decomposition of an **ES** complex. Therefore, in general terms k_{cat} can be considered as a first order rate constant that is a function of the first order rate constants for decomposition or transformation of all complexes formed between the biocatalyst and a substrate, reaction intermediate or even product during a catalytic process.

8.2.3 Determination of k_{cat} and K_m

Parameters k_{cat}, V_{max} and K_m may be determined from real catalytic rate data by means of a **single** or **double reciprocal plot**. The single reciprocal plot or **Hanes plot** is based upon the following algebraically rearranged version of Equation (8.8):

$$\frac{[S]}{\nu} = \frac{K_m}{V_{max}} + \frac{[S]}{V_{max}} \tag{8.12}$$

When initial rate data, ν, and initial substrate concentration data, [S], are plotted with Equation (8.12), data should fit to a straight line (Figure 8.25). The dotted lines illustrate typical error distortion associated with this plot, emphasising that most systematic error is encountered at very low values of ν and [S] only. The double reciprocal plot or **Lineweaver-Burk plot** has tended to be used more widely than the Hanes plot. However, this popularity is a little misplaced. A Lineweaver-Burk plot is based upon this alternate rearranged version of Equation (8.8):

$$\frac{1}{\nu} = \frac{K_m}{V_{max}} \cdot \frac{1}{[S]} + \frac{1}{V_{max}} \tag{8.13}$$

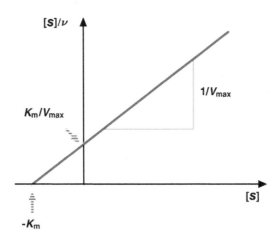

Figure 8.25   Linear, single-reciprocal Hanes plot.

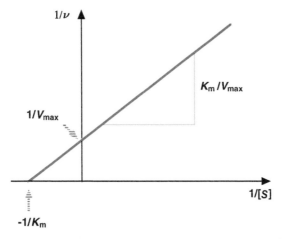

Figure 8.26   Linear, double-reciprocal Lineweaver-Burk plot.

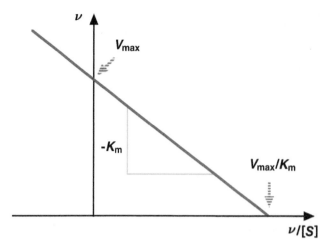

Figure 8.27　Linear, single-reciprocal Eadie-Hofstee plot.

In this case, initial rate and concentration data also give a straight-line plot, but one suffering from gross systematic error distortion with increasingly low values of ν and $[S]$ (Figure 8.26). For this reason, we would not recommend the use of Lineweaver-Burk plots as a rule unless for the multisubstrate kinetics experiments. Another version of kinetic data presentation is the **Eadie-Hofstee plot** which is a single reciprocal plot that is based upon the following rearranged version of Equation (8.8):

$$\nu = V_{max} - K_m \cdot \frac{\nu}{[S]}\qquad(8.14)$$

In this case, initial rate and concentration data also give a straight-line plot, but one suffering from gross systematic error distortion with increasingly low values of ν and $[S]$ (Figure 8.27).

8.2.4 Effect of steady state inhibitors

Biocatalyst inhibitors **I** are "substrate-like" molecules that interact with a given biocatalyst and interfere with the progress of biocatalysis. Inhibitors usually act in one of three ways, either by **competitive inhibition, non-competitive inhibition** or **uncompetitive inhibition**. The mode of inhibition is different in each case and as a result a different steady state kinetic scheme is required to account for each mode of inhibition. Consequently, each mode of inhibition is characterised by a different steady state kinetic equation that gives rise to a different graphical output of ν versus $[S]$ data, as we will show in the next section. These substantial differences in graphical output can be used to diagnose the type of inhibition if unknown.

8.2.4.1 Competitive inhibition

When inhibitor **I** acts as a **competitive inhibitor** of the biocatalyst, then the inhibitor acts by binding to the active site of the biocatalyst in direct competition with the substrate. If we look at inhibition through the lens of the Uni Uni kinetic scheme used to derive the Michaelis-Menten Equation (8.8), then the presence of the competitive inhibitor modifies the kinetic scheme in the following way (Scheme 8.2). Applying a Briggs-Haldane steady state approach to this kinetic scheme yields the following steady state kinetic equation:

$$\nu = \frac{k_2[E]_o \cdot [S]}{K_m\left(1+[I]\big/K_I\right)+[S]}\qquad(8.15)$$

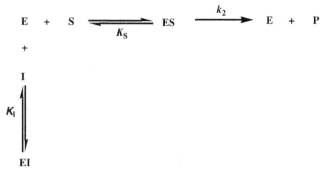

Scheme 8.2

where K_I is an **equilibrium dissociation constant** for the biocatalyst-inhibitor complex, **EI**, that defines the strength of interaction between biocatalyst and inhibitor. Equation (8.15) can be rearranged into a double reciprocal form as follows:

$$\frac{1}{\nu} = \frac{K_m\left(1+\dfrac{[I]}{K_I}\right)}{V_{max}} \cdot \frac{1}{[S]} + \frac{1}{V_{max}} \tag{8.16}$$

In the event that a particular inhibitor **I** acts by competitive inhibition, then plots of $1/\nu$ versus $1/[S]$ data obtained at different fixed concentrations of $[I]$ should obey Equation (8.16) and take on the visual appearance shown in Figure 8.28. This graphical depiction of $1/\nu$ versus $1/[S]$ data not only confirms the existence of competitive inhibition but allows for the determination of K_I as well.

8.2.4.2 Non-competitive inhibition

If **I** acts as a **non-competitive inhibitor** of the biocatalyst, then the inhibitor acts by binding not to the active site but to an **allosteric site** (i.e. alternate, non-overlapping, non-active site binding region) present in both the free biocatalyst and the biocatalyst-substrate complex. The classical non-competitive inhibitor has no direct effect upon substrate binding and vice versa. However, the resulting **ESI** complex is catalytically inactive. Hence, if we repeat the analysis outlined in Section 8.2.4.1 then the appropriate kinetic scheme becomes as follows (Scheme 8.3), and the modified steady state equation becomes:

$$\nu = \frac{\left\{k_2[E]_o \middle/ \left(1+\dfrac{[I]}{K_I}\right)\right\} \cdot [S]}{K_m + [S]} \tag{8.17}$$

Once again, Equation (8.17) can be rearranged into a double reciprocal form as follows:

$$\frac{1}{\nu} = \frac{K_m}{V_{max}}\left(1+\frac{[I]}{K_I}\right) \cdot \frac{1}{[S]} + \frac{1}{V_{max}}\left(1+\frac{[I]}{K_I}\right) \tag{8.18}$$

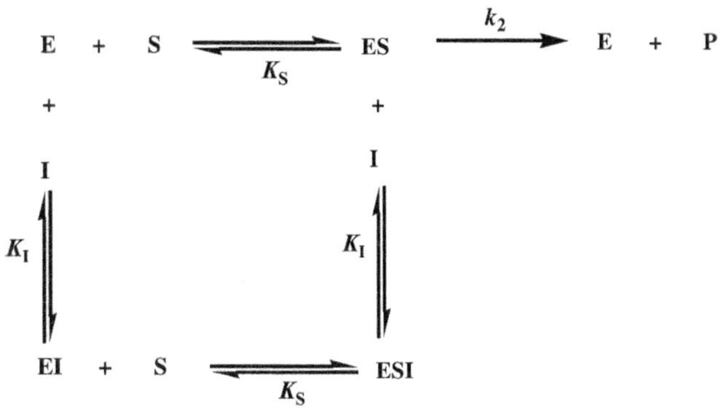

Scheme 8.3

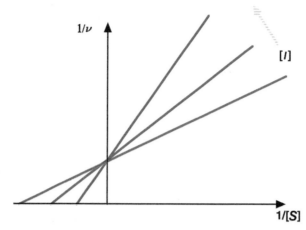

Figure 8.28   Successive Lineweaver-Burk plots showing competitive inhibition. These are typical for a biocatalyst that operates mechanistically through the simplest **Uni Uni kinetic scheme** (Scheme 8.1).

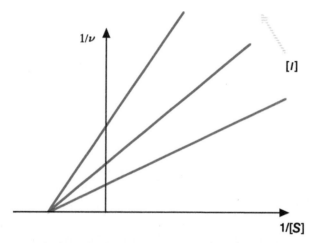

Figure 8.29   Successive Lineweaver-Burk plots showing non-competitive inhibition. These are also typical for a biocatalyst that operates mechanistically through the simplest **Uni Uni kinetic scheme** (Scheme 8.1).

In the event that a particular inhibitor **I** acts by non-competitive inhibition, then plots of $1/\nu$ versus $1/[S]$ data obtained at different fixed concentrations of **[I]** should obey Equation (8.18) and take on the visual appearance shown (Figure 8.29). Clearly, when **I** acts as a non-competitive inhibitor then the only practical effect is to reduce the maximal effective rate of catalysis, V_{max}.

8.2.4.3   Uncompetitive inhibition

If **I** acts as an uncompetitive inhibitor of the biocatalyst, then the inhibitor acts by binding to an allosteric site revealed only in the biocatalyst-substrate complex. In this case the kinetic scheme becomes modified again (Scheme 8.4), and the modified steady state equation becomes:

$$\nu = \frac{\left\{ k_2 [E]_o \Big/ \left(1 + {[I]}\Big/{K_I} \right) \right\} \cdot [S]}{\left\{ K_m \Big/ \left(1 + {[I]}\Big/{K_I} \right) \right\} + [S]} \qquad (8.19)$$

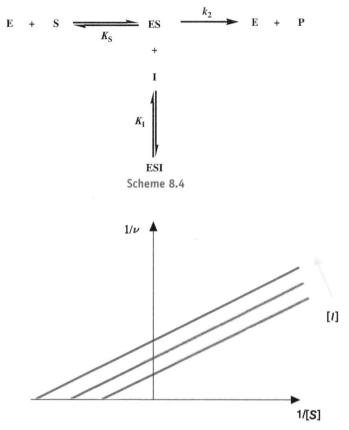

Scheme 8.4

Figure 8.30 **Successive Lineweaver-Burk plots showing uncompetitive inhibition.** Once more, these are typical for a biocatalyst that operates mechanistically through the simplest **Uni Uni kinetic scheme** (Scheme 8.1).

that rearranges to double reciprocal form:

$$\frac{1}{\nu} = \frac{K_m}{V_{max}} \cdot \frac{1}{[S]} + \frac{1}{V_{max}}\left(1 + \frac{[I]}{K_I}\right) \tag{8.20}$$

When a particular inhibitor **I** acts by uncompetitive inhibition, then plots of **1/ν** versus **1/[S]** data obtained at different fixed concentrations of **[I]** should obey Equation (8.20) and take on the visual appearance shown in Figure 8.30. In this case, when **I** acts as a uncompetitive inhibitor then there are practical reductions in both the maximal effective rate of catalysis, V_{max} and K_m, in equal proportion.

8.2.5 Applicability of Michaelis-Menten equation

The Michaelis-Menten Equation (8.8) and the irreversible Uni Uni kinetic scheme (Scheme 8.1) are only really applicable to an **irreversible** biocatalytic process involving a **single substrate** interacting with a biocatalyst that comprises a single catalytic site. Hence, with reference to the biocatalyst examples given in Section 8.1, Equation (8.8), the Uni Uni kinetic scheme is only really directly applicable to the steady state kinetic analysis of TIM biocatalysis (Figure 8.1, Table 8.1). Furthermore, even this statement is only valid with the proviso that all biocatalytic initial rate values are determined in the absence of product. Similarly, the Uni Uni kinetic schemes for competitive, non-competitive and uncompetitive inhibition are only really applicable directly for the steady state kinetic analysis for the inhibition of TIM (Table 8.1). Therefore, why are Equation (8.8) and the irreversible Uni Uni kinetic scheme apparently used so widely for the steady state analysis of many different biocatalytic processes? The main reason for this is that Equation (8.8) is simple to use and measured k_{cat} and K_m parameters can be easily interpreted. There is only a necessity to adapt catalysis conditions such that steady state kinetic measurements of initial rates are made under pseudo-irreversible Uni Uni kinetic conditions where all but one substrate are in substantial excess and no product is present. Furthermore, where a biocatalyst has more than one catalytic site, steady state analyses should be performed under conditions where only one of the sites is active at any one time, or else where the catalytic sites are at least able to operate completely independently of each other. In the latter case, measured values of k_{cat} numbers should then be divided through by the number of catalytic sites to provide a value per catalytic site, otherwise known as the turnover number per catalytic site.

Hence steady state kinetic analyses of RNAse A, lysozyme and CA (Figures 8.4, 8.6 and 8.9) can easily be accommodated within Equation (8.8) and the irreversible Uni Uni kinetic scheme to a reasonable approximation, since water is obviously an abundant substrate. The same could be said to be true of steady state kinetic analyses of bovine pancreatic α-chymotrypsin catalysed polypeptide hydrolysis and AChE catalysed acetylcholine hydrolysis (Figures 8.3 and 8.10). Even reactions catalysed by PLP-dependent AlaR, GadB, and AspAT (Figure 8.5), or those assisted by CAT and LysU (Figures 8.7 and 8.11), may be accommodated within Equation (8.8) and the Uni Uni kinetic scheme given appropriate steady state kinetic conditions. However, while such an approach may have the virtue of simplicity, more careful and intensive steady state kinetic analyses with more complex kinetic schemes has the power to demonstrate and/or confirm mechanisms in biocatalysis and should be performed. So, let us take a more detailed look at more complex kinetic schemes.

8.2.6 Multiple substrate/product steady state kinetics

In this section, we shall begin to see how the Briggs-Haldane steady state approach can be enlarged to derive steady state kinetics equations appropriate to more complex kinetic schemes. In doing this, there will be some pleasant surprises in that the form of these new steady state kinetic equations will follow the form of Michaelis-Menten Equation (8.8) with a few adaptations, not unlike those seen in the Uni Uni steady state kinetic scheme adapted to fit the presence of inhibitors (see Section 8.2.4).

8.2.6.1 Multiple catalytic sites, non-cooperative Uni Uni kinetic scheme

Initially, let us look at the situation of a single substrate biocatalyst with two independent catalytic sites where each site is capable of binding substrate and catalysing the formation of product. This scenario can be analysed by means of the Briggs-Haldane steady state approach with reference to the extended Uni Uni kinetic scheme, once again assuming irreversibility (Scheme 8.5). By analogy to the treatment in Section 8.2.1, we can generate by inspection two important Equations (8.21) and (8.22) from the scheme:

$$\nu = k_2\left[ES\right] + k_2\left[SE\right] + 2k_2\left[SES\right] \tag{8.21}$$

$$\left[E\right]_o = \left[E\right] + \left[ES\right] + \left[SE\right] + \left[SES\right] \tag{8.22}$$

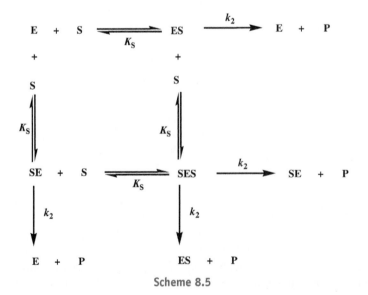

Scheme 8.5

Dividing Equation (8.21) through by 8.22 gives:

$$\frac{\nu}{[E]_o} = \frac{k_2[ES] + k_2[SE] + 2k_2[SES]}{[E] + [ES] + [SE] + [SES]} \tag{8.23}$$

Alternate equilibrium dissociation equations (see Chapter 7) can then be derived on the basis of stasis for constant K_S such as Equation (8.24) for example that defines K_S as:

$$K_S = \frac{[E] \cdot [S]}{[ES]} = \frac{[SE] \cdot [S]}{[SES]} \tag{8.24}$$

Thereafter, these equations can be rearranged to give solutions for $[ES]$, $[SE]$ and $[SES]$ that may be substituted into Equation (8.23) to give us the complete steady state kinetics Equation (8.25) after algebraic manipulation:

$$\nu = \frac{2k_2[E]_o \cdot [S]}{K_S + [S]} \tag{8.25}$$

This is essentially identical with the form of the Michaelis-Menten Equation (8.8), although the meaning of the corresponding biocatalytic parameters is slightly modified. The kinetic scheme upon which this derivation is based is clearly limited to a single substrate biocatalyst that operates with two independent catalytic sites. Such a biocatalyst could be monomeric with two catalytic sites or else homodimeric with one catalytic site per subunit. With reference to the biocatalyst examples described in Section 8.1, the chemical biology reader should be able to see that Equation (8.25) and the Uni Uni kinetic scheme for two catalytic sites seems appropriate for the steady state kinetic analysis of homodimeric *B. stearothermophilus* AlaR biocatalysis (Figure 8.5, Table 8.1). Furthermore, we might easily extrapolate from Equation (8.25) to a four independent catalytic sites Equation (8.26):

$$\nu = \frac{4k_2[E]_o \cdot [S]}{K_S + [S]} \tag{8.26}$$

In this case, Equation (8.26) and the irreversible Uni Uni kinetic scheme for four catalytic sites is appropriate for the steady state kinetic analysis of homotetrameric, human mitochondrial MnSOD biocatalysis (Figure 8.8, Table 8.1) with one catalytic site containing one manganese ion per subunit. These sites are not only independent, but each turnover of a catalytic site involves one substrate superoxide radical being transformed to only one of two possible redox products, depending upon the oxidation state of the manganese ion involved.

8.2.6.2 Multiple catalytic sites, cooperative Uni Uni kinetic scheme and Hill equation

Where biocatalysts possess more than one catalytic site and these sites are distributed over several associated subunits, then the catalysis at one catalytic site is much more likely to be dependent upon catalysis taking place at the other sites as well. Dependency can either be positive or negative. If positive, then catalysis at the one catalytic site should boost catalysis at the other sites in the biocatalyst. If negative, then catalysis at the one catalytic site should attenuate catalysis at the other sites even to the limit at which none of the other sites are able to function at all. The positive effect is known as **positive cooperativity** and the negative effect as **negative cooperativity**. The same concept has been discussed in Chapter 7 with respect to ligand binding to receptors with multiple binding sites.

Now we shall consider explicitly the situation of a single substrate biocatalyst with two catalytic sites each capable of binding substrate and catalysing the formation of product, but where the efficiency of substrate binding at the one site is dependent upon substrate binding at the other. The relevant extended Uni Uni kinetic scheme is as shown, assuming irreversibility again (Scheme 8.6). In fact, the overall rate Equation (8.23) still applies, although, owing to the presence of αK_S, the substitutions for $[ES]$, $[SE]$ and $[SES]$ are different. Equations can be derived for equilibrium dissociation constants K_S and αK_S (assuming that $\alpha > 1$), for example Equation (8.27) defining αK_S:

$$\alpha K_S = \frac{[ES] \cdot [S]}{[SES]} = \frac{[SE] \cdot [S]}{[SES]} \tag{8.27}$$

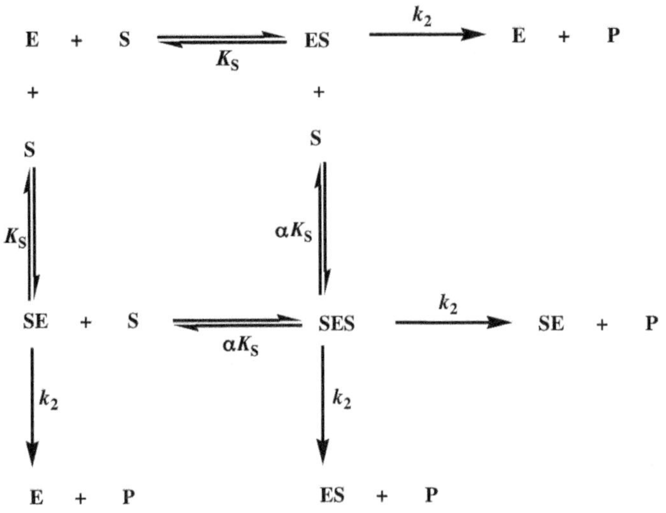

$$E + S \;\underset{K_S}{\overset{}{\rightleftharpoons}}\; ES \;\xrightarrow{\;k_2\;}\; E + P$$

Scheme 8.6

Thereafter, these equations can be rearranged to give solutions for [ES], [SE] and [SES] such as Equation (8.28) that defines [SES]:

$$[SES] = \frac{[E]\cdot[S]^2}{\alpha K_S^2} \tag{8.28}$$

all of which are substituted into Equation (8.23) to give complete Equation (8.29) after some arithmetic manipulation:

$$\frac{\nu}{[E]_o} = \frac{2k_2\left(\dfrac{[S]}{K_S} + \dfrac{[S]^2}{\alpha K_S^2}\right)}{1 + \dfrac{2[S]}{K_S} + \dfrac{[S]^2}{\alpha K_S^2}} \tag{8.29}$$

This equation embodies all the features of positive cooperative binding of substrate and the effect of this upon biocatalytic rate. Given the power terms, this equation is decidedly *not* identical with the Michaelis-Menten Equation (8.8). However, an alternative and equally important biocatalytic equation can be derived from Equation (8.29) with a little more work and a few simple assumptions. First, let us assume that cooperative binding is sufficiently positive that terms in [S] are eliminated from Equation (8.29). Second, let us define V_{max} according to Equation (8.30):

$$V_{max} = 2k_2 \cdot [E]_o \tag{8.30}$$

Hence, if we simplify Equation (8.29) with the assumption and substitute in from Equation (8.30), then the result is power Equation (8.31):

$$\frac{\nu}{V_{max}} = \frac{[S]^2 \big/ \alpha K_S^2}{1 + \left([S]^2 \big/ \alpha K_S^2\right)} = \frac{[S]^2}{\alpha K_S^2 + [S]^2} \tag{8.31}$$

This Equation (8.31) is applicable only to a biocatalyst with two distinct but strongly binding-dependent active sites. A more general form can be deduced from 8.31 that applies to any biocatalyst with *n*-distinct but strongly binding-dependent active sites (where $n \geq 2$). This general form is given by Equation (8.32):

$$\frac{\nu}{V_{max}} = \frac{[S]^n}{K' + [S]^n} \tag{8.32}$$

Equation (8.32) is known as the **Hill equation**. This equation is almost the same in form as the original Michaelis-Menten equation except for the presence of power terms. These power terms ensure that a plot of ν against $[S]$ no longer fits a rectangular hyperbolic function but instead fits a **sigmoidal function** whose dimensions are influenced strongly by the value of n and also K' (Figure 8.31). **The Hill equation and associated sigmoidal graphical depictions are fundamental to cooperative biocatalysis.** Note that although the saturating hyperbolic function of non-cooperative biocatalysis has been abolished in cooperative biocatalysis, the alternative sigmoidal functions must still tend to saturation since biocatalysis must always involve saturation at high values of $[S]$.

In common with the Michaelis-Menten equation, there are linearised forms of the Hill equation of which the most widely used form is given by Equation (8.33):

$$\log\left(\frac{\nu}{V_{max} - \nu}\right) = n \cdot \log[S] - \log K' \qquad (8.33)$$

that derives by simple arithmetic manipulation from Equation (8.32). Where unambiguous cooperative biocatalysis is taking place, then a plot of values of $\log\left(\nu/V_{max} - \nu\right)$ versus $\log[S]$ should render a straight-line of slope n (Figure 8.32).

In principle, all single substrate multi-subunit/multi-catalytic site biocatalysts may be subject to positive cooperativity in catalysis under certain conditions according to Equation (8.33). The single substrate multi-subunit/multi-catalytic site biocatalyst is almost designed for this opportunity. However, as we have seen already, this opportunity need not be taken advantage of for a given single substrate biocatalyst operating in a given catalytic niche. Also, positive cooperativity

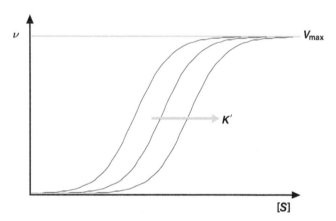

Figure 8.31 Classical sigmoidal rate curves indicative of biocatalysis involving strong positive cooperativity in the catalytic mechanism. Curves move to the right as constant, K', increases.

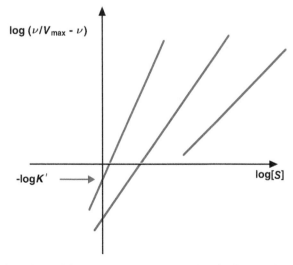

Figure 8.32 Linear kinetic Hill plots derived from sigmoidal data. As already illustrated (Figure 8.31). Gradients and intercepts define values of n and K' respectively.

does not necessarily have to be a subject to substrate binding. For instance, pH changes may induce cooperative shifts in catalytic behaviour of a single substrate biocatalyst such as the homohexameric *E. coli* GadB enzyme responsible for α-decarboxylation of glutamate to give carbon dioxide and **γ-amino butyric acid (GABA)** (Figure 8.5, Table 8.1). In this case, Equation (8.33) needs to be modified so that $[H^+]$ replaces $[S]$ to reflect the actual single participant in this positive cooperative process. The effect of increasing $[H^+]$ is to promote a rapid and substantial conformational change within the homohexameric enzyme structure leading to transition from inactive form (**T**-state) of the enzyme at neutral pH to the highly active form (**R**-state) at lower pH (<5.2).

Clearly, Equation (8.32) does not hold strictly for multiple substrate multi-subunit/multi-catalytic site biocatalysts. However, catalysis conditions might be adapted such that steady state kinetic measurements of initial rates are made under pseudo-irreversible Uni Uni kinetic conditions where all but one substrate are in substantial excess and no product is present. In this way, positive cooperativity involving multiple substrate binding at multiple dependent catalytic sites may still be analysed to a reasonable approximation using Equation (8.32).

Please note that in general, cooperativity in catalysis may just as well be negative as positive. For instance, homodimeric *E. coli* LysU has two catalytic sites (one per monomer), but only one is involved in catalysis at any one time since multiple substrate binding at the one binding site suppresses multiple substrate binding at the other in a process of negative cooperativity. The catalytic mechanism of LysU is still complex, involving as it does multiple substrates (see below), but at least only one catalytic site need be considered (Figure 8.11, Table 8.1). In addition, the chemical biology reader should also be aware that whether single or multiple substrate, multi-subunit/multi-catalytic site biocatalysts may be subject to the binding of **allosteric activators** or **allosteric modulators** as well, namely molecules that bind in discrete binding sites separate from catalytic sites, thereby stimulating or reducing catalytic activity. In the case of GadB enzyme, Cl⁻ and other halide ions act as allosteric activators that not only stimulate GadB catalysis but also promote the positive cooperative effect of $[H^+]$ on GadB conformational change and catalysis by encouraging the transition from **T** to **R** state to take place at higher pH (6.0–5.5).

8.2.6.3 Uni Bi kinetic scheme

The next level of complexity is to review the situation of a single substrate biocatalyst with with a single catalytic site that is responsible for more than one product-forming/release step (a **multiple product** situation). This scenario will be analysed by means of the Briggs-Haldane steady state approach with reference to the indicated **Uni Bi kinetic scheme**, where Uni refers to one substrate and Bi to the evolution of two products. Irreversibility is also assumed (Scheme 8.7). By analogy with the previous treatments above, we may take Equation (8.1) and divide through by Equation (8.34):

$$[E]_o = [E] + [ES] + [ES'] \tag{8.34}$$

This results in the following:

$$\frac{\nu}{[E]_o} = \frac{k_2[ES]}{[E] + [ES] + [ES']} \tag{8.35}$$

Finding solutions to $[E]$ and $[ES']$ requires that stasis exists, that $[E]$ and $[ES']$ are constant with time and that the rates of formation and decomposition of **ES** and **ES′** are equal, giving rise to the the following two equations:

$$k_1[E][S] = (k_2 + k_{-1})[ES] \tag{8.36}$$

$$[ES'] = (k_2 / k_3)[ES] \tag{8.37}$$

$$E + S \underset{k_{-1}}{\overset{k_1}{\rightleftharpoons}} ES \overset{k_2}{\longrightarrow} ES' \overset{k_3}{\longrightarrow} E + P_B$$
$$+$$
$$P_A$$

Scheme 8.7

By using Equation (8.37) and rearranging Equation (8.36) to give an expression for $[E]$, we can then substitute for $[E]$ and $[ES']$ in Equation (8.35) and cancel out $[ES]$ terms thereby giving intermediate equation:

$$\frac{\nu}{[E]_o} = \frac{k_2}{\left(\dfrac{k_2+k_{-1}}{k_1[S]}\right) + \left(\dfrac{k_2+k_3}{k_3}\right)} \tag{8.38}$$

With further algebraic manipulation to convert this into the form of the Michaelis-Menten Equation (8.8), we arrive at Equation (8.39):

$$\nu = \frac{\left(\dfrac{k_2 k_3}{k_2+k_3}\right)[E]_o[S]}{\left\{\dfrac{k_3(k_2+k_{-1})}{k_1(k_2+k_3)}\right\} + [S]} \tag{8.39}$$

where the collection of rate constants in the numerator is equivalent to the Michaelis-Menten k_{cat} and the collection of rate constants in the denominator is the equivalent of K_m. Once again, Equation (8.39) is essentially identical with the form of the Michaelis-Menten Equation (8.8). The kinetic scheme upon which this derivation is based is clearly limited to a single substrate biocatalyst that generates two products from a single active site. With reference to the biocatalyst examples described in Section 8.1, the chemical biology reader should be able to surmise that Equation (8.39) and the Uni Bi kinetic scheme should be applicable to the steady state kinetic analysis for carbonic acid heterolysis into carbon dioxide and water catalysed by monomeric CA1 from human erythrocytes (Figure 8.9, Table 8.1) under conditions where initial rates can be measured with no product present.

8.2.6.4 Ordered Bi Uni kinetic scheme

Now we shift from multiple product to a **multiple substrate** scenario involving a biocatalyst with a single catalytic site that binds a leading substrate, S_A, and then a following substrate, S_B. Any such **bisubstrate reaction** is usually known as a having a **sequential** or **single-displacement** reaction mechanism on the basis that substrates bind first, react and then release product in a consistent manner as part of the standard catalytic cycle. This particular scenario may be analysed by means of the Briggs-Haldane steady state approach with reference to an **ordered Bi Uni kinetic scheme** where Bi refers to the involvement of two substrates and Uni to the evolution of only a single product by an irreversible catalytic process (Scheme 8.8). Once again, by analogy to the treatment in Section 8.2.1, we can generate Equations (8.40) and (8.41):

$$\nu = k_3[ES_A S_B] \tag{8.40}$$

$$[E]_o = [E] + [ES_A] + [ES_A S_B] \tag{8.41}$$

that combine easily to produce Equation (8.42):

$$\frac{\nu}{[E]_o} = \frac{k_3[ES_A S_B]}{[E] + [ES_A] + [ES_A S_B]} \tag{8.42}$$

As before, equilibrium equations can then be derived for constants K_A and K_B. Thereafter, these equations can be rearranged to give solutions for $[ES_A]$ and $[ES_A S_B]$ that may be substituted into Equation (8.42), giving us the complete steady state kinetics Equation (8.43):

$$\nu = \frac{k_3 \cdot [E]_o[S_A][S_B]}{[S_A][S_B] + K_B[S_A] + K_A K_B} \tag{8.43}$$

At constant $[S_A]$ then Equation (8.43) reduces to Equation (8.44) after dividing right-hand side by $[S_A]$ and gathering terms:

$$\nu = \frac{k_3 \cdot [E]_o[S_B]}{K_B(1 + K_A / [S_A]) + [S_B]} \tag{8.44}$$

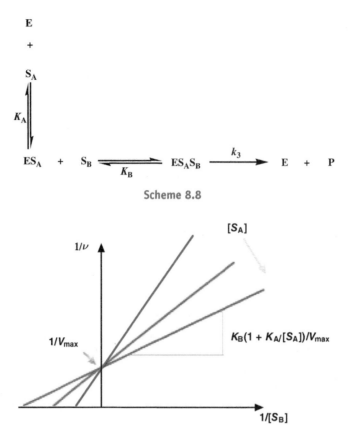

Scheme 8.8

Figure 8.33 Successive Lineweaver-Burk plots typical for an ordered Bi Uni kinetic scheme. The kinetic reaction scheme is also illustrated (Scheme 8.8).

which the chemical biology reader should recognise as the classic form of the Michaelis-Menten Equation (8.8) once again. If we now substitute in for V_{max} according to Equation (8.45):

$$V_{max} = k_3 \cdot [E]_o \qquad (8.45)$$

and rearrange the resulting equation into a double reciprocal form, then we end up with Equation (8.46):

$$\frac{1}{\nu} = \frac{K_B \left(1 + K_A / [S_A]\right)}{V_{max}} \cdot \frac{1}{[S_B]} + \frac{1}{V_{max}} \qquad (8.46)$$

In the event that the kinetic mechanism of a given biocatalyst of interest obeys the irreversible, single catalytic site, ordered Bi Uni mechanism, then plots of actual $1/\nu$ versus $1/[S_B]$ data obtained at different fixed initial concentrations of $[S_A]$ will also all obey Equation (8.46) and take on the visual appearance shown (Figure 8.33). Clearly, such linear graphical data would be sufficient to allow all constants to be determined. It is hoped that the chemical biology reader can recognise the relationship between Equation (8.46) and Equation (8.13) that defines the simple Lineweaver-Burk plot derived from the original Michaelis-Menten equation for the linear analysis of kinetic data. With reference to the biocatalyst examples described in Section 8.1, the chemical biology reader may be able to guess that Equation (8.46) and the ordered Bi Uni kinetic scheme should be applicable to the steady state kinetic analysis for carbonic acid synthesis from carbon dioxide and water catalysed by monomeric CA1 from human erythrocytes (Figure 8.9, Table 8.1) when initial rates are measured with no product present. In this situation, water is the leading substrate, S_A, and carbon dioxide the following substrate, S_B.

8.2.7 Multiple substrate/product King-Altman kinetics

A more generally applicable way of providing steady state kinetics equations for multiple substrate/product kinetic schemes is to use the **King-Altman derivation approach**. We shall examine the principles behind this with reference

$$\text{E} + \text{S} \; \overset{k_1}{\underset{k_{-1}}{\rightleftharpoons}} \; \text{ES} \; \overset{k_2}{\underset{k_{-2}}{\rightleftharpoons}} \; \text{EP} \; \overset{k_3}{\underset{k_{-3}}{\rightleftharpoons}} \; \text{E} + \text{P}$$

Scheme 8.9

to the simplest **single catalytic site, reversible Uni Uni kinetic scheme** that is shown (Scheme 8.9). This reversible scheme is actually most appropriate for the complete steady state kinetic analysis of chicken TIM (Figure 8.1, Table 8.1). According to the King-Altman treatment, we must begin as usual by stating equations that represent the rate of product formation, ν. In this case, they are essentially modified versions of Equations (8.1) and (8.35) as follows:

$$\nu = k_2[ES] - k_{-2}[EP] \tag{8.47}$$

$$\frac{\nu}{[E]_o} = \frac{k_2[ES] - k_{-2}[EP]}{[E] + [ES] + [EP]} \tag{8.48}$$

However, instead of trying to derive explicit expressions for [E], [ES] and [EP] and then substituting back into Equation (8.48), the King-Altman approach seeks first to construct a vector equivalent diagram linking all the biocatalyst species together (Figure 8.34). Expressions for [E], [ES] and [EP] are then derived by considering all the alternative pathways to any given species, starting from one or other or both of the other biocatalyst species (Figure 8.34). The result is Equation (8.49) to Equation (8.51):

$$[E] = k_{-2}k_{-1} + k_{-1}k_3 + k_2k_3 \tag{8.49}$$

$$[ES] = k_{-2}k_1[S] + k_3k_1[S] + k_{-3}[P]k_{-2} \tag{8.50}$$

$$[EP] = k_1[S]k_2 + k_{-1}k_{-3}[P] + k_2k_{-3}[P] \tag{8.51}$$

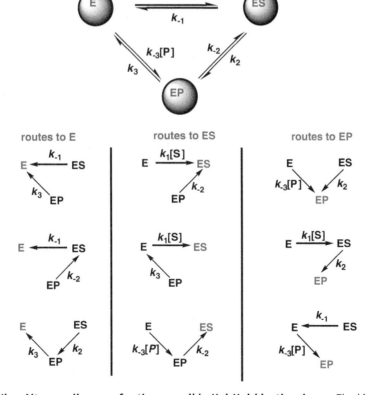

Figure 8.34 **Full set of King-Altmann diagrams for the reversible Uni Uni kinetic scheme.** The kinetic reaction scheme is also illustrated (Scheme 8.9).

These can all be substituted back into Equation (8.48) and after the gathering (extensive) and subtracting of terms, where relevant, we end up with a relatively simple result in Equation (8.52):

$$\frac{\nu}{[E]_0} = \frac{k_1 k_2 k_3 [S] - k_{-1} k_{-2} k_{-3} [P]}{(k_{-2} k_{-1} + k_{-1} k_3 + k_2 k_3) + k_1 (k_{-2} + k_2 + k_3)[S] + k_{-3}(k_{-1} + k_{-2} + k_2)[P]}$$

(8.52)

8.2.7.1 Reversible Uni Uni kinetic scheme

Unfortunately, the form of Equation (8.52) is a little way off the form of the Michaelis-Menten equation. For this reason, the King-Altman approach is usually supplemented by an approach developed by **Cleland**. The **Cleland approach** seeks to group kinetic rate constants together into **numbers** (*num*), **coefficients** (*Coef*) and **constants** (*const*) that themselves can be collectively defined as experimental steady state kinetic parameters equivalent to k_{cat}, V_{max} and K_m of the original Michaelis-Menten equation. After such substitutions, the result is that equations may be algebraically manipulated to reproduce the form of the Michaelis-Menten Equation (8.8). Use of the Cleland approach is illustrated as follows.

First, we will need to rewrite Equation (8.52) in the so-called Cleland form, giving Equation (8.53):

$$\nu = \frac{num_1 [S] - num_2 [P]}{const + Coef_S [S] - Coef_P [P]}$$

(8.53)

Second, we will need to define steady state kinetic parameters as follows:

$$K_{eq} = \frac{num_1}{num_2} = \frac{k_1 k_2 k_3}{k_{-1} k_{-2} k_{-3}}$$

(8.54)

$$V_{maxf} = \frac{num_1}{Coef_S}$$

(8.55)

$$V_{maxb} = \frac{num_2}{Coef_P}$$

(8.56)

$$K_{mS} = \frac{const}{Coef_S}$$

(8.57)

$$K_{mP} = \frac{const}{Coef_P}$$

(8.58)

then substitute all these into Equation (8.53) after algebraic manipulation of the right-hand side by multiplication with the term ($num_2/[Coef_S \times Coef_P]$) in order to facilitate the in-substitution of all the steady state kinetic parameters given by Equations (8.54) to (8.58). The initial result is that Equation (8.53) is now transformed into Equation (8.59):

$$\nu = \frac{V_{maxf} V_{maxb} ([S] - [P]/K_{eq})}{K_{mS} V_{maxb} + V_{maxb}[S] + V_{maxf} \cdot [P]/K_{eq}}$$

(8.59)

Furthermore, algebraic manipulation of Equations (8.54) to (8.58) results in the following equation for K_{eq}:

$$K_{eq} = \frac{V_{maxf} K_{mP}}{V_{maxb} K_{mS}}$$

(8.60)

This can be substituted into the denominator of Equation (8.59), allowing us to transform 8.59 substantially to give a properly useful steady state kinetic Equation (8.61):

$$\frac{\nu}{V_{maxf}} = \frac{[S] - [P]/K_{eq}}{K_{mS}(1 + [P]/K_{mP}) + [S]}$$

(8.61)

This equation now expresses initial net forward rate for the conversion of substrate into product in the presence of significant concentrations of both [S] and [P]. Any relationship that relates any equilibrium constant to kinetic rate constants is known as a **Haldane relationship**. Equation (8.60) is a typical Haldane relationship. Therefore, the operation using Equation (8.60) to simplify Equation (8.59) to give Equation (8.61) is known as a **Haldane simplification**. Clearly, if there is no significant [P] then Equation (8.61) reduces to the following:

$$\frac{\nu}{V_{maxf}} = \frac{[S]}{K_{mS} + [S]} \tag{8.62}$$

which now possesses the simple form of the Michaelis-Menten Equation (8.8) once more, although the meaning of the corresponding biocatalytic parameters is obviously different. Equations (8.55) and (8.57) define V_{maxf} and K_{mS} respectively such that by in-substitution from Equation (8.52) we end up with:

$$V_{maxf} = \frac{k_1 k_2 k_3 [E]_o}{k_1 (k_{-2} + k_2 + k_3)} \tag{8.63}$$

$$K_{mS} = \frac{(k_{-2}k_{-1} + k_{-1}k_3 + k_2 k_3)}{k_1 (k_{-2} + k_2 + k_3)} \tag{8.64}$$

Provided that the kinetic mechanism of a given biocatalyst of interest follows the reversible Uni Uni mechanism, then plots of actual $1/\nu$ versus $1/[S]$ data obtained in the presence of negligible [P] will be fitted by the double reciprocal version of Equation (8.62):

$$\frac{1}{\nu} = \frac{K_{mS}}{V_{maxf}} \cdot \frac{1}{[S]} + \frac{1}{V_{maxf}} \tag{8.65}$$

and take on the visual appearance shown (Figure 8.35). However, there is little likelihood of being able to determine individual microscopic rate constants from this graphical representation of rate data even with Equations (8.63) and (8.64) to help. Instead, assistance from pre-steady-state kinetic analyses is essential to identify some or all of the microscopic rate constants that make up the steady state kinetic parameters V_{maxf} and K_{mS}. Equation (8.65) is almost indistinguishable from Equation (8.8), but for the identities of the kinetic components. This equivalence proves that whether biocatalyst reactions are irreversible or reversible, provided that steady state analyses are performed in which initial rates are determined in the absence of product, then irreversible kinetic schemes are really quite adequate for almost all steady state kinetic analyses of biocatalysis.

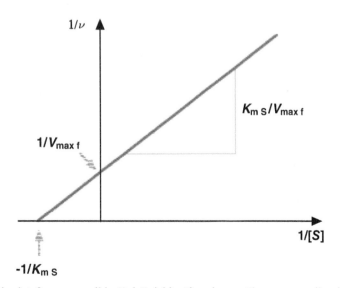

Figure 8.35 Lineweaver-Burk plot for a reversible Uni Uni kinetic scheme. The corresponding kinetic reaction scheme is also illustrated (Scheme 8.9).

The King-Altman approach described here can be summarised as a process in which an "original rate equation" (such as Equation (8.52)) is customarily developed, converted into a coefficient form (such as Equation (8.53)) and from there simplified to "steady state kinetic forms" (e.g. Equations (8.61) and (8.62)) by algebraic manipulation and Haldane simplification. This King-Altman approach is an approach that can be generalised for the derivation of most steady state kinetic equations based upon most complex kinetic schemes. Clearly these derivations can be substantial, but we shall not bother to reproduce these here except to cover a few important examples of particular relevance to the biocatalyst examples described in Section 8.1.

8.2.7.2 Ordered Bi Bi kinetic scheme

The first of these important examples is another multiple substrate/multiple product scenario. In this instance a biocatalyst with a single catalytic site binds a leading substrate, S_A, then a following substrate, S_B, thereby setting up a process of chemical transformation. Upon completion of the transformation, a leading product, P_P, is then released, succeeded by a following product, P_Q. Steady state kinetic equations are derived by means of the King-Altman approach with reference to the **ordered Bi Bi kinetic scheme** that also assumes catalytic irreversibility (Scheme 8.10). This scheme corresponds to a typical bisubstrate reaction with a sequential or single-displacement mechanism. The corresponding King-Altman diagram is illustrated (Figure 8.36). From this diagram, a rate equation may be derived that accounts for the initial net forward rate for the conversion of substrates into products under conditions where product concentrations are negligible:

$$\frac{\nu}{V_{\text{maxf}}} = \frac{[S_A][S_B]}{K_{iS_A}K_{mS_B} + K_{mS_B}[S_A] + K_{mS_A}[S_B] + [S_A][S_B]} \tag{8.66}$$

when $[S_B]$ is constant, the Equation (8.66) reduces to:

$$\frac{\nu}{V_{\text{maxf}}} = \frac{[S_A]}{K_{mS_A}\left\{1 + \left(\dfrac{K_{iS_A}K_{mS_B}}{K_{mS_A}[S_B]}\right)\right\} + [S_A]\left(1 + \dfrac{K_{mS_B}}{[S_B]}\right)} \tag{8.67}$$

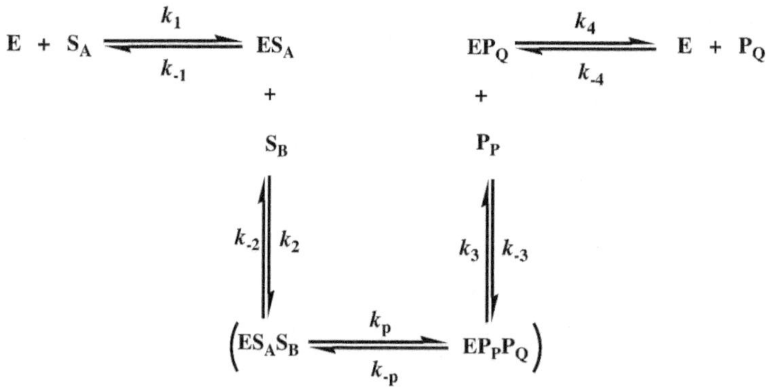

Scheme 8.10

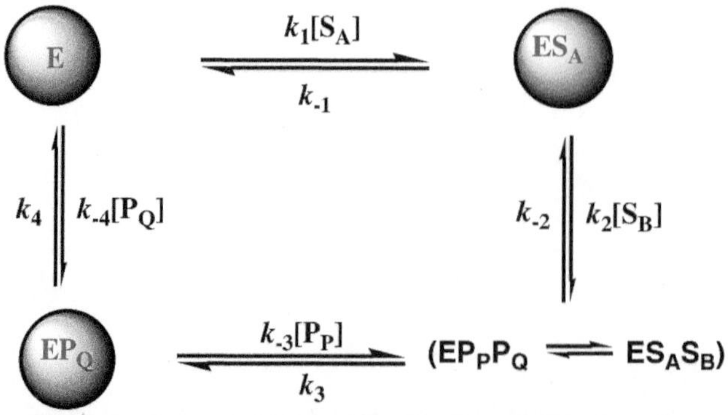

Figure 8.36 **King-Altmann diagram for the ordered Bi Bi kinetic scheme.** The kinetic reaction scheme is also illustrated (Scheme 8.10).

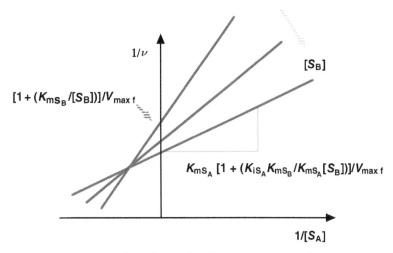

Figure 8.37 Set of Lineweaver-Burk plots for the ordered Bi Bi kinetic scheme. The kinetic corresponding reaction scheme is also illustrated (Scheme 8.10). Plot gradient decreases with increasing initial levels of substrate [S_B].

which is then rearranged into a double reciprocal form as follows:

$$\frac{1}{\nu} = \frac{K_{mS_A}}{V_{maxf}}\left\{1 + \left(K_{iS_A}K_{mS_B}\Big/K_{mS_A}[S_B]\right)\right\} \cdot \frac{1}{[S_A]} + \frac{1}{V_{maxf}}\left(1 + K_{mS_B}\Big/[S_B]\right) \tag{8.68}$$

Provided that the kinetic mechanism of a biocatalyst of interest obeys the irreversible, single catalytic site, ordered Bi Bi kinetic scheme, then plots of actual $1/\nu$ versus $1/[S_A]$ data obtained in the presence of various fixed concentrations of the S_B substrate, $[S_B]$, should obey the double reciprocal Equation (8.68) giving the putative graphical output shown in Figure 8.37.

Equation (8.68) and this ordered Bi Bi kinetic scheme are applicable for the steady state kinetic analyses of several of the biocatalysts outlined in Section 8.1. These include, porcine mitochondrial MDH and *E. coli* CAT (Figures 8.2 and 8.7, Table 8.1). In the case of MDH, the leading substrate, S_A, is NAD$^+$ and the following substrate, S_B, is **oxalate**. MDH is homodimeric but both catalytic sites are completely independent. Clearly, V_{maxf} values derived from Equation (8.68) can be converted into k_{cat} values for MDH that must be divided through by two in order to derive the correct turnover number per catalytic site. With respect to CAT, only one catalytic site binds both substrates, but the mechanism may be random, in which case either chloramphenicol or **acetyl-CoA** bind first. Finally, under appropriate conditions, the ordered Bi Bi kinetic scheme is also applicable for steady state kinetic analysis of the first catalytic step (step 1) of homodimeric LysU catalysed Ap$_4$A formation (Figure 8.11, Table 8.1). LysU step 1 is highly ordered since the leading substrate, L-lysine-Mg^{2+}, must bind to both catalytic sites first in order to organise the catalytic domains of LysU prior to the binding of the following substrate, ATP, to only one out of the two available catalytic sites, hence leading to the catalysis of **lysyl-adenylate intermediate** formation in only one of the two catalytic sites at any one time.

8.2.7.3 Ping-pong Bi Bi kinetic scheme

A bisubstrate reaction that invokes a situation in which one or more products are released before all substrates have been bound to catalytic site(s) is said to have a double-displacement or ping-pong reaction mechanism. The corresponding **ping-pong Bi Bi kinetic scheme**, assuming catalytic irreversibility (Scheme 8.11), is the simplest summary of this scenario. In such a ping-pong kinetic scheme, the biocatalyst itself with a single catalytic site is also temporarily altered (usually covalently) as part of the biocatalytic mechanism from a standard state represented by E to an altered state defined by F, prior to return to the E state. Once again, the King-Altman diagram is illustrated (Figure 8.38), and deriving from this the rate equation that expresses the initial net forward rate for the conversion of substrates into products in the presence of negligible concentrations of products:

$$\frac{\nu}{V_{maxf}} = \frac{[S_A][S_B]}{K_{mS_B}[S_A] + K_{mS_A}[S_B] + [S_A][S_B]} \tag{8.69}$$

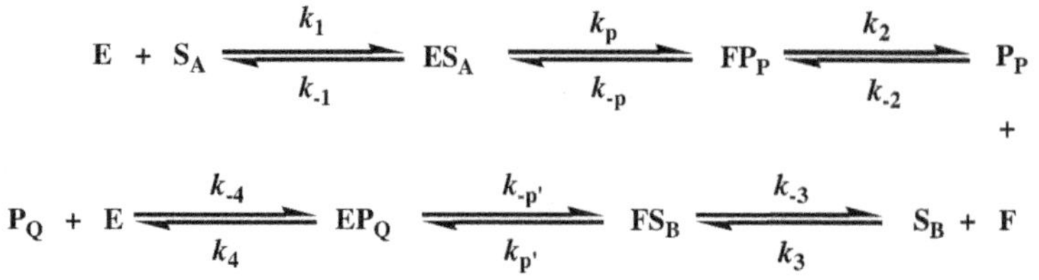

$$E \; + \; S_A \underset{k_{-1}}{\overset{k_1}{\rightleftharpoons}} ES_A \underset{k_{-p}}{\overset{k_p}{\rightleftharpoons}} FP_P \underset{k_{-2}}{\overset{k_2}{\rightleftharpoons}} P_P$$

$$+$$

$$P_Q \; + \; E \underset{k_4}{\overset{k_{-4}}{\rightleftharpoons}} EP_Q \underset{k_{p'}}{\overset{k_{-p'}}{\rightleftharpoons}} FS_B \underset{k_3}{\overset{k_{-3}}{\rightleftharpoons}} S_B \; + \; F$$

Scheme 8.11

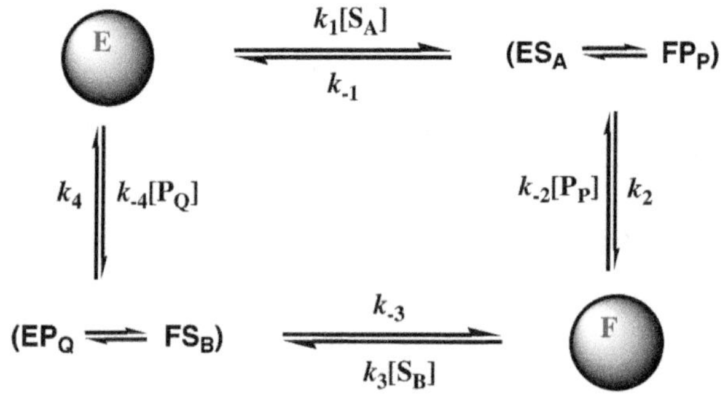

Figure 8.38 **King-Altmann diagram for the ping-pong Bi Bi kinetic scheme.** The kinetic reaction scheme is also illustrated (Scheme 8.11).

When $[S_B]$ is constant, the Equation (8.69) reduces to:

$$\frac{\nu}{V_{\text{maxf}}} = \frac{[S_A]}{K_{mS_A} + [S_A]\left(1 + {K_{mS_B}}\big/{[S_B]}\right)} \tag{8.70}$$

which relates to the double reciprocal form:

$$\frac{1}{\nu} = \frac{K_{mS_A}}{V_{\text{maxf}}} \cdot \frac{1}{[S_A]} + \frac{1}{V_{\text{maxf}}}\left(1 + {K_{mS_B}}\big/{[S_B]}\right) \tag{8.71}$$

As before, provided that the kinetic mechanism of the given biocatalyst of interest brings about biocatalysis through the ping-pong Bi Bi mechanism, then plots of actual $1/\nu$ versus $1/[S_A]$ data obtained in the presence of various fixed concentrations of the S_B substrate, $[S_B]$, should be fitted closely by Equation (8.71), resulting in the parallel graphical output shown (Figure 8.39). Equation (8.71) and this ping-pong Bi Bi kinetic scheme are applicable for the steady state kinetic analyses of many of our example biocatalysts, including homodimeric *E. coli* AspAT (Figure 8.5, Table 8.1), together with the monomeric hydrolytic enzymes bovine pancreatic α-chymotrypsin (Figure 8.3, Table 8.1) and *T. californica* AChE (Figure 8.10, Table 8.1), which both require an acyl-enzyme covalent intermediate (F state) as an essential part of their catalytic cycles. Similarly, Equation (8.71) and this ping-pong Bi Bi kinetic scheme are also applicable for the steady state kinetic analyses of monomeric hydrolytic enzymes bovine pancreatic RNAse A (Figure 8.4, Table 8.1) and hen egg-white lysozyme (Figure 8.6, Table 8.1) since they both involve a non-covalent enzyme-intermediate complex (F state) as an integral part of their catalytic cycles. Please note that although *E. coli* AspAT is homodimeric and there are substantial conformational changes taking place in subunits during the catalytic cycle, evidence for catalytic site dependence and cooperativity in catalysis has not been observed so far, even though the two active sites do not appear to be exactly equivalent. Hence both catalytic sites are essentially independent of each other, making this ping-pong Bi Bi kinetic scheme applicable for each individual AspAT catalytic site.

8.2.7.4 *Ordered Ter Bi and Ter Ter kinetic schemes*

The kinetic scheme corresponding with the **ordered Ter Bi kinetic scheme** is illustrated (Scheme 8.12), which gives the following initial net forward rate for the conversion of substrates into products in the presence of negligible concentrations of products:

$$\frac{\nu}{V_{\text{maxf}}} = \frac{[S_A][S_B][S_C]}{\begin{array}{c} K_{iS_A}K_{iS_B}K_{mS_C} + K_{iS_B}K_{mS_C}[S_A] + K_{iS_A}K_{mS_B}[S_C] + K_{mS_C}[S_A][S_B] + \\ K_{mS_B}[S_A][S_C] + K_{mS_A}[S_B][S_C] + [S_A][S_B][S_C] \end{array}} \tag{8.72}$$

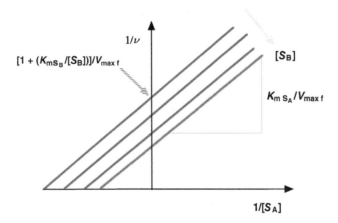

Figure 8.39 Lineweaver-Burk Plots for the ping-pong Bi Bi kinetic scheme. The corresponding kinetic reaction scheme is also illustrated (Scheme 8.11).

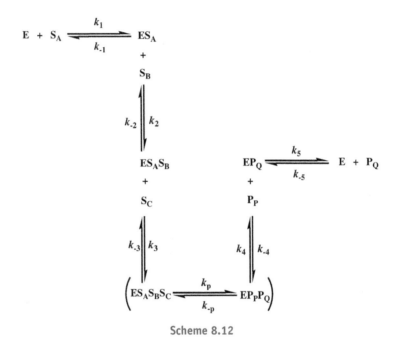

Scheme 8.12

When $[S_B]$ and $[S_C]$ are constant, Equation (8.72) rearranges to:

$$\frac{\nu}{V_{maxf}} = \frac{[S_A]}{K_{mS_A}\left(1 + \dfrac{K_{iS_A}K_{iS_B}K_{mS_C}}{K_{mS_A}[S_B][S_C]} + \dfrac{K_{iS_A}K_{mS_B}}{K_{mS_A}[S_B]}\right) + [S_A]\left(1 + \dfrac{K_{iS_B}K_{mS_C}}{[S_B][S_C]} + \dfrac{K_{mS_B}}{[S_B]} + \dfrac{K_{mS_C}}{[S_C]}\right)} \tag{8.73}$$

which relates to the following double reciprocal:

$$\frac{1}{\nu} = \frac{K_{mS_A}}{V_{maxf}}\left(1 + \frac{K_{iS_A}K_{iS_B}K_{mS_C}}{K_{mS_A}[S_B][S_C]} + \frac{K_{iS_A}K_{mS_B}}{K_{mS_A}[S_B]}\right)\frac{1}{[S_A]} + \frac{1}{V_{maxf}}\left(1 + \frac{K_{iS_B}K_{mS_C}}{[S_B][S_C]} + \frac{K_{mS_B}}{[S_B]} + \frac{K_{mS_C}}{[S_C]}\right) \tag{8.74}$$

Assuming that the ordered Ter Bi mechanism applies, then plots of actual $1/\nu$ versus $1/[S_A]$ data obtained in the presence of various fixed concentrations of the S_B and S_C substrates, $[S_B]$ and $[S_C]$, should be fitted by Equation (8.74), resulting in the off-axis graphical output shown in Figure 8.40. The scheme corresponding with the **ordered Ter Ter kinetic scheme** is also shown (Scheme 8.13). Fortunately, the initial net forward rate for the conversion of substrates into products in the presence of negligible concentrations of products obeys the same steady state rate equations as the Ter Bi kinetic scheme.

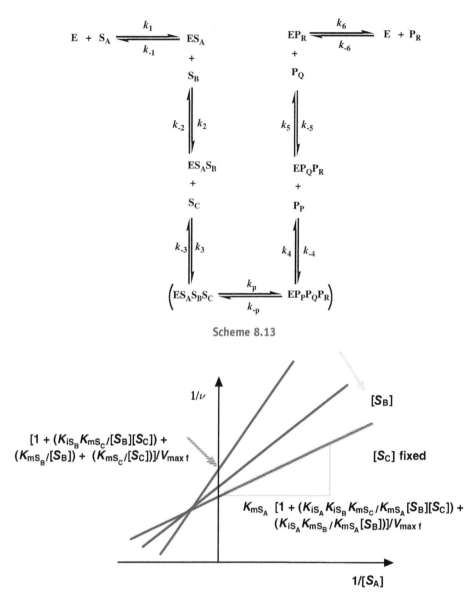

Scheme 8.13

Figure 8.40 Lineweaver-Burk Plots for the ordered Ter Bi kinetic scheme. The kinetic reaction scheme is also illustrated (Scheme 8.12).

Interestingly, Equation (8.74) and the irreversible, single catalytic site, Ter Ter kinetic scheme may also be applicable for the steady state kinetic analysis of the overall conversion of two molecules of ATP to Ap$_4$A and pyrophosphate, catalysed by LysU with the assistance of third substrate L-lysine-Mg^{2+} (Figure 8.11, Table 8.1). The combination of ATP, first nucleotide substrate, with L-lysine-Mg^{2+} to give intermediate lysyl-adenylate represents the formation of a transient non-covalent enzyme-intermediate complex. Pyrophosphate product now dissociates from catalytic site allowing for ATP, second nucleotide substrate, to enter in its place and form Ap$_4$A by in-line displacement of L-lysine from the lysyl-adenylate intermediate. This is not a perfect fit with the ordered Ter Ter kinetic scheme since there are some elements of ping-pong kinetics but the Ter Ter kinetic scheme is workable on the assumption that enzyme associated lysyl-adenylate intermediate is indeed transient.

Unsurprisingly, the complexity of steady state kinetics equations starts to tell above the Ter Ter kinetic scheme level. Moreover, this complexity is probably superfluous to further understanding of biocatalysis. Essentially, the equations and kinetic schemes described in this chapter thus far are more than adequate for the steady state analysis of most biological catalytic events. Moreover, steady state kinetics, whilst being a useful measure of "surface" biocatalyst activity, is unable to penetrate the realities of biocatalyst associated chemical transformations, the rapid binding of substrate and then the release of product. For a more detailed analysis of chemical events on a given biocatalyst, other faster techniques are required. In particular, pre-steady-state kinetics.

8.3 Pre-steady-state kinetics

The arena of **pre-steady-state kinetics**, also known as **single turnover kinetics**, begins under conditions where bio-catalysis is being observed in the presence of a large amount of biocatalyst under rapid mixing ($\ll$ sec), non-equilibrium conditions. This is no longer steady state kinetics. Therefore, complex kinetic schemes and complex steady state kinetics equations derived from steady state analyses are no longer required. In fact, kinetic schemes become very simple since we are usually only observing state-to-state changes (either in the biocatalyst itself or in substrate/product bound to the biocatalyst under investigation) that essentially all obey first order or pseudo first order kinetics. For this reason, an appreciation of first order kinetics is an essential prerequisite for understanding pre-steady-state kinetics.

Pre-steady-state analyses are used primarily to reach a much more detailed understanding surrounding the mechanistic behaviour of a given biocatalysis. In order to achieve this, such analyses are always performed under non-equilibrium rapid mixing conditions and involve only a single catalytic turnover. **Relaxation towards equilibrium** is then observed by continuous monitoring of an appropriate physical property or spectroscopic signal (usually fluorescence intensity) over a time interval sufficient for a change in state (either in the biocatalyst itself or in substrate/product bound to the biocatalyst) to take place. Relaxation data allows catalytic effects to be correlated with actual physical and chemical properties of the biocatalyst or substrate/product. Therefore, pre-steady-state kinetic analyses can open up the precise process by which a given biocatalyst promotes inter-conversions between substrate(s) and product(s), including substrate binding and product release steps, with precise microscopic rate constants. However, in spite of this utility such analyses require large amounts of biocatalyst with some relatively sophisticated fast-kinetics apparatus and so are not performed routinely without a great deal of background preparation.

8.3.1 Pre-steady-state bioassays

Pre-steady-state bioassays are performed by linking up some form of **rapid mixing device** to a spectrometer that is usually a spectrofluorimeter calibrated to detect changes in fluorescence signal intensity at a defined wavelength over a fixed time period (approximately 4–200 ms). There are three main types of rapid mixing device, namely **rapid mixing continuous flow**, **stopped flow** and **rapid quench**. In the first (Figure 8.41), two syringes are individually loaded with samples of biocatalyst, substrate, product or even biocatalyst-substrate species that are "fired" together into a mixing chamber and then expelled down a flow-tube at constant flow rate (approximately $10\,\text{ms}^{-1}$) such that the distance along the tube is proportional to "age" (i.e. time). For instance, given the indicated flow rate, at 1 cm from the mixing chamber the solution is 1 ms "old". The power of this method is that the output from rapid mixing may be observed even at an age as recent as $10\,\mu s$.

In the case of the stopped flow device (Figure 8.42), two compressed syringes are set up to express small volumes (50–200 µl) at any one time. These syringes are individually loaded with samples as above and small volumes of these samples are then fired together into a mixing chamber and expelled down a flow-tube at constant flow rate (approximately $10\,\text{m s}^{-1}$) prior to a mechanical stop. A fixed detector at a distance of 1 cm from the mixing chamber is then positioned to observe solution that is initially 1 ms "old" at the end of continuous flow. Compression converts continuous into static flow so that spectroscopic and/or physical changes in solution are then monitored normally *in situ* in full view of the fixed detector. Note that the first 1 ms is lost and represents the **dead time of mixing** of the apparatus before which nothing can be observed. Finally, the rapid quench device (Figure 8.43), consists similarly of two compressed syringes set up to express small volumes into a mixing chamber that links through a flow-tube into a second mixing chamber where a third syringe is on hand to introduce an agent such as strong acid that prevents any further reaction from taking place, by acting as a chemical quenching agent. Such a device has a dead time of mixing of 4–5 ms and a window of observation extending from 100 to 150 ms only. Ultimately, stopped flow devices are much the most popular and versatile devices for pre-steady-state analyses.

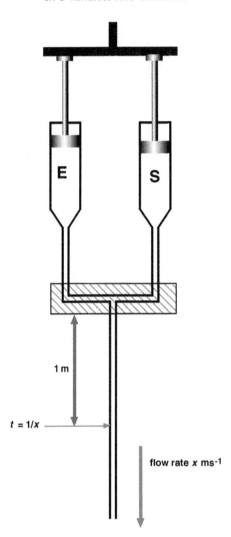

Figure 8.41 Rapid mixing continuous flow device. Biocatalyst, **E**, and substrate, **S**, are combined in a **mixing zone** (hatched area) and the mixture ejected along a common outlet tube. Distance from the mixing zone determines **time, t**, from mixing. Spectroscopic monitoring of mixture as a function of distance generates first order **relaxation curves** for analysis.

8.3.2 First order pre-steady state kinetics

When there is an irreversible conversion of a species **A** to **B** that follows the given kinetic scheme (Scheme 8.14 (a)), then the rate of disappearance of **A** follows the Equation (8.75):

$$-\mathrm{d}[A]\Big/_{\mathrm{d}t} = k_1[A] \tag{8.75}$$

Reaction rate depends cleanly on the concentration of only one chemical species and so is said to obey **first order kinetics**. In a similar way, if a reaction does depend upon the concentration of two chemical species but one species is in such large excess that its concentration is effectively constant during the course of the reaction, then the reaction rate will depend effectively on the concentration of only one species again and is said to obey **pseudo first order kinetics**. Equation (8.75) integrates to give Equation (8.76):

$$[A]_t = [A]_o \exp(-k_1 t) \tag{8.76}$$

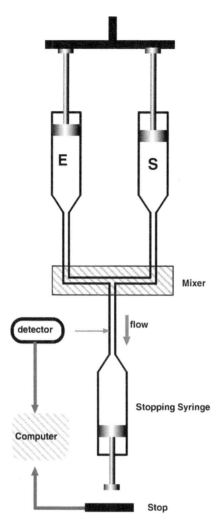

Figure 8.42 Stopped flow device. Biocatalyst, **E**, and substrate, **S**, are combined in a mixing zone (hatched area) and the mixture ejected along a common outlet tube activating a stopping syringe that provokes detection by the detector. Spectroscopic monitoring of mixture as a function of **time, *t***, then generates first order **relaxation curves** for analysis.

where $[A]_o$ is the **initial concentration** of species **A** at the beginning of reaction, and $[A]_t$ is the **concentration** of species **A** at **time *t***. Note the exponential dependence of $[A]_t$ with time. Given the fact that $[B]_t$, the **concentration** of species **B** at **time *t***, is related simply to $[A]_t$ by the relationship 8.77:

$$[A]_t = [A]_o - [B]_t \tag{8.77}$$

then Equation (8.76) can be rewritten for $[B]_t$ as shown:

$$[B]_t = [A]_o \left\{ 1 - \exp\left(-k_1 t\right) \right\} \tag{8.78}$$

that represents an inverted exponential dependence. In the case of both Equations (8.76) and (8.78) the term k_1 is the **first order rate constant** for the conversion of species **A** to **B**, and the term $[A]_o$ behaves as an **amplitude** that scales the dimensions of the exponential together with k_1 (Figure 8.44).

Now for the reversible conversion of a species **A** to **B** that follows an alternate kinetic scheme (Scheme 8.14 (b)). In this instance, matters could become more complicated. However, if we just consider the situation in which pure **A** is allowed to convert to **B** until equilibrium is established, then the rate of approach to equilibrium will still involve an exponential dependence, although modified with respect to Equations (8.76) and (8.78). A complete derivation of these modified equations is not necessary here, but the reader should be aware that there must be differences in the way rate constants

Scheme 8.14

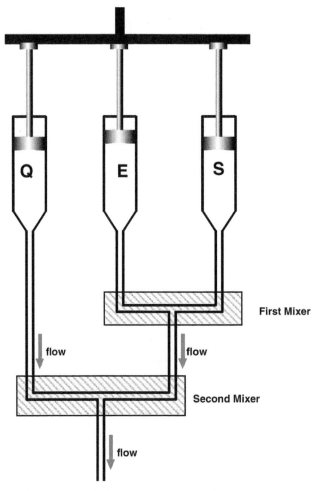

Figure 8.43 Rapid quench device. Biocatalyst, **E**, and substrate, **S**, are combined in a mixing zone (hatched area) and the mixture ejected along a common outlet tube in order to be combined with quencher, **Q**, in another mixing zone. Different mixing times correlate with different reaction **times**, *t*, pre-quenching. Spectroscopic monitoring as a function of *t* gives first order **relaxation curves** for analysis.

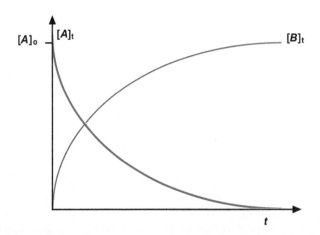

Figure 8.44 Graphical depiction of committed first order reaction. Variation of concentration of species **A** and **B** with time where the conversion from **A** to **B** obeys first order kinetics and is irreversible.

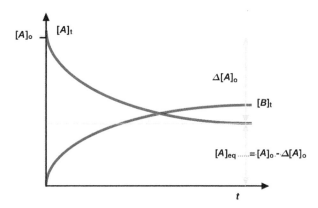

Figure 8.45 **Graphical depiction of first order reaction converging to equilibrium.** Variation of concentration of species **A** and **B** with time where the conversion from **A** to **B** obeys first order kinetics, but is reversible and as such converges to equilibrium.

appear and also differences in the amplitude to accommodate the existence of equilibrium. Hence the equilibrium version of Equation (8.76) is as follows:

$$[A]_t = \Delta[A]_o \exp\{-(k_1 + k_{-1})t\} + [A]_{eq} \tag{8.79}$$

where $\Delta[A]_o$ is the **total change in concentration** as species **A** is converted to **B** starting from initial concentration $[A]_o$ and converging at **final equilibrium concentration** $[A]_{eq}$. $\Delta[A]_o$ represents the correct amplitude in this case (Figure 8.45), and is given explicitly by the following Equation (8.80):

$$\Delta[A]_o = [A]_o \left(\frac{k_1}{k_1 + k_{-1}} \right) \tag{8.80}$$

We can rewrite Equations (8.79) and (8.80) in terms of product **B** as follows:

$$[B]_t = \Delta[B]_o \exp\{-(k_1 + k_{-1})t\} + [B]_{eq} \tag{8.81}$$

$$\Delta[B]_o = [B]_o \left(\frac{k_{-1}}{k_1 + k_{-1}} \right) \tag{8.82}$$

Obviously, Equations (8.81) and (8.82) apply to the reverse scenario in which pure **B** is allowed to convert to **A** starting from **initial concentration** $[B]_o$ and converging at **final equilibrium concentration** $[B]_{eq}$. In other words, the exact opposite of the scenario that Equations (8.79) and (8.80) apply to. Now let us return to pre-steady-state kinetic analysis.

8.3.3 Further pre-steady-state equations

Consider the following Uni Uni biocatalysis kinetic scheme that involves two biocatalyst-substrate species intermediates that are generated prior to product formation and release (Scheme 8.15). Most of the steps involve the interconversions of biocatalyst-substrate or biocatalyst-product species and hence microscopic inter-conversion rates will depend only upon single species concentrations. In other words, these inter-conversions must obey first order kinetics. Similarly, if **S** and **E** are combined under conditions where $[S] \gg [E]$ such that $[S]$ is effectively unchanged during reaction, then the microscopic inter-conversion rate from **E** to **ES** will once again depend effectively only upon $[E]$, such that this inter-conversion obeys pseudo first order kinetics. Accordingly, every step/inter-conversion shown in the biocatalysis reaction scheme here obeys and could be analysed by appropriate variations of Equation (8.79).

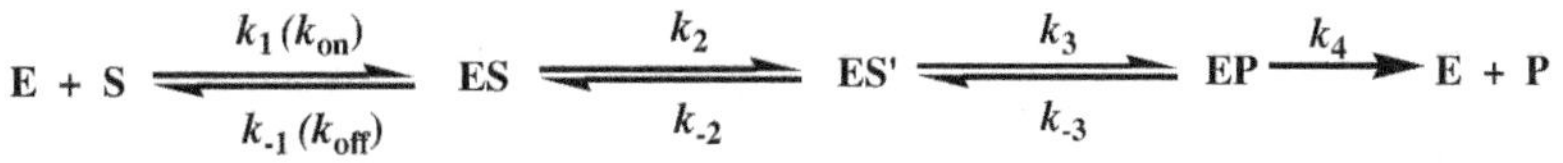

$$\mathbf{E + S} \underset{k_{-1}\,(k_{off})}{\overset{k_1\,(k_{on})}{\rightleftharpoons}} \mathbf{ES} \underset{k_{-2}}{\overset{k_2}{\rightleftharpoons}} \mathbf{ES'} \underset{k_{-3}}{\overset{k_3}{\rightleftharpoons}} \mathbf{EP} \overset{k_4}{\longrightarrow} \mathbf{E + P}$$

Scheme 8.15

In practice, if we are to analyse biocatalysis under conditions in which pure substrate and/or product are followed to chemical equilibrium in the presence of pure biocatalyst, then rapid mixing devices are essential (see Section 8.3.1). Using such devices, a large quantity of biocatalyst ($[E] \cong$ mM-level) may be combined (ms timescale) with a significant excess of substrate(s) and/or product(s) ($[S]$ and/or $[P] \gg$ mM) and the relaxation to chemical equilibrium followed over the ms–sec timescale. Relaxation to equilibrium may then be followed through changes with time in an appropriate physical property or spectroscopic signature (frequently fluorescence output) that is linked with mechanistic activity involving the biocatalyst. The plot of **signal S_t** against time results in a **relaxation curve**. This is typically analysed using a multi-exponential Equation (8.83), derived in form from Equation (8.79):

$$S_t = at + b\sum_i C_i \cdot \exp(-k_i t) \tag{8.83}$$

where a is the **slope of the signal baseline**, b the value of the **baseline** at **time t**, C_i the **amplitude** for the i-th mechanistic step and k_i the corresponding **rate constant**. In effect, one rapid mixing experiment could provide sufficient information from the evolution of S_t with time t to determine the microscopic rate constants and amplitude factors for each mechanistic step of a biocatalyst pathway. Values of k_i are not true microscopic rate constant values since each k_i term corresponds to a sum of rate constants (see Equations (8.79) and (8.81)). However, actual microscopic rate constant values may be assigned given knowledge of corresponding values of C_i, since amplitudes also contain rate constant terms (see Equations (8.80) and (8.82)).

In practice, one rapid mixing experiment is rarely sufficient, if ever, to obtain all the information required on steps in a pathway, including microscopic rate constants. Instead, a range of relaxation curves must usually be acquired over a substantial range of $[E]_o$, $[S]$ and/or $[P]$ values to confirm measured values of C_i and k_i. Furthermore, individual steps in the biocatalyst pathway may need to be isolated for more extensive steady state analysis by careful choice of pre-steady-state conditions, the use of substrate and/or product analogues, and changes in the physical property or spectroscopic signature used to monitor relaxation. For instance, the selection of a substrate analogue able to bind to biocatalyst **E** but not able to proceed to the **ES′** state would allow for precise pre-steady-state analysis of the first binding equilibrium step without interference from the later steps. Experiments of this type ensure that a significant range of pre-steady-state kinetic experiments may be devised to tease out the mechanistic detail of different steps in a biocatalyst pathway and/or determine individual microscopic rate constants in that pathway.

8.4 Theories of biocatalysis

So far in this chapter, the chemical biology reader has been introduced to examples of biocatalysts, kinetics assays, steady state kinetic analysis as a means to probe basic mechanisms and to pre-steady-state kinetic analysis as a means to measure rates of on-catalyst events. In order to complete this survey of biocatalysis, we now need to consider those factors that make biocatalysis possible. In other words, how do biocatalysts achieve the catalytic rate enhancements that they do? This is a simple question that in reality needs to be answered in many different ways according to the biocatalyst concerned. For certain, there are general principles that underpin the operation of all biocatalysts, but there again other principles are employed more selectively. Several classical theories of catalysis have been developed over time that include the concepts of **intramolecular catalysis**, **"orbital steering"**, **general acid-base catalysis**, **electrophilic catalysis** and **nucleophilic catalysis**. Such classical theories are useful starting points in our quest to understand how biocatalysts are able to effect biocatalysis with such efficiency.

8.4.1 Intramolecular catalysis and stereo-control in catalysis

Intramolecular catalysis is a fundamental concept in biocatalysis. In essence, some catalysis is made possible just by binding substrates in close proximity to each other in an **active** or **catalytic site(s)**. The same is true of binding substrates in close proximity to active site functional groups that participate in the catalytic process (Figure 8.46). Binding in proximity within catalytic site(s) enhances the contact frequency between substrates and/or functional groups to a level significantly in excess of that possible in free solution. The physical consequence of this is enhanced reaction rates. This effect is known as **intramolecular catalysis**. Intramolecular catalysis can be quantified by saying that the binding of substrates and/or functional groups in close proximity enhances contact frequencies in a manner

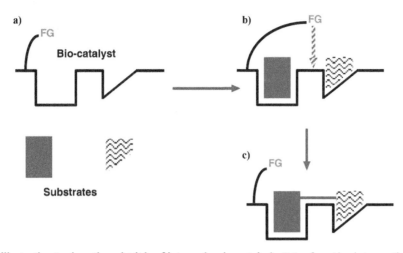

Figure 8.46 Schematic illustration to show the principle of intramolecular catalysis. Rate of reaction between the two substrates is enabled by binding and close proximity in an **active** or **catalytic site** region (**a**). **Neighbouring group participation** involving neighbouring functional group (**FG**) assistance provides additional rate enhancement (**anchimeric assistance**) (**b**) for bond construction (or indeed destruction, if so required) (**c**). Anchimeric assistance represents a general term for those types of catalytic assistance described in Sections 8.4.4 and 8.4.5.

equivalent to increasing the **effective concentration** of each substrate and/or functional group in free solution to the point that identical contact frequencies would be made possible. The consequence of such simple binding on reaction rates is impressive. Intramolecular catalysis can account for between 10^2 to 10^3-fold of the total rate enhancement in biocatalysis that typically varies anywhere from 10^5 to 10^9-fold depending upon the nature of the chemical reactions involved and the biocatalyst.

Another consequence of the simple binding of substrates to catalytic site(s) is that biocatalysts can exert strong effects on chirality during reactions. Biocatalysts such as enzymes are themselves chiral and so, too, are their active or catalytic site(s). Consequently, where substrates are chiral then **ES** complexes binding different enantiomers are themselves diastereomeric with respect to each other, involving different binding modes and different binding energies. Frequently, this effect ensures that biocatalysts choose only certain select substrate enantiomers for biocatalysis. Where substrates are achiral, then the **ES** complex is still chiral and reactions involving achiral substrates frequently give rise to new chiral centres. In addition, since substrates typically remain in association with the biocatalyst during catalytic conversion into products, then the transfer of chiral functional groups frequently takes place with either retention or inversion of absolute configuration, depending upon the nature of the chemical reactions involved and the biocatalyst (Figure 8.47).

8.4.2 "Orbital steering"

The concept of "orbital steering" is a direct successor to intramolecular catalysis. According this concept, rate enhancements conferred by simple binding can be significantly enhanced if the frontier orbitals responsible for reactivity can also be oriented and brought into optimal alignment. Such optimal alignments would then enhance reaction rates by increasing the frequency of productive contacts between substrates and/or functional groups. Both intramolecular catalysis and "orbital steering" effects provide a general background to biocatalysis and may make contributions towards the rate enhancement of most biocatalytic processes.

8.4.3 Induced-fit and strain

Strain is an old concept that has been used in organic chemistry since the beginning of stereo-electronic explanations for reactivity. Strain is connected with the consequences of introducing unnatural bond angles and dihedral angles into small molecules, either by design or by coercion that leads to an increase in molecular potential energy and hence chemical potential for reactivity. According to one theory of biocatalysis, a biocatalyst structure is rigid and the substrate must undergo a process of **induced-fit** to the relevant active site for optimal binding. Induced-fit implies the introduction of strain into the substrate molecule, thereby preparing the bound substrate in the biocatalyst-ground state **ES** complex to enter the biocatalyst-transition state complex, **ES‡**. Therefore, the active or catalytic site of a given rigid biocatalyst structure

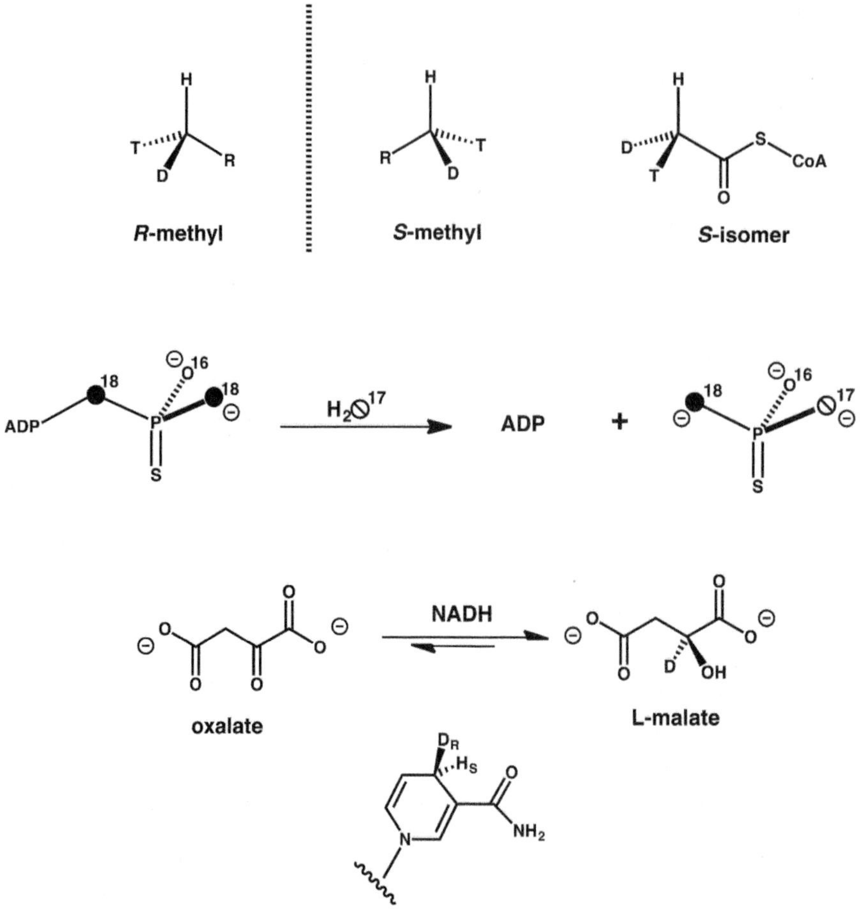

Figure 8.47 Stereochemistry in biocatalysis. Stereochemical changes in biocatalysis may be followed with chiral substrates to reveal stereochemical changes during catalysis. Substrates include: **chiral acetyl-CoA** and phosphate (that undergo **inversion** or **retention of configuration** with reaction), and **chiral NADH(D)** designed to demonstrate whether prochiral (**proR** or **proS**) are employed in reduction. Enzyme mMDH (see Figure 8.2) uses the proR hydrogen as illustrated by deuterium (D) transfer to the product (bottom). **D: deuterium; T: tritium.**

is envisaged to be completely complementary to the reaction transition state, hence biocatalyst-transition state interactions are said to be strain free. However, evidence is accumulating to show that biocatalyst structure should not be regarded as rigid at all. Rather substrate-biocatalyst interactions are just as likely to involve induced fit of the biocatalyst structure to accommodate optimal binding of "rigid substrate" within an active or catalytic site. Therefore, strain may play a role in promoting rate enhancements of some reactions under the influence of a biocatalyst, but the role should be subtle at best.

8.4.4 General acid-base catalysis

General acid-base catalysis is an essential part of biocatalysis. In order to appreciate the principles of rate enhancement through general acid-base catalysis, we need to begin with the two **Brønsted equations** that separately describe general base and general acid catalysis. These equations relate rate of reaction to the equilibrium that exists between a given base **B** and its conjugate acid **BH⁺** as defined by the **base equilibrium ionisation constant, K_d^B**:

$$K_d^B = \frac{[B][H^+]}{[BH^+]} \tag{8.84}$$

By definition:

$$pK_d^B = -\log K_d^B \tag{8.85}$$

General base catalysis is said to occur when the **measured rate**, k_B, of a given chemical reaction changes according to the nature of the base, **B**, used to catalyse the reaction. Quite frequently, the measured rate, k_B varies systematically with the strength of the various bases used to catalyse the reaction according to the **first Brønsted Equation** (8.86):

$$\log k_B = C + \beta\, pK_d^B \tag{8.86}$$

where C is a **reaction constant** and β is the **Brønsted β-value**. The value of β provides a measure of the sensitivity of reaction rate to pK_d^B values that in turn define the relative basic strengths of the different bases used to catalyse the reaction. The higher is the pK_d^B value, the greater is the basic strength of the corresponding base **B** and the faster the reaction. Brønsted β-values vary from 0 to 1, but are typically measured in the 0.3–0.6 region.

In a similar way, **general acid catalysis** is said to occur when the **measured rate**, k_A, of a given chemical reaction changes according to the nature of the acid, **A**, used to catalyse the reaction. Quite frequently, the measured rate, k_A, varies systematically with the strength of the various acids used to catalyse the reaction according to the **second Brønsted Equation** (8.87):

$$\log k_A = C - \alpha\, pK_d^A \tag{8.87}$$

where C is a reaction constant once again and α is the **Brønsted α-value**. The value of α provides a measure of the sensitivity of reaction rate to pK_d^A values that in turn define the relative acid strengths of the different acids used to catalyse the reaction. The lower the pK_d^A value, the greater is the acid strength of the corresponding base, **A**, and the faster the reaction. Brønsted α-values vary within a similar range to Brønsted β-values. For the sake of completeness, note that the equilibrium that exists between a given acid, **AH**, and its conjugate base, **A⁻**, is defined by the **acid equilibrium ionisation constant**, K_d^A:

$$K_d^A = \frac{\left[A^-\right]\left[H^+\right]}{\left[HA\right]} \tag{8.88}$$

8.4.4.1 Dixon-Webb log plots

Clearly the effect of pH upon biocatalytic rates can be indirect or direct. In indirect terms, conformational and other structural changes (e.g. changes in oligomerisation state) induced by changes in solution pH can indirectly perturb biocatalysis rates. However, if the ionisation state of given functional group(s) is in fact critical to efficient biocatalysis, then the effect of pH on biocatalysis rates will be direct. In this case, the concepts introduced in the previous section can be used to quantify the effect of pH as follows. Almost the simplest case is to consider a scenario in which the ionisation state of two functional groups (one acidic designated **group 1**, and one basic designated **group 2**) is critical to efficient catalysis by the simplest irreversible, single catalytic site, Uni Uni kinetic scheme (see Section 8.2.1). Hence, assuming that biocatalysis is not possible unless the ionisation state is completely correct, then the Uni Uni kinetic scheme may be integrated with a series of protonation/deprotonation ionisation equilibria as shown (Scheme 8.16). Now we can use the same kinetic treatment that was used previously in this chapter to derive a meaningful kinetic equation or two. By analogy to the treatment in Section 8.2.1, we can generate two initial Equations (8.89) and (8.90):

$$\nu = k_2 \left[E^n S\right] \tag{8.89}$$

$$\left[E\right]_o = \left[E^n\right] + \left[E^{n+1}\right] + \left[E^{n-1}\right] + \left[E^n S\right] + \left[E^{n+1} S\right] + \left[E^{n-1} S\right] \tag{8.90}$$

that combine smoothly to produce Equation (8.91):

$$\frac{\nu}{\left[E\right]_o} = \frac{k_2 \left[E^n S\right]}{\left[E^n\right] + \left[E^{n+1}\right] + \left[E^{n-1}\right] + \left[E^n S\right] + \left[E^{n+1} S\right] + \left[E^{n-1} S\right]} \tag{8.91}$$

Different equilibrium dissociation constant and equilibrium ionisation constant equations may then be derived with reference to the kinetic scheme (Scheme 8.16) on the basis of stasis, such as Equation (8.92) that defines an

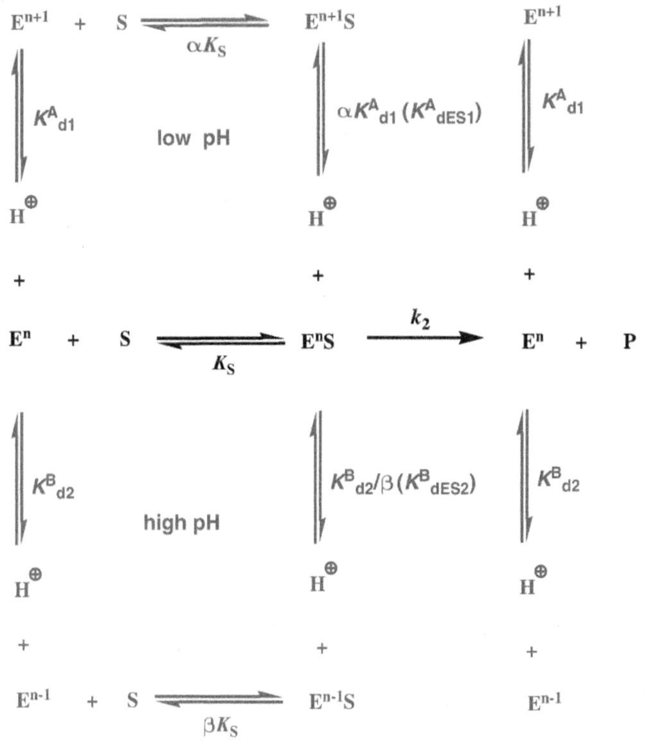

Scheme 8.16

equilibrium dissociation constant αK_S, or Equation (8.93) that defines the **equilibrium ionisation constant K_{d1}^A** for acidic functional group 1, or 8.94 that defines the **equilibrium ionisation constant K_{d2}^B** for basic functional group 2:

$$\alpha K_S = \frac{\left[E^{n+1}\right]\left[S\right]}{\left[E^{n+1}S\right]} \tag{8.92}$$

$$K_{d1}^A = \frac{\left[E^n\right]\left[H^+\right]}{\left[E^{n+1}\right]} \tag{8.93}$$

$$K_{d2}^B = \frac{\left[E^{n-1}\right]\left[H^+\right]}{\left[E^n\right]} \tag{8.94}$$

Thereafter, these equations and others can be rearranged to give solutions for all the enzyme species, including $[E^{n+1}]$ and $[E^{n+1}S]$ for instance. These solutions are then substituted back into Equation (8.91) giving us a complete steady state kinetics/ionisation equilibrium Equation (8.95):

$$\frac{\nu}{k_2[E]_o} = \frac{[S]}{K_S\left(1 + \frac{\left[H^+\right]}{K_{d1}^A} + \frac{K_{d2}^B}{\left[H^+\right]}\right) + [S]\left(1 + \frac{\left[H^+\right]}{\alpha K_{d1}^A} + \frac{K_{d2}^B}{\beta\left[H^+\right]}\right)} \tag{8.95}$$

This equation is still relatively complicated if concise. Therefore, we will have to apply some "boundary conditions" in order to reduce this further in order to obtain useful information. Let us assume that $[S] \gg [K_S]$ (i.e. k_{cat} conditions), then making the usual substitution for V_{max} (see Section 8.2.1), we arrive at Equation (8.96):

$$\frac{V_{max,app[H+]}}{V_{max}} = \frac{1}{\left(1 + \frac{\left[H^+\right]}{K_{dES1}^A} + \frac{K_{dES2}^B}{\left[H^+\right]}\right)} \tag{8.96}$$

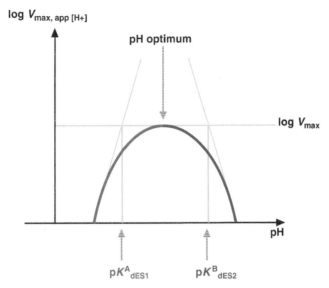

Figure 8.48   Schematic illustration of Dixon-Webb log plots. Plot due to low pH Equation (8.98) is shown (**red**) and the plot due to high pH Equation (8.99) is also shown (**blue**). On the combined plot, optimal pH position is shown, along with log values of **ES** equilibrium ionisation constants, **pK^A_{dES1}** and **pK^B_{dES2}**.

where ν is now defined and substituted for by $V_{\text{max, app[H+]}}$, the **optimal catalytic rate at a given pH**, and αK^A_{d1} and K^B_{d2}/β are substituted for by K^A_{dES1} and K^B_{dES2} respectively, corresponding to the **ionisation equilibrium constants** for functional group 1 and group 2 respectively, modified by the presence of substrate bound to the biocatalyst active site (see Scheme 8.16). The logarithmic form of 8.96 is now given by Equation (8.97):

$$\log\frac{V_{\text{max,app[H+]}}}{V_{\text{max}}} = -\log\left(1+\frac{[H^+]}{K^A_{dES1}}+\frac{K^B_{dES2}}{[H^+]}\right) \tag{8.97}$$

This is the famous **Dixon-Webb equation.** Further simplification is now possible if we take the two extremes of low pH (high $[H^+]$), and high pH (low $[H^+]$). The low pH form of Equation (8.97) is Equation (8.98) and the high pH form is Equation (8.99):

$$\log\frac{V_{\text{max,app[H+]}}}{V_{\text{max}}} = -\log[H^+] + \log K^A_{dES1} = \text{pH} - \text{p}K^A_{dES1} \tag{8.98}$$

$$\log\frac{V_{\text{max,app[H+]}}}{V_{\text{max}}} = \log[H^+] - \log K^B_{dES2} = -\text{pH} + \text{p}K^B_{dES2} \tag{8.99}$$

Both equations link catalytic rate to general pH ($-\log[H^+]$) and the specific **pK^A_{dES1}** ($-\log K^A_{dES1}$) of the acidic functional group 1 or the specific **pK^B_{dES2}** ($-\log K^B_{dES2}$) of the basic functional group 2. Both equations plotted together produce a classic bell-shaped curve (Figure 8.48), where optimal rate is set to fall at a value of pH that is a compromise between the joint requirements for acidic functional group 1 to be **unprotonated** and for basic functional group 2 to be **protonated** in order for optimal biocatalysis to take place. The classic bell-shaped curve is a fusion of two **Dixon-Webb log plots** and is the classical representation for the dependence of biocatalytic rate on pH for a **diprotic mechanism.**

8.4.5 Electrophilic and nucleophilic catalysis

Both electrophilic and nucleophilic catalysis are exceedingly common components of biocatalytic mechanisms, involving as they do the formation of reactive, metastable covalent intermediates. For instance, metal ions are frequently used by biocatalysts to enhance reaction rates by coordinating to reactive functional groups in substrates and then stabilising those anionic charges formed post-reaction. Other examples of electrophilic catalysis at work in biocatalytic mechanisms involve the covalent use of non-peptidic cofactors that promote enzyme catalysed reaction pathways by acting as electrophilic reactants (electron sink) that trap substrates covalently in the form of high energy reactive intermediates which themselves then react before releasing the cofactor for a following round of catalytic activity.

Nucleophilic catalysis is also an exceedingly common component of biocatalytic mechanisms. In addition to substrate coordination, metals may also ionise water and in the process provide a source of extremely nucleophilic hydroxyl ions that are considerably more reactive than water. Alternatively, enzyme active sites have a particularly rich source of nucleophilic functional groups donated by such residues as **serine, threonine, cysteine, aspartic acid, glutamic acid**, lysine, **histidine** and **tyrosine** (see Chapter 1). These groups all have the propensity to trap appropriate substrates covalently under the right conditions in order to form high energy reactive intermediates which themselves then react before releasing the functional group for a following round of catalytic activity.

8.4.6 Mechanisms of biocatalysis by selected biocatalysts

With reference to the examples of biocatalysts in Section 8.1, this is a good moment to pause and consider the known mechanisms of these biocatalysts. Intramolecular catalysis, and to some extent "orbital steering", almost certainly make mechanistic contributions to rate enhancements brought about by all the biocatalyst examples mentioned in this chapter. Contributions from induced fit and strain are harder to assess. By contrast, general acid-base catalysis appear to be employed in a highly selective manner by some biocatalysts to bring about catalytic rate enhancements. The same is true of electrophilic and nucleophilic catalysis. Bovine pancreatic RNAse A and hen egg-white lysozyme are almost archetypical biocatalysts that employ general acid-base catalysis with classical diprotic mechanism that may be analysed by Dixon-Webb plots (Figure 8.49). Chicken TIM, too, makes use of general acid-base catalysis for rate acceleration (Figure 8.49). Both bovine pancreatic α-chymotrypsin and *T. californica* AChE make use of general acid-base catalysis and nucleophilic catalysis involving the formation of reactive acyl-enzyme intermediates (Figure 8.50). By contrast, *E. coli* AspAT, *E. coli* GadB and *B. stearothermophilus* AlaR all employ general acid-base catalysis used in combination with electrophilic catalysis made possible through the provision of the cofactor pyridoxal phosphate (PLP) (Figure 8.51). Finally, ribozymes, human mitochondrial MnSOD, human erythrocyte CA1 and even *E. coli* LysU, make substantial use of metal ion centred electrophilic catalysis to make mechanistic contributions to rate enhancements (Figure 8.52). Having said this, the role of Zn^{2+} in *E. coli* LysU catalysis remains a little controversial. In the illustrated mechanism, the contribution of Zn^{2+} is in N7–N7 chelation, increasing the effective concentration of incoming ATP with respect to the lysyl-adenylate intermediate. Such a chelate contribution could be considered electrophilic catalysis in one sense, although Zn^{2+} may also have a role to coordinate oxygen attached to the α-phosphate of the lysyl-adenylate intermediate in order to encourage nucleophilic attack by incoming ATP as well as or instead of N7–N7 chelation (Figure 8.52).

8.4.7 Transition state stabilisation and biocatalysis

A chapter on biocatalysis is not really complete without introducing the idea of transition state theory and the concept of transition state stabilisation as the most fundamental means in biocatalysis for biocatalysts to effect rate enhancements.

8.4.7.1 Basic transition state concepts

In order for a reaction to take place between two species, **A** and **B**, both reactants need to meet in a productive collision, as a result of which mutual orientations are appropriate for a reaction to occur. Frequently, collisions will be unproductive, leaving the reactants to separate and await the opportunity for a future productive collision. Collisions, both productive and unproductive, can be appreciated by means of a 3D **potential energy surface** that depicts the mutual potential energy of two reactants as a function of all possible trajectories of approach and separation (Figure 8.53). As a rule, productive collisions are associated with the trajectory, or pathway, of minimum potential energy on the potential energy surface. This is known as the **reaction coordinate**. The point of highest potential energy along a reaction coordinate is typically a **saddle-point** or **col** in the potential energy surface. This represents the point of highest potential energy in a productive collision between two reactants and corresponds with the point of maximum molecular reorganisation appropriate for product formation.

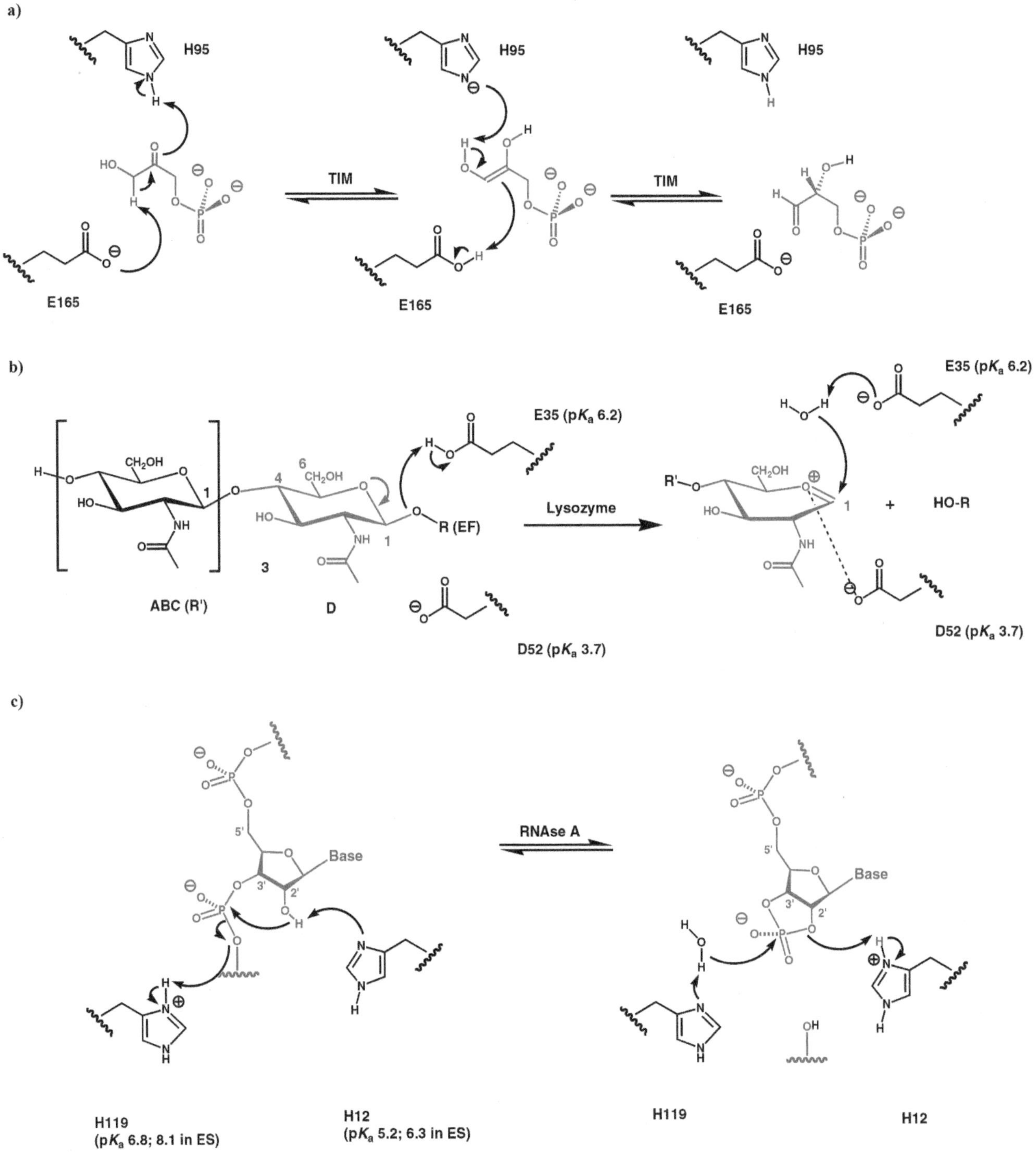

a)

b)

c)

Figure 8.49 Mechanisms of three enzymes that utilise **general acid-base catalysis** as part of their mechanistic paths to successful biocatalysis. (**a**) TIM; (**b**) Lysozyme; (**c**) RNAse A. In all cases substrates are shown in red. Lone pair donor amino acid residues are general bases, lone pair acceptor amino acid residues are general acids. Note that pK_a (a commonly used term) is the equivalent of pK_d^A or pK_d^B (as written in this book) as appropriate for an acidic or basic functional group.

Transition state theory or **absolute reaction rate theory** is built upon these ideas of a potential energy surface and reaction coordinate to account for reactivity. The theory seeks to understand and appreciate reactivity in terms of the structure and behaviour of reaction transition states. A **transition state** is defined as a transient, unstable species that is found at the energy maximum along a given reaction coordinate (i.e. at the saddle-point in a given potential energy surface) through which reactants must pass for successful conversion to product (Figure 8.53). Given this definition, a

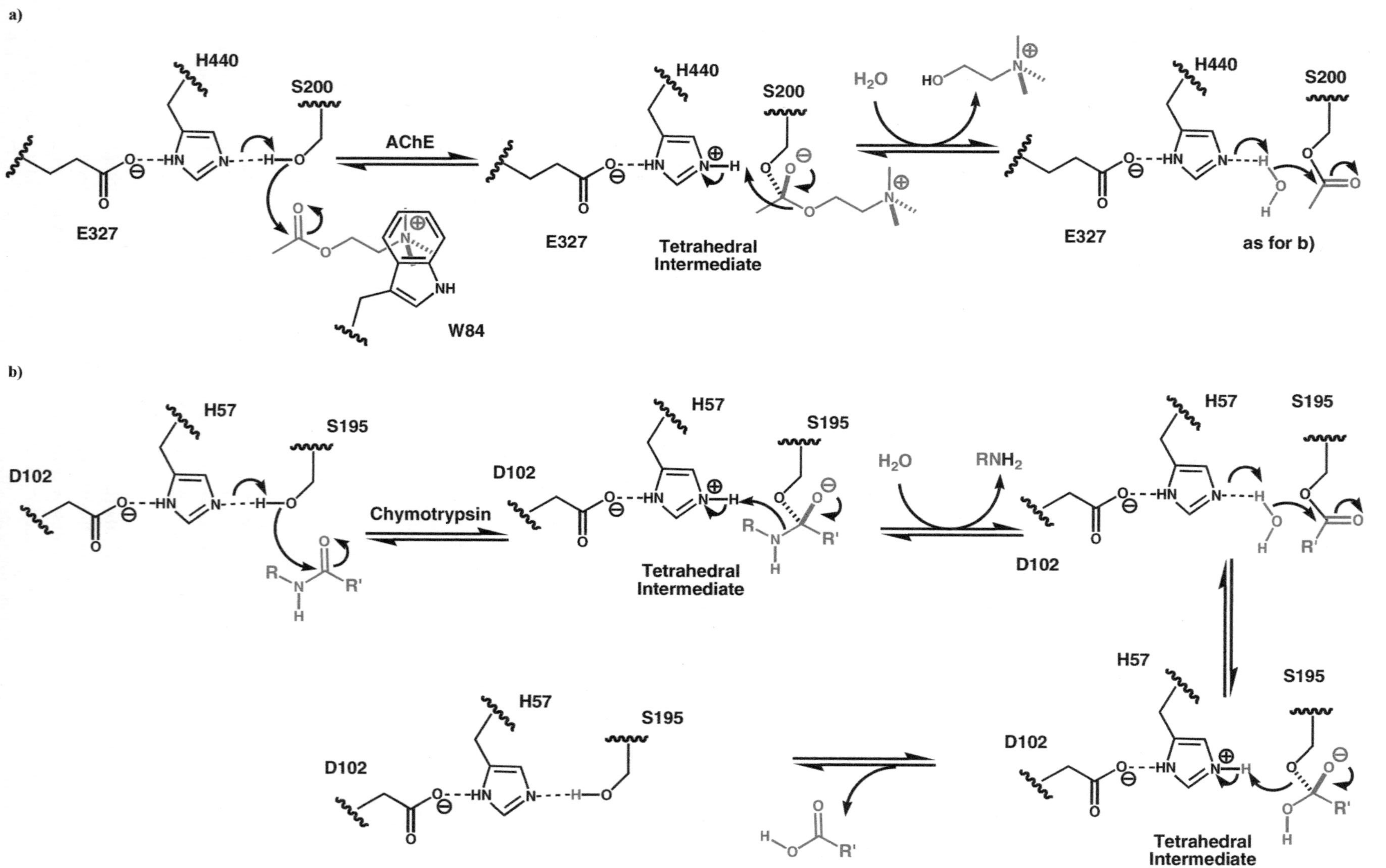

Figure 8.50 Mechanisms for two biocatalysts that employ nucleophilic catalysis for biocatalysis. (a) AChE; (b) α–chymotrypsin. All substrates are shown in red. Nucleophilic catalysis is brought about by appropriate amino acid residues that possess nucleophilic side chains for routine nucleophilic catalysis operations.

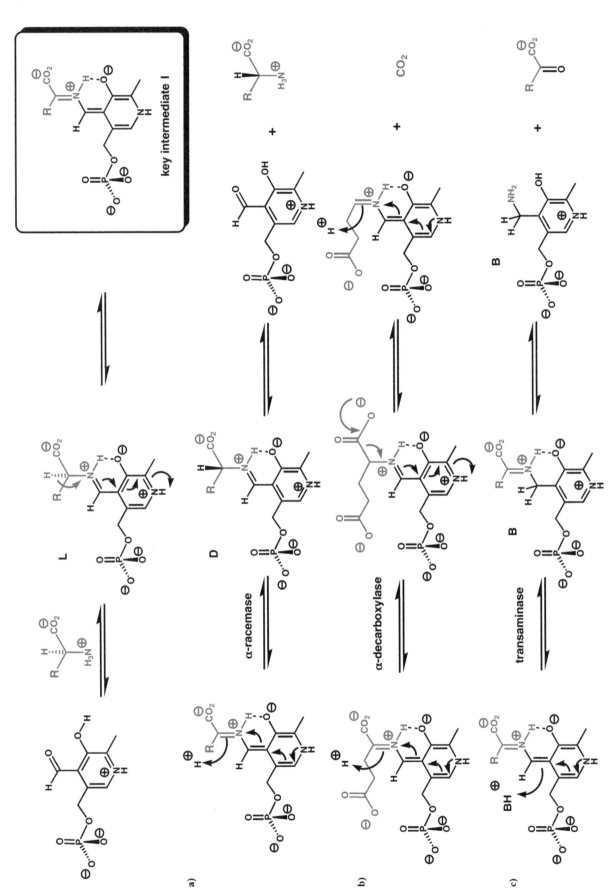

key intermediate I

α-racemase

α-decarboxylase

transaminase

a)

b)

c)

Figure 8.51 Mechanisms of three types of biocatalysts that employ electrophilic catalysis with PLP for biocatalysis. (a) AlaR; (b) GadB; (c) AspAT. All substrates are shown in red.

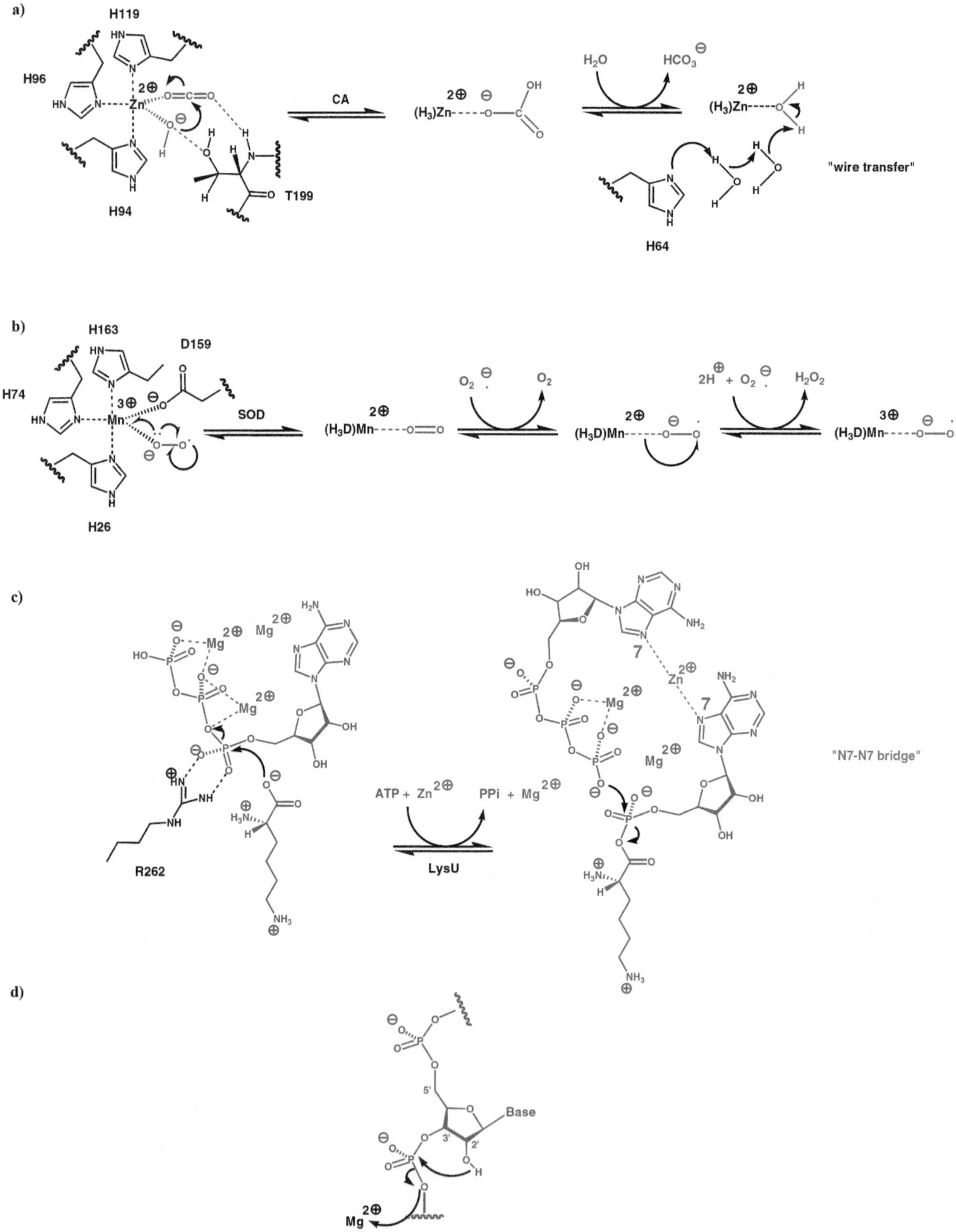

Figure 8.52 Mechanisms of four biocatalysts that employ different modes of electrophilic catalysis for biocatalysis. (a) CA; **(b)** SOD; **(c)** LysU; **(d)** ribozymes. Substrates are shown in red. Electrophilic catalysis is brought about in all cases by the use of a critical metal ion such as Zn^{2+}, Mn^{2+} and/or Mg^{2+}.

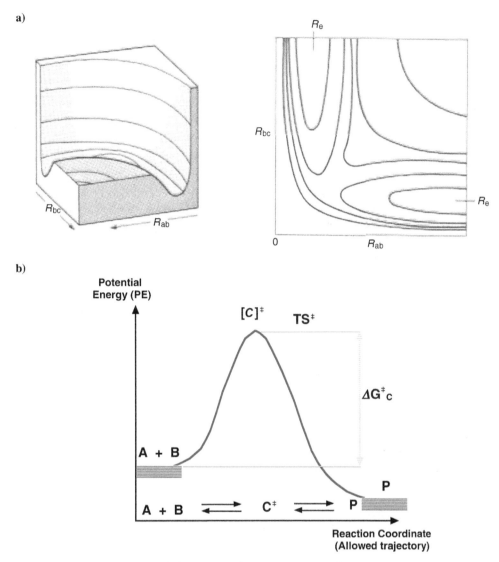

Figure 8.53 Schematic illustration of potential energy surface, reaction coordinate and transition state. (a) 3D depiction of potential energy surface (left) and contour depiction (right); (b) Conversion of species A and B to P, illustrating the transition state [C]‡ and free energy of activation $\Delta G^{\ddagger}_{C}$. Illustrations (a) and (b) from Atkins, 1995, Figures 27.15 and 27.16, Oxford University Press.

transition state can only be formed by reactants if they approach each other along the reaction coordinate. Any other trajectory will not do. In other words, we can redefine a productive collision as one able to generate a transition state and an unproductive collision as one that is unable to generate a transition state. The structure of a given transition state is typically thought to comprise a number of fragile, purely transient high energy bonds that **vibrate through the saddle-point at a frequency,** ν_{TS}, in a manner unconstrained by the usual restoring forces found in ground state covalent bonds (see Chapter 4). While the existence of transition states is still the subject of some debate, the idea that there is some form of "activated complex" formed between reactants that represents a "gateway to product formation" has been very appealing to chemists for some time and is certainly a powerful aid to understanding biocatalysis as well, as we shall see.

In relating transition state structure and behaviour to reactivity, transition state theory relies on two key assumptions. These are that the absolute rate of reaction depends upon the rate of decomposition of the transition state, and that any transition state is in quasi-equilibrium with its corresponding reactants. Hence, if we consider the most rudimentary of reaction schemes (Figure 8.53) (Scheme 8.17), then a simple rate equation may be devised in which rate of formation of product, **P**, obeys Equation (8.100):

$$d[P]\Big/{dt} = k^{\ddagger}_{C}\left[C^{\ddagger}\right]$$

(8.100)

$$A + B \rightleftharpoons \overset{K_C^{\ddagger}}{} C \rightleftharpoons C^{\ddagger} \overset{k_C^{\ddagger}}{\longrightarrow} P$$

Scheme 8.17

The **transition state forward decomposition rate constant** $k_C^{\ddagger}$ can be further defined as the product of two terms:

$$k_C^{\ddagger} = \kappa \cdot \nu_{TS} \tag{8.101}$$

where κ is the **transmission coefficient** that represents the mole fraction of transition states formed that progress towards product instead of reverting to reactants. This constant of proportionality acts as a correction for factors such as quantum mechanical tunnelling (see Chapter 6) and solvent friction effects (see Chapter 7). According to transition state theory, the **microscopic rate constant** of reaction k_p is then defined as the product of $k_C^{\ddagger}$ and the **quasi-equilibrium association constant** $K_C^{\ddagger}$ as shown:

$$k_P = \kappa \cdot \nu_{TS} K_C^{\ddagger} \tag{8.102}$$

Customarily, an equilibrium concentration is defined as a ratio of bulk concentrations; however, according to statistical thermodynamics, an equilibrium constant can also be written in terms of **partition functions** that quantify the relative populations of molecules in different states. Hence, the term $K_C^{\ddagger}$ is usually written out as:

$$K_C^{\ddagger} = \frac{q^{\ddagger}}{q^A q^B} \cdot \exp\left(-E_o \big/ kT\right) \tag{8.103}$$

where the q terms are the relevant partition functions, E_o is the **difference between transition state and reactant ground state zero-point energies**, and k is the standard **Boltzmann constant**. The partition function $q^{\ddagger}$ is usually replaced by a product term in which the vibrational motion of the transition state along the reaction coordinate that promotes product formation is separately factored out giving:

$$q^{\ddagger} = q^{\neq} \frac{kT}{h\nu_{TS}} \tag{8.104}$$

Hence, substituting Equation (8.104) back into 8.103 gives:

$$K_C^{\ddagger} = \frac{kT}{h\nu_{TS}} \frac{q^{\neq}}{q^A q^B} \cdot \exp\left(-E_o \big/ kT\right) \tag{8.105}$$

Further substituting Equation (8.105) back into Equation (8.102) results in:

$$k_p = \kappa \cdot \frac{kT}{h} \frac{q^{\neq}}{q^A q^B} \cdot \exp\left(-E_o \big/ kT\right) \tag{8.106}$$

Finally, the quasi-equilibrium statistical thermodynamics part of Equation (8.106) may be substituted for by the standard expression relating equilibrium constant with a standard Gibbs free energy change for the equilibrium (see Chapter 7) with the result that Equation (8.106) may be rewritten as:

$$k_p = \kappa \cdot \frac{kT}{h} \exp\left(-\Delta G_o^{\ddagger} \big/ RT\right) \tag{8.107}$$

where the term $\Delta G_o^{\ddagger}$ is actually a quasi-Gibbs free energy term that is known as the **standard free energy of activation** and represents the quasi-thermodynamic barrier to reaction.

8.4.7.2 Binding energy in biocatalysis

The concepts of reaction coordinate, transition state and free energy of activation are central to reaching a fundamental understanding of chemical reactivity and also the effects of biocatalysis upon chemical reactivity. Haldane was first to propose that biocatalysts use the energy released from substrate binding to increase reaction rates by distorting the

$$E \; + \; S \; \underset{K_S}{\rightleftharpoons} \; ES \; \overset{K_{ES}^{\ddagger}}{\rightleftharpoons} \; ES^{\ddagger} \; \overset{k_{ES}^{\ddagger}}{\longrightarrow} \; P$$

Scheme 8.18

structures of their substrates towards those of their products, and in so doing allowing transition states to make more favourable contacts with the biocatalyst than substrates. Pauling went on to suggest that the structure of a biocatalyst should be more properly complementary to a transition state than a substrate in order to optimise the use of binding energy for catalysis. In any event, both suggestions have at their heart the central idea that the biocatalyst principally acts to catalyse a given reaction by using binding energy to reduce free energy of activation, leading to increases in reaction rate (see Equation (8.108)).

Nowadays, we can be even more explicit. Let us return to a simple Uni Uni kinetic scheme (Scheme 8.18). If the rate of conversion of **ES** to **P** is slow, then K_m reduces to K_S the equilibrium dissociation constant for **ES** complex. Assuming also that $[S] \gg K_m$ then the binding equilibrium is very much in favour of **ES** formation (i.e. **ES** formation is energetically favoured). Together, this allows us to draw a simple reaction coordinate diagram representing the minimum energy pathway for the combination of substrate with biocatalyst, followed by the formation of a biocatalyst-transition state complex, **ES‡**, from the ground state **ES** complex (Figure 8.54). Note that the **free energy of activation** for the biocatalyst assisted reaction from free **E** and **S** is denoted $\Delta G_{ES}^{\ddagger}$ and that this is essentially equivalent to $\Delta G_0^{\ddagger}$ that governs the measured rate constant of reaction k_p according to Equation (8.107). $\Delta G_{ES}^{\ddagger}$ is also involved in the following sum:

$$\Delta G_T^{\ddagger} = \Delta G_{ES}^{\ddagger} - \Delta G_S \tag{8.108}$$

where ΔG_S is the **free energy of association for substrate binding** linked to the corresponding equilibrium association constant ($1/K_S$; NB K_S is an equilibrium dissociation constant), and $\Delta G_T^{\ddagger}$ is the **free energy of activation** for the direct conversion of ground state **ES** into **ES‡** that is associated with a **pseudo-equilibrium association constant** $K_T^{\ddagger}$.

Conditions where $[S] \gg K_m$ are often known as the "k_{cat} conditions" of catalysis. When $[S] \gg K_m$ there is no free biocatalyst **E** at all, hence the first equilibrium effectively disappears and the measured rate of reaction, k_p in Equation (8.106) becomes identified directly with the first order Michaelis-Menten term, k_{cat}, according to Equations (8.1) and (8.9). In a similar way $\Delta G_0^{\ddagger}$ in Equation (8.107) is now no longer identified with $\Delta G_{ES}^{\ddagger}$, but becomes identified directly with $\Delta G_T^{\ddagger}$ by definition. If both **ES** and **ES‡** are bound equally more effectively by a uniform value ΔG_R (Figure 8.55), then K_S and by implication K_m are reduced but there is no change in k_{cat}. If **ES** is bound to an extent ΔG_R more effectively

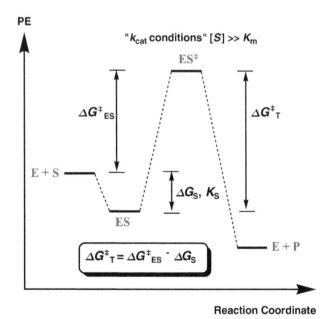

Figure 8.54 **Reaction coordinate diagram for the simplest Uni Uni kinetic scheme.** The corresponding kinetic reaction scheme for biocatalysis is shown (Scheme 8.18). This reaction coordinate diagram is drawn under the assumption that excess substrate is present so that k_{cat} **conditions** prevail. Two types of free energy of activation exist, from **E + S** ($\Delta G_{ES}^{\ddagger}$) and from **ES** ($\Delta G_T^{\ddagger}$).

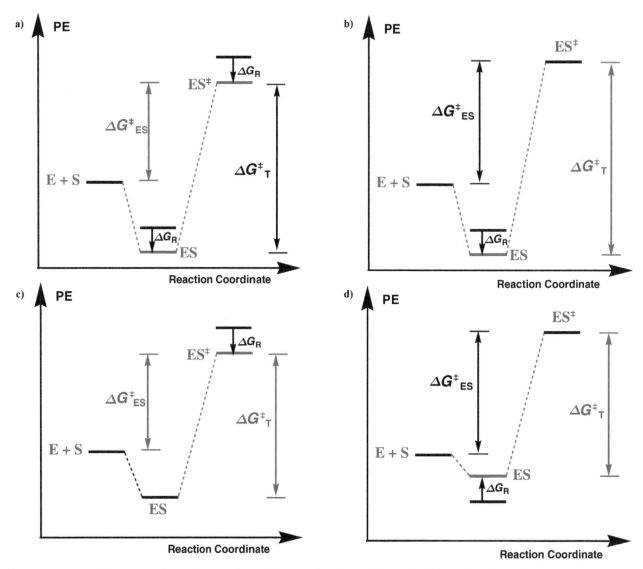

Figure 8.55 **Set of four related reaction coordinate diagrams for the simplest Uni Uni kinetic scheme.** The corresponding kinetic reaction scheme for biocatalysis is shown (Scheme 8.18). (**a**) Illustrates the neutral effect of **uniform stabilisation** of ES and ES‡ by $\Delta G_{R'}$; (**b**) The deleterious effect of **differential stabilisation** of ES by $\Delta G_{R'}$; (**c**) The optimally beneficial effect of **differential stabilisation** of ES‡ by $\Delta G_{R'}$; (**d**) The partially beneficial effect of **differential destabilisation** of ES by ΔG_{R}.

than ES‡ then K_{m} is reduced but there is a fall in k_{cat}, too, since $\Delta G_{T}^{\ddagger}$ must increase (Figure 8.55). That is to say that there is an actual reduction in catalytic enhancement. In fact, an increase in catalytic enhancement can only occur if ES‡ is bound to an extent ΔG_{R} more effectively than ES, or if ES is bound to an extent ΔG_{R} less effectively than ES‡ (Figure 8.55). In the former case, K_{m} remains unchanged but k_{cat} is increased, while in the latter case K_{m} increases but k_{cat} is simultaneously increased. In both of these scenarios where differential binding interactions between substrate and transition state are able to bring about catalytic enhancement, then this is known as **catalysis of an elementary step**.

By contrast, conditions where $K_{m} \gg [S]$ are often known as the "k_{cat}/K_{m} conditions" of catalysis. When $K_{m} \gg [S]$ then the binding equilibrium is very much less in favour of ES formation (i.e. ES formation is much less energetically favoured) and free biocatalyst E will be generated. Hence the first equilibrium reappears and the measured rate of reaction, k_{p}, in Equation (8.107) becomes identified directly with the second-order Michaelis-Menten term k_{cat}/K_{m} according to Equation (8.11). Furthermore, $\Delta G_{0}^{\ddagger}$ in Equation (8.107) becomes identified with $\Delta G_{ES}^{\ddagger}$ once more. The relevant reaction coordinate diagram is illustrated (Figure 8.56). When both ES and ES‡ are bound equally more effectively by a uniform value, ΔG_{R}, then K_{s} and by implication K_{m} decreases but k_{cat}/K_{m} increases since $\Delta G_{ES}^{\ddagger}$ is reduced, implying catalytic enhancement (Figure 8.57). When ES is bound to an extent ΔG_{R} more effectively than ES‡, K_{m} decreases but k_{cat}/K_{m} remains unchanged (Figure 8.57). If we scrutinise the other two alternatives, then when ES‡ is bound to an extent

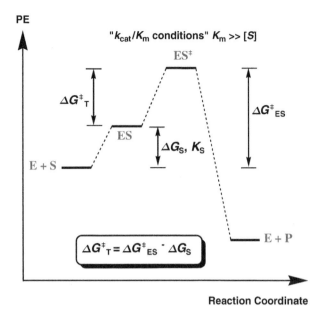

Figure 8.56 An alternative reaction coordinate diagram for the simplest Uni Uni kinetic scheme. The corresponding kinetic reaction scheme for biocatalysis is shown (Scheme 8.18). This alternative reaction coordinate diagram is drawn under the assumption that substrate is limited so that k_{cat}/K_m **conditions** prevail.

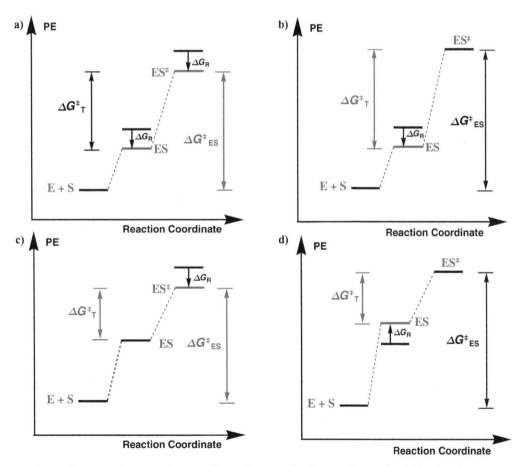

Figure 8.57 An alternative set of four reaction coordinate diagrams for the simplest Uni Uni kinetic scheme. The corresponding kinetic reaction scheme for biocatalysis is shown (Scheme 8.18). (**a**) Illustrates the partially beneficial effect of **uniform stabilisation** of **ES** and **ES‡** by ΔG_R; (**b**) The neutral effect of **differential stabilisation** of **ES** by ΔG_R; (**c**) The optimally beneficial effect of **differential stabilisation** of **ES‡** by ΔG_R; (**d**) The neutral effect of **differential destabilisation** of **ES** by ΔG_R.

ΔG_R more effectively than **ES**, K_m remains unchanged but k_{cat}/K_m is increased again. Finally, if **ES** is bound to an extent ΔG_R less effectively than **ES‡**, then K_m increases but k_{cat}/K_m remains constant. A quick comparison between the reaction coordinate analyses for k_{cat} and k_{cat}/K_m conditions (Figures 8.55 and 8.57), shows that only differential binding of the transition state is sufficient to render catalytic enhancement under both conditions of catalysis. Hence, binding energy released through the binding of a substrate or transition state to a biocatalyst can be used in a variety of ways to bring about catalytic enhancement but the only consistent way to achieve catalytic enhancement under all operating conditions of a biocatalyst is through differential binding of the transition state in preference to binding of the substrate. This is then the purest condition for catalysis of an elementary step, and represents the primary means by which any biocatalyst may bring about reaction catalysis according to transition state theory.

8.4.8 "Perfect biocatalyst" theory

A single biocatalyst is said to reach perfection when k_{cat}/K_m reaches a value of 10^8–10^9 M^{-1}s^{-1}. At this value, the biocatalyst must be catalysing multi-step reactions at the diffusion limit when catalytic rate becomes a matter purely of encounter efficiency between substrate and biocatalyst, or else controlled release of product. At the same time, flux through the chemical inter-conversion pathways must be optimal and cannot be increased. In order to achieve all this, free energy barriers between different chemical species in a biocatalysis pathway must be equivalent in magnitude to each other as well as to the physical free energy barriers associated with either substrate binding or product release. Accordingly, internal equilibrium constants governing the concentrations of different biocatalyst species (comprising bound substrate/product) must also be almost identical in order to avoid accumulation of any one intermediate over another.

In practical terms, taking a simple Uni Uni kinetic scheme (Scheme 8.19), the relative proportions of biocatalyst species **ES** and **EP** are critical in order to establish catalytic perfection. According to **Brønsted-type rate-equilibrium relationships** (Section 8.4.4), the relative ratio of **ES** to **EP** will have a direct effect upon barrier heights to the transition state **ES‡** in both directions and hence the catalysis of an elementary step. Therefore, when a biocatalyst is operating in the presence of equilibrium concentrations of substrate **S** and product **P**, then the optimal internal equilibrium constant K_{int} should be 1 (i.e. **ES** and **EP** are iso-energetic) to ensure that the conversion between **S** and **P** is diffusion limited. When operating far from equilibrium, K_{int} is adjusted so that the rate of chemical transformation matches the rate of release of product **P**, ensuring that only the rate of release of product **P** determines the rate of catalysis and not the chemical transformation step. The protein biocatalyst TIM is notable for having catalytic characteristics close to perfection. The appearance of a rate determining product release step and a value of $K_{int} \approx 1$ are illustrated by the known TIM reaction coordinate diagram linking free substrate with released product (Figure 8.58). Most other biocatalysts are not perfect but are usually adequate biocatalysts for the biological niche in which they are required to operate.

8.4.9 Linear free energy relationships

Linear free energy relationships derive from so-called Brønsted-type rate-equilibrium relationships and represent a potentially useful tool for unlocking the topography of the reaction coordinate or "**energy landscape**" linking a sequence of equilibrated biocatalyst species of the type described above (Section 8.4.8). The critical idea is encapsulated in the following Equation (8.109):

$$k_n = A K_{eq,n}^{\beta} \tag{8.109}$$

where k_n is the **rate constant** for a given conversion step of interest between adjacent biocatalyst species in a catalytic pathway. $K_{eq,n}$ represents the **related equilibrium constant** defining the concentration relationship between either two ground state species or a ground state and a transition state species (see Section 8.4.7). The power term β is known as a **Brønsted coefficient** or "**beta value**". In the event that k_n varies according to substrate/product structure or even according to variations in biocatalyst structure (such as changes in amino acid sequence in an enzyme), then there is always the possibility that the linked variations of k_n and $K_{eq,n}$ values will obey the following linear relationship:

$$\log(k_n) = const + \beta \log(K_{eq,n}) \tag{8.110}$$

$$E + S \rightleftharpoons ES \rightleftharpoons EP \rightleftharpoons E + P$$

Scheme 8.19

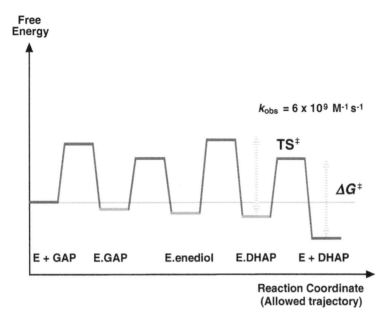

Figure 8.58   Reaction coordinate diagram for triose phosphate isomerase. This represents a near perfect energy landscape pathway allowing for near perfect 1:1:1 stoichiometric equilibrium between all enzyme-bound species optimal for flux through from one enzyme-bound species to another. Enzyme turnover rate k_{obs} is at the diffusion limit, the rate determining step is the association of DHAP with the TIM catalytic site (Figure 8.1), hence chemistry is not rate limiting. Therefore, TIM is considered a perfect enzyme. For TIM enzyme assay see Figure 8.17; for TIM enzyme mechanism see Figure 8.49. Illustration adapted from Knowles, 1991, Figure 2.

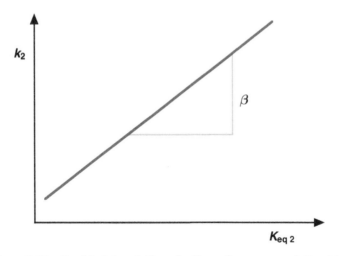

Figure 8.59   Graphical description of a linear free energy relationship.

The implications of linearity can be explained quite simply. We consider the following catalytic pathway (Scheme 8.20), in which **E** corresponds with an enzyme. Assume that values of $K_{eq,2}$ and k_2 have been measured by pre-steady-state kinetic analyses for the wild-type enzyme **E**, together with the values for a variety of point mutant enzymes wherein the amino acid residue change takes place at one residue position and most changes influence catalytic rate, k_2, at least to some extent. Provided that a plot of values of $K_{eq,2}$ and k_2 is linear, then the gradient β may be determined for the conversion of **ES** to **EP** (Figure 8.59). Taking into account the standard relationship between a rate constant such as k_2 and the standard free energy of activation, see Equation (8.89), then we could say that β represents the fractional change in binding energy for the conversion of **ES** to **EP**. For instance, if β is +0.8, then 80% of the binding energy is realised in the transition state, **ES**‡, for the conversion of **ES** to **EP** and as a result **EP** also closely resembles **ES**‡ in energy and structure. More specifically, we

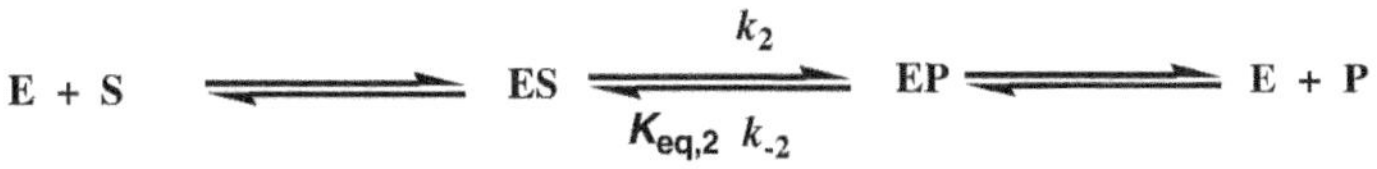

$$\text{E + S} \xrightleftharpoons{\hspace{1.5cm}} \text{ES} \underset{K_{eq,2}\ \ k_{-2}}{\overset{k_2}{\rightleftharpoons}} \text{EP} \xrightleftharpoons{\hspace{1.5cm}} \text{E + P}$$

Scheme 8.20

could say that 80% of the binding energy generated by interactions between the specific amino acid residue position and substrate is realised in binding the transition state. Clearly, the maximum possible value for β is +1.0. Under these circumstances, 100% of the binding energy generated by interactions between the specific amino acid residue position and substrate is realised in binding the transition state only and hence interactions between the specific amino acid residue position and the substrate ground state are negligible. This is almost the perfect situation for catalysis of an elementary step.

8.5 Electron transfer

Traditionally in electron transfer reactions, there is an **electron donor species**, **D**, and an **electron acceptor species**, **A**. **D** and **A** are usually brought together by diffusion, with the assistance of long-range electrostatic forces (see Chapter 7). Thereafter, short-range forces ensure molecular recognition and binding in order to optimise electron transfer rates through the formation of a **specific encounter complex** (**DA**). Electron transfer then takes place prior to separation.

8.5.1 Electron transfer kinetics

The corresponding kinetic scheme is as illustrated (Scheme 8.21). This kinetic scheme bears some similarity to the simplest kinetic scheme and hence simplest Briggs-Haldane steady state kinetics treatment can usefully apply on the assumption that the donor species, **D**, is in excess (i.e. $[D] \gg [A]$) and so is constant during the progress of the reaction. In that case, we can assume that acceptor species, **A**, behaves in an equivalent manner to a biocatalyst substrate and donor species, **D**, to the biocatalyst itself at a **fixed total concentration** of $[D]_o$. Hence, Equation (8.6) neatly transforms into Equation (8.111):

$$\frac{\nu}{[D]_o} = \frac{k_{ET} \cdot [A]}{\left(k_{ET} + k_{-1}\big/k_1\right) + [A]} \tag{8.111}$$

where the term k_2 in Equation (8.6) is replaced by the equivalent term k_{ET} in Equation (8.111) that is the **rate constant for the actual electron transfer step**, about which more later. Term $\nu /[D]_o$ represents the rate of the total electron transfer process described by the kinetic scheme above. Equation (8.111) neatly rearranges to:

$$\frac{\nu}{[D]_o} = \frac{k_1 k_{ET} \cdot [A]}{k_{ET} + k_{-1} + k_1 [A]} \tag{8.112}$$

At the extreme where $k_{ET} \gg k_{-1} + k_1[A]$ then Equation (8.112) reduces to Equation (8.113):

$$\frac{\nu}{[D]_o} = k_1 [A] \tag{8.113}$$

This equation corresponds to the situation when the rate of the total electron transfer process is said to be at the diffusion limit, and so is as perfect as possible. At another extreme where $k_{-1} \gg k_{ET} + k_1[A]$ then Equation (8.112) reduces to Equation (8.114):

$$\frac{\nu}{[D]_o} = \frac{k_1 k_{ET} \cdot [A]}{k_{-1}} = K_{a,DA} k_{ET} \cdot [A] \tag{8.114}$$

where $K_{a,DA}$ is **equilibrium association constant** (k_1/k_{-1}) that governs the association binding equilibrium between acceptor species, **A**, and donor species, **D**. At this extreme, the rate of the total electron transfer process is said to be activation controlled, with a reaction pre-equilibrium. That is, the reaction rate depends partly upon the rate of the actual electron transfer step that is associated with a free energy of activation and the equilibrium association between acceptor species, **A**, and donor species, **D**.

$$D + A \underset{k_{-1}}{\overset{k_1}{\rightleftharpoons}} (DA) \xrightarrow{k_{ET}} (D^{\oplus} A^{\ominus}) \longrightarrow D^{\oplus} + A^{\ominus}$$

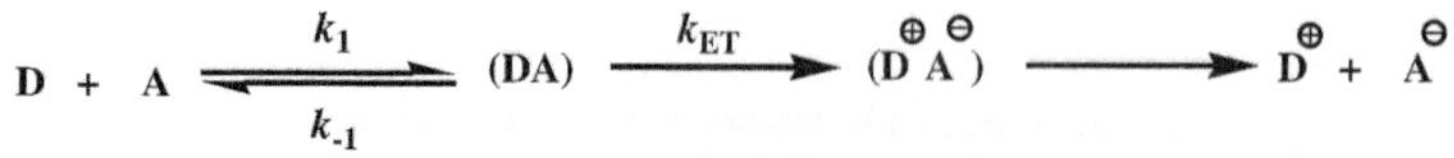

Scheme 8.21

8.5.2 Electron transfer step

The actual electron transfer step is enormously complex since such a process goes right to the heart of the quantum mechanical description of atoms and molecules. Therefore, we are going to keep the discussion as simple as possible here. For instance, in chemical biology we only need be concerned by electron transfer processes involving transfers within or between proteins in which the actual electron transfer processes take place between redox-active prosthetic groups and/or redox-active amino acid residues (usually tyrosine). There are many redox-active prosthetic groups, varying from the haem system of cytochrome c (see Chapter 1) through to varieties of iron-sulfur clusters, blue copper complexes and flavins (see Chapter 4). Multiple assemblies of redox-active prosthetic groups are also found embedded in the enormous protein assemblies that make up the photosynthesis photosystems in many plants.

The great paradox of electron transfer processes involving redox-active prosthetic groups embedded in proteins is that these electron transfer processes are able to take place without any one redox-active prosthetic group/amino acid residue making direct contact one with another. In fact, the central theme of biological electron transfer is that the redox-active centres are "insulated" one from another by "polypeptide walls" that are usually several Å thick. Electron transfer takes place without the formation of intermediate protein excited states so that there is no sense in which polypeptide walls facilitate electron transfer by acting as electronic conductors. Instead, these electron transfer processes involve primarily long-range electron tunnelling (see Chapter 6) between localised donor and acceptor redox-active centres. Under these circumstances, the rate constant k_{ET} for the actual electron transfer step is then defined by the following Equation (8.115) known as the **Fermi equation**:

$$k_{ET} = \frac{2\pi}{\hbar}\left|T_{DA}\right|^2 (FC) \tag{8.115}$$

where the modulus squared $\left|T_{DA}\right|^2$ is the equivalent of the square of an electronic wave function and is a measure of frontier orbital overlap between linked redox-active donor and acceptor centres, while **(FC)** is known as the **Franck-Condon term** and is a measure of overlap between the nuclear wave functions of linked donor and acceptor centres. Note how the equation separates the dependency of the electron transport rate constant into electronic and nuclear overlap terms. This apparently complex equation may be simplified by the following proportionality:

$$\left|T_{DA}\right|^2 \propto \exp\left(-\beta_{ET}\,R_{ET}\right) \tag{8.116}$$

that converts into the following Equation (8.117) by amalgamation with Equation (8.115):

$$\log\left(k_{ET}\right) = C - \beta_{ET}\,R_{ET} \tag{8.117}$$

where R_{ET} is the **edge-to-edge distance** between redox-active donor and acceptor groups, while β_{ET} is known as "**beta value**" for electron transfer. When $\log(k_{ET})$ values measured for electron transfers between redox-active donor and acceptor groups embedded in proteins are plotted as a function of the R_{ET} distance between them, then the result is a straight-line function over a separation range of 20 Å, encompassing values of k_{ET} distributed over 12 orders of magnitude (Figure 8.60). The resulting beta value is 1.4 Å^{-1} a little more than the value of 1.2 Å^{-1} found for electron transfers through glassy solvents. In other words, all polypeptide walls of whatever thickness whether comprised of one or more polypeptide backbones behave in a similar way to a glassy solvent. Undeniably a beautifully simple approximation.

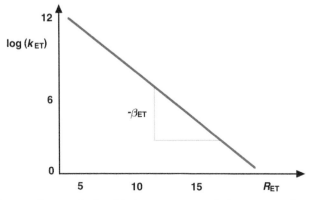

Figure 8.60 **Illustration of inverse linear relationship between rates of electron transfer and edge-to-edge distances between redox-active donor and acceptor groups.**

9
Mass Spectrometry and Proteomics

9.1 Mass spectrometry in chemical biology

Mass spectrometry is a technique with origins in the physical sciences that fast became one of the most powerful techniques available for probing biological systems. Advances in mass spectrometry made possible by the advent of **soft ionisation techniques** have enabled detailed investigations into the primary structures of biological macromolecules and amphiphilic lipids. Indeed, mass spectrometry could be seen as the pre-eminent technique for determining the primary structures of polypeptides, nucleic acids and complex carbohydrates owing to the possibility of obtaining molecular mass and sequence information using minute quantities of material (fmol–pmol, even amol levels). Such information can be obtained without even the need for recombinant technologies (see Chapter 3). Hence, although 3D information is not directly available by mass spectrometry, impressive amounts of primary structural information are available that can be integrated with other pieces of information to try and provide useful insights into biological macromolecular structure without the need for analyses of the type presented in Chapters 5 and 6. However, nowadays, mass spectrometry is also known to be essentially indispensable to the field of **proteomics**, defined as the study of all the interactions and implied functions of all the expressed proteins in a genome.

Chemical biology and multidisciplinary thinking have very much driven the development of proteomics from the development of suitable mass spectrometric methods for large-scale protein identification, to the development of suitable chromatographic systems to map out protein networks (Figure 9.1). In other words, mass spectrometry gives us not only primary structures but has the capacity to define complex, functional, 3D networks of interacting proteins and polypeptides within cells as well. Mass spectrometry quite simply lifts us from the opportunity to make detailed investigations into the structures and functions of certain biological macromolecules of interest, to the opportunity to study interactions in multi-protein complexes and beyond to networks of protein-protein interactions that determine not just molecular function but also cellular function. Hence, there should be no doubt that any discussion on mass spectrometry should follow Chapters 7 and 8 that deal with individualised functions for biological macromolecules of interest. In this chapter we present mass spectrometry to enable the move from individual to collective molecular structure and function analyses that must surely be an ultimate goal for chemical biology research.

Essentials of Chemical Biology: Structures and Dynamics of Biological Macromolecules In Vitro *and* In Vivo, Second Edition. Andrew D. Miller and Julian A. Tanner.
© 2024 John Wiley & Sons, Inc. Published 2024 by John Wiley & Sons, Inc.
Companion Website: www.wiley.com/go/miller/essentialschembiol2

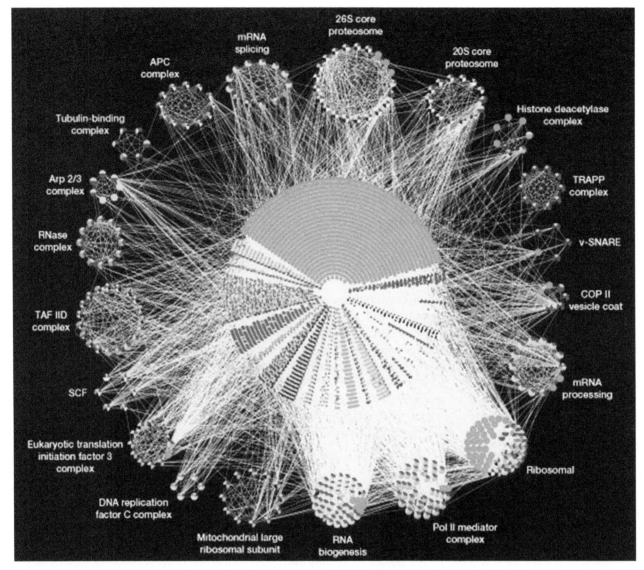

Figure 9.1 **Visualisation of combined datasets of protein–protein interaction in yeast.** A total of 14,000 physical interactions obtained from the **GRID database** were represented with the **Osprey network visualisation** system. Each edge in the graph represents an interaction between nodes, which are coloured according to **gene ontology (GO)** functional annotation. Highly connected complexes within the data set, shown at the perimeter of the central mass, are built from nodes that share at least three interactions within other complex members. The complete graph contains 4543 nodes of 6000 proteins encoded by the **yeast genome**, 12,843 interactions and an average connectivity of 2.82 per node. The 20 highly connected complexes contain 340 genes, 1835 connections and an average connectivity of 5.39 (Illustration from Tyers and Mann, 2003, Figure 2, Springer Nature).

9.2 Key principles in mass spectrometry

Mass spectrometry involves the ionisation of individual molecules in the gas/vapour phase, followed by their differentiation and detection according to **mass/charge (m/z) ratio**. Typically, z is unity so the ratio is equivalent to mass. Accordingly, mass spectrometry was used primarily to determine molecular weight precisely from the identification of the **molecular ion ($[M]^+$)**. Fortuitously, ions in gas phase are able to undergo decomposition reactions (spontaneous) leading to **fragment ions ($[M\text{-}x]^+$)** that could be used to identify elements of structure. In instrumental terms, a mass spectrometer consists of three basic components: an **ionisation** or **ion source**, a mass analyser and the detector (Figure 9.2). The molecules must first be passed into the gas phase as ionic species by the ion source so that their flight may be manipulated. The flight takes place under high vacuum so that the ions are able to avoid colliding or interacting with other species. A **mass analyser** separates the ions for detection according to their m/z ratio. The

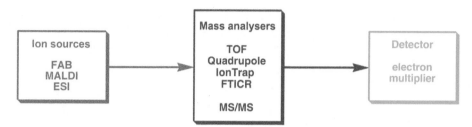

Figure 9.2   The basic components of a mass spectrometer. All mass spectrometers consist of an **ion source** linked to a **mass analyser**, then to an *m/z* **detector**. The important ion sources and mass analysers for biological mass spectrometry are listed. There are many other potential ion sources and mass analysers used generally in mass spectrometry, but only those indicated are of use in the analysis of biological macromolecules and amphiphilic lipids, and also in proteomics: **FAB: fast atom bombardment; MALDI: matrix-assisted laser desorption and ionisation; ESI: electrospray ionisation; TOF: time of flight; FTICR: Fourier transform ion cyclotron resonance; MS/MS: tandem mass spectrometry.**

detector measures both the abundance (a function of the signal) and *m/z* (a function of physical property and/or time) of detected ions.

Mass spectrometry has a long history in chemistry, using classic methods of ionisation such as **electron** or **chemical ionisation** to analyse the molecular weights of small molecules although neither technique was actually applicable for the mass analysis of biological macromolecules or lipid amphiphiles. However, the development of **fast atom bombardment (FAB) mass spectrometry** in the early 1980s brought the mass analysis of oligo- and polypeptides into the sphere of mass spectrometry. Thereafter, the development of **matrix assisted laser desorption ionisation (MALDI)** and **electrospray ionisation (ESI) mass spectrometry**, set up extensive possibilities for the mass analysis of nearly all biological macromolecules and lipid amphiphiles. Hence, the focus of the chemical biology reader is naturally drawn to these three main soft ionisation techniques and how these are then linked with appropriate mass analyser and detector units to derive mass spectrometers appropriate for biological macromolecule and amphiphilic lipid characterisation.

9.2.1 Ionisation sources

9.2.1.1 Traditional techniques of ionisation

Electron ionisation (EI) is the most widespread technique of ionisation used for mass spectrometry in synthetic chemistry. EI consists of two steps: first, a sample of **analyte molecules** of interest is passed into the **vapour phase**; then second, bombarded with a stream of electrons. Should the energy of the colliding electrons exceed the ionisation energy of a given analyte molecule then an electron will be displaced, resulting in a positively charged **molecular ion** $[M]^+$ (radical cation). Since the mass of an electron is negligible, then the observed mass of $[M]^+$ must correlate with the molecular mass. Excess collisional energy encourages spontaneous **fragmentation** and **fragment ion** $[M\text{-}x]^+$ formation, from which information the structure of the analyte molecule of interest may be deduced at least in part. EI remains a useful technique for the analysis of small molecules (<1000 Da) but has some significant drawbacks. First, fragmentation is often so widespread that the molecular ion cannot be observed. Second, neither thermally labile compounds nor non-volatile compounds can be analysed by this ionisation technique that is too high energy (i.e. **hard**) for mass analysis of either biological macromolecules or lipid amphiphiles.

Chemical ionisation (CI) was developed as a technique to complement EI wherein the ionisation method was not quite so hard. In this technique a **reagent gas** at a partial pressure significantly greater than that of a sample of analyte molecules of interest in the vapour phase, is ionised with a beam of electrons (Figure 9.3). The ionised reagent gas then combines with analyte molecules of interest to form stable **molecular ion adducts**: the *m/z* ratio may be measured and the adduct mass accounted for during analysis. A number of different reagent gases may be used for this purpose such as hydrogen, methane, water, methanol, ethanol or ammonia. The extent of fragmentation can be controlled by the choice of reagent gas to a certain degree. Although CI has some advantages over EI in that the fragmentation is more controlled (although EI would hold the advantage of being able to provide more detailed structural information), this ionisation technique remains a hard technique that has similar drawbacks to EI regarding non-volatile compounds, so is only suitable for the analysis of compounds <1000 Da.

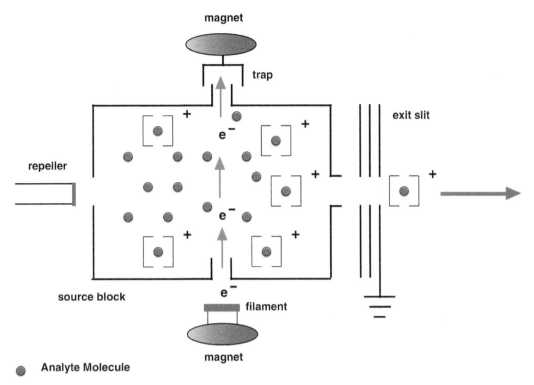

Figure 9.3   Illustration of chemical ionisation for mass spectrometry. A **reagent gas** is present in the source block and is initially ionised by a beam of electrons. Ionised gas then interacts with sample **analyte molecules** also in the vapour phase to form stable **molecular ion adducts** that are repelled into the mass analyser. The mass of the reagent gas must be corrected for during mass analysis. The combined process is known as **chemical ionisation (CI)**.

9.2.1.2 Desorption ionisation techniques: FAB and MALDI

Virtually all desorption ionisation methods share the same approach to ionisation: a sample of analyte molecules of interest is combined with/dissolved in a matrix before being bombarded with high-energy particles or light, resulting in the production of a high concentration of solvated gas-phase neutral or ionic species just above the point of impact. After the impact of the beam, vaporisation occurs before thermal decomposition so that ions are able to "escape" the impact region before fragmentation begins. The original desorption ionisation techniques included **field desorption** (**FD**) and **plasma desorption**. In FD, ions were desorbed directly into the vapour phase by means of an extremely strong electric field. The field lowers the barrier to removal of an electron from the molecule by **quantum-mechanical tunneling**, leaving a radical cation molecular ion $[M]^+$. Although FD spectrometry was successful for the mass analysis of non-volatile compounds, the technique was not particularly popular due to difficulties experienced in making reproducible emitter electrodes to enable ionisation. The need for softer, more versatile ionisation techniques was obvious.

The first of these versatile soft ionisation techniques to be developed was the technique of FAB that is also a desorption technique. FAB makes use of a fast stream of argon or xenon atoms to strike a sample of analyte molecules of interest dissolved in a **liquid matrix** (typically **glycerol**) (Figure 9.4). This fast stream of inert gas atoms is created by electron bombardment of inert gas atoms to generate ions that are accelerated through a potential difference. The fast-moving ions are then converted into atoms by a cloud of excess neutral gas atoms that neutralise these ions by an electron transfer process (residual ions are deflected by a potential field before they are able to strike the matrix). Impact of the fast-moving large atoms with the matrix causes a number of ions to be sputtered from the sample, including the molecular ions $[M + H]^+$ or $[M-H]^-$. If FAB is run in the **positive ion mode**, then the protonated molecular ion and cationic fragment ions are picked up and mass analysed. If the **negative ion mode** is used, then only the deprotonated molecular ion and anionic fragment ions will be mass analysed. In the positive ion mode, **Na$^+$** and **K$^+$** molecular ion adducts (i.e. $[M + Na]^+$ or $[M + K]^+$) are also frequently seen.

Irrespective of mode, fragmentation is also observed that is usually well balanced with respect to molecular ion intensity, so FAB mass spectrometry can also be used to gain insight into the structure of analyte molecules of interest. One of the main reasons for this is that FAB mass spectra can be acquired over timescales of minutes, during which time molecular ion (or molecular ion adduct) and fragment ion intensities can be sustained and even increase. The reason for this is the nature

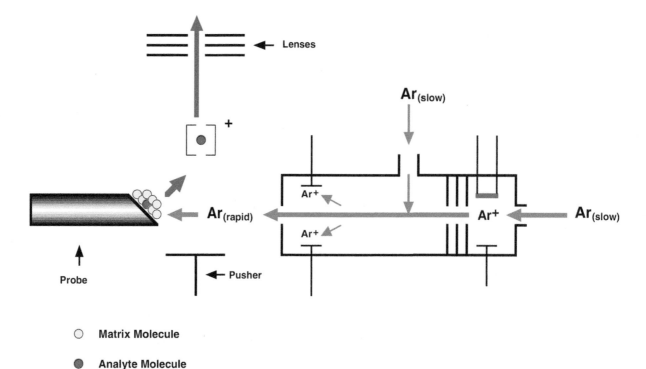

Figure 9.4   Illustration of fast atom bombardment for mass spectrometry. A stream of inert atoms (argon or xenon) strike a probe target on which analyte molecules of interest are dissolved in a **liquid matrix** (frequently **glycerol**). Electron bombardment of a stream of inert atoms (**argon [Ar]** atoms in this case) leads to the formation of ions that are then accelerated by a potential difference. Following electron transfer from other inert atoms, a stream of fast atoms is created that acts by **fast atom bombardment** (**FAB**) to sputter molecular and fragment ion species from the liquid matrix into the vapour phase for mass analysis.

of the matrix. The matrix plays an important role in FAB mass spectrometry, absorbing the excess energy of the atomic bombardment, and replenishing the matrix surface with analyte molecules of interest during the collision process. Viscous, nonvolatile liquids such as glycerol are favoured FAB mass spectral matrices, but occasionally other matrices are useful. FAB mass spectrometry has proved very useful for the mass analysis of amphiphilic lipids and smaller biological macromolecules (or fragments thereof) <4000 Da, comprising short polysaccharides, short polypeptides and oligodeoxy-/oligonucleotides.

MALDI represents an important technical development on FAB. In MALDI, a sample of analyte molecules of interest is dispersed in a **crystalline** not liquid **matrix** and the beam of fast atoms is replaced by laser irradiation. The analyte is pre-mixed with a matrix material and crystals are then placed on a solid surface (Figure 9.5). Nanosecond laser pulses sublime the matrix, together with analyte molecules of interest. Thereafter, the matrix absorbs most of the laser energy, preventing excessive breakdown of the analyte during sublimation and subsequent ionisation, so ensuring a significant molecular ion population in the vapour phase. The crystalline matrix plays two other important supporting roles in MALDI. First, the matrix actually solvates analyte molecules, thereby reducing aggregation and promoting molecular ion formation in the vapour phase. Second, matrices have been identified with specific characteristics for specific applications (Figure 9.6).

The most commonly used matrix for the mass analysis of oligo-/polypeptides and for peptide fingerprinting is **α-cyano-4-hydroxycinnamic acid (CHCA)**. **Sinapinic acid (SA)** is often used as the matrix of choice for the mass analysis of proteins, **hydroxypicolinic acid (HPA)** for oligonucleotides and oligodeoxynucleotides, **di-2,5-hydroxybenzoic acid (DHB)** for oligosaccharides and **2-hydroxy-5-methoxybenzoic acid (HMB)** for lipids. MALDI is sufficiently soft and versatile as an ionisation technique to allow for the observation of the molecular ions of even very large biological macromolecules, approaching 10^6 Da, when using an appropriate version of MALDI mass spectrometry.

9.2.1.3 Spray ionisation techniques: thermospray and electrospray

Spray ionisation techniques solve many of the problems of how to pass liquid solutions into the gas phase. As these techniques all take samples from the liquid phase, they are the ones most easily coupled to a liquid chromatography system. Unlike desorption techniques described above, these techniques take place at ambient pressure rather than

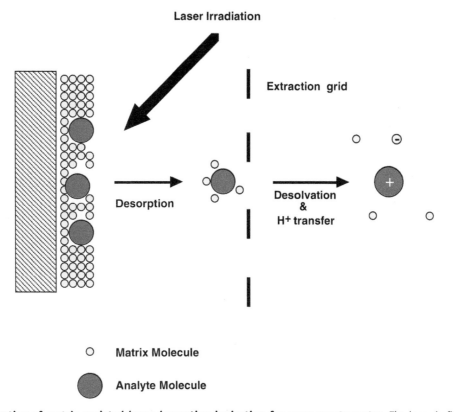

Figure 9.5 **Illustration of matrix assisted laser desorption ionisation for mass spectrometry.** The laser is fired at a sample of analyte molecules of interest admixed with a **crystalline matrix** that readily absorbs laser energy and allows the sputtering of analyte molecules into the vapour phase in association with matrix molecules of solvation. Desolvation and proton transfer leads to the formation of naked **single-charged molecular ion species**, ready for mass analysis. The combined process is known as **matrix assisted laser desorption ionisation (MALDI).**

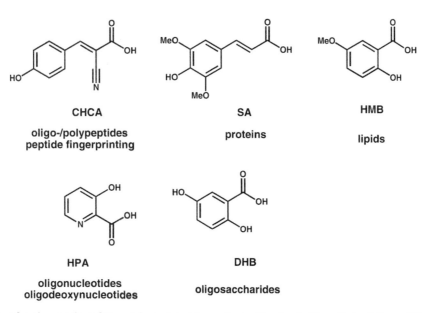

Figure 9.6 **Summary of main matrices for matrix assisted laser desorption ionisation.** Each of these different matrices are used typically for optimal mass analysis of certain types of biological analytes only, and not for others.

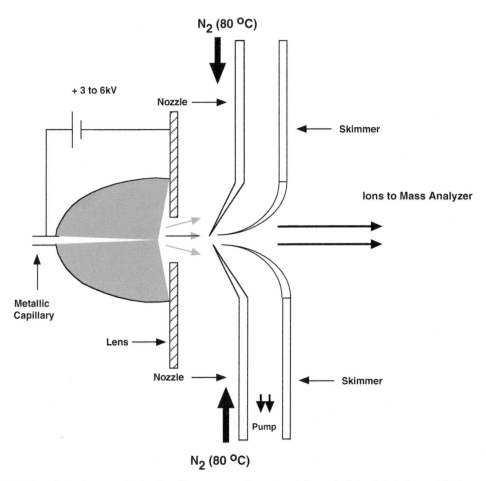

Figure 9.7   Illustration of electrospray ionisation for mass spectrometry. A "spray" of droplets is formed that evaporates until the destabilising electrostatic forces cause the droplets to "explode", releasing **multi-charged molecular ion species** for mass analysis. The combined process is known as **electrospray ionisation (ESI)**.

in a vacuum. One of the first spray ionisation techniques to be reported was **thermospray ionisation (TI)**. This technique is best suited to volatile compounds <1000 Da, so has no role for biological macromolecule of amphiphilic lipid mass analysis, but the technique is illustrative. TI involves passing a solution containing analyte molecules of interest through a heated capillary, the rapid expansion and cooling of the liquid results in the ejection, from the end of a capillary, of a superheated mist containing a mixture of tiny droplets together with analyte molecules in the vapour phase. Sprayed droplets tend to become charged, allowing molecular ions to develop for onward mass analysis. Thermospray is a useful ionisation technique for the analysis of small oligopeptides, dinucleotides and dideoxynucleotides, small oligosaccharides and other small organic molecules.

ESI is the large molecule upgrade of TI. ESI was developed as an elaboration of the **atmospheric pressure chemical ionisation (APCI)** technique. In the case of ESI, a solution containing analyte molecules of interest is passed through a capillary to which is applied a potential difference of 3–4 kV, producing a significant electrostatic field around the tip of the capillary that aids in molecular ion generation (Figure 9.7). As the solution leaves the end of the capillary, the electrostatic field enables the generation of a "spray" of droplets. Two forces then determine the stability of a given droplet. First, droplet surface tension acts to maintain droplet integrity. Second, electrostatic forces in analyte molecules act to disintegrate the droplet. Hence, as droplets evaporate, destructive electrostatic forces increase until droplets "explode", releasing molecular ions and fragment ions for mass analysis. Unusually in the case of the ESI technique, analyte molecular ions may be multiply charged. For instance, biological macromolecules such as proteins frequently produce molecular ion series of the form $[M + zH]^{z+}$ or $[M\text{-}zH]^{z\text{-}}$. As with FAB so with ESI, both positive and negative ion modes are available using this technique, therefore the cationic molecular ion series can be observed in the positive ion mode and the anionic molecular ion series in the negative ion mode respectively. A main advantage of multiple ion series is that while actual values of m/z are within range of the mass analysis, true molecular ion weights can be orders of magnitude greater. Hence, ESI opens up the possibility for mass analysis of very large biological macromolecules indeed, of at least 10^6 Da in molecular weight.

9.2.2 Mass analysers in mass spectrometry

Soft ionisation devices for FAB, MALDI and ESI are the front end of the modern mass spectrometer used by the chemical biology researcher. Clearly these devices need to be associated with mass analysers for the resolution and subsequent detection of molecular and fragment ions. The traditional EI mass spectrometer relied upon the use of different combinations of electric and magnetic sectors to perturb ion motion *in vacuo* through arc trajectories that were a function of ion mass, velocity and net charge. Such mass analysers were large and unwieldy, making mass analysis instrumentation highly specialised.

Nowadays, the provision of soft ionisation devices has opened the door for much less complex mass analysers that considerably reduce instrument specialisation in favour of more complex experiments. There are four major types of mass analysers of most importance to the chemical biology reader. These are the **time of flight** (**TOF**), **quadrupole**, **ion trap** and **Fourier transform ion cyclotron resonance** (**FTICR**) analysers. These mass analyser devices are linked to a soft ionisation device either singly or in tandem in certain optimal combinations as described in the following sections.

9.2.2.1 Time of flight (TOF) mass analysers

TOF mass analysis is based on the simple principle that molecular and fragment ions are separated as a function of ion velocities. Ions are all accelerated with the same electrostatic potential and hence acquire the same total kinetic energy. Accordingly, after acceleration to a constant kinetic energy (equivalent to zV, where V is the **accelerating electrostatic potential**), **ions travel at a velocity**, ν_z, that is related to m/z according to Equation (9.1):

$$\nu_z = \sqrt{\frac{2zV}{m}} \qquad (9.1)$$

Therefore, ions are accelerated to velocities proportional to the inverse square roots of their m/z values. Post acceleration, ions are allowed to travel in a **field-free flight tube** of **length** L_z (typically over 1 m in length), before striking the detector. Of course, lighter ions have a greater velocity and reach the detector in less time than heavier ions. The **time to detector**, t_z, is quantified according to Equation (9.2):

$$t_z = \frac{L_z}{\nu_z} = L_z \cdot \sqrt{\frac{m}{2zV}} \qquad (9.2)$$

illustrating that the t_z value of a given ion is proportional to the square root of the corresponding m/z value. Obviously, optimal resolution in ion detection requires that all accelerated ions are coordinated to enter the field free flight tube at the same time. For this reason, **TOF mass analysers** are best matched with MALDI ionisation devices, although it

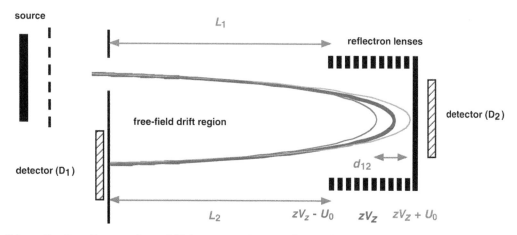

Figure 9.8 **Schematic of a reflectron time of flight mass analyser.** Reflectron lenses act as an electrostatic mirror to both increase the effective length of the flight path, but also to compensate for ion kinetic energy variations, resulting in higher mass accuracy relative to purely linear **TOF mass analysers**. Consequently, linear TOF mass analysers will become increasingly obsolete.

is possible to couple continuous ionisation techniques such as ESI to TOF if care is taken in the design of the way ions are delivered into the flight tube.

Many of the first TOF mass analysers consisted of linear flight tubes, but space can now be saved with a **reflectron** device that uses multiple ion reflection to increase the effective value of L_z and hence the resolution of mass analysis (Figure 9.8). A reflectron is in effect an electrostatic mirror consisting of a series of electrical lenses, which not only improves resolution by increasing the effective value of L_z, but also does so by equalising out small variations in ion kinetic energy, should they exist, due to coupling between laser energy and ion kinetic energies (Figure 9.8). Where a reflectron device is used, then Equation (9.2) should be modified to Equation (9.3):

$$t_z = (L_1 + L_2 + d_{12}) \cdot \sqrt{\frac{m}{2zV}} \qquad (9.3)$$

where the terms L_1, L_2 and d_{12} correspond to the illustrated **distance elements in the reflectron** device (Figure 9.8). Please note that another method for improving MALDI-TOF resolution is by using **delayed extraction**, where the electrical charge for acceleration is applied shortly after the laser shot onto the solid surface for MALDI.

9.2.2.2 Quadrupole mass analysers

Quadrupole mass analysers are said to work on a filtration principle. They consist of four cylindrical rods organised in an orthogonal array (Figure 9.9). A quadrupole field is created by applying an oscillating current at a set frequency to one opposing pair of rods, and repeating the same to the other opposing pair of rods, ensuring that the current oscillation is 180° (π) out of phase with respect to the current experienced by the first opposing pair of rods. In so doing, an oscillating electric field is generated between the rods. By convention the field z-axis is parallel to the rods and located equidistant from each rod of the orthogonal array such that the **field potential** $\Phi_{x,y}$ at any point between the rods in any arbitrary (x, y) plane perpendicular to the rods is described according to Equation (9.4):

$$\Phi_{x,y} = \Phi_o \frac{x^2 - y^2}{r_o^2} \qquad (9.4)$$

where x and y are distances from a given z-axis in the arbitrary (x, y) plane, and r_o is the **field radius between the rods** (i.e. the perpendicular distance from the z-axis to the centre-line of each rod of the orthogonal array). The term Φ_o corresponds to the **maximum possible field potential**. Equation (9.4) indicates that the field potential will decline to zero if x and y are either zero or otherwise equivalent in magnitude. Hence if an ion were able to travel from the

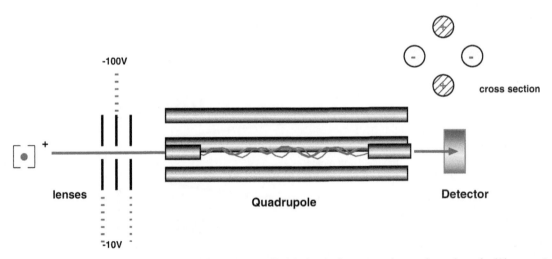

Figure 9.9 Schematic of a quadrupole mass analyser. Four cylindrical rods form an orthogonal **quadrupole (Q)** array. One of the rods has been cut away in the diagram to illustrate the complex trajectories of analyte ions trapped between the electrodes. For a given potential cycling between opposing rods, a molecular ion with a certain weight, *m/z*, will be enabled to follow a **zero field trajectory** to the detector. Changes in potential, change the *m/z* values of detectable analyte ions.

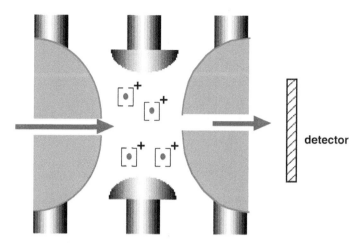

Figure 9.10 **Schematic of an ion trap mass analyser.** Ion traps work on a similar principle to quadrupole analysers, but here the trap acts as a 3D capture box that releases ions when the potential and **m/z** are such that the analyte ion has a field-free path to the detector.

ionisation device and follow a trajectory of zero field potential along or parallel to the z-axis of the orthogonal array in the mass analyser, then the ion would be able to reach a detector at the other end. Hence, the link between Equation (9.4) and **m/z** can be described in the following terms. For each ion type with a given value of **m/z** there becomes a corresponding value of Φ_0 when interactions between ion type and external quadrupole field are such as to enable this ion type to follow a **zero field trajectory** to the detector. Increasing values of Φ_0 are obtained using increasing current amplitudes. Therefore, a full mass spectrum may be obtained by sweeping the current amplitude (frequency is kept constant) of the oscillating applied current that creates the quadrupolar field between the rods. In so doing, different values of Φ_0 are generated, enabling ions of different **m/z** to traverse the quadrupole on a stable trajectory in order to reach the detector at the other end. The filtration principle is now clear. Only ions with a value of **m/z** appropriate for the given value of Φ_0 are able to reach the detector, otherwise all other ions are "filtered out".

9.2.2.3 Ion trap mass analysers

Ion trap analysers use a similar principle to quadrupole mass analysers, but employ a system of entrance, exit and endcap electrodes together with a ring electrode that surrounds the **trap cavity** (Figure 9.10). As with quadrupole so with ion trap, for each ion type with a given value of **m/z** there becomes a corresponding value of Φ_0 when interactions between ion type and external quadrupole field are such as to enable the trapping of ion type of interest within the analyser prior to release for detection. Ion traps are relatively inexpensive, quite sensitive and robust, so therefore are fairly widespread, despite being less accurate than TOF and quadrupole mass analysers.

9.2.2.4 Fourier transform ion cyclotron resonance and orbitrap mass analysers

FTICR represents an alternative type of trap mass analyser that requires superconducting magnets and high vacuum (Figure 9.11). Ions are constrained spatially by both electric and magnetic fields, and move in **circular orbits**. The technique uses a three-step process. First, following release from a given ion source, ions are trapped within a small potential in a 3D cell. Second, an excitation pulse is applied so that ions with a precessional frequency matching the excitation pulse absorb the excitation pulse's energy. Third, ions are detected by measuring image current induced when ions move in close proximity to receiving plates. The method displays excellent sensitivity and mass accuracy, but these mass analysers are expensive, and are better suited to molecular than fragment ion analysis. The **orbitrap** mass analyser is the latest analyser, with origins in FTICR, which is now frequently used in biological mass spectrometry. Orbitraps remove many of the limitations of FTICR; there is no need for a superconducting magnet, and there can be a wide dynamic range of detection. Orbitraps consist of outer electrodes in the shape of cups facing each other, comprising a spindle-like central electrode. When a voltage is applied between outer and central electrodes then a linear electric field is set up such that oscillations are purely harmonic. Ions may then be injected into the space between the electrodes so that when voltage parameters are applied correctly the ions enter into orbit within the trap.

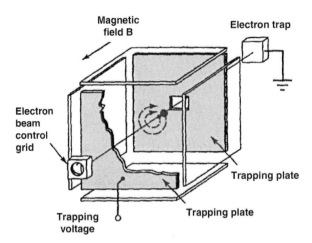

Figure 9.11 **Schematic of a Fourier transform ion cyclotron resonance mass analyser.** Ions are constrained in circular orbits by electric and magnet fields generated by superconducting magnet before selective detection. These **Fourier transform ion cyclotron resonance (FTICR)** mass analysers have the highest sensitivity and accuracy of any mass analysers presently available (illustration from Dass, 2001, Wiley).

Outer electrodes are used as receiver plates for image current detection, which can be processed by **Fourier transform** into mass spectra.

9.2.2.5 Tandem mass analysers

Tandem mass spectrometry (MS/MS), results from the coupling together of two consecutive mass analysers in a mass spectrometer so as to obtain further information regarding the sample. Typically, there are three stages to tandem mass analysis:

1. **Mass selection** of a user-defined **parent** or **precursor ion**

2. **Fragmentation** of the parent ion to form **daughter** or **product ion**, followed by

3. **Mass analysis** of the product ions

Fragmentation is typically achieved through the impact of neutral gas atoms, such as by using a stream of argon or helium for **collision induced dissociation (CID)**. The **triple-quadrupole (TQ)** is one of the most popular tandem mass analysers. In this instance, the mass analyser consists of three sectors. The first and last sectors are normal quadrupole mass analysers, but the central sector constitutes a **collision cell** where CID can take place. The first sector is normally chosen to transmit a particular parent ion (or precursor ion) (such as a molecular or fragment ion of interest), which then enters the collision cell and undergoes controlled fragmentation into daughter ions (or product ions) that are subsequently scanned or mass analysed in the third sector (a **product ion scan**). Alternatively, a specific daughter fragment ion (or product ion) of interest can be targeted in the third sector, and the first sector used to scan the parent ions (or precursor ions) responsible for the selected daughter ion targeted in the third sector (a **precursor ion scan**).

Post-source decay (PSD) is another mass analyser technique that is incorporated into some reflectron TOF instruments. **Metastable dissociation** refers to the dissociation of ions during flight in a time of flight mass analyser. In a linear TOF mass analyser, metastable dissociation would make no difference to the time of arrival of the molecular or fragment ions at the detector. However, with **PSD mass analysis**, a parent ion can be selected by monitoring the linear detector. Then the reflectron is switched off just as the parent ion is passing through. A second detector then records the daughter fragment ions from metastable dissociation of the selected parent ion. Typically, the reflectron voltage is stepped so that a wide range of daughter fragment ions may be observed from as many parent ions as possible.

As the chemical biology reader might expect in this ever growing world of mass spectrometry, more orthogonal tandem mass analyser combinations are now becoming more and more widespread as well, including hybrid

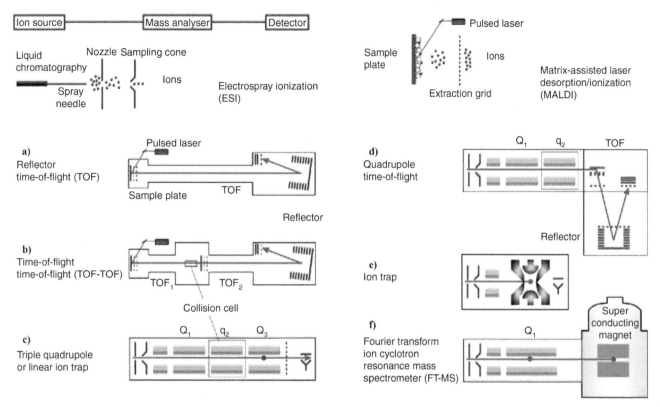

Figure 9.12 **Summary of some major use combinations of ion sources and mass analysers in use in biological mass spectrometry today.** (a) **MALDI-TOF**; (b) **MALDI-TOF/TOF**; (c) **ESI-TQ**; (d) **ESI-Q-TOF**; (e) **ESI-ion trap**; (f) **ESI-Q-FTICR**. The combination between MALDI ion source and TOF is becoming standard, as is the combination of ESI ion source and Q mass analyser. The combinations above also split into **MS systems** (a, e, f), and **MS/MS systems** (b, c, d). MS/MS systems are themselves becoming more popular due to **product ion scanning** (illustration after Aebersold and Mann, 2003, Figure 2, Springer Nature).

quadrupole-TOF (**Q-TOF**) instruments. Using this device combination, parent ions are easily selected using the quadrupole filter, fragmented in a collision cell, after which a TOF mass analyser may be used for accurate resolution of daughter ions prior to detection. The configuration of some major use combinations of ion sources, mass analysers and detectors are illustrated (Figure 9.12).

9.3 Structural analysis of biological macromolecules and lipids by mass spectrometry

The mass analysis of biological macromolecules and amphiphilic lipids can be very sophisticated given the range of possible mass spectrometers available, including the range of ionisation techniques. Clearly the data from the mass analyser depends heavily on the ionisation method used. ESI can give a matrix of multiply charged ions, particularly when molecular ions are very heavy, whereas MALDI will typically result in single molecular ions. Fragment ions should be seen in both cases, but tandem MS/MS mass analysis will be more helpful if fragment ion analysis is required.

9.3.1 Analysis of individual peptides by mass spectrometry

With respect to oligo-/polypeptides, the parent molecular ion may be identified by most FAB, MALDI and ESI mass spectrometry. Besides detection of the molecular ion, a great deal of primary structure information can be gained from studying the mass spectra of an individual oligo-/polypeptide. Oligo-/polypeptides have standard fragmentation patterns in the positive ion mode based upon charge separation in individual peptide links and amino acid residues.

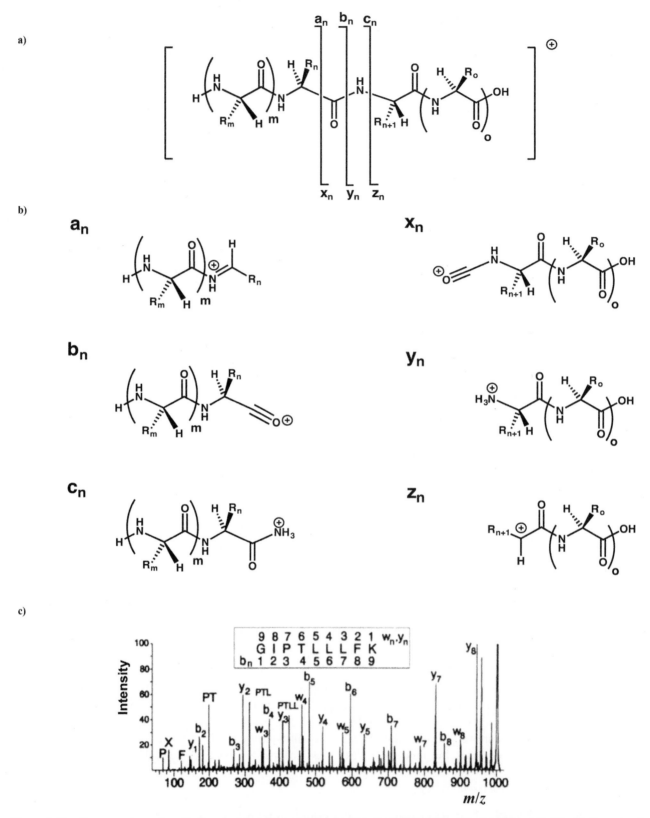

Figure 9.13 Mass spectrometry of oligo-/polypeptides. (a) Origin of main fragmentation ions as illustrated (**b**); (**c**) Example of FAB MS/MS mass spectrum of indicated oligopeptide. The main fragment ions visible are the **b**ₙ series (*N*-terminal fragment ions) and the **y**ₙ series (*C*-terminal fragment ions). Mass differences between consecutive **b**ₙ series fragment ions define oligopeptide primary structure (*N→C* direction); mass differences between consecutive **y**ₙ **series fragment ions** define oligopeptide primary structure (*C→N* direction) (illustration (c) from Biemann, 1988, Wiley).

These fragmentations give rise to a_n, b_n and c_n **series fragment ions** corresponding to the situation where fragmentation takes place around the peptide link of n-th and ($n + 1$)-th amino acid residues, starting from the *N*-terminus, so creating defined **N-terminal fragment ions**; then there are the x_n, y_n, z_n **series fragment ions** where fragmentation takes place around the peptide link of n-th and ($n + 1$)-th amino acid residues, starting from the *C*-terminus, hence creating defined **C-terminal fragment ions** (Figure 9.13). These fragment ions are readily seen in sequence in FAB mass spectra of oligo-/polypeptides (particularly b_n and y_n fragment ions) allowing the oligo-/polypeptide amino acid residue sequences to be determined from the mass differences between consecutive, sequential fragment ion peaks. Typically, the molecular ion $[M + H]^+$ is most abundant, followed by fragment ions adequate to identify 5–6 amino acid residues in sequence, moving in either the *C*- or *N*-terminal direction from the initial point of fragmentation according to the fragment ion series followed. In this way, a sequence length of 10–12 amino acid residues can be identified in total. This approach has proved excellent for sequence mapping small signal peptides found in minute quantities in mammalian brains or oligopeptide toxins excreted in similarly small quantities from sources such as *Xenopus* (frog) skin.

Polypeptide sequencing becomes possible using fragment ions and the technique of **peptide ladder sequencing**. According to this technique, a parent molecular ion is identified and then the *N*- or *C*-terminal amino acid residue are removed before the new parent molecular ion is found. The mass difference between original and new parent molecular ion identify the original *N*- or *C*-terminal amino acid residue. This process can be continued for as long as desired to build up a polypeptide sequence. *N*-terminal residues are removed sequentially either by **aminopeptidase** enzymes, specific for peptide link hydrolysis at the *N*-terminus of polypeptides, or vapour phase acid hydrolysis. *C*-terminal residues are removed by carboxypeptidase enzymes such **carboxypeptidase Y** that is specific for peptide link hydrolysis at the *C*-terminus of polypeptides. Such peptide ladder sequencing is laborious and has been largely superseded by mass spectrometry with a tandem MS/MS mass analyser system making use of product ion scanning. In this instance, a single tandem MS/MS mass analysis of a pure polypeptide of interest can be sufficient to reveal the entire b_n and y_n fragment ion series. In addition, tandem mass spectrometers are often coupled with software to ease the time-consuming procedure of assigning observed peaks to expected b_n and y_n daughter ions.

9.3.2 Analysis of proteins by mass spectrometry

Proteins may be analysed using two main approaches by mass spectrometry: the molecular mass may be measured for the whole protein and/or the protein may be broken into smaller peptides that may be individually analysed, providing insight into the structure of the parent protein.

9.3.2.1 Protein molecular weight determination

Using either ESI or MALDI mass spectrometry the molecular weight determination of the protein may be determined to an accuracy of ±0.01% (Figure 9.14). This information can provide initial clues as to any deviation from the expected sequence, or potential types of modification to the protein post translation. Furthermore, the softness of MALDI or ESI ionisation ensures that even non-covalent complexes between a protein and cognate ligand or substrate

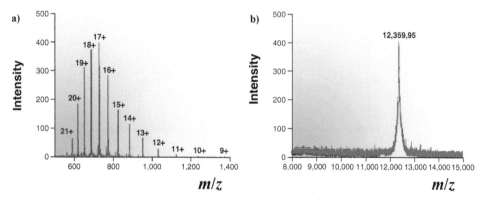

Figure 9.14 **Comparison of protein mass analysis data from mass spectrometry.** (**a**) ESI produces a family of multi-charged molecular ion species, whereas (**b**) MALDI typically generates single-charged molecular ion species (illustration from Glish and Vachet, 2003, Figure 2, Springer Nature).

can be observed. Post-translational modifications frequently include glycosylation (covalent addition of heteroglycan oligo-/polysaccharides to amino or hydroxyl functional groups) and phosphorylation of serine, threonine and/or tyrosine amino acid residues.

9.3.2.2 Gel-based isolation and digestion of a protein for mass spectrometry

The real power in using mass spectrometry is analysis of minute amounts (pmol) of proteins isolated from biological sources. In this case the stage is to digest the protein "in gel" with an **endopeptidase**, such as **trypsin**, into oligo-/polypeptide fragments that can be individually isolated and sequenced by ESI or MALDI tandem mass spectrometry. Sequence mapping of all of these oligo-/polypeptide fragments allows for at least partial reconstruction of the entire protein amino acid residue sequence. Complete sequences may then be determined by repeating the process again with another endopeptidase such as **chymotrypsin** to derive another endopeptidase sequence map. By comparing both endopeptidase sequence maps all the oligo-/polypeptide fragments should then interlock together in order to define a unique protein amino acid residue sequence.

Gel-based isolation of a protein followed by digestion with trypsin and identification of the protein from tryptic fragment analysis is now a standard approach in proteomics (Figure 9.15). Gel separated proteins (**1D** and **2D gel electrophoresis**) are initially stained by modified **silver staining methods** that omit **glutaraldehyde**, and employ a rapid **destaining** that makes use of **iron ferricyanide**, together with **sodium thiosulfate**. After staining, intact protein bands are cut from the gel, destained (an optional step involves cysteine blocking with **iodoacetamide**), then treated with a peptidase/proteinase solution to fragment the protein into resolvable oligo-/polypeptides with distinct

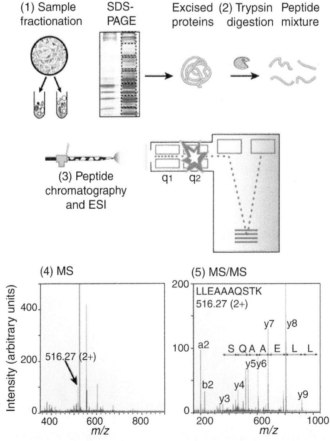

Figure 9.15 Typical mass spectrometric experiment for protein identification/characterisation. (1) Proteins are fractionated by chromatography, separated by **1D polyacrylamide gel electrophoresis (1D PAGE)**, then excised from gel. **(2)** The protein of interest is digested into peptide fragments, that are then **(3)** identified by **ESI-Q-TOF-MS/MS**. **(4)** The first dimension involves molecular ion analysis for **peptide mass finger-printing** (q1, MS only). **(5)** Tandem MS/MS is used when protein identification is not unambiguous, in which case **parent molecular ions** are activated by **collision induced dissociation (CID)** (q2) and **daughter (product) ions** are characterised (TOF) according to the technique of **product ion scanning** (illustration from Aebersold and Mann, 2003, Springer Nature).

m/z values that can be mass analysed. The range of proteinases is as follows. First and foremost, as noted above, there is trypsin that catalyses peptide link hydrolysis on the *C*-terminal side of arginine and lysine residues. Then there is chymotrypsin that catalyses peptide link hydrolysis on the *C*-terminal side of tyrosine, phenylalanine and trypto-phan residues, with occasional cleavage at lysine residues. Alternatives include: **V8 protease** (also known as **glutamyl endoproteinase I** or **Glu-C**) that cleaves peptide links on the *C*-terminal side of glutamic and aspartic acid residues; **endopeptidase Lys-C** that cleaves on the *C*-terminal side of lysine residues; **endopeptidase Arg-C** (also known as clostripain) that cleaves on the *C*-terminal side of arginine residues and **endopeptidase Asp-N** that cleaves on the *N*-terminal side of aspartic acid residues. Chemicals such as **cyanogen bromide** may also be used to chemically cleave on the carboxylic side of methionine residues, converting the methionine to a homoserine, which can cyclise to a lactone.

9.3.2.3 Peptide mass fingerprinting for protein identification

MALDI-TOF mass spectrometry is frequently used alone to identify proteins by a technique called **peptide mass fin-gerprinting** (also known as peptide mapping or peptide-mass mapping), although similar approaches can be taken using ESI mass spectrometry. A protein digest (prepared as in Section 9.3.2.2) is spotted or combined with the matrix (usually CHCA) onto the MALDI probe. The experimentally observed peptide masses are then compared to a cal-culated list of proteolytic masses generated from gene databases. Peptide mass fingerprinting is most useful when working with proteins from sequenced species to enable this matching to be carried out. Typically, only 20% of a pro-tein sequence needs to be observed in order to unequivocally identify a protein. In fact, rarely more than 60% of any protein amino acid sequence is fitted to the databases; therefore peptide mass fingerprinting is remarkably efficient. Should peptide mass fingerprinting prove inadequate to identify a protein unambiguously, then tandem mass spec-trometry may be required to further narrow the list of candidate proteins.

9.3.2.4 Tandem mass spectrometry for protein identification

Tandem mass spectrometry provides a most definitive approach to protein identification. For tandem mass spectrom-etry, the work up of proteins – typically by gel-based isolation and digestion – is exactly the same as for peptide mass fingerprinting. Also, the "first dimension" of mass spectrometry also gives information practically identical to finger-printing. However, the real power of tandem mass spectrometry is in product ion scanning. The molecular ions of each individual oligo-/polypeptide in a proteinase digest can be individually selected as a parent ion in the first sector of tandem MS/MS mass analyser, allowing the corresponding b_n and y_n daughter ions to be generated, resolved and assigned in the second and third sectors of the mass analyser (Figure 9.15). Fragment ion determination like this rep-resents the "second dimension" of mass spectrometry. The combination of peptide mass fingerprinting and specific sequence analysis should be sufficient to identify the protein unambiguously provided that the protein derives from a sequenced organism. If this is not possible, then complete sequencing becomes necessary using more protein and different endopeptidase enzymes.

9.3.3 Analysis of oligonucleotides by mass spectrometry

With respect to oligo-/oligodeoxynucleotides, the parent molecular ion may be identified by most MALDI and ESI mass spectrometry. In addition, besides detection of the molecular ion, primary structure information can be gained from studying the mass spectra of an individual oligo-/polydeoxynucleotide as with oligo-/polypeptides. With regard to ESI mass spectrometry, negative ion mode tends to give better results than positive ion mode, yielding a sequence of multiply charged molecular ions $[M\text{-}zH]^{z\text{-}}$. Metal cations must be excluded from oligo-/oligodeoxynucleotide samples, since metal cation adducts suppress signal to noise and complicate spectra. By contrast, MALDI mass spectrometry has proved more useful for primary structure analysis. Once again, it is important to take steps to avoid metal cationic adduct formation. The recommended matrix when working with oligo-/oligodeoxynucleotides is **hydroxypicolinic acid** (**HPA**). In a comparable way to oligo-/polypeptides, oligo-/oligodeoxynucleotides have standard fragmentation patterns in the negative ion mode based upon charge separation in individual phosphodiester links. These fragmentations then take place around the phosphodiester link between the *n*-th and (*n* + 1)-th nucleotide/deoxynucleotide residues, so creating defined **5′-series fragment** or **3′-series fragment anions** (Figure 9.16), that start from the 5′-terminus or 3′-terminus respectively. These fragment ions are readily seen in sequence in MALDI mass spectrometry, allowing the oligo-/oligode-oxynucleotide residue sequences to be determined from the mass differences between consecutive, sequential fragment

364   CH 9 MASS SPECTROMETRY AND PROTEOMICS

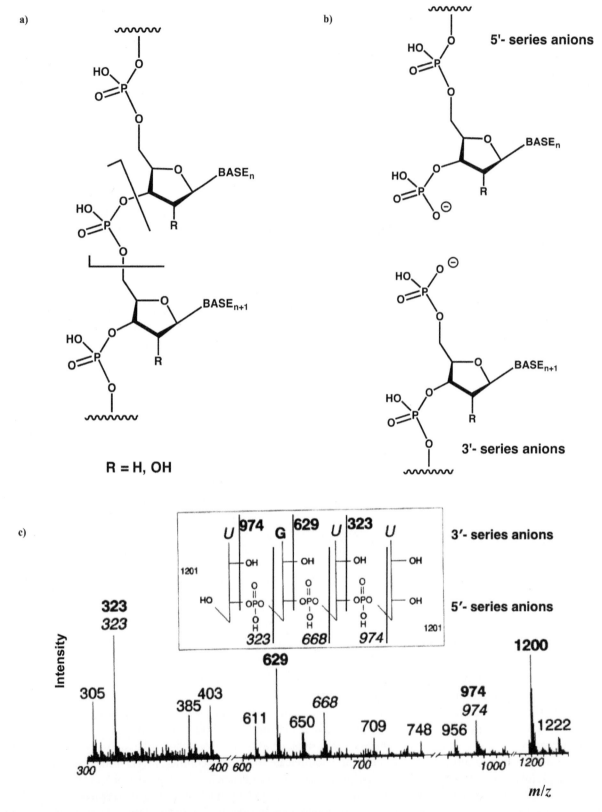

Figure 9.16 Mass spectrometry of oligonucleotides/oligodeoxynucleotides. (a) Origin of main fragmentation ions as illustrated (**b**); (**c**) Example of negative FAB mass spectrum of indicated oligonucleotide. The main fragment ions visible are the **5′-series fragment anions** and the **3′-series fragment anions.** Mass differences between consecutive 5′-series fragment anions define oligo-nucleotide primary structure (5′→3′ direction); mass differences between consecutive 3′-series fragment anions define oligonucleotide primary structure (3′→5′ direction) (illustration (c) from Grotjahn, 1986, Wiley).

anion peaks. Typically, the molecular ion [*M-H*]⁻ is most abundant, followed by fragment anions. Of course, sequence analysis is potentially much enhanced by MALDI tandem mass spectrometry again, in the same way that polypeptide and protein sequencing is also much enhanced using product ion scanning. Alternatively, **ladder sequencing** can be done in an equivalent manner to peptide ladder sequencing (see Section 9.3.1), using exonuclease enzymes to selectively hydrolyse the phosphodiester link nearest the 5′-/3′-terminus. However, this is laborious. In the event, neither product ion scanning nor ladder sequencing compete very well with gel or capillary electrophoresis based ddNTP approaches to sequencing (see Chapter 3).

Although the sensitivity and efficacy of MALDI mass spectrometry is not so high as that of gel-based approaches, mass spectrometry does allow identification of nucleic acid changes more specifically. Indeed, the main application of mass spectrometry of nucleic acids is in the identification of **single-nucleotide polymorphisms (SNPs)**, using mass spectrometry as a platform for genomics studies. Single-nucleotide polymorphisms are single-base variations in the genome that occur at a significant frequency (>1%) in the human population, typically appearing once on average every 1000 bases in the human genome. Typically, biotin-tagged ddNTPs are used for extension, together with primers that anneal immediately next to the polymorphic sites. The terminated products are captured by streptavidin protein binding, and MALDI-TOF mass spectrometry can be used to identify polymorphic variation by the combined mass of the product.

9.3.4 Analysis of carbohydrates and glycoproteins by mass spectrometry

In contrast with nucleic acids, MALDI and ESI mass spectrometry come into their own with oligo-/polysaccharide sequencing particularly where heteroglycans are involved. Oligo-/polysaccharides have standard fragmentation patterns in the positive ion mode based upon charge separation in individual glycosidic links and within monosaccharide (glycosyl) residues. These fragmentations give rise to **A, B** and **C series fragment ions** corresponding to the situation where fragmentation takes place around the glycosidic link of *n*-th and (*n* + 1)-th glycosyl residues, starting from a non-reducing end or terminus, so creating defined **non-reducing end fragment ions**; then there are the **X, Y, Z series fragment ions** where fragmentation takes place around the glycosidic link of *n*-th and (*n* + 1)-th glycosyl residues, starting from the reducing end or terminus, hence creating defined **reducing end fragment ions** (Figure 9.17). As with the oligo-/polypeptide sequencing, **B** and **Y** series fragment ions are important. So, too, are the **A** and **X** series fragment ions that characterise actual fragmentations within the (*n* + 1)-th and *n*-th glycosyl residues, respectively. These are critical to demonstrate the presence and location of multiple oligo-/polysaccharide-chain branching from a single monosaccharide residue. Consequently, careful analysis of fragment ion series by MALDI or ESI mass spectrometry and tandem mass spectrometry become very powerful tools for the primary structure characterisation of complex heteroglycans, even isolated in very small quantities (pmol) from biological sources such as from glycoproteins.

The heteroglycans associated with glycoproteins can generally be characterised in a series of steps. First, the molecular weight of the entire glycoprotein can be measured, typically by ESI mass spectrometry. Peptide maps of the glycoprotein can then be compared before and after treatment with an endoglycosidase enzyme, for instance **peptide-*N*4-(*N*-acetyl-β-glucosaminyl)asparagine amidase** (**PNGase**) that can be used to cleave the bond between asparagines amino acid residues and 2-amino-2-deoxyglucopyranose residues. Otherwise, NaBH₄/NaOH is sufficient to release heteroglycans by reductive cleavage. Comparative peptide maps allow for the identification of glycopeptides, and their precise site of glycosylation within the amino acid chain can be determined by tandem mass spectrometry. Otherwise the heteroglycan(s) can be resolved by chromatography from the glycoprotein and separately sequence characterised as indicated above.

9.3.5 Analysis of lipids by mass spectrometry

FAB, MALDI and ESI mass spectrometry are both suitable to observe molecular ions of amphiphilic lipids. This molecular ion is normally more than enough to determine structure, since molecular ions are unique between biologically important acylglycerols, phospholipids and cholesterol. Should more information be required, lipids are able to fragment about ester bonds, yielding fragment ions that are best visualised in negative ion mode. These include fatty acid fragment ions that may further fragment by carboxyl group dehydration to ω-decomposition to reveal fatty acid alkyl chain length and homologation (Figure 9.18).

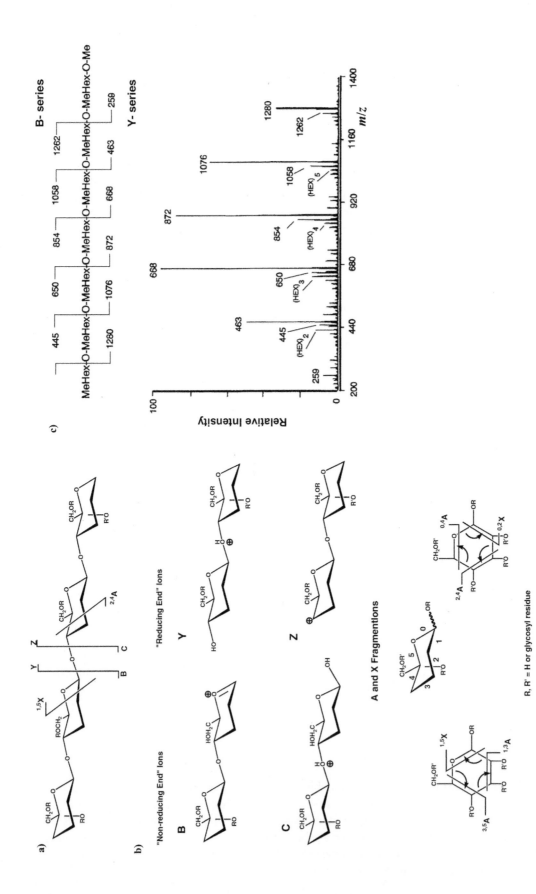

Figure 9.17 Mass spectrometry of oligosaccharides. (a) Origin of main fragmentation ions as illustrated **(b)**; **(c)** Example of ESI MS/MS mass spectrum of indicated methylated oligosaccharide. The main fragment ions visible are the **B** series **(non-reducing end fragment ions)**, and the **Y** series **(reducing end fragment ions)**. Mass differences between consecutive **B series fragment ions** define oligosaccharide primary structure (non-reducing to reducing end direction); mass differences between consecutive **Y series fragment ions** define oligosaccharide primary structure (reducing to non-reducing end direction) (illustration (c) from Reinhold et al., 1995, Figure 7, American Chemical Society).

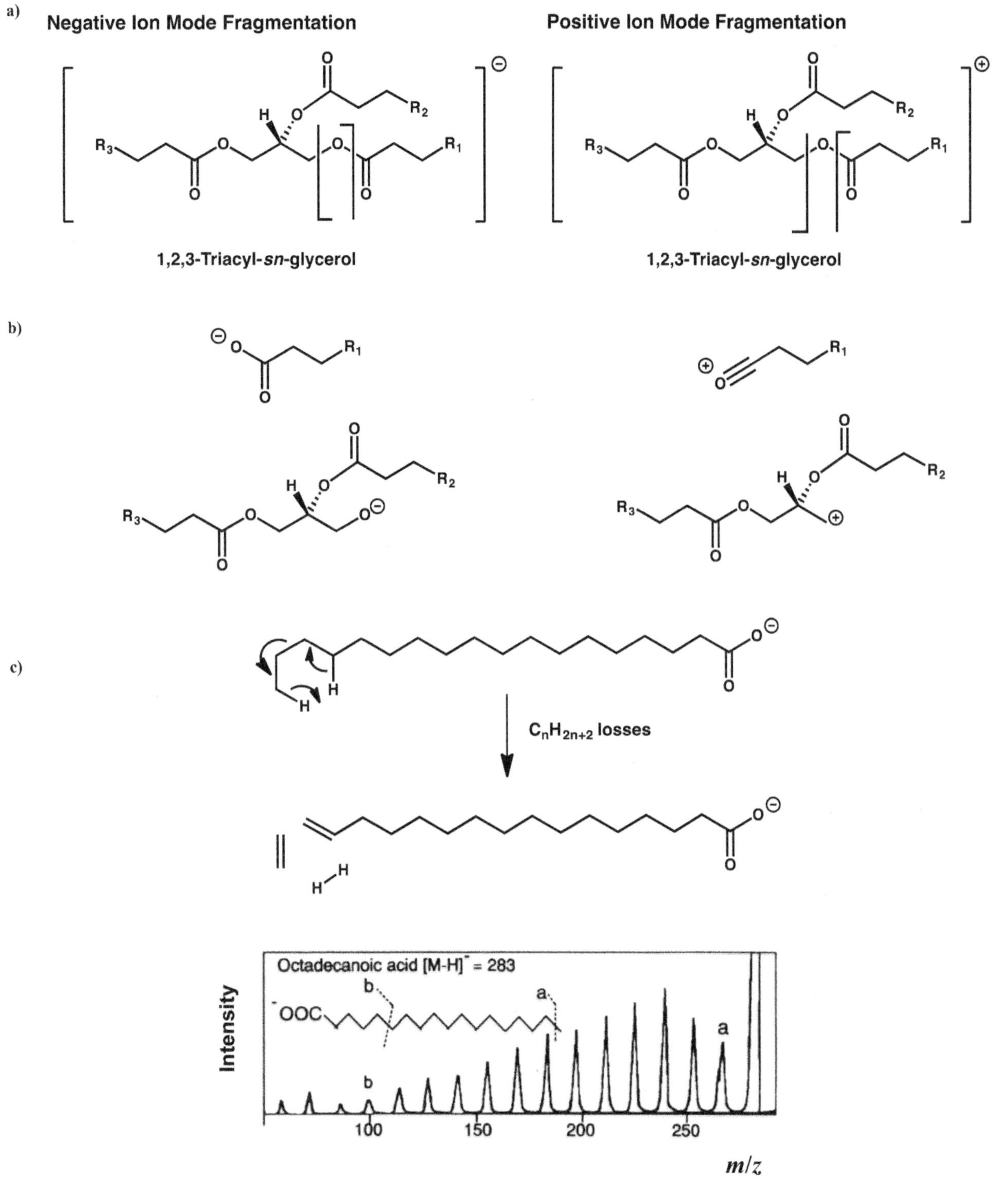

Figure 9.18 **Mass spectrometry of glycerides, phospholipids and fatty acids.** (a) Origin of main fragmentation ions as illustrated (b); (c) Example of negative ion FAB MS/MS CID mass spectrum of indicated fatty acid. The main fragment ions visible are homologation fragment anions. Mass differences between consecutive fragment anions define fatty acid CH_2-length ($C\omega \rightarrow C\alpha$ direction) (illustration (c) adapted from Jensen *et al.*, 1985, Figure 1).

9.4 The challenge of proteomics

Studies on individual biological macromolecules and macromolecular lipid assemblies have done much to explain the inner workings of cells and organisms. This will almost certainly continue to be the case. Much can be gained by removing cellular components, then seeking to characterise their structures and functions without complication from the "background noise" of other cellular components. Having said this, the "background noise" must be every bit as much associated with the function and even structure of the cellular components of interest. Therefore, chemical biology research must seek to follow not only the reductionist philosophy of chemistry but also the inclusionist philosophies of biology and medicine. In this respect, mass spectrometry through the study and practice of **proteomics** has an enormously powerful role "to enable the move from individual to collective molecular structure and function analyses" as stated previously. This is especially true of studies involving cellular proteins since there is a growing appreciation that proteins do not always function in isolation, but are also frequent participants in larger multiprotein complexes. In other words, quaternary association of catalytic and functional units is far more widespread than previously assumed. The implications of this realisation remain to be established.

9.4.1 Early developments in proteomics

Attempts to describe systems of proteins on a large scale were initially outlined during the 1970s with the development of **2D polyacrylamide gel electrophoresis (2D PAGE)** as a method to visualise sets of proteins (Figure 9.19). This method involved separation of protein mixtures by pI in the first dimension and then by molecular mass in the second dimension, resulting in a 2D gel that may then be stained and visualised as a complex pattern. Experiments were typically conducted comparatively, for example two tissues were compared, one diseased and one normal. Patterns were then probed to check for changes in protein spot pattern. The problem with this original approach was that it was very difficult to identify proteins from spots. Chemical degradation could be attempted to sequence the protein, but still the protein could only be identified if the protein was already known and sequenced. These difficulties really held 2D PAGE back during the 1980s and this situation was only resolved later on given access to large gene databases together with soft-ionisation mass spectrometry techniques to identify peptides from proteins digested "in gel".

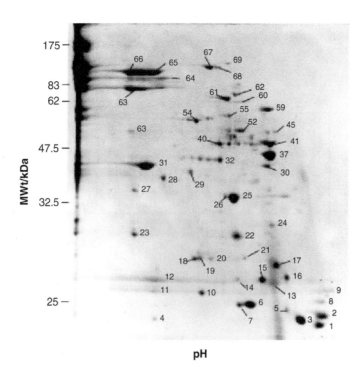

Figure 9.19   2D polyacrylamide gel electrophoresis of protein mixtures. Relative abundances of proteins may be visualised by **2D polyacrylamide gel electrophoresis (2D PAGE)**, then gel staining after which individual spots can be excised and identified by mass spectrometry. (Pandey and Mann, 2000, Springer Nature).

9.4.2 Using 2D PAGE with mass spectrometry

The combination of 2D PAGE and mass spectrometry nowadays defines the traditional approach towards proteomics. 2D PAGE separates proteins according to orthogonal properties, gels are stained, then spots cut from the gel and identified by mass spectrometry using sequence information from gene databases and peptide mass fingerprinting (Figure 9.20). The use of these techniques alone has allowed types of "cataloguing" experiments to be performed wherein patterns on 2D gels can be examined and spots on gel identified. These approaches have had most impact when examining systems somewhat simpler than the whole cell situation, for instance the compositions of subcellular organelles or multiprotein complexes. Alternatively, comparative experiments may be performed using **pattern-matching algorithms** to compare 2D PAGE results in an attempt to make out the differences, for example in protein profile between diseased and normal tissue, in an attempt to uncover disease biomarkers.

However, there are significant problems with using the combination of 2D PAGE and mass spectrometry. First, only the high abundance proteins are identified on 2D gels; it has been estimated that the number of genes expressed in a human cell type could exceed 10,000. If post-translational modification were also considered then the numbers of distinct polypeptides even within a single cell could run into the millions. On a 2D gel several hundred spots may be separately analysed, perhaps increasing a few-fold with specialised techniques of subcellular fractionation or zoom gels that cover narrower pH ranges. In addition, many membrane and basic proteins may not be separated on 2D gels at all. As a result, when one examines the databases that catalogue proteins in certain tissues, the bulk of proteins are the abundant proteins such as cytoskeletal proteins, chaperones and matrix-associated proteins. Many low-abundance proteins, such as signalling proteins, transmembrane receptors and the like are very under-represented. Hence the push for a more quantitative proteomics became imperative.

9.4.3 Isotope-coded affinity tags

One of the main problems to overcome in order to realise quantitative proteomics is that mass spectrometry is not strictly a quantitative technique with respect to molecular abundance. One approach to overcome this problem is to perform comparative experiments in parallel, where one sample is labelled with an isotope (such as ^{18}O or ^{2}H) and the other is unlabelled. Mass differences between labelled and unlabelled samples may be observed in a mass spectrometer and the relative ratios quantified (Figure 9.21). Obviously, this method of quantification relies on equivalent mass "expression" by

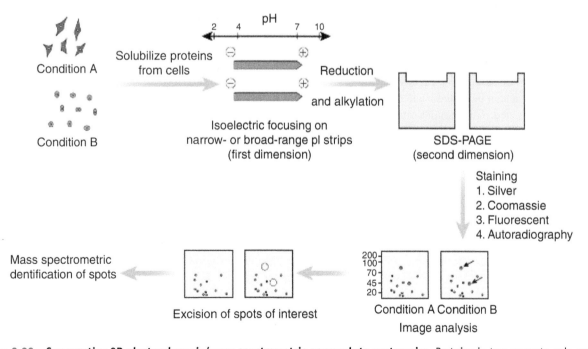

Figure 9.20 Comparative 2D electrophoresis/mass spectrometric approach to proteomics. Proteins in two separate gels can be compared, then differentially expressed spots identified and protein identities confirmed by mass spectrometry (illustration from Pandey and Mann, 2000, Figure 3, Springer Nature).

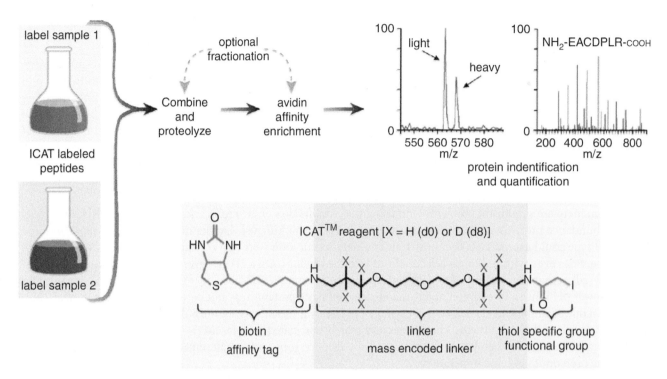

Figure 9.21 **Isotope coded affinity tagging for comparative proteomics.** Two pools of peptides can be differentially labelled with tags of different masses, then the pools combined, enriched by **avidin affinity column purification** (avidin has a very high affinity for biotin) and relative abundance of particular peptides can be directly compared by mass spectrometry (illustration from Patterson and Aebersold, 2003, Figure 5, Springer Nature).

both samples in spite of the isotopic labelling of the one and not the other. In other words, in spite of the labelling, both samples should behave identically chemically. The simplest way to effect labelling is to use **isotope-coded affinity tags (ICATs)**. According to this technique, two identical populations of proteins are maintained under different conditions (e.g. temperature, pH, ionic strength), but tagged differently on conclusion with ICATs. Tagging is site-specific (e.g. on sulfhydryl, amine groups or phosphate esters) and the ICAT tagging agent also possesses an affinity tag. Typically, the heavier ICAT has a mass difference of 8 relative to the lighter ICAT owing to the substitution of eight ^{1}H hydrogen atoms for eight ^{2}H deuterium (D) atoms. Both protein populations are then pooled post-tagging and digested into oligo-/poly-peptides prior to affinity purification of the affinity tagged oligo-/polypeptides of the combined digest. Tagged oligo-/polypeptides are then compared directly by tandem mass spectrometry, looking for quantitative differences in molecular ion abundance that can be linked back to differences in the conditions under which the two identical populations of proteins were maintained. This remains one of the best approaches to quantitative and comparative proteomics.

9.4.4 Deciphering protein networks

One of the most notable early successes of proteomics has been in providing a protein-protein interaction map for yeast cells. A major approach to building such maps, has been to use the genetic approach of the **yeast two-hybrid system**. Yeast two-hybrid system experiments are based on the idea that some **transcriptional activators** can be separated into two proteins: one involved in **DNA binding** and the other involved in **transcriptional activation** (Figure 9.22). Transcriptional activation then only takes place when the two proteins are tethered to one another. The yeast protein GAL4 is one such transcriptional activator that can be separated into its **GAL4 DNA-binding domain** (**GAL4-BD**) and transcriptional **activation domain** (**GAL4-AD**). Given this, potential interactions between two different proteins of interest, coded for by two different **open reading frames** (**ORFs**), may then be tested by preparing two GAL4 fusion proteins wherein one of the proteins of interest is fused to **GAL4-BD** and the other to **GAL4-AD**. If the two different proteins of interest are able to interact, then **GAL4-BD** and **GAL4-AD** will become tethered, so enabling reporter gene transcription. If not, then transcription is not enabled (Figure 9.22).

There are two approaches to performing these experiments on a large scale: either using an array approach or a library approach. In an array approach, a living array of haploid yeast (**ORF-AD**) transformants (the **prey**) is designed, each

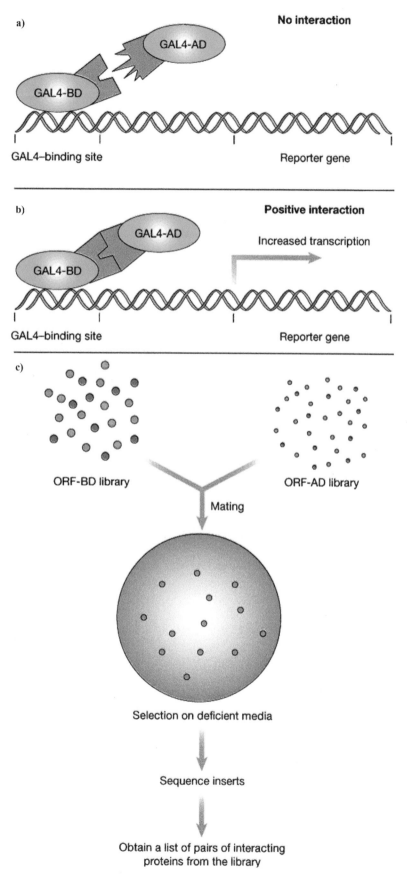

Figure 9.22   The yeast two-hybrid system. Different **open reading frames (ORFs)** are fused to a **GAL4 DNA-binding domain (GAL4-BD)** or an **activation domain (GAL4-AD)**. If the encoded proteins interact then a reporter gene is transcribed and translated, allowing for identification of protein-protein interactions. (Illustration from Pandey and Mann, 2000, Figure 7, Springer Nature).

expressing different proteins, coded for by different corresponding ORFs, fused to an activation domain like GAL4-AD. Another yeast (**ORF-BD**) transformant (the **bait**) is prepared expressing a single protein of interest, coded for by a single alternate ORF, fused with a DNA-binding domain like GAL4-BD. When the bait transformant is mated with all the transformants on the array, diploid yeast strains are produced but only those expressing proper molecular partners to the single protein of interest from the ORFs will show transcriptional activation. Since the identities of ORFs in the array are known, together with their array addresses, then molecular partners to our single protein of interest should be identified with ease. In a library approach, all the prey transformants are pooled and fused to the bait transformant. According to this method, transcriptional activation is associated with a survival trait (e.g. antibiotic resistance) so that the only surviving diploid yeast strains are those expressing proper molecular partners to the single protein of interest from the ORFs. Following this, molecular partners of the protein of interest can be identified separately by gene sequencing of the "survival" ORFs.

Tandem affinity purification (TAP) is another method for probing protein-protein interactions that is able to characterise multiprotein complexes rather than just the binary partners, as illustrated with the yeast two-hybrid system. TAP is an extension of protein tagging; two tags are used so as to enable specific purification of molecular partners interacting specific bait protein of interest. Two different protease sites are also included: one between the two tags (**tag 1**) and one between the two tags and the specific bait protein of interest (**tag 2**). Initially, an extract (which includes the TAP-tagged bait protein of interest) is passed over an affinity column for tag 1 allowing tag 1 to bind. The column is then washed and tag 1 is protease cleaved from tag 2, allowing the protein of interest and interacting protein partners to elute from the affinity column in association with each other. This enriched protein mixture is subsequently passed over a second affinity column for tag 2, allowing tag 2 to bind. This time, the bait protein of interest is protease cleaved from the tag 2, allowing protein of interest and molecular partners to elute together, free of non-specific binding interactions. The bait protein of interest and partners can then be identified by 2D gel electrophoresis and mass spectrometry.

Protein correlation profiling (PCP) is a newer approach based on co-elution of proteins under non-denaturing conditions. This technique works on the assumption that when proteins co-elute under different chromatographic separations, then the proteins are likely to be together in a complex. Cross-linking can also be employed, particularly for membrane protein complexes where solubility becomes an issue outside the membrane environment. Such approaches are useful in creating reference proteome sets, but remain quite challenging for assessing protein dynamics in cells.

Spatial proteomics is another recent approach for determining binding interactions through affinity labelling. One approach for spatial proteomics is using **ascorbic acid peroxidase** (**APEX**) protein tags that generate hydroxyl radicals from hydrogen peroxide in the presence of phenoxy biotin so as to biotinylate all proteins positioned within 30 Å. These biotinylated proteins can then be enriched and analysed by mass spectrometry. Similarly, the **BioID approach** uses the biotin ligase enzyme **BirA** to biotinylate proteins in close proximity to the BirA-fusion protein. Finally, another variation on this theme is tagging proteins with the Cas9 protein (see Chapter 12). Thereafter, the use of a so-called **single guide RNA** (**sgRNA**) strand brings Cas9 labelled proteins into interaction with specific gene loci proximal to transcription factors and histone proteins which control the corresponding gene expression. These transcription factors and histones can then be enriched to understand histone modifications in these different genome loci.

9.4.5 The challenge of membrane proteins in proteomics

It has been estimated that 20–30% of proteins encoded by the human genome are membrane proteins. Clearly, this makes membrane proteins a vital class to understand and, furthermore, more than 70% of drug targets are membrane proteins. A significant problem in mass spectrometry is that a far smaller number of membrane proteins are identified than are known by other methods to be present in any particular sample. The observation of membrane proteins by standard 2D gels is hindered by solubility. First, these proteins are not solubilised in the buffer (for the first dimension). Second, solubilised membrane proteins will all too easily precipitate again at their isoelectric point pI. One approach to solving the initial problem is solubilising membrane proteins in either organic solvents or non-ionic/zwitterionic detergents. The pI problem may be avoided by just using 1D gels. However, even if membrane proteins are clearly resolved within a gel, further problems are encountered during the digestion steps because oligo-/polypeptides in digests tend to be particularly hydrophobic (Figure 9.23).

An alternative to using gels is **shotgun proteomics** [i.e. *in situ* digestion of complex mixtures of proteins followed by **2D liquid chromatography mass spectrometry** (**2D LC-MS**)] to identify proteins in particular samples (Figure 9.24). Again, initial solubilisation is an important factor; organic acids and solvents, as well as various detergents, may be used. Proteolysis may be carried out with enzymes (e.g. trypsin, which is even functional in 60 % methanol) or chemically (e.g. cyanogen bromide). An alternative approach is to cleave the exposed soluble domains

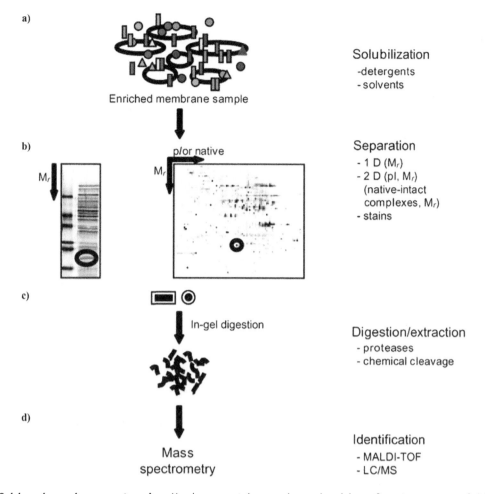

a) **Solubilization**
- detergents
- solvents

Enriched membrane sample

b) **Separation**
- 1 D (M$_r$)
- 2 D (pI, M$_r$)
  (native-intact complexes, M$_r$)
- stains

pI or native

M$_r$

c) **Digestion/extraction**
- proteases
- chemical cleavage

In-gel digestion

d) **Identification**
- MALDI-TOF
- LC/MS

Mass spectrometry

Figure 9.23 Gel-based membrane proteomics. Membrane proteins can be analysed by a four-step process of (**a**) solubilisation, (**b**) separation, (**c**) digestion/extraction and finally (**d**) identification by mass spectrometry (illustration from Wu and Yates, 2003, Figure 1, Springer Nature).

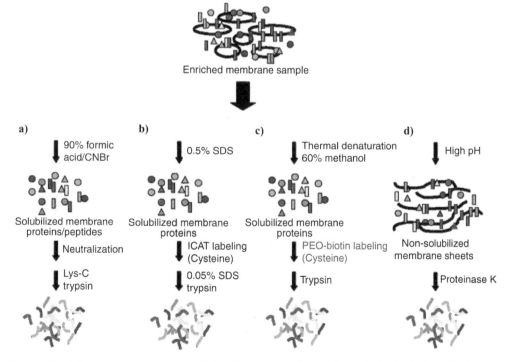

Enriched membrane sample

a)
90% formic acid/CNBr

Solubilized membrane proteins/peptides

Neutralization

Lys-C trypsin

b)
0.5% SDS

Solubilized membrane proteins

ICAT labeling (Cysteine)

0.05% SDS trypsin

c)
Thermal denaturation 60% methanol

Solubilized membrane proteins

PEO-biotin labeling (Cysteine)

Trypsin

d)
High pH

Non-solubilized membrane sheets

Proteinase K

Figure 9.24 Four methods to prepare membrane proteins for mass spectrometric analysis. Analysis can be by shotgun proteomics or by conventional mass spectrometry (illustration from Wu and Yates, 2003, Figure 2, Springer Nature).

using a non-specific protease such as **proteinase K**, then use LC-MS to sequence the resultant peptides and employ protein mass fingerprinting to identify proteins. To summarise, membrane protein proteomics remains a challenge, but these approaches will undoubtedly become more widespread, especially considering the pharmaceutical interest in membrane proteins.

9.4.6 Proteomics and post-translational modifications

Post-translational modifications (PTMs) are vitally important in modifying protein structure and function, especially in eukaryotes. On a large scale, researchers are faced with a separate challenge. Normally in proteomics experiments, proteins can be isolated, identified by protein mass fingerprinting and combined with product ion scanning if necessary. Unfortunately, **PTM analysis** requires a much more thorough analysis of the entire amino acid sequence. This is further complicated by the fact that modifications are not typically homogeneous, and a wide range of modifications can be found on a single type of protein of interest. The role of PTMs in cells remains a huge area of research, widely open to the chemical biology reader to explore, since not even biologists understand structural and functional consequences.

9.4.6.1 Comprehensive PTM analysis of a single protein

The first stage in a comprehensive PTM analysis of a protein of interest is purification. Depending on the initial level of purification, this can be most conveniently achieved by gel electrophoresis, either 1D or 2D gel. Often electrophoresis may give clues to the nature of modification, for example **phosphorylation** of a protein may result in spots separated horizontally on a 2D gel due to differential pI values between phosphorylated and non-phosphorylated states of the protein of interest. A useful first stage in the process is to accurately identify the molecular ion, comparisons between the expected and measured mass can give initial information about which modifications may be present. After gaining information on the intact protein, detailed information regarding site(s) of modification may be obtained by analysing peptides obtained after enzymatic or chemical degradation. A typical combination would be to use trypsin in one experiment and then to use endoproteinase Asp-N or Glu-C in a second experiment. Modified oligo-/polypeptides in digests can be identified by unexpected molecular ions and then analysed by product ion scanning to determine the amino acid residues at which PTMs have taken place. **Acetylation** (often on arginine amino acid residues) is easiest to identify. Phosphorylation can be either stable or labile (phosphotyrosine tends to be stable, phosphothreonine less so, whilst phosphoserine tends to be labile), but can usually be seen. Unfortunately, **sulfation** and **glycosylation** modification sites can be much more labile and less easy to identify by tandem mass spectrometry.

9.4.6.2 PTM analysis of protein populations

The PTM mapping of protein populations requires a different approach to the single protein approach. For instance, once a protein sample mixture has been resolved by 2D gel electrophoresis, then phosphorylated proteins can be identified by **Western-blot** using **anti-phospho-amino acid antibodies**. Individual anti-phosphoserine, anti-phosphothreonine or anti-phosphotyrosine antibodies may also be used. Another approach is to specifically target and purify a subset of proteins from a complex mixture by affinity column purification. For instance, anti-phospho-amino acid antibodies coupled to a resin (e.g. agarose beads) may be used to affinity purify phospho-proteins from complex protein mixtures. The enriched population of proteins may then be eluted from the column, resolved by 2D-electrophoresis and individual proteins identified with phosphorylation sites defined by tandem mass spectrometry.

Gels may be eliminated by shotgun proteomics, but specific protein modifications of interest need to reviewed and resolved by some form of chromatography prior to mass spectrometry. For example, phosphopeptides may be specifically purified before mass spectrometry, using **immobilised-metal affinity chromatography** (IMAC) (see Chapter 2). Both iron (as Fe^{3+}) and gallium (as Ga^{3+}) have proved useful in metal ion affinity column purification of phosphorylated peptides prior to mass spectrometry. Their sequences and site(s) of phosphorylation can then be identified with ease by tandem mass spectrometry, and subsequently related back to the identities of the parent proteins from which these daughter digest peptides derived. Significant progress continues to be made in PTM analysis and this continues as an important area in proteomics.

9.5 Genomics: assigning function to genes and proteins

Gene sequences deriving from the ORFs for many organisms are already available, therefore the bottleneck in biology has moved from gene sequencing to gene function and protein structure/function. Protein network studies by yeast, two-hybrid and proteomics screens are useful steps forward (all described in this chapter), but these need to be complemented and supplemented by other approaches to obtain information on proteins and other biological macromolecules of interest.

9.5.1 Protein microarrays

Protein microarray technology has emerged from the more established technique of DNA microarray technology. Microarrays are simply a technique where individual molecules such as oligo-/oligodeoxynucleotides, proteins, antibodies and small molecules can be separately spotted onto a surface such as a glass slide and then the surface is analysed for specific functions or activities (Figure 9.25). **DNA microarrays** can now be used at densities of tens of thousands of different oligonucleotide probes per square centimetre and can be used for genetic analysis as well as expression analysis at the mRNA level. As previously discussed, mRNA levels do not strictly correlate with protein levels and therefore **protein microarrays** have also been developed to probe proteomes in a microarray format. One significant problem with protein arrays relative to oligonucleotide arrays is stability: it is important to array the proteins without denaturing, but maintaining a reasonable density for facile detection of activity. The surfaces are therefore modified and engineered to prevent evaporation and drying by the use of sophisticated pads, films, nanowells and microfluidic channels. Arrays used in proteomics take two forms: analytical arrays and functional arrays.

9.5.1.1 Analytical arrays

Analytical arrays typically use molecules that are able to capture and bind to proteins from a complex protein mixture. For instance, antibodies, antigens, carbohydrates or small molecules are spotted onto a derivatised surface. The typical approach used is that two protein populations from different biological states are compared: one set is labelled with a red fluorescent dye, the other labelled with a green fluorescent dye, mixed and then incubated with the chips. Each spot in the array is then analysed for the relative quantities of the two dyes. This information can lead to insights regarding structural and functional differences between protein populations maintained in two different biological states prior to experiment. Nevertheless, there remain significant problems with the general stability of protein microarrays.

9.5.1.2 Functional arrays

Functional arrays use individually purified or synthesised proteins or peptides that are arrayed onto a suitable surface. The microarray can then be probed for binding partners of many different types. For instance, proteins, nucleic acids, drugs or other small molecules could be passed over the surface to identify macromolecular binding interactions. This technique has growing importance in drug discovery and drug validation research programs. In fact, functional arrays for many organisms are close to completion now. For instance, the near complete proteome of yeast has now been assembled into microarray format. The method used was to His-tag all ORFs (see Chapter 2) so that all expressed yeast proteins may be spotted onto Ni-coated slides.

9.5.2 Biochemical genomics

Another approach is to assign not just binding interactions but catalytic function to proteins in a large-scale manner too. Once again, work in yeast has demonstrated the proof of principle. The principle of the approach is as follows:

1. Construction of several thousand yeast strains by standard genetic recombination methods, each bearing a plasmid expressing a different **glutathione S-transferase (GST)**-ORF fusion under the control of an inducible promoter.

2. Pooling of extracts from a convenient number of yeast strains (e.g. 100) so that each of the pools may be assayed side by side for a specific activity.

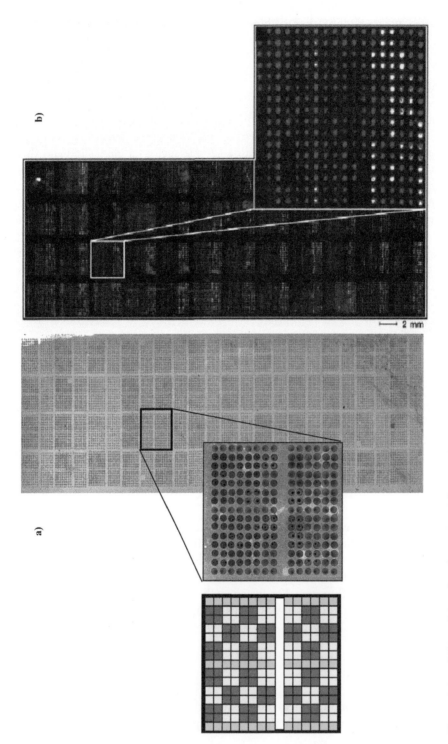

Figure 9.25 **DNA and protein microarrays. (a) DNA microarray** comprising 10,080 features and 1959 **plasmid DNA (pDNA)** clones for probing. Schematic detail indicates layout arrangement for each clone analysed as clusters of $n = 4$ between guide tracks of **green fluorescent protein (GFP)** expressing pDNA (illustration adapted from Palmer *et al.*, 2006, Springer Nature, CC BY 2.0); **(b) Protein microarray** comprising 6566 Proteins are arrayed on a microscope slide, such arrays have a diverse number of applications in globally understanding proteome function (illustration from MacBeath, 2002, Springer Nature).

3. Purification of the GST-ORFs from the pooled strains so that each pool contains approximately 100 GST-tagged proteins.

4. Each pool is assayed for an enzymatic activity; if a specific pool is found to have activity then it is possible to create smaller and smaller sub-pools within the pool, eventually narrowing down the search until the ORF encoding the activity is pinpointed.

As biochemical genomics uses catalytic activity rather than solely binding activity the sensitivity of the technique can be controlled by the time of incubation of the enzyme reaction, thereby dramatically increasing sensitivity compared to most binding assays. As proteins are tested within pools initially rather than individually it can also provide information about catalytic activities that might be dependent on complexes of proteins that would be missed when working with the protein in isolation. The drawbacks of the method are that the libraries bias against larger proteins, and any overexpressed GST-ORF fusions that retard growth during propagation of the yeast strain will also be underrepresented in the pool.

9.5.3 Chemical genomics

In recent times, libraries of chemical compounds have been instituted due to the advent of combinatorial chemistry. The quality of libraries is highly variable and often of doubtful utility, but the approach is here to stay. Such libraries can be built in a more controlled way by **diversity-orientated synthesis** that seeks to create chemical diversity in a less randomised manner, underpinned by greater synthetic robustness. The benefits of libraries prepared with better chemical control is then the opportunity to screen the library against cells, looking either for any reproducible phenotypic change or looking for one phenotypic characteristic in particular. Once a reproducible phenotype is found, synthesis of a pure compound can then definitively match novel small molecule with novel biological function. This **chemical genomics** approach clearly opens the possibility of identifying novel leads for pharmaceutical investigation against cancer, virus infection, etc (see Chapter 11). In addition, chemical genomics should certainly contribute in the discovery of the next generation of chemical tools for the investigation of structure and function in chemistry and biology (see Chapter 11).

9.5.4 Structural genomics

Structural genomics aims to take large-scale approaches at the protein structural level to compile a dictionary of protein structure from the level of the fold to the domain, to the protein and to its quaternary interactions. Although it shares a philosophy with that of the hugely successful sequencing projects of the 1990s, it will be a far more challenging task to solve protein structure on a large scale due to great variation in protein structure compared to the conservative properties of nucleic acids. Rapid developments are taking place both in X-ray crystallography and NMR spectroscopy to facilitate high-throughput structural analysis by partial automation of various stages of each process.

9.5.5 Perspectives on the future of proteomics with genomics

Proteomics continues to be very much driven by new technologies, both from the perspective of hardware, namely development of newer generations of mass spectrometers, and from the methodologies perspective, for applications in biological systems. Nowadays, proteomics is making essential contributions in mapping protein interaction pathways in cells in the same way that genomic expression data did previously. Proteomics is also opening up the whole issue of PTMs and epigenetic (i.e. non-genome-based) control of structure and function in cells and organisms. Hence there is no question that proteomics is playing a greater and greater role in both improving our understanding of basic biology, and in developing new methodologies for advanced diagnostic applications in medicine.

10
Molecular Selection and Evolution

10.1 Chemical biology and the origins of life

The origins of life and existence are uniquely interesting. There can be no more significant scientific or philosophical question than what is/was the origin of life and existence. Leaving aside all but scientific approaches to this question, this chapter aims to introduce the chemical biology reader to a rudimentary theory for the origins of life as seen from the perspective of chemical biology and chemistry. Seeing that biological macromolecules and assemblies, as well as cofactors and prosthetic groups, are the primary building blocks of living organisms, then there can be no more logical a starting point. Once our rudimentary theory is in place, a number of useful processes will be described that have drawn inspiration from these ideas and principles.

10.1.1 Order from complexity

The simplest description of a living organism is that the organism consists of a system of chemicals that has the capacity to catalyse its own reproduction. Given this view, the root property of life is the achievement of autocatalytic closure among a collection of molecular species. Alone, each molecular species is "dead"; jointly the collective system of molecules is "alive". This concept is very compelling for the chemical biologist since biocatalysis no longer just sustains life but is also a key requirement for the origins of life. However, catalysis alone cannot be enough to originate life. All living systems "eat"; they take in matter and energy in order to sustain and reproduce themselves. Therefore, living systems must be the equivalent of **closed autocatalytic** but **open (non-equilibrium) thermodynamic systems**. But even supposing that life can originate from autocatalytic closure among a collection of molecular species, how does this view of life account for the astonishing order that characterises living systems and seems so much at variance with the second law of thermodynamics? In Kauffman's view, this order arises directly from **molecular diversity** and complexity in an open thermodynamic system. That is to say that **self-organisation** arises naturally as a direct consequence of molecular diversity and complexity in an open thermodynamic system (i.e. "order for free").

10.1.2 Evolution from the molecular level

Amongst biologists, discussions about the origins of life have been focused by arguments of **natural selection**. There has been an apparent tendency to ignore self-organisation. Most likely this is because there is a fundamental difficulty in recognising how living systems may be governed simultaneously by two sources of order, self-organisation born of molecular diversity and complexity, and the "forces" of natural selection. However, if we are to understand the origin

Essentials of Chemical Biology: Structures and Dynamics of Biological Macromolecules In Vitro *and* In Vivo, Second Edition. Andrew D. Miller and Julian A. Tanner.
© 2024 John Wiley & Sons, Inc. Published 2024 by John Wiley & Sons, Inc.
Companion Website: www.wiley.com/go/miller/essentialschembiol2

of life properly, a final theory of biology must allow for the commingling of self-organisation and selection processes as an expression of an even deeper order. Surely, as Kauffman argues, self-organisation precedes natural selection. For natural selection to be able to operate there should be two main criteria satisfied:

1. The existence of compartmentalisation or the means of compartmentalisation.

2. The existence of low energy, self-organised molecule states ("robust systems") that provide a foundation upon which to exercise the forces of natural selection to create and mould living organisms.

Compartmentalisation is a key part of the self-organisation process, if for nothing else than to sustain spatial integrity and prevent dilution of the reacting molecules that comprise each living system. The ability of lipids to self-assemble into lipid bilayers and more complex mesophases is well established, and has long been accepted to form a key part of compartmentalisation in biology (see Chapter 1). **Stable, low energy, self-organised states** ("robust systems") such as DNA, RNA and even folded proteins (including their prosthetic groups and cofactors) intuitively provide a sound foundation upon which to exercise the forces of natural selection to create and mould living organisms (see Chapter 1). For the forces of natural selection to have an effect, random structural variations (**mutations**) must exist in the primary structure of all such robust systems (namely base variations in nucleic acids and amino acid residue sequence variations in proteins) that represent a molecular pool of diversity upon which **selection pressures** can act. Natural selection pressures should then select for that mutation or group of mutations belonging to a given robust system that are most beneficial for the existence of the self-assembled living system. For instance, if enhanced catalytic efficiency is required, then the forces of "natural selection" acting under the influence of a given set of metabolic selection pressures should select for that amino acid residue mutation or group of mutations that promote the necessary enhancements in catalytic efficiency of a key enzyme or group of enzymes. This is a form of **molecular evolution**.

Molecular evolution is the putative engine of macro-evolution and is thought to involve primarily successions of single (point) mutations leading to ever more enhanced robust systems that in turn promote the well-being and competitiveness of the self-assembled living system. Such molecular evolution by a succession of single point mutations seems to be a satisfying way to define the molecular basis of natural selection. Hence, if we combine this with the idea that natural selection is preceded by self-organisation, then we have an integrated, holistic **molecular hypothesis for the origin of life**. In other words, the origin of life should be explained as follows. First, self-organisation exists as a direct consequence of molecular complexity and diversity in order to provide the means to compartmentalise a "living" system of reacting molecules. Second, self-organisation results in stable, low-energy, "robust" systems such as DNA, RNA and folded proteins that can be moulded under pressure of natural selection by a succession of single point mutations so as to promote catalysis or other functions necessary to originate and promulgate living organisms. Reproduction should then be the final stage, resting upon a process of autocatalytic replication of robust systems once the enclosed chemical environment allows such reactivity to take place, converting closed autocatalytic "living" systems of reacting molecules into sustainable living organisms.

10.1.3 Chemical self-organisation from complexity

If self-organisation exists as a direct consequence of molecular complexity and diversity, then such a hypothesis must be tested experimentally as well as by theoretical mathematical treatments. In other words, chemical biology should be able to provide clear evidence for molecular diversity and complexity leading to lifelike "order for free", through well-designed **pre-biotic chemistry** experiments. The main objective of pre-biotic chemistry is to try to generate by experimental means, chemical structures that accurately presage the biological macromolecules of life, including their cofactors and prosthetic groups, by reaction from pre-biotic "starting materials". In so doing, credible molecular links can be established from these pre-biotic "starting materials" to recognisable robust systems and hence "living" organisms. There are three key early experiments that provide the foundations for pre-biotic chemistry, namely the **Miller**, **Oro** and **Fischer experiments** (Figure 10.1). In the Miller experiment, proteinogenic amino acids were generated spontaneously by the electrical discharge through anaerobic mixtures containing methane, ammonia, hydrogen and water vapour. In the Oro experiment, adenine was generated through the heating of aqueous ammonium cyanide. Finally, in the Fischer experiment, formaldehyde and water generated sugars in the presence of calcium hydroxide (**formose reaction**).

Although all three types of experiment are important, they have now been superseded by studies involving "starting mixtures" of compounds known to be resident in interstellar gas clouds, or else "starting mixtures" of compounds identified from amongst gas phase products generated in methane/nitrogen mixtures post electrical discharge (Figure 10.2).

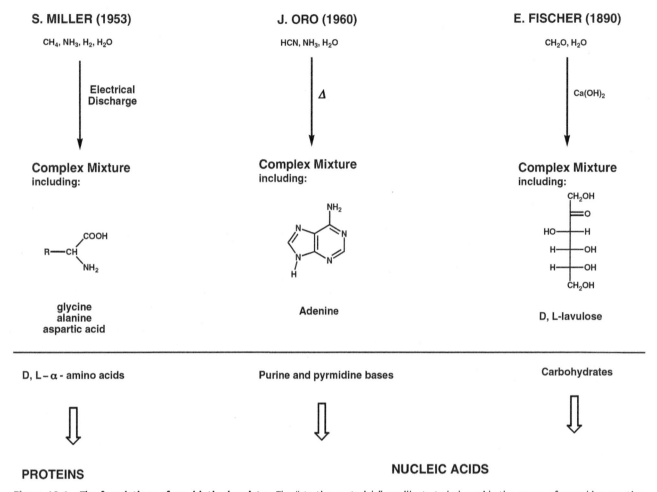

Figure 10.1 The foundations of pre-biotic chemistry. The "starting materials" are illustrated alongside the means of provoking reaction and a summary of main elemental compounds produced.

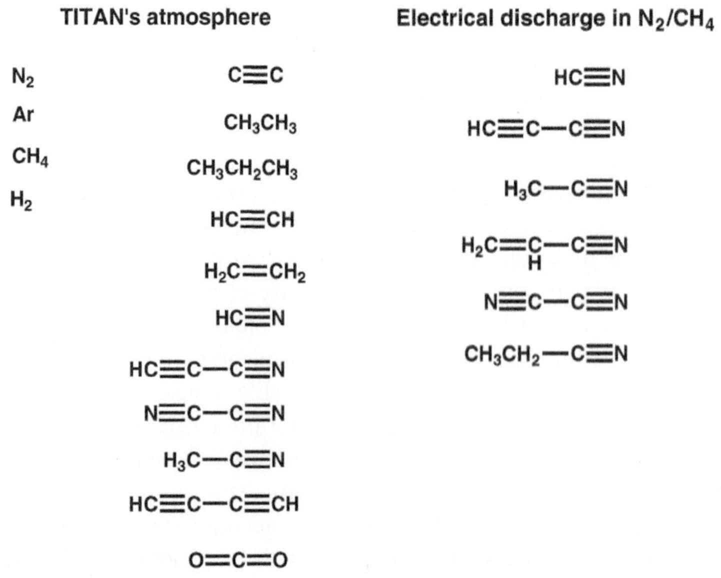

Figure 10.2 Titan's atmosphere. The atmosphere of this satellite moon of the planet Saturn comprises "starting materials" for **pre-biotic chemistry**, compared with the compounds that result from electrical discharge in an atmosphere of N_2 and CH_4. Note the extraordinary compatibility between these lists.

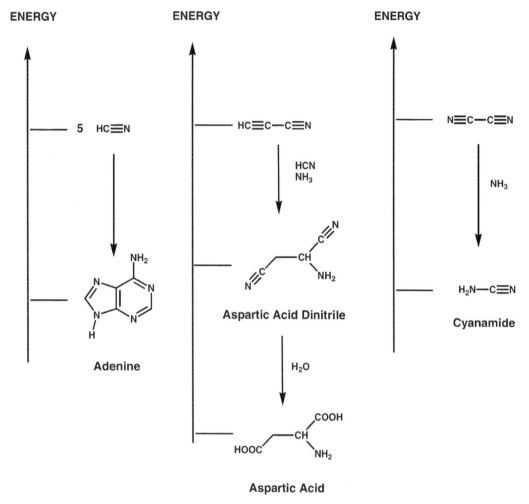

Figure 10.3 Key principle of pre-biotic chemistry. Triple-bond "starting materials" are high energy and have significant chemical potential for onward conversion into elemental compounds that form basic building blocks for biological macromolecule formation.

Molecular compositions are impressively similar in both cases. Clearly water/oxygen is not present, but this could be introduced at a later, more appropriate stage. Significantly, there are many triple-bond-containing "starting materials" that are chemical-energy-storage molecules (in an anaerobic environment) and which certainly could have the chemical potential for substantial forward reactions (Figure 10.3). In some cases, these forward reactions have actually been studied and certain key structures have been identified that form routinely from the pool of anaerobic, interstellar/electrical discharge "starting materials". These are known as **elementary structures** (from a chemical point of view). A molecular theory of the origins of life demands that anaerobic equivalents of amino acids, bases, prosthetic groups and cofactors should all be elementary structures able to form from lists of triple-bond compounds, by any one of a number of energetically acceptable pathways. A growing body of evidence appears to suggest that this is indeed the case.

10.1.4 Origins of biological macromolecules of life

There is a reasonable argument to suggest that protein prosthetic groups and cofactors should arise in advance of proteins and enzymes since these are often directly responsible for the chemical reactivity of a given protein supported by the surrounding polypeptide infrastructure. For this reason, prosthetic groups and cofactors are sometimes referred to as "**molecular fossils**". A good example of a molecular fossil is **uroporphyrinogen III** that is the main precursor to **protoporphyrin IX** found in **myoglobin**, **haemoglobin** and **cytochrome c** proteins (see Chapter 1). The anaerobic version of uroporphyrinogen III is **cyano-uroporphyrinogen III** that can originate from **cyano-aspartic acid** and **cyano-glycine** (α-amino nitriles) via a number of steps involving heat or acid catalysis (Figure 10.4). Cyano-uroporphyrinogen III is therefore a classic elementary structure. There are four main isomers of cyano-uroporphyrinogen,

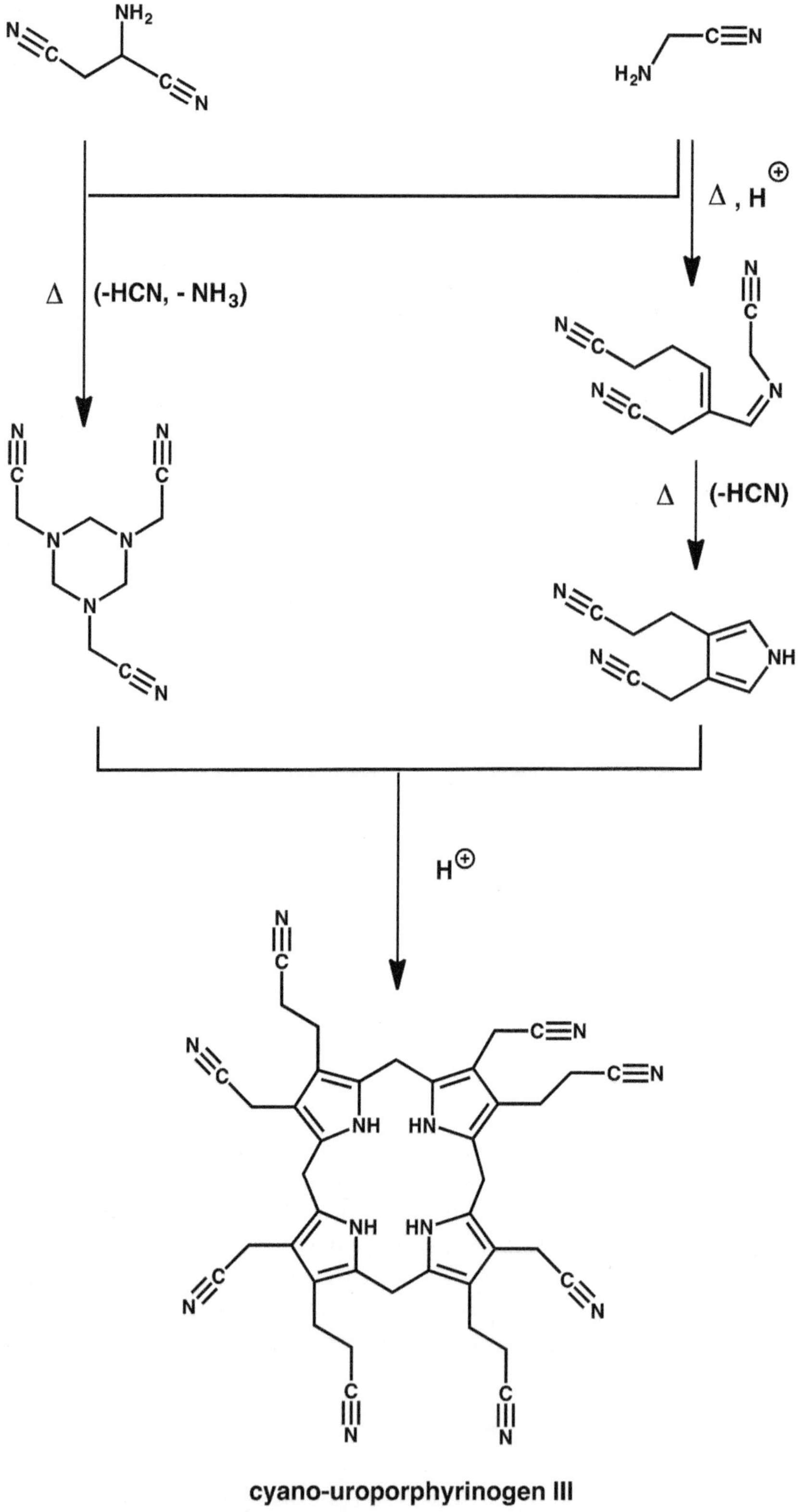

cyano-uroporphyrinogen III

Figure 10.4 Important transformations in pre-biotic chemistry. Putative pre-biotic chemistry route of conversion of **cyano-aspartic acid** (also known as **aspartic acid dinitrile**, see Figure 10.3) **cyano-glycine** (also known as **cyanamide**, see Figure 10.3) (α-amino nitriles) into **cyano-uroporphyrinogen III**, the anaerobic equivalent to the key biosynthetic intermediate **uroporphyrinogen III** that is required for the subsequent formation of cofactor **protoporphyrin IX** found in **myoglobin**, **haemoglobin** and **cytochrome c.** Uroporphyrinogen III also gives rise to many porphyrin-related pigments employed in respiration and photosynthesis.

but even with this relatively crude level of pre-biotic chemistry, the type III isomer still dominates the other three and readily yields protoporphyrin IX after oxidation and hydrolysis. The main DNA/RNA purine and pyrimidine bases are also elementary structures. The formation of anaerobic versions of each of these from such simple chemical starting points is even more impressive, as shown in Figure 10.5. Please note that putative intermediates formed along the way even suggest pre-biotic routes to anaerobic forms of the redox active cofactors/prosthetic groups as well. Therefore, these may also be regarded as elementary structures.

Mapping putative pathway(s) from such anaerobic precursors to fully functional proteins or nucleic acids is not straightforward. In particular, routes to DNA need some careful consideration. The formose reaction gives complex product mixtures and low levels of racemic 5-carbon ribose, a fact that suggests that the Gilbert concept of a pre-biotic "**RNA world**" is unlikely to be correct, in spite of the appearance of catalytic RNA (ribozymes, see Chapter 8). One of the conundrums yet to be resolved in pre-biotic chemistry is the apparent preference for hexose over pentose sugars. Given this preference, why are DNA and RNA constructed from a pentose sugar and not hexose sugars? The answer may lie with the discovery that hexose-DNA (**homo-DNA**) (Figure 10.6) is able to form anti-parallel helices without rigorous application of the **Watson-Crick base pairing rules**, to such an extent that this critical basis of helical cohesion does not apply. Consequently, autocatalytic molecular replication of homo-DNA could not lead to the propagation of base sequence encoded genetic information. By contrast, pentose-DNA does obey Watson-Crick base pairing rules rigorously and therefore offers the best means for the transmission of base sequence encoded genetic information, coincident with autocatalytic molecular replication. From this discovery now flows a **chemical aetiology** of nucleic acid structure (Figure 10.6). In effect, the molecular evolution of nucleic acids requires a convergence upon pentose-DNA and RNA via hexose-nucleic acid "evolutionary intermediates" (Figure 10.6).

10.2 Molecular breeding: natural selection acting on self-organisation

Breeding is essentially a man-made process of natural selection. Mankind has used breeding to select for desired traits in other species for many millennia. Crops have been bred to improve the yield and quality of their produce, yeasts for baking and fermentation, livestock for their increased nutritional value and increased docility, horses for their speed and strength, and cats and dogs for their companionship. For many thousands of years, all that was understood was that cross-breeding of individual organisms showing the desired characteristics often led to the accumulation of desired characteristics in their progeny and subsequent generations. This approach used so successfully for many thousands of years is entirely empirical: function and utility were improved without any real understanding of the underlying principles behind the improvements. Since the pioneering work of Mendel, we are nowadays aware of genes that "code for traits" and how breeding seeks to modify or mutate genes, and the way in which these genes are controlled and/or regulated, in order to "code for alternative traits" that are considered desirable (or which may offer some selective advantage) in the biological niche required by the breeder.

As yet, the entire process of breeding cannot be replicated *de novo* in the laboratory at the molecular level, but those aspects concerning the modification or mutation of genes leading to protein mutations certainly can be. This might be thought of as the basis of **molecular breeding** in its simplest form. One theory of natural selection acting at the molecular level is that protein structure is the most open class of biological macromolecule to moulding and adaptation of function by successive mutations. Therefore, protein mutations are central to the process of organismal adaptation under selection pressure. The provision of a pool of protein mutations is then a mainstay requirement for the evolutionary process at the molecular level. However, there is some divergence of opinion between those that maintain that routes to improvements or new functions always involve a sequential evolutionary series of protein species each differing from its neighbouring species, before and after, by a single point mutation, and those that allow for multiple mutations in each evolutionary series of protein species as well. In either event, the road to molecular breeding of improved or new protein functions becomes clear. Provided that a form of artificial selection pressure can be applied to an appropriately created pool of single or multiple mutations, then mutant proteins can be identified and isolated in principle, with improved or new functions according to the desires of the molecular breeder.

Typically, the usual approach in chemistry is to gain an understanding of structure and mechanism to provide insights that enable a reasoned approach to improved or altered binding or reactivities. Most of synthetic organic chemistry has used this approach to build an impressive corpus of knowledge that couples structure and reaction mechanism. This philosophy has also been extended to rational protein design: knowledge of structure and mechanism followed by reasoned site-directed mutagenesis can, for example, improve the catalytic efficiency of an enzyme. The **molecular breeding approach** is the obverse. In principle, molecular breeding allows for the discovery of improvements or alterations in biocatalytic activities (or binding activities) of proteins, RNA (and even now DNA) without much pre-knowledge of

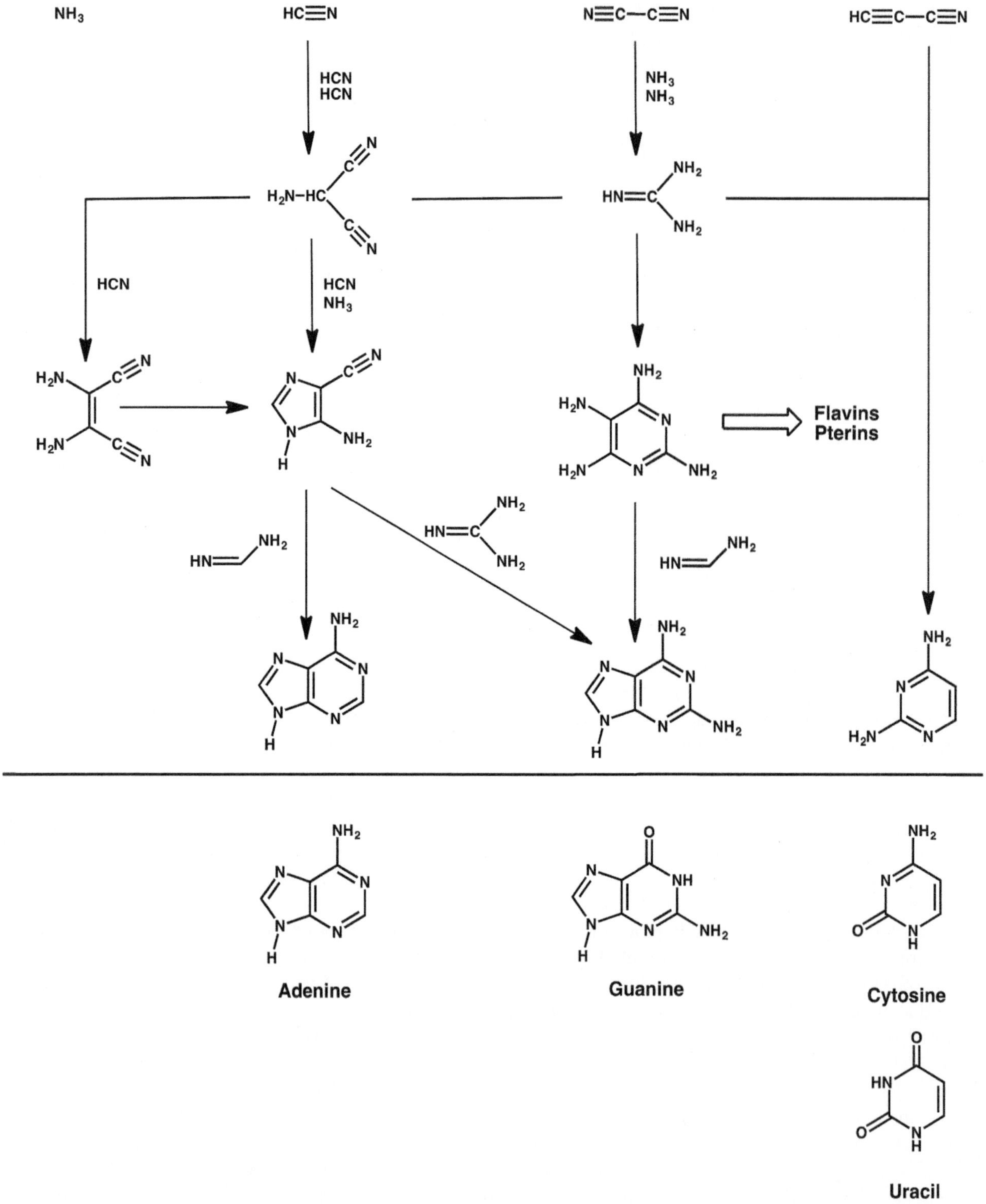

Figure 10.5 Pre-biotic chemistry routes towards DNA and RNA. Putative pre-biotic chemistry routes lead to the formation of anaerobic equivalents of the main purine and pyrimidine bases found in DNA and RNA. Intermediates formed also suggest pathways to anaerobic equivalents of redox active cofactors.

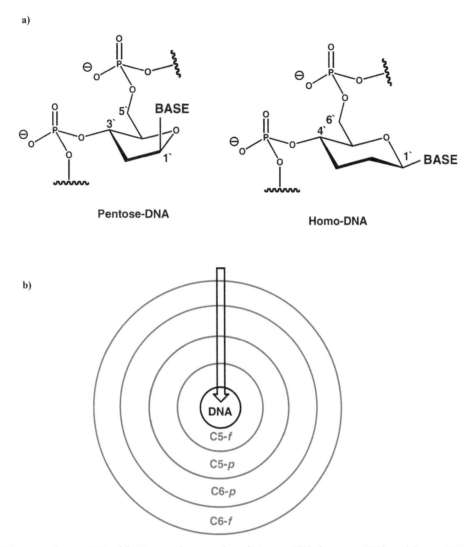

Figure 10.6 DNA versus homo-DNA. (a) Structural comparison between **DNA** (pentose-DNA) and **homo-DNA**. Homo-DNA forms double helices in which **Watson-Crick base pairing** rules are not rigorously retained. **(b)** Diagram to illustrate the aetiology of nucleic acid polymer formation. The **formose reaction** generates very little and few pentose (**C5**) sugars, hence pre-biotic nucleic acids are presumed to be made initially from hexose (**C6**) sugars first in furanose (*f*) and then pyranose (*p*) cyclic conformations, prior to selection of **C5-*p*** and then **C5-*f*** rings that give nucleic acid double helices which both require and enable Watson-Crick base pairs.

structure and mechanism. Hence, the problems involved in molecular breeding are quite different to the usual problems encountered in chemistry. How do we generate molecular diversity and then select for specific desired traits in an iterative process? Although not yet orthodoxy this molecular breeding approach is having some practical and academic success. Practically, evolved proteins such as improved lipases and evolved modified green fluorescent protein are already in the marketplace. Academically, *in vitro* **evolution** is providing insight into the evolutionary process itself and studies of evolution of self-replicating systems may even be providing further insights into the molecular origins of life.

10.3 Directed evolution of protein function

One of the simplest approaches to molecular breeding is to begin with a gene for an enzyme that performs a certain activity, then introduce **variation** or molecular diversity at the genetic level into the gene by a number of methods, including chemical mutagenesis, error-prone **error-prone polymerase chain reaction** (**error-prone PCR**), gene shuffling or incremental truncation. Virtually all basic molecular breeding approaches start with a protein that has a desired trait (at least to a small extent). Then the aim of the experiment *in vitro* is to improve or evolve that trait. Such a molecular breeding experiment

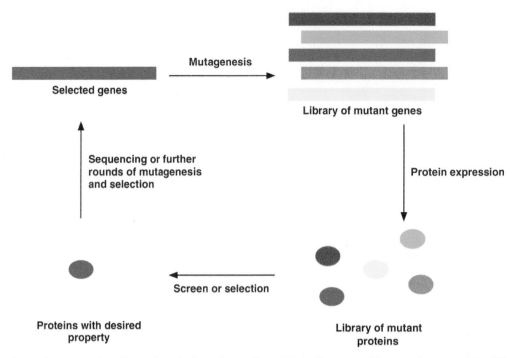

Figure 10.7   General strategy for directed evolution of protein activity. The gene of interest is mutated by different methods and the resultant **mutant gene library** is translated to a **library of protein mutants** or variants. A **selection screen** is then used to identify protein mutants or variants with improved or novel characteristics. The corresponding mutant genes coding for the most improved or novel protein variants are then reselected for further rounds (**iterations**) of mutation and selection screening to allow for the identification of further improved protein variants. This iterative procedure of **directed evolution** continues as long as required.

may be called **directed evolution** of protein function. Mutant genes are generated and then protein variants are expressed and selected for by a specific assay designed to provide the artificial selection pressure. The most improved mutant protein variant is then identified and then the whole process is repeated again, starting from the gene for the improved mutant protein variant (Figure 10.7).

10.3.1 Random mutagenesis and PCR

A useful method for the generation of molecular diversity starting from a given gene is to use a method of **random mutagenesis** by error-prone PCR (Figure 10.8). Using error-prone PCR, the mutation rate must be carefully tuned; beneficial mutations are rare but deleterious mutations are common. If the mutation rate is too high, then all beneficial mutations might be inactivated by secondary deleterious mutations. Conversely, if the mutation rate is too low then too few beneficial mutations will be observed. However, this method has significant drawbacks in that usually only one advantageous mutation will take place per selection event. Even by using sequential random mutagenesis, it is a relatively slow process to evolve a protein, and many useful mutations may be discarded (whose usefulness might only be uncovered through the presence of a second mutation), and other negative mutations are likely to linger.

There are many ways of experimentally performing error-prone PCR. The standard PCR enzyme *Taq* **polymerase** is remarkably accurate, so different approaches are required to increase the error rate, such as using Mn^{2+} in place of Mg^{2+} and varying the levels of dNTPs away from equimolar amounts. However, it is very difficult to eliminate **bias** from an error-prone PCR experiment. For instance, the characteristics of the polymerase mean that some mutations are more likely to occur than others (**error bias**). There are also problems in the nature of the genetic code, as some amino acids are coded for by single codons, whereas others are coded by many different codons. This means the likelihood of substituting for a particular amino acid within a particular codon by mutation varies from amino acid to amino acid (**codon bias**). A further source of bias is in the iterative nature of PCR itself; any mutation that occurs early in PCR is likely to occur in a significant percentage of the final DNA sequences (**amplification bias**). An alternative approach to error-prone PCR is to use **chemical mutagenesis**, where chemical mutagens are used to introduce less biased variations into the gene coding the original protein of interest. Another approach is to use **mutator strains**. These are bacterial strains that have defects

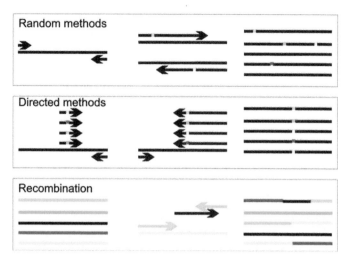

Figure 10.8 **Methods for the randomisation of DNA sequences using PCR.** Each line in the left column represents DNA duplex and each arrow a primer. The 5'-sense primer initiates strand copying using the complementary strand as a template and the 5'-complementary primer initiates strand copying of the complementary strand using the sense strand as a template. **Random methods**, such as **error-prone PCR**, introduce random changes at positions throughout the gene sequence. **Directed methods** will introduce random changes at a specific position(s) defined by lesions in the primers themselves. **Recombination methods** such as **family shuffling** start from a **library of homologous genes** from related species that can be artificially fragmented by DNase I treatment and recombined by PCR (without added primers) leading to **chimeric genes** that code for **chimeric proteins**. This represents a high level of **molecular diversity** upon which to exercise selection screening looking for novel or improved characteristics.

in the DNA repair pathways, resulting in a far higher mutation rate than usual in the DNA carried in these strains. A significant problem with both these approaches is that there is little control over the extent of mutation. Mutations will be found in the chromosome as well as in the construct harbouring the intended target gene. In addition, when using mutator strains, there is no simple control of the rate of mutation. Accordingly, error-prone PCR has generally superseded both of these methods as the preferred method to bring about random mutagenesis due to the greater control of mutagenesis rate and position that this technique is able to deliver on.

10.3.2 Mutagenesis and DNA shuffling

A great step forward was made with the development of DNA shuffling as a combinatorial approach for searching "sequence space". DNA shuffling consists of mixing together similar sequences harbouring mutations (Figure 10.8). **DNA shuffling**, or **sexual PCR**, solves the problem concerning the loss of potentially useful mutations in using error-prone PCR, mentioned above. Performing DNA shuffling involves a number of steps. First, **random mutations** are introduced into a sequence using error-prone PCR. These random mutations are initially screened to evaluate those that are apparently beneficial. Second, these beneficial mutants are then mixed together and gene sequences are randomly fragmented using the enzyme DNAse I, then recombined by performing a PCR reaction in the absence of added primer. Consequently, each beneficial mutant is embedded in a mixture of recombined genes at different locations in the gene sequence. Furthermore, new mutations might also be generated during this stage. Finally, primers are added before a final cycle of PCR to obtain a full-length PCR product. An extension of DNA shuffling approach is to start from a **library of homologous genes** that were originally from related species. This approach has been termed **family shuffling**, and it is possible to generate interesting chimeric genes and proteins that incorporate features of many of the parent genes into a single gene coding for a single polypeptide. In many ways, DNA shuffling is much closer than error-prone PCR to the events that give rise in nature to the pool of random mutations upon which the forces of natural selection operate for the evolution of improved protein activities or new functions.

10.3.3 Oligodeoxynucleotide cassette mutagenesis

Where more limited randomness is either desired or required, then selected sequences of the gene coding for the protein of interest may be targeted by oligonucleotide cassette mutagenesis. Typically, a short region of the gene, 10–30 bases in length, is selected for mutagenesis and primers may be chosen for focused error-prone PCR within that short region, after which the amplified region is recombined back into the wild-type gene replacing the original short region PCR. Alternatively, multiple "site directed mutagenesis" experiments can be performed (see Chapter 3), using **oligodeoxynucleotide mutagenesis primers** ("dirty oligos") that are "**spiked**" at each position of the primer with a small percentage of the three alternative nucleotides to the correct, complementary Watson-Crick base pair nucleotide at each position. "Spiking" is designed to ensure that each mutant gene generated at the conclusion of the mutagenesis experiment will possess only one or possibly two base mutations per mutant gene. Accordingly, the "spiking" should ensure that as close as feasible every possible single point amino acid residue mutation may be sampled in a mutant library across the entire range of the chosen short region within the original protein of interest. This process represents **random** or **saturation mutagenesis** in the purest sense (Figure 10.9). One the other hand, mutagenesis primers may instead be designed to promote only single point mutations of single amino acid residues of a similar type, class or identity. There is in fact a huge range of possibilities using cassette mutagenesis for the creation of molecular diversity within a limited region of structure within a given protein of interest.

10.3.4 Screening strategies

The number of protein variants that can be screened in a molecular breeding experiment is often limited by the speed and convenience of the assay for the activity that is desired. Therefore, the screen should be well designed so that there is an easy way of identifying improved proteins when they arise. For instance, if improved fluorescent properties are required in a molecular breeding experiment then the screen only needs to measure gains in protein fluorescence above a set threshold. Other objectives of a molecular breeding experiment may be to change the conditions under which biocatalysis take place, for example in non-aqueous solvents, under different conditions of pH or at increased temperatures. In these cases, simple screens can also be devised to identify improved proteins. Where altered enzyme specificity is required, the design of effective screens becomes more challenging and especially so if *de novo* biocatalytic activity is required. The following worked example should serve to demonstrate how a **screening process** integrates with the mutagenesis steps in a molecular breeding experiment to lead to the identification of improved proteins with the desired functions.

The trials and difficulties of the methodologies that lie behind a directed protein evolution experiment linking molecular diversity with effective screening, can best be touched on by example. **Asymmetric catalysis** is the ability of a chiral catalyst to catalyse an achiral substrate to a chiral product with a bias for the formation of one of the enantiomers. There is a huge demand for chiral products throughout the chemical industry, in particular for pharmaceuticals, where it is often the case that for chiral drugs only one of the enantiomers is pharmaceutically active and in the most infamous example of thalidomide the other enantiomer can actually be harmful. Despite the demand, asymmetric catalysis remains a significant challenge using conventional organic chemistry. One approach to meeting the challenge is in the use of directed evolution of a carefully selected enzyme to evolve the requisite, desired enantioselectivity. There are two main challenges in realising such a molecule breeding experiment:

1. Designing methods to introduce molecular diversity that are likely to be beneficial.

2. Designing a simple high-throughput screen to identifying improved proteins.

In the following case, we shall look at the molecular breeding of the **lipase** from the bacterium *Pseudomonas aeruginosa*. This particular molecular breeding experiment was the first example of the directed evolution of an enantioselective enzyme. The wild-type enzyme catalyses the hydrolysis of esters to carboxylic acids (Figure 10.10), and shows very little enantioselectivity: only 2% enantiomeric excess biased towards the (S) configuration.

In the original experiment, molecular diversity was introduced by error-prone PCR, used under conditions designed to ensure that only one or possibly two base mutations were introduced per mutant gene in an attempt to ensure that every possible single point amino acid residue mutation could be sampled at least across the entire primary structure of the original protein of interest (Figure 10.11). The numbers start to speak for themselves. The lipase of interest consists of 285 amino acids. Each amino acid could be mutated to any one of the 19 other natural amino acids, so therefore there can be 5415 possible single point amino acid residue mutants produced. This figure rises where double point mutants are concerned to 15 million possible unique variations, onwards to 52 billion possible variations when unique triple point mutants are envisaged. Such numbers are normal for protein mutant libraries and reflect the size of the challenge for efficient and effective screening. **High-throughput screening** (HTS) is indispensable

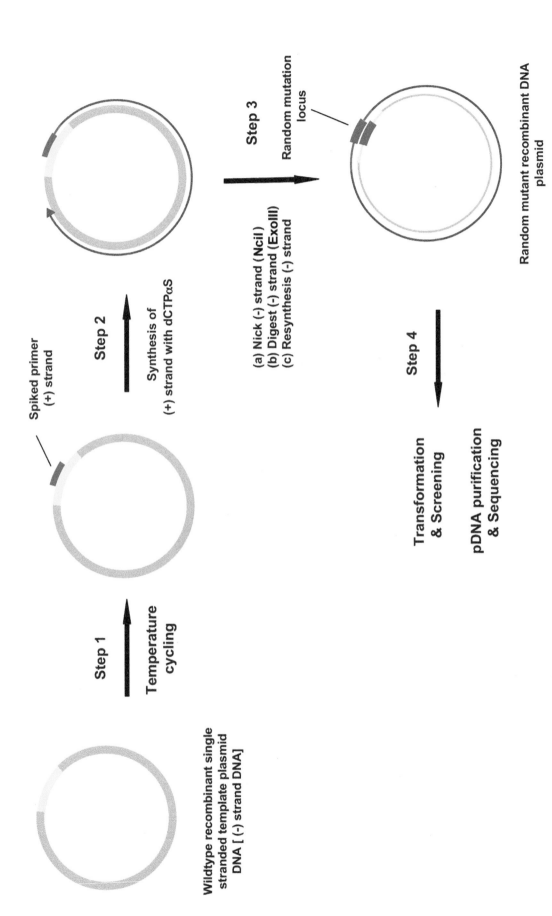

Figure 10.9   Non-PCR random mutagenesis. Overview of a **random site-directed mutagenesis method**. Obtain template DNA [ssDNA (–) strand]. Step 1: anneal **spiked oligodeoxy-nucleotide mutagenesis primer (blue)** with desired mutation range; Step 2: extend and incorporate mutagenic primer into new (+) strand containing dCTPαS; Step 3: digest the template DNA (–) strand with **NciI** and **ExoIII**, then resynthesise (–) strand to include mutations into (–) strand as well; Step 4: transform **E. coli** with random mutant recombinant DNA plasmid ready for selection, DNA purification and sequence identification of spiked mutant recombinant DNAs.

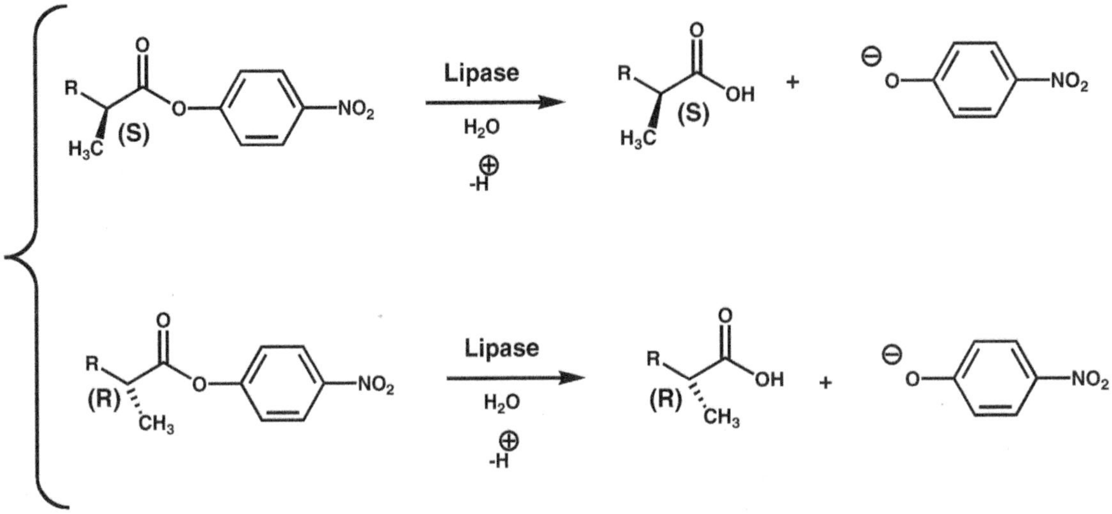

Figure 10.10 Directed evolution of enantioselectivity. A lipase from *Pseudomonas aeruginosa* was chosen for directed evolution of enantioselectivity.

given such molecular diversity. In the case of this lipase example, the assay used was a simple UV-visible spectroscopy assay monitoring effective enzymatic hydrolysis in terms of the appearance of the *p*-nitrophenolate anion (detected by absorbance at A_{405}). Each mutant protein was tested in a pairwise assay, testing hydrolysis of both the (S)- and the (R)-substrates. The relative rates of hydrolysis gave a measure of the enantiomeric selectivity (the **enantiomeric excess, *ee*** in %) as given by Equation (10.1):

$$ee = \left(\frac{[R]-[S]}{[R]+[S]}\right)\cdot 100 \tag{10.1}$$

where [*R*] and [*S*] are the molar concentrations or the (R)- and (S)-products respectively.

In the initial screening, around 1000 mutant enzymes were generated from the first (and subsequent rounds) of mutation, and of these approximately 1% showed an increased enantioselectivity, up to 31% *ee*, a significant improvement over the wild-type enzyme. This mutant enzyme was then subjected to three further rounds of mutation and selection, after which a new mutant was identified able to deliver on an enantioselectivity of up to 81% *ee* (corresponding to an **E value of enantioselectivity** of 11.3). Saturation mutagenesis was then performed at local "hotspots", designated in the primary sequence where amino acid residue changes were thought most likely to be beneficial, leading to mutant enzyme competent to produce an E value of 25. Further improvements to this efficacy were obtained by a new error-prone PCR step modified to introduce around three mutations per gene per generation, followed by DNA shuffling, resulting in a mutant protein with an E value of 32 post selection. Thereafter, a modified form of DNA shuffling named **combinatorial multiple-cassette mutagenesis** resulted in the identification of a significantly improved mutant with the ability to deliver on an E value of >52.

10.4 Directed evolution of nucleic acids

The evolution of nucleic acids is a more direct process when compared to the evolution of proteins. For directed protein evolution, the full cycle of transcription and translation is required to take place, and this generally means that stages of the process generally have to be performed using a living system to translate the protein. Selection and evolution of nucleic acids can, however, be routinely carried out using an entirely chemical *in vitro* process: ***in vitro* molecular breeding**. The binding strengths of nucleic acids can approach those of proteins, though the catalytic repertoire of nucleic acids is significantly reduced. In this section, we shall describe approaches for the selection and evolution of both RNA and DNA for both binding and catalysis.

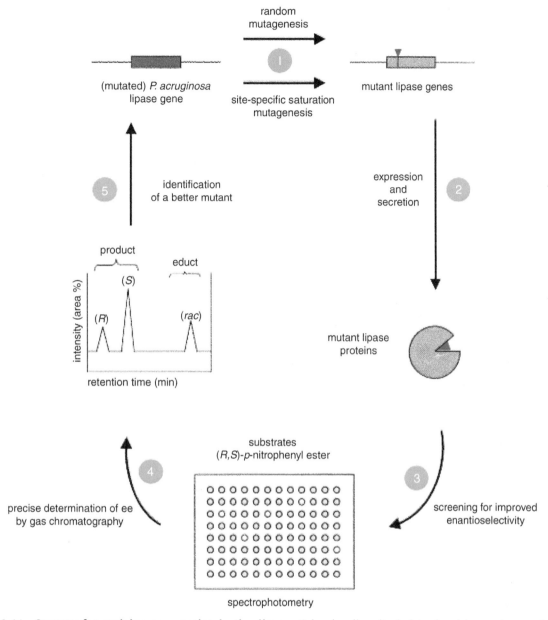

Figure 10.11 Strategy for evolving an enantioselective lipase. Molecular diversity is introduced by **random mutagenesis** of the lipase gene and site-specific **random** or **saturation mutagenesis.** Mutant genes resulting are segregated and separately expressed in individual bacterial clones. Protein product is secreted into extracellular medium and a spectrophotometric assay used to assay for nitrophenolate production from **(R)** and **(S) substrates** by action of active lipase enzymes (see Figure 10.10). Relative quantities of **(R)** and **(S)** product are quantified by gas chromatography and where enantioselectivity is also discovered then the corresponding gene is harvested for another round of mutagenesis and selection (illustration from Liebeton *et al.*, 2000, Figure 5, Elsevier).

10.4.1 Aptamers

Nucleic acids can play roles far beyond merely harbouring the coding information for proteins. Single-stranded nucleic acids can fold into intricate structures capable of molecular recognition and even catalysis. 3D structures are specified by the primary structure, namely the deoxynucleotide (or nucleotide for RNA) sequence (5′→3′) by analogy to the situation in which the amino acid residue sequence determines the 3D structures of polypeptides. In nature, **transfer RNAs (tRNAs)** use their 3D shape for molecular recognition, while some **ribosomal RNAs (rRNAs)** are able to catalyse crucial steps even within the protein synthetic pathways themselves.

From a technology perspective, a massive pool of nucleic acids (either RNA, DNA, or modified nucleotides) that provides for the required molecular diversity, represents a huge library from which to select, amplify and identify novel binding partners (**aptamers**) or even novel biocatalysts (**aptazymes**). This type of molecular breeding is known as the **systematic evolution of ligands by exponential enrichment** (**SELEX**) process (Figure 10.12). Using the SELEX process, it is possible to evolve naïve nucleic acids until they are capable of binding specifically to proteins, peptides, cells, nucleic acids and even other small molecules. Nucleic acids capable of catalysing chemical reactions may also be evolved. SELEX may be used to select from both RNA and DNA libraries. RNA has greater potential structural diversity than DNA; hence affinities for a potential target should be generally higher overall. DNA has less structural diversity than RNA; hence affinities for a potential target should be generally lower overall, albeit with a higher degree of overall intrinsic stability. SELEX steps are outlined below.

10.4.1.1 Design and construction of a polydeoxynucleotide/polynucleotide library

A polydeoxynucleotide library is synthesised for molecular breeding by standard solid phase methodologies, except that **randomised single-stranded (ss) DNA(pre-template DNA)** is synthesised with a 3′-flanking region of known sequence that can be recognised by a primer in a PCR reaction. Randomisation is achieved by using an approximately equal mix of the four phosphoramidite building blocks during solid phase synthesis, weighted according to differences in coupling efficiencies in order to ensure lack of deoxynucleotide residue bias in the final library. When a fully randomised polydeoxynucleotide is prepared in this way then the molecular diversity is equivalent to 4^n, where n is the number of randomised nucleotide residue positions, so leading to a total library of 10^{15} different polydeoxynucleotides, assuming that n is 25. If n is increased to 30, then the molecular diversity increases to a library of 10^{18}. Beyond this, molecular diversity becomes unmanageable. For instance, if a stretch of 100 random positions were to be studied, then the molecular diversity would approach the number of elementary particles in the universe. Therefore 25–30 randomised positions are used in preference. After the pre-template DNA is synthesised, a primer is annealed to the known sequence region and complementary strands are synthesised using the **Klenow fragment** of DNA polymerase I, yielding **randomised double-stranded (ds) DNA (template DNA)**.

In nature, natural RNA-protein recognition sites comprise 15–25 nucleotide residues, suggesting that a high-affinity binding polynucleotide should also be 25–30 nucleotide residues in length too. Therefore, an appropriate **RNA library** may be generated by direct transcription from template DNA by means of **T7 RNA polymerase** operating from a **T7 promoter** element inserted during synthesis of the template DNA. Often modified nucleotides are used in place of the natural ones to confer extra stability on RNA molecules produced. This is especially important in the design of aptamers for therapeutic purposes. The most common approach is to modify β-D-ribofuranose rings attached to pyrimidine bases with 2′-F or 2′-NH$_2$, modifications that confer resistance to most RNAases. The 5-position of uracil base may also be modified. Fortuitously, the T7 RNA polymerase is able to tolerate the insertion of modified bases and β-D-ribofuranose rings at the 2′-position reasonably well.

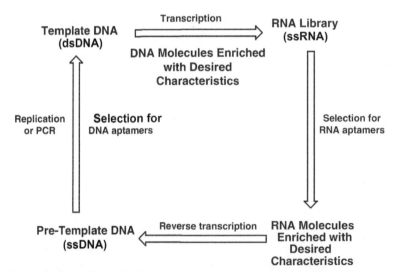

Figure 10.12   A general scheme for the *in vitro* selection and evolution of RNA aptamers. This is otherwise known as **SELEX** (systematic evolution of ligands by exponential enrichment).

10.4.1.2 Partition, amplification and iteration

The next stage in the molecular breeding of functional DNA molecules from template DNA should be simple screening in order to distinguish those DNA molecules that are able to perform the required task from those that are not. For instance, if DNA aptamers are desired then the target ligand could be immobilised on a column and **affinity chromatography** used: nucleic acids that do bind to the column would bind to the target ligand, whilst those that do not would pass through. An alternative that is often used in SELEX experiments is to immobilise the target on a **nitrocellulose filter**. Partition is the most critical variable of a SELEX experiment, it is important to be able to efficiently cut down the initial complexity of 10^{15}–10^{18} sequences in a template DNA library to a manageable number of sequences in as few rounds of screening as possible. Part of this process is also to prevent or minimise the possibility of non-specific binding of nucleic acids to column or filter media giving rise to false positives. Several rounds of screening may be carried out to refine the pool of high-affinity DNA binders to the target ligand of choice before proceeding to sequence characterisation.

In the case of RNA library screening for RNA aptamers, the process involves first the isolation of functional mRNA that survives the selection procedure, and then this selected pool of mRNA is converted into next-generation pre-template DNA by reverse transcription and thereafter into next-generation template DNA by means of PCR (this is equivalent to the cDNA synthesis described previously, see Chapter 3). Transcription by means of T7 RNA polymerase is subsequently performed so that RNA selection may then begin again, starting from this reduced, next-generation mRNA library of "positive hits". Selection then proceeds through further complete rounds of target ligand binding and release until high-affinity RNA binders to the target ligand of choice are found. If required, next-generation template DNA can be introduced into a cloning vector and individual clones may be sequenced (see Chapter 3).

Where the identification of either DNA or RNA aptamers is required, alignment of either set of sequences post several rounds of screening should lead to the identification of critical consensus sequences, required for the desired aptameric activity. Often this results in **families of aptamers**. Competition experiments may then be performed to see whether these families compete for the same target. The aptamer families selected can then be refined in a number of ways. First, **truncation** may be used to determine the minimum length aptamer capable of binding to the target with high affinity. Sometimes possible truncations may be clear from the alignments of the various sequences; truncations can be easily designed and tested using standard molecular biology techniques. Alternatively, digestions from either end may be attempted, then separated against the target to define the exact length required for high-affinity binding. Several structures of aptamers in complexation with their targets have been solved (Figure 10.13 and Figure 10.14).

10.4.1.3 Applications of aptamers

Many parallels may be seen between **therapeutic aptamers** and **therapeutic antibodies**. Although therapeutic antibodies have perhaps a more established reputation, aptamers compete well with antibodies in the following respects:

1. Aptamers are selected in an entirely *in vitro* process that allows for specificity and affinity to be tightly controlled, and allows selection against toxic or non-immunogenic targets.

2. Aptamers generally bind more strongly to a target than an antibody and are able to disrupt protein-protein interactions.

3. Aptamers show little immunogenicity and no toxicity, and can be delivered sub-cutaneously.

4. Aptamers are easily synthesised, readily producible in bulk and easily stored as lyophilised powders.

Similarly to antibodies, aptamers are able to bind to virtually any target with an affinity in the nM to mM range. In order to play a therapeutic role, they must bind tightly and inhibit a specific biomolecule function by binding tightly whilst showing no harmful side effects. The first aptamer tested in a therapeutic context was a DNA aptamer against **thrombin** that was able to prolong blood clot time *in vitro*. Aptamers are also playing increasingly important roles in molecular diagnostics, for example they can be linked to fluorescence tags and be converted into beacons to signify the presence of a key disease marker at very low levels of detection.

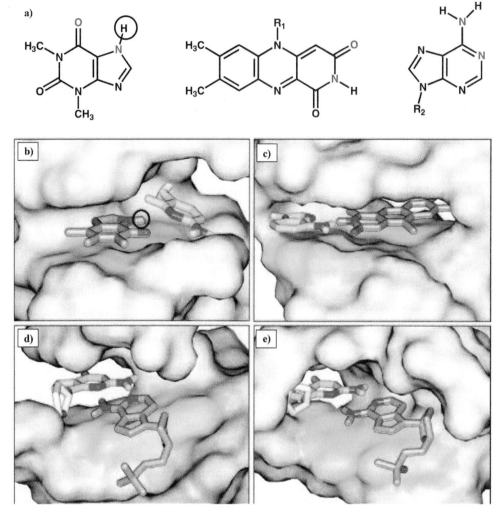

Figure 10.13   SELEX enables enhanced ligand binding. Shown are a set of small molecules [panel (**a**)] and used to select for powerful RNA aptamers [(**b**)–(**e**)]. Small molecules are rendered in **stick representations** and aptamers are depicted using **Van der Waals surface** representation (illustrations from Hermann and Patel, 2000, Figure 1, American Association for the Advancement of Science).

10.4.2 Selection of catalytic RNA

There are two general steps to evolving a catalytic RNA. First, from a pool of sequence variants there must be a screening protocol that separates active RNA with the desired trait from others that do not. Second, the genome of the survivors must be copied and amplified, via DNA intermediates, prior to the next round of selection. The process is iterative and heavily reliant on the ease of nucleic acid copying by PCR amplification. Thereafter, there are two methods of selection, direct or indirect. In a **direct selection**, there is chemical transformation of the catalytic RNA during the selection step. In an **indirect selection**, an RNA aptamer approach is taken, but RNA aptamers are identified and evolved to bind to a transition state in the reaction that is desired to be catalysed (this indirect method is similar to the strategy used to develop catalytic antibodies, see Section 10.5).

In **direct selection**, one needs a method to separate catalytic vs non-catalytic RNA. There are three general approaches:

1. **Differential migration** by **polyacrylamide gel electrophoresis (PAGE)**

2. **Primer-binding site tagging** technique

3. **Affinity probe tagging** technique

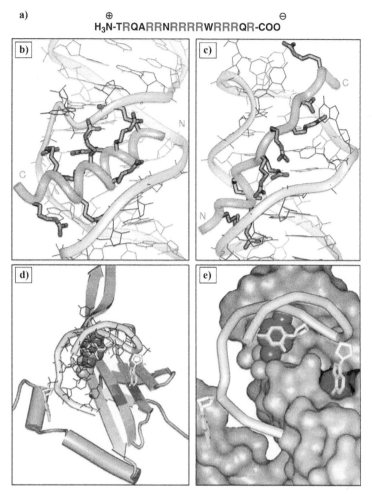

Figure 10.14 **Molecular recognition of peptides and proteins by nucleic acid aptamers**. A peptide is shown [panel (**a**)], alongside its selected nucleic acid aptamers [(**b**) and (**c**)]; binding of proteins by nucleic acid aptamers [(**d**) and (**e**)] is also shown. Aptamers are rendered throughout in **ribbon display structure** representation (**grey**); peptide (**yellow-brown**) is shown by **CA stick display** in (b) and (c) with amino acid residue side chains in **stick representations**; protein (**yellow-brown**) is shown as **schematic display structure** (d) and in a **Van der Waals surface** representation (e). Protein binding pockets are illustrated in **red** (illustrations from Hermann and Patel, 2000, Figure 4, American Association for the Advancement of Science).

In differential migration, the catalytic RNA itself is the product of the reaction and therefore may be selected. For example, the target may be self-cleavage of RNA or the ligation of RNA that is already covalently linked to the catalytic region elsewhere. Clearly, if a reaction takes place then the result should have a different electrophoretic mobility to the situation where no reaction takes places. Hence, PAGE may be used to separate cleanly reacting from non-reacting scenarios. Alternatively, the primer-binding site tagging technique is useful where, for example, a ligation reaction is being analysed for. In this case, successful ligation introduces a stretch of RNA sequence onto the catalytic RNA itself that could double as a PCR primer site. Consequently, successful PCR amplification will only be possible if the ligation reaction has succeeded in the first place. Therefore, only RNA that is properly catalytic can and will be amplified by PCR amplification. Finally, affinity probe tagging is useful for the identification of catalytic RNA that is able to catalyse other chemical reactions. The chemical reaction of interest should be designed to result in the RNA self-tagging with an affinity probe such as biotin. Hence, correctly catalytic RNA may then be simply separated from non-catalytic RNA by streptavidin protein affinity chromatography.

A number of attempts have been made to evolve catalytic RNAs or ribozymes from **natural ribozymes** and also from pools of random sequence RNA. Natural ribozymes have the advantage that a natural active site already exists and so can be evolved. Such approaches have had some success in improving or altering catalytic properties. Otherwise, ribozymes that have been evolved so far include catalytic RNA nucleases, ligases and polynucleotide kinases. Similar approaches have also been employed to evolve ribozymes that catalyse reactions at carbon bonds, such as functional group transfers, including *N*-alkylation (e.g. using the 2-amino functional group of guanosine base to displace iodide from iodoacetamide) and S-alkylation (e.g. using 5′-phosphorothioate to displace bromide from an *N*-bromacetyl peptide). In general, the structure of catalytic RNAs that catalyse reactions at carbon bonds tends to be more complex than those that catalyse reactions at phosphodiester bonds, reflecting the more complex conformations required to bind non-nucleotide substrates. Overall, natural ribozymes turn out to be much more efficient catalysts than evolved ribozymes, so there is much room for further research in this area.

10.4.3 DNA aptamers and catalytic DNAs

Catalytic DNA is made possible because single-stranded DNA can adopt complex tertiary structures, in a similar way to RNA, although unlike RNA no DNA-based catalysts have yet been found in nature. Both DNA aptamers and **DNA catalysts (deoxyribozymes)** can be evolved by molecular breeding experiments equivalent to those used to evolve RNA aptamer and ribozymes. However, DNA polydeoxynucleotide sequences are less flexible and accommodating than corresponding RNA polynucleotide sequences owing to the absence of a 2′-hydroxyl group in each sugar ring of each deoxynucleotide residue. Nevertheless, DNA aptamers and deoxyribozymes are significantly more stable to most nucleases than unmodified RNA aptamer and ribozymes. The possibility of deoxyribozyme catalysts has reignited the debate about the molecular origins of life and reopened the RNA world concept to further scrutiny.

The first single-stranded DNA aptamers identified were found to bind thrombin with a K_d value of 25–200 nM. The central core (15–17 residues in length) was found to assume a **G-quadruplex structure (G4)**, a very common structural motif in both DNA and RNA aptamers (Figure 10.15). DNA aptamers that bind other proteins have since been found to have several other defining structural features, including bulges, stem-loops and pseudoknots. DNA aptamers have also been enriched against small molecules. For instance, a DNA aptamer has been evolved to selectively bind ATP with a 6 mM affinity. DNA aptamers have also been enriched that bind to a variety of porphyrin rings. Curiously, where RNA aptamers have been evolved to bind the same target ligands, they show no sequence homology with DNA aptamers at all. In other words, DNA and RNA aptamers are in fact structurally diverse, although functionally equivalent.

Going beyond DNA aptamers, catalytic DNA, or **deoxyribozymes**, have also been evolved to cleave RNA or DNA, metalate porphyrins, cleave *N*-glycosidic bonds or ligate DNA. The majority of deoxyribozymes isolated

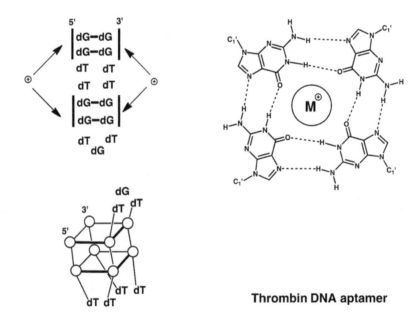

Thrombin DNA aptamer

Figure 10.15 Structural order in DNA aptamers. Some DNA aptamers possess higher order structural elements based upon the planar **G-tetrad** structure (on right) which is found in many DNA aptamers. Solid bars indicate regions of **G-quadruplex structure** formation by G-tetrad stacking (on left). The ion in the **thrombin DNA aptamer** (illustrated) is typically Na⁺ or K⁺. G-tetrads can stack to form G-quadruplex structures wherein phosphodiester backbones are either parallel or anti-parallel with respect to each other. G-quadruplex structure motifs are also identified in special regions of eukaryotic chromosomes known as **telomeres**.

catalyse RNA cleavage by phosphoester transfer, usually in the presence of a metal cofactor. The general approach for discovery of a cleaving deoxyribozyme is as follows. An initial population of ssDNA is synthesised that incorporates a primer target site, a target sequence, a randomised sequence, a second 3′-primer target site and a 5′-end biotin label for simple immobilisation. This multifunctional polydeoxynucleotide is first immobilised on a streptavidin protein matrix, then non-binding DNA molecules are washed away. Those immobilised DNA molecules are then incubated under suitable reaction conditions, importantly incorporating potential deoxyribozyme metal cofactors. Thereafter, successful catalytic DNA molecules that self-cleave from the column are washed clear and then amplified using PCR and a biotin-labelled primer. The resulting dsDNA from PCR amplification is then rebound to the streptavidin matrix. The process is repeated until the catalytic activity of the deoxyribozyme reaches a plateau. At this stage the evolved deoxyribozyme may then be eluted, sequenced and the different species aligned to define the catalytic motif. In order to evolve deoxyribozymes against other fissile bonds or links, the screening procedure is actually the same, although the appropriate target cleavage site should be included in the initial multifunctional polydeoxynucleotide.

The process of discovering a deoxyribozyme capable of ligation of RNA is an excellent example of catalytic DNA development. The process was begun with an initial multifunctional polydeoxynucleotide designed to possess two regions complementary to two arbitrary RNA substrates (Figure 10.16). Between these two arms was positioned a

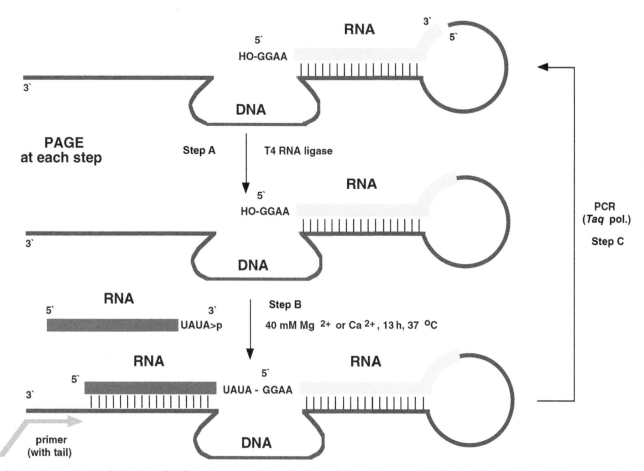

Figure 10.16 Deoxyribozyme selection. An approach to selection of deoxyribozymes that ligate RNA. DNA template is represented by dark grey line. The DNA loop contains random sequence variations to allow for the possibility of catalysis. **Step A: right-hand side RNA (yellow)** is ligated to a DNA template with a 5′-terminal overhang; **Step B: left-hand side RNA (red)** is Watson-Crick base pair associated with the same DNA template but with a 3′-terminal overhang. A ligation reaction is then promoted. Where ligation is possible, the PAGE electrophoretic mobility of the product will differ from the unligated situation; **Step C:** PCR with *Taq* polymerase (*Taq* pol.) may then be used to determine the DNA sequence(s) responsible for RNA ligation catalysis. Further rounds of maturation are also possible (illustration adapted from Flynn-Charlebois *et al.*, 2003, Figure 1B).

randomised sequence wherein ligase activity was to be evolved and enriched. Otherwise, a few bases of the RNA were left unpaired to minimise steric hindrance with respect to DNA. In the first step of selection, **right-hand side RNA** was annealed to the DNA and ligated using T4 RNA ligase enzyme. PAGE was then used to separate out and purify any DNA/RNA duplex formed. In the second step, the **left-hand side RNA** incorporating a 2′, 3′-cyclic phosphate was added. This second RNA was expected to anneal by Watson-Crick base pairing to the available DNA binding site to set up the opportunity for ligation. Any such potential ligation reactions were also enabled through the incorporation of potential deoxyribozyme metal cofactors again. Successful ligation was diagnosed by a change in electrophoretic mobility on polyacrylamide gels of the ligated product as compared to the mobility of the non-ligated, uncatalysed situation. In this way, catalytic DNA was gradually separated from non-catalytic DNA. This whole process was actually repeated in an iterative manner with multiple cycles until most of the pool of DNA remaining was observed to have some ligase activity.

10.5 Catalytic antibodies

Many parallels can be drawn between the use of aptamers and the use of antibodies. Of course, the critical structural difference is that aptamers are based on polynucleotide or polydeoxynucleotide chains, whilst antibodies are based on polypeptide chains and in particular the immunoglobulin fold. A vertebrate immune system typically has the capacity to biosynthesise in immune cells known as B-cells, at least 10^{10} structurally distinct antibodies, each with a specific affinity to foreign materials known as antigens (see Chapter 7). Antibody production comprises part of the **humoral immune response** mounted against invading pathogens or other foreign agents. If a chemical ligand (**hapten**) is coupled to a **carrier protein** such as **keyhole limpet haemocyanin (KLH)** or **bovine serum albumin (BSA)** then the bioconjugation product can be very immunogenic with respect to the humoral immune response. Injection of such a product into a mouse or rabbit creates a substantial blood-borne response that is reflected by huge B-cell production in their spleens. Hence, these animals are sacrificed and their spleens extracted for B-cell populations that are fused to cancer cell lines such as myeloma cells, resulting in **immortalised B-cell lines (hybridomas)**. Hybridomas are intended to act as antibody cell factories, and each hybridoma should be able to produce a unique antibody specific to a different antigen. Where hybridomas are responsible for the production of anti-hapten antibodies, they are identified by cell population screening, isolated and cloned by cell culturing. Each hybridoma so cloned should be responsible for the production of one single type of antibody (**monoclonal antibody [MAb]**). Typically, monoclonal antibodies must be subject to final screening to identify that population of monoclonal antibodies optimal for hapten binding.

According to the most advanced theories of enzyme catalysis, **differential stabilisation** of reaction transition states relative to ground states is the most effective way to effect catalysis irrespective of whether there is an excess or lack of substrate available for biocatalysis (see Chapter 8; Figure 10.17). Therefore, in order for an antibody to act as an enzyme, the antibody should no longer possess an optimal affinity in binding to a ground state molecule such as an antigen, but to a reaction transition state. In principle then, any monoclonal antibody raised against a **transition state analogue** should be able to act as a biocatalyst for the corresponding reaction (Figure 10.17). The design of a transition state analogue requires an element of chemical surrogacy and imagination. For instance, phosphonate appears to be a useful surrogate for the rate determining tetrahedral transition state formed by the addition of water to an ester, prior to the second alcohol elimination step. Indeed, the phosphonate group is not only mono-charged but the P-O bond (1.52 Å) is intermediate in length between the C-O bond found in the actual tetrahedral intermediate and the presumed C-O bond length projected in the transition state. Other examples of transition state analogues and catalytic antibody catalysis are shown in Figure 10.18. Catalytic antibodies may affect biocatalysis of biomolecular reactions such as transamidation also (Figure 10.19). A word of caution though, as with natural ribozymes so with natural enzymes, on the whole catalytic antibodies do not affect biocatalysis with the same efficiency as natural enzymes. Indeed, they usually fall a long way short. Hence, in order to improve catalytic antibody mediated biocatalysis, alternative, additional theories of catalysis have been employed, including **general acid-base**, **nucleophilic** and/or **electrophilic catalysis** (see Chapter 8; Figure 10.20). Advances in the area of catalytic antibodies continue to be made.

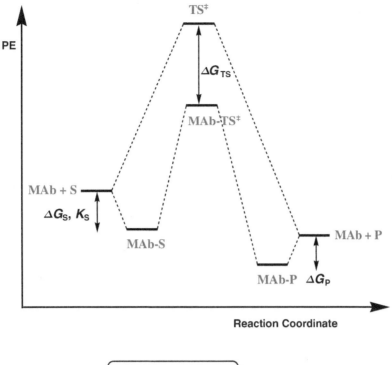

Figure 10.17 Illustration of catalytic antibody catalysed chemical conversions. Catalytic **monoclonal antibody (MAb)** mediated conversion of substrate **S** to product **P** is illustrated versus the uncatalysed situation. Catalysis by antibody is made possible by **differential stabilisation** through binding of the rate determining step transition state for the reaction relative to **S** and/or **P**. This is also a fundamental for enzyme-based reaction catalysis as well (see Chapter 8).

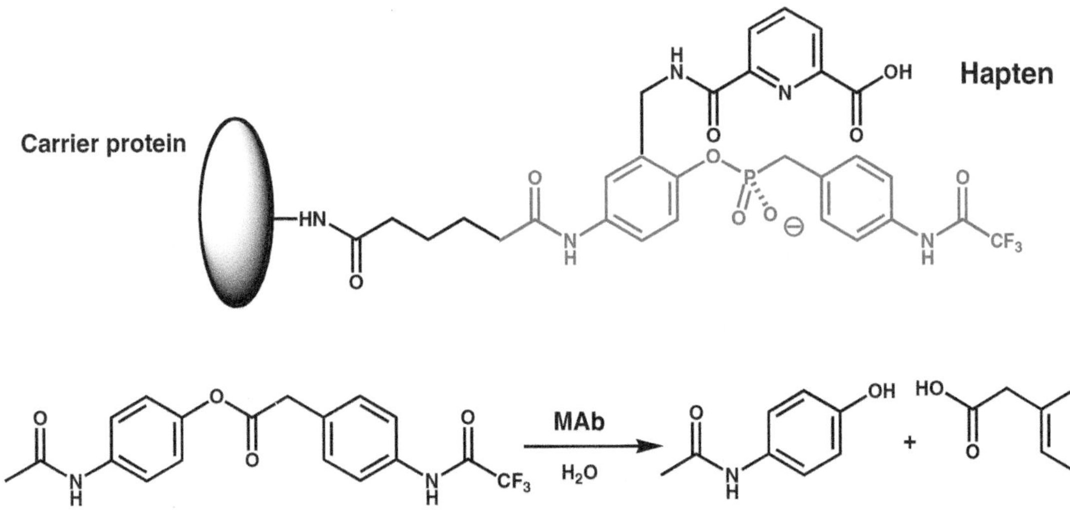

Figure 10.18 Catalytic antibody mediated ester hydrolysis. An appropriate catalytic MAb is generated through immunoreaction with the illustrated **hapten** linked to an appropriate **carrier protein**. The part of the hapten related to the substrate structure is shown in red. The phosphonate link is used as a **transition state analogue** of the rate determining step transition state leading to the key tetrahedral intermediate of ester hydrolysis. The catalytic MAb should optimally bind the rate determining step transition state relative to substrate or products in order to effect maximum catalytic effect.

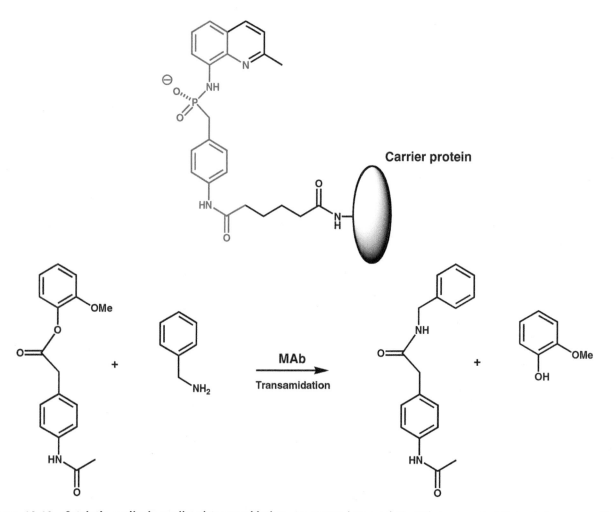

Figure 10.19 **Catalytic antibody mediated transamidation.** An appropriate catalytic MAb is generated through immunoreaction with the illustrated hapten linked to an appropriate carrier protein. The part of the hapten related to the substrate structure is shown in red. The phosphonate link is used as a transition state analogue of the rate determining step transition state leading to the key tetrahedral intermediate. The catalytic MAb should optimally bind the rate determining step transition state relative to the substrates or products in order to effect maximum catalytic effect.

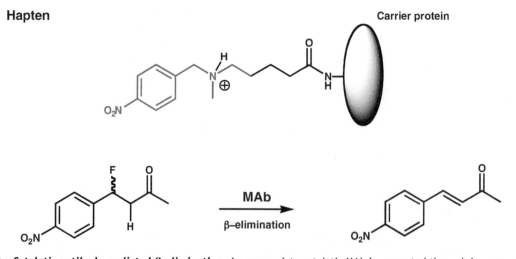

Figure 10.20 **Catalytic antibody mediated β-elimination.** An appropriate catalytic MAb is generated through immunoreaction with the illustrated hapten linked to an appropriate carrier protein. The part of the hapten related to the substrate structure is shown in red. The tertiary amine functional group is introduced to encourage the generation of catalytic antibodies with complementary bases/ nucleophiles in the vicinity of the substrate α-proton when substrate binds to a catalytic Mab, in order to facilitate catalysis of E2 elimination by a combination of **general base catalysis**, as well as the more usual differential stabilisation of the rate determining step transition state through binding. Enzymes usually employ more than just differential stabilisation of rate determining step transition states in order to effect catalysis but also employ several other physical "tricks" as well (see Chapter 8).

11

Chemical Biology of Cells

11.1 General introduction

In this chapter, we shall look at how chemical biology, as a multidisciplinary subject that aims to understand the way biology works at the molecular level inside cells, can be used to map cellular networks of interacting molecules and suggest new drug candidates for the treatment of patients.

11.2 Array technologies, microfluidics and miniaturisation

Arrays, miniaturisation and microfluidics are at the forefront of new methods to understand molecular mechanisms in biology at the level of individual molecules, at subcellular levels and at the cellular level. Critically, the emergence and inclusion of these technologies has vastly improved the range and quantity of data available to understand problems at the chemistry/biology interface more comprehensively than at any other time previously.

11.2.1 Arrays from past to present

Microarrays were developed as a platform for **high-throughput (HT) analysis** of gene expression since the 1990s (see Chapter 9). Since then, microarrays have been used in many cell biology and chemical biology applications, aided by microarrayer technologies for spotting high-density arrays on glass microscope slides. Following this, microfabrication techniques, such as **soft microlithography**, have been introduced and used to create small-molecule chemical microarrays, suitable for HT analysis (Figure 11.1). Such small-molecule chemical microarrays are defined as monolithic, flat surfaces that bear a systematic arrangement (spatially addressable) of probe sites (usually 1000 to 100,000) and each contains a small molecule (e.g. peptide, carbohydrate, drug-like molecule, natural product) that is immobilised either covalently or through non-specific adsorption. The potential value of such microarrays will be discussed below in the context of HT screening in chemical genomics.

11.2.2 Micropatterned and microfluidic devices

Micropatterned devices arise from **soft microlithography** techniques that have led to the creation of numerous structures integrating complex biochemicals or cells as required. Figure 11.2 illustrates an example of how this works. In

Essentials of Chemical Biology: Structures and Dynamics of Biological Macromolecules In Vitro *and* In Vivo, Second Edition. Andrew D. Miller and Julian A. Tanner.
© 2024 John Wiley & Sons, Inc. Published 2024 by John Wiley & Sons, Inc.
Companion Website: www.wiley.com/go/miller/essentialschembiol2

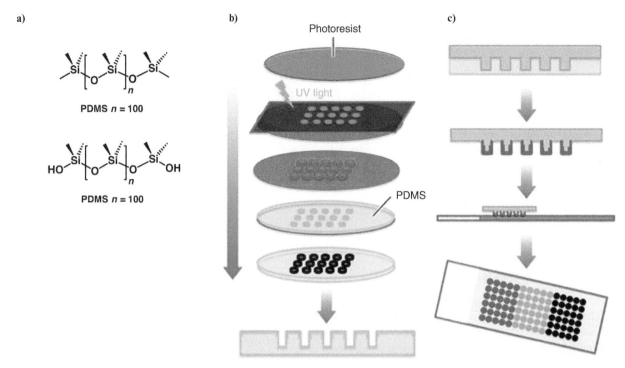

Figure 11.1 Soft microlithography for microarray preparations. Microfabricated **polydimethylsiloxane** (**PDMS**) moulds/stamps with mm scale patterns are used in soft microlithographic techniques as pattern transfer agents to modify bio-surfaces (microcontact printing) and regulate fluid flow (microfluidics): (**a**) PDMS structures; (**b**) PDMS moulds are fabricated by an initial lithography step that patterns photoresist onto a silicon wafer. Next, PDMS is cured on top of the patterned silicon wafer to create a soft or elastomeric, micropatterned mould. The mould can then be removed and used directly as a microwell platform; (**c**) The mould may be used alternatively in PDMS stamps by replica moulding, which can further transfer the patterns to culture surfaces by microcontact printing. (Taken from Figure 2 of Ashton *et al.*, 2011, *Journal of Annual Reviews*.)

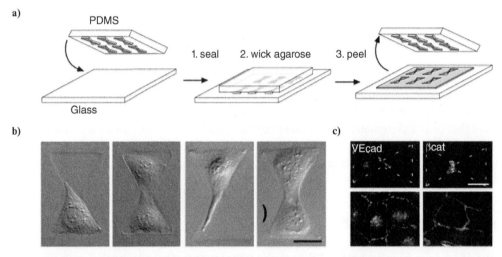

Figure 11.2 Effect of cell-cell contact on proliferation. (**a**) Schematic outline of method used to pattern substrates to control cell spreading and cell-cell contact simultaneously; (**b**) Differential interference contrast images of single cells or pairs of cells in agarose wells of 750 μm^2/half (left two images) and 1000 μm^2/half (right two images). Bovine **pulmonary artery** and **adrenal microvascular endothelial cells** were G_0-synchronised and cultured on arrays of wells for 24 h and fixed for analysis. Cells distributed randomly as single cells and pairs of cells in the wells; (**c**) Immunofluorescence images of pairs of cells in wells of 750 μm^2/half (top images) or monolayers (bottom images) stained for **vascular endothelial-cadherin (VEcad)** or β-**catenin (betacat)**. Both VE-cadherin and β-catenin specifically localised to the zone of contact. Broken lines (white) indicate the borders of the wells. (Figures taken from Nelson and Chen, *Journal of Cell*, 2003, the Company of Biologists.)

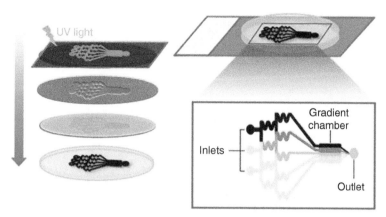

Figure 11.3 Soft lithography for microfluidics. In this case, PDMS moulds (see Figure 11.1) can be used to synthesise microfluidic devices, that are used to generate microscale gradients of soluble factors. (Taken from Figure 2 of Ashton *et al.*, 2011, *Journal of Annual Reviews.*)

this case, stamps of **polydimethylsiloxane (PDMS)** were cast on a silicon master with 20 μm-thick, raised "bow-tie" features created by soft microlithography. The stamp was positioned over a glass surface and the gaps between filled with agarose. Removal of the stamp revealed bow-tie shaped wells with agarose walls and glass bases. These glass bases were then coated with the protein **fibronectin**, after which ovine pulmonary artery or adrenal microvascular endothelial cell suspensions were plated over the substrate and retained in the fibronectin-coated wells. Each bow-tie well was designed to enable the attachment and spreading of either one cell or a pair in close cell-cell contact. Cell-cell contacts thus formed were found to comprise intercellular adherens junctions containing the proteins **vascular endothelial-cadherin** and β-**catenin**. In this scenario, cell-cell contact interactions were decoupled from cell density and cell spreading using the micropatterned surface. In doing this, cell-cell contact was shown to increase or decrease the overall proliferation rate of cells by differentially shifting the balance between cellular proliferation and anti-proliferation signals, as mediated through the cell cytoskeleton. Such an observation demonstrates the complex interplay of mechanical and chemical signals with which cells must navigate around their local microenvironment.

The ability to fabricate micropatterned features sized appropriately to generate clonal populations or single-cell arrays is uniquely convenient using soft microlithography. This technique is expected to be increasingly utilised in understanding many different cell-cell interactions and individual cell susceptibilities. In addition, micropatterned devices can be used with HT chemical genetics screens that require execution using single-cell or low-cell number arrays. Unlike features created by microarrayers, soft lithography patterned features can facilitate the synthesis of micropatterned surfaces for HT biological studies at the single-cell level owing to the μm scale of operations. Such micropatterning enables detailed simultaneous studies on the impacts of multiple microenvironmental properties on cell behaviour. Furthermore, microfluidic devices can also be prepared by soft microlithographic techniques, leading to the creation of numerous microfluidic structures that enable highly controlled studies with complex biochemical mixtures and/or cells as required (Figure 11.3). These microfluidic designs are becoming increasingly more sophisticated, allowing for the creation of spatial gradients and transient soluble factor exposure regimens in HT. Massively parallel, miniaturised cell culture platforms can be used for elaborate HT cell chemical genetics or stem cell applications (as discussed further below).

11.3 Chemical genetics and potential new therapeutics

The terms **chemical genomics** and **chemical genetics** are often used interchangeably. However, chemical genomics is really a broader term describing different types of large-scale whole genome approaches used for drug discovery *in vivo*. Such approaches include large-scale screening of compound libraries for bioactivity against a specific cellular target/phenotype. Chemical genomics might be described as the systematic search for small-molecule modulators of all functional proteins coded for by all gene products, and the characterisation of corresponding phenotypes at the cellular level. On the other hand, chemical genetics refers to the systematic assessment of the impact of genetic variance in cells on drug activities as observed by screening, particularly **high-throughput screening (HTS)**.

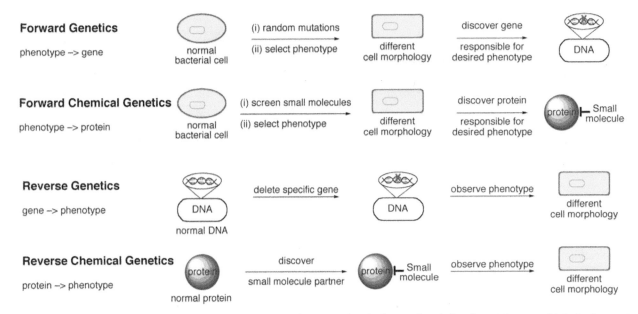

Figure 11.4   **Forward and reverse genetics with forward and reverse chemical genetics.** (Taken from Scheme 1 of D.R. Spring, 2005, Royal Society of Chemistry.)

One way to identify the function of a protein is to perturb its function. Genetically, gene function is modulated through a mutation and then phenotypes (physiological effects) are observed. Genetics can be divided into **forward genetics** (built on random mutations followed by phenotypic screening and gene identification) and **reverse genetics** (built on defined gene mutations and phenotype characterisation). So, genetics in the "forward" direction is from phenotype to gene; in the "reverse" direction it is from gene to phenotype (Figure 11.4). **Chemical genetics** can be divided similarly. Chemical genetics uses small molecules to modulate protein function. Hence, **forward chemical genetics** implies the use of small molecules to perturb protein function(s) leading to observed drug-related phenotype(s), in a biological system under investigation, followed by the identification of the gene(s) that code for those proteins with perturbed function(s). **Reverse chemical genetics** implies the application of small molecules which perturb known target protein function(s) and which may then be screened in a biological system under investigation for observable drug-related phenotypic effect(s). So, chemical genetics in the "forward" direction is from "phenotypes" to proteins then genes; in "reverse" it is from protein targets to "phenotypes". In both forward and reverse chemical genetics, the identification of small molecules followed by detailed biological investigations is required.

11.3.1 Early chemical genetics

In forward chemical genetics, collections of small molecules can be used in an HT screen to select the small molecule that gives a desired phenotype. Major pharmaceutical companies have proprietary compound collections consisting of around a million small molecules each. Structurally diverse compound collections can also be obtained from nature, or **diversity-oriented synthesis** (**DOS**). Nature has provided natural products that are undoubtedly diverse and complex structurally; however, there are disadvantages with their use. For example, natural products are often not easily available from nature as single compounds; therefore they are screened as mixtures, making it difficult to identify the active constituent(s). Furthermore, natural products are usually only found in low abundance in nature. Therefore, their chemical derivatisation is challenging from the synthetic chemistry point of view. On the other hand, DOS mimics nature and can enable the efficient, simultaneous synthesis of numerous structurally complex and diverse compounds. One well-established approach for DOS is **split-pool synthesis** (Figure 11.5). Such DOS strategies look very promising at one level, but details matter and chemical reactions need to take place efficiently with near quantitative conversion for such library syntheses to yield meaningful results. Any synthetic chemist will know that this is far from a given. Also, synthetic strategies are not always obvious and remain a significant challenge to implementation by synthetic chemistry.

Still, there can be some delightful surprises. For example, several complex natural products are known that interfere with cell cycle and might be used as potential anti-cancer drugs (Figure 11.6). These variously act on cell scaffold pro-

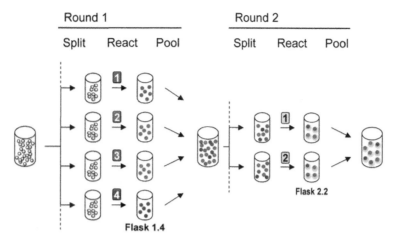

Figure 11.5 Diversity orientated split-pool synthesis. The synthesis starts with beads each having multiple copies (e.g. approximately 6×10^{16} copies per 550 μm polystyrene bead) of a starting core small molecule attached to it. The beads are then split out in round 1 into any number of different reaction flasks (four here). Analogous chemical reactions are then used to attach different building blocks to the small molecules in each flask. The beads from each flask are then pooled together and split back out into a second set of reaction flasks (two here). As in round 1, analogous chemistry is used in each reaction flask in round 2 to add a second building block to the small molecules on each bead. The beads from each flask are then pooled again and two rounds of split-pool synthesis have been completed. In this example, eight possible small-molecule variants on individual beads are prepared in six chemical steps. The key strength of split-pool synthesis is that small molecules containing all possible combinations of building blocks are prepared in just a few chemical steps. For example, if ten reaction flasks are used in four rounds involving 40 different building blocks then 10,000 (10^4) small molecules are prepared in just 40 chemical steps. At the end of the synthesis the beads are physically separated from each other and independently treated with a reagent that cleaves the small molecules from the bead for analytical and biological testing. (Taken from Figure 3 of Ward *et al.*, 2002, John Wiley & Sons.)

teins known as **tubulin** or **actin**. Tubulin modulators include **taxol, colchicine, vinblastine, vincristine, epothilone, eleutherobin** and **discodermolide**. Actin modulators include **jasplakinolide, latrunculin B** and **cytochalasin B**. In one forward chemical genetic experiment, a small-molecule library was screened for control of cell cycle in tissue culture cells. This led to the discovery of a synthetic compound known as **monastrol** (Figure 11.6) that promoted a remarkable reorganisation of the mitotic spindle even though considerably simpler in chemical shape and composition than the aforementioned natural products. In a comparable way, natural products are known potentially to inhibit key cancer inducing protein-protein interactions in cells, such as that between **T cell factor (Tcf)**/β-catenin protein complex. Interestingly, following HTS of a 7000-strong natural products library, the best inhibitor turned out to be the simplest of bicyclic heterocycles with no chiral centres, **PKF118-310** (Figure 11.6). Another example of note has come from studying small-molecule modulators that might inhibit **leukocyte function associated antigen-1 (LFA-1)** binding interactions with **intracellular adhesion molecule-1 (ICAM-1)**, as a means to inhibit both inflammatory and immune responses. HTS of compound libraries led to the identification of the **cinnamide** (IC_{50}, 6 nM) (Figure 11.7). Further investigation of the inhibition mechanism revealed that the cinnamide does not directly inhibit the protein-protein interactions by binding to the protein-protein interface. The likely mode of action is by binding in a pocket of LFA-1, thereby preventing LFA-1 from adopting a conformation appropriate for ICAM-1 binding. Allosteric regulation of protein-protein interactions avoids the problem of binding a small molecule to a large, flat surface and is a useful alternate way to modulate protein-protein interactions by means of small molecules. Interestingly, when inhibitors of LFA-1/ICAM-1 were developed by means of rational design to bind to the LFA-1/ICAM-1 protein-protein binding interface, then the resulting drug lead was the illustrated **thiophene** (IC_{50}, 1.4 nM) (Figure 11.7) that is surprisingly similar to the previous cinnamide derived by HTS.

The HTS of most small-molecule libraries is typically carried out using 384- and/or 1536-well microtitre plates that support the growth of mammalian cells in spite of their small volume (20–70 μl per well for 384-well plates, 2–10 μl for 1536-well plates). Pin array devices can then be used to transfer nanolitre to microlitre volumes of liquid between multi-well plates, often using robotic systems to pick up aliquots of solutions each containing different small molecules from the given library in solution. Thereafter, outcomes/well can be studied in plate-reader format, using absorbance, luminescence, radioactivity, scintillation proximity or various fluorescence-based assays as a readout as appropriate. Alternatively, if a specific antibody is available for measuring a protein or process of interest, a versatile high HTS detection is available called "cytoblot". On the other hand, for phenotype-based screens, which may involve the analysis of complex phenomena such as embryonic development, changes in cell morphology or changes in intracellular trafficking patterns,

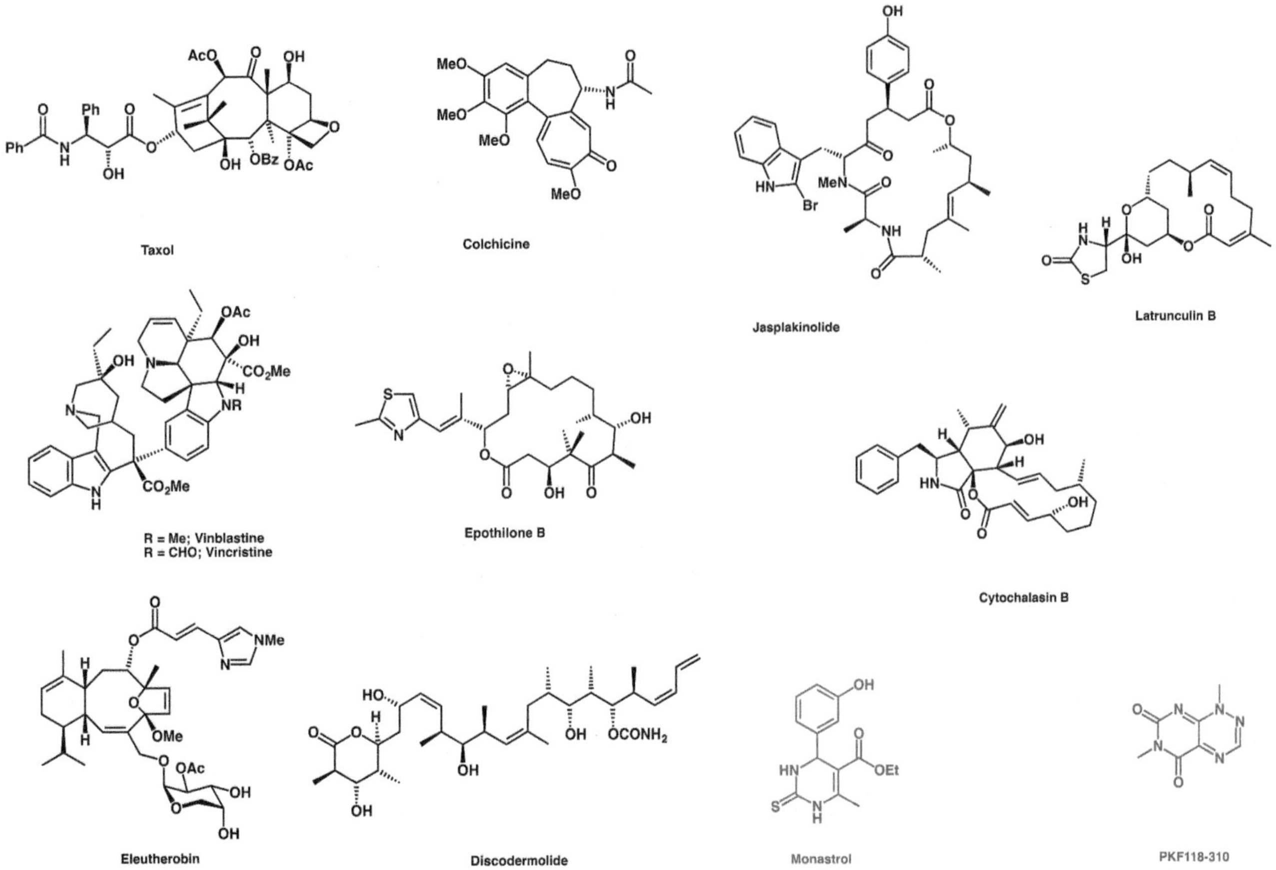

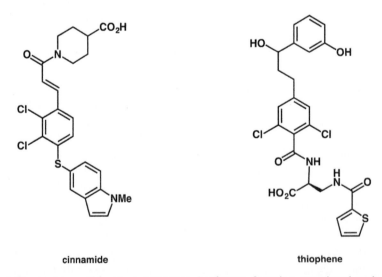

Taxol

Colchicine

Jasplakinolide

Latrunculin B

R = Me; Vinblastine
R = CHO; Vincristine

Epothilone B

Cytochalasin B

Eleutherobin

Discodermolide

Monastrol

PKF118-310

Figure 11.6 **Chemical modifiers of cell cycle. Monastrol** and **PKF118-310** were discovered by **high-throughput screening (HTS)** of chemical libraries.

cinnamide

thiophene

Figure 11.7 **Controlling inflammatory and immune responses. Leukocyte function-associated antigen-1 (LFA-1)/intercellular adhesion molecule-1 (ICAM-1)** binding interaction inhibitors were identified in two ways. The **cinnamide** (left) was identified by HTS, the **thiophene** (right) by "rational design" based upon a structural knowledge of the **leukocyte function associated antigen-1/ intracellular adhesion molecule-1 (LFA-1/ICAM-1)** interface. (Adapted from Winn *et al.*, *Journal of Medicinal Chemistry*, 2001 and Gadek *et al.*, *Science*, 2002.)

the inspection of images collected by light (transmitted or fluorescence) microscopy is typically the only readout available and is often rich in information. To assist in this, special microtitre plates with thin polystyrene or glass bottoms are available. Also, handling and image collection can be automated, either by custom modification of existing microscope equipment or with automated microscopes specifically designed for this purpose.

A notable example of HTS of compound libraries as a function of phenotypes can be found in **zebrafish** research. Zebrafish possess discrete organs that are very similar to human organs, and they are transparent in their early stages of life, so that all organ development can be visualised during development. Also, zebrafish are only small vertebrates, so embryos can be grown in a single well of a 384-well microtitre plate within a few days. Moreover, a pair of adults can lay routinely hundreds of fertilised eggs every day, thereby making it possible to HTS large numbers of small-molecule libraries (Figure 11.8). Accordingly, forward genetic screens on zebrafish have identified thousands of small molecules that affect the development of every organ. One example being **31N3** that inhibits the development of otoliths that are small, bony structures that are attached to bundles of hair cells in the zebrafish ear (Figure 11.8). In the presence of 31N3, zebrafish embryos between 14 to 26 h post-fertilisation were found to have lost their balance and often swam on their sides or upside down.

In the case of reverse chemical genetics, the protein target is essentially pre-selected and a small-molecule binding partner is required, which could be identified though protein-binding screens. An HT method for the identification of protein-binding partners is represented by small-molecule microarrays, or **chemical microarrays** (Figure 11.9). Protein binding partners in the chemical array may then be detected by incubating the microarray probe sites with the protein of interest which is labelled for spatio-local detection (with a fluorophore, for example) after "washing" of the microarray to remove non-specifically bound protein. Deconvolution of the array identifies small molecules capable of specific protein binding. By using such techniques as resonant mirror biosensing (see Section 7.3.5), the protein small-molecule binding events can be detected directly on microarray, thereby avoiding the need for protein labelling. Once small-molecule binders are identified, then they should be applied to cellular phenotypic assays to determine observable phenotypic effect(s) that result from perturbation of protein function by small-molecule protein binding.

The exploitation of small-molecule microarrays for reverse chemical genetics has been nicely illustrated with the following example. DOS was used to generate around 4000 structurally-complex and structurally-diverse small molecules that were printed onto a microscope slide to make a small-molecule microarray. This microarray was then challenged with **fluorescently-labelled Ure2p**, and **uretupamine** was detected as a small-molecule binder (Figure 11.10). Phosphoprotein Ure2p is a central gene repressor protein involved in metabolism. Experiments with uretupamine showed that Ure2p binding only affected a subset of functions controlled by Ure2p, which is important since selective modulation like this would not be seen in traditional genetics where the URE2 gene would simply be deleted. This example highlights the flexibility of chemical genetics in deciphering the individual functions of multifunctional proteins.

Systematic identifications of proteins of interest are difficult. So, too, is determining the **mechanism of action** (MoA) that follows from small-molecule protein interactions. The identifications of proteins of interest can make use of small

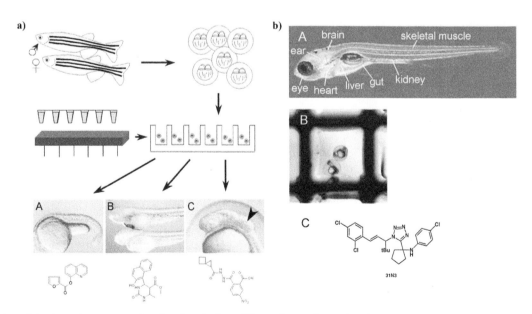

Figure 11.8 **High-throughput screening for chemical modifiers of vertebrate development.** (a) Screening for chemical modifiers of vertebrate development adult **zebrafish**, which lay hundreds of fertilised eggs each morning. Embryos are arrayed in 384-well assay plates and compounds from small-molecule libraries are added to each well. Embryos are allowed to develop and are screened visually for developmental defects. Examples of specific developmental phenotypes include elongation of the notochord (A), absence of blood (B, untreated, upper; treated, lower), and loss of a single otolith in the ear (C); (b) (A) Zebrafish larvae six days post fertilisation possess most of the tissues and organs of the fully developed vertebrate, (B) Zebrafish embryos in the wells of a standard 384-well plate, (C) Structure of **31N3**. (Figures taken from MacRae *et al.*, 2003, Elsevier.)

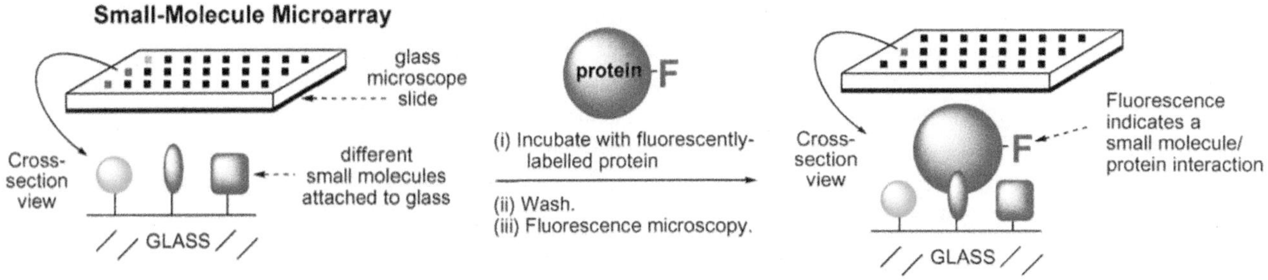

Small-Molecule Microarray

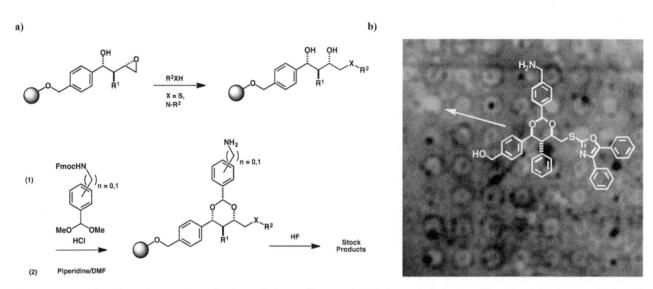

Figure 11.9 **Small molecule microarray application.** Different small molecules can be covalently attached to glass slides and probed with fluorescent labelled proteins requiring a small-molecule partner. After the slides are washed, to remove non-specific interactions, they can be scanned for spots of fluorescence, indicating a protein small-molecule interaction, and thereby detecting potential protein binders. (Taken from Scheme 2 of D.R. Spring, 2005, Royal Society of Chemistry.)

Figure 11.10 **Diversity orientated synthesis to find specific protein binder.** 1,3-dioxane small-molecule library synthesis and identification of **uretupamine** (right-hand structure). (**a**) **Diversity orientated synthesis** (**DOS**) leading to uretupamine and other library members; (**b**) Expanded view of 64 compound spots on the 3780-member small-molecule microarray (800 spots/cm). **Cy5-labelled Ure2p** was passed over a microarray of the 1,3-dioxane small-molecule library and the resulting slide was washed three times and scanned for fluorescence. The spot corresponding to uretupamine is shown. (Figures taken from Kuruvilla *et al.*, 2002, Springer Nature.)

molecules as "bait" to trap and hence label a target protein of interest. Typically, the small-molecule bait contains a reactive functional group that may be induced to form a covalent link with the protein of interest after specific, non-covalent binding has taken place. The result is to "tag" the protein of interest for later identification by purification and then characterisation (Figure 11.11). In some cases, small-molecule-protein interactions will lead spontaneously to covalent interactions, but mostly a bio-conjugation, cross-linking step is required, making use of photoactivation with ultraviolet-light irradiation, for example. Purification of the tagged protein will depend upon the biophysical characteristics of the tag. Where the tag is radioactive or fluorescent, then tagged protein may be isolated by chromatography and characterised by mass spectrometry, for example. Where the tag is a high affinity ligand, such as **biotin**, then the protein of interest may be purified by affinity chromatography. Alternatively, the small-molecule "bait" might itself be attached directly to a solid phase matrix via a linker. Hence, protein extracts can be eluted through a column containing the solid phase matrix, with immobilised small molecule, such that the protein of interest will bind tightly and selectively, while unbound and non-specifically bound proteins can be washed free.

A good example of this latter kind of approach has been described in studies with a stable analogue of the natural product **di-adenosine 5′,5‴-P¹, P⁴-tetraphosphate (Ap₄A)** (see Chapter 8). This stable analogue, known as β, β'-**methylene di-adenosine 5′,5‴-P¹, P⁴-tetraphosphate (AppCH₂ppA)**, was found to possess some very interesting biological properties against chronic pain and epilepsy in key biological models *in vivo*. Therefore, AppCH₂ppA was tagged with a biotin ligand to act as "bait" for the identification and isolation of intracellular **Ap₄A binding proteins** by **magnetic bio-panning** (Figure 11.12). According to this, probes are bound to streptavidin-coated magnetic beads via their biotin-tags, then

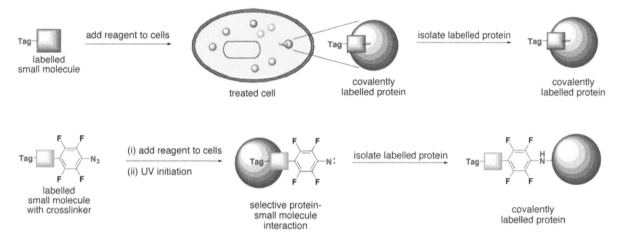

Figure 11.11 Identification of protein binders for reverse chemical genetics. A target protein becomes covalently labelled (tagged) by using a small-molecule tag that attaches itself covalently to the target protein by covalent crosslinking. The tag may alternatively bind target protein non-covalently only to undergo unmasking (e.g. photochemical unmasking of azide to nitrene) to reveal a highly reactive functional group for follow-on covalent crosslinking to the target protein. (Taken from Scheme 3 of D.R. Spring, 2005, Royal Society of Chemistry.)

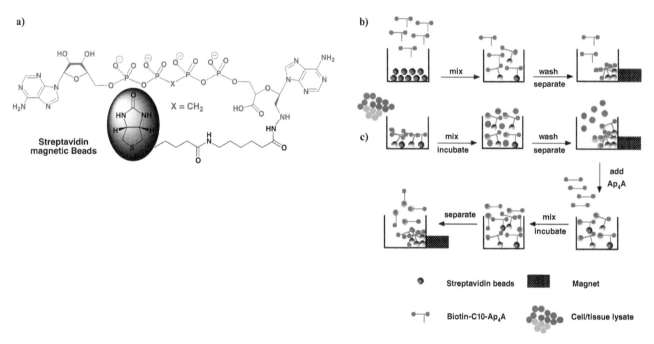

Figure 11.12 Molecular bio-panning for intracellular diadenosine-5′, 5‴-P^1, P^4-tetraphosphate binding proteins. This was performed using: (**a**) **Biotin-C10-AppCH$_2$ppA probes** that bind to streptavidin-coated magnetic beads; (**b**) Probes (red) are mixed with beads, washed and bead-bound probes are separated from bead-free probes by means of a magnet; (**c**) Bead-bound probes are then combined with cell lysates to allow intracellular **Ap$_4$A binding proteins** to bind, unbound/non-specifically bound proteins are removed by a wash step, then binding proteins are released by incubation and washing with free **Ap$_4$A**, prior to liquid chromatography elution and MS identification. (Figures taken from Azhar *et al.*, 2014, Elsevier; Guo *et al.*, 2011, with permission of Elsevier.)

mixed with cell lysates of interest to enable the binding of intracellular Ap$_4$A binding proteins to the probes. The pool of magnetic beads can then be aggregated by means of an external magnet, after which unbound/non-specifically bound proteins may be removed by washing. Finally, specifically probe-bound proteins are eluted by application of an excess of free AppCH$_2$ppA. Thereafter LC-MS analysis can be employed to identify any potential intracellular Ap$_4$A binding proteins. While this approach has been successful, such target identification studies require a strong affinity between protein and small-molecule partner, which is far from guaranteed. Also, such protein target identification methods, require the chemical derivatisation of small-molecule ligands, which can be time-consuming and difficult to achieve without impairing small-molecule protein-binding affinities.

11.3.2 More recent chemical genetics

More recent chemical genetic studies with chemical libraries routinely extend from studies with microbes all the way to human cells. Genetic variation ranges from highly controlled (in microbes) to natural (in human), such that at its most powerful, chemical genetics can now make use of genome-wide libraries of mutant cells of interest as well, comprising either **loss-of-function** (**LoF**, by knockdown or knockout) or **gain-of-function** (**GoF**, overexpression) mutations in every chromosome expressed gene. These cellular libraries are then either pooled or arrayed in single microarray format and then exposed to chemical libraries (Figure 11.13). At its simplest, the output of such HTS experiments is a set of comparative **growth inhibition curves** or **fitness profiles** that register cell growth in hundreds to thousands of different mutant cells as a function of different concentrations of chemical library components. Comparative cell growth inhibition curve data represent quantitative as well as qualitative chemical phenotypes for each of these chemical library components (Figure 11.14). Furthermore, by bringing chemical and cellular libraries together in one HTS experiment, the principal outcome is that both forward (phenotypes to protein targets then genes) and reverse (genes to protein targets then phenotypes) chemical genetic data can be acquired simultaneously. The sum total of chemical phenotypes obtained may then be interpreted in the light of the genetic backgrounds of the library cells concerned, leading to potential drug identification, the characterisation of MoAs, mapping of drug influx, efflux, identification of potential drug resistance mechanisms, and an understanding of potential drug synergies all together.

Crucially, nowadays, although microbial chemical genetics screens have concentrated on measuring bulk chemical phenotypes, such as comparative cell growth fitness profiles, other chemical phenotypes may be derived, for instance, by observing for potential chemical impacts on cellular developmental processes, DNA uptake and cell lysis. In addition, the quantification of single-cell chemical phenotypes and population behaviours across mutant cell libraries is also possible with current advances in HTS with microscopy. In such cases, cell markers and classifiers of drug responses can provide further insights into the biological activities of a given drug in the cell. Single-cell readouts and multi-parametric phenotyping analysis are more common in chemical genetics in human cell lines.

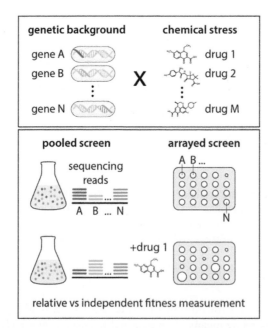

Figure 11.13   Basic concepts and approaches in chemical genetics. Modern chemical-genetic approaches are based on combinations of **genetic** and **chemical perturbations**. The fitness of genome-wide libraries of **gain-of-function (GoF)** and **loss-of-function (LoF)** mutations is assessed upon exposure to large numbers of chemical entities. Mutant libraries can be pooled or arrayed. In pooled screens, barcoded mutants compete among each other after exposure to a certain drug and their relative abundance is measured by barcode sequencing. In arrayed screens, mutants are ordered and their fitness or additional macroscopic phenotypes can be assessed independently. (Taken from Cacace et al., 2017, Figure 1, Elsevier, CC BY-4.0.)

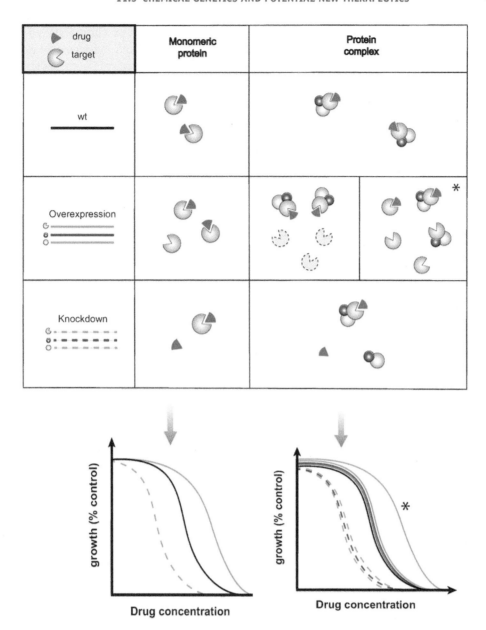

Figure 11.14 Gene-dosage perturbations reveal drug target. Gene dosage perturbations lead to different insights on a drug **mechanism of action (MoA)** depending on the nature of the drug target. In the case of a monomeric protein target, its overexpression determines a right-shift in the growth inhibition curve of the drug (i.e. higher drug concentrations are needed to produce the same growth inhibition). However, if the drug target belongs to a protein complex, then a curve shift in comparison to wild-type is evident only if the drug can still bind to the target protein and the conjugate is also present/functionally active without involving other members of the protein complex (*). If the drug target is functional only as part of the protein complex, then overexpression does not yield any evident change in the effect of the drug. So, too, overexpression of protein complex co-members does not change the growth inhibition curve. On the other hand, knockdown of a protein target can cause detectable changes in growth inhibition curves (left-shift), irrespective of whether the drug target is a monomeric protein or belongs to a protein complex. In addition, knockdown of protein complex co-members will also cause similar shifts in growth inhibition curves. (Taken from Cacace *et al.*, 2017, Figure 2, Elsevier, CC BY-4.0.)

11.3.2.1 Chemical genetics in MoA identification

There are two primary ways in which chemical genetics can be used to map drug targets. First, by using cellular libraries in which the levels of essential genes, which typically express the protein targets of small-molecule drugs, may be modulated. In this case, when a given target gene is downregulated, then the mutant cells concerned often become more drug sensitive (since less drug is required for titrating the cellular target). Obviously, the opposite holds true for target gene overexpression (see Figure 11.14). In cells or organisms that are diploid, heterozygous deletion mutant libraries can be used to reduce, to "downregulate" in effect, essential genes. Such libraries give rise to **haplo-insufficiency profiling** (**HIP**), that was successfully used to map cellular drug targets in yeast. In cells or organisms that are already haploid, in bacteria for example, increasing gene levels is technically simpler, hence target gene overexpression is preferred to identify target genes and hence protein targets of drugs. Both approaches have caveats, since genes that confer direct or indirect resistance to a given drug could be confused as target genes coding for protein targets. Also, given mutant cells may experience more profound changes in their cellular networks than just a single, simple change in the expressed level of one essential gene. Nevertheless, knockdown and overexpression approaches can now be used successfully to identify drug targets in human cell lines.

Second, drug targets can be inferred from chemical genetics data by comparing drug signatures. A drug signature comprises the compiled quantitative growth inhibition curves or fitness profiles for each cellular mutant within a genome-wide deletion library (all non-essential genes) in the presence of the drug. Drugs with similar signatures are likely to share cellular targets and/or cytotoxicity mechanisms. This guilt-by-association approach becomes more powerful when more drugs are profiled, so revealing "chemogenomic" signatures that correlate to known MoAs. However, drug signatures are also driven by pathways controlling intracellular drug concentrations as much as they depend on pathways related to drug MoA or its cytotoxic effects to the cell. Thus, machine-learning algorithms can be used to recognise the chemical genetic interactions that are reflective of a given drug MoA. These chemogenomic signatures can be further refined using multi-parametric analysis of microscopy images, increasing the resolution of MoA identification.

11.3.2.2 Chemical genetics in drug resistance and drug-drug interactions

Chemical genetics data are influenced by drug detoxification mechanisms and the ways in which cells import drugs from outside or export them from the inside. Indeed, in yeast, up to 12% of the genome confers multi-drug resistance, whereas only a few dozen genes have similar pleiotropic roles in *E. coli*, implying that prokaryotes have more diverse and/or redundant drug resistance mechanisms. Curiously, in bacteria many drug transporters and pumps are cryptic; although they have the capacity to help the organism survive drug treatments, they do not sense the presence of drugs and so can remain minimally expressed even when drugs are present. Such a suboptimal expression of drug transporters underlines the high capacity that bacteria possess for intrinsic antibiotic resistance. In addition, chemical genetics can be used to assess the level of cross-resistance and collateral sensitivity between drugs (see Figure 11.15), if mutations lead to resistance (or sensitivity) or make one cell more resistant to a given drug but another more sensitive. Chemical genetics data also measure the contribution of every non-essential gene towards drug resistance. Evaluating cross-resistance patterns can facilitate an understanding of drug resistance, and reveal the means to mitigate or even reverse drug resistance.

When combined, drugs can synergise, antagonise or even mask each other's effects (Figure 11.16a). The combined effects of anti-infectives (antibiotics or antifungals), and the combination use of anti-infectives with possible adjuvants, have been studied by chemical genetics, but there is much more. In the clinic, drugs are combined to exploit synergistic activities, increasing the potency, widening the spectrum of action and reducing doses and side effects of individual drugs. Most combination uses of drugs arise from empirical studies and *ad hoc* testing, though drug-drug interactions are also seen as a potential regulatory problem. Chemical genetics has a real role to play in understanding the underlying mechanisms of drug-drug interactions. In this case, drugs are profiled alone or in combination across a library to identify the small fraction of genetic backgrounds in which the drug-drug interactions are no longer detectable anymore (Figure 11.16b). The corresponding genetic background and the genes concerned reflect molecular pathways and processes behind drug-drug interactions of interest. Although drug-drug interactions are generally robust to genetic backgrounds, cells often require just single mutations to neutralise or even reverse such interactions, for example converting the interaction from synergistic to antagonistic. Accordingly, drug-drug interactions within or across species do not appear robust or strongly conserved. Still, chemical genetics profiles have been used to predict drug-drug interaction outcomes. Drugs eliciting similar chemical genetic signatures not only often target the same cellular processes, but are also more likely to interact synergistically. This characteristic can be used to predict synergies and potentially beneficial drug-drug interactions going forward.

11.4 Chemical cellular dynamics

Much fundamental cellular research is now focused on probing the dynamic characteristics of gene regulation and signalling networks in response to changes in local cellular environments. Chemical cellular dynamics is the study and understanding of these dynamic characteristics. Significantly, most dynamic characteristics of cells are under the control of feedback systems that monitor conditions in and around a given cell and respond by modulating biochemical

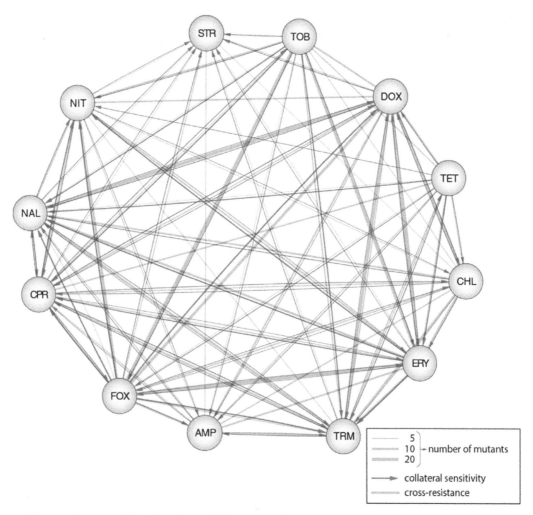

Figure 11.15 Cross-resistance and collateral sensitivity maps from chemical genetics data. In the map, grey edges depict **cross-resistance** (i.e. gene-drug phenotypic events where mutants are either significantly more susceptible (negative) or more resistant (positive) to both drugs), red edges depict directional **collateral sensitivity**-mutants which make cell more resistant to drug A, but more sensitive to drug B, or vice versa. Edge thickness denotes number of mutants. Data used to create network come from published chemical genetics data and selected drugs are based on overlaps with previous cross-resistance studies. (Taken from Cacace *et al.*, 2017, Figure 3, Elsevier, CC BY-4.0.)

pathways leading to changes in the dynamic characteristics of cells. In cellular dynamics studies, mechanisms of modulation may be exposed by deliberately introducing a precise change to a cultured cell's environment and observing for cellular subsystem responses.

Lately, such studies are much enhanced using microfluidic techniques that expose cells to microfluidic "switching flows" which permit bulk changes in cellular surroundings at a faster rate than with traditional perfusion techniques. The small dimensions of microfluidic channels lead to conditions of laminar flow which can be exploited to situate two dissimilar fluid streams parallel to one another in a channel. Such co-flowing streams can be used to concurrently treat distinct regions of cells, clusters of cells or embryos with multiple fluid environments that differ in temperature or chemical composition. Furthermore, **hydrodynamic focusing** can be used to deliver solutes selectively to subpopulations of cells by constraining one fluid stream tightly between two others. Also, the application of **microperfusion** helps expose cells to the very minutest chemical perturbations.

Critically, cells can be cultured in microfluidic devices, permitting long-term monitoring in carefully controlled environments (see Figures 11.17 and 11.18). According to this example, **NIH 3T3 fibroblast cells**, cultured in a microfluidics device, can be made to respond uniformly to artificially generated pulses of **ionomycin**, a calcium ionophore, both during and shortly after such pulses. Experimentally speaking, these pulses of ionomycin cause rapid entry of ionomycin into fibroblasts to mobilise internal calcium stores, leading to rapid rises in free intracellular $[Ca^{2+}]$ that are visualised directly using **Fluo-4 AM**, a membrane-permeable, fluorescence indicator of intracellular free Ca^{2+}

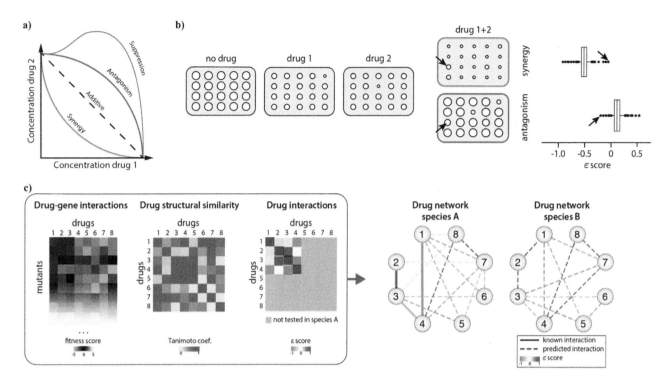

Figure 11.16   Chemical genetics facilitate the mechanistic dissection of drug action and prediction of drug-drug interactions.
(a) **Isobologram** illustrating different cases of drug interactions. Synergistic, antagonistic and suppressive interactions are represented as phenotypic deviations from an expected additive effect; (b) Drug synergies or antagonisms can be profiled in a genome-wide library of mutants. A ε-**score** assessing the drug-drug interaction is calculated for every mutant, and corresponds to the difference between observed and expected cell growth behaviour in the presence of two drugs. Although the vast majority of mutants exhibit wild-type growth behaviour, in some mutants drug-drug interaction promotes such growth behaviour (arrowed data). These mutants reflect the molecular mechanisms that drug-drug interactions depend upon; (c) Compilations of chemical genetics, drug structural similarity, and known drug-drug interaction data can be integrated to help map out the extent of drug-drug interaction networks. Such networks can be extrapolated to phylogenetically related species (denoted A and B above). (Taken from Cacace *et al.,* 2017, Figure 4, Elsevier, CC BY-4.0.)

ions. Quite clearly, ionomycin-stimulated fibroblast cells in all cases are strongly fluorescent for just over as long as ionomycin stimulation persists. In effect, the calcium signalling dynamics of cultivated fibroblast cells are actively controlled by artificially generated pulses of ionomycin, an exogeneous chemical signal generator, under the control of pressure differences between two microfluidics inlets. In other words, controlled pressure differences make possible actual spatiotemporal modulations in cellular behaviour mediated via the exogenous chemical signal ionomycin. Going forward from such an example, the goal must be to achieve more detailed chemical control of cultured cells, using microfluidics systems. However, microfluidics techniques still only allow for limited variations in extracellular environments continuously over time. Therefore, significant progress remains to be made before more sophisticated exogenous signal-mediated cellular behaviour can be studied using microfluidics.

Another example of real-time observation of cellular dynamics has been reported using **iriomoteolide-3a (irio-3a)** and analogues. Despite its biological profile, the cellular targets of iriomoteolide-3a, a novel 15-membered macrolide isolated from *Amphidinium sp.,* and analogues were unknown until a small library of non-natural iriomoteolide-3a analogues was prepared and then shown to target actin proteins (Figure 11.19). The importance of this is that the two major constituents of the cytoskeleton in eukaryotic cells are the proteins actin and tubulin. Tubulin has been long recognised as a target for anti-cancer therapies, with cytoskeleton interacting drugs such as paclitaxel or epothilones (see Figure 11.6). Actin has also been recognised as a potentially important anti-cancer target given the utmost importance of the **actin cytoskeleton** to a wide variety of cellular processes, ranging from cell shape and locomotion to cell division, cell adhesion and cell transport (endo- and exocytosis). The actin cytoskeleton *per se* actually involves a dynamic interchange between **monomeric G-actin** and **polymeric F-actin filaments** under the control of secondary proteins, such as **profilin**, that acts to "sequester" monomeric G-actin, so inhibiting addition to F-actin filaments. The illustrated data (Figure 11.19) and corroborating data demonstrated that irio-3a and analogues, at sub-μM concentrations, are able to inhibit cell migration and induce severe morphological changes in actin cytoskeletons, through stabilisation of F-actin filaments, inhibition of F-actin depolymerisation and the enhancement of G-actin polymerisation into F-actin filaments.

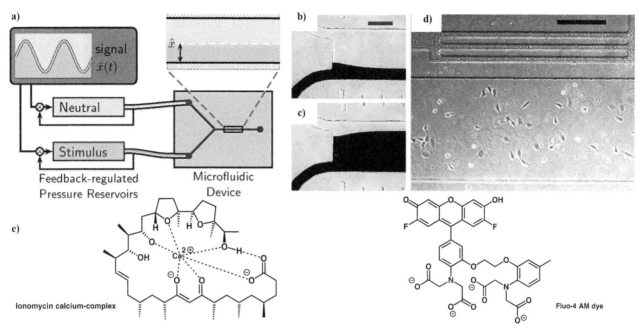

Figure 11.17 **Microfluidics device to deliver controlled, artificially generated pulses of an external, chemical signal generator to cultivated cells.** (a) Two fluid reservoirs (**Neutral** or **Stimulus**) are linked to two separate inlet channels, of a microfluidic device, both of equal fluid resistance. These merge to form a single outlet channel where cultivated cells are located. Each inlet channel conducts either the Neutral or Stimulus fluid to the outlet channel where a laminar flow interface is established. The static pressures of the two fluid reservoirs (**Neutral** and **Stimulus**) are controlled by a pair of feedback pressure regulators according to commands received from a supervisory controller. In (**b**) and (**c**) note how the interface may be shifted laterally (up or down) depending on fluid pressure differences between the two inlets (scale bar 250 μm). Pressure differences applied over time enable the laminar interface to shift in a controlled, even periodic manner (see Figure 11.18). In (**d**) locations of **NIH 3T3 fibroblast cells** in the outlet channel are shown, positionally defined by "latitude" (location across the channel width, bottom to top) and "longitude" (distance from the start of the outlet channel) (scale bar 250 μm). Neutral fluid comprises: neutral phosphate buffer with 1 μM $CaCl_2$. Stimulus fluid comprises: neutral phosphate buffer with 1 μM $CaCl_2$ and 1 μM **ionomycin**; (**e**) Structure of ionomycin calcium ion complex (left) and fluorescent dye **Fluo-4 AM** (right). This is loaded into fibroblasts to bind free intracellular calcium ions released from intracellular storage, by ionomycin stimulation, allowing fibroblasts to become fluorescent during ionomycin stimulation. (Figure adapted from Kuczenski *et al.*, 2009, Figure 1, *PLOS*, CC BY-4.0; (b–d) Kuczenski *et al.*, 2009, PLOS ONE.)

11.5 Chemical biology and *in vivo* cell connectomics

In order to investigate and probe at niche microenvironments that control cell behaviour *in vivo*, genetic cell-labelling techniques are now being developed where hundreds of different hues can be generated by stochastic and combinatorial expression of a few spectrally distinct fluorescent proteins. Most significantly to date are new genetic cell-labelling techniques. According to these, unique colour profiles can be used as cellular identification tags for multiple applications, such as tracing axons through a nervous system, following individual cells during development or analysing cell lineage. In recent years, this genetic cell-labelling technique and other combinatorial expression strategies have expanded from the mouse nervous system to other model organisms and a wide variety of tissues. Particularly exciting is the adaptation of **Brainbow** in parsing out complex cellular relationships during organogenesis.

11.5.1 Brainbow connectomics

Green fluorescent protein (GFP), and its engineered transgenic variants, namely **red fluorescent protein (RFP)** and **blue fluorescent protein (BFP)** are the basis of *in vivo* **Brainbow connectomics**. The background strategy relies on the fact that the three visual primary colours red (R), green (G) and blue (B) can be combined in any proportions to generate all colours in the visual spectrum. Brainbow achieves the same effect by combining three or four distinctly **coloured fluorescent proteins (FPs)**, by including recombinant **yellow fluorescent protein (YFP)**, and expressing them in different ratios within each cell of a given integrated network. The resulting colour combinations are unique to each Brainbow-expressing cell and can therefore serve as cellular identification tags that can be visualised by light

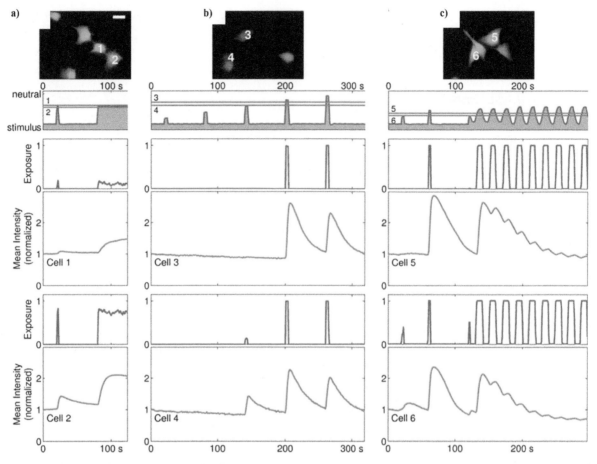

Figure 11.18 Controlling cell responses in microfluidics device. This follows on from Figure 11.17. Here we observe the responses of NIH 3T3 fibroblast cells to different, artificially generated pulses of ionomycin exposure (scale bar 25 μm) such that in: (**a**) Two cells (**1** and **2**) differentially exposed to a short ionomycin pulse (from a short interface fluctuation) followed by a step-change in exposure (from a longer term interface fluctuation). Because of their different "latitudes", Cell **1** experiences a sub-maximal exposure, while Cell **2** experiences a near-maximal exposure, reflected by differences in fluorescence intensity output from both cells. In (**b**) two cells (**3** and **4**) are exposed to short ionomycin pulses of increasing amplitude. Cell **3** experiences one less pulse than Cell **4** due to its higher "latitude" and so exhibits one less peak of fluorescence. In (**c**) Two cells (**5** and **6**) are exposed to short ionophore pulses followed by periodic stimulation. Cell **5** experiences one less pulse than Cell **6** due to its higher "latitude". All these data demonstrate time- and space-controlled intracellular stimulation by ionomycin of cultivated fibroblast cells. (Figure adapted from Figure 4 of Kuczenski *et al.*, 2009, *PLOS ONE*.)

microscope. Many different Brainbow and Brainbow-like strategies are now used, based on **Cre-recombinase**-mediated DNA excision or inversion (Figure 11.20).

Cre-recombinase is a tyrosine recombinase (38 kDa) from the P1 bacteriophage that catalyses site-specific recombination between two DNA recognition sites (***lox*P sites**). Each *lox*P recognition site is 34 bp in length and comprises two 13 bp palindromic sequences either side of an 8 bp spacer region. The products of **Cre-recombination** are dependent on location and relative orientation of the two ***lox*P recognition sites** (Figure 11.20A). Where one DNA species contains two matching *lox*P sites, the intervening DNA sequence is said to be "floxed". Should the two matching *lox*P sites be orientated in the same direction, then the floxed DNA will be excised as a **covalently closed circular DNA** (**cccDNA**). Should the two matching *lox*P sites be orientated in the opposite direction, then the floxed DNA will be inverted. Cre-recombinase needs neither **adenosine 5'-triphosphate** (**ATP**) nor co-protein to function fully. Note that where two separate DNA species contain matching *lox*P sites, then these can undergo fusion as a result of Cre-recombinase action.

Cre-recombinase is a well-used tool in molecular biology with a unique ability to operate efficiently in a wide range of cellular environments (mammalian, plant, bacteria and yeast). Hence Cre-recombinase activities may be observed in anything from individual cells up to transgenic organisms. In addition, Cre-recombinase is able to tolerate up to five mismatches in each of the two palindromic sequences that comprise a *lox*P recognition site, without loss of recombinase function. On the other hand, Cre-recombinase is intolerant of mismatches in the 8 bp spacer region; this core sequence is the so-called cleavage site and its asymmetry defines the direction of the *lox*P site. The importance of this is that, on

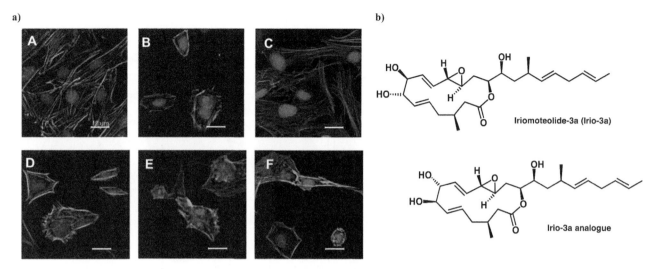

Figure 11.19 **Effect of Irio-3a on murine NIH/3T3 fibroblast growth as seen by fluorescence micrographs of the actin cytoskeleton.** In: (**a**) Actin cytoskeleton is stained with **fluorescein isothiocyanate labelled phalloidin (FITC-phalloidin)** (green) and nuclei with **2-(4-amidinophenyl)-6-indolecarbamidine hydrochloride (DAPI) (blue)**; panel (A) control fibroblasts; panel (B) fibroblasts incubated with **iriomoteolide (irio-3a)** at 250 nM for 2 h and panel (C) for 8 h showing complete recovery of their normal morphology and microfilament organisation; in panels (D) and (E), fibroblasts are incubated with irio-3a at 1 and 4 μM respectively for 2 h; in panel (F) cells are incubated with irio-3a analogue at 10 μM for 8 h; (**b**) Structures of irio-3a and analogue. (Unzue *et al.,* 2018, adapted from Royal Society of Chemistry.)

a statistical level, the 34-bp consensus *lox*P site is not expected to be present in mammalian genomes, but the mismatch tolerance ensures that functional Cre-recombinase recognition sites can be found in both human and mouse genomes. These cryptic (or pseudo) *lox* sites, can support Cre-recombinase mediated recombination at high efficiency.

The role of Cre-recombinase in Brainbow, begins with **excision-based Brainbow 1.0** (Figure 11.20B). According this methodology, genes coding for GFP, RFP and BFP are linked in a single transgene (that may or may not genome integrate) along with two matching pairs of *lox* sites (*lox*2272 and *lox*P) that are independently recognised by Cre-recombinase. Cre-recombinase is introduced to target cells for labelling in the form of another expressible transgene to enable recombination. Before Cre-recombination, only the first colour in the array is expressed (termed the "default" colour). Following recombination, each of the three FP genes will be separately expressed, resulting in a three-colour transgenic cell population. This strategy can be expanded to four FP genes by utilising a third matched pair of *lox* sites. Thereafter, in the case of **inversion-based Brainbow 2.0** (Figure 11.20C), two pairs of matching *lox* sites are positioned such that they face each other. Hence, Cre-recombinase inverts (or "flips") the floxed DNA as opposed to excising it. In this strategy, two different FP genes are aligned in head-to-head orientations such that Cre-recombinase mediates inversion, leading to the expression of only one of these two FP genes at a time, resulting in a two-colour transgenic cell population. With **Brainbow 2.1** (Figure 11.20D), excision and inversion can take place with four different FP genes within a single cell mediated by added Cre-recombinase, resulting in four-colour transgenic cell populations.

A more recent amplification of this concept is **combinatorial Brainbow** that makes use of combinatorial expression of multiple FP genes from Brainbow expression cassettes. Fundamentally if each cell in a target population contains only one copy of a three-colour construct (e.g. from Brainbow 1.0), this would result in a three-colour transgenic cell population after Cre-recombinase action (Figure 11.20B). However, more complex multicolour expression results when multiple Brainbow expression cassettes are present in each cell. With more than one copy of a cassette present, then Cre-recombinase can act randomly on each copy and collectively generate a multi-colour population of transgenic cells (Figures 11.20E and F). In practice, up to one hundred different cellular pigments can be generated, wherein each pigment acts like a cellular "bar code" for cell tracing (e.g. following cell movement or neurite growth) and lineage analysis (where colour is used to distinguish cell populations derived from different progenitors). In contrast, **TIE-DYE** involves multiple single-FP transgenes that are expressed simultaneously under control of a promoter that is subject to stochastic Cre-recombinase excisions. Therefore, each transgene is stochastically expressed to create a cell population of transgenic cells with combinatorial and diverse hues (Figure 11.20G).

Clearly, Brainbow labelling of transgenic cells can be achieved by means of an impressive variety of transgenic techniques. However, arguably the most important aspect of Brainbow labelling is that this is a genuine genetic labelling technique and the result of DNA recombination events are inheritable. Consequently, once an initial pool of progenitor cells has been created with a certain colour or colours, then clonal progeny will result that are labelled in a way that

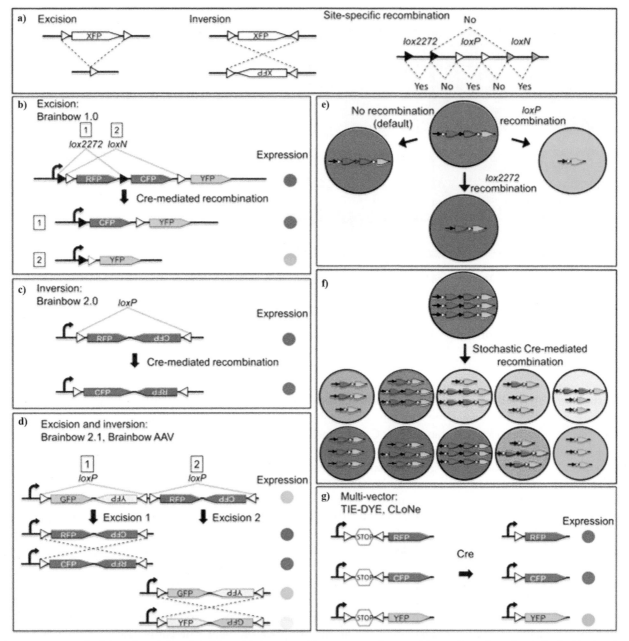

Figure 11.20   Principles of Brainbow labelling. (a) Cre-recombinase can perform excision or inversion of DNA flanked by *lox* sites (triangles), depending on orientation of *lox* sites. Different sites such as *lox*2272 (black triangle), *lox*P (white triangle) and *lox*N (grey triangle) function identically but are incompatible with each other; **(b) Excision-based Brainbow 1.0**, genes for **coloured fluorescent proteins** (**FPs**) are flanked by two pairs of matched *lox* sites. In the absence of recombination, **RFP** is expressed. Recombination results in excision expression of either **CFP** (event 1) or **YFP** (event 2); **(c) Inversion-based Brainbow 2.0**, FP gene expression exchanges between RFP and CFP by DNA inversion; **(d)** In **Brainbow 2.1**, excision leads to selection of either the GFP/YFP pair or the RFP/CFP pair. DNA inversion then decides which FP of either pair is expressed; thereafter in **(e)** only the first FP gene in a gene array is expressed, then in a cell population with a single Brainbow transgene per cell, cells can be RFP+ (no recombination, i.e. "default"), CFP+ or YFP+; **(f) Combinatorial Brainbow**, multiple copies of a Brainbow transgene are present in a cell, such that each copy recombines independently. Three copies of the Brainbow transgene then generates ten distinct colours and more copies will generate even greater colour diversity; **(g) TIE-DYE**, combinatorial multicolour labelling makes use of multiple vectors, each carrying a single FP gene. As the expression of each FP gene is stochastic, the colour profile within each cell is different. (Taken from Figure 1 of Weissman and Pan, 2015, Oxford University Press.)

reflects cellular lineage (Figure 11.21). In other words, all cells within a clone will have the same colour. Given this, Brainbow labelling has superb potential applications for observing cellular growth and differentiation in complex multicellular organisms, and shedding light on the complex cellular anatomy found within multicellular organisms that is otherwise very difficult to determine by other means.

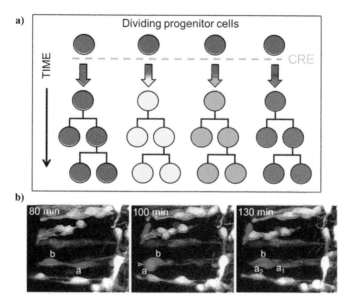

Figure 11.21 Brainbow for clonal analysis. (a) A uniform population of dividing progenitor cells becomes multicolour upon **Cre-recombination**. Following recombination, each dividing cell produces progeny that share its unique colour, thus colour coding the resulting clones; **(b)** This type of Brainbow labelling was used *in vivo* to follow dividing radial progenitor cells in the chick spinal cord over time. Over a period of 50 mins, one member of the blue clone (cell **a**) divides, producing two daughter cells (**a₁** and **a₂**). (Adapted from Figure 2 of Weissman and Pan, 2015, Oxford University Press.)

11.5.2 Brainbow applications

Brainbow labelling was primarily intended to label individual cells of a given cell type so that they become colour differentiated. As the name implies, this cell-colouring approach was first used for the labelling of brain cells, allowing for the functional analyses of individual neuronal cells within a population (as opposed to population behaviour). Critically, this technique allowed for the following of individual cells over time and space, as well as for tracing projections in the nervous system. Following this, Brainbow labelling techniques have been applied to several different tissues and model organisms such as **mouse, rat, chick**, zebrafish, **fruit fly** and **plants**. These applications have included both **germline transgenic** approaches and **somatic labelling** approaches (i.e. non-germline transgenic) (see Tables 11.1 and 11.2). Initially, Brainbow labelling was launched with mouse transgenic neuronal cell lines in order to study, in particular, neuronal development in the mouse brain *in vivo*. Thereafter, as applications of interest have evolved, additional mouse transgenic cell lines were developed, such as **R26-Confetti** and **R26-Rainbow** (involving gene integration/knock-in to the ROSA26 chromosomal

Table 11.1 Applications of Brainbow and variants based on transgenic cells lines.

Organism	Latin name	Promoter	Transgenic cell lines
Mouse	*Mus musculus*	Neuronal	Brainbow 1.0/1.1/2.0/2.1
			Brainbow 3.0/3.1/3.2
			Flpbow, Autobow
		Ubiquitous	R26-Confetti, R26-Rainbow
			Rainbow, MAGIC, UBow
Zebrafish	*Danio rerio*	Gal4 inducible	Brainbow, Zebrabow
		Ubiquitous	PriZm, Zebrabow
Fruit fly	*Drosophila melanogaster*	Gal4 inducible	dBrainbow, Flybow 1.0/1.1/2.0, LOLLIbow
		Ubiquitous	TIE-DYE
Plant	*Arabidopsis thaliana*	Ubiquitous	Brother of Brainbow (BOB)

Table 11.2 Applications of Brainbow and variants based on somatic labelling approaches.

Transgenesis method	Transgene	Organism	Application
DNA injection	Brainbow	Zebrafish	Cell/axon labelling
Electroporation	Brainbow	Mouse, chick	Cell/axon labelling and lineage analysis
	CLoNe		
	MAGIC		
Lentivirus	RGB LeGO	Mouse, culture cells	Lineage analysis
	LeGO with DNA bar code		
AAV	Brainbow AAV	Mouse	Cell/axon labelling

locus), and non-knock-in cell lines such as **Rainbow**, **Cytbow** and **Nucbow**. Moving away from mouse, Brainbow labelling has been made possible in zebrafish using the variants **Zebrabow** and **PriZm**, in the fruit fly using **Flybow**, **LOLLIbow** (with a light-activated Cre-recombinase) and TIE-DYE, and, finally, in plants using **Brother of Brainbow** (**BOB**). Importantly, these newer variants of Brainbow labelling may also involve the replacement of Cre-*lox* excision and inversion strategies. For example, the **Flpbow** variant makes use of Flp-recombinase and pairs of matching FRT recognition sites, which are functionally equivalent to Cre-recombinase and *lox* sites, respectively (see Table 11.1).

Where somatic approaches have been used to enact Brainbow labelling, non-germline cells have been rendered transgenic by direct DNA injection methods or using electroporation and viral nucleic acid delivery systems (see Chapter 13) (see Table 11.2). Both the germline and somatic approaches are suitable for short-term Brainbow cell labelling experiments and long-term lineage analyses. Germline approaches give more homogenous expression. So it is easier to produce consistent labelling density and colour diversity across different animals. In contrast, labelling density and colour diversity are often more variable with somatic labelling. On the other hand, somatic labelling is applicable to a wide range of models, allowing for direct cross-species comparisons and applications in organisms for which it is difficult to generate transgenic lines. Furthermore, these methods do not require the time and costs required for generating and maintaining Brainbow transgenic lines. In conclusion to this section, recorded images taken in different organisms beautifully demonstrate how Brainbow and variants might contribute to structural and functional studies of cell populations *in vivo* (Figure 11.22).

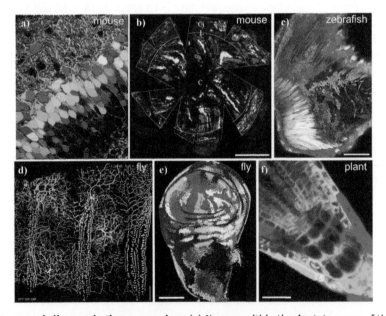

Figure 11.22 Brainbow transgenic lines and other approaches. (a) Neurons within the **dentate gyrus** of the **Brainbow mouse hippocampus** (line L; Image by T. Weissman and J. Lichtman); **(b)** Radial clones of cells in the **mouse cornea** from Di Girolamo *et al.*, 2014, included with permission from Wiley, Copyright ©2014 AlphaMed Press; **(c)** Pectoral fin in **"zebrabow" zebrafish**, from Pan *et al.*, 2013; **(d) Sensory neurons** in the ventrolateral body wall of a Drosophila **LOLLIbow** larva, adapted from Boulina *et al.*, 2013, with permission from Elsevier; **(e) Wing-imaginal disc** in TIE-DYE *Drosophila*, adapted from Worley *et al.*, 2013, with permission from Elsevier; **(f)** Cells in *Arabidopsis thaliana* **root meristem** labelled using the Brother of Brainbow system from Wachsman *et al.*, 2011. Image is copyrighted by the American Society of Plant Biologists and is reprinted with permission. (Adapted from Figure 3 of Weissman and Pan, 2015, Oxford University Press.)

12

Chemical Biology of Stem Cells to Tissue Engineering

12.1 General introduction

Advanced therapeutics such as somatic cell therapy, tissue engineering and gene therapy are the result of multidisciplinary problem-driven research involving several fields, including nanotechnology and cellular and molecular biotechnology. In this chapter, we shall look at how chemical biology approaches are being used to understand stem cell biology and how this field opens up possibilities for cell therapies. We also examine the emergence of tissue engineering and the possibilities for regenerative medicine for wound healing and disease treatment.

12.2 Chemical stem cell biology

Stem cells are the cells of early embryogenesis and biological development. They maintain a primitive state (self-renewal) and then differentiate upon signal into one or more specialised lineages (potency). In 1981, **mouse embryonic stem cells (mESCs)** were successfully cultured and demonstrated to have pluripotency, the capacity to generate all cell types of the adult organism. Thereafter, in 1998, **human embryonic stem cells (hESCs)** were successfully derived from blastocyst-stage embryos. Most recently, **induced pluripotent stem cells (iPSCs)** with properties similar to **embryonic stem (ES)** cells were generated by the overexpression of four transcription factors that can collectively drive a differentiated cell back to a pluripotent state. Importantly, **induced pluripotent stem (iPS)** cells may bypass potential ethical challenges associated with hESC research because they are not derived from human embryos. Stem cells are still present in fully adult bodies in numerous tissues, including the brain, muscle, adipose/fat and tissues of the haematopoietic system. However, such adult stem cells are multipotent, or capable of differentiating into multiple cell types, but not all, typically only those appropriate for their local tissue location.

12.2.1 Stem cell regulation

12.2.1.1 Biochemical regulation

In vivo, stem cells are regulated by specialised microenvironments that present them with numerous regulatory signals, soluble signalling molecules, biophysical cues, cell-**extracellular matrix (ECM)** interactions and cell-cell contacts that

Essentials of Chemical Biology: Structures and Dynamics of Biological Macromolecules In Vitro *and* In Vivo, Second Edition. Andrew D. Miller and Julian A. Tanner.
© 2024 John Wiley & Sons, Inc. Published 2024 by John Wiley & Sons, Inc.
Companion Website: www.wiley.com/go/miller/essentialschembiol2

are collectively referred to as the stem cell niche. Within the niche, biochemical regulation is achieved by ECM components, soluble factors or cell-surface factors. Soluble signalling molecules include **Wnt proteins, insulin** and **fibroblast growth factors (FGFs), transforming growth factors (TGFs)** and **cytokines**. In the ECM, **laminin, fibronectin** and **collagen** are important ligands. Each of these proteins is highly intricate and often exhibits both multiple isoforms with numerous cellular receptor binding motifs per isoform. Accordingly, the biochemical information an ECM molecule is conveying to a given stem cell may not be very clear. Interestingly, ES cells and iPS cells interact well with laminin-511, suggesting that this adhesion, mediated by $\alpha_6\beta_1$ integrin proteins, is sufficient to maintain pluripotency.

On the other hand, **pluripotent stem (PS)** cells interact well with vitronectin, as mediated by $\alpha_v\beta_5$ integrin proteins, or with vitronectin- and bone sialoprotein-derived peptides. Hence, pluripotency of stem cells should involve at least two main ECM proteins and a couple of different adhesion receptors. Soluble signalling molecules also interact with ECM proteins. FGFs, TGFs and many other soluble signalling molecules have ECM-binding domains. Therefore, interactions between soluble signalling molecules and stem cells will also be indirectly affected by ECM interactions. However, in addition to the identities of biochemical factors and their specific effects on stem cells, the contextual presentation of these moieties, including potential immobilisation and spatial organisation on scaffolds or particles, also impacts on cell behaviour.

12.2.1.2 Biophysical regulation

In vivo, stem cell niches are incredibly diverse biochemically; however, there are many accompanying differences in the biophysical properties of niches. Most apparent are differences in stiffnesses and topographies of different tissues, as well as the forces imparted during the natural motions of organisms, including joint bending, muscle contraction, compressive impact and strain on tissues, and pulsatile flow of the circulatory system. Even early in development and embryogenesis, significant forces are generated during cell adhesion and migration. Accordingly, biophysical niche properties such as stiffness also regulate stem cell behaviours. In addition to static biophysical properties such as stiffness, microenvironments can also impart dynamic forces on stem cells. Indeed, data clearly support that stiffness controls stem cell behaviour. Otherwise, dynamic forces, such as shear flow and strain also control stem cell behaviour, including differentiation. In addition to mechanical properties such as stiffness, shear and strain, other biophysical properties such as topographical organisation matter, for example the fibrous structures of ECM proteins and the pores in bone marrow are also critical in regulating stem cell behaviour.

12.2.2 Controlling stem cell regulation by biochemical and biophysical means

Clearly, stem cell niches are biochemically and biophysically complex, hence the design and development of systems to explore their structure-function relationships are challenging. Furthermore, niches are highly variable, as stem cells reside in different tissues during all stages of development, from germ layer segregation during embryogenesis and tissue formation during development to declining niche properties in aged tissues. In each niche, ECM macromolecules and resident cells interact in unique ways to shape biochemical properties, such as the identities of natural and synthetic ligands and their spatial/architectural presentation, as well as its biophysical properties, such as modulus, topography, dimensionality and shear/strain.

Nevertheless, several chemical biology strategies can now be used to investigate mechanisms by which the niche controls stem cell behaviour, including the use of HTS technologies to identify soluble signalling molecules (chemical genetics style), ECM components and combinations that modulate stem cell self-renewal and differentiation. Such basic information can then be translated into biomimetic or synthetic microenvironments that control cell behaviour, both at laboratory and bioprocess scale (Figure 12.1). Fundamentally, there are currently two experimental approaches for determining the factors that regulate stem cell fates:

1. **Candidate approach**: seeks to investigate one or more factors likely to have an effect on cell function. This approach is limited to known factors and studies typically explore only a relatively small set of candidates. This approach can be laborious and can require relatively large amounts of materials.
2. **Library approach**: seeks to use HTS to explore the impact of numerous known and unknown factors as well as complex combinations of factors analogous to those encountered in endogenous stem cell niches.

Library approaches, for example using differentiated ES cells on arrayed combinations of ECM proteins, soluble growth factors and recombinant cell adhesion molecules (spotted on cell adhesion-resistant, acrylamide hydrogel-coated slides)

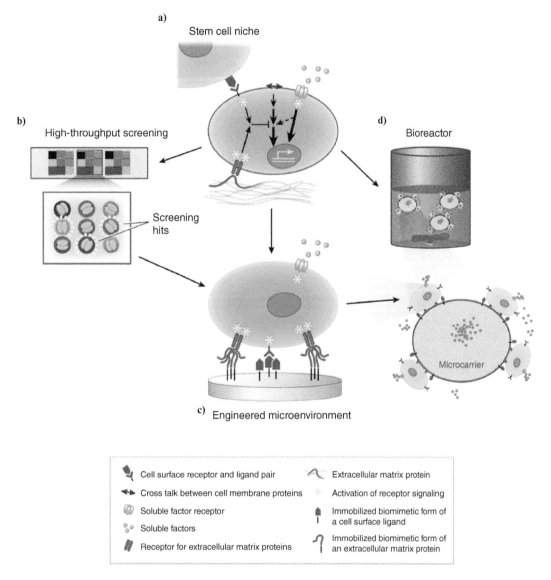

Figure 12.1 Stem cell biology and therapeutics. (a) Stem cells process both biochemical and biophysical signals from adjacent cells, the extracellular matrix and the soluble medium in their biological niches. The complex signal transduction and genetic networks (black and grey arrows inside cell, respectively) that process these microenvironmental signals regulate self-renewal, death or differentiation behaviours that can be mathematically modelled to facilitate an understanding of stem cell biology; (**b**) **High-throughput screening** (**HTS**) technologies, such as seeding stem cells on arrays of micropatterned extracellular matrix proteins, or synthetic polymers, promote the discovery of regulatory factors; (**c**) Factors are applied in engineering synthetic microenvironments to study and control stem cell behaviour *ex vivo*; (**d**) Knowledge gained about stem cell biology and microenvironmental factors from modelling and the use of engineered microenvironments will facilitate the design of bioreactors for large-scale and clinical-grade stem cell therapies (taken from Figure 1 of Ashton et al., 2011, *Annual Reviews*).

have proven a potent way to explore the combinatorial effect of specific microenvironmental factors on stem cell fates. In this case, culture surfaces composed of collagen I and fibronectin vastly enhanced early hepatic differentiation of **embryonic stem cells** (**ESCs**). Collagen I, collagen IV, fibronectin and laminin are indispensable for the extended culture of ESCs.

As an example of the candidate approach, an automated cell culture platform was constructed around a **polydimethylsiloxane** (**PDMS**) microfluidic device (see Chapter 11) with 96 individually addressable culture chambers (60 nL each). Individual addressability offers the opportunity for HTS, by allowing for the customisation of distinct culture environments from chamber to chamber, as a function of time. Fortunately, PDMS is biocompatible and highly permeable to CO_2 and O_2, thereby guaranteeing rapid exchange of these gases between the atmosphere surrounding the device and the medium within each culture chambers (Figure 12.2). This was applied to study the osteogenic differentiation of **human primary mesenchymal stem cells** (**hMSCs**), which are bone marrow stromal cells (non-haematopoietic stem

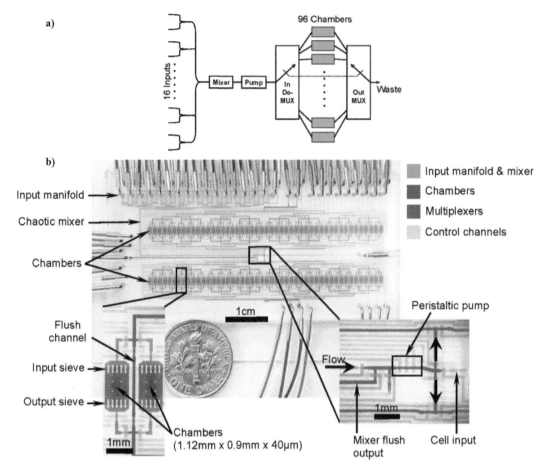

Figure 12.2 Design of a cell culture chip. (a) Simplified schematic diagram of the fluidic path in the chip (MUX: multiplexer); **(b)** Annotated photograph of a chip with the channels filled with coloured water to indicate different parts of the device. In left inset, two culture chambers can be seen, with a MUX flush channel in between them. In the right inset, the MUX is shown with peristaltic pump, a waste output for flushing the mixer, and a cell input line (Adapted from Figure 1 of Gómez-Sjöberg et al., 2007, American Chemical Society).

cell population of the bone marrow). Through investigation of a range of osteogenic media (containing ascorbic acid, β-glycerophosphate and dexamethasone), and exposure periods (0–168 h), it was determined that 4 d was sufficient to induce osteogenic differentiation (Figure 12.3).

As an alternative to microfluidics, two examples using microcontact printed arrays will serve to demonstrate the candidate approach linked to biophysical parameters. In the first instance, microcontact printed arrays of cell-adhesive islands (20 μm diameter) were created for the formation of clonal microarrays, arrays of clonal cell populations derived from individual stem cells, appropriate for HTS analysis of genetic libraries from gain (cDNA) or loss (RNAi) of function (Figure 12.4). As a proof of principle, clonal microarrays have been used to demonstrate that overexpression of akt1 gene increased the proliferation of **neural progenitor cells (NPCs)** compared to the situation without akt1 gene overexpression (Figure 12.4). In another example of microcontact printing, high-density arrays consisting of fibronectin islands of various dimensions were prepared by using PDMS as a support, this time, for fibronectin microprinting. This was done in order to generate well-defined arrays of fibronectin "islands" of different surface areas to which individual hMSCs could attach (one cell per island) and spread to different degrees depending on island sizes (Figure 12.5). In a stem cell local environment change, there can be many cues when cells are grown in different ways. With increasing density, cell adhesion and cell spreading decrease, while cell-cell contact and paracrine (hormonal) signalling increase. The intention here was to create an artificial stem cell micro-environment to determine the impact of cell spreading, namely cell shape (cross-sectional area), on stem cell commitment in the absence of cell-cell communication. Analysis of stem cell fate on these arrays under various differentiation conditions clearly demonstrated that size matters. Greater hMSC cell spreading clearly promotes **osteogenic differentiation** into osteoblasts, while less hMSC spreading promotes **adipogenic differentiation** into adipocytes (Figures 12.5a and 12.5b). In this instance, changes in cell shape may be transduced into a regulatory signal by several structures in the cell, including the actin cytoskeleton itself. Mechanistically speaking, in well-spread hMSCs, so-called

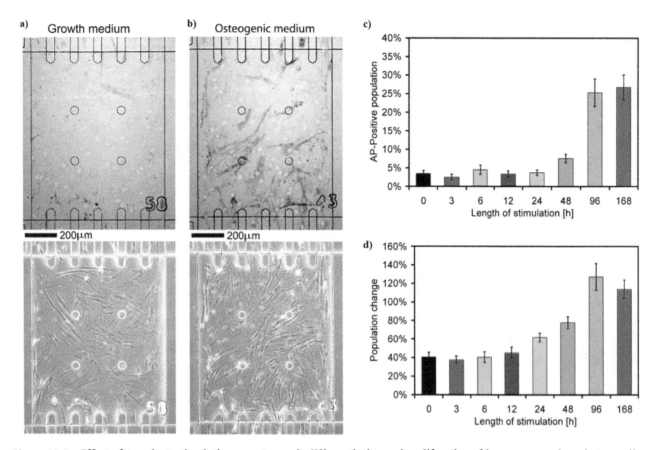

Figure 12.3 **Effect of transient stimulation on osteogenic differentiation and proliferation of human mesenchymal stem cells.** Culture chambers were treated with fibronectin for cell adhesion and then **human mesenchymal stem cells** (**hMSCs**) seeded at 6000 cells/cm² for osteogenesis. After growth periods (see below) chambers were fixed with 4% (w/w) paraformaldehyde, then subject to **alkaline phosphatase** (**AP**) cellular staining, and **2-(4-amidinophenyl)-6-indolecarbamidine hydrochloride** (**DAPI**) nuclear staining in preparation for microscopy analysis. Cell proliferation was assessed by counting cells at the beginning and end. Osteogenic differentiation was assessed by counting the percentage of cells staining positively for AP activity (AP is a bone matrix enzyme whose expression is upregulated during early osteogenic differentiation and is routinely used as a marker of osteogenesis). Shown are bright field and phase contrast microscopic observations (false-colour DAPI image superimposed) of representative chambers of: (**a**) A control chamber fed only with growth medium for 168 h (initial cell number is 59 and the final cell number is 83); (**b**) An osteogenesis chamber, treated for 96 h with osteogenic medium (initial cell number is 52 and the final cell number is 160); (**c**) Average fraction of cells that have differentiated in all osteogenesis chambers (as indicated by the number of cells that stain positive for AP activity); (**d**) Average relative change in the number of cells in all osteogenesis chambers. Error bars correspond to the standard error of the means (*n* = 12) (Adapted from Figure 2 of Gómez-Sjöberg et al., 2007, American Chemical Society).

"stress fibres" are more prominent than in less well spread hMSCs. The physical stresses that are the result of cell spreading then appear to activate a protein called **RhoA GTPase**, leading to upregulation of a **p160-Rho-associated coiled-coil kinase** (**ROCK**) protein that in turn appears to direct hMSC commitment towards osteogenic differentiation.

In a companion set of experiments, cell shape was also found to be a key regulator of **transforming growth factor β3** (**TGFβ3**) induced hMSC differentiation to **myogenic smooth muscle cell** (**SMC**) versus **chondrogenic fates**. Those hMSCs on larger fibronectin "islands", subject to greater cell spreading, appear triggered towards **myogenic differentiation** into SMC fates. On the other hand, hMSCs on small fibronectin "islands", subject to minimal cell spreading, appear triggered to continue into a chondrogenic fate (Figure 12.5c). Mechanistically speaking, TGFβ3 activates **Rac1** (a small GTPase protein, like RhoA, above), so that when hMSCs are well spread there is an increase in expression of *N*-cadherin (a protein involved in cell-matrix adhesion and linked with the cell actin cytoskeleton), leading to the upregulation of myogenic/SMC genes. When hMSCs are less well spread, TGFβ3 fails to activate Rac1 signalling and upregulates chondrogenic gene expression instead, leading to a chondrocyte fate.

Moving on to an impressive example of the library approach, a polymer library was created from 22 well-defined polyethylene/propylene glycol-acrylamide monomers admixed together in various different ratios (Figure 12.6a). Monomers were spot printed in microarray format in a humid argon atmosphere on an epoxy monolayer-coated glass slide that was

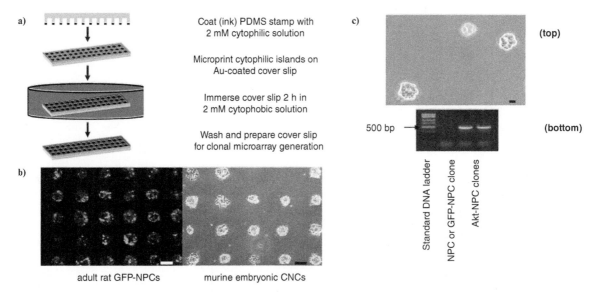

adult rat GFP-NPCs   murine embryonic CNCs

Figure 12.4   Formation of neurosphere microarrays. (a) Schematic of procedure used to generate micropatterned substrates starting with PDMS elastomeric stamps. Stamp tip surfaces were coated with a 2 mM solution (cytophilic) of **11-mercapto-undecanoic acid** dissolved in ethanol (EtOH) dried under a nitrogen stream and then brought in contact for 30 s with a glass cover slip pre-coated with 25 Å of titanium followed by 125 Å of gold using an electron beam evaporator. The cover slip was then immersed into a 2 mM solution (cytophobic) of **tri(ethylene glycol)-undecanethiol** that forms a self-assembled monolayer in the remaining non-printed regions. After a 2 h incubation period, the chemically modified gold cover slips were rinsed with EtOH ready for generating clonal microarrays; **(b)** Post cell seeding and growth, patterned array of neurospheres of adult rat hippocampal **neural progenitor cells expressing green fluorescent protein (GFP-NPCs)** and mouse embryonic **cortical neural cells (CNCs)** (scale bar 100 μm); **(c)** (top) Proof of concept experiment where GFP-NPCs and **akt1 overexpressing NPCs (Akt-NPCs)** were seeded side by side and cultured. Post four days' culturing with 1 ng/ml FGF-2, GFP-NPC clonal populations were seen adjacent to larger unlabelled NPC clonal populations (scale bar 10 μm); (bottom) DNA agarose gel from clonal populations showing that larger unlabelled NPC clonal populations are Akt-NPCs overexpressing akt1, consistent with the fact that akt1 signalling is necessary for robust NPC proliferation. The smaller GFP-clonal populations did not overexpress akt1, so, too, some smaller unlabelled NPC clonal populations were identified that were also found not to overexpress akt1 (Adapted from Figures 1 and 4 of Ashton et al., 2007, Oxford University Press).

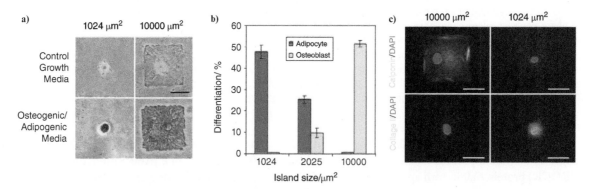

Figure 12.5   Cell shape drives human mesenchymal stem cell commitment. A PDMS support was microprinted with fibronectin to generate fibronectin "islands" (cytophilic regions) surrounded by regions blocked off to cellular growth with the non-adhesive, **pluronic F108** (cytophobic regions). Next, hMSCs were plated onto these islands at a density of one single per island, allowing hMSCs to spread to different degrees depending on the size of the islands (1024 or 10,000 μm²). **(a)** Brightfield images of hMSCs plated onto small (1024 μm²) or large (10,000 μm²) fibronectin "islands" then cultured for one week in standard control growth or mixed (osteogenic/adipogenic) media, followed by staining. Lipids stain red, as a marker for hMSC to adipocyte differentiation, while alkaline phosphatase (AP) stains blue, as a marker for hMSC to osteoblast differentiation (scale bar 50 μm); **(b)** Percentage differentiation is shown of hMSCs plated on 1024, 2025 or 10,000 μm² fibroblast islands after one week of culture in mixed (osteogenic/adipogenic) media (taken from Figure 12.3 of McBeath et al., 2004, *Developmental Cell*); **(c)** Fluorescence confocal microscopy images are shown of hMSCs plated onto small (1024 μm²) or large (10,000 μm²) fibronectin "islands" then cultured for one week in **transforming growth factor β3 (TGFβ3)** media, followed by staining. DAPI stains all cell nuclei blue, Calponin stains green, as a marker of hMSC myogenic differentiation into **smooth muscle cells (SMCs)**, and collagen II stains green, as a marker for hMSC chondrogenic differentiation into **chondrocytes** (scale bar 50 μm) (Adapted from Figure 1 of Gao et al., 2010, Oxford University Press).

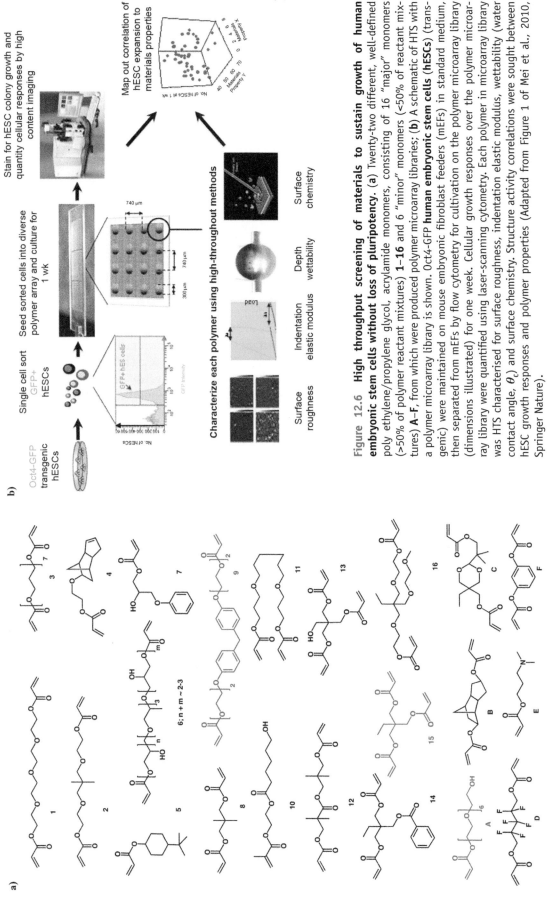

Figure 12.6 High throughput screening of materials to sustain growth of human embryonic stem cells without loss of pluripotency. (a) Twenty-two different, well-defined poly ethylene/propylene glycol, acrylamide monomers, consisting of 16 "major" monomers (>50% of polymer reactant mixtures) **1–16** and 6 "minor" monomers (<50% of reactant mixtures) **A–F**, from which were produced polymer microarray libraries; **(b)** A schematic of HTS with a polymer microarray library is shown. Oct4-GFP **human embryonic stem cells (hESCs)** (transgenic) were maintained on mouse embryonic fibroblast feeders (mEFs) in standard medium, then separated from mEFs by flow cytometry for cultivation on the polymer microarray library (dimensions illustrated) for one week. Cellular growth responses over the polymer microarray library were quantified using laser-scanning cytometry. Each polymer in microarray library was HTS characterised for surface roughness, indentation elastic modulus, wettability (water contact angle, θ_c) and surface chemistry. Structure activity correlations were sought between hESC growth responses and polymer properties (Adapted from Figure 1 of Mei et al., 2010, Springer Nature).

first dip-coated in 4% (v/v) poly-(hydroxyethyl methacrylate). Once done, each microarray monomer spot was polymerised via 10 s exposure to long-wave UV (365 nm), dried at <50 mτ pressure for at least 7 d, then coated with 20% serum or with other proteins (e.g. human or bovine serum albumin, human vitronectin). Thereafter, polymer microarray libraries, so created, were subject to HTS in order to determine those synthetic polymers to which proteins adsorbed most ably to sustain growth of hESCs without loss of pluripotency (Figure 12.6b). Using HT surface characterisation techniques (e.g. atomic force microscopy, water contact angle and time-of-flight MS/MS), structure activity correlations were sought between the physico-chemical properties of library polymers and their abilities to sustain hESC growth without loss in pluripotency, resulting in the identification of two hit polymers. One was prepared from 100% monomer **9**, and the other **15A-30%** (prepared from 70% monomer **15** and 30% monomer **A**) (Figure 12.6a). The leading combination for optimal clonal expansion of hESCs was found to be 15A-30% that adsorbed vitronectin from the serum containing culture medium.

The example above clearly demonstrates how variations in the physicochemical properties of natural and synthetic substrate materials can have powerful impacts on niche microenvironments that promote stem cell growth and control stem cell fates. Leading the way here are synthetic materials such as **polyacrylamide** and **polyethylene glycol** (**PEG**) that do provide several advantages over natural materials. For example, a library of ultrathin polyacrylamide gels has been prepared with different levels of elasticity (in 2D: $100–10^5$ Pa), in common with actual tissue elasticity values, through variations of three parameters, namely cross-linking, gel thickness and collagen-I deposition. As a result, hMSCs cultured on such gels appear to differentiate into specialised cell types of widely different cellular morphologies according to medium elasticity (Figure 12.7). Therefore, there is a great deal to be gained by making and testing a wide range of different synthetic substrate materials to mimic natural systems, such as the fibrous structure of ECM proteins and the pores in bone marrow, in order to control stem cell growth and differentiation. Whilst soft microlithography, for micropatterning and microfluidics with micro-syringe deposition, can be part of this process, another technique is also emerging with great utility for the creation of biomimetic materials that are also biocompatible. This technique is known as **electrospinning**. The basic principles of electrospinning are illustrated in Figure 12.8a. An interesting recent use of electrospinning has been to prepare biocompat-

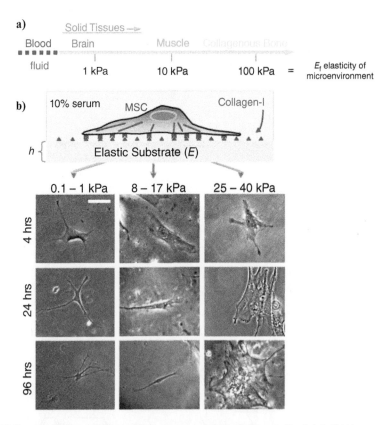

Figure 12.7   Tissue elasticity and differentiation of human mesenchymal stem cells. (a) Solid tissues exhibit a range of stiffness, as measured by their elastic modulus, E; (b) Ultrathin polyacrylamide gels are prepared for *in vitro* cell cultivation adapted for control of E through cross-linking and control of thickness (~70–100 μm in thickness), plus control of cell adhesion with covalently attached collagen-I at 0.25–1 μg/cm². Naive hMSCs are initially small and round, but develop increasingly branched, spindle or polygonal shapes when grown on different respective matrices in the range typical of E_{brain} (0.1–1 kPa), E_{muscle} (8–17 kPa), or $E_{collagenous}$ (25–40 kPa) (scale bar is 20 μm) (Adapted from Engler et al., 2006, with permission of Elsevier).

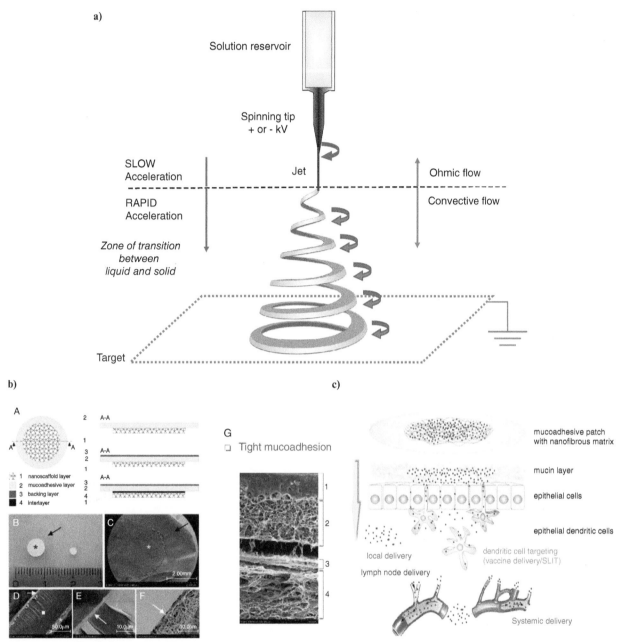

Figure 12.8 **Electrospinning process. (a)** A schematic of this fibre production technique, which uses electric fields to create charged threads of synthetic or natural polymer that may be drawn out into uniform fibres with nm-dimensions (of a few 100 nm). This technique relies on the fact that a droplet of polymer solution, formed on a tip in the presence of a sufficiently high electric field, will become charged to the point that electrostatic repulsion exceeds droplet surface tension. A liquid "eruption" then takes place, forming a jet that elongates as it dries in flight leaving behind charged polymer fibres that are themselves further elongated and thinned into uniform nanofibres. These are eventually deposited on a target under the influence of convective flow, so producing nanofibrous materials suitable for many potential applications such as stem cell ECMs, nanoparticle adsorption, etc; **(b) Needle-free, sublingual and buccal nanodelivery patch technologies.** In panel (A) the general design of the patch technology is shown as three variants of a multi-layered **nanofibrous mucoadhesive film.** In panel (B) patch dimensions are illustrated: nanoscaffold/nanofibrous layer (asterisk), mucoadhesive layer (arrowed). In panel (C) is a scanning electron micrograph from a patch bottom surface, while panels (D) and (E) are cross-sectional, cryo-scanning electron micrographs wherein nanoscaffold layer (asterisk) and mucoadhesive layer (square) are identified with the backing layer (by the head of the white arrows). Panel (F) shows the nanoscaffold layer in close up detail (right of view), the interface with the mucoadhesive layer is indicated (white arrow) (taken from Figure 12.2, Mašek et al., 2017, with permission of Elsevier). In panel (G) is a scanning electron micrograph of the nanoscaffold layer (2) in intimate contact with mucin (3) mucosal surface tissue (4) (Mašek et al., 2017, with permission of Elsevier); **(c)** A schematic to demonstrate how the nanoscaffold layer is intended to adsorb nanoparticles, virus particles and alike, then release these through the mucosal surface by "mass action delivery" for mucosal immune cell surveillance ("mucosal vaccination"), followed by their subsequent distribution to the systemic circulation, via available lymph or blood vessels (Adapted from Figure 1, Mašek et al., 2017, Elsevier).

ible/biomimetic materials for use in the design and creation of **needle-free, sublingual and buccal nanodelivery patch technologies** intended to enable mucosal vaccination (Figures 12.8b and 12.8c). Electrospinning is used to prepare a nanofibrous layer that adsorbs biomaterials such as vaccination nanoparticles (see Chapter 13) that are then transferred by mass action across mucosal membranes to which the nanofibrous layer is placed in contact by an additional mucoadhesive layer.

12.2.3 Controlling stem cell regulation by genetic means

With unlimited self-renewal capacity, hESCs are the benchmark for cells of value to regenerative medicine and disease modelling. However, given ethical concerns with using hESCs, the discovery of patient-specific iPSCs was exceptionally important. Yet, to realise the full potential of iPSCs, disease causing genes typically need to be corrected or modified prior to applications. A completely orthogonal way to achieve this in stem cells is now possible using some of the latest genetic engineering tools. These derive from so-called **Class 2, clustered, regularly interspaced, short palindromic repeat (CRISPR)** systems. Originally studied as part of an adaptive immune system in bacteria, CRISPR has become rapidly adopted as a popular genome engineering approach tool in association with **CRISPR-associated endonuclease** protein (**Cas protein**, typically Cas9 from *Streptococcus pyogenes*). Cas9 is a nuclease capable of creating targeted **double-strand breaks (DSBs)** when directed to a given DNA locus by means of CRISPR and a **guide RNA (gRNA or sgRNA)** (Figure 12.9). An sgRNA is a short stretch of synthetic RNA composed of a scaffold sequence region at its 3′-end for binding to Cas9 and a user-defined **protospacer sequence** (20–22 nts) region at its 5′-end, for binding to the AS strand of a selected CRISPR protospacer target in DNA (which is obviously complementary to the protospacer target sequence in DNA). The mechanism is shown giving rise to highly selective DSBs (Figure 12.9). These DSBs are then repaired by one of two general repair pathways, either the efficient but error-prone **non-homologous end joining (NHEJ)** pathway, or the less efficient but high-fidelity **homology directed repair (HDR)** pathway.

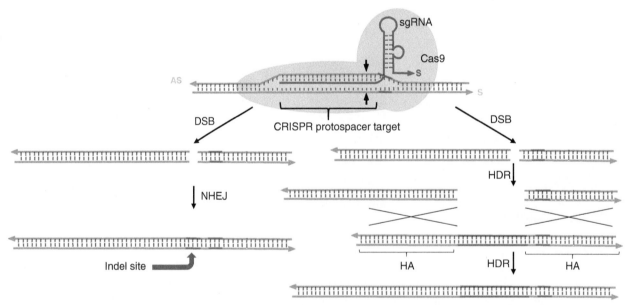

Figure 12.9 Mechanism of class 2, clustered, regularly interspaced, short palindromic repeat/Cas9 gene editing system. A ribonucleoprotein complex is first formed between **Cas9** and an **sgRNA** co-factor. This complexation begins with interactions between the sgRNA scaffold sequence region and surface-exposed positively-charged grooves on Cas9. Thereafter, a conformational change occurs, enabling the sgRNA protospacer region to zip-bind with the AS strand of its **Class 2, clustered, regularly interspaced, short palindromic repeat (CRISPR) protospacer target sequence** (in a 3′→5′ direction). Initial binding involves interactions between the seed sequence (first 8–10 nts of sgRNA) with the 3′-end of the corresponding AS strand of the CRISPR protospacer target sequence. This CRISPR protospacer target is located immediately adjacent to a **protospacer motif (PAM)** (typically 5′-NGG in the DNA S strand) (**blue stranded region** in figure). Assuming seed sequence/AS strand complementarity, the sgRNA will then continue to anneal to completion, thereby enabling the Cas9 functional endonuclease domains (RuvC and HNH) to undergo a second conformational change that causes a **double-strand break (DSB)** (~3–4 nts upstream of the PAM sequence). DSBs are repaired by one of two general repair pathways: (1) the efficient but error-prone **non-homologous end joining (NHEJ)** that frequently causes **small nucleotide insertions** or **deletions** (**indels**) (left: **short purple strand region**); (2) the less efficient but high-fidelity **homology directed repair (HDR)** (right: **long purple strand region**). Targeted DSB repairs are the basis, then, of CRISPR/Cas 9 engineering of genomes. Other abbreviations: **homology arm: HA.**

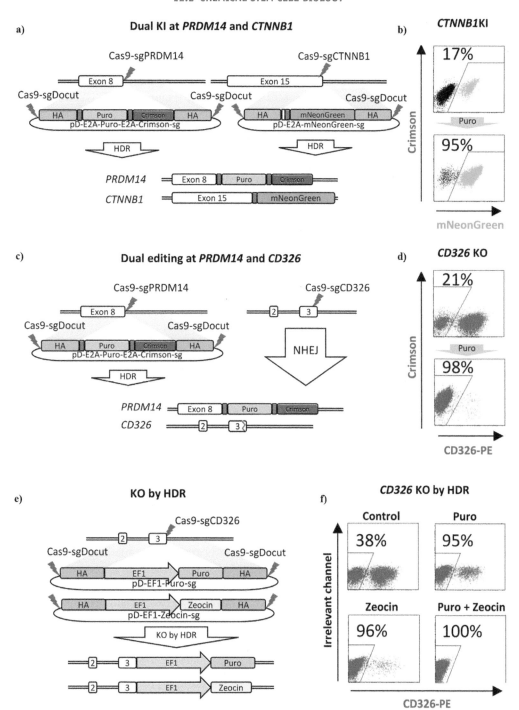

Figure 12.10 **CRISPR/Cas9 engineering strategies with knock in or knock out gene editing.** (a) HDR-mediated dual **knock-in (KI)** at PRDM14 and CTNNB1 sites in **induced pluripotent stem cell (iPSC)** genomes. Cas9, sgRNA and donor pDNAs were transferred into iPSCs by electroporation. HDR KI of the E2A-Puro-E2A-Crimson donor cassette at the PRDM14 site allows for puromycin selection to enrich for iPSCs with HDR editing at CTNNB1; (b) Co-enrichment of PRDM14/CTNNB1 double HDR-edited iPSCs was achieved by single selection, using added puromycin (1 g/ml) (added two days after electroporation). KI efficiency (mNeonGreen-positive) at CTNNB1 site was determined by **fluorescence-activated cell sorting (FACS)**; (c) Editing at PRDM14 by HDR KI and at CD326 by NHEJ **knockout (KO)**. Similar procedure to (a); (d) Co-enrichment of PRDM14/CD326 HDR/NHEJ-edited iPSCs was achieved by single selection using added puromycin (1g/ml) (added 2 d after electroporation). NHEJ KO efficiency at CD326 site (CD326-PE-negative) was determined by FACS; (e) Schematic for dual biallelic HDR KO-editing. Two HDR pDNA donors (pD-EF1-Puro-sg and pD-EF1-Zeocin-sg) were designed to insert puromycin or zeocin resistance genes at CD326, leading to biallelic disruption of the open reading frame. All Cas9, sgRNA and donor pDNAs were transferred into iPSCs by electroporation, followed by single or double selection; (f) Co-enrichment of HDR KO gene edited iPSCs was achieved with single or double selection. Single selections were performed using puromycin (1 g/ml) or zeocin (100 g/ml) and double selection using both (always added two days after electroporation). HDR KO efficiency at CD326 site (CD326-PE-negative) was determined by FACS (taken from Figure 6, Li et al., 2018, Oxford University Press).

The NHEJ repair pathway is the most active repair mechanism, but frequently causes small nucleotide **insertions or deletions (indels)** at the DSB site. In most cases, indels result in frameshift mutations leading to premature stop codons within the targeted gene, and may result in amino acid deletions, insertions and/or protein loss of function at the level of translation. Accordingly, the NHEJ repair pathway can be used to disrupt the open reading frame of a gene and generate a **knockout (KO)** allele to a target gene bearing the selected protospacer target sequence (Figure 12.9). This editing approach is relatively efficient and has been widely used in genetic engineering and functional genomics research. On the other hand, the HDR pathway can be used to integrate a donor DNA sequence that is flanked by **homology arms (HAs,** 300–600 bp up to 10 kbps in length) into the DSB site in order to create a precise DNA deletion, substitution or insertion, leading to the correction of pathologic genes or else the targeted **knock-in (KI)** integration of a new gene or DNA fragment of interest. As always with such technologies, the efficiency of use can vary widely with cell types and methods. The inefficiency in editing human iP-SCs is largely due to low cell viability after manipulation. To precisely edit iPSCs, the CRISPR components Cas9 and sgRNA should be introduced to iPSCs in the form of pDNA expression cassettes, together with a pDNA donor template (pDonor), that contains a donor DNA sequence of interest flanked by HAs. The most efficient means of such pDNA delivery *in vitro* is electroporation (an electric shock technique to temporarily open up the membranes of cells, at 60–70% confluence, to heterologous nucleic acid entry), which does result in functional delivery but can induce massive cell death. Means are now being found to improve on this situation and it is very likely that CRISPR/Cas9 will become a powerful future tool for the regulation and engineering of iPSCs and other stem cell types (Figure 12.10).

12.2.4 Stem cell modelling

In spite of undoubted experimental successes, overall the nonintuitive agonistic and antagonistic cross talk between ECM protein and growth factor signalling pathways, as revealed from multifactorial microarray data, is limited. Computational modelling approaches are also required to understand and make sense of underlying signalling mechanisms. Furthermore, although HTS methods will facilitate the discovery of new regulators of stem cell fate, the complexity of microarray data requires a deeper understanding of the corresponding molecular and signalling mechanisms to translate these discoveries into future therapeutics. What is clear is that stem cell fates are governed by complex intracellular signalling networks that process input signals from the cell surface and relay those signals to the nucleus. These signalling cascades may contain nonlinear components, such as signal amplification, oscillation, feedforward or feedback loops and cross talk between multiple pathways (Figure 12.1). Once inside the nucleus, signal processing continues with circuits of transcription factors that control the expression of one another and of genes that regulate fate choice. Hence in actuality there is a complex, nonlinear, multilevel cascade that is difficult to even begin to really investigate and understand without the aid of systems-level analysis and mathematical tools.

Multiple classes of computational models have been used to analyse stem cell signal processing networks, including deterministic, stochastic and attractor state models that highlight knowledge gaps and drive further experimentation. Complementary approaches include statistical methods such as **Bayesian networks** and **principal components analysis (PCA)/partial least squares (PLS)** regression designed to data mine large "-omic" data sets (e.g. from transcriptomics, genomics, proteomics, kinomics, etc.) to identify genes and modules whose behaviours are correlated, thereby offering mechanistic hypotheses that can be further tested to deepen an understanding of these complex systems.

12.2.4.1 Deterministic modelling

Deterministic models define molecular interactions among microenvironmental inputs and intracellular signalling networks as mass action expressions in steady state and the outputs of models are time trajectories of the concentrations of network constituents. Such models utilise and require detailed knowledge of most constituent molecular interactions, including the appropriate kinetic and binding constants. Because such data can be limiting, often owing to a lack of measured physical constants, the estimation of these constants from analogous systems is often required. Additionally, these models assume that reactants are abundant and thus use sets of continuous ordinary or partial differential equation formulations, which are often nonlinear and thus can only be solved numerically.

Deterministic models have highlighted intriguing and unintuitive network behaviours in stem cell systems already. For example, stem cells execute all-or-nothing fate decisions in response to microenvironmental cues. One network characteristic that could mediate such decision-making is **bi-stability**, in which a change in input parameter results in an all-or-nothing binary change with respect to an output parameter. Such bi-stable networks also exhibit hysteresis (system memory), making them resistant to noise in the input signal and thus avoiding rapid or indecisive switching between cell states, even when levels of input signal(s) are close to the threshold for switching between states.

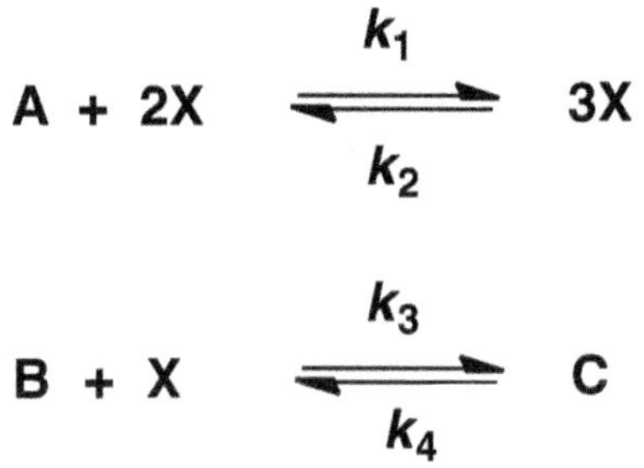

$$\mathbf{A\ +\ 2X} \quad \underset{k_2}{\overset{k_1}{\rightleftharpoons}} \quad \mathbf{3X}$$

$$\mathbf{B\ +\ X} \quad \underset{k_4}{\overset{k_3}{\rightleftharpoons}} \quad \mathbf{C}$$

Scheme 12.1 **Nonlinear chemical reaction scheme**

A way of understanding bi-stability can come from considering more closely what happens to dynamics of biochemical reactions when small volumes are involved (i.e. as within cells). In this case, the dynamics of biochemical reaction systems in a small volume (i.e. mesoscopic) are considered in terms of a stochastic, discrete-state, continuous-time formulations, called **chemical master equations** (**CMEs**). A given CME is similar in concept to a wavefunction in quantum mechanics, except that instead of representing a probability density function for an electron, the CME maps out a probability landscape that characterises the dynamic behaviour of the corresponding biochemical reaction system which links through to macroscopic determinism. This can be illustrated with respect to one simple system of non-linear chemical reactions (Scheme 12.1). The traditional macroscopic kinetics of this system obey the following mass action Equation (12.1), in effect, an **ordinary, deterministic differential equation** (**ODE**):

$$\frac{d[X]}{dt} = k_1[A][X]^2 + k_4[C] - k_2[X]^3 - k_3[B][X] \tag{12.1}$$

which becomes modified by replacing molarity (mol l^{-3}) with moles of each reactant/product. This then gives the following Equation (12.2):

$$\frac{dX}{dt} = k_1 A \frac{X^2}{V^2} + k_4 C - k_2 \frac{X^3}{V^2} - k_3 B \frac{X}{V} \tag{12.2}$$

which can be further adapted to give Equation (12.3):

$$\frac{dX}{dt} = k_1 \frac{AX(X-1)}{V^2} + k_4 C - k_2 \frac{(X-1)X(X+1)}{V^2} - k_3 \frac{B(X+1)}{V} \tag{12.3}$$

that reduces further by grouping the terms, as follows:

$$\frac{dX}{dt} = v_X - \omega_X \tag{12.4}$$

and thence by analogy to the molecular version of the same:

$$\frac{dP(x,t)}{dt} = v_x - \omega_x \tag{12.5}$$

which can be summed over time, *t*, to give:

$$P(x) = C_0 \sum_{j=0}^{j=x-1} v_j - \omega_j \tag{12.6}$$

where C_0 is a normalisation constant such that $P(x)$ is 1. By taking natural logs of both sides we end up with Equation (12.7):

$$\ln P(x) = \sum_{j=0}^{j=x-1} \ln \frac{v_j}{\omega_j} + C_1 \tag{12.7}$$

where C_1 is a natural log of C_0. This can be turned into an integral approximation as shown:

$$\ln P(x) \approx \int_0^x \ln\frac{\upsilon_x}{\omega_x}\,dx + C_1 \tag{12.8}$$

that is rearranged to give Equation (12.9):

$$\Phi(x) = -\int_0^x \ln\frac{\upsilon_x}{\omega_x}\,dx \tag{12.9}$$

where the term $\Phi(x)$ is the "landscape" for the given nonlinear chemical reaction system (Scheme 12.1), which links back to the ODE (Equation 12.1). Finally, since $[X]$ is defined by X/V or x/V, where X is mole and x is molecular number, then a stationary probability distribution function can be defined for a given system volume, V, as shown:

$$f([X]) \approx \exp\{-V\Phi(x/V)\} \tag{12.10}$$

A solution for this function will demonstrate two alternate peaks of distribution of $[X]$ that demonstrate the bi-stability of the nonlinear chemical reaction system in small cellular volumes. The implication in stem cell systems is that one peak of distribution could correlate with a defined cell fate decision and the other with a defined cell decision to retain current status.

By definition, a CME is the differential equation for the probability distribution appropriate for a given nonlinear chemical reaction system. In this context, Equations (12.8), (12.9) and (12.10) reflect different "versions" of the CME for the given nonlinear chemical reaction system (Scheme 12.1). Please note that probability distributions are just the beginning. These need to be combined with stochastic trajectory information that might be supplied from **particle-state-tracking** (PST) or **particle-number-tracking** (PNT) analyses for the chemical reaction system of interest. Indeed, in analysing a CME model, the twin approaches of CME and trajectory analysis complement each other. More generally, for a real understanding of a biochemical reaction system, and biochemical networks in turn, accurate solutions to multiple CMEs will be required. In this way the topological features of probabilistic landscapes, such as basins of attraction, craters, peaks and saddle points, may be quantitated in terms of their widths, breadth and depths. Such data in turn may then become linked with actual biological implications such as stem cell fates and epigenetic modifications of cell behaviour.

12.2.4.2 Other modelling approaches

Deterministic models assume that system states are uniquely determined by fixed parameters. However, many biological systems such as cells are characterised by slow biochemical reactions and/or low concentrations of reactants, resulting in a greater influence of fluctuations or noise (stochasticity) on signalling behaviour. Hence some stem cell fates may be the result of stochastic processes. In fact, ESCs may exist in multiple metastable states, that are interchangeable. However, molecular stochastic simulations to investigate potential multistate behaviour of gene networks and biochemical reactions are yet to see widespread use. Similarly, there is a second alternative to deterministic models, known as attractor state models, which rely on the assumption that systems will always converge towards a stable state (states) according to mathematical convergence formulae or algorithms. Attractor state models are also yet to see widespread use, but could be useful in understanding the networks underlying stem cell fate choice, particularly in iPSCs as they revert to ESC-like states.

Finally, there are thoroughly statistical models; these are considered useful for studying cell responses to the microenvironment provided the responsible signalling pathways are well understood. In many instances a researcher is faced with analysing a large "-omic" data set with little or no knowledge of the critical signalling network(s). **Bayesian networks** (BNs) can help "map" causal relationships between measured quantities in a large data set. In order to find a BN model that best fits a data set well, the start is a probability function, $P(\text{Data} \mid \text{Model})$, that converges towards the unitary value of 1 for best fit. Such a function is defined to act on a data set with m entries, wherein each entry is made up of n discrete variables defined as i. The number of states that each variable can attain is r_i (Figure 12.11). In terms of a BN analysis, each variable, i, represents a network node linked to other **nodes** by arrows to represent apparently causal connections between nodes. Each node, i, is networked with **parent nodes**, π_i, that adopt a total of q_i possible combinations of states

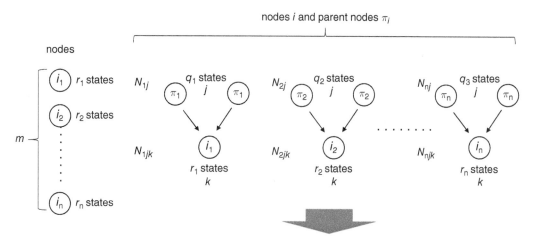

Figure 12.11 **Summary for Bayesian network terms.** These **Bayesian network (BN)** terms are for a data set of m entries where each entry comprises n of **nodes**, i. These nodes, i, in k possible r_i states are in causal relationships with **parent nodes**, π_i, in j possible q_i states. Causality is determined by (1) the total number of instances, N_{ijk}, when each node, i, is in a particular r_i state, k, while parent nodes, π_i, are in a particular q_i state, j, and by (2) the total number of instances, N_{ij}, when parent nodes, π_i, are in a particular q_i state, j.

(Figure 12.11). Finally, N_{ijk} represents the total number of instances where node, i, is in a state, k, while parent nodes, π_i, are in a state, j. Similarly, N_{ij} represents the total number of instances where parent nodes, π_i, are in a state, j. Given these terms, a closed form expression for $P(\textbf{Data} \mid \textbf{Model})$ can be written as Equation (12.11):

$$P\big(\textbf{Data}|\textbf{Model}\big)=\prod_{i=1}^{n}\prod_{j=1}^{q_i}\frac{\big(r_i-1\big)!}{\big(N_{ij}+r_i-1\big)!}\prod_{k=1}^{r_i}N_{ijk}! \tag{12.11}$$

In effect, this equation represents the product of the probability of observing nodes, i, in a particular state, k, given parent nodes, π_i, that are in some state, j. Combinations of parent nodes that are more informative, or predictive, of the nodes will have a higher probability. Biologically, each node, i, is a value that can be measured, such as the phosphorylation level of a particular protein or the rate of cell differentiation. Arrows between nodes indicate apparent causal relationships uncovered within the data (Figure 12.11). However, arrows between nodes do not differentiate among positive, negative or more complicated interactions, but only indicate a directional relationship between variables.

The procedure for finding a BN model that best fits a data set requires the generation of a wide range of possible networks subject to certain constraints (e.g. no more than three parent nodes, π_i, can be linked with a given node, i, some i nodes (cues) should have no π_i parent nodes or some π_i parent nodes (responses) should have no links to any i nodes), followed by analysis for best fit. This analysis results in a graphical map (Figure 12.12) representing the likelihood of finding a species in a particular state given the states of the surrounding species. At times the resulting network yields results that would have been difficult to uncover through typical reductionist experimental approaches, and these results can drive more experimentation. This approach has been used to study mESC proteomics data focused on the detection of protein phosphorylation in response to varying levels of **fibroblast growth factor 4 (FGF4)**, cytokine **leukaemia inhibitory factor (LIF)**, plus laminin and fibronectin of the ECM. Although no assumptions were made about the structure of the underlying signalling network responsible for transmitting signals from microenvironmental cues, the BN analysis highlighted the importance of the **extracellular signal-regulated kinase (ERK)**, **mitogen-activated protein kinase (MAPK)**, **LIF/Janus kinase (Jak)/signal transducer and activator of transcription 3 (STAT3)** pathways, all in good agreement with prior knowledge of key mESC signalling networks. In addition, the importance of **adducin-α** and **protein kinase Cε (PKCε)** in mESC differentiation was highlighted and subsequently confirmed by experiment for the first time (Figure 12.12).

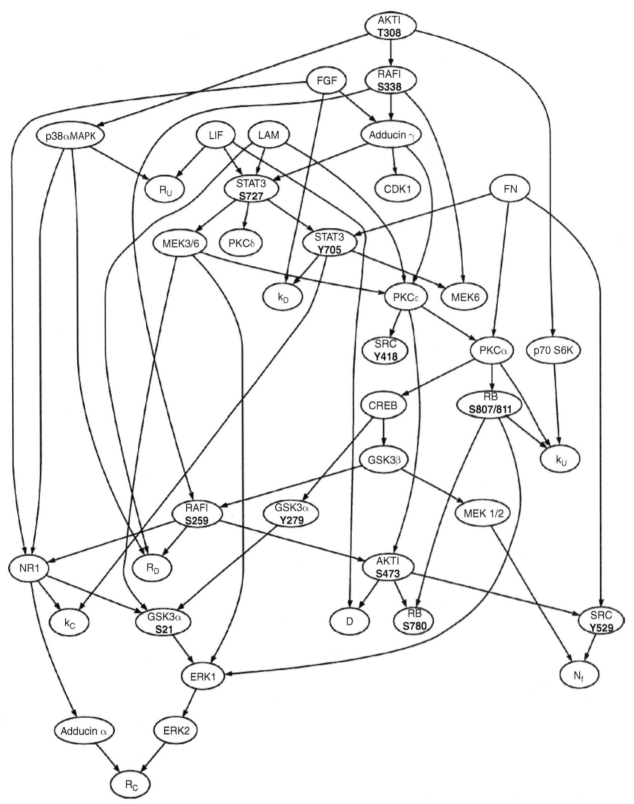

Figure 12.12   Best fit Bayesian network scheme to model signalling events leading to mouse embryonic stem cells differentiation.
Each node, *i*, represents a different signal protein and arrows connecting nodes represent apparent causal relationships. This scheme is analogous to a cellular biochemical pathway in **mouse embryonic stem cells (mESCs)**. In this case, including the suggestion of relationships not identified by traditional biochemical means. Highlights of this scheme are the dominant involvement of **extracellular signal-regulated kinase (ERK)**, **mitogen-activated protein kinase (MAPK)** and **leukaemia inhibitory factor (LIF)/Janus Kinase (Jak)/signal transducer and activator of transcription 3 (STAT3)** protein combinations. In addition, the importance of **adducin-α** and **protein kinase Cε (PKCε)** were seen for the first time (taken from Figure 3 of Woolf et al., 2005, Oxford University Press).

12.3 The road to cell therapies

As detailed above (see Section 12.2.2), chemical biology research into materials can lead to quantitative control of certain regulatory stem cell features in a modular manner, thereby enabling the step-by-step mastery of individual properties that impact on stem cell function in complex stem cell microenvironments. This knowledge can then be applied to the development of large-scale and clinical-grade bioreactors for the discovery of bulk structure-function relationships and for the bulk scale production of stem cells of interest. Thereafter, the priority objective must be to liberate the biomedical potential of stem cells for the treatment of tissue(s) degraded by disease or trauma. In principle, this can be done *in vivo* by means of advanced stem cell-driven tissue replacement therapies that make use of both stem cell self-renewal and the capability of stem cells for controlled differentiation *in vivo* into key cell types required in multiple adult tissues.

12.3.1 The need for bioreactors

For clinical applications, approximately 10^9 cells are apparently required to regenerate one cardiac patient's tissue after a myocardial infarction or to overcome lack of insulin sensitivity in one 70 kg diabetic patient. Given that standard suspension bioreactors can produce cultures of 10^6–10^7 cells ml^{-1}, then culture volumes of hundreds of mL^{-1} L must be available for each cell therapy patient, assuming that corresponding therapeutic cell populations remain appropriately homogeneous after culturing. Accordingly, bioprocesses need to be designed capable of producing therapeutic cell populations at this scale for numerous patients in a cost effective, pathogen-free and reproducible manner. Moreover, materials used for stem cell culture and differentiation should be fully defined and produced via synthetic or recombinant means, for example no feeder cell layers, conditioned media or animal or human-derived serum or proteins. In addition, all operational stem cell bioreactor parameters should be fully controlled, for example dissolved oxygen, pH and agitation-induced shear. This is very important. ESCs in the developing embryo and in the adult brain function at oxygen levels much lower than those of standard culture conditions, and oxygen levels are known to regulate stem cell proliferation and differentiation *in vitro*.

12.3.2 Practical bioreactors

Stirred suspension bioreactors (**SSBs**) are the traditional workhorse of the biomanufacturing industry (Figure 12.13), but single-phase SSB cultures, which generally contain only culture media and cells, are arguably insufficient to deliver on the necessary level of control over ESC or iPSC production, given the inherent heterogeneity of local microenvironments inside cultured cell aggregates, for example formation of **embryoid bodies** (**EBs**). Importantly, ESCs and iPSCs together, which can be defined collectively as **pluripotent stem cells** (**PSCs**), are prone to some degree of apoptosis when cultured as single cells under standard conditions. Also, EBs that are formed make an exterior epithelial-like cell layer with tight cell-cell junctions and deposits and an exterior basal lamina. As a result, PSCs in EBs undergo spontaneous and relatively uncontrolled differentiation into cell derivatives of the three embryonic germ layers (endo-, meso- and ectoderm), such as neural, haematopoietic, endothelial, cardiac or pancreatic cells. Finally, 30–50% of any cell cultures are lost during subculturing, which must be performed at least weekly to limit the development of larger cell aggregates (>500 μm in diameter) that result in spontaneous cell differentiation and promote cell death due to limited oxygen and nutrient diffusion.

Apoptosis may be reduced by including ROCK inhibitors (see Section 12.2.2) during PSC culturing. Other molecular interventions such as the inhibition of ERK and **glycogen synthase kinase** (**GSK**) 3β plus stimulation with LIF and **Forskolin** (2i/LIF/FK) (see Section 12.2.4.2) help to define PSC ground state for self-renewal and thereby facilitate single-cell seeding and expansion of PSCs in culture. These methods represent a significant advance toward the development of SSBs for large-scale production of PSCs. Further culture improvements may also be possible with two-phase SSB systems that make use of cylindrical or spherical microcarriers within a given bioreactor. Two-phase microcarrier bioreactors achieve high culture cell densities, for example 10^6 cells ml^{-1}, while maintaining the pluripotent state of the stem cells. However, the translational potential of such protocols may be limited due to the need to coat microcarriers with ECM proteins such as collagen to promote cell adhesion. Similar to single-phase SSBs, two-phase microcarrier SSBs also suffer from loss of significant cell numbers during subculturing.

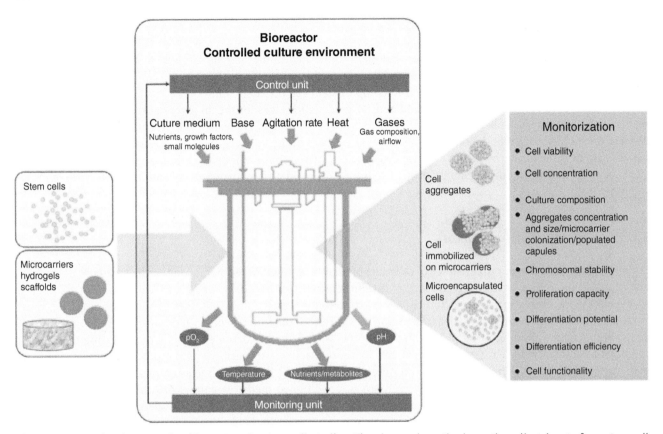

Figure 12.13 Stirred suspension bioreactors in stem cell studies. The above schematic shows the salient inputs for a stem cell **stirred suspension bioreactor (SSB)**. **Pluripotent stem cells (PSCs)** are grown either as individual cells or else in hydrogel microcarriers as a means to avoid the formation of **embryoid bodies (EBs)** (cell aggregates) forming during culturing. The formation of EBs leads to spontaneous and relatively uncontrolled differentiation of PSCs into cell derivatives of the three embryonic germ layers. This is but one main problem during culturing. Other aspects that need monitoring are described (right-hand side). The goal is to avoid the contamination of PSC cultures with differentiated cells or debris from cellular apoptosis.

As an alternative to cell adhesion to the exterior of solid carriers, cell microencapsulation within hydrogels can be used to physically isolate proliferating cell clusters from the bioreactor's fluidic environment (Figure 12.13). Hydrogels used are so far based on **hyaluronic acid**, **agarose**, **alginate** or **alginate-poly-L-lysine**. To varying extents, cultures prepared with all these hydrogels limit EB formation and promote the pluripotent state of encapsulated PSCs. Nevertheless, regardless of the technique, hydrogel pore size, porosity and mechanical properties will have to be engineered carefully to achieve the desired PSC growth profiles.

12.3.3 Practical cell therapies

Although the practicalities of cell therapies are still immense, a number of cell therapies are now approved for clinical use, particularly for the treatment of cancer. For example, processes and procedures have been developed for the isolation, by centrifugation, of autologous (patient-derived) **antigen presenting cells (APCs)** and **dendritic cells** of the innate immune system. After this, isolated cells are cultivated before activation *ex vivo* by means of exposure to a recombinant antigen protein, then returned to the same patient as an immune modulator or vaccine (PROVENGE®) to treat asymptomatic or minimally symptomatic metastatic castrate resistant (hormone refractory) **prostate cancer**. This is a personalised cell therapy treatment that works in effect by "programming" the immune system of individual patients to seek out that cancer and destroy as if it were foreign tissue.

In two other cases, other immune cells known as T-cells are involved. In the first, processes and procedures were developed for the isolation of autologous (patient-derived) T-cells of the adaptive immune response. After this, isolated cells are cultivated and then subject to virus-mediated genetic modification *ex vivo* to express the chimaeric antigen receptor (CAR) (with 4-1BB co-stimulatory domain), then returned to the same patient as an immune modulated transgenic "CAR-T" cell line (KYMRIAH®) that targets leukaemia cells for destruction in cases of B-cell precursor **acute lymphoblastic leukaemia (ALL)**, **diffuse large B-cell lymphoma (DLBCL)**, and/or **high grade B-cell lymphoma** and DLBCL arising from follicular lymphoma. In the second, autologous T-cells were also isolated, cultivated and genetically modified to become an alternative CD19-directed "CAR-T" cell line (YESCARTA®) that is returned to the same patient for treatment of **primary mediastinal large B-cell lymphoma**, high grade B-cell lymphoma and/or DLBCL. There will undoubtedly be much more to come.

12.4 Tissue engineering

Tissue engineering can be said to have evolved out of the development of biomaterials, and is concerned with making combinations of artificial bio-inspired scaffolds, cells and biologically active molecules into functional tissues. The simplest form of tissue engineering could be considered cell engineering for cell therapies. In this instance, the focus is the creation of appropriate artificial ECMs from proteins to synthetic plastics. With the artificial ECM in hand, cells with or without a requisite "cocktail" of growth factors can be introduced together and may self-assemble into artificial tissue. Currently, tissue engineering plays a relatively small part in patient treatment even though this represents a potentially very powerful form of advanced therapeutic approach. Primarily, the procedures are still experimental in nature and very costly. Accordingly, tissue engineering is now expanding, growing beyond therapeutic applications to embrace non-therapeutic applications, including the development of tissue models for fundamental pre-clinical disease studies, using tissues as biosensors to detect biological or chemical threat agents and the creation of tissue chips for fast toxicology analyses.

12.4.1 Biomaterials to tissue engineering

In spite of diversification, the primary current goal of tissue engineering remains to assemble functional combinations able to restore, maintain or improve tissues or whole organs damaged by disease or trauma. In this context, promising results have come from artificially engineered human tissue that can be created using autologous (self-derived) cells and an artificial collagen "scaffold" to help guide the growth of new tissue that has been created by stripping cells from a donor organ and using the scaffold. This process has been used to bioengineer heart, liver, lung and kidney tissue. In the case of kidney tissue, kidney scaffolds have been seeded with epithelial and endothelial cells. The resulting tissue was found to clear metabolites, reabsorb nutrients and produce urine in animal models. However, there is currently a very real limit in that artificially engineered tissues larger than 200 μm in dimensions cannot survive owing to a lack of vasculature. Therefore, there is much room for innovation in the design and creation of biomaterials for this purpose. Nevertheless, this approach holds out the genuine possibility for the routine generation of pre-existing scaffolds from human tissue discarded after surgery, and autologous cells (direct from the patient), so leading to the creation of artificial tissues and customised organs that may survive immune system rejection.

By way of a reality check though, it should be noted that the only tissue engineering combinations approved for clinical use currently are artificial skin and cartilage, with only limited patient applications. For example, allogeneic (non-self-derived) **keratinocyte** and **fibroblast cells** cultured in bovine collagen (artificial ECM) (GINTUIT®) are used to assist with wound healing in the treatment of some dental conditions in patients. By definition, keratinocytes are the most common type of skin cells that make **keratin**, a fibrous protein that provides strength to skin, hair and nails. Fibroblasts are common cells that derive from hMSCs and are found in connective tissue where they synthesise the ECM including collagen. Dermal fibroblasts found within the skin are responsible for generating connective tissue and allowing the skin to recover from injury. In a similar way, cultured **chondrocytes** cultured on porcine collagen (artificial ECM) (MACI®) can be used for repair of single or multiple symptomatic, full-thickness cartilage defects of the knee with or without bone involvement in adults. MACI is an autologous (self-derived) cellularised scaffold product. Chondrocytes, by definition, are the only cells found in healthy cartilage. They produce and maintain the cartilaginous matrix, which consists mainly of collagen and proteoglycans. Clearly this is just the start and tissue engineering potentially promises so much more with the production of different ECMs (see for example Figure 12.14).

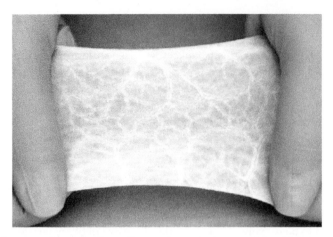

Figure 12.14 A biomaterial from porcine intestine. When moistened, the material, which is called SIS, is flexible and easy to handle, so may be used for wound healing in humans (Stephen Badylak, University of Pittsburgh).

12.4.2 Tissue engineering and regenerative medicine

Regenerative medicine is a broad field that includes tissue engineering, but also merges with self-healing and self-regeneration concepts where the body is stimulated, with or without the addition of foreign biological material, to recreate cells and rebuild body tissues and organs *in situ*. The terms tissue engineering and regenerative medicine are used interchangeably, but they should not be. Regenerative medicine relies on the basic principle that healthy cells, particularly stem cells, could be used to help replace diseased or otherwise damaged tissues in such areas as **spinal cord injuries**, **type 1 diabetes**, **Parkinson's disease**, **amyotrophic lateral sclerosis**, **Alzheimer's disease**, **heart disease**, **stroke**, **burns**, **cancer** and **osteoarthritis**.

For several years now, the understanding that stem cell fates can be controlled externally has been a central pillar in the design and creation of stem cell therapy approaches for the treatment of chronic diseases and for regenerative medicine. Most especially, it should now be apparent that different types of growth medium and different types of artificial scaffolds, involving different types of confined spaces, will determine stem cell fates through modulation of very specific gene networks therein. Once a deeper understanding of such processes has been reached and sufficient stem cells can be cultured reproducibly and efficiently enough, then there is every likelihood that the controlled transformation of autologous stem cells into mature tissue will become possible, ready for potential transplant into the given patient concerned.

In practice, the only clear opportunities with stem cells in regenerative medicine, have been with applications of **human haematopoietic stem cells** (**hHPCs**) that originate from cord blood removed from neonatal umbilical cords. These hHPCs are used to treat disorders of the haematopoietic system, which is the bodily system involved in the production and maturation of **red blood cells**, **white blood cells** and **platelets**, consisting in adult mammals of the **bone marrow**, **thymus**, **lymph nodes** and other **lymphoid organs/tissues**. **Stem cell therapies** in patients using hHPCs are for haematopoietic and immunologic reconstitution/regeneration of haematopoietic system deficiencies that are either inherited, acquired or result from myeloablative treatment (high-dose chemo-/radiotherapy that kills cells in the bone marrow). Otherwise stem cell therapies and application in regenerative medicine are yet to achieve widespread use in clinic, although thousands of stem cell clinical trials have taken place over the last decade or more. Still, stem cell studies are already leading to a much-enhanced understanding of basic developmental biology and how diseases occur, through the observation of how stem cells mature into bones, heart muscle, nerves and other body organs/tissues. So too, stem cell colonies can also be used in drug testing, particularly for adverse reactions to drug use. Stem cell therapies and regenerative medicine offer an enormous opportunity yet for chemical biology research and development too.

13

Chemical Biology, Nanomedicine and Advanced Therapeutics

13.1 General introduction

Advanced therapeutics are the result of multidisciplinary problem-driven research involving several fields such as nanotechnology and cellular and molecular biotechnology. In this chapter, we shall see how chemical biology informs a process that begins in nanotechnology and culminates in the emerging fields of nanomedicine and advanced therapeutics, in particular in gene therapy.

13.2 The chemical biology approach to gene therapy

In its primary iteration, chemical biology seeks to understand the way biology works at the molecular level. In its secondary iteration, chemical biology seeks to understand the way biology works at the nanomolecular level as well, and this is where the chemical biology approach to gene therapy begins.

13.2.1 Nanotechnology to nanomedicine

A fundamental of biology is that most significance in biology, if it is not of molecular consequence, is of nanomolecular consequence. This implies structure and matter at dimensions of approximately 1–250 nm. This is the scale at which legions of multimolecular "machines" operate in cells. As such, this is the scale that chemical biologists should understand in order to appreciate properly cell structures and functions, cell behaviour as a whole, and hence whole-body physiologies, medicine and disease. **Nanotechnology** involves imaging, measuring and modelling, but primarily the synthesis and manipulation of matter on the nanomolecular scale. Clearly, nanostructured products that are derived can be used in biology where they interact primarily with similarly sized **nanostructures** outside cells, on cell surfaces and in intracellular environments. Indeed, although it can be said that all the truly significant mechanisms inside, outside and between cells are fundamentally molecular in character, they take place in the context of nanostructure interactions. Therefore, the link from the molecular to the nanoscale is central to a proper understanding and exploitation of molecular mechanisms in biology.

Unfortunately, the major weakness of current nanoscale studies is the paucity of characterisation methods available for the study in general of structures and functions of nanostructures inside cells. Indeed, there appears a substantial analytical gap between those techniques suitable for detailed molecular studies (as documented in Chapters 3–6 of this

Essentials of Chemical Biology: Structures and Dynamics of Biological Macromolecules In Vitro *and* In Vivo, Second Edition. Andrew D. Miller and Julian A. Tanner.
© 2024 John Wiley & Sons, Inc. Published 2024 by John Wiley & Sons, Inc.
Companion Website: www.wiley.com/go/miller/essentialschembiol2

textbook) and those techniques suitable for detailed nanomolecular studies. Recent advances in **confocal microscopy** and atomic force microscopy, for example, are enabling nanomolecular scale insights into cell behaviour, but there is ample need and opportunity for new techniques to bridge the gap from the molecular to nanomolecular scale, particularly where dynamic interaction and real time behaviour studies are required.

Nanomedicine is the application of nanotechnology to medicine. Nanomedicine aims to overcome unmet needs in disease management and treatment through interventions on the nanoscale that correlate with the operational scale of biological macromolecules inside cells. Although widely applicable for the diagnosis and treatment of many diseases, nanomedicine has progressed most in research directed at the diagnosis and treatment of cancer. Today, researchers are constantly developing new nanomaterials, nanodevices and nanoparticles with different applications in mind. Of particular interest here are nanoparticles that are genuine particles (approximately 50–250 nm in dimension). These nanoparticles are intended to facilitate the functional delivery of **active pharmaceutical ingredients** (**APIs**) to disease-target cells for treatment (see this chapter) and/or of imaging agents to disease-target cells for diagnosis (see Chapter 14). Where advanced therapeutics are concerned, APIs are therapeutic nucleic acids intended for use in gene therapy strategies for disease treatment. Ultimately, nanoparticle studies are now opening up the possibility of genuine **precision therapeutic approaches** (**PTAs**) for the treatment of chronic diseases that are made possible by the unique combinations of physical, biophysical and chemical properties that can be combined within nanostructures of all origins, but most importantly biocompatible nanoparticles.

13.2.2 Advanced therapeutics: gene therapy

Gene therapy can be defined as the use of therapeutic nucleic acids as medicines that exploit genetic level mechanism(s) involving either gene modification, replacement, replication and/or transcription to treat disease. Given this, a fundamental of gene therapy that is axiomatic to success is that the practical realisation of gene therapies in the clinic requires functional delivery of therapeutic nucleic acids to targets in the body by means of "**vectors**". Vectors provide a "means of carriage" for therapeutic nucleic acids from their site(s) of administration to their desired site(s) of action in cells of interest within a target organ(s) of choice. There is no question that the key rate limiting step in the development of most promising gene therapy strategies for disease treatment today is identification of the most appropriate vector. Viral vectors (Figure 13.1) have proved to be the most popular means of carriage in clinical trials, and physical non-viral approaches (such as **direct injection**, **biolistics** (**gene gun**) or **electroporation**) (Figure 13.2) used in conjunction with naked **plasmid DNA** (**pDNA**). However, the approach closest to the heart of chemical biology is that of using synthetic nucleic acid delivery systems that are biocompatible nanostructures, derived from nanotechnology, designed and created for the express purpose of targeted delivery of therapeutic nucleic acid APIs from site(s) of administration to desired target site(s).

Overall, clinical trials that have made use of synthetic nucleic acid delivery systems used to represent at least 10% of the total number of trials by September 2005. However, this figure dropped to less than 5% of the total of 2409 gene

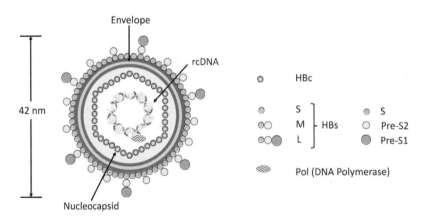

Figure 13.1   Viral "vectors". These vectors are transgenic variants of virus nanoparticles adapted for the delivery of "foreign" nucleic acids to cells. The illustration is of **hepatitis B virus** (**HBV**) is intended to emphasise the elegant simplicity and reproducibility of virus structure that has seduced many researchers into believing that these professional nucleic acid delivery systems could be adapted with ease for gene therapy applications, to the exclusion of all other types of delivery systems, either physical or synthetic (taken from Figure 1 of Duraisamy et al., 2020, *Viruses*).

a)

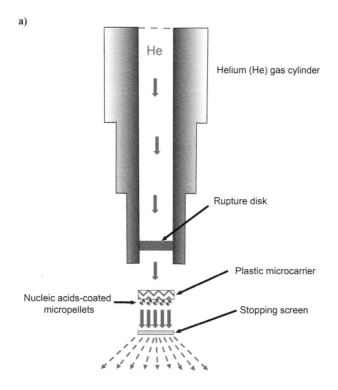

High velocity micropellets distributed over area
for tissue/cell penetration and functional nucleic acid delivery

b)

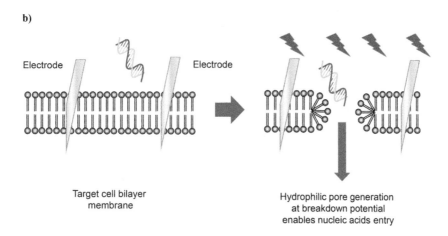

Figure 13.2 **Physical methods for functional delivery of nucleic acids.** (a) The schematic shows a **gene gun** that works by kinetic acceleration of nucleic acid-coated micropellets for distribution over and then penetration of an exposed target tissue target; (**b**) A schematic for **electroporation.** This technique requires approximately 20 ms electrical pulses from an electrode cluster at an electric field strength of 50 V cm⁻¹, representing a minimum membrane breakdown potential of −0.7 V that enables entry of nucleic acids into target tissues and cells.

therapy clinical trials reported by October 2016, including only one ongoing phase 3 trial. Collectively, clinical trial data continue to suggest that viral nucleic acid delivery systems may be highly effective for nucleic acid delivery, but can still suffer from a variety of problems, including immunogenicity, toxicity and oncogenicity. By contrast, synthetic nucleic acid delivery systems should be much less affected by immunogenicity, toxicity and oncogenicity, but they are comparatively inefficient in mediating functional delivery of associated therapeutic nucleic acids. Well over a decade ago, there was a consensus view that synthetic nucleic acid delivery systems could be widely taken up in clinical gene therapy settings once engineered for improved efficacy. Unfortunately, today the field of synthetic nucleic acid delivery systems still carries the same burden of innovation. In other words, there still remains a critical requirement for newer and better synthetic nucleic acid delivery systems sufficiently functional for routine use in clinical gene therapy settings.

13.2.3 Gene therapy strategies

The strategy for any given gene therapy depends upon the disease concerned and the nature of the therapeutic nucleic acids delivered. For instance, where monogenic disorders are concerned, caused by a single defective gene, a typical strategy has been to carry out gene supplementation and/or gene replacement. In other words, supplementing and/or replacing an existing, non-/partially functional mutant gene with a normal wild-type gene able to express sufficient wild-type protein for therapeutic correction in diseased cells. In the case of cancers, gene therapy strategies have been typically designed using gene addition strategies to promote cellular destruction either by introducing genes for immuno-stimulation (cytokine and/or antigen genes) or for programmed cell death/necrosis (tumour suppressor, replication inhibitor or suicide genes). Gene **knockdown (KD)** represents an alternative form of gene therapy strategy, particularly for the treatment of degenerative diseases and viral infections. In virtually all early gene therapy strategies that relied on synthetic nucleic acid delivery systems, the therapeutic nucleic acids of interest were pDNA constructs.

13.2.4 First applications of synthetic nanoparticles in gene therapy strategies

Since their inception in the late 1980s, a variety of **cationic liposomes/micelles** and **cationic polymers** have been prepared and combined with therapeutic nucleic acids to form early version **lipid-based nanoparticles (LNPs)** or **polymer-based nanoparticles (PNPs)** respectively, that were then used to mediate functional delivery of therapeutic nucleic acids to target cells *in vitro*, *ex vivo* and to some extent *in vivo*. Early version LNPs were prepared by lipid self-assembly from synthetic and natural lipid components. Chief amongst these synthetic components were cationic amphiphiles (known as **cytofectins**) (Figure 13.3) that were used typically in combination with naturally available, charge neutral lipid components such as **dioleoyl L-α-phosphatidylethanolamine (DOPE)** or **cholesterol (Chol)** (Figure 13.4). In contrast, early version PNPs were prepared from cationic polymers such as **poly L-lysine (pLL)**, **polyethylenimine (PEI)**, **polyamidoamine (PAMAM)**, **dendrimers** and **chitosan** (Figure 13.5). In formulating early version LNPs, cationic liposomes/micelles were typically prepared first by self-assembly of lipid components in aqueous solution by two primary methods, **thin-film evaporation** and **reversed-phase evaporation** (Figure 13.6), followed by polishing with **sonication** (Figure 13.7) and **extrusion** (Figure 13.8).

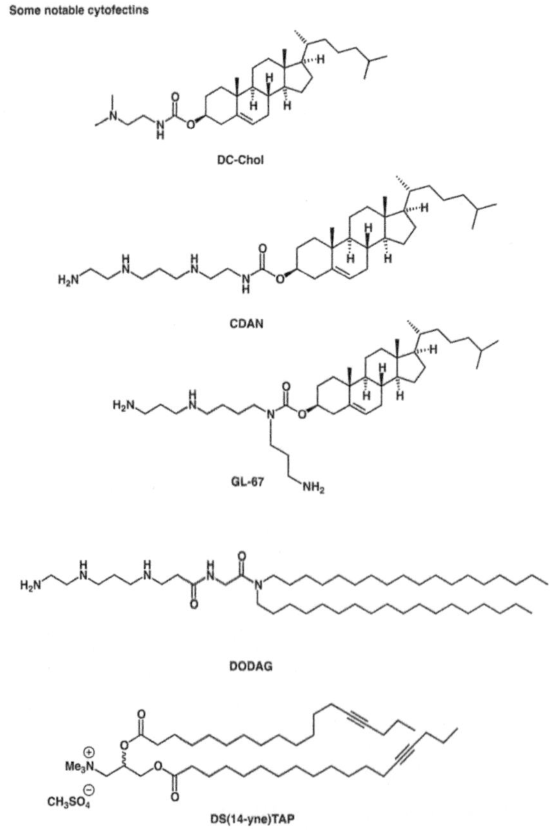

Figure 13.3   **Notable cytofectins used in early version LNPs.** Glossary of lipid names for Chapter 13 is included on page xvi.

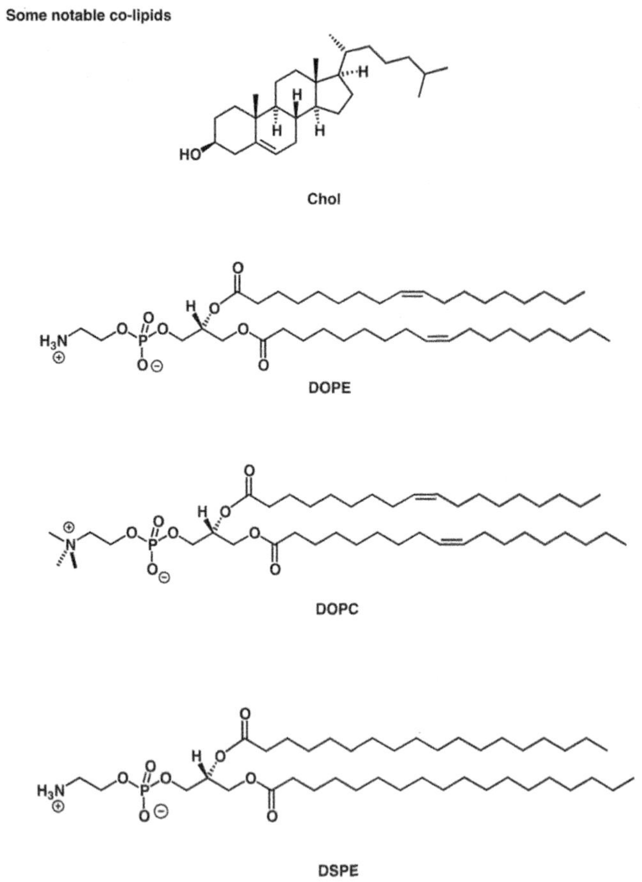

Some notable co-lipids

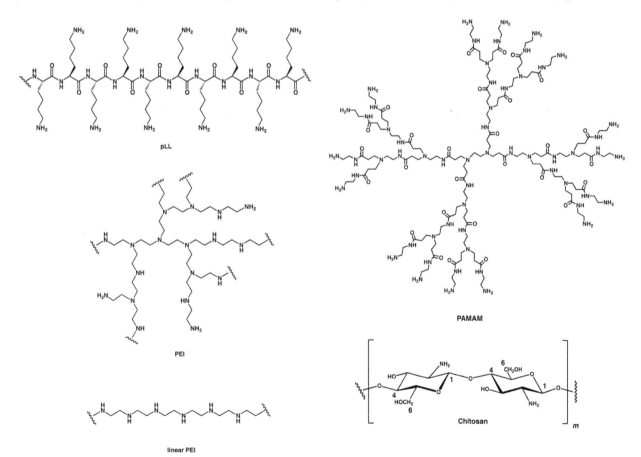

Chol

DOPE

DOPC

DSPE

Figure 13.4 **Notable co-lipids used in early version LNPs.** Glossary of lipid names for Chapter 13 is included on page xvi.

pLL

PEI

PAMAM

linear PEI

Chitosan

Figure 13.5 **Notable cationic polymers used in early version PNPs.**

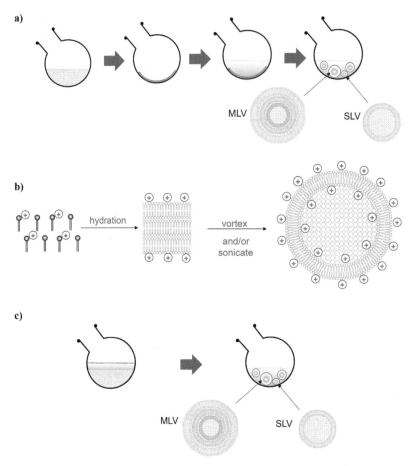

Figure 13.6 Cationic liposome preparation methods. (a) **Thin film hydration.** According to this technique lipids dissolved (1–10 mg mL^{-1} type concentrations) in organic solvent are allowed to form into a thin film on the walls of a glass vessel (flask) by the standard organic chemistry technique of rotary evaporation. After vacuum drying to remove residual organic solvent, the film is allowed to hydrate slowly in buffer solution (pH 4–9) until separated from the walls of the flask. Thereafter, a mixture of **multilamellar vesicles (MLVs)** and **single lamellar vesicles (SLVs)** are found in suspension; (**b**) An alternative schematic view of cationic liposome SLV preparation; (**c**) **Reversed-phase evaporation.** According to this technique, lipids dissolved in organic solvent (1 mg mL^{-1} type concentrations) are mixed in a glass vessel with an equal volume of buffer solution (pH 4–9), then the organic solvent is selectively removed by rotary evaporator. The aqueous solution left contains SLV suspensions, also typically centred around 100–200 nm in diameter. The problem with this method is the presence of trace organic solvents in aqueous solution which is anathema to use in biological experiments, given their general cytotoxicity.

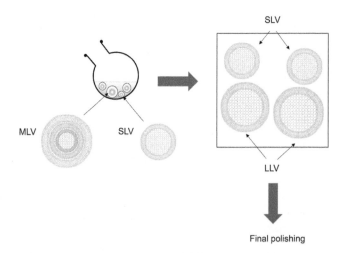

Figure 13.7 Multilamellar vesicles to single lamellar vesicle polishing. This is achieved by **bath sonication** at around 40 °C, in order to create mixtures of LLVs and SLVs of around 100–200 nm in diameter.

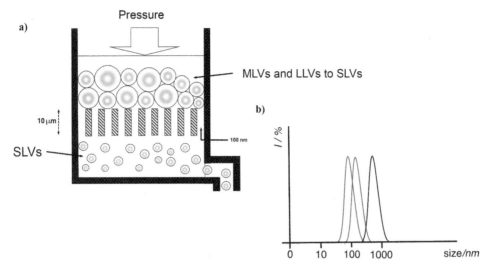

Figure 13.8 Medium/high pressure extrusion. (a) A schematic of high or medium pressure extrusion, that follows bath sonication (see Figure 13.7), to "polish" these suspensions from SLVs from MLVs or LLVs. The above two methods by resuspensions are a very useful means of moving larger SLV aggregates (>1000 nm), formed because of colloidal instability, and by reducing SLV polydispersity through filter extrusion (typically filter pores are 50 nm, 100 nm or 200 nm in diameter). These enable the formation of preparations of moderately monodisperse liposomal SLVs in solution; **(b)** Data obtained from **dynamic light scattering (DLS)** involving a monochromatic laser (see Figure 13.33).

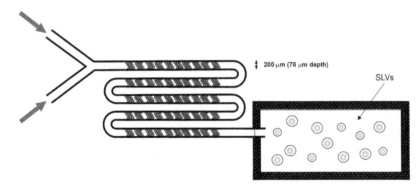

Figure 13.9 Microfluidics for rapid kinetic mixing of lipid components. This kinetic mixing technique can give very well-defined cationic liposomes/micelles. The most recent technique involves microfluidic mixing, wherein lipids in ethanol solution are rapidly mixed with buffer in a "T-junction" arrangement following by kinetic mixing and self-association during turbulent flow through microfluidic channels etched with "herring bone" patterns. By controlling the ratio of ethanol to buffer, lipid concentrations and mixing rates, highly monodisperse SLVs may be produced anywhere from 40 up to 150 nm in diameter.

The formulation of early version PNPs was carried out in similar ways. Most recently, cationic liposomes/micelles have been formulated from lipid components using **microfluidics** although all lipid components need to be ethanol soluble for this approach to be effective (Figure 13.9).

In mechanistic terms, both cationic liposomes/micelles and cationic polymers are polycationic entities that were intended to make multiple, energetically favourable electrostatic interactions with the polyanionic phosphodiester backbones of pDNAs of all possible sizes, molecular weights, configurations and topologies. These polyvalent interactions led to the "collapse" and condensation of pDNAs into early version LNP or PNP structures (Figure 13.10) that can mediate functional delivery of entrapped or encapsulated pDNAs to target cells. At the cellular level, this intracellular delivery process involves **endocytosis** (Figure 13.11), followed by some form of endosomolytic process (involving osmotic shock and/or fusogenic events) (Figure 13.12) that enables the release of free pDNA into a target cell cytoplasm. Following this, released pDNA needs to traffic into the cell nucleus in order to initiate its therapeutic function there. DNA constructs of all types, not just pDNA, all need to act similarly. On the other hand, where RNA constructs are concerned, these need only to remain in the target cell cytoplasm to carry out their therapeutic functions (Figures 13.11 and 13.12). This mechanism of functional delivery has been deeply inefficient to the extent that ≪1% of delivered pDNAs, and other pDNA constructs,

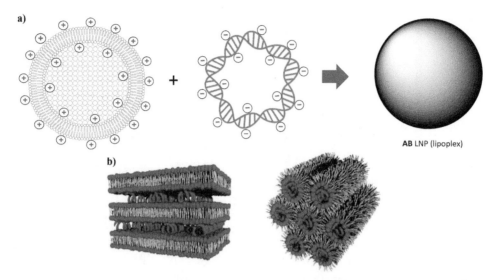

Figure 13.10   Early version LNP formulation. (a) The simplest way to achieve early version LNP formulation is rapid vortex mixing of cationic lipid containing SLVs with nucleic acids of interest. There will be variations, but early version LNPs will formulate by the addition of nucleic acids of interest (**A**) to a suspension of these SLVs (**B**) under rapid vortex mixing conditions, such that a final lipid/nucleic acid ratio of 3:1 up to 6:1 w/w is approached over several mins of mixing. In doing this, the SLV structure disappears. In its place (**b**), a dense multilamellar lipid/nucleic acid nanoparticle is formed wherein nucleic acids are located between stacks of lipid bilayers (left) or within hexagonal tubes (right). Unfortunately, early version LNP, may be quite polydisperse, especially should the cationic lipid positive to nucleic acid negative charge ratio approach 1. For this reason, microfluidic mixing procedures can be used (see Figure 13.9), particularly where formation of early version LNPs requires fast kinetic mixing (in µs-ms range), for instance when the mol fraction of cationic/ionisable lipids used is low, giving rise to positive/negative charge ratios of close to 1.

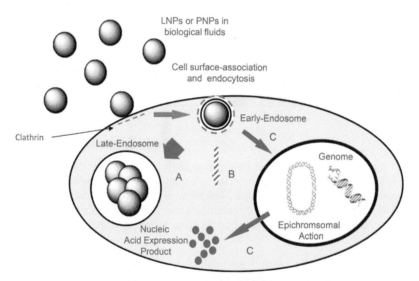

Figure 13.11   Mechanism of functional nucleic acid delivery to a target cell *in vitro* and *in vivo*. This assumes delivery is mediated by early version LNPs or PNPs. The most effective mechanism of cell entry is by clathrin coated pit **endocytosis**. This is followed either by aggregation of early endosomes into late endosomes (**pathway A**, dead-end entry pathway), or by **endosomolysis** that releases therapeutic nucleic acids into the target cell cytoplasm (see Figure 13.12). RNA then **traffics** to its site of action in the cytoplasm (**pathway B**); DNA **traffics** into the nucleus for epichromosomal expression, after which protein expression is returned to the target cell cytoplasm (**pathway C**). Pathway A dominates, pathway C is ≪1% efficient. Accordingly, functional delivery of therapeutic DNA constructs within a target cell is a very wasteful process. Pathway B has proved markedly less inefficient in recent times, although there is still plenty of room for improvement (taken from Figure 14.15 of Escriou *et al.*, 2014, World Scientific Publishing).

are actually able to carry out their intended therapeutic functions due to miss-trafficking and weaknesses in intracellular delivery. The functional delivery of RNA constructs has been less troubled, as will become clear later on in this chapter.

In general, efforts over the years to move early version LNPs and PNPs into the clinic, formulated with DNA constructs, have proved only partially successfully due to many problems encountered. For instance, most early version PNPs

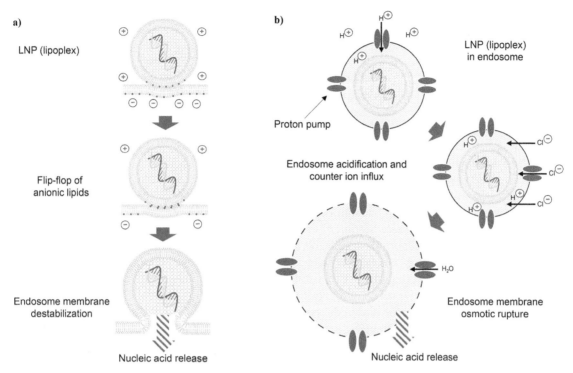

Figure 13.12 **Mechanisms of endosomolysis in cells *in vitro* and *in vivo*.** (a) Illustrates the **fusogenic mechanism** in which cationic, early version LNPs (selected cationic cytofectin head groups, in blue) contact cell membranes (anionic lipid head groups, in red) causing "flip-flop" of anionic lipids, so bringing their anionic head groups into intimate electrostatic contact with cytofectin head groups. These interfacial interactions trigger endosomolysis by membrane **fusogenesis** and hence nucleic acid release into cell cytoplasm; (b) Illustrates the **osmotic shock mechanism** in which endosome acidification post endocytosis of charged, early version LNPs increases endosome ionic strength, leading to osmotic rupture of the endosome membrane and indirect nucleic acid release from endosomes. This mechanism is applicable to early version PNPs too.

were too cytotoxic to be of widespread use *in vivo*, let alone in clinic. On the other hand, although much more biocompatible, early version LNPs were found difficult to formulate reproducibly on kg scales, and unstable to long-term storage (>weeks). In addition, these early versions LNPs were also significantly unstable with respect to aggregation in biological fluids and prone to undesirable interactions with host immune systems. Overall, the intended therapeutic benefits did not emerge in clinic because LNP-mediated functional delivery of therapeutic nucleic acids turned out to be just too inefficient *in vivo* and in clinic too. The scale of the inefficiency can be demonstrated in the following way. Most effective gene therapy experiments in animal models of disease with simple LNPs have typically involved pDNA doses in the 1–2 mg kg^{-1} region if not more. The level of delivery inefficiency then becomes clear from a simple calculation. Assuming that a 5 kbp pDNA is being delivered, then each microgram will contain 2×10^{11} pDNA molecules, and yet each target cell only requires a few pDNA copies (approximately 1–10 per cell nucleus) to produce a therapeutic effect.

13.2.5 Improving on the therapeutic nucleic acid APIs

As noted in the previous section, most early gene therapy research using synthetic nucleic acid delivery systems, was conducted using pDNA therapeutic nucleic acid constructs. Since then, the popularity of pDNA has declined and other therapeutic nucleic acid systems have started to emerge for two main reasons. First, the delivery of pDNA constructs (typically 4–7 kbp) to cell nuclei *in vivo*, using synthetic nucleic acid delivery systems, is very problematic and very inefficient. Second, pDNAs once in nuclei are typically subject to **plasmid silencing**, namely a decline in pDNA expression levels to background, 7–14 d post-delivery. This can be averted in pDNA using such elements as **scaffold/matrix-attachment regions** (**S/MARs**). However, highly inefficient delivery to cells and to their nuclei remains a tremendous problem.

Alternative DNA expression systems for use in gene therapy have included artificial chromosomes (50 kbp–1 Mbp). These were constructed to express genes in an epi-chromosomal manner supported by all the main features of a chromosome (such as the centromere) so that they may operate as pseudo-chromosomes. Depending upon the source of sequences and genes, **bacterial artificial chromosomes (BACs)**, **P1-derived artificial chromosomes (PACs)**, **yeast artificial chromosomes (YACs)**, **mammalian artificial chromosomes (MACs)** and **human artificial chromosomes (huACs)** have all been designed and created. Otherwise, gene therapy strategies have also been developed based on site-specific integration of small to large segments of DNA into host cell chromosomes. This can be achieved using mammalian **transposons**, namely stretches of DNA (either linear or circular) that are capable of insertion into chromosomal DNA at defined sites with the assistance of a transposase enzyme. For example, the **Sleeping Beauty transposon** with transgene can be embedded into mammalian chromosomal DNA, leading to long-term transgene expression (months). Regrettably, transposons may integrate into too many sites in chromosomal DNA and hence may be cancer-inducing (oncogenic).

Most recently, direct gene editing technologies have been coming to the fore with the capability to mediate gene addition, gene supplementation and/or gene replacement strategies for gene therapy, all based on gene **knock-in (KI)** approaches mediated by the direct gene editing technology concerned. These gene editing approaches can also be turned to complete gene **knockout (KO)** approaches as well, which further increases options for gene therapy strategies. There are currently three major gene editing technologies of note. These are **Class 2, clustered, regularly interspaced, short palindromic repeat (CRISPR)/Cas9** (see Section 12.2.3), **zinc finger nucleases (ZFNs)** and **transcription activator-like effector nucleases (TALENs)** (Figure 13.13). Clearly, these are all essentially proteins so their use requires functional delivery of all necessary proteins and cofactors to target cells *in vivo*. Making use of synthetic nucleic acid delivery systems, this can be achieved by delivering pDNAs expressing the relevant proteins and cofactors (sgRNA in the case of CRISPR/Cas9, see Figure 12.9). However, given the difficulty of achieving functional pDNA delivery *in vivo*, there is likely to be a substantial interest going forward in delivering stable *in vivo* mRNA expression systems instead.

This brings us to RNA-based gene therapies. There was initially little attention given to RNA delivery until the emergence of **non-coding RNAs (ncRNAs)** at the beginning of the century. First and foremost, **small interfering RNAs (siRNAs)** are now widely considered powerful therapeutic nucleic acids, with great current and future potential in gene KD strategies for gene therapy, very applicable for the treatment of cancers, degenerative diseases and viral infections. Most siRNAs known and used are siRNA duplexes (typically comprising a 19–23 bp antiparallel A-form double helix) (Figure 13.14). Synthetically generated siRNAs can be delivered to cells, using, for example, LNPs to interact with the intracellular **RNA induced silencing complex (RISC)** (Figure 13.15). The **Ago2 protein** of RISC acts to separate sense

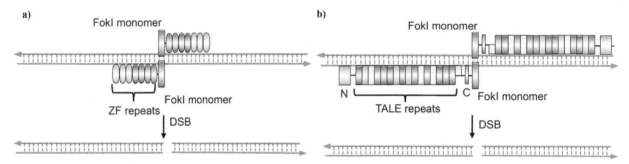

Figure 13.13   Schematic representations of gene editing technologies. (a) Cys$_2$His$_2$ **zinc fingers (ZFs)** are DNA-binding domains that each recognise approximately three bps of DNA. Alteration of a small number of amino acid residues in or near an α-helix within this domain can lead to changes in DNA-binding specificity. Engineered zinc fingers can be joined together into more extended arrays capable of recognising longer DNA sequences. A large number of zinc finger arrays engineered can be fused to a non-specific nuclease domain from the type IIS **FokI restriction enzyme** to create **zinc finger nucleases (ZFNs)**. The FokI nuclease functions as a dimer and therefore two zinc finger arrays must be designed for each target site. Most recent ZFN pairs contain complementary obligate heterodimeric FokI domains. A ZFN pair are shown to produce a highly selective **double-strand break (DSB)** that is followed by **non-homologous end joining (NHEJ)** or **homology directed repair (HDR)** as appropriate (see Figure 12.9); (b) **Transcription activator-like effector nucleases (TALENS)** have rapidly emerged as an alternative to ZFNs. TALENs are similar to ZFNs and comprise a non-specific FokI nuclease domain fused to a customisable DNA-binding domain. The fundamental building blocks used to create the DNA-binding domain of TALENs are highly conserved repeats derived from naturally occurring **transcription activator-like effectors (TALEs)** encoded for by *Xanthomonas* proteobacteria. DNA binding by TALEs is mediated by arrays of highly conserved 33–35 amino acid residue repeats flanked by additional TALE-derived domains at the amino- and carboxy-terminal ends of a given array. Individual repeats bind a single deoxynucleotide residue of DNA as determined by the identities of two hypervariable residues typically found at amino acid residue positions 12 and 13 in each TALE repeat. TALE repeats with hypervariable residues N and N (**green**) recognise dG nucleotide residues, N and I (**yellow**) recognise dA nucleotide residues, H and D (**purple**) recognise dC nucleotide residues, and N and G (**red**) recognise dT nucleotide residues, respectively. Once again, the FokI nuclease functions as a dimer and therefore two TALENs must be designed for each target site. A TALEN pair may then produce a highly selective DSB, followed by an NHEJ or HDR event as appropriate (see Figure 12.9) (taken from Figure 6 of Duraisamy et al., 2020, *Viruses*).

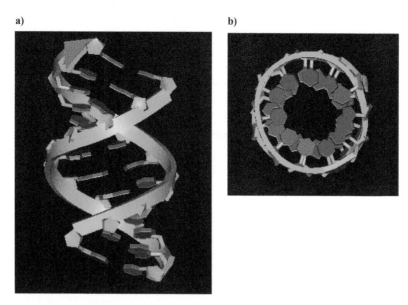

Figure 13.14 **A-form small interfering RNA duplex. Small interfering RNA (siRNA)** is a high molecular weight, biopharmaceutical agent that is metabolically unstable and too hydrophilic for cell entry without assistance. This consists of two anti-parallel aligned phosphodiester strands (in **green**) stabilised by complementary Watson-Crick bps (purine bases in **red**, pyrimidine bases in **blue**) (see Section 1.4.5) with unpaired overhangs consisting of two nucleotide residues each at each 3′-terminus. (**a**) Side view; (**b**) Top view to emphasise the A-form double helix (duplex) structure. The duplex typically comprises 19–21 bp, up to 23 bp, in a typical siRNA construct.

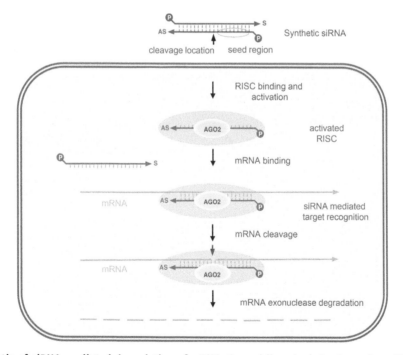

Figure 13.15 **Schematic of siRNA mediated degradation of mRNA.** General linearised structure of an siRNA is shown with **sense strand (S)** (in **red**) and complementary **antisense strand (AS)** (in **blue**). The S strand harbours a nucleotide residue sequence that matches with corresponding sequence in a target mRNA. The AS strand is complementary, but is the functional strand, with a 5′-seed region (circled) and "cleavage location" (**short black arrow**) shown at the phosphodiester link between nucleotide residues 10–11. After cell entry, **RNA induced silencing complex (RISC)** asymmetrically separates siRNA duplex (starting from the 3′-end of the S strand), then sequesters the AS strand (by **Ago2 protein**) for activation. Watson-Crick base pair binding of target mRNA is then followed by Mg²⁺-ion assisted phosphodiester cleavage between mRNA residues complimentary to AS strand residues 10–11, followed by more extensive hydrolytic cleavage involving exonuclease degradation of mRNA fragments. RISC bearing AS strand continues with subsequent rounds of hydrolytic degradation of target mRNAs (taken from Figure 4 of Duraisamy *et al.*, 2020, *Viruses*).

(S) and antisense (AS) strands from each other and typically adopts the AS over the S strand as a "template" with which to bind a target mRNA for destruction. A target mRNA is singled out for destruction through siRNA-mediated target recognition, provided by the AS strand on RISC, that makes complementary Watson-Crick base pair interactions with a corresponding region in this target mRNA (Figure 13.15).

Nowadays, there are defined rule sets for the identification of siRNAs, with optimal base-sequences for RISC-mediated destruction of a target mRNA of interest. Furthermore, these sequences may then be screened by means of high-end bioinformatics analyses (such as the siDIRECT analysis) ensuring that they have a minimal likelihood of cross-reactivity with off target mRNA sequences and hence little likelihood of eliciting undesirable cytotoxicities and even immunogenic reactions. Most recently, siRNAs have been designed with chemical modifications to the sequence (see Figure 13.16). Such modifications enhance intracellular stability of siRNAs with respect to intracellular nucleases, but also improve RISC interactions.

Nowadays, even mRNA has begun to feature with synthetic nucleic acid delivery systems for various applications, particularly in vaccination. A key element that has made this possible has been recent improvements in the synthesis of mRNA by *in vitro* transcription (IVT) using a T7 polymerase. Transcription also involves replacing **uridine 5′-triphosphate (UTP)** with **m¹ψTP** (Figure 13.17), where uridine is replaced by **1-methylpseudouridine (m¹ψ)** (see Chapter 1). The resulting mRNA must then be capped by m⁷G nucleoside using a 2′-*O*-methyltransferase. Then it can be purified by standard chromatography techniques prior to storage at −20 °C. The inclusion of pseudo-uridine is interesting since this makes mRNA not only much more translatable, but immunologically silent. The use of non-immunogenic mRNA is crucial because a series of innate immune receptors (TLR3, TLR7, TLR8, RIG-I, MDA5, NOD2, PKR and others) recognise mRNA, resulting in the release of type I interferons, activation of interferon-inducible genes and inhibition of translation. Such m¹ψ- modified mRNAs have already been shown to be promising for use in vaccinations against HIV, Zika virus and influenza virus infections in animal models of these human infections.

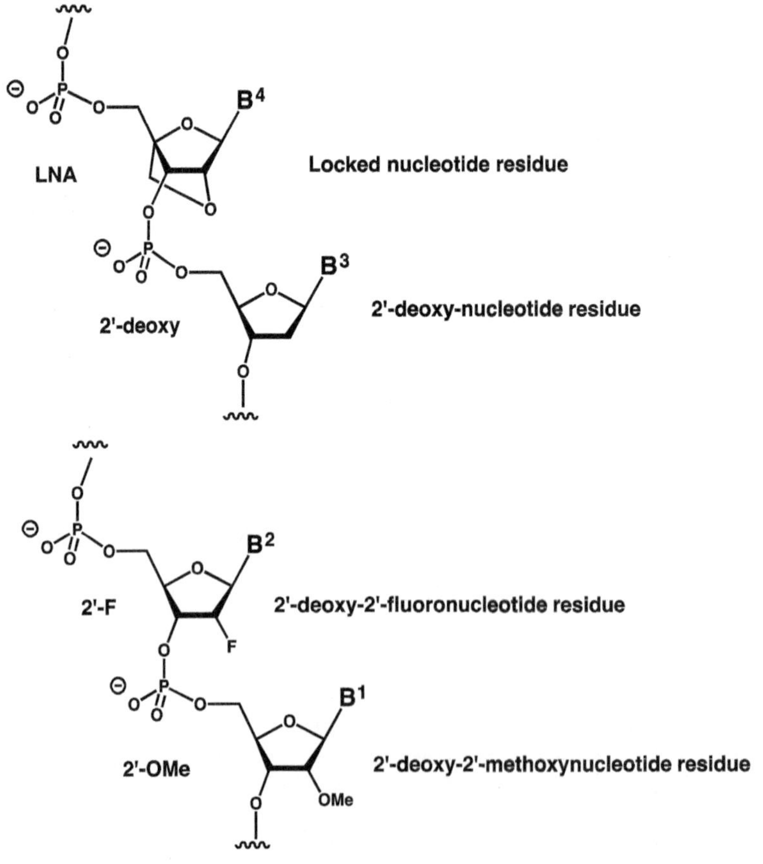

Figure 13.16   Frequent chemical modifications in backbones of small interfering RNAs.

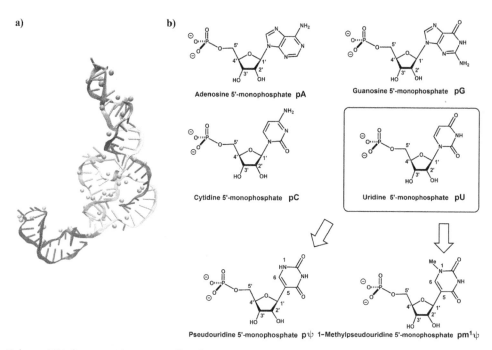

Figure 13.17 Using mRNA for gene therapy applications. For many years this has been thought impossible. (**a**) Illustrates a typical mRNA molecule (ladder schematic) containing Watson-Crick base paired regions, and informal paired regions (loops and bulges). Mg^{2+} ions are sequestered throughout. Such nucleic acid constructs have been thought too metabolically labile and too immunogenic to be used in gene therapy applications; (**b**) A key way to resolve this problem has been the use of *in vitro* **cell free translation** (**IVT**) wherein pU nucleotide residues are replaced with **pψ** or **pm¹ψ**.

13.2.6 The ABCD nanoparticle concept

Following clinical experience with early version LNPs and PNPs, the obvious imperative was for the design and creation of significantly more biocompatible, clinically appropriate nanoparticles. This led to the **ABCD** nanoparticle design concept for self-assembly nanoparticles that are those most likely to mediate successful and effective functional delivery of therapeutic nucleic acids to target cells *in vivo*. The details of this structural paradigm are illustrated in Figure 13.18. Implicit to this design concept is:

1. That nanoparticle formulations can and should be regular and uniform (with diameters typically of approximately 100 nm).

2. That multiple **ABC/ABCD** nanoparticle variants can be formulated by self-assembly from toolkits of purpose designed chemical components, equipped with different functionalities to address different biological barriers (biobarriers) on the critical path to efficient therapeutic nucleic acid delivery to target (see below).

3. That multiple **ABC/ABCD** nanoparticle variants are biocompatible and so potentially appropriate for clinical use, whilst core **AB** nanoparticles (equivalent to early version LNPs, described above) are not so and only potentially useful as synthetic nucleic acid delivery systems for *in vitro* and *ex vivo* use, or for very local/regional use *in vivo*. Much of this biocompatibility derives from the **C**-layer and can be accounted for by the **Derjaguin, Landau, Verwey and Overbeek (DVLO) theory** (Figure 13.19).

Overall, the nanoscale of **ABCD** nanoparticles enables each nanoparticle to present an impressive range of different functionalities and capabilities with which to address specific nucleic acid delivery requirements (see below). Hence, we should be able to anticipate the future design and creation of tailor-made nanoparticles intended for individual nucleic acid delivery applications in individual patients. Accordingly, **ABC/ABCD** nanoparticles have the potential

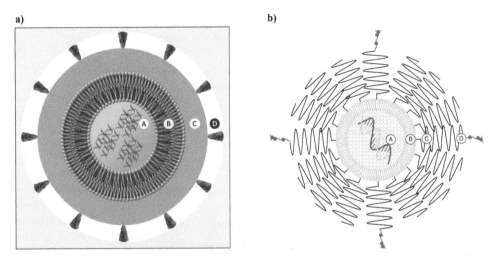

Figure 13.18   The ABCD nanoparticle concept. A structural paradigm for *in vivo* functional synthetic nucleic acid and small molecule drug delivery systems. Two different schematic views are shown, in (**a**) and (**b**). According to this structural paradigm, **ABCD** nanoparticles are multi-component, multi-layer synthetic systems that contain at their core **active pharmaceutical ingredients** (**APIs**) (**A**-component, e.g. pDNA, siRNA, mRNA, or drugs) entrapped or encapsulated by compaction/association agents (**B**-components, e.g. lipids or synthetic polymers, etc.) selected for their ability to condense **A**-components into nanoparticles for functional delivery to target cells *in vitro*. These **AB** nanoparticles are in effect equivalent to the early version LNPs or PNPs described above. **AB** nanoparticles are upgraded for functional delivery *in vivo* by the addition of a polymeric surface coat (**C**-layer; primary **C**-component, most often **polyethylene glycol [PEG]**) that confers **ABC** nanoparticles with colloidal stability in biological fluids (Figure 13.19), plus additional "stealth" properties, namely resistance to the **reticuloendothelial system** (**RES**) and partial immuno-protection towards other immune system responses. Finally, functional delivery *in vivo* may be enhanced with a biological targeting layer (**D**-layer; primary **D**-components, *bona fide* biological receptor-specific targeting ligands), giving rise to **ABCD** nanoparticles with cell target specificity.

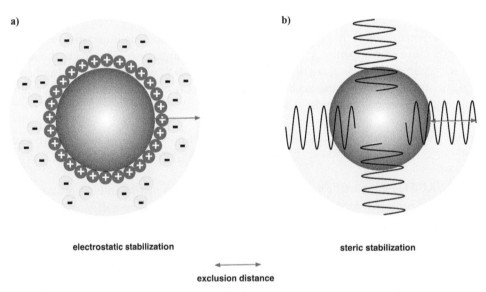

Figure 13.19   Derjaguin, Landau, Verwey and Overbeek (DVLO) theory. In simple terms, this is a theory about colloidal stability of nanometric and other particle assemblies in solution. (**a**) A particle with cationic charge can be stabilised with respect to aggregation in solution by a "wall" of anionic counter ions; (**b**) A particle of any charge status can be stabilised sterically by surface presentation of extended polymer chains (e.g. linear PEGs). In either (a) or (b) an exclusion distance is established with other particles.

to be a potent platform technology for precision therapeutic approaches (PTAs) for disease treatment in future (see Chapter 14). In closing, it should be noted that the **ABCD** nanoparticle design concept is API independent and can be adapted in principle for the functional delivery of all main classes of therapeutic nucleic acid APIs, and for the functional delivery of small molecule drugs as well (see Section 13.4).

13.2.7 Formulation of ABC/ABCD nanoparticles for *in vivo* and clinical use

The realisation of actual **ABC** or **ABCD** nanoparticles for therapeutic nucleic acid delivery is dependent on the type and nature of the nucleic acids to be delivered. Where LNPs are concerned, the **B**-components are lipids. Where PNPs are concerned, alternative **B**-components might be organic polymers, proteins or peptides respectively. Hybrid situations are also possible. For example, if pDNA is to be delivered (as primary **A**-component), then a 19 amino acid residue μ, mu-peptide (found in the core of the adenovirus) has been found to pre-condense pDNA as uniform mu:pDNA (MD) particles that can be further encapsulated within cationic liposomes to form non-aggregating stable, core liposome:mu:pDNA (LMD)-**AB** nanoparticles. In comparison, much lower molecular weight therapeutic nucleic acids such as siRNA may be combined successfully with cationic liposomes alone by rapid vortex procedure giving rise to uniform core siRNA-**AB** nanoparticles.

LNPs of the **ABC/ABCD** nanoparticle type are prepared most simply by direct self-assembly from lipids and PEG-lipid components in solution, in a process known as **pre-modification self-assembly** (Figure 13.20). According to this approach, PEGylated cationic SLVs may be prepared in the first instance, by thin-film evaporation or reverse-phase evaporation (Figure 13.6), followed by polishing with sonication (Figure 13.7) and extrusion (Figure 13.8). The nucleic acid API, quite likely siRNA today, is then combined with these PEG-SLVs in a slow, careful stepwise process (with vortex-mixing) over several minutes, until an appropriate lipid/nucleic acid w/w ratio is reached, resulting in complete LNP **ABC** or **ABCD** nanoparticle formation. In doing this, PEG-SLV structures disappear, only to be replaced by a dense multilamellar lipid/nucleic acid core **AB** nanoparticle wherein nucleic acids are located between stacks of lipid bilayers, while PEG molecules (**C**-layer moieties) project from the exterior surface and create the nanoparticle **C**-layer. In an alternative to this **thermodynamic mixing** process, LNPs of the **ABC** nanoparticle type might also be prepared using microfluidics in a **kinetic mixing** process that seeks to combine ethanolic solutions of lipids with nucleic acid APIs rapidly (ms timescales) in aqueous buffer solution (Figure 13.9).

LNPs of the **ABC** and **ABCD** nanoparticle type can also be prepared by **post-modification self-assembly** (Figure 13.21). In the case of **ABCD** nanoparticles, PEG-SLVs are prepared with PEG-lipids and ligand-PEG-lipid conjugates, wherein biological receptor-specific targeting ligands (**D**-layer moieties) are directly conjugated to PEG and so able to

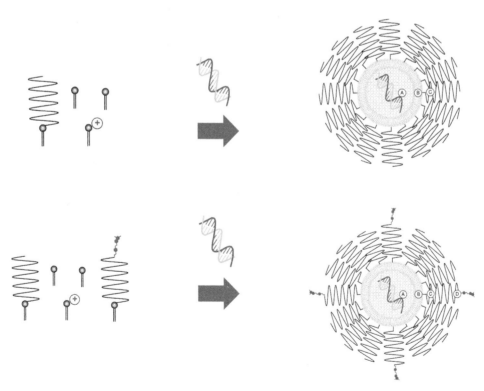

Figure 13.20 **LNPs of the ABC and ABCD nanoparticle type assembled with small interfering RNA by pre-modification self-assembly.** According to this process, all lipids (left-hand side) (including ligand-PEG-lipids, PEG-lipids and cytofectin [cationic/ionisable lipid]) are used to prepare PEGylated cationic liposomes (see Figures 13.6, 13.7, 13.8, 13.9) that are combined with siRNA either simultaneously (**kinetic mixing**) or in a separate step (**thermodynamic mixing**), leading to siRNA-**ABC** LNPs (top right-hand side) and siRNAs-**ABCD** LNPs (bottom right-hand side).

project into solution beyond the nanoparticle **C**-layer. Otherwise, the desire for better control of **ABCD** nanoparticle preparation using a multi-stage **pre-modification post-modification self-assembly** approach might be employed (Figure 13.22). Finally, LNPs of the **ABC/ABCD** nanoparticle type might be prepared by **pre-modification post-coupling self-assembly** approaches, although these are relatively new and need further development (Figure 13.23). Importantly though, these pre-modification, post-coupling, self-assembly approaches are particularly interesting given their use of highly regio- and chemo-selective **click chemistry** reactions.

Almost without exception, linear PEG is used as a preferred **C**-component, with a molecular weight of 2000–5000 Da. Linear PEG^{2000} confers sufficient stealth/biocompatibility characteristics to most **ABC/ABCD** type nanoparticles in the bloodstream, PEG^{5000} for lung applications (see Section 13.2.10). On the whole, LNPs of the **ABC/ABCD** nanoparticle type make use of a variety of PEG^{2000}-lipids and ligand-PEG^{2000}-lipid conjugates to ensure *in vivo* biocompatibility. Gratifyingly, several LNPs of the **ABC** nanoparticle type have made it for systemic, clinical use (see Section 13.2.8). Finally, LNPs of the

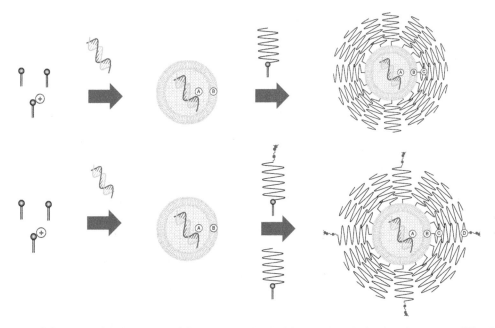

Figure 13.21   LNPs of the ABC and ABCD nanoparticle type assembled with small interfering RNA by post-modification self-assembly. According to this process, **B**-component lipids (left-hand side) (including cytofectin [cationic/ionisable lipid]]) are used to prepare core siRNA-**AB** LNPs (middle) (see Figures 13.6, 13.7, 13.8, 13.9) either by simultaneous combination with siRNA (kinetic mixing) or in separate steps (thermodynamic mixing). siRNA-**ABC** LNP formulation then requires an additional incubation step with PEG-lipids (top right-hand side). siRNA-**ABCD** LNP formulation requires an additional incubation step with ligand-PEG-lipids and PEG-lipids (bottom right-hand side).

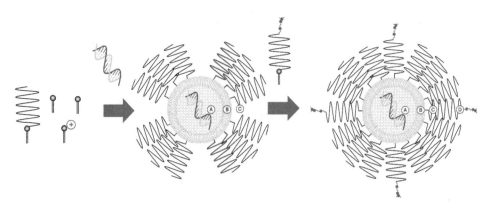

Figure 13.22   LNPs of the ABCD nanoparticle type assembled with small interfering RNA by pre-modification post-modification self-assembly. According to this process, all lipids (left-hand side) (including PEG-lipids and cytofectin [cationic/ionisable lipid]) are used to prepare siRNA-**ABC** LNPs (middle) (see Figures 13.6, 13.7, 13.8, 13.9) that are combined with siRNA either simultaneously (kinetic mixing) or in a separate step (thermodynamic mixing). Thereafter, siRNA-**ABCD** LNP formulation requires an additional incubation step with ligand-PEG-lipids (right-hand side).

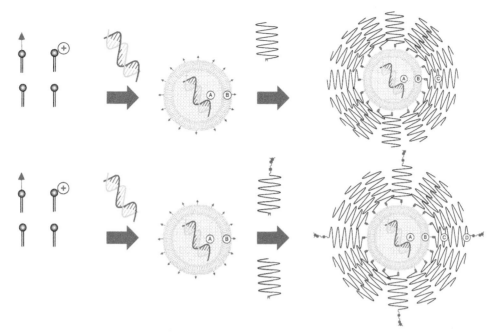

Figure 13.23 **LNPs of the ABC and ABCD nanoparticle type assembled with small interfering RNA by pre-modification post-coupling self-assembly.** According to this process, **B**-component lipids (left-hand side) (including cytofectin [cationic/ionisable lipid] and **click chemistry** lipid [**blue arrow**]) are used to prepare click chemistry-enabled core siRNA-**AB** LNPs (middle) (see Figures 13.6, 13.7, 13.8, 13.9) either by simultaneous combination with siRNA (kinetic mixing) or in separate steps (thermodynamic mixing). siRNA-**ABC** LNP formulation then requires an additional incubation step with click-complementary PEGs (top right-hand side). siRNA-**ABCD** LNP formulation requires additional incubation step with click-complementary ligand-PEGs (bottom right-hand side).

ABC or **ABCD** nanoparticle type can be modified by the inclusion of imaging agents, hence may be easily introduced to enable the monitoring of real time nanoparticle trafficking, nanoparticle biodistribution and API pharmacokinetic behaviour all *in vivo* (see Chapter 14). Currently, the extent to which different imaging modalities will have a substantive role to play in clinically appropriate synthetic nucleic acid delivery systems still remains to be seen.

13.2.8 ABC nanoparticles in clinic

Undoubtedly, LNPs of the **ABC** nanoparticle type are the primary class of synthetic nucleic acid delivery vehicles that have shown realistic potential in gene therapy studies in clinic. Early version LNPs of the **ABC** nanoparticle type were prepared with cystic fibrosis (CF) gene therapy studies in mind, using PEGylated cationic liposomes, GL-67/DOPE/PEG^{5000}-DMPE (Figures 13.3, 13.4 and 13.24) formulated with pDNA expressing wild-type cystic fibrosis transmembrane conductance regulator (CFTR) protein, for CF gene therapy by gene supplementation. In Phase 1 clinical trials, use of these LNPs resulted in some correction of the typical CF abnormalities seen in CF patients. LNPs of the **ABC** nanoparticle type have since been used in a range of important systemic anti-cancer, anti-infectivity and hypercholesterolaemia studies *in vivo* (Table 13.1; Figure 13.24). Now, at least one LNP of the **ABC** nanoparticle type has passed Phase 3 clinical trials and is market registered as a genuine gene therapy product known as ONPATTRO® (Table 13.2; Figure 13.25). In addition, by the end of 2022 at least two further LNPs of the **ABC** nanoparticle type, similar to ONPATTRO®, were marketed as mRNA-gene therapy vaccination products against the Covid-19 global pandemic infection.

Without doubt, all this is very promising, but all the signs are that the majority of current LNPs of the **ABC** nanoparticle type will have limited applications in clinic, however well designed, because LNP-mediated functional delivery of therapeutic nucleic acids still remains inefficient. This weakness places potential patients at risk from undesirable side effects. In general, improvements of up to 10^2-fold would be desirable in the efficiency of synthetic nucleic acid delivery system processes from points of administration to target cells. Similarly, improvements of 10^2-fold are also needed in the process of delivery from target cell surfaces to intracellular mechanistic sites of action. Therefore, overall efficiency improvements of at least 10^4-fold are essential before synthetic nucleic acid delivery system mediated gene therapy can become routinely viable.

Table 13.1   Notable LNPs of the ABC nanoparticle type used *in vivo*. Included are lipids, formulations and ratios. For chemical structures of component lipids noted please refer to Figures 13.4 and 13.24. Glossary of lipid names for Chapter 13 is included on page xvi.

LNP system	Lipids	Ratios	Companies involved
SPLP (Pro-1) Tumour tropic	DODMA/DSPC/Chol/ PEG^{2000}-DSG pDNA	25:20:45:10 (m/m/m/m) lipid:pDNA ~ 12:1 (w/w)	Protiva
SNALP (TKM-HBV) Liver tropic	DLinDMA/DSPC/Chol/ PEG^{2000}-C-DMA siRNA	30:20:48:2 (m/m/m/m) lipid:siRNA ~ 12:1 (w/w)	Sirna/Protiva
SNALP (TKM-ApoB) Liver tropic	DLinDMA/DSPC/Chol/ PEG^{2000}-C-DMA siRNA	40:10:48:2 (m/m/m/m) lipid:siRNA ~ 10:1 (w/w)	Alnylam/Protiva
SNALP (ALN-VSP02) Liver tropic	DLinDMA/DPPC/Chol/ PEG^{2000}-C-DMA siRNA	57.1:7.1:34.3:1.4 (m/m/m/m) lipid:siRNA ~ 6:1 (w/w)	Alnylam
SNALP (ALN-TTR01) Liver tropic	DLin-KC2-DMA/DPPC/ Chol/PEG^{2000}-C-DMA siRNA	57.1:7.1:34.3:1.4 (m/m/m/m) lipid:siRNA ~ 6:1 (w/w)	Alnylam/Tekmira
SNALP (ALN-TTR02) Liver tropic	DLin-MC3-DMA/DSPC/ Chol/PEG^{2000}-C-DMG siRNA	50:10:38.5:1.5 (m/m/m/m) lipid:siRNA ~ 6:1 (w/w)	Alnylam
Atuplex (Atu027) Tumour tropic	AtuFECT01/DPhyPE/ PEG^{2000}-DSPE siRNA	50:49:1 (m/m/m) lipid:siRNA ~ 7:1 (w/w)	Silence Therapeutics

13.2.9 Designing future ABC/ABCD nanoparticles

Designing the future of LNPs will rely upon a proper appreciation of the problems that currently stand in the way of progress. Logically speaking, LNPs and other synthetic nucleic acid delivery systems are unlikely to see widespread application in clinical gene therapy trials until the biological barriers to successful delivery and the biophysical parameters most appropriate to overcome these barriers are known. A new biophysical framework of understanding is required to link nanoparticle biophysical properties to *in vivo* delivery outcomes. This has led to early efforts to correlate the four key attributes of drug-delivery processes (4Ts of delivery, see below) with the four general structural parameters of nanoparticles (4Ss of nanotechnology). The 4Ts are as follows:

- **Trafficking**: the collective process by which APIs are carried from a point of administration to disease-target cells (extracellular) and then to sites of action within cells post-target cell entry (intracellular).
- **Triggerability:** design characteristic(s) of a nanoparticle that promote nanoparticle stability in biological fluids from a point of administration to a disease target site, followed by triggered API release at the disease target site caused either by:
  - intrinsic changes in local conditions (at or within target cells) or
  - extrinsic changes in local conditions resulting from the application of an exogenous stimulus/stimuli.
- **Targeting**: the process by which nanoparticles can be encouraged to approach and enter disease target cells according to a natural nanoparticle tropism (passive targeting), and/or by means of surface attached *bona fide* biological receptor-specific ligands (**D**-components) that mediate close-range, receptor-mediated, specific uptake (active targeting) of nanoparticles into target disease cell populations.
- **Timing**: an appropriate dosing-regime to optimise the correlation between dosage and phenotypic outcomes. Structurally and functionally optimised nanoparticles should enable minimal dosing for maximum effect, ideal for the treatment of chronic diseases.

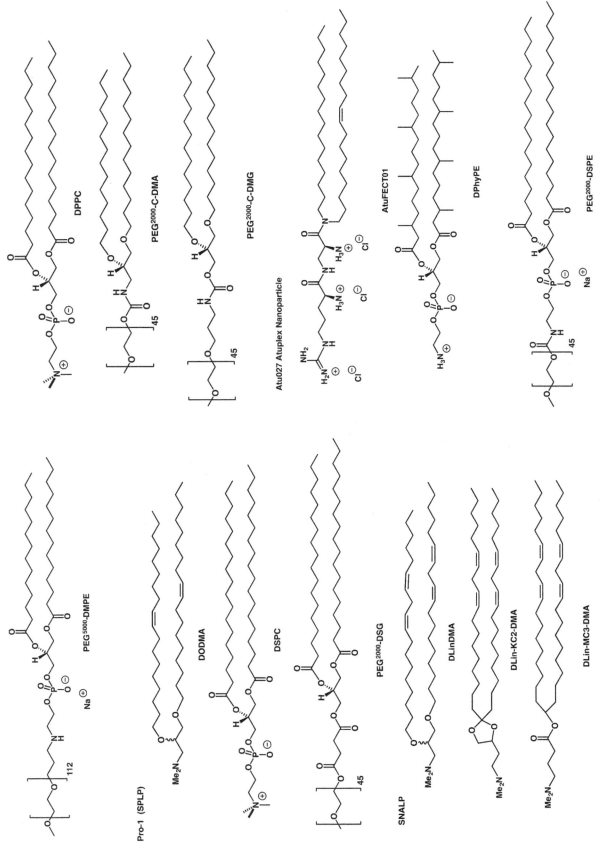

Figure 13.24 Lipid components for LNPs of the ABC nanoparticle type under development in clinic. Glossary of lipid names for Chapter 13 is included on page xvi.

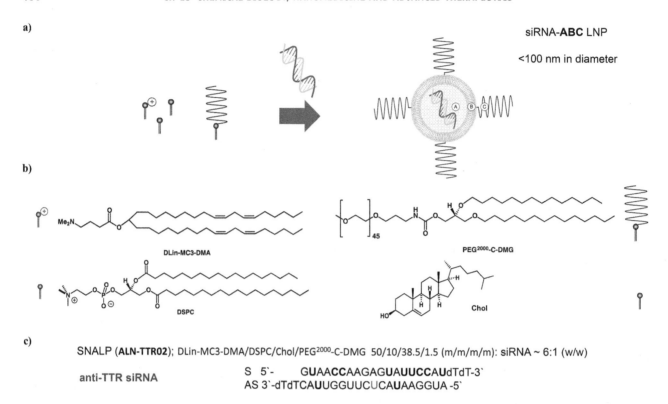

a)

siRNA-**ABC** LNP

<100 nm in diameter

b)

DLin-MC3-DMA

PEG²⁰⁰⁰-C-DMG

DSPC

Chol

c)

SNALP (**ALN-TTR02**); DLin-MC3-DMA/DSPC/Chol/PEG²⁰⁰⁰-C-DMG 50/10/38.5/1.5 (m/m/m/m): siRNA ~ 6:1 (w/w)

anti-TTR siRNA

S 5`-          GUAA**CC**AAGAGUA**UUCC**AUdTdT-3`
AS 3`-dTdTCAUUGGUUC**U**CAUAAGGUA -5`

Figure 13.25 ONPATTRO® patisiran (ALN-TTR02). This marketed siRNA-**ABC** LNP system targets **transthyretin (TTR)-mediated amyloidosis.** The formulation (**a**) of these siRNA-**ABC** LNPs results from a combination of siRNA-API anti-TTR siRNA and lipid self-assembly, involving one cationic/ionisable, two neutral and one PEGylated lipid, as shown in (**b**). In (**c**) the mol fractions of each lipid component are indicated, as is the approximate ratio of lipid to siRNA in the fully formulated siRNA-**ABC** LNP. The structure of anti-TTR siRNA is chemically modified. Chemical modifications are indicated as follows: dN are 2′-deoxynucleotide residues (dT: 2′-deoxythymidine nucleotide residue); **bold letters** are 2′-OMe nucleotide residues. The **red letter** denotes the residue position (10) of the putative mRNA cleavage in AS that acts as a **guide strand** for RNA silencing, while the complementary S is a **passenger strand.**

Table 13.2 LNP ABC-type mediated delivery of RNAi effectors in clinic.

LNP system	Company	Clinical indication	Trial phase
Atu027	Silence Therapeutics	Pancreatic adenocarcinoma	Phase 2
ALN-VSP02	Alnylam	Hepatocellular carcinoma (HCC)	Phase 1
ALN-TTR01	Alnylam	TTR amyloidosis	Phase 1
ALN-TTR02	Alnylam	TTR amyloidosis	Marketed product – ONPATTRO®
ALN-PCS	Alnylam	Hypercholesterolaemia	Phase 1
TKM-ApoB	Tekmira	Hypercholesterolaemia	Pre-phase 1
TKM-PLK1	Arbutus	Hepatocellular carcinoma (HCC)	Phase 1/2
TKM-Ebola	Arbutus	Ebola virus infection	Phase 2
ARB-1467 (TKM-HBV)	Arbutus	HBV infection	Phase 2

Of these four drug-delivery attributes, **timing** is a function of the overall efficacy of nanoparticle-mediated delivery and the overall efficacy of the API for therapy. Efficient **trafficking** is the primary attribute required by any API delivery nanoparticle to navigate from a point of administration to a site of therapeutic action overcoming bio-barriers and other obstacles to functional delivery according to disease biology and pathology. Thereafter, **triggerability** and **targeting** are the primary attributes that can be used to improve trafficking and timing. Accordingly, triggerability and targeting deserve a little more discussion.

The need for triggerability arises from an inherent design paradox in **ABC/ABCD** type nanoparticles. The **C**-layer provides for nanoparticle stealth/biocompatibility. However, the presence of the **C**-layer is typically inhibitory or refractory to functional API delivery once target cells are reached. Therefore, in order not to significantly impair local API release and functional delivery to intracellular sites of action, triggered release of **C**-layer should take place just prior to or during cell entry. In the case of LNPs of the **ABC/ABCD** nanoparticle type, triggered release of the **C**-layer may be accomplished by changes in local pH during endocytosis (Figure 13.26), changes in local enzyme

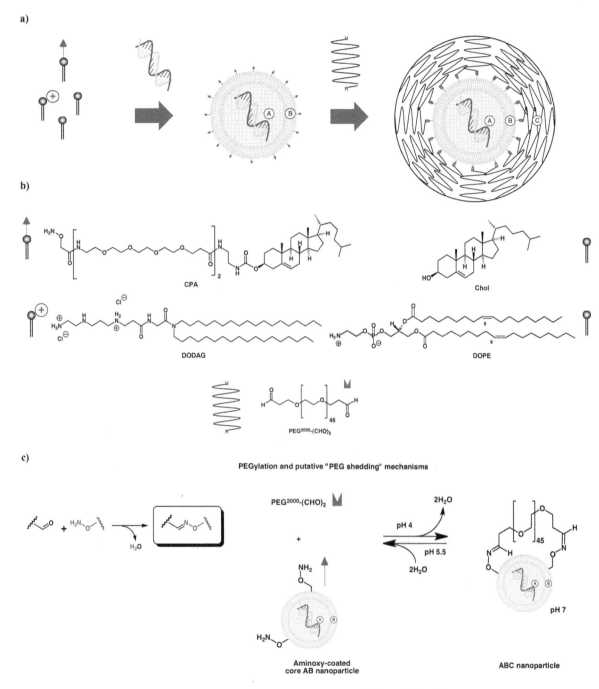

Figure 13.26   pH-triggered LNPs of the ABC nanoparticle type, assembled with small interfering RNA by pre-modification, post-coupling self-assembly. (a) Core **AB** LNPs with siRNA are formulated with lipids mounting a **click chemistry donor** moiety (**aminoxy**, denoted by **blue arrow**); in a second step, pH-triggered siRNA-**ABC** LNPs are prepared by **post-coupling** at pH 4 of donor moieties with **the click chemistry acceptor** moieties (**aldehyde**, denoted by **red "crown" shape**) of **PEG²⁰⁰⁰-dipropionaldehyde**; (b) Relevant siRNA-**ABC** LNP lipid structures and PEG²⁰⁰⁰-dipropionaldehyde; (c) The mechanism of click chemistry post-coupling leading to **oxime** formation (left). In the other reaction scheme is shown (right) how oxime formation takes place at optimal pH 4, leading to formation of pH-triggered siRNA-**ABC** LNPs. Thereafter, these oxime linkages are stable at pH 7, but subject to cleavage at pH 5.5 (acidic endosome pH). Therefore, after pH-triggered siRNA-**ABC** LNPs enter target cells by endocytosis, oxime hydrolysis begins as the endosome pH drops, so leading to pH-triggered "PEG shedding" and re-exposure of core siRNA-**AB** LNPs. Subsequent lipid-lipid interactions should then cause endosomolysis and enable functional delivery of entrapped siRNAs to the cytoplasm of target cells (see Figure 13.12) (adapted from Miller, 2013, Figure 10).

levels (Figure 13.27) or redox conditions (Figure 13.28). In addition, the modulation of nanoparticle half-life $t_{1/2}$ can be employed (Figure 13.29). Thermally triggered release nanoparticles have also been devised, although with small molecule drug delivery in mind (see Chapter 14).

The need for targeting arises since this is a key way to maximise the efficiency of local drug delivery to disease target cells. First, **ABC** and **ABCD** nanoparticles themselves have their own natural tendencies to partition to certain regions

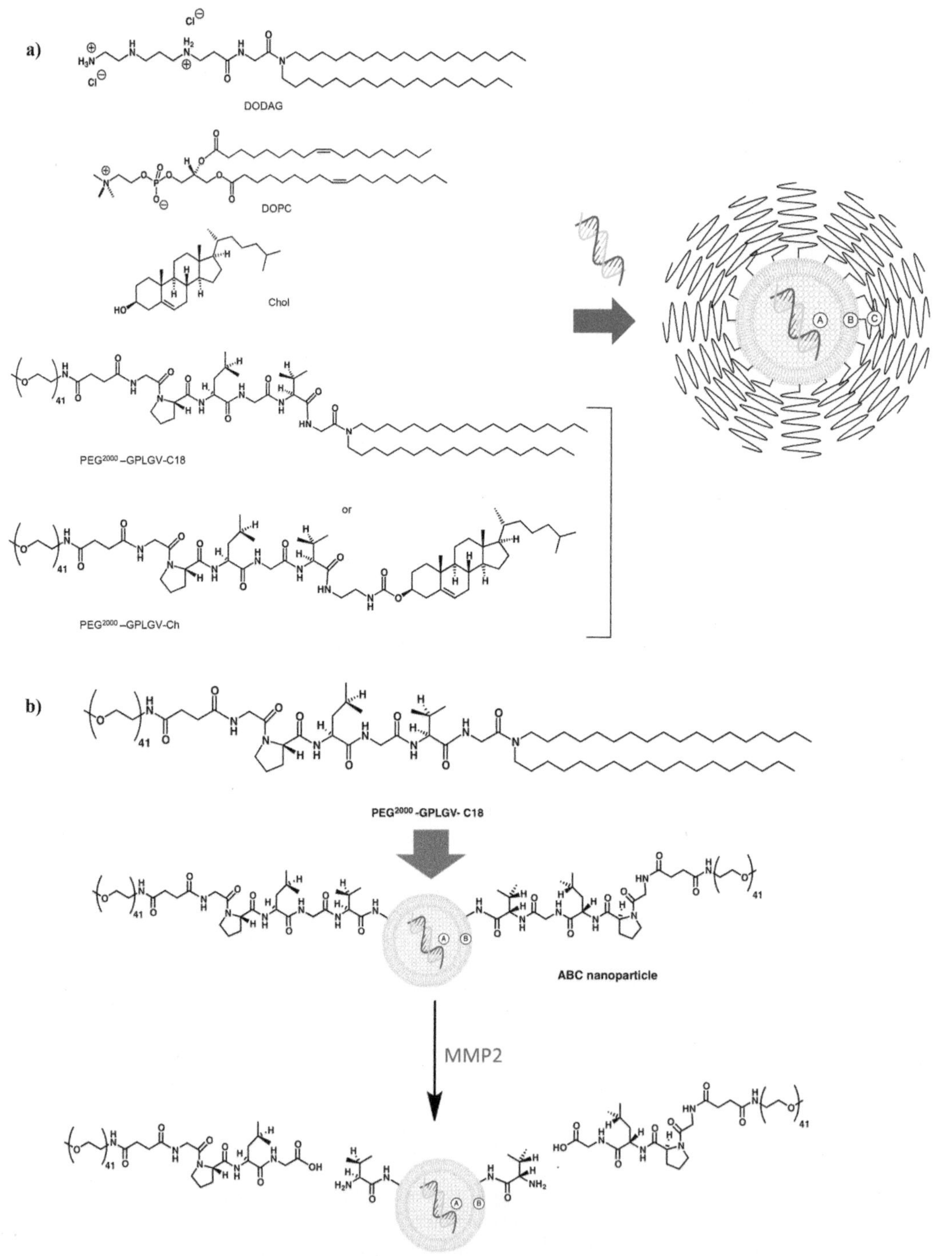

Figure 13.27 Enzyme-triggered LNPs of the ABC nanoparticle type assembled with small interfering RNA. (a) Schematic for preparation of enzyme-triggered siRNA-**ABC** LNPs by **pre-modification self-assembly**. PEGylated cationic liposomes are prepared first, using one or other of the two indicated **PEG-peptidyl lipids**, followed by rapid vortex mixing with siRNA; (**b**) Schematic to illustrate how enzyme-triggered siRNA-**ABC** LNPs are stable at pH 7, but are subject to **matrix metalloproteinase (MMP-2)** mediated enzyme-triggered "PEG shedding" for extracellular re-exposure of core siRNA-**AB** LNPs within tumour target cell matrices. Core siRNA-**AB** LNPs may then freely enter target cells by endocytosis. Subsequent lipid-lipid interactions then cause endosomolysis and enable functional delivery of entrapped siRNAs to the cytoplasm of target cells (see Figure 13.12) (adapted from Miller, 2013, Figure 10).

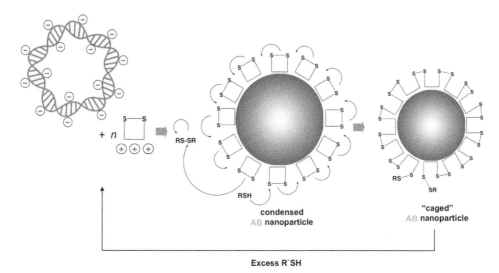

Figure 13.28 **Redox-triggered core pDNA-AB PNPs for lung delivery.** The scheme above shows how core pDNA-**AB** PNPs are prepared by pre-modification self-assembly. Initially, pDNA is condensed with **disulfide polyamine monomers**, then becomes redox "caged" by oxidation. Redox caged pDNA-**AB** PNPs are stable at pH 7 in oxidising conditions, but in extracellular lung environments with reducing excipients present, such as **N-acetyl cysteine** (**NAC**), these nanoparticles are triggered to destabilise *en route* to lung target cells. At target, redox destabilised nanoparticles promote functional pDNA delivery by means of endocytosis, endosomolysis and trafficking to target cell nuclei (adapted from Miller, 2013, Figure 10).

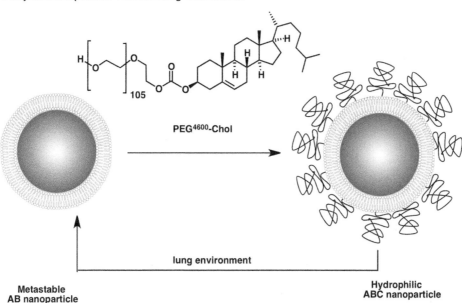

Figure 13.29 **Half-life triggered LNPs of the ABC nanoparticle type assembled with pDNA for topical lung** delivery. In this scheme, pDNA-**ABC** LNPs are formulated by **post-modification self-assembly**. Initially, core pDNA-**AB** LNPs are prepared by combining cationic liposomes followed by rapid vortex mixing with pDNA. Thereafter pDNA-**ABC** LNPs are generated by incubation with **PEG^{4600}-Chol** lipids. These pDNA-**ABC** LNPs are stable during storage at pH 7, but undergo spontaneous "PEG shedding" in extracellular lung environments leading to gradual exposure of core pDNA-**AB** LNPs that may then freely enter lung target cells by endocytosis. Subsequent lipid-lipid interactions then cause endosomolysis and trafficking to target cell nuclei (see Figure 13.12) (Miller, 2013, Figure 10, Taylor and Francis).

over other regions *in vivo*. This passive targeting may be associated with, for example, the **enhanced permeability and retention (EPR)** effect (Figure 13.30). Thereafter, **ABCD** nanoparticles with *bona fide* receptor-specific targeting ligands (**D**-layer components) attached to their outer surfaces are primed to enter disease target cells by receptor-mediated uptake (active targeting). Contrary to popular expectation, receptor-mediated uptake requires a tremendous amount of optimisation, and there are surprisingly few ligands available that actually assist functional delivery *in vivo*. A useful current list of ligands currently includes: **transferrin (Tf), anti-TfR antibody fragment (scFv$_{anti-TfR}$), arginine-glycine-aspar-**

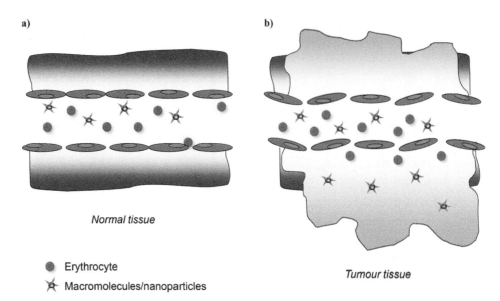

Figure 13.30 **Enhanced permeability and retention effect.** The **enhanced permeability and retention (EPR)** effect applies in certain locations, such as the liver, in tumours (as shown) and at sites of inflammation, all due to fenestrations (large-sized gaps, >100 nm in dimension) that exist between endothelial cells in the local vasculature. (**a**) Normal tissue, fenestrations do not appear. However, when fenestrations do appear, in (**b**), these enable circulating nanoparticles to partition out of the blood pool and gain access to the corresponding tissue locations beyond. Accumulation of nanoparticles in tissue by EPR is a key mechanism for passive targeting of nanoparticles (adapted from Miller, 2013, Figure 5, Taylor and Francis).

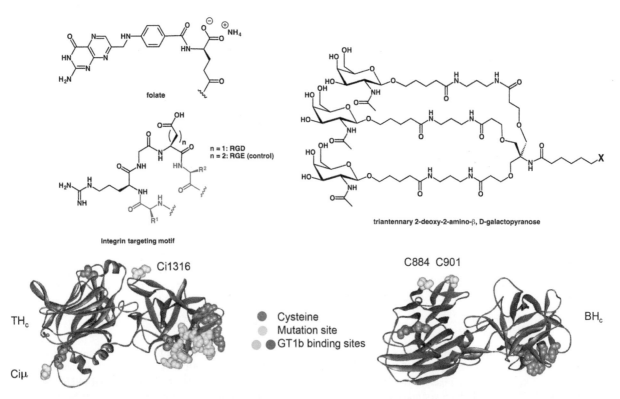

Figure 13.31 *Bona fide* **biological receptor-specific targeting ligands.** The number of genuinely functional receptor-targeting ligands available that function equally well *in vitro* and *in vivo* is arguably very small. Representative examples are illustrated. First, **folate** binds specifically to folate receptors found on many tumour cell surfaces; second, **arginine-glycine-aspartate (RGD) peptide moieties** bind specifically to different cell surface integrins depending on the flanking amino acid residues (in red) and third, **the triantennary *N*-acetylgalactosamine ligand** was designed to bind to the asialoglycoprotein receptor located on hepatocyte cell surfaces. Finally, are illustrated **tetanus toxin C fragment (TH$_c$)** and **botulinum toxin C fragment (BH$_c$)** that both bind specifically to the oligosaccharides of **ganglioside GT1b** in neurons.

tate (**RGD**) integrin receptor targeting peptides, folate, galactose/N-acetyl-galactose, **urokinase plasminogen activator receptor U11 peptide (uPAR-U11)** and recombinant **tetanus toxin or botulinum toxin C fragments (TH$_c$ and BH$_c$,** respectively) (Figure 13.31). Without optimisation, targeting ligands can easily end up promoting non-specific enhanced cell uptake processes that assist delivery without being functional by receptor-independent means. Nevertheless, when they are optimised for use, biological receptor-specific target ligands really can improve local API delivery and overcome the C-layer paradox, even without recourse to triggerability. Clearly there is a substantial unmet need to identify further biological receptor-specific targeting ligands able to promote genuine functional nanoparticle-mediated delivery *in vivo*.

13.2.10 The 4Ss of nanotechnology; impact on 4Ts of delivery

The 4Ss of nanotechnology are as follows:

- **Shape**: meaning nanoparticle shape (e.g. spherical or non-spherical) and degree of homogeneity.
- **Size**: meaning main nanoparticle dimensions.
- **Surface**: meaning overall impacts of surface hydrophobicity versus hydrophilicity, charged versus neutral.
- **Structure**: meaning the structures of molecular components, plus their impact on nanomolecular structure and nanoparticle mediated functional delivery.

Where **shape** is concerned, LNPs that are spherical and monodisperse (i.e. homogenous in formation) appear best for current delivery priorities.

Where **size** is concerned, LNPs of various sizes appear appropriate, depending on the location of the target cells in the organ(s) of interest. Regarding LNPs of the **ABC/ABCD** nanoparticle type, a 100 nm diameter size, or just under, seems best for nanoparticle mediated trafficking to liver *in vivo* and to tumours. Where topical delivery of nucleic acids to lung surfaces is concerned, sizes of ≫100 nm in diameter seem more appropriate. Appropriate sizes for trafficking and functional delivery to other organs of potential interest, like brain, spleen and pancreas, for example, remain to be established and need to be the subject of future research.

Where **surface** is concerned, variations in nanoparticle surfaces can impact substantially on trafficking and targeting outcomes. In the case of LNPs of the **ABC/ABCD** nanoparticle type, nanoparticle surface hydrophilicity and residual surface charge appear critical parameters for functional delivery. The degree of hydrophilicity is determined as a function of PEG polymer molecule weight and the extent of polymer branching, and the percentage by mol (mol%) of PEG-lipid with respect to the total lipids used in any particular LNP assembly. LNPs can be rendered essentially inert in serum conditions (for at least 3–4 h) and will traffic comfortably to liver cells (hepatocytes) post-i.v. administration assuming a 100 nm diameter particle size, a ζ potential of near zero and a surface cover of up to 5 mol% of a linear PEG^{2000}. Higher levels (6–20 mol%) of linear PEG^{2000} appear beneficial where tumour delivery is desired over liver delivery. Where topical lung delivery is concerned (via intra nasal [i.n.]-administration), a > 100 nm nanoparticle diameter, negative ζ potentials and at least 14 mol% surface coverage with a large, linear PEG^{4600} moiety, appears appropriate for successful functional delivery of nucleic acids to lung epithelial cell surfaces. Once again, appropriate surface characteristics for functional delivery to other organs of potential interest remain to be established and need to be the subject of future research.

Otherwise, surface factors also synergise with size and shape factors to impact significantly on biological receptor-mediated targeting. Current guidelines suggest that LNPs of the **ABCD** nanoparticle type should be as spherical and as monodisperse as possible at or below 100 nm in diameter. Nanoparticle ζ potentials should approach zero, and ligand surface coverage should be typically at or around 1–2 mol% of total lipid mol (on a case-by-case basis). Moreover, all *bona fide* ligands need to be surface attached in correct orientation(s) and conformation(s) for optimal interactions with receptors. Unsurprisingly perhaps, correct orientations and conformations usually result from case-by-case optimisations.

Finally, where **structure** is concerned, the unrivalled capacity of lipids to self-assemble into LNPs is fundamental to the design and creation of functional LNPs of the **ABC/ABCD** nanoparticle type. Using organic chemistry expertise, LNPs should be improved with lipid components that possess different molecular shapes and structures to enhance key LNP properties such as biological fluid stability, biocompatibility, triggerability and optimal trafficking/targeting characteristics (Figure 13.32).

It should now be clear that studies to correlate the 4Ss of nanotechnology to the 4Ts of delivery are well underway, but will require considerably more future effort. Once done, the current over-emphasis on synthetic nucleic acid delivery system mediated delivery to liver, tumours and lung can be expanded to other body organs, opening up substantial new potential opportunities for ncRNA therapeutics and other approaches to gene therapy. Furthermore, once such an understanding has been reached, then functional delivery may be further improved with toolkits of chemical components designed to impact logically on the 4Ts of delivery and so completely optimise drug delivery mediated by synthetic nucleic acid delivery systems.

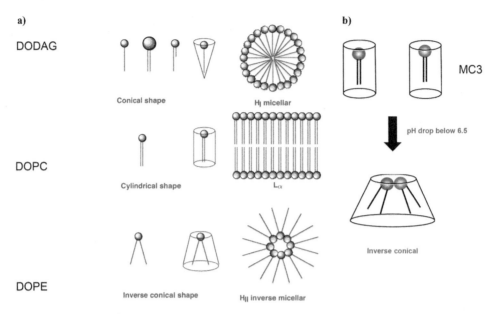

Figure 13.32   Lipid shape and potential impact on LNP behaviour. (a) Lipids are characterised in terms of the their volumetric 3D shapes (see Figures 13.27 and 13.4 for DODAG, DOPC and DOPE chemical structures). Lipids with a large polar head group region or asymmetric chain region (comprising hydrophobic single hydrophobic chains or chains of widely different length) have a **conical** shape that tends towards **fluid hexagonal H$_I$ (micellar) mesophase** structures; lipids with a small polar head group region or high volume asymmetric region (i.e. through inclusion of *cis*- double bonds) have an **inverse conical** shape that tends towards **fluid hexagonal H$_{II}$ (inverse micellar) mesophase** structures; finally, lipids with a polar head group region and symmetric chain region in volumetric balance have a **cylindrical** shape, and tend towards **fluid lamellar L$_\alpha$ mesophase** structures; LNPs comprise all three types of lipid for tuning LNP stability for storage, for biocompatibility in biological fluids and for functional delivery after uptake into target cells. DOPE, for example, aids endosomolysis (the inverse conical shape promotes membrane fusion, fusogenesis, and rupture) (see Figure 13.12); **(b)** A schematic that indicates the proposed capability of **MC3 (DLin-MC3-DMA)** and **DLin-KC2-DMA** cytofectins (see Figure 13.24), with single amine polar head group regions of low p*K*a values (<6.5), to pair at low endosome pH values with anionic lipids, forming inverse conical shapes that act like DOPE to aid endosomolysis (note above) (Miller, 2013, Figure 7 and 9, Taylor and Francis).

13.2.11 Future prospects for gene therapy enabled by LNP nanomedicine

Gene therapy research has evolved considerably since the first clinical trials, moving from *ex vivo* and local administration to systemic administration. Today, the main challenge for gene therapy remains the delivery problem. Therapeutic nucleic acids will always have to be formulated with a delivery system in such a way that somatic and local pharmacokinetics can guarantee the delivery of the therapeutic nucleic acid to the target cells of interest within the organ(s) of choice. Currently, there are three gene therapy products registered for use in the clinic, based upon viral delivery systems. There are two recombinant **adeno-associated virus (AAV)** based viral delivery system products. One, an AAV-9 based product (ZOLGENSMA®), is directed at the treatment of paediatric patients under two years of age suffering from **spinal muscular atrophy (SMA)**. The second, an AAV-2 based product (LUXTURNA®) is for the treatment of patients with confirmed biallelic *RPE65* mutation-associated **retinal dystrophy**. In addition, there is a **herpes simplex virus 1 (HSV-1)** based product (IMLYGIC®) for the treatment of **melanoma** by melanoma cell-specific oncolytic mechanisms, coupled with melanoma cell-specific immunotherapy. These compare with the one synthetic nucleic acid delivery system-based product (ONPATTRO®) mentioned above.

These products are important milestones and underline the growing promise of gene therapy to meet many unmet medical needs. In principle, synthetic nucleic acid delivery systems are able to mediate the delivery of all types of nucleic acids. Therefore, synthetic nucleic acid delivery systems have the potential to deliver on the widest possible range of gene therapy strategies going forward, once delivery efficacy is significantly improved to appropriate levels. This means that there is every incentive to create newer and better synthetic nucleic acid delivery systems with synthetic designs informed by a detailed and thorough knowledge of the relevant biological barriers that stand in the way of efficient, extracellular nanoparticle trafficking from the selected site of administration to the disease target site of interest, followed by local, highly efficient functional API delivery. Such future nanoparticle designs will require development by means of pre-clinical studies, with the best possible, most accurate biological models, such as the use of primary tissue samples and humanised animal disease models.

Overall, gene therapy *per se* is complicated and the promise is yet to be properly fulfilled. Nevertheless, all in good time, gene therapies mediated by synthetic nucleic acid delivery systems can emerge with all the necessary capabilities to become a dominant feature of future gene therapy treatments. And even if this is not quite the case, one thing should be clear, there is certainly a place for synthetic nucleic acid delivery systems in the future of gene therapy.

13.3 Biophysical characterisation of LNPs

The most direct ways of observing nanoparticle dimensions in solution are **atomic force microscopy** (**AFM**) and **electron microscopy** (**EM**) (in transmission scanning modes and cryo-electron microscopy modes). However, there are simpler, more direct techniques that are frequently used to great effect.

13.3.1 Dynamic light scattering

Nanoparticle dimensions can be measured in solution by the technique of **dynamic light scattering** (**DLS**), also known as **photon correlation spectroscopy** (**PCS**). Typically, when light passes through a transparent solution with suspended particles that are small (<250 nm) compared to the **wavelength** λ (~400 nm), then light scatters in all directions as a result of productive "collisions" with the suspended particles or **scatterers**. This is known as **Rayleigh scattering**. Importantly, Rayleigh scattering intensity fluctuates over time, even when the incident light is monochromatic laser light. Such intensity fluctuations are caused by the Brownian motion of scatterers, a motion which alters the relative spatial distribution of scatterers as a function of time. Intensity fluctuations are further modulated by constructive or destructive interference between light scattered from neighbouring and other surrounding scatterers. Hence, the intensity of Raleigh scattering can be said to fluctuate over time as a function of dynamic changes in the relative spatial distribution of scatterers. The opportunity given by such dynamic intensity fluctuations is that these may be analysed by a **photon autocorrelation function** (**ACF**).

An ACF analysis is a mathematical approach for determining patterns in periodic data. In essence here, autocorrelation compares intensity data at a certain time with data after a fixed **delay time**, τ. Assuming τ is short, then the correlation between intensity data should be high, since scatterers have barely moved. However, as τ increases, the quality of the correlation declines until the delay time is such that there is no correlation between intensity data at initial time and after the delay time. Accordingly, provided that scatterers are reasonably **monodisperse**, namely of roughly equivalent dimensions in solution, then the decline in correlation between intensity data as delay time increases can be fitted to an exponential function, Equation (13.1):

$$g^1(q;\tau) = \exp(-\Gamma\tau) \tag{13.1}$$

where $g^1(q;\tau)$ is a so-called first order autocorrelation function dependent on the properties of **wave vector**, q, the delay time, τ, and **the decay constant** Γ. The decay constant is related to the dynamic behaviour of scatterers as follows in Equation (13.2):

$$\Gamma = q^2 \cdot D_{NM} \tag{13.2}$$

where D_{NM} is a **nanomolecular diffusion constant** appropriate for the temperature and pressure of observation. Finally, the wave vector is defined as follows:

$$q = \frac{4\pi \cdot n_R}{\lambda} \operatorname{Sin}\left(\frac{\theta}{2}\right) \tag{13.3}$$

where n_R is the **refractive index of the medium** and θ a selected **Raleigh scattering angle**. In the case of any nanomolecular particle scatterers suspended in solution, there is an optimal θ value with which to derive the best intensity data with time for autocorrelation. Clearly the most important potential outcome of autocorrelation is size determination of nanomolecular particle scatterers. Fortunately, once a decay constant Γ is determined, Equation (13.2) enables the determination of D_{NM} for the monodisperse nanomolecular scatterers under investigation, from which value a **spherical macromolecular radius**, r_{sph}, or **hydrodynamic radius** may be determined through the Stokes and Einstein-Sutherland equations (see

Chapter 7, Equations 7.2, 7.5, and 7.14), assuming that the monodisperse nanomolecular scatterers under investigation are spherical in shape, which is a reasonable assumption for LNPs, for example. Furthermore, in the event that nanomolecular scatterers under investigation are **polydisperse**, namely of nonequivalent dimensions in solution, then the autocorrelation function (as defined in Equation 13.1) may be redefined as the sum of as many different exponential decays as there are distinct populations, *n*, of different size scatterers present in the sample under investigation (Equation 13.4):

$$g^1(q;\tau) = \sum_{i=1}^{n} G_i(\Gamma_i) \cdot \exp(-\Gamma_i \tau) = \int G(\Gamma) \cdot \exp(-\Gamma \tau) \tag{13.4}$$

which in principle can also be used with Equation (13.2) to determine D_{NM} values for distinct populations, *n*, of different size scatterers, followed by their hydrodynamic radii.

A typical experimental set up for studying DLS is illustrated in Figure 13.33. A monochromatic light source, usually a laser, is shot through a polariser and into a sample. The scattered light then goes through a second polariser where it is collected by a photomultiplier. The result is a **speckle pattern** image that comprises dark areas (destructive interference) and bright speckles (constructive interference). This process is repeated at short delay time intervals and speckle patterns are subject to ACF analysis in order to determine the decay in speckle pattern correlations with time and arrive at Γ decay constant values. The polarisers can be set up in two geometrical configurations. One is a vertical/vertical (VV) geometry, where the second polariser allows light through that vibrates in the same plane as the primary polariser. In vertical/horizontal (VH) geometry the second polariser allows light through that vibrates in a plane 90° to light transmitted by the primary polariser.

A high-quality DLS analysis should always be performed at several scattering angles (multi-angle DLS). This becomes even more important when scatterers are polydisperse with an unknown particle size distribution. At certain angles the scattering intensity from certain populations of scatterers will completely overwhelm the weak scattering from others, thus making them invisible to any data analysis at this given angle. Therefore, the presence of irregular solution contaminants (e.g. dust and other particulate artefacts) could easily distort the pattern over time of Raleigh scattering and intensity fluctuations by *bona fide* scatterers. Accordingly, DLS measurements should only be made with samples of suspended particles that have been prefiltered or centrifuged. Otherwise, DLS analyses are best achieved under conditions where scatterers do not interact with each other significantly either through collision or through electrostatic interactions involving ions. Therefore, DLS measurements are better performed when scatterer-scatterer interactions can be suppressed by concentration dilution and at low ionic strengths in order to collapse the size of electrical double layers that exist around positively or negatively charged nanoparticulate scatterers (see Section 13.3.2).

Finally, what happens when scatterers themselves diverge from being spherical? Clearly, this question is not so important as far as LNPs and other types of nanoparticulates are concerned, since these are frequently genuinely spherical in suspension. However, if scatterers are non-spherical, like random coil polymers, then standard DLS analyses with autocorrelation functions will still treat scatterers as spherical, such that any hydrodynamic radius values obtained are those of "equivalent sphere" scat-

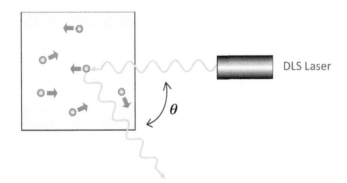

Figure 13.33 Dynamic light scattering involving a laser. The technique of dynamic light scattering (DLS) involves irradiation of particles in solution with a polarised monochromatic laser light source (~400 nm). Light is then subject to **Raleigh scattering** in all directions from particles (if <250 nm in diameter), but collected at fixed angles, θ, by photomultiplier after passing through a second polariser. The fluctuation of scattered light intensity, from Brownian motion of particles as a function of time, is analysed mathematically by a **photon autocorrelation function (ACF)** in order to determine good estimates of particle size distribution and diffusion behaviour (as explained in the text). DLS may also be known as **photon correlation spectroscopy (PCS)** or **quasi-elastic light scattering (QELS)**. Good particle sizing data relies on good sample preparation involving filtration or centrifugation to remove dust and artefacts from the solution that would otherwise distort data outputs.

terers that move in the same manner in suspension as the corresponding real, non-spherical scatterers. Finally, it is also useful to point out that nanoparticulate scatterer dimensions will by default include other solutes or solvent molecules that associate with a given scatterer in solution, since these obviously contribute towards the measured hydrodynamic radii. So, for example, when **ABC/ABCD** type LNP samples are under analysis by DLS, PEG **C**-layer surfaces can be observed and will contribute fully towards measured nanoparticle hydrodynamic radii. By contrast, such PEG **C**-layers are typically invisible to electron microscopy (since this technique is unable to visualise such PEG **C**-layers due to poor contrast imaging).

13.3.2 Measuring nanoparticulate zeta-potentials

Nanoparticle charges in suspension can be determined by a **zeta-sizer** that represents a combination technique of electrophoresis and laser light scattering to measure nanoparticle electrophoretic mobilities. Interactions of ions in solution with positively charged nanoparticles are illustrated (Figure 13.34). Critically, positively charged nanoparticles can be expected to be covered by a **Stern layer** of counter anions (see capillary electrophoresis, see Section 7.3.4), surrounded by an organised **electrical double layer** consisting of cations and further counter anions. This gives way to a **slipping plane** at which point formal cation and anion structures merge with bulk ionic solution. The electrical potential at the beginning of the slipping plane is known as the **ζ-potential**. This represents the *de facto* electrical potential experienced by other nanoparticles, ions, molecular structures, etc. in close proximity to positively charged nanoparticles in suspension in solution. Where the interactions of ions in solution with negatively charged nanoparticles are concerned, the situation is essentially identical with that for positively charged nanoparticles, except that the Stern layer is now comprised of counter cations and measured ζ-potentials should be negative, not positive.

The measurement of ζ-potential values relies on the ability of charged nanoparticles to migrate in suspension in solution under the influence of an applied electric field, while under the observation of visible laser light applied orthogonally

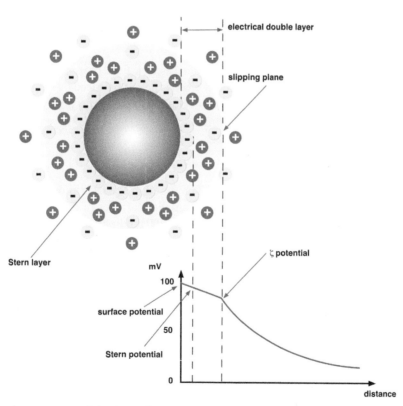

Figure 13.34 **Defining the zeta potential values of nanoparticles.** A cationic nanoparticle in solution is surrounded by a **Stern layer** of essentially immobile counter anions (the reverse is true with an anion nanoparticle, not illustrated). Around this is generated an informally organised **electrical double layer**, comprised of solution anions and cations, which terminates at a **slipping plane**, after which the electrostatic field emanating from the nanoparticle is too weak to induce significant order. The nanoparticle **zeta** (**ζ**)-**potential** (±mV) represents the electrostatic potential in solution generated by the nanoparticle at the slipping plane, which is equivalent to the nanoparticle surface potential modulated by Stern and electrical double layer.

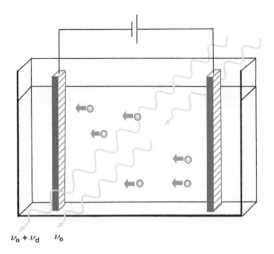

$F(\kappa a)$ - Henry's function:

❑ Hückel limit ~ >1.0 (smaller NPs, dilute or non-aqueous)
❑ Smoluchowski limit ~ <1.5 (~ 200nm nanoparticles, mM electrolyte)

Figure 13.35   Zeta potential measurement by zeta-analyser. The ζ-potential analysis involves irradiation of particles in solution with a polarised monochromatic laser light source (~400 nm) in the presence of a mobilising electric field. Charged particles (illustrated are cationic nanoparticles) travel towards the electrode of opposite charge. When the attractive force of the electric field is balanced by viscous drag on particles, then particles travel at a constant velocity defined by their **electrophoretic mobility**, μ_e. Values of μ_e are measured from interactions of incident laser light (orthogonal to the plane of electrophoretic mobility) and forward re-transmission to detector. Particle interactions change the frequency of incident light from ν_o to $\nu_o + \nu_d$ where ν_d is the apparent change (**Doppler shift**) in the frequency of transmitted light due to particle electrophoretic mobility. This apparent change can be measured and inter-polated to give values of μ_e from which values of ζ-potentials may be determined directly (see Equation 13.6), assuming certain fixed values of the **Henry function** (see above).

to the plain of nanoparticle migration (Figure 13.35). In an applied electric field, charged nanoparticles travel towards the electrode of opposite charge. When the attractive force of the electric field is balanced by the viscous drag force acting on each nanoparticle, then nanoparticles will travel at a constant velocity that relates to their **electrophoretic mobility**, μ_e, as defined by the Henry Equation (13.5):

$$\mu_e = \frac{2\varepsilon\zeta \cdot F(\kappa a)}{3\eta} \tag{13.5}$$

where ε is **permittivity** of the solution, and η is **viscosity**. The function $F(\kappa a)$ is known as **Henry's function**, which limits with a value of 1.0 at the **Hückel limit** (appropriate for smaller particles in dilute or non-aqueous solution), or 1.5 at the **Smoluchowski limit** (appropriate for nanoparticles of ~ 200 nm in dimension and up to mM concentrations of electrolyte). The determination of μ_e values for nanoparticles moving at constant velocity in an applied electric field is then made by either **laser doppler velocimetry** or **phase analysis light scattering** (**PALS**), of which the second method is most applicable for slower moving nanoparticles. For nanoparticles, Equation (14.5) can be rewritten in the form of the **Helmholtz-Smoluchowski equation** that gives ζ-potential values as follows directly in Equation (13.6):

$$\zeta = \frac{3\eta \cdot \mu_e}{2\varepsilon \cdot F(\kappa a)} \tag{13.6}$$

13.3.3 Measuring nanoparticle properties in complex solutions

One of the main problems for nanoparticle characterisation by either of the two methods mentioned above is that both methods are limited to nanoparticle measurements in non-biological media. Accordingly, the best that can be hoped for is to use DLS techniques for the characterisation of nanoparticle colloidal instability with respect to aggregation. Aggregates (>1 mm in dimension) can be expected to form from nanoparticles when electrostatic charge neutralisation removes electrostatic forces of repulsion. In addition, nanoparticles will aggregate with the surface adsorption of biological fluid components such as serum proteins. The PEG (**C**-layer) described is intended to obviate this tendency, hence the reason why this is known as a biocompatibility layer, but nanoparticle aggregation remains a risk. Stability in biological fluids (>80% serum equivalent) is in fact a central requirement for any LNP to have any widespread use in clinic as a synthetic nucleic acid delivery system.

13.4 Applications of LNPs with small molecule drugs

While this chapter has largely focused on LNPs in gene therapy applications, there have been substantial applications of small molecule drug-loaded LNPs. Indeed, LNPs loaded with small molecule drugs have proven quite effective as drug delivery systems in clinic, particularly where the delivery of cytotoxic, anti-cancer drugs is concerned. Such small molecule drug-loaded LNPs are typically formulated directly from liposomes by pre-modification, self-assembly (see Chapter 14). Selected structural lipids self-assemble into neutral or anionic liposomes that are typically approximately 100 nm in diameter and consist of a lipid bilayer surrounding an aqueous cavity. This cavity can be used to entrap water-soluble drugs in an enclosed volume, resulting in a drug-**AB** nanoparticle system. The first reported LNPs of this type were designed to improve the pharmacokinetics and biodistribution of the anthracycline drug **doxorubicin** (see Chapter 14). Doxorubicin is a potent anti-cancer agent but is cardiotoxic. Therefore, in order to minimise cardiotoxicity, doxorubicin was encapsulated in anionic liposomes giving anionic doxorubicin drug-**AB** nanoparticles that enabled improved drug accumulation in tumours and increased anti-tumour activity, while diminishing side effects from cardiotoxicity. This nanoparticle formulation has since been used efficiently in clinic for the treatment of ovarian and breast cancer.

Thereafter, the product DOXIL® was devised corresponding to a drug-**ABC** LNP, comprising PEGylated liposomes with encapsulated doxorubicin. These DOXIL® drug-**ABC** LNPs (also known as PEGylated drug-nanoparticles) were designed to improve drug pharmacokinetics and reduce toxicity further by maximising **reticuloendothelial system (RES)** avoidance, making use of the PEG layer to reduce uptake by RES macrophages of the **mononuclear phagocyte system (MPS)**. Since then, other LNP drug delivery systems have emerged, such as MARQIBO®, ALOCREST® and BRAKIVA® for the delivery of the anti-cancer drugs, **vincristine**, **vinorelbine** and **topotecan**, respectively. Unfortunately, the clinical impact of drug-loaded LNPs has been relatively limited in spite of their potential for re-purposing the drugs concerned and giving them new life as APIs. A primary reason for this is that nanoparticle biodistribution and drug pharmacokinetic behaviour has not proved as understandable, controllable or as beneficial as originally anticipated.

14

Chemical Biology and Advanced Diagnostics Leading to Precision Therapeutic Approaches

14.1 General introduction

Advanced diagnostics are the result of multidisciplinary problem-driven research that involves the fields of nanotechnology, and cellular and molecular biotechnology, linked to diagnostic imaging modalities such as **magnetic resonance imaging (MRI)**, **computed tomography (CT)** and **nuclear medicine imaging**, which includes **single-photon emission computed tomography (SPECT)** and **positron emission tomography (PET)**. In this chapter, we shall look at how chemical biology helps to turn this knowledge into advanced diagnostics with potentially profound ways to help manage the treatment of patients.

14.2 MRI basic principles leading to diagnostic applications

MRI is an imaging technique that relies on the analysis of the ^1H-NMR behaviour of water molecules, analysed by MRI scanner in different tissue environments, as a function of whole body (x, y, z) Cartesian coordinates (Figure 14.1). The resulting high-resolution images are typically presented as 2D sectional grey-scale images where volumes of highest water **relaxivities** (see Equation 14.1) appear as lighter areas, and volumes of lower water relaxivities appear as darker areas (Figure 14.1).

$$S_1 = S_0 \left[1 - \exp(-xT_1) \right] \tag{14.1}$$

$$R_1[C] = 1/T_1[C] + \alpha_1 \cdot C \tag{14.2}$$

where S_0 is the **resting state bulk magnetisation signal** along the z-axis for a given "pool" of equivalent water molecules *in vitro* or *in vivo*. Otherwise, S_1 is the **bulk magnetisation signal observed along the z-axis** as a function of time, T_1, the spin-lattice relaxation time constant and R_1 is the corresponding **concentration dependent water relaxivity rate**. Both α_1 and C are constants of proportionality in Equation (14.2). Importantly, while MRI whole body scanners are routine clinical tools, MRI scans are of limited value in managing disease treatment unless tissue

Essentials of Chemical Biology: Structures and Dynamics of Biological Macromolecules In Vitro *and* In Vivo, Second Edition. Andrew D. Miller and Julian A. Tanner.
© 2024 John Wiley & Sons, Inc. Published 2024 by John Wiley & Sons, Inc.
Companion Website: www.wiley.com/go/miller/essentialschembiol2

a)

b)

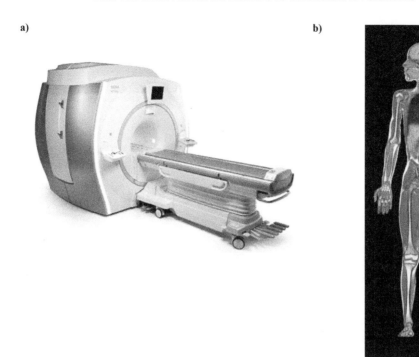

Figure 14.1 MRI in the clinic. (a) Image of standard clinical MRI machine with a flat-bed device with which to transfer a patient to the centre of the magnet for MRI scanning; (b) Whole body MRI scan shown as a 2D "slice". Internal organ resolutions hover around 4–5 µm. (Simon Fraser/Science Source; AroPhoto/Adobe Stock).

pathologies can be clearly identified beyond reasonable doubt at high-resolution. This is not always easy, hence the need to use contrast agents.

14.2.1 Small molecule positive contrast agents

Small molecule (<500 Da) contrast agents initially enhance MR image contrast definition and improve on organ visibility. Localities where contrast agents accumulate are in effect "painted", giving enhanced visibility (Figure 14.2). Mainstream small molecule **positive contrast agents** today are, for example, **Magnevist**™ (Bayer) and **Gadovist**™ (Bayer). All such agents act through the temporary coordination and release of water molecules to the central rare earth metal gadolinium (III) ion. While coordinated, a water molecule is subject to paramagnetic relaxation such that the molecular T_1 time constant is reduced (Figure 14.3). Provided that sufficient water molecules from a given "pool" of equivalent bulk water molecules are processed in this way, then the T_1 time constant of this pool of bulk water is reduced and the corresponding R_1 relaxivity rate increases according to Equations (14.1) and (14.2). In consequence, this pool of bulk water resonates with greater intensity than neighbouring pools of water not subject to paramagnetic relaxation and, as such, appears as a bright field zone in MRI scan. In the event that positive contrast agents accumulate in tissue pathologies, such as tumours, then MRI becomes an ideal technique to identify the locations and regional dimensions of such tissue pathologies, anywhere in the human body, including in the brain. Small molecule contrast agents are unable to achieve this convincingly, consequently there is room for new solutions from nanomedicine.

14.2.2 Imaging nanoparticles

In terms of cancer nanomedicine, there is one factor that is very much in favour of imaging nanoparticle use. Nanoparticles administered in the bloodstream (i.v. administration) are able to accumulate in certain pathologies, such as tumours, due to the **enhanced permeability and retention (EPR)** effect (see Figure 13.30). The EPR mediated accumulation of nanoparticles in tumours apparently takes place due to the presence of highly permeable blood vessels in tumours with large fenestrations (>100 nm in size), a result of rapid, defective angiogenesis. In addition,

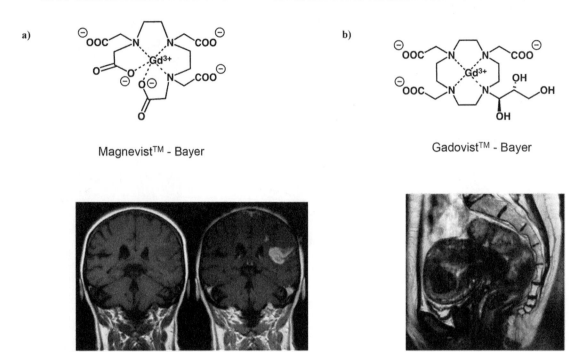

Figure 14.2   MRI contrast agent assisted imaging. (a) Small molecule gadolinium (III) chelate (**Magnevist™**) (top), used routinely as a positive contrast agent to image **blood brain barrier** (**BBB**) lesions (bottom) BY-SA 3.0 (Hellerhoff, Wikipedia Commons/CC); (**b**) Small molecule gadolinium (III) chelate (**Gadovist™**) (top), used routinely as a **positive contract agent** to image uterine fibroids (bottom).

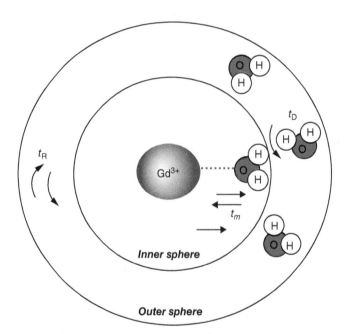

Figure 14.3   Mechanism of gadolinium (III) ion mediated paramagnetic relaxation of water molecules. Schematic demonstrates how inner sphere water molecule chelation to a paramagnetic gadolinium (III) ion results in enhanced relaxation of ^1H-resonance signals of water (β- to α-spin states) by magnetisation transfer over time, t_M. This enhanced relaxation results in higher local water ^1H-resonance signal intensities, resulting in bright-field images in a clinical MRI scan wherever gadolinium (III) ions are accumulated (temporarily or otherwise).

tumours are characterised by dysfunctional lymphatic drainage that helps the retention of nanoparticles in the tumour for long enough to enable local nanoparticle disintegration in the vicinity of tumour cells. The phenomenon has been used widely to explain the relative efficiency of nanoparticle and macromolecular drug accumulation in tumours.

Imaging nanoparticles are best described in terms of the **ABCD** nanoparticle paradigm (see Section 13.2.6), where the A-component is now an imaging contrast agent rather than a traditional **active pharmaceutical ingredient** (**API**). In one good example, small molecule positive contrast agent Gadovist™ was effectively converted into a gadolinium (III)-lipid (Gd.DOTA.DSA) lipid by the attachment of two long-chain, hydrophobic fatty acids. Thereafter, this gadolinium lipid was formulated into **Gd-imaging lipid-based nanoparticles** of the **ABC** nanoparticle type (Gd-**ABC** imaging LNPs) by a pre-modification self-assembly process (Figure 14.4). Similarly, this gadolinium (III)-lipid was formulated into Gd-imaging LNPs of the **ABCD** nanoparticle type (Gd-**ABCD** imaging LNPs), too, using an equivalent pre-modification, self-assembly process (Figure 14.5). In this case, the first receptor-specific targeting ligand (**D**-component) of choice was **folate**, which targets **folate receptors** on the surfaces of, for example, ovarian, breast or prostate cancer tumour cells. The detection of such cancer cells by positive contrast MRI may also be possible using X-ray radiation treatment, owing to the powerful X-ray scattering properties of gadolinium (III) ions internalised into cells. Hence, these Gd-**ABC/D** imaging LNPs are potentially excellent clinical nanotechnology tools for the early detection and diagnosis of primary and metastatic cancer lesions. It remains to be seen how effective these are at enabling cancer treatment as cancer clinical trials are performed.

Otherwise, a small number of **polymer-based nanoparticle** (**PNP**) **systems** have also been prepared with chelated gadolinium (III) ions to act as positive contrast agents for tumour detection in clinic, as shown (Figure 14.6). On the other hand, chelated gadolinium (III) ions might be replaced instead by an encapsulated **superparamagnetic iron oxide** (**SPIO**) core particle (**A**-component), lipid coated to confer biological compatibility and function. So, too, inorganic "hard" **ultra-small-superparamagnetic iron oxide particles** (**USPIOs**) (FERUMOXTRAN-10®) can be used directly *in vivo* without the need for lipid coating. Irrespective, these particular nanoparticles are able to act as MRI negative contrast agents, the exact opposite of chelated gadolinium (III) ions. The reason for this is that SPIOs enable paramagnetic relaxation of water molecules by influencing water ^1H resonance spin-spin relaxation, T_2, time constants instead of ^1H resonance spin-lattice relaxation, T_1, time constants. In consequence, any pool of bulk water subject to SPIO paramagnetic relaxation, resonates with much lower intensity than neighbouring pools of water, not exposed to the contrast agent, and as such appears as a dark field zone in MRI scan. As with positive contrast agents, in the event that negative contrast agents accumulate in tissue pathologies, such as tumours, then MRI can also be used to identify the locations and regional dimensions of such tissue pathologies, anywhere in the human body, including in the brain. Indeed, SPIO systems are currently already in use as diagnostic tools for pre-operative stage(s) of cancer.

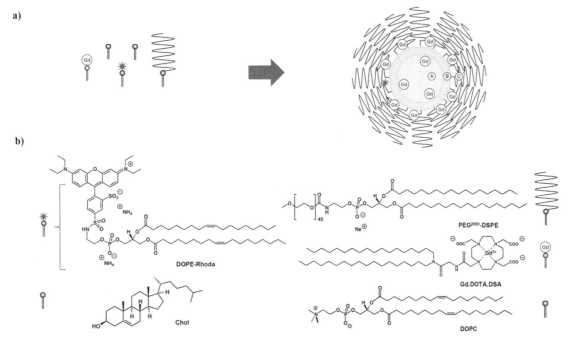

Figure 14.4 Gd-imaging LNPs of the ABC nanoparticle type as MRI positive contrast agent for *in vivo* labelling of tumours by passive targeting. (a) Gd-**ABC** imaging LNPs were prepared by **pre-modification self-assembly** from lipids Gd.DOTA.DSA/DOPC/Chol/ PEG²⁰⁰⁰-DSPE/DOPE-Rhoda in the ratio 30:32:30:7:1 (m/m/m/m/m). Size: ~100 nm; net **ζ-potential** neutral; (b) Illustrated lipids, PEG-lipid, plus gadolinium (III) containing **Gd**-lipid and fluorescent (**red star**) lipids labelled with rhodamine are shown. Glossary of lipid names for Chapter 14 is included on page xvi.

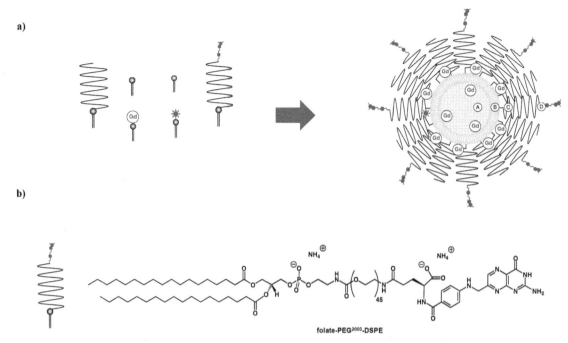

Figure 14.5   Gd-imaging LNPs of the ABCD nanoparticle type as MRI positive contrast agent for *in vivo* labelling of tumours using active folate receptor mediated-targeting of tumour cells. Many different types of tumour cells present folate receptors, for example ovarian and breast cancer cells. (**a**) Gd-**ABCD** imaging LNPs were prepared by pre-modification self-assembly from lipids Gd.DOTA.DSA/DOPC/Chol/PEG²⁰⁰⁰-DSPE/DOPE-Rhoda/folate-PEG²⁰⁰⁰-DSPE in the ratio 30:32:30:4:1:3 (m/m/m/m/m/m). Size: ~100 nm; net ζ-potential neutral; (**b**) The **folate-PEG-lipid** conjugate for receptor specific targeting is shown. Other lipids involved are as shown in Figure 14.4. Glossary of lipid names for Chapter 14 is included on page xvi.

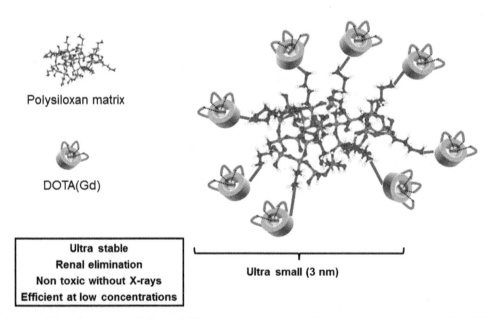

Figure 14.6   Ultra small imaging nanoparticle as positive contrast agent. The illustrated agent is known as **AGuIX™ – NH TherAguix** that is in clinical development for brain tumour imaging and radiation therapy. This imaging nanoparticle crosses the blood brain barrier (BBB) and appears to accumulate in brain tumours, after which radiation therapy is used to ablate tumours on the basis that gadolinium (III) ions are powerful heavy metal ion scattering centres for cytotoxic radiation and so can be used to enhance the ability of incident radiation to destroy tissue pathology in the vicinity of these scattering centres.

14.3 PET/CT and SPECT fundamentals

PET/CT and SPECT are two closely related whole body imaging techniques based around the emission of high energy radiation emission from centres in the body that have accumulated **tracer agents**.

14.3.1 Tracer agents

These are molecular agents intended to track to disease pathology in the hope and expectation of revealing pathologically damaged tissue and/or defective function. In the case of PET/CT, an appropriate tracer agent contains a highly unstable, high energy radioactive isotope such as fluorine-18, that seeks to decay to oxygen-18 with the release of **positron**, $\beta +$, subatomic particles (Figure 14.7a). By contrast, in SPECT, the tracer agent contains high energy radioactive isotopes such as iodine-123 or technetium-99m that decay to telurium-123 and ruthenium-99 respectively with the release of high energy γ-photons (Figure 14.7b and 14.7c). PET/CT and SPECT images then result by superimposing whole body (x, y, z) Cartesian coordinate radiation data with whole body imaging data from CT scanning. Specific examples are given in the cases of cancer and neurological disease detection (Figure 14.8).

In the case of PET/CT, **2-deoxy-2-fluoro-D-glucopyranose (FDG)** is a very popular tracer agent for applications in detection of brain cancers and **Alzheimer's disease (AD)** (Figure 14.8a). FDG is taken up in metabolically active tissues (e.g. tumours) or reflects areas of metabolic decline relative to controls (e.g. in AD disease regions). As a glucose analogue, FDG will also readily transfer across the **blood brain barrier (BBB)** into metabolically active deep brain regions or away from metabolically inactive deep brain regions. In the case of SPECT, tracers can offer more functional variety. For example, the technetium (99mTc) tracer **Sestamibi** provides for real-time images of dynamic heart behaviour (Figure 14.8b), whilst the iodide tracer **Ioflupane** accumulates in deep brain **Parkinson's disease (PD)** lesions (Figure 14.8c).

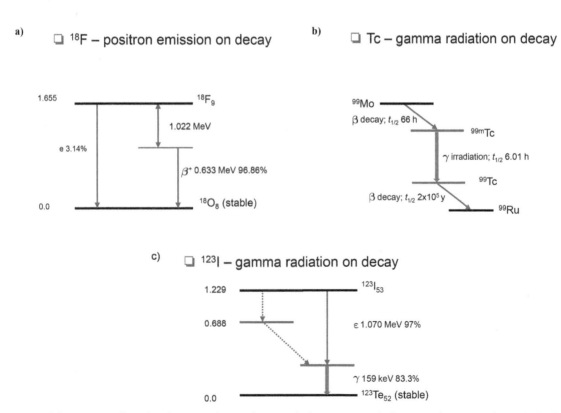

Figure 14.7 High energy radioactive isotopes for positron emission tomography/computed tomography and single-photon emission computed tomography. (a) Positron emission tomography/computed tomography (PET/CT) requires positron emission, which can, for example, be released through decay of fluorine isotope ^{18}F to oxygen isotope ^{18}O. In (**b**) and (**c**) **single-photon emission computed tomography (SPECT)** single γ-ray photon emission, which can, for example, be released through energetic decay of molybdenum isotope ^{99}**Mo** (via the intermediate technetium isotope ^{99m}**Tc**) or iodine isotope ^{123}**I** respectively.

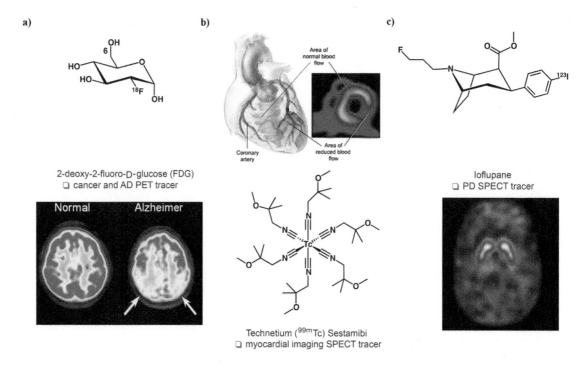

a) 2-deoxy-2-fluoro-D-glucose (FDG)
☐ cancer and AD PET tracer

Technetium (^{99m}Tc) Sestamibi
☐ myocardial imaging SPECT tracer

c) Ioflupane
☐ PD SPECT tracer

Figure 14.8 **Advanced imaging of body organs in disease states.** In (**a**) **PET** tracer **FDG** is illustrated and a standard data set showing how this tracer can be used in conjunction with **CT** to illuminate zones of pathological brain damage due to **Alzheimer's disease** (**AD**) (see yellow arrows). In (**b**) and (**c**) SPECT tracers **Sestamibi** and **Ioflupane** are illustrated and standard data sets are presented to show how both are used respectively to study dynamic blood flow in the heart impaired by heart disease and pathological brain damage from **Parkinson's disease** (**PD**).

An interesting variant of Sestamibi can be bioconjugated to a cyclic arginine-glycine-aspartate peptide that acts as a primary ligand specific for binding to biological $\alpha_{IIb}\beta_3$-**integrin** receptors on platelets involved in blood clot formation. Accordingly, i.v. administration of this bioconjugate should result in "targeted" accumulation *in vivo* in regions where biological $\alpha_{IIb}\beta_3$-integrin receptors are accumulated, hence allowing for the 3D-imaging of **deep vein thrombosis** (**DVT**) anywhere in a body (Figure 14.9). The advantage of such tracer studies is their biological specificity and sensitivity. The primary disadvantage is the cost of maintaining a PET/CT or SPECT facility that requires a cyclotron particle accelerator close at hand to generate the high-energy radioactive isotopes for PET/CT and SPECT that possess half-lives of only minutes to hours.

14.3.2 Nanomolecular tracer agents

LNPs have also been described for PET/CT and SPECT radionuclide delivery to tumour lesions. Typically, these consist of a central liposome that entraps a radionuclide of interest by analogy to drug-**AB/C** nanoparticles and whose surface may be modified by receptor targeting ligands (**D-components**). LNPs of this type have been used to entrap the chelate ^{111}**In-diethylenetriamine-pentaacetic acid** (^{111}**In-DTPA**). These LNPs were found useful for the imaging of solid tumours, particularly those of the head and neck. Moreover, once delivered, the radionuclide itself was able to act not only as a SPECT imaging agent but also as a therapeutic agent to destroy tumour mass by radiation according to the principles of nuclear medicine. This concept of combining imaging with therapy is known as **theranostics**, and will be discussed further as the basis of **precision therapeutic approaches** (**PTAs**) for the treatment of chronic diseases (see Section 14.5).

14.4 Understanding how to control nanoparticle biodistribution behaviour *in vivo*

One of the greatest challenges facing nanomedicine today is the control of nanoparticle biodistribution *in vivo*. Assuming that the nanoparticle is some form of delivery system for an API, either drug or imaging system, then some control of delivery is essential. However, the reality is that the research focus of too much nanomedicine has been on creating

a)

b)

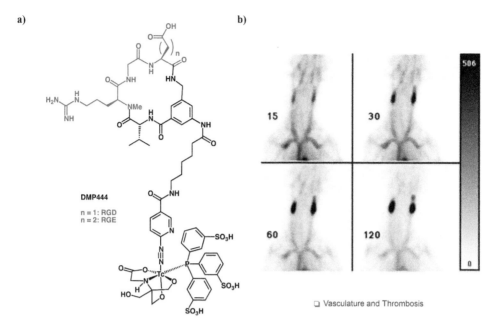

Figure 14.9 Targeted single-photon emission computed assisted imaging. This is made possible by using **arginine-glycine-aspartate** (**RGD**) biological receptor-specific targeting ligands specific to $\alpha_{IIb}\beta_3$-**integrin** receptors. These are found on platelets involved in blood clot formation from **deep vein thrombosis** (DVT). (**a**) Illustrates an RGD conjugate **DMP444** (with RGE control, where E corresponds to glutamate); (**b**) Thromboid lesions in dog vasculature detected over 15, 30, 60 and 120 mins post i.v. infusion of DMP444. DMP444 entered Phase 2 clinical trials for detection of DVT-related human clot formation. (Zhou *et al.*, 2011, Ivyspring International Publisher).

ever more diverse materials to act as API vectors, many of which unsurprisingly turn out to be bioincompatible anyway. Accordingly, there has been little real understanding at the structure-activity level to correlate nanoparticle structural properties with favourable nanoparticle-mediated delivery outcomes.

14.4.1 Designing future theranostic ABC/ABCD nanoparticles

Designing the future of synthetic nucleic acid delivery systems will rely upon a proper appreciation of the problems that currently stand in the way of progress. Logically speaking, synthetic nucleic acid delivery systems are unlikely to see widespread application in clinical gene therapy trials until the biological barriers to delivery success and the biophysical parameters that are most appropriate to overcoming these barriers are mapped out. In practice this means that new biophysical frameworks of understanding are required to link nanoparticle biophysical properties to *in vivo* delivery outcomes. One key approach that has been developed is to seek to correlate the 4T attributes of drug-delivery processes with the general 4S structural parameters of nanoparticles (see Section 13.2.9). Such studies can be made most conveniently using systematic variations of Gd-**ABC** and Gd-**ABCD** imaging LNPs (Figures 14.4 and 14.5). Such studies also open the door to the design and creation of **theranostic lipid-based nanoparticles** equipped with one or more imaging agent, loaded with API(s) for disease treatment and potentially equipped with cell-surface, receptor-specific targeting ligands (Figure 14.10).

14.5 Theranostics

Theranostics is the study of image-associated or image guided therapy which typically makes use of different types of **theranostic nanoparticles** (**TNPs**) (i.e. *thera*py + diag*nostic*). Initial TNPs have been monomodal or bimodal imaging systems. Such TNPs are intended to facilitate the functional delivery of APIs to disease target sites *in vivo*, but also to enable the simultaneous diagnostic/real-time imaging of nanoparticle biodistribution and API pharmacokinetics in order to follow and understand the delivery process in more detail.

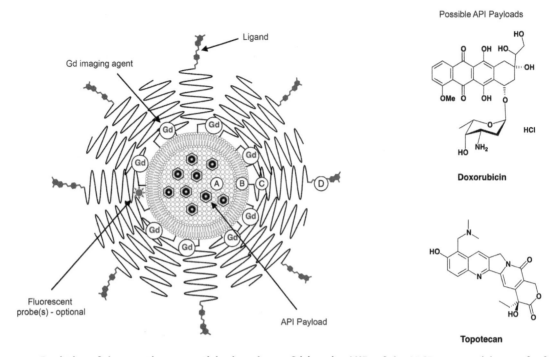

Figure 14.10   **Depiction of theranostic nanoparticles based upon Gd-imaging LNPs of the ABCD nanoparticle type for functional delivery of small molecule drugs to tumours *in vivo*.** The illustrated **theranostic nanoparticles** (**TNPs**) should be prepared by loading Gd-**ABCD** imaging LNPs, previously prepared by standard pre-modification self-assembly process (Figure 14.5), with **active pharmaceutical ingredient** (**API**) payloads such as anti-cancer drugs **doxorubicin** or **topotocan** (right). The resulting drug-**ABC** TNPs should act as a MRI positive contrast agent (given the inclusion of gadolinium (III), **Gd**-lipid) and may also be equipped with optional **near infrared fluorescence** (**NIRF**) (**red star**) lipid probe for NIR fluorescence detection; the indicated receptor-specific targeting ligand is folate.

14.5.1 Multimodal "hard" TNPs

SPIO nanoparticles have proved a sound "template" on which to construct multimodal TNPs able to support imaging involving at least two different imaging techniques such as fluorescence and MRI. For example, dextran-coated SPIO nanoparticles bearing a Cy5.5 **near infrared** (**NIR**) probe and anti-cancer drug **doxorubicin**, have been shown to deliver doxorubicin to cancer cells for treatment and enable real-time imaging of nanoparticle biodistribution and drug pharmacokinetic behaviour, alongside the observation of substantial phenotypic (pharmacodynamic) reductions in tumour size. In addition, bimodal imaging siRNA-**ABC** nanoparticles were realised by coupling siRNA effectors to the dextran coat alongside the Cy5.5 NIR probe. The beauty of enabling bimodal imaging by MRI and NIRF, is that both imaging modalities are available in clinic so that real-time/diagnostic imaging of functional delivery of siRNA effectors to target cells should be possible in future in clinic.

14.5.2 Multimodal "soft" TNPs

What was achieved with inorganic "hard" iron oxide nanoparticles can also be realised using LNPs. For instance, a multimodal imaging theranostic siRNA-**ABC** TNP system was recently described that had been assembled by the stepwise formulation of PEGylated cationic liposomes (prepared using Gd.DOTA.DSA and DOPE-Rhoda amongst other lipids), followed by the encapsulation of **Alexa Fluor 488**-labelled anti-survivin siRNA (Figure 14.11). These multimodal imaging TNPs were found to mediate functional delivery of siRNA to tumours, giving rise to a significant phenotypic (pharmacodynamic) reductions in tumour sizes relative to controls, while at the same time nanoparticle biodistribution (DOPE-Rhoda fluorescence plus MRI) and siRNA pharmacokinetic behaviour (Alexa Fluor 488 fluorescence) could be observed by means of simultaneous real-time imaging. In addition, a $t_{1/2}$-triggered, theranostic pDNA-**ABC** TNP system has been reported using the LMD design principle (see above), but with an expansive level of surface PEGylation (provided by inclusion of 23 mol% of PEG^{2000}-lipid in formulation) to promote a high level of tumour tropism post *in vivo* i.v. administration.

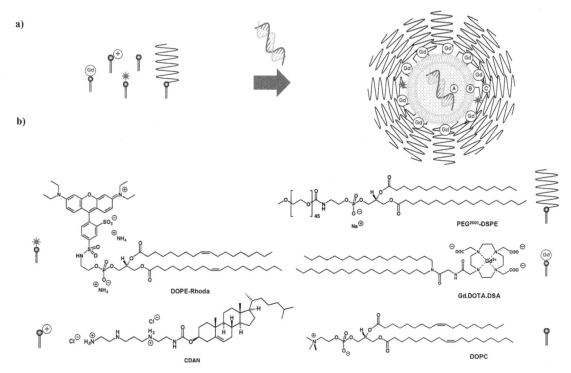

Figure 14.11 **Half-life triggered theranostic nanoparticles based on Gd-imaging LNPs of the ABC nanoparticle type for functional delivery of siRNAs to tumours _in vivo_.** (a) The siRNA-**ABC** TNPs were prepared by combining Gd-**ABC** imaging LNPs, previously prepared by a standard pre-modification self-assembly process, with anti-cancer siRNAs by vortex mixing; (b) The lipid/PEG-lipid composition was Gd.DOTA.DSA/DOPC/CDAN/PEG^{2000}-DSPE/DOPE-Rhoda in the ratio 30:31:31:7.5:0.5 (m/m/m/m/m). The final lipid:siRNA ratio was ~12:1 (w/w) and the final lipid:Gd ratio ~19:1 (w/w). Glossary of lipid names for Chapter 14 is included on page xvi.

This principle of multimodal imaging TNPs for cancer imaging and therapy is certain to grow in importance in both preclinical and clinical studies. Multimodal imaging TNPs should offer substantial benefits for cancer diagnosis and therapy going forward, but only in combination with further advances in nanoparticle platform delivery technologies. What might these advances be and how might they be implemented? As far as imaging nanoparticles are concerned, provided that an _in situ_ diagnosis is all that is required, then current imaging nanoparticle technologies that already accumulate in cancer lesions may well be sufficient. However, for **precision medicine** to really take off, the detection of cancer disease specific biomarkers _in vivo_ is really required. In order to achieve this, considerable attention may well have to be paid to the appropriate design and selection of ligands (**D**-components) for the biological targeting layer (or **D**-layer). Furthermore, TNPs would certainly benefit from triggerability, as described in Chapter 13. Much of previous work on this topic has revolved around change(s) in local endogenous conditions, hence the development of appropriate exogenous stimuli looks to be a real growth area for the future. In principle, all **ABC** and **ABCD** TNPs could be triggered for controlled release through interaction with external stimuli such as light, ultrasound, radiofrequency and thermal radiation from defined sources. Appropriate **ABC** or **ABCD** TNPs of most types should be prepared with ease by pre-modification, self-assembly (Figure 14.12); post-modification, self-assembly (Figure 14.13); pre-modification post-modification self-assembly (Figure 14.14) or by pre-modification, post-coupling, self-assembly approaches (Figures 14.15 and 14.16). The latter processes make use of click chemistry reactions (see also Chapter 13).

14.5.2 Towards precision therapeutic approaches for treatment

Beginning with "soft" organic nanoparticles, **thermally triggered** drug-**ABC** LNPs (now known as THERMODOX®) were described a few years ago based upon DOXIL®. These THERMODOX® nanoparticles were formulated with lipids to encapsulate doxorubicin within thermosensitive, nanoparticle lipid bilayer membranes. Owing to the inclusion of one _lyso_-phospholipid, LNP membranes appeared to become unexpectedly porous, at induced temperatures above 37 °C, allowing for substantial local controlled release of drug. In clinical studies, THERMODOX® was

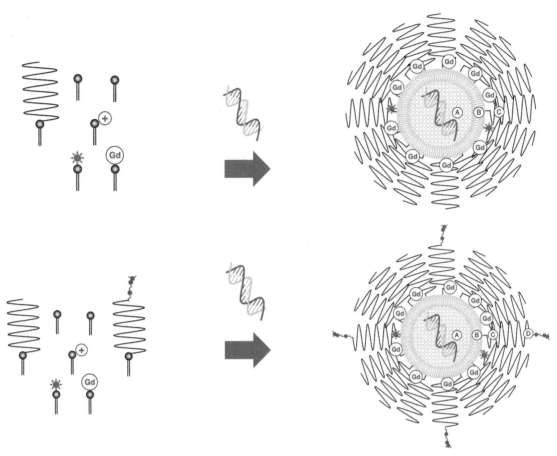

Figure 14.12 **Pre-modification self-assembly of theranostic nanoparticles based on Gd-imaging LNPs of the ABC and ABCD nanoparticle type for functional delivery of siRNAs.** According to this process, siRNA-**ABC** TNPs might be prepared by combining the lipids needed for Gd-**ABC** or Gd-**ABCD** imaging LNP preparation (left-hand side) (including gadolinium (III) lipids [**Gd**-lipids] and fluorescent [**red star**] lipids) with siRNA either simultaneously (kinetic mixing) or in separate steps (thermodynamic mixing), leading to the formation of siRNA-**ABC** TNPs deriving from Gd-**ABC** imaging LNPs (top right-hand side) or siRNA-**ABCD** TNPs deriving from Gd-**ABCD** imaging LNPs, respectively (bottom right-hand side).

administered i.v. in combination with **radiofrequency ablation** (**RFA**) of tumour tissue. In this case, the RFA acted as an exogenous stimulus to induce local tissue hyperthermia (39.5–42 °C) that simultaneously acted as a thermal trigger to enable controlled release of encapsulated doxorubicin from the central aqueous cavity of THERMODOX® nanoparticles. This is the first time that thermally triggered drug-**ABC** LNPs were devised and used in clinical trials. Following this, a most compelling translation of thermally triggered drug-**ABC** LNPs into TNPs was recently achieved with reference to the Gd-**ABC** imaging LNPs described above (Figures 14.4). By introducing THERMODOX® LNP concepts, thermally triggered drug-**ABC** TNPs were derived (Figure 14.17). These thermally triggered drug-**ABC** TNPs might also be described as thermal trig-anostic drug-**ABC** nanoparticle systems.

In parallel with "hard" nanoparticle systems, gold nanoshells, prepared from a thin metal gold shell surrounding a silica core, were shown to have tunable surface plasmon resonances able to absorb NIR radiation from a bespoke laser source in order to damage local cancer tissue by induction of local hyperthermia. Other nanoshell variants have been created with a 10 nm iron oxide layer (MRI negative contrast agent) coating a silica core. These were then upgraded by surface modification with a **NIRF** probe (indocyanine green dye), anti-HER2 antibodies and PEG polymers for biocompatibility. As a result, HER-2 antibody-mediated targeted delivery of these nanoshells to tumour cells could be followed by real-time/diagnostic bimodal MRI/NIRF imaging, after which tumour cells were destroyed by photo-thermal ablation therapy caused by local hyperthermia induced by NIRF laser irradiation. Importantly, these nanoshells were also sufficiently paramagnetic that an external magnetic field could be used to help optimise their accumulation at tumour target sites prior to the application of **photo-thermal ablation therapy**. These nanoshells might be described as targeted trig-anostic drug-**ABCD** nanoparticle systems (or **targeted TNPs**).

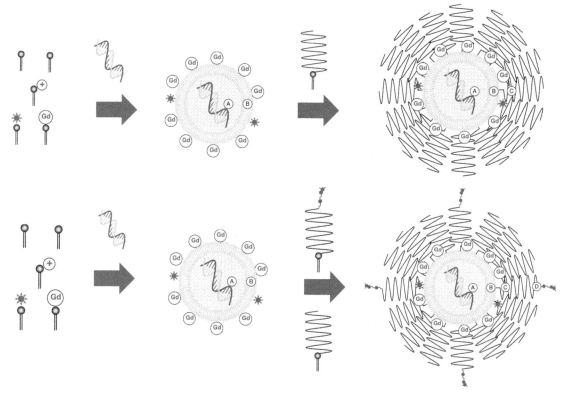

Figure 14.13   Post-modification self-assembly of theranostic nanoparticles based on Gd-imaging LNPs of the ABC and ABCD nanoparticle type for functional delivery of siRNAs. According to this process, **B**-component lipids (left-hand side) (including gadolinium (III) lipids [**Gd**-lipids] and fluorescent [**red star**] lipids) might be used to prepare core siRNA-**AB** TNPs (middle), either by simultaneous combination with siRNA (kinetic mixing) or in separate steps (thermodynamic mixing). Final formulation of siRNA-**ABC** TNPs deriving from Gd-**ABC** imaging LNPs would then require an additional incubation step with PEG-lipids (top right-hand side). Final formulation of siRNA-**ABCD** TNPs deriving from Gd-**ABCD** imaging LNPs would require an additional incubation step with ligand-PEG-lipids and PEG-lipids (bottom right-hand side).

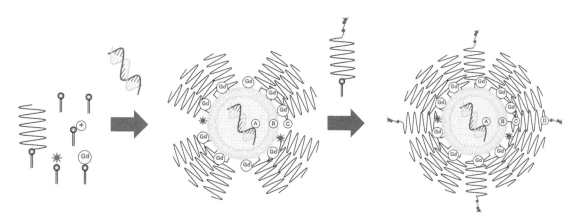

Figure 14.14   Pre-modification post-modification self-assembly of theranostic nanoparticles based on Gd-imaging LNPs of the ABCD nanoparticle type for functional delivery of siRNAs. According to this process, siRNA-**ABC** TNPs deriving from Gd-**ABC** imaging LNPs might be prepared by combining all lipids needed for Gd-**ABC** imaging LNPs preparation (left-hand side) (including gadolinium (III) lipids [**Gd**-lipids] and fluorescent [**red star**] lipids) with siRNA either simultaneously (kinetic mixing) or in separate steps (thermodynamic mixing). Thereafter, final formulation of siRNA-**ABCD** TNPs deriving from Gd-**ABCD** imaging LNPs would require an additional incubation step with ligand-PEG-lipids (right-hand side).

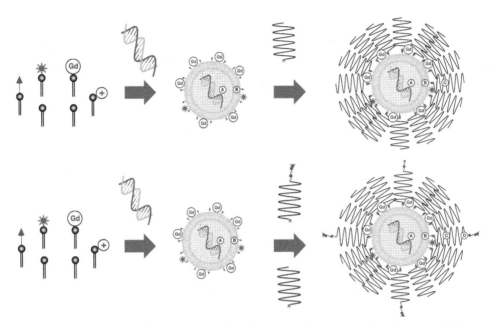

Figure 14.15 **Pre-modification post-coupling self-assembly of theranostic nanoparticles based on Gd-imaging LNPs of the ABC and ABCD nanoparticle type for functional delivery of siRNAs.** According to this process, **B**-component lipids (left-hand side) (including gadolinium (III) lipids [**Gd**-lipids], fluorescent [**red star**] lipids and **click chemistry** [**blue arrow**] lipids) might be used to prepare click chemistry-enabled core siRNA-**AB** TNPs (middle) either by simultaneous combination with siRNA (kinetic mixing) or in separate steps (thermodynamic mixing). Final formulation of siRNA-**ABC** TNPs deriving from Gd-**ABC** imaging LNPs would then require an additional incubation step with **click-complementary PEGs** (top right-hand side). Final formulation of siRNA-**ABCD** TNPs deriving from Gd-**ABCD** imaging LNPs would require an additional incubation step with **click-complementary ligand-PEGs** (and potentially with click-complementary PEGs too) (bottom right-hand side).

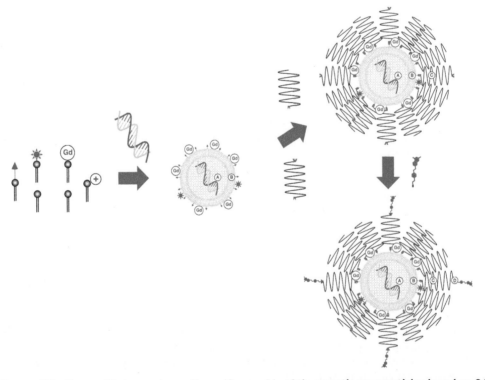

Figure 14.16 **Pre-modification multi-step post-coupling self-assembly of theranostic nanoparticles based on Gd-imaging LNPs of the ABCD nanoparticle type for functional delivery of siRNAs.** As for the left-hand side of Figure 14.15, except that final formulation of siRNA-**ABCD** TNPs deriving from Gd-**ABCD** imaging LNPs would require an incubation step with click-complementary and enabled PEGs (top right-hand side), then an additional incubation step with **click-complementary ligands** (bottom right-hand side).

14.5.3 Devising precision therapeutic approaches for treatment

Since their creation, thermally triggered drug-**ABC** TNPs (Figure 14.17) have led to some potentially very important realisations. One very good example has emerged from pharmacokinetic and pharmacodynamic studies of thermally triggered drug-**ABC** TNPs in a xenograft ovarian cancer model. A key aspect of these studies was to investigate the potential pharmacokinetic and pharmacodynamic effects of **focused ultrasound** (**FUS**) irradiation of tumour xenograft tissues (Figure 14.18). Results were dramatic. Data demonstrated that FUS irradiation of selected tumours resulted in massively enhanced NIRF labelling and magnetic resonance positive contrast imaging of irradiated tumours, compared to the situation with non-irradiated controls, wherein no accumulation of any kind was observed (Figure 14.19). This accumulation of label also correlated with near quantitative drug-delivery into irradiated tumour lesions with the potential for highly selective and potent anti-tumour effects (Figure 14.20). The clear implication was that whatever FUS was doing in physical terms to irradiated tissue was apparently sufficient to guide thermally triggered drug-**ABC** TNPs quantitively from the blood pool, post i.v. injection, into irradiated tumour lesions.

In comparison, if imaging LNPs are introduced to mice by i.v. injection, then they will circulate for an extended period (several hours) and will partially partition into tumour tissue from the blood pool by EPR effect (Figure 14.18). This process is typically very inefficient, and 2–3% at best of an administered dose will appear in tumour tissue, as opposed to elsewhere in the body (Figure 14.18). This is a substantial difference to the situation observed post application of pulsed FUS (Figures 14.19 and 14.20). Indeed, FUS irradiation appears to exert a transformative effect on the irradiated tissue that would appear to be caused by FUS-induction of sub-lethal hyperthermia in irradiated tissue, which changes tumour tissue permeability reversibly from typical EPR levels to **hyperpermeability and retention** (**HPR**) levels. Importantly, this accumulation of label is not only coupled with near quantitative drug delivery into irradiated tumour lesions, but also with very highly selective and potent anti-tumour effects (Figure 14.20). The implication of all the above is that FUS irradiation of tumour tissue has the potential to alter anti-cancer drug pharmacokinetic and pharmacodynamic behaviour

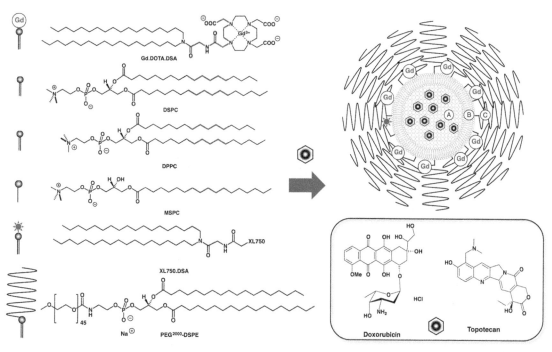

Figure 14.17 **Thermally triggered theranostic nanoparticles based on Gd-imaging LNPs of the ABC nanoparticle type for functional anti-cancer drug delivery to tumours *in vivo*.** In this scheme, **thermally triggered** drug-**ABC** TNPs were prepared by initial preparation of thermally triggered Gd-**ABC** imaging LNPs by pre-modification self-assembly, using PEG-lipid, NIR-fluorescent (**red star**) and gadolinium (III) (**Gd-**) lipids, plus two phospholipids and a *lyso-phospholipid*, MSPC. Final formulation of thermally triggered drug- **ABC** TNPs was then accomplished by encapsulation of anti-cancer drugs (inset, doxorubicin or topotecan). The presence of MSPC weakens the thermal stability of the thermally triggered drug-**ABC** TNPs such that full drug release can be attained in mins by raising ambient temperature from 37 to 42°C. The final composition of the thermally triggered drug-**ABC** TNPs was Gd.DOTA.DSA/ DPPC/DSPC/MSPC/PEG²⁰⁰⁰-DSPE/XL750.DSA using natural and synthetic lipids in the ratio 30:54:5:5:6:0.05 (m/m/m/m/m/m). The final lipid:topotecan ratio was ~59:1 (w/w) whilst the final lipid:doxorubicin ratio was also ~59:1 (w/w). Glossary of lipid names for Chapter 14 is included on page xvi.

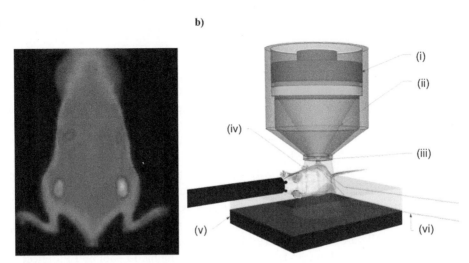

Figure 14.18   Near infrared imaging *in vivo* and focused ultrasound. (a) Typical biodistribution of thermally triggered drug-**ABC** TNPs (monitored by NIRF, see Figures 14.10 and 14.17) several h post i.v. delivery to a mouse. The mouse has two flanking xenograft tumours which are both equally labelled by TNP extravasation caused by the **enhanced permeability and retention (EPR)** effect. Otherwise, considerable label is spread elsewhere (Centelles *et al.*, 2018, Elsevier); **(b)** Schematic of a device used for the application of **focused ultrasound (FUS)** irradiation to one out of the two xenograft tumours. An FUS transducer (i) produces an FUS beam (ii) that is focused on a xenograft tumour (iii) on the right-flank of a mouse (iv). The mouse is positioned on a stage (v) while under sedation. Local tissue temperature rises are monitored by thermocouple (vi).

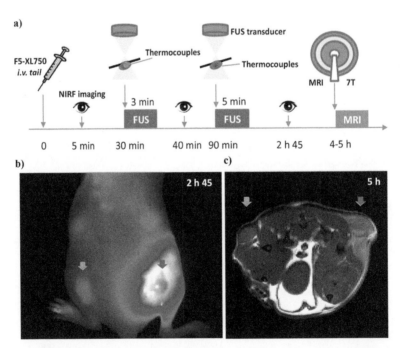

Figure 14.19   Experimental preclinical data in support of a precision therapeutic approach for the treatment of cancer using focused ultrasound. Data shown from an extended FUS experiment with thermally triggered drug-**ABC** TNPs (see Figure 14.18) monitored by NIRF and MRI. **(a)** Schematic diagram documents experimental timings post i.v. administration of thermally triggered drug-**ABC** TNPs, including two FUS bursts, that expose one tumour (on the right flank, **red arrow**) to temporary sublethal heat shock hyperthermia, but not the control tumour (on the left flank, **blue arrow**); **(b)** Imaging data show the effect of two FUS bursts on the accumulation of NIRF into the irradiated tumour (on the right flank, **red arrow**) after 2 h 45 mins, in comparison to the situation with the non-irradiated tumour (on the left flank, **blue arrow**); **(c)** Imaging data show the effect of two FUS bursts on the accumulation of thermally triggered drug-**ABC** TNPs into the irradiated tumour (on the right flank, red arrow) after 5 h, in comparison to the situation with the non-irradiated tumour (on the left flank, **blue arrow**). All data apparently demonstrate that FUS irradiation is sufficient to induce i.v. administered thermally triggered drug-TNPs, with entrapped drug, to accumulate from the blood pool into the irradiated tumour alone and apparently hardly anywhere else in the animal body. (Centelles *et al.*, 2018, Elsevier).

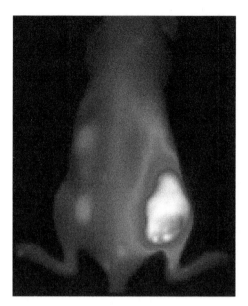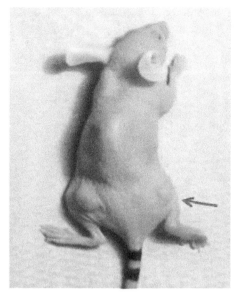

Figure 14.20   Impact of focused ultrasound irradiation on changes tissue permeability. NIRF image (left) of a mouse at the end of the experiment shown above (Figures 14.18 and 14.19) and a corresponding photograph (right). The irradiated tumour (right flank) has accumulated much of the NIRF signal and has been essentially eliminated. The non-irradiated tumour exhibits no NIRF signal and still grows.

dramatically by substantially modifying the biodistribution behaviour of i.v.-delivered drug-delivery nanoparticles (e.g. LNPs, TNPs, etc.). This has led to the proposal for a hypothetical precision therapeutic approach (PTA) for the treatment of cancer in animal models of cancer (Figure 14.21).

Following on from this, the key question to ask about such a hypothesis is how this might be implemented in clinic? Fortunately, this is much less problematic than it might appear. In a clinical setting, the imaging modality of choice is MRI. Therefore, were a clinical PTA set up for the treatment of cancer based on the above preclinical data, then there would be a need for clinical MRI facilities with integrated FUS installations. Fortunately, there is already in clinic a technique known as **magnetic resonance guided focused ultrasound** (**MRgFUS**), which is a pre-existing clinical technique. Bespoke MRgFUS installations have been in use for over a decade, with clinical MRI scanners and bespoke FUS devices

Identification

Using TNPs

Guidance

Image-guided focused ultrasound (IgFUS)
to facilitate delivery of TNPs

Confirmation

Using TNPs

Figure 14.21   Outline of a precision therapeutic approach for cancer treatment in animal model of human cancer. Successful preclinical implementation should improve pharmacokinetics of API delivery, but also pharmacodynamics, even of advanced APIs such as siRNA (see Chapter 13). All this is made possible owing to the creation of TNPs fit for purpose.

for the controlled oblation of fibroid lesions *in utero*, for example. Accordingly, we have been suggesting a potentially viable clinical PTA for cancer treatment as follows:

- **Identification** of tumour lesions (primary or metastatic), achieved using a combination of MRI with imaging LNPs and/or drug-delivery TNPs acting as tumour selective, MRI positive contrast agents.

- **Guidance** of drug-delivery TNPs to these tumour lesions, achieved by the application of MRgFUS to direct TNPs to accumulate in selected tumour lesions.

- **Confirmation** of therapeutic effects, achieved following accumulation of drug-delivery TNPs into the cells of tumour lesions. MRI contrast images of these lesions would be possible to view for as long as tumour cells remain viable.

If such a PTA for the treatment of cancer could be implemented, then our data above suggest that cancer chemotherapy could be transformed in clinic from a crude, frontline approach to cancer treatment, dominated by cytotoxic side effects, into an approach to treatment where anti-cancer drugs are targeted precisely to tumour lesions and nowhere else in the body. Such a PTA for cancer treatment could be expected to dominate the future multi-billion chemotherapy market, leading to substantial reductions in future anti-cancer drug use costs and in future costs of post-treatment, healthcare management of chemotherapy patients. From a patient point of view, chemotherapy would cease to be an object of fear, with long hospital stays and weeks of recovery time, but would become a simple outpatient activity based around clinical MRgFUS installations. MRgFUS-based PTAs for the treatment of other chronic diseases could be similarly transformative in the future.

15

DNA Nanotechnology

15.1 Background

Watson-Crick base pairing in DNA has many unexpected benefits, not least the opportunity to program DNA structure and dynamics on the nanoscale to produce an astonishing range of new nanostructures with many potential uses and applications in fields such as diagnostics, drug delivery, biomolecular structure determination and synthetic biology, amongst others. These advances have been made possible by two clear advances, namely: (1) in the quantitative understanding of DNA thermodynamics and (2) in the bulk preparation of **single-stranded** (**ss**) **DNA**. Fundamentally, DNA self-assembly can be considered as a "bottom-up" assembly process driven by the autonomous interactions of small 2′-deoxyoligonucleotide "fragments" coming together. This allows for finer structural resolution than is possible with a "top-down" assembly process. Another advantage of bottom-up assembly is massive parallelism: billions of DNA sequences can be assembled autonomously in parallel. Hence **DNA nanotechnology** has become an unparalleled tool for nanostructure creation.

15.1.1 From Holliday junctions to double-crossover and paranemic DNA

DNA nanotechnology emerged in the 1980s from Ned Seeman's work on branched junctions, based on the underlying idea that branched junctions might be used to scaffold the 3D organisation of proteins or other macromolecules for X-ray crystallisation. Such thinking led to four-way DNA junction assembly via four 2′-deoxyoligonucleotides that could be assembled in a quadrilateral manner (Figure 15.1a) and then into lattices to aid the assembly of proteins for crystallisation (Figure 15.1b). The result of such thinking was the DNA cube, produced by the self-assembly of six separate DNA strands (Figure 15.2). Thereafter, a major breakthrough in DNA nanotechnology was made possible by going beyond simple Holliday-junction-like junctions (see Chapter 1) into the development of **double-crossover** (**DX**) **motifs** involving two DNA helices linked through strand exchange (Figure 15.3a). Such DX motifs promote geometric rigidity and stability and so were seen potentially as useful for the creation of more extensive DNA nanostructures with higher levels of connectivity and topology. Hence DX motifs were further extended to **triple-crossover** (**TX**) **motifs** (Figure 15.3b), culminating in **paranemic crossover** (**PX**) **motifs** (Figure 15.3c and 15.3d). PX motifs comprise two parallel DNA strands, with their phosphodiester backbones orientated in the same 5′→3′ direction, engaged in base-pair hydrogen bonding with two partially complementary antiparallel DNA strands; the whole takes on the appearance of two parallel DNA double helices linked together by alternating strand segments engaged in complete inter-helical, base-pair hydrogen bonding and strand segments engaged in partial intra-helical base-pair hydrogen bonding (Figure 15.4). Each half-turn of the PX-DNA motif possesses a **wide major groove separation** (**W**) and a **narrow minor groove separation** (**N**) flanking the central dyad axis of the motif. The particular rigidity and stability of a PX motif renders this a significant building block for the construction of a wide range of DNA nanostructures.

Essentials of Chemical Biology: Structures and Dynamics of Biological Macromolecules In Vitro *and* In Vivo, Second Edition. Andrew D. Miller and Julian A. Tanner.
© 2024 John Wiley & Sons, Inc. Published 2024 by John Wiley & Sons, Inc.
Companion Website: www.wiley.com/go/miller/essentialschembiol2

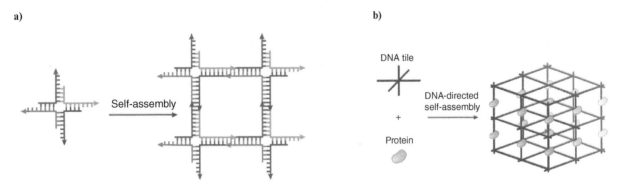

Figure 15.1 DNA four-way junction and DNA scaffold. (a) Self-assembly of a DNA four-way junction through quadrilateral annealing of four 2′-deoxyoligonucleotide strands; **(b)** DNA scaffold to assist protein crystallisation. (N.C. Seeman and H.F. Sleiman, 2017, Springer Nature.)

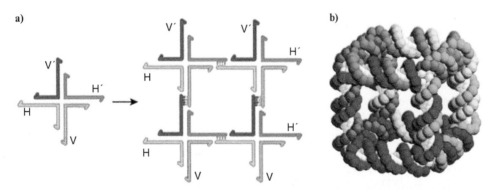

Figure 15.2 DNA cube. (a) A **DNA branched junction** has strands with overhangs H and H′ and overhangs V and V′ which are able to base-pair hydrogen bonds with one another; **(b)** The ligated molecules interconnect to form a DNA cube wherein the structure consists of six interlocked strands. (N.C. Seeman, 2003, Springer Nature.)

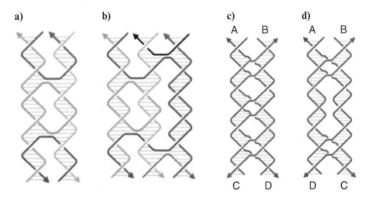

Figure 15.3 Crossover DNA motifs. (a) Double-crossover (DX) motif from double reciprocal exchange; **(b) Triple-crossover (TX)** from two successive double reciprocal exchanges with three helical domains; **(c) Paranemic crossover (PX)** DNA where two double helices in proximity exchange strands; **(d) JX2 DNA**, a variant of PX DNA that lacks crossovers. (X. Wang *et al.*, 2019, adapted from American Chemical Society)

15.1.2 Scaffolded DNA origami

In 2006, there was a breakthrough in DNA nanotechnology from Paul Rothemund in the form of the **scaffolded DNA origami** strategy. According to this strategy, a single long poly-2′-deoxynucleotide scaffold strand, comprising several thousand 2′-deoxynucleotide residues, could be encouraged to fold into desired, patterned structures with the

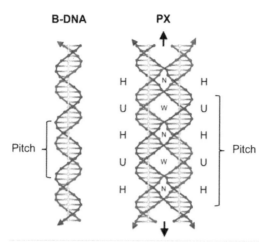

Figure 15.4   A comparison of B-form with paranemic crossover DNA. In the PX DNA motif, the major grooves are indicated by **W**, while minor narrow grooves are indicated by **N** moving along the dyad axis indicated by the arrow. The **U regions** comprise strand segments engaged in partial intra-helical base-pair hydrogen bonding, while the **H regions** comprise strand segments engaged in complete inter-helical base-pair hydrogen bonding. (X. Wang *et al.,* 2019, adapted from American Chemical Society).

aid of multiple complementary **staple strands** (Figure 15.5a). Initially, a **DNA scaffold strand** was obtained from the circular genomic ssDNA of the **M13mp18 virus** (7249 nt residues in length) subject to a simple restriction digest. The linear DNA scaffold strand so formed was then guided into defined folding pathways by the use of staple strands that act to create crossover motifs that also minimise and balance molecular twist strain. For each folding pathway, a 100-fold excess of 200–250 of selected staple strands were mixed and annealed to the same linear DNA scaffold. The results of a few selected folding pathways are shown, with false colouring to enhance the image and emphasise the pathway itself (Figure 15.5b). Scaffolded DNA origami was a breakthrough concept in DNA nanostructure design for three reasons: (1) the DNA scaffold strand does not need to be optimised to avoid any secondary structure, prior to nanostructure assembly; (2) the assembly process is facile, involving only simple mixing of the single-stranded scaffold with a large excess of the staple strands, without any need for equimolar mixing and (3) the approach is relatively inexpensive and simple to implement so that the scaffolded DNA origami method can be adapted easily for numerous applications.

15.1.3 Single-stranded tile (SST) assembly

An alternative approach, without the need for the linear DNA scaffold strand, is **single-stranded tile assembly** or **"brick" assembly**. In this case, **DNA single-strand tiles** (**SSTs**), or "bricks", are prepared with four modular domains that are partially complementary to each other and so form crossover motifs via complementary base-pair hydrogen bonding. The result is a DNA lattice that is akin to a brick wall in appearance (Figure 15.6a). A desired shape may then be assembled from SSTs using a one-pot annealing method to bring together hundreds or thousands of SSTs into association with each other on a molecular canvas for visualisation by **atomic force microscopy** (**AFM**) (Figure 15.6b). This SST assembly approach carries the inbuilt advantage that desired shapes can be designed *de novo* and assembled from SSTs without the need for any linear DNA scaffold strand. Also, in the same way as with the scaffolded DNA origami method, there is no need for careful relative control of SST stoichiometries.

15.2 Three-dimensional DNA nanostructures

Following the initial form of the cube (Figure 15.2), the design of 3D DNA nanostructures has surged in recent years. For instance, there has been a consistent effort to promote the hierarchical self-assembly of polyhedra into larger 3D structures by making use of many copies of identical elements (e.g. three-point star motifs or SSTs). The impact of this

a)

b)

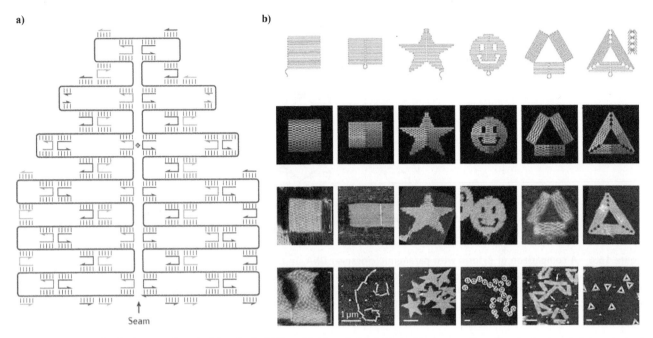

Figure 15.5 Scaffolded DNA origami. (a) A single **DNA scaffold strand** is folded with dozens of smaller staple strands into a desired shape; **(b)** Top row shows the folding pathways for various shapes with dangling curves and loops showing unfolded sequence. The second row shows the bend of helices at crossovers where the helices touch and move away from crossover motifs as helical segments bend apart. The colour indicates the base-pair index along the folding path such that **red** is the first base and **purple** is the 7000th base. The third and fourth rows show **atomic force microscopy** (**AFM**) images. Scale bar for second column from the left is 1 µm and 100 nm for the third to sixth columns. (P.W.K. Rothemund *et al.*, 2006, adapted from Springer Nature.)

a)

b)

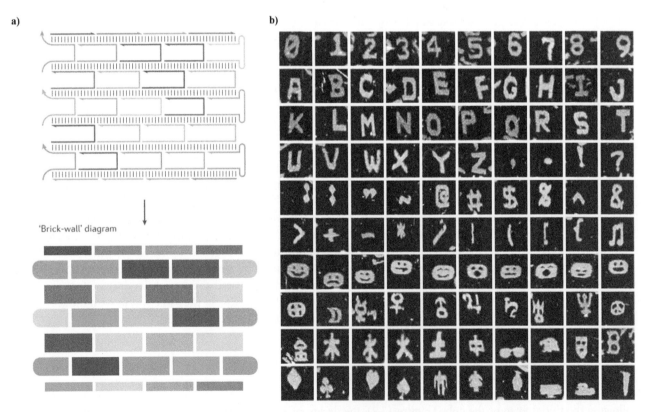

Figure 15.6 Single-stranded tile assembly. (a) Each DNA **single- stranded tile** (**SST**) consists of four domains designed such that distinct SST will arrange into a DNA lattice akin to forming a brick wall; **(b)** Use of a molecular canvas including constituent pixels and excluded pixels to design arbitrary shapes of any design observed by AFM. (N.C. Seeman and H.F. Sleiman, 2017, adapted from Springer Nature.)

approach is shown, making use of three-point star motifs (Figure 15.7a) that are able to assemble into tetrahedra, dodeca-hedra and "buckyballs", as observed and characterised by **cryo-electron microscopy (cryo-EM)** (Figure 15.7b). The scaf-folded DNA origami approach in particular has been found especially adept at exploring 3D, for example when staple strands have the ability to enable evenly spaced crossovers between non-adjacent sections of a linear DNA scaffold strand (Figure 15.8a). These crossovers give rise to multi-layer crossover structures (Figure 15.8b and 15.8c) that may all be visu-alised by **transmission electron microscopy (transmission EM)** with impressive varieties of shape and dense molecular packing (Figure 15.8d and 15.8e). Furthermore, once staple strands can be designed to enable asymmetric crossovers between non-adjacent sections of a linear DNA scaffold strand, such that the distances between crossovers increase along the length of the DNA scaffold strand, then complex curvatures can be induced, giving rise to concentric DNA ring nanostructures instead (Figure 15.9a). In this respect, one of the recent great achievements of this process has been the design of a **nanoflask** consisting of multiple concentric DNA ring nanostructures (Figure 15.9b). This scaffolded DNA origami approach has since been further extended for the preparation of a wide variety of other nanostructures with high curvature, including spherical and ellipsoidal shells, all fully characterised by AFM or transmission EM.

A further extension of the scaffolded DNA origami approach has been made possible using digital 3D mesh com-putational approaches to map out the theoretical pathway for a linear DNA scaffold strand and the location of staple strands to stabilise the pathway, thereby allowing actual polygonal structures to come into being. The design method is amenable to "one-click" 3D printing wherein polygonal shapes can be drawn initially, then appropriate DNA sequences can be determined by the software to order for the realisation of the polygonal design of interest in reality. Real 3D mesh-directed structures from this computational approach are open in conformation with a single helix per polygon edge and a hollow centre (Figure 15.10), quite different from the dense structures described using more conventional scaffolded DNA origami (Figure 15.8). These 3D mesh-directed structures appear to be remarkably stable in physiological buffers, such as **phosphate buffered saline (PBS)**, and as such do not require the typically high cation counterion concentrations needed to stabilise more conventional scaffolded DNA origami nanostructures. Hence, these 3D mesh-directed struc-tures have been used for direct applications in cell culture assays or *in vivo*. Furthermore, designs can be impressively complex and structures such as a polygonal rabbit have even been produced (Figure 15.10l and 15.10k) and visualised by transmission EM.

An alternative wireframe approach has also been described that uses multi-arm junctions with carefully controlled angles, wherein a threading sequence scaffold strand is routed by staple stands to create a larger integrated polygonal structure (Figure 15.11). The approach starts from a desired target pattern that is treated as a planar graph where all edges

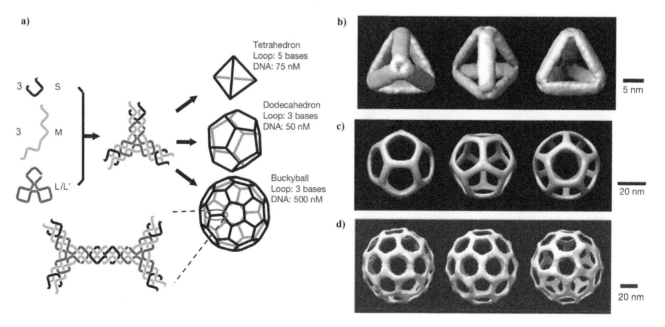

Figure 15.7   Self-assembly of DNA polyhedral. (**a**) Symmetric three-point star motifs assemble into tiles which then assemble into the larger nanostructures in a one-pot reaction; (**b**) Reconstructed **cryo-EM** images of DNA tetrahedron; (**c**) DNA dodecahedron; (**d**) DNA "buckyball". (Y. He *et al.*, 2008, adapted from Springer Nature.)

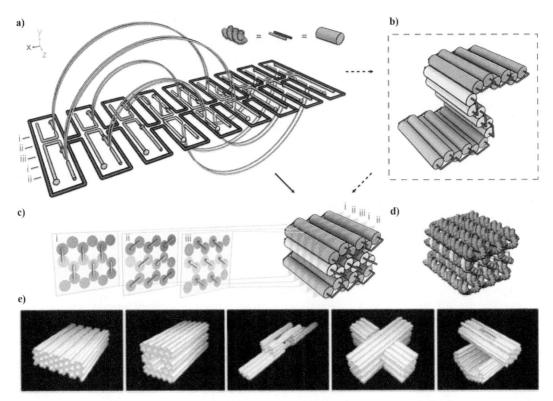

Figure 15.8   Three-dimensional origami. (a) The DNA scaffold strand is shown in dark grey with staple strands in **orange**, **blue** and **white**; (b) Cylinders represent double helices in a "half-rolled up" intermediate; (c) A cross section shows the hexagonal honeycomb arrangement with the staple strand crossovers indicated; (d) Molecular view of layered honeycomb origami structure; (e) Various structures that may be derived from the honeycomb design including (left to right) a monolith, a square nut, a railed bridge, a slotted cross and a stacked cross. (S.M. Douglas et al., 2009, adapted from Springer Nature.)

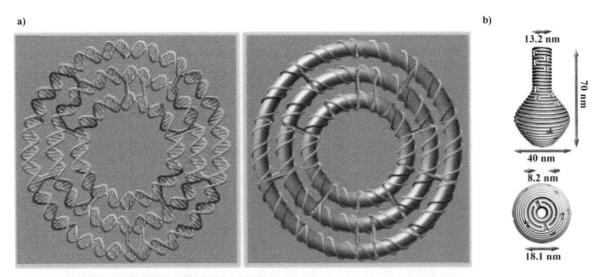

Figure 15.9   Scaffolded DNA origami with complex curvature in 3D space. (a) Bending of the DNA scaffold strand is facilitated through asymmetric distances between crossover motifs such that the distance between crossover motifs is greater in transitioning from inner to outer rings. Helical and cylindrical views of a three-ring concentric structure are shown; (b) Schematic representation of a nanoflask designed through scaffolded DNA origami methodologies with complex 3D curvature. (D. Han *et al.*, 2011, American Association for the Advancement of Science.)

are said to comprise of double helices formed from crossover interactions with staple strands, that are used with effect to mediate crossovers between non-adjacent sections of the linear DNA scaffold strand so as to link polygon edges where appropriate (Figure 15.11b). This approach was particularly effective in quasi-crystalline structure design (Figure 15.11c) and for the creation of 3D polyhedra from 2D structures by direct analogy to traditional paper origami (Figure 15.11d).

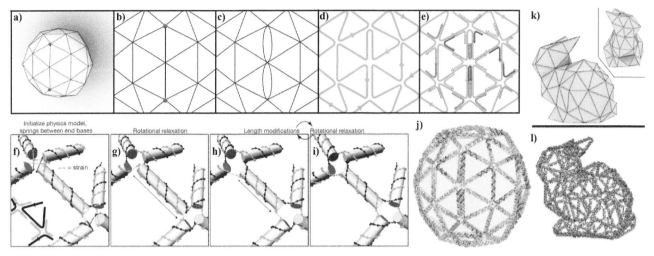

Figure 15.10 Scaffolded DNA origami meshes designed by scaffold routing. (a) A 3D mesh is initially drawn using computer software; (b) Odd-degree vertices are paired; (c) Double edges are introduced; (d) Then an algorithm routes the scaffold (in blue); (e) Thereafter staple strands are designed to follow the scaffold routing; (f, g, h, i) A physical model is used to relax and evenly distribute strain across the design; (j) Final design of structure; (k) Mesh design of rabbit; (l) Now rendered in DNA. (E. Benson et al., 2015, Springer Nature.)

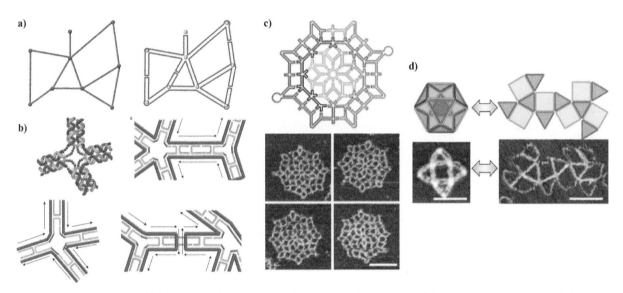

Figure 15.11 Wireframe scaffolded DNA origami with multi-arm junction vertices. (a) An arbitrary structure with lines in **grey** and vertices in **blue**; (b) DNA helical model of a 4xf junction with line model with the DNA scaffold strand shown (in **dark blue**) and staple strands (in **grey** and **cyan**), arrows indicate 5′→3′ orientation of the DNA scaffold strand; (c) An eight-fold quasicrystalline structure design and observation by AFM (d) 3D folded structure showing initial 2D origami format then reconfigured into 3D polyhedra also as observed by AFM. (F. Zhang *et al.*, 2015, Springer Nature.)

Efforts to expand the SST method into 3D have also been rather successful (Figure 15.12). In this case, each SST consists of 32 2′-deoxynucleotide residues which can make crossover motifs with up to four neighbouring SSTs by complementary eight base-pair hydrogen bonding interactions. In effect, each SST comprises four crossover domains or segments of 8 2′-deoxynucleotide residues each: two head domains and two tail domains (Figure 15.12a). Given this, SSTs may then interconnect with a 90° dihedral angle by hybridisation (Figure 15.12b) to form extended 3D structures (Figure 15.12c), such as the 6 × 6 cuboid structure (Figure 15.12d). This cuboid structure may then be adapted to other 3D shapes (Figure 15.12e and 15.12f), wherein each SST block might be omitted at will, enabling the formation of a wide array of different block-like structures (Figure 15.12g) for characterisation by transmission EM.

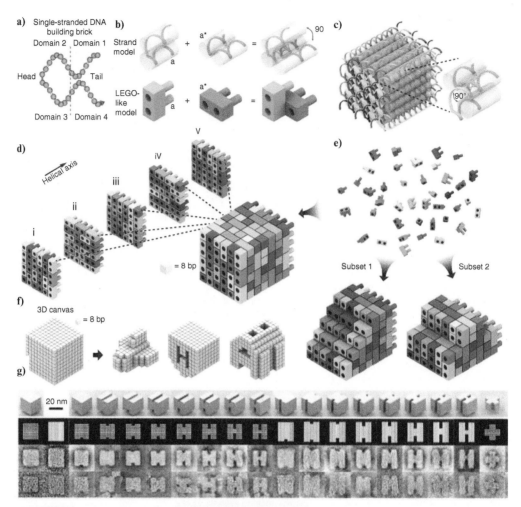

Figure 15.12 "Lego" DNA brick assembly of 3D structures. (a) Each SST consists of 32 2′-deoxynucleotide residues with four 8 2′-deoxynucleotide residue domains: two head domains and two tail domains; (b) Two SSTs come together at a dihedral angle of 90°C with the sequence of the 8 2′-deoxynucleotide residue arms dictating specificity of interactions; (c) A helical model of an 8 × 8 cuboid structure built up from the SSTs; (d) SSTs come together in alternating horizontally and vertically aligned planes, with different sequences indicated by different colours; (e) The entire pool forms a cuboid structure, but when particular SSTs are omitted then differ-ent subsets create different structures; (f) The canvas allows for the design of a variety of shapes; (g) Shapes observed by transmission EM. (Y. Ke *et al.*, 2012, American Association for the Advancement of Science.)

15.3 Dynamic DNA nanostructures

The move to responsive dynamic from static DNA nanostructures is fascinating. In this respect, a key first step was the development of **DNA tweezers**, a simple small DNA nanostructure within which the DNA structure itself provided the energy source for reversible motion (Figure 15.13). The DNA tweezer design was based on three 2′-deoxynucleo-tides, two of which have floppy overhang sequences (Figure 15.13a). The tweezers were labelled with a FRET pair of **tetrachlorofluorescein** (**TET**) and **tetramethylrhodamine** (**TAMRA**). TET is the initial fluorophore, whilst TAMRA is the FRET partner with a fluorescence activity responsive to the distance between the two moieties (see Chapter 4). As a result, FRET allows for observation of the reversible closing and opening of the DNA tweezers in real time.

In a nutshell, the operation of the DNA tweezers was shown to be driven by two complementary 2′-deoxyoligonucleo-tide "**fuel strands**", F and $\overline{\text{F}}$. The addition of F was used to enable initial strand rehybridisation within the DNA tweezer, such that hybridisation of F with both floppy overhang sequences encouraged "closing" of the tweezers, so bringing TET and TAMRA into close proximity for the observation of FRET effects (Figure 15.13b). "Opening" of the tweezers was then made possible by the addition of $\overline{\text{F}}$ that sequesters F into **double-stranded DNA** (**dsDNA**) "**waste product**" F-$\overline{\text{F}}$ so releasing the tweezers to return to their original state, wherein TET and TAMRA are no longer in sufficient proximity

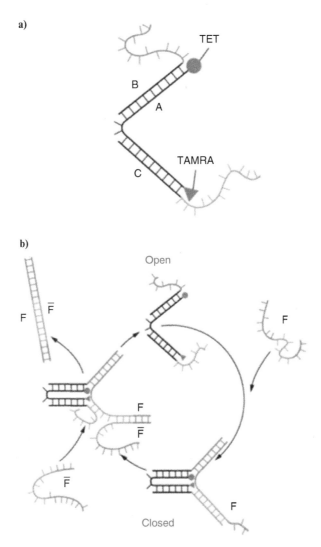

Figure 15.13   DNA molecular tweezers activated by strand displacement. (a) **DNA tweezers** are formed from three 2′-deoxyoli-gonucleotides; (b) **Fuel strand F** is added and hybridises with the free dangling ends of the tweezers so that the **TET** fluorophore and FRET partner **TAMRA** come into proximity. When **fuel strand F̄**, the complement of F, is added then F is sequestered as **double-stranded DNA (dsDNA) "waste product"** F-F̄ allowing the DNA tweezers to return to their original state. (B. Yurke *et al.*, 2000, Springer Nature.)

for the observation of FRET effects. The complete cycle of events required that the "fuel strands" were highly specific to the particular sequences of the DNA tweezer in order to achieve coordinated motion of this molecular machine. Subsequently, such ideas have been applied to facilitate the physical locomotion of many other DNA nanostructures.

Inspired by the DNA tweezers, efforts were then made to simulate the bipedal stepping motion of kinesin along microtubules, a major example of physical locomotion in biology. The result was the design of a progressive **bipedal DNA nanomotor** designed for locomotion by a trailing to leading "foot" mechanism (Figure 15.14). The bipedal DNA nanomotor was set up to move over a DNA 2′-deoxyoligonucleotide residue track (laid down in analogy to a microtubule surface) with short 2′-deoxyoligonucleotide branches for transient attachment of the bipedal DNA nanomotor in the form of a double-stranded **DNA walker** (Figure 15.14a). The double-stranded DNA walker was designed with two overhangs ("feet"). To begin the process of bipedal "walking", the trailing overhang was initially hybridised to a first 2′-deoxyoligonucleotide attachment fuel strand that was also designed to hybridise simultaneously the trailing overhang of the DNA walker with the first branch of the **DNA track** (Figure 15.14b). Thereafter, a second 2′-deoxyoligonucleotide attachment fuel strand was used to hybridise the leading overhang of the DNA walker with the next branch on the DNA track (Figure 15.14c). Following this, the first "step" of the DNA walker was made possible by introduction of a first 2′-deoxyoligonucleotide detachment fuel strand to sequester the first attachment fuel strand as a dsDNA "waste product" (Figure 15.14d). The next "step" of the walk was then made possible by supplying more of the first

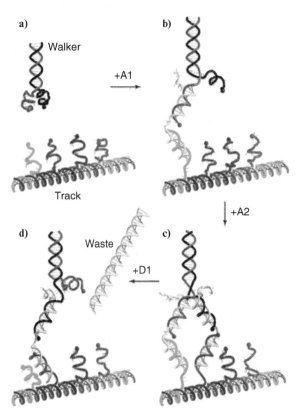

Figure 15.14 **Locomotive DNA walker.** (a) Unbound walker shown separately from the **DNA track**; (b) First attachment fuel strand (**A1, yellow**) is added, enabling hybridisation to the first **green** branch of the track; (c) Second attachment fuel strand (**A2, pink**) is added, enabling hybridisation with the **purple** branch of the track; (d) First detachment fuel strand (**D1, white**) is added, sequestering A1 as a dsDNA "waste product" A1–D1 and locomotive progression along the track. (J.S. Shin and N.A. Pierce, 2004, American Chemical Society.)

attachment fuel strand to hybridise the trailing overhang of the DNA walker to the following branch of the DNA track, followed by the introduction of a second detachment fuel strand to release the leading overhang of the DNA walker and sequester the second attachment fuel strand as a second dsDNA "waste product". By repeating this cycle, the DNA walker could literally "walk" one step at a time the length of the DNA track with a 5 nm step size. This step size may be smaller than the 8 nm stride of kinesin along a microtubule. However, these distances would be easily tunable through redesign of the track scaffold.

Similarly inspired by the DNA tweezers, another example of biomimetic physical locomotion was the implementation of a multi-step organic synthetic chemistry process (Figure 15.15) involving a series of amine acylation reactions intended to mimic peptide/protein synthesis on the ribosome (Figure 15.15). The approach closely mimicked concepts of ribosomal translation, making use once again of a **DNA track** (analogous to mRNA) with three different binding sites (C1, C2 and C3; analogous to mRNA codons) intended to bind three different substrates (S1, S2, S3). Associated enzymatic system comprising a **DNAzyme** was also involved, integrated into a DNA walker (W) once again. Substrates S1, S2 and S3 each consisted of N-hydroxysuccinimidyl esters with 2′-deoxyoligonucleotide strands attached (analogous to tRNA anticodon loops) separated from each other by ribonucleotide residue pairs whose fissile phosphodiester links were introduced for hydrolysis by the DNAzyme mentioned above. The whole reaction sequence process was initiated by the binding of an initiator strand S0 to an "initiation site" on the DNA track. Once done, S1, S2 and S3 were invited to bind to their respective "codon" sites on the DNA track (Figure 15.15a). Following this, W was then added, which bound first to S0 and then to a D1 sequence in S1, via its complementary D1′ sequence, so enabling W to translocate from binding S0 to S1. At this point, W's DNAzyme cleaved the fissile phosphodiester link of S1, while its terminal amino functional group performed a nucleophilic attack on the N-hydroxysuccinimidyl ester of S1 to complete the first aminoacylation reaction. Thereafter, the D1′ sequence of W then bound a complementary D1 sequence in S2, so enabling W

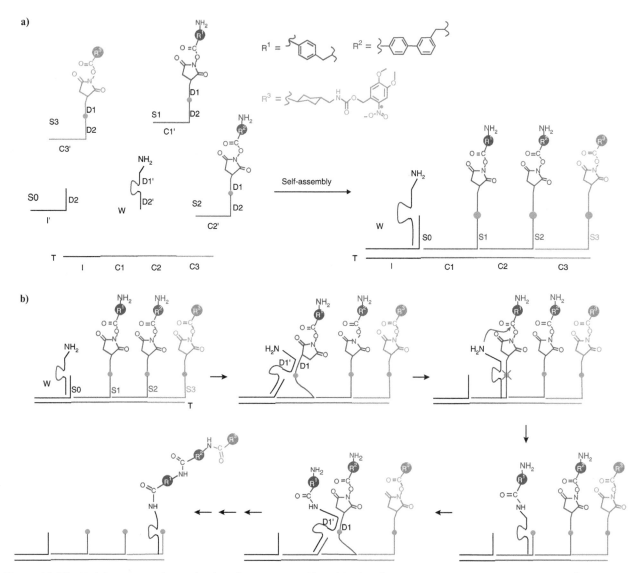

Figure 15.15 Multistep organic synthesis using a DNA walker. (a) The ribosome mimetic system comprises six 2′-deoxyoligo-nucleotide strands: a DNA track, an initiator strand S0, a DNA walker strand W, with **DNAzyme** inbuilt, and three substrate strands, S1, S2 and S3. The green dot represents the site of two ribonucleotide residues with a fissile phosphodiester link for cleavage by the DNAzyme (purple region); **(b)** After self-assembly, the reaction proceeds via a multistep process of hybridisation, phosphodiester link cleavage and aminoacylation to generate a triamide product. (Y. He and D.R. Liu, 2010, Springer Nature.)

to translocate from binding S1 to S2, and setting up W's DNAzyme to hydrolyse the phosphodiester link of S2 while its terminal amino functional group effected a second aminoacylation reaction. A final round of translocation of W to S3 concluded with W bound to S3 while its terminal amino functional group effected a third and final aminoacylation reaction, thereby bringing the multi-step series of aminoacylation reactions to an end (Figure 15.15b). This entire one-pot, three-step reaction process was found to occur spontaneously without any external intervention, at a single temperature and at a single pH. Finally, reaction intermediates and products were characterised successfully by mass spectrometry, and subsequently a variety of different products were generated by interchanging R functional groups associated with S1, S2 and S3. The beauty of this whole approach is that a single DNA mechanical device was able to perform multiple synthetic chemistry activities under mild conditions without the need for external assistance, arguably given the favourable energetics of RNA phosphodiester link cleavage.

Following the examples above, the reader should be confident that many new dynamic DNA nanostructures can and will be designed for different applications going forward, especially to mediate efficient, mild organic chemistry reactions.

15.4 Biomedical applications of DNA nanostructures

A key later development of DNA nanotechnology has been the drive towards potential use of DNA nanostructures in diagnostic and therapeutic applications. One seminal breakthrough in this regard has been the development of logic-gated **DNA "nanorobots"** for the functional delivery of therapeutics to specific cells (Figure 15.16). These typically comprise a scaffolded DNA origami nanostructure which encloses a therapeutic payload such as a protein, antibody fragments or gold nanoparticles. The DNA nanostructure is gated or triggered for therapeutic agent release by **an aptamer lock mechanism** which makes use of a surface **DNA aptamer**, specific for a given **target cell antigen**, hybridised initially to a complementary strand (Figure 15.16b). In the absence of the target cell antigen (e.g. a certain target cell specific receptor), the DNA nanostructure is stable. In the presence of the target cell antigen, the surface DNA aptamer prefers to bind to the antigen and in the process releases its complementary strand, so triggering the DNA nanostructure to open and release the therapeutic agent (Figure 15.16b and 15.16c). Thus far, DNA nanorobots have been shown to mediate functional delivery of therapeutic agents to target cells in tissue culture *in vitro*, but not *in vivo*.

One other example of a DNA nanostructure created with therapeutic applications in mind has been prepared with the **class 2, clustered, regularly interspaced, short palindromic repeat (CRISPR)/Cas9** for gene editing system (Figure 15.17) (see Chapter 12). **Rolling circle amplification (RCA)** was used with palindromic sequences to drive the synthesis and self-assembly of a DNA nanostructure known as a "**nanoclew**", with clew originating from the word for a 'ball of thread'. The nanoclew sequence was designed to be partially complementary to **guide RNA (sgRNA)** to facilitate

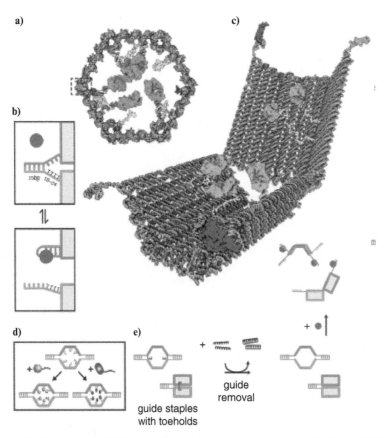

Figure 15.16   An aptamer-gated nanorobot. (a) Cross section view of nanostructure showing the scaffolded DNA origami nanostructure (grey) carrying a **therapeutic protein payload (pink)** held by linker sequences **(yellow)**; **(b)** Nanostructure delivery is mediated by the **aptamer lock mechanism**. The **DNA aptamer (blue)** is initially hybridised to a complementary strand **(orange)**. When a **target cell antigen (red)** comes into view, the aptamer preferentially binds to the antigen, so opening of the nanostructure; **(c)** View of opened nanorobot showing where the target cell antigen **(red)** is bound and locations of a protein therapeutic payload **(pink)**; **(d)** Gold nanoparticles **(yellow)** and antibody Fab fragments **(pink)** can also be loaded inside the closed **DNA nanorobot**; **(e)** Guide staples bear toeholds to help assemble the DNA nanorobot in readiness for the **aptamer lock mechanism**. (S.M. Douglas *et al.*, 2012, American Association for the Advancement of Science.)

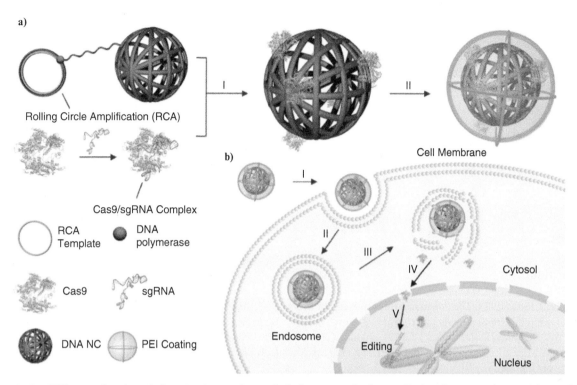

Figure 15.17 **DNA nanoclew-based class 2, clustered, regularly interspaced, short palindromic repeat (CRISPR)/Cas9 delivery system.** (a) The **nanoclew (red)** is synthesised by **rolling circle amplification (RCA)** from a DNA template then loading with Cas9/sgRNA and encapsulated in **polyethyleneimine (PEI)**; (b) The nanoclew enters target cells by endocytosis followed by PEI-assisted endoso-molysis that then enables Cas9/sgRNA to be transported to the cell nucleus for gene editing. (W. Sun *et al.,* 2015, John Wiley & Sons.)

encapsulation of the Cas9/sgRNA complex (Figure 15.17a). Thereafter the whole complex was coated with the cationic polymer **polyethylenimine (PEI)** to help cellular uptake and induce endosomal escape (Figure 15.17b). In this case, the nanoclew was shown to mediate functional delivering of CRISPR/Cas9 to the nuclei of target cells *in vitro*, and then to target cells in xenograft tumours *in vivo*, following direct intra-tumoural injection.

Other biomedical applications of DNA nanostructures are emerging all the time now and the interested reader is pointed to the general reading lists to be up to date with the latest state of the art. Without doubt, DNA nanotechnology has a firm foundation and it looks likely that this field will continue to expand by making use of alternatives to standard 2′-deoxynucleotide residues to enable preparation of ever more diverse DNA nanostructures. Hence, the field is certain to have a significant impact in the realisation of novel therapeutics, diagnostics and imaging agents in years to come.

Bibliography

Chapter 1

General Reading

Aspinall, G.O. (ed.) (1982). *The Polysaccharides*, 1. New York, NY, USA: Academic Press.

Aspinall, G.O. (ed.) (1983). *The Polysaccharides*, 2. New York, NY, USA: Academic Press.

Atkins, P.W., De Paula, J. and Keller, J. (2019). *Physical Chemistry*, 11e. Oxford, UK: Oxford University Press.

Blackburn, G.M., Egli, M., Gait, M.J. and Watts, J.K. (2022). *Nucleic Acids in Chemistry and Biology*, 4e. Cambridge, UK: Royal Society of Chemistry.

Brändén, C.-I. and Tooze, J. (1999). *Introduction to Protein Structure*, 2e. New York, NY, USA: Garland Publishing.

Cevc, G. (ed.) (2007). *Phospholipids Handbook*, 2e. New York, NY, USA: Taylor & Francis.

Creighton, T.E. (1993). *Proteins: Structures and Molecular Principles*, 2e. New York, NY, USA: W.H. Freeman & Co.

Fersht, A.R. (2017). *Structure and Mechanism in Protein Science: A Guide to Enzyme Catalysis and Protein Folding*, 4e. Hackensack, NJ, USA: World Scientific Publishing.

Kennedy, J.F. and White, C.A. (1983). *Bioactive Carbohydrates in Chemistry, Biochemistry and Biology*. Chichester, UK: Ellis Horwood Ltd.

Lesk, A.M. (1991). *Protein Architecture A Practical Approach*. Oxford, UK: IRL Press at Oxford University Press.

Rees, D.A. (1977). *Outline Studies in Biology Polysaccharide Shapes*. London, UK: Chapman & Hall.

Sinden, R.R. (1994). *DNA Structure and Function*. San Diego, CA, USA: Academic Press.

Small, D.M. (1986). *Handbook of Lipid Research-4: The Physical Chemistry of Lipids*. New York, NY, USA: Plenum Press.

Sternberg, M.J.E. (Ed) (1996). *Protein Structure Prediction A Practical Approach*. Oxford, UK: IRL Press at Oxford University Press

Voet, D., Voet, J.G. and Pratt, C.W. (1999), *Fundamentals of Biochemistry*, New York, NY, USA: John Wiley & Sons.

Voet, D., Voet, J.G. and Pratt, C.W. (2016). *Fundamentals of Biochemistry: Life at the Molecular Level*, 5e. New York, NY, USA: John Wiley & Sons.

Extra Reading

Arnott, S., Fulmer, A., Scott, W.E., Dea, I.C., Moorhouse, R. and Rees, D.A. (1974). The agarose double helix and its function in agarose gel structure. *J. Mol. Biol.* 90: 269–284 (pdb: **1aga**).

Arnott, S., Scott, W.E., Rees, D.A. and McNab, C.G. (1974). Iota-carrageenan: molecular structure and packing of polysaccharide double helices in oriented fibres of divalent cation salts. *J. Mol. Biol.* 90: 253–267 (pdb: **1car**).

Banner, D.W., Bloomer, A., Petsko, G.A., Phillips, D.C. and Wilson, I.A. (1976). Atomic coordinates for triose phosphate isomerase from chicken muscle. *Biochem. Biophys. Res. Commun.* 72: 146–155 (pdb: **1tim**).

Brown, R.S., Dewan, J.C. and Klug, A. (1985). Crystallographic and biochemical investigation of the lead (II)-catalyzed hydrolysis of yeast phenylalanine tRNA. *Biochemistry* 24: 4785–4801 (pdb: **1tn2**).

Essentials of Chemical Biology: Structures and Dynamics of Biological Macromolecules In Vitro *and* In Vivo, Second Edition. Andrew D. Miller and Julian A. Tanner.

© 2024 John Wiley & Sons, Inc. Published 2024 by John Wiley & Sons, Inc.

Companion Website: www.wiley.com/go/miller/essentialschembiol2

Bushnell, G.W., Louie, G.V. and Brayer, G.D. (1990). High-resolution three-dimensional structure of horse heart cytochrome c. *J. Mol. Biol.* 214: 585–595 (pdb: **1hrc**).

Diamond, R. (1974). Real-space refinement of the structure of hen egg-white lysozyme. *J. Mol. Biol.* 82: 371–391 (pdb: **6lyz**).

Drew, H.R., Wing, R.M., Takano, T., Broka, C., Tanaka, S., Itakura, K. *et al.* (1981). Structure of a B-DNA dodecamer: conformation and dynamics. *Proc. Natl. Acad. Sci. USA* 78: 2179–2183 (pdb: **1bna**).

Egli, M., Minasov, G., Tereshko, V., Pallan, P.S., Teplova, M., Inamati, G.B. *et al.* (2005). Probing the influence of stereoelectronic effects on the biophysical properties of oligonucleotides: comprehensive analysis of the RNA affinity, nuclease resistance and crystal structure of ten 2′-O-ribonucleic acid modifications. *Biochemistry* 44: 9045–9057 (pdb: **1wv5**).

Frier, J.A. and Perutz, M.F. (1977). Structure of human foetal deoxyhaemoglobin. *J. Mol. Biol.* 112: 97–112 (pdb:**1fdh**).

Govil, G. (1976). Conformational structure of polynucleotides around O-P bonds; refined parameters for CPF calculations. *Biopolymers* 15: 2303–2307.

Kannan, K.K., Ramanadham, M. and Jones, T.A. (1984). Structure, refinement and function of carbonic anhydrase isozymes: refinement of human carbonic anhydrase I. *Ann. N.Y. Acad. Sci.* 429: 49–60 (pdb: **2cab**).

Kramer, R.Z., Bella, J., Mayville, P., Brodsky, B. and Berman, H.M. (1999). Sequence dependent conformational variations of collagen triple-helical structure. *Nat. Struct. Biol.* 6: 454–457 (pdb: **1bkv**).

Lionetti, C., Guanziroli, M.G., Frigerio, F., Ascenzi, P. and Bolognesi, M. (1991). X-ray crystal structure of the ferric sperm whale myoglobin: imidazole complex at 2.0 Å resolution. *J. Mol. Biol.* 217: 409–412 (pdb: **1mbi**).

Pan, B., Ban, C., Wahl, M.C. and Sundaralingam, M. (1997). Crystal structure of d(GCGCGCG) with 5′-overhang G residues. *Biophys. J.* 73: 1553–1561 (pdb: **331d**).

Rees, D.A. and Smith, P.J.C. (1975). Polysaccharide conformation. Part IX. Monte-Carlo calculation of conformational energies for disaccharides and comparison with experiment. *J. Chem. Soc.-Perkin Trans.* 2: 836–840.

Seddon, J.M. (1990). Structure of the inverted hexagonal (H_{II}) phase and non-lamellar transitions of lipids. *Biochim. Biophys. Acta* 1031: 1–69.

Tame, J.R. and Vallone, B. (2000). The structures of deoxy human haemoglobin and the mutant Hb Tyrα42His at 120 K. *Acta Crystallogr., Sect. D* 56: 805–811 (pdb: **1a3n**).

Chapter 2

General Reading

PerSeptive Biosystems (1996). *The Busy Researcher's Guide to Biomolecule Chromatography.* Framingham, MA, USA: PerSeptive Biosystems Inc.

Chan, W.C. and White, P.D. (eds.) (2000). *FMOC Solid Phase Peptide Synthesis: A Practical Approach.* Oxford, UK: Oxford University Press.

Gunstone, F. (1996). *Fatty Acid and Lipid Chemistry,* 1e. Glasgow, UK: Blackie Academic & Professional.

Hecht, S.M. (ed.) (1999). *Bioorganic Chemistry: Carbohydrates.* Oxford, UK: Oxford University Press.

Herdewijn, P. (ed.) (2005). *Oligonucleotide Synthesis: Methods and Applications.* Totowa, NJ, USA: Humana Press.

Robyt, J.F. (1998). *Essentials of Carbohydrate Chemistry.* New York, NY, USA: Springer-Verlag.

Seeberger, P.H. (ed.) (2001). *Solid Support Oligosaccharide Synthesis and Combinatorial Carbohydrate Libraries.* New York, NY, USA: John Wiley & Sons.

Voet, D. and Voet, J.G. (1995). *Biochemistry,* 2e, New York, NY, USA: John Wiley & Sons.

Voet, D., Voet, J.G. and Pratt, C.W. (2016). *Fundamentals of Biochemistry: Life at the Molecular Level,* 5e. New York, NY, USA: John Wiley & Sons.

Extra Reading

Bhurruth-Alcor, Y., Røst, T.H., Jorgensen, M.R., Müller, R.M., Skorve, J., Berge, R.K. *et al.*(2010). Novel phospholipid analogues of pan-PPAR activator tetradecylthioacetic acid are more PPARα selective. *Bioorg. Med. Chem. Lett.* 20: 1252–1255.

Carmona, S., Jorgensen, M.R., Kolli, S., Crowther, C., Salazar, F.H., Marion, P.L. *et al.* (2009). Controlling HBV replication *in vivo* by intravenous administration of triggered PEGylated siRNA nanoparticles. *Mol. Pharmaceutics* 6: 706–717.

Chabner, B.A. and Longo, D.L. (eds.) (2005). *Cancer Chemotherapy and Biotherapy: Principles and Practice,* 4e. Philadelphia: Lippincott Williams & Wilkins.

Effenberg, R., Turánek-Knötigová, P., Zyka, D., Čelechovská, H., Mašek, J., Bartheldyová, E. *et al.* (2017). Nonpyrogenic molecular adjuvants based on norAbu-muramyldipeptide and norAbu-glucosaminyl muramyldipeptide: synthesis, molecular mechanisms of action, biological activities *in vitro* and *in vivo. J. Med. Chem.* 60: 7745–7763.

Fedurco, M., Romieu, A., Williams, S., Lawrence, I. and Turcatti, G. (2006). BTA, a novel reagent for DNA attachment on glass and efficient generation of solid-phase amplified DNA colonies. *Nucleic Acids Res.* 34: e22.

Fletcher, S., Ahmad, A., Perouzel, E., Heron, A., Miller, A.D. and Jorgensen, M.R. (2006a). *In vivo* studies of dialkynoyl analogues of DOTAP demonstrate improved gene transfer efficiency of cationic liposomes in mouse lung. *J. Med. Chem.* 49: 349–357.

Fletcher, S., Ahmad, A., Perouzel, E., Jorgensen, M.R. and Miller, A.D. (2006b). A dialkynoyl analogue of DOPE improves gene transfer efficiency of lower-charged, cationic lipoplexes. *Org. Biomol. Chem.* 4: 196–199.

Fletcher, S., Ahmad, A., Price, W.S., Jorgensen, M.R. and Miller, A.D. (2008). Physical characterization of CDAN/DOPE-analogue lipoplexes accounts for enhanced gene delivery. *ChemBioChem* 9: 455–463.

Jorgensen, M.R., Røst, T., Bhurruth-Alcor, Y.A.H., Bohoy, P., Guisado, C., Kostarelos, K. *et al.* (2009). Development of novel lipids for the treatment of the metabolic syndrome and diabetes type II. *J. Med. Chem.* 52: 1172–1179.

Koeller, K.M. and Wong, C.-H. (2000). Synthesis of complex carbohydrates and glycoconjugates: enzyme-based and programmable one-pot strategies. *Chem. Rev.* 100: 4485–4493.

Kolli, S., Wong, S.-P., Harbottle, R., Johnston, B., Thanou, M. and Miller, A.D. (2013). pH-triggered nanoparticle mediated delivery of siRNA to liver cells *in vitro* and *in vivo*. *Bioconjugate Chem.* 24: 314–332.

Ledvina, M., Ježek, J., Šaman, D., Vaisar, T. and Hříbalová, V. (1994). Synthesis of O-[2-acetamido-2-deoxy-6-O-stearoyl- and -6-O-(2-tetradecylhexadecanoyl)-β-D-glucopyranosyl]-(1→4)-N-acetylnormuramoyl-L-α-aminobutanoyl-D-isoglutamine, lipophilic disaccharide analogs of MDP. *Carbohydr. Res.* 251: 269–284.

Lund, J., Stensrud, C., Rajender, R., Bohov, P., Thoresen, G.H., Berge, R.K. *et al.* (2016). The molecular structure of thio-ether fatty acids influences PPAR-dependent regulation of lipid metabolism. *Bioorg. Med. Chem.* 24: 1191–1203.

Mével, M., Kamaly, N., Carmona, S., Oliver, M.H., Jorgensen, M.R., Crowther, C. *et al.* (2010). DODAG; a versatile new cationic lipid that mediates efficient delivery of pDNA and siRNA. *J. Control. Rel.* 143: 222–232.

Nielsen, P.E., Hansen, J.B. and Buchardt, O. (1984). Photochemical cross-linking of protein and DNA in chromatin. 1. Synthesis and application of a photosensitive cleavable derivative of 9-aminoacridine with 2 photoprobes connected through a disulfide-containing linker. *Biochem. J.* 223: 519–526.

Seeberger, P.H. and Haase, W.-C. (2000). Solid-phase oligosaccharide synthesis and combinatorial carbohydrate libraries. *Chem. Rev.* 100: 4349–4393.

Takahara, P.M., Frederick, C.A. and Lippard, S.J. (1996). Crystal structure of the anticancer drug cisplatin bound to duplex DNA. *J. Am. Chem. Soc.* 118: 12309–12321.

Theoclitou, M.-E., Wittung, E.P.L., Hindley, A.D., El-Thaher, T.S.H. and Miller, A.D. (1996). Characterisation of stress protein LysU. Enzymic synthesis of diadenosine 5', 5'''-P^1, P^4-tetraphosphate (Ap$_4$A) analogues by LysU. *J. Chem. Soc., Perkin Trans.* 1: 2009–2019.

Wright, M., Azhar, M.A., Kamal, A. and Miller, A.D. (2014). Syntheses of stable, synthetic diadenosine polyphosphate analogues using recombinant histidine-tagged lysyl tRNA synthetase (LysU). *Bioorg. Med. Chem. Lett.* 24: 2346–2352.

Wright, M. and Miller, A.D. (2013). Quantification of diadenosine polyphosphates in blood plasma using a tandem boronate affinity-ion exchange chromatography system. *Anal. Biochem.* 432: 103–105.

Wright, M., Tanner, J.A. and Miller, A.D. (2003). Quantitative single-step purification of dinucleoside polyphosphates. *Anal. Biochem.* 316: 135–138.

Yingyuad, P., Mével, M., Prata, C., Furegati, S., Kontogiorgis, C., Thanou, M. *et al.* (2013). Enzyme-triggered PEGylated pDNA-nanoparticles for controlled release of pDNA in tumours. *Bioconjugate Chem.* 24: 343–362.

Chapter 3

General Reading

Alberts, B. (2022). *Molecular Biology of the Cell*, 7e. New York, NY, USA: W. W. Norton & Co.

Fersht, A.R. (2017). *Structure and Mechanism in Protein Science: A Guide to Enzyme Catalysis and Protein Folding*, 4e. Hackensack, NJ, USA: World Scientific Publishing.

Green, M.R. and Sambrook, J. (2012). *Molecular Cloning: A Laboratory Manual*, 4e. Cold Spring Harbor, NY, USA: Cold Spring Harbor Laboratory Press.

Sinden, R.R. (1994). *DNA Structure and Function*. San Diego, CA, USA: Academic Press.

Voet, D. and Voet, J.G. (1995). *Biochemistry*, 2e, New York, NY, USA: John Wiley & Sons.

Voet, D., Voet, J.G. and Pratt, C.W. (1999), *Fundamentals of Biochemistry*, New York, NY, USA: John Wiley & Sons.

Voet, D., Voet, J.G. and Pratt, C.W. (2016). *Fundamentals of Biochemistry: Life at the Molecular Level*, 5e. New York, NY, USA: John Wiley & Sons.

Watson, J.D., Baker, T.A., Bell, S.P., Gann, A., Levine, M. and Losick, R. (2014). *Molecular Biology of the Gene*, 7e. New York, NY, USA: Pearson Education.

Extra Reading

Abdul-Wahab, M.F., Homma, T., Wright, M., Olerenshaw, D., Dafforn, T.R., Nagata, K. *et al.* (2013). The pH sensitivity of murine heat shock protein 47 (HSP47) binding to collagen is affected by mutations in the breach histidine cluster. *J. Biol. Chem.* 288: 4452–4461.

Amarasinghe, S.L., Su, S., Dong, X., Zappia, L., Ritchie, M.E. and Gouil, Q. (2020). Opportunities and challenges in long-read sequencing data analysis. *Genome Biol.* 21: 30.

Chen, X., Boonyalai, N., Lau, C., Thipayang, S., Xu, Y., Wright, M. *et al.* (2013). Multiple catalytic activities of *Escherichia coli* lysyl-tRNA synthetase (LysU) are dissected by site directed mutagenesis. *FEBS J.* 280: 102–114.

Lah, M.S., Dixon, M.M., Pattridge, K.A., Stallings, W.C., Fee, J.A. and Ludwig, M.L. (1995). Structure-function in *Escherichia coli* iron superoxide dismutase: comparisons with the manganese enzyme from *Thermus thermophilus*. *Biochemistry* 34: 1646–1660 (pdb: **1isa**).

Pascal, J.M., O'Brien, P.J., Tomkinson, A.E. and Ellenberger, T. (2004). Human DNA ligase I completely encircles and partially unwinds nicked DNA. *Nature* 432: 473–478 (pdb: **1x9n**).

Perona, J.J. and Martin, A.M. (1997). Conformational transitions and structural deformability of EcoRV endonuclease revealed by crystallographic analysis. *J. Mol. Biol.* 273: 207–225 (pdb: **1az0**).

Tong, Y., Jorgensen, T.S., Whitford, C.M., Weber, T. and Lee, S.Y. (2021). A versatile genetic engineering toolkit for *E. coli* based on CRISPR-prime editing. *Nat. Commun.* 12: 5206.

Chapter 4

General Reading

Cantor, C.R. and Schimmel, P.R. (1980a). *Biophysical Chemistry Part II. Techniques for the Study of Biological Structure and Function*. San Francisco, CA, USA: W. H. Freeman & Co.

Cantor, C.R. and Schimmel, P.R. (1980b). *Biophysical Chemistry Part III. The Behaviour of Biological Macromolecules*. San Francisco, CA, USA: W. H. Freeman & Co.

Fleming, I. and Williams, D.H. (2019). *Spectroscopic Methods in Organic Chemistry*, 7e. Cham, Switzerland: Springer.

Haugland, R.P., Gregory, J.D., Spence, M.T.Z. and Johnson, I.D. (2002). *Handbook of Fluorescent Probes and Research Products*, 9e. Eugene, OR, USA: Molecular Probes.

Kaim, W., Schwederski, B. and Klein, A. (2014). *Bioinorganic Chemistry: Inorganic Elements in the Chemistry of Life: An Introduction and Guide*, 2e. Chichester, UK: John Wiley & Sons.

Serdyuk, I.N., Zaccai, N.R. and Zaccai, G. (2017). *Methods in Molecular Biophysics, Structure Dynamics and Function*, 2e. Cambridge, UK: Cambridge University Press.

Walker, J.M. (ed.) (2009). *The Protein Protocols Handbook*, 3e. New York, NY, USA: Humana Press.

Extra Reading

Abdul-Wahab, M.F., Homma, T., Wright, M., Olerenshaw, D., Dafforn, T.R., Nagata, K. *et al.* (2013). The pH sensitivity of murine heat shock protein 47 (HSP47) binding to collagen is affected by mutations in the breach histidine cluster. *J. Biol. Chem.* 288: 4452–4461.

Brejc, K., Sixma, T.K., Kitts, P.A., Kain, S.R., Tsien, R.Y., Ormo, M. *et al.* (1997). Structural basis for dual excitation and photoisomerization of the *Aequorea victoria* green fluorescent protein. *Proc. Natl. Acad. Sci. USA* 94: 2306–2311 (pdb: **1emb**).

Cohen, B.E., McAnaney, T.B., Park, E.S., Jan, Y.N., Boxer, S.G. and Jan, L.Y. (2002). Probing protein electrostatics with a synthetic fluorescent amino acid. *Science* 296: 1700–1703.

Dafforn, T.R., Della, M. and Miller, A.D. (2001). The molecular interactions of heat shock protein 47 (Hsp47) and their implications for collagen biosynthesis. *J. Biol. Chem.* 276: 49310–49319.

El-Thaher, T.S.H., Drake, A.F., Yokota, S., Nakai, A., Nagata, K. and Miller, A.D. (1996). The pH-dependent, ATP-independent interaction of collagen specific serpin/stress protein HSP47. *Prot. Pept. Lett.* 3: 1–8.

Griffin, B.A., Adams, S.R. and Tsien, R.Y. (1998). Specific covalent labeling of recombinant protein molecules inside live cells. *Science* 281: 269–272.

Keller, M., Harbottle, R.P., Perouzel, E., Colin, M., Shah, I., Rahim, A. *et al.* (2003). Nuclear localisation sequence templated nonviral gene delivery vectors: investigation of intracellular trafficking events, of LMD and LD vector systems. *ChemBioChem* 4: 286–298.

Lilley, D.M.J. and Wilson, T.J. (2000). Fluorescence resonance energy transfer as a structural tool for nucleic acids. *Curr. Opin. Chem. Biol.* 4: 507–517.

Palmer, E. and Freeman, T. (2004). Investigation into the use of C- and N-terminal GFP fusion proteins for sub-cellular localization studies using reverse transfection microarrays. *Compar. Funct. Genom.* 5 (4): 342–353.

Preuss, M., Tecle, M., Shah, I., Matthews, D.A. and Miller, A.D. (2003). Comparison between the interactions of adenovirus-derived peptides with plasmid DNA and their role in gene delivery mediated by liposome-peptide-DNA virus-like nanoparticles. *Org. Biomol. Chem.* 1: 2430–2438.

Schwinn, M.K., Machleidt, T., Zimmerman, K., Eggers, C.T., Dixon, A.S., Hurst, R. *et al.* (2018). CRISPR-mediated tagging of endogenous proteins with a luminescent peptide. *ACS Chem. Biol.* 13: 467–474.

Tanner, J.A., Wright, M., Christie, E.M., Preuss, M.K. and Miller, A.D. (2006). Investigation into the interactions between diadenosine 5′, 5′′′-P^1, P^4-tetraphosphate and two proteins: molecular chaperone GroEL and cAMP receptor protein. *Biochemistry* 45: 3095–3106.

Tecle, M., Preuss, M. and Miller, A.D. (2003). Kinetic study of DNA condensation by cationic peptides used in nonviral gene therapy: analogy of DNA condensation to protein folding. *Biochemistry* 42: 10343–10347.

Weiss, S. (1999). Fluorescence spectroscopy of single biomolecules. *Science* 283: 1676–1683.

Wright, M. and Miller, A.D. (2004). Synthesis of novel fluorescent-labelled dinucleoside polyphosphates, Bioorg. *Med. Chem. Lett.* 14: 2813–2816.

Wright, M. and Miller, A.D. (2006). Novel fluorescent labelled affinity probes for diadenosine 5′, 5′′′-P^1, P^4-tetraphosphate (Ap_4A)-binding studies. *Bioorg. Med. Chem. Lett.* 16: 943–948.

Chapter 5

General Reading

Atkins, P.W., De Paula, J. and Keller, J. (2019). *Physical Chemistry*, 11e. Oxford, UK: Oxford University Press.

Evans, J.N.S. (1995). *Biomolecular NMR Spectroscopy*. Oxford, UK: Oxford University Press.

Leach, A.R. and Modelling, M. (2001). *Principles and Applications*, 2e. Harlow, UK: Prentice Hall.

Serdyuk, I.N., Zaccai, N.R. and Zaccai, G. (2017). *Methods in Molecular Biophysics, Structure Dynamics and Function*, 2e. Cambridge, UK: Cambridge University Press.

Wüthrich, K. (1986). *NMR of Proteins and Nucleic Acids*. New York, NY, USA: John Wiley & Sons.

Extra Reading

Alderson, T.R. and Kay, L.E. (2021). NMR spectroscopy captures the essential role of dynamics in regulating biomolecular function. *Cell* 184: 577–595.

Clore, G.M., Bax, A., Driscoll, P.C., Wingfield, P.T. and Gronenborn, A.M. (1990). Assignment of the side-chain ^1H and ^{13}C resonances of Interleukin-1β using double-resonance and triple-resonance heteronuclear 3-dimensional NMR-spectroscopy. *Biochemistry* 29: 8172–8184.

Clore, G.M., Wingfield, P.T. and Gronenborn, A.M. (1991). High-resolution three-dimensional structure of interleukin 1β in solution by three- and four-dimensional nuclear magnetic resonance spectroscopy. *Biochemistry* 30: 2315–2323 (pdb: **7i1b**).

Finzel, B.C., Clancy, L.L., Holland, D.R., Muchmore, S.W., Watenpaugh, K.D. and Einspahr, H.M. (1989). Crystal structure of recombinant human interleukin-1β at 2.0 Å resolution. *J. Mol. Biol.* 209: 779–791 (pdb: **1i1b**).

Homans, S.W. (1990). Oligosaccharide conformations – application of NMR and energy calculations. *Prog. NMR Spec.* 22: 55–81.

Ikura, M., Bax, A., Clore, G.M. and Gronenborn, A.M. (1990). Detection of nuclear overhauser effects between degenerate amide proton resonances by heteronuclear 3-dimensional nuclear-magnetic-resonance spectroscopy. *J. Am. Chem. Soc.* 112: 9020–9022.

Jacobs, R.E. and Oldfield, E. (1980). NMR of membranes. *Prog. NMR Spec.* 14: 113–136.

Kay, L.E., Ikura, M. and Bax, A. (1991). The design and optimization of complex NMR experiments: application to a triple-resonance pulse scheme correlating H_a, NH and ^{15}N chemical-shifts in ^{15}N-^{13}C "-labeled" proteins. *J. Magn. Reson.* 91: 84–92.

Patel, D.J., Shapiro, L. and Hare, D. (1987). DNA and RNA – NMR-studies of conformations and dynamics in solution. *Quat. Rev. Biophys.* 20: 35–112.

Pullen, J.R., Dalmaris, J., Serapian, S.A. and Miller, A.D. (2013). Assessing the preferred solution conformation of an interacting sense-antisense (complementary) peptide pair. *Bioorg. Med. Chem. Lett.* 23: 496–502.

Reid, B.R. (1987). Sequence-specific assignments and their use in NMR-studies of DNA-structure. *Quat. Rev. Biophys.* 20: 1–34.

Seelig, J. and MacDonald, P.M. (1987). Phospholipids and proteins in biological-membranes - ^2H- NMR as a method to study structure, dynamics and interactions. *Acc. Chem. Res.* 20: 211–228.

van Halbeck, H. and Poppe, L. (1992). Conformation and dynamics of glycoprotein oligosaccharides as studied by ^1H-NMR spectroscopy. *Magn. Reson. Chem.* Special issue, 30: S74–S86.

Wijmenga, S.S. and van Buuren, B.N.M. (1998). The use of NMR methods for conformational studies of nucleic acids. *Prog. NMR Spec.* 32: 287–387.

Chapter 6

General Reading

Drenth, J. (1994). *Principles of Protein X-ray Crystallography*, New York, NY, USA: Springer-Verlag.

Drenth, J. (2011). *Principles of Protein X-ray Crystallography*, 3e. New York, NY, USA: Springer.

Egerton, R.F. (2016). *Physical Principles of Electron Microscopy: An Introduction to TEM, SEM and AEM*, 2e. Cham, Switzerland: Springer.

Henderson, R. (2015). Overview and future of single particle electron cryomicroscopy. Arch. *Biochem. Biophys.* 581: 19–24.

Lippard, S.J. and Berg, J.M. (1994). *Principles of Bioinorganic Chemistry*. Mill Valley, CA, USA: University Science Books.

McRee, D.E. and David, P.R. (1999). *Practical Protein Crystallography*, 2nd. San Diego, CA, USA: Academic Press.

Wiesendanger, R. (1994). *Scanning Probe Microscopy and Spectroscopy*. Cambridge, UK: Cambridge University Press.

Extra Reading

Agbandje, M., Parrish, C.R. and Rossmann, M.G. (1995). The structure of parvoviruses. *Semin. Virol.* 6: 299–309.

Ando, T. (2019). High-speed atomic force microscopy. *Curr. Opin. Chem. Biol.* 51: 105–112.

Blundell, T.L. and Chaplin, A.K. (2021). The resolution revolution in X-ray diffraction, Cryo-EM and other technologies. *Prog. Biophys. Mol. Biol.* 160: 2–4.

Chen, C.H., Clegg, D.O. and Hansma, H.G. (1998). Structures and dynamic motion of laminin-1 as observed by atomic force microscopy. *Biochemistry* 37: 8262–8267.

Conway, J.F., Trus, B.L., Body, F.P., Newcomb, W.W., Brown, J.C. and Steven, A.C. (1996). Visualization of three-dimensional density maps reconstructed from cryoelectron micrographs of viral capsids. *J. Struct. Biol.* 116: 200–208.

Desogus, G., Todone, F., Brick, P. and Onesti, S. (2000). Active site of lysyl-tRNA synthetase: structural studies of the adenylation reaction. *Biochemistry* 39: 8418–8425 (pdbs: **1e1t, 1e22 & 1e24**).

Guckenberger, R., Heim, M., Cevc, G., Knapp, H.F., Wiegrabe, W. and Hillebrand, A. (1994). Scanning-tunneling-microscopy of insulators and biological specimens based on lateral conductivity of ultrathin water films. *Science* 266: 1538–1540.

Hughes, S.J., Tanner, J.A., Hindley, A.D., Miller, A.D. and Gould, I.R. (2003). Functional asymmetry in the lysyl-tRNA synthetase explored by molecular dynamics, free energy calculations and experiment. *BMC Struct. Biol.* 3: 5.

Hughes, S.J., Tanner, J.A., Miller, A.D. and Gould, I.R. (2006). Molecular dynamics simulations of LysRS: an asymmetric state. *Proteins* 62: 649–662.

Jones, H., Dalmaris, J., Wright, M., Steinke, J.H.G. and Miller, A.D. (2006a). Hydrogel polymer appears to mimic the performance of the GroEL/GroES molecular chaperone machine. *Org. Biomol. Chem.* 4: 2568–2574.

Jones, H., Preuss, M., Wright, M. and Miller, A.D. (2006b). The mechanism of GroEL/GroES folding/refolding of protein substrates revisited. *Org. Biomol. Chem.* 4: 1223–1235.

Muller, D.J., Janovjak, H., Lehto, T., Kuerschner, L. and Anderson, K. (2002). Observing structure, function and assembly of single proteins by AFM. *Prog. Biophys. Mol. Biol.* 79: 1–43.

Onesti, S., Miller, A.D. and Brick, P. (1995). The crystal structure of the lysyl-tRNA synthetase (LysU) from *Escherichia coli*. *Structure* 3: 163–176 (pdb: **1lyl**).

Ranson, N.A., Clare, D.K., Farr, G.W., Houldershaw, D., Horwich, A.L. and Saibil, H.R. (2006). Allosteric signaling of ATP hydrolysis in GroEL-GroES complexes. *Nat. Struct. Mol. Biol.* 13: 147–152 (fitted coordinates, Cryo-EM) (pdbs: **2c7c & 2c7d**).

Ranson, N.A., Farr, G.W., Roseman, A.M., Gowen, B., Fenton, W.A., Horwich, A.L. et al. (2001). ATP-bound states of GroEL captured by cryo-electron microscopy. *Cell* 107: 869–879 (pdbs: **1gr5 & 1gru**).

Ranson, N.A., White, H.E. and Saibil, H.R. (1998). Chaperonins. *Biochem. J.* 333: 233–242.

Roseman, A.M., Chen, S., White, H., Braig, K. and Saibil, H.R. (1996). The chaperonin ATPase cycle: mechanism of allosteric switching and movements of substrate-binding domains in GroEL. *Cell* 87: 241–251.

Saibil, H.R. (2022). Cryo-EM in molecular and cellular biology. *Mol. Cell* 82: 274–284.

Smith, C.M., Kohler, R.J., Barho, E., El-Thaher, T.S.H., Preuss, M. and Miller, A.D. (1999). Characterisation of Cpn60 (GroEL) bound cytochrome c: the passive role of molecular chaperones in assisting folding/refolding of proteins. *J. Chem. Soc., Perkin Trans.* 2: 1537–1546.

Wang, J. and Boisvert, D.C. (2003). Structural basis for GroEL-assisted protein folding from the crystal structure of (GroEL-KMgATP)$_{14}$ at 2.0 Å resolution. *J. Mol. Biol.* 327: 843–855.

Wang, J. and Chen, L. (2003). Domain motions in GroEL upon binding of an oligopeptide. *J. Mol. Biol.* 334: 489–499.

Wikoff, W.R., Tsai, C.J., Wang, G.J., Baker, T.S. and Johnson, J.E. (1997). The structure of cucumber mosaic virus: cryoelectron microscopy, X-ray crystallography and sequence analysis. *Virology* 232: 91–97.

Xu, Z., Horwich, A.L. and Sigler, P.B. (1997). The crystal structure of the asymmetric GroEL-GroES-(ADP)$_7$ chaperonin complex. *Nature* 388: 741–750 (pdb: **1aon**).

Chapter 7

General Reading

Bendall, D.S. (ed.) (1996). *Protein Electron Transfer*. Oxford, UK: BIOS Scientific Publishers.

Cantor, C.R. and Schimmel, P.R. (1980). *Biophysical Chemistry Part II. Techniques for the Study of Biological Structure and Function*. San Francisco, CA, USA: W.H. Freeman & Co.

Creighton, T.E. (ed.) (1997). *Protein Function: A Practical Approach*, 2e. Oxford, UK: IRL Press.

Houslay, M.D. and Stanley, K.K. (1982). *Dynamics of Biological Membranes*. Chichester, UK: John Wiley & Sons.

Ladbury, J.E. and Doyle, M.L. (eds.) (2004). *Biocalorimetry 2: Applications of Calorimetry in the Biological Sciences*. Chichester, UK: John Wiley & Sons.

Leach, A.R. and Modelling, M. (2001). *Principles and Applications*, 2e. Harlow, UK: Prentice Hall.

Pain, R.H. (ed.) (2000). *Mechanisms of Protein Folding*, 2e. Oxford, UK: Oxford University Press.

Voet, D. and Voet, J. G. (1995). *Biochemistry*, 2e. New York, NY, USA: John Wiley & Sons.

Voet, D., Voet, J. G. and Pratt, C. W. (1999), *Fundamentals of Biochemistry*, New York, NY, USA: John Wiley & Sons.

Voet, D., Voet, J.G. and Pratt, C.W. (2016). *Fundamentals of Biochemistry: Life at the Molecular Level*, 5e. New York, NY, USA: John Wiley & Sons.

Extra Reading

Abdul-Wahab, M.F., Homma, T., Wright, M., Olerenshaw, D., Dafforn, T.R., Nagata, K. *et al.* (2013). The pH sensitivity of murine heat shock protein 47 (HSP47) binding to collagen is affected by mutations in the breach histidine cluster. *J. Biol. Chem.* 288: 4452–4461.

Bhakoo, A., Raynes, J.G., Heal, J.R., Keller, M. and Miller, A.D. (2004). De-novo design of complementary (antisense) peptide mini-receptor inhibitor of interleukin 18 (IL-18). *Mol. Immunol.* 41: 1217–1224.

Boonyalai, N., Pullen, J.R., Abdul-Wahab, M.F., Wright, M. and Miller, A.D. (2013). *Escherichia coli* LysU is a potential surrogate for human lysyl tRNA synthetase in interactions with the C-terminal domain of HIV-1 capsid protein. *Org. Biomol. Chem.* 11: 612–620.

Chen, X., Boonyalai, N., Lau, C., Thipayang, S., Xu, Y., Wright, M. *et al.* (2013). Multiple catalytic activities of *Escherichia coli* lysyl-tRNA synthetase (LysU) are dissected by site directed mutagenesis. *FEBS J.* 280: 102–114.

Dafforn, T.R., Della, M. and Miller, A.D. (2001). The molecular interactions of heat shock protein 47 (Hsp47) and their implications for collagen biosynthesis. *J. Biol. Chem.* 276: 49310–49319.

Davids, J.W., El-Bakri, A., Heal, J., Christie, G., Roberts, G.W., Raynes, J.G. *et al.* (1997). Design of antisense (complementary) peptides as selective inhibitors of cytokine interleukin-1. *Angew. Chem. Intl. Ed.* 36: 962–967.

Desogus, G., Todone, F., Brick, P. and Onesti, S. (2000). Active site of lysyl-tRNA synthetase: structural studies of the adenylation reaction. *Biochemistry* 39: 8418–8425 (pdbs: **1e1t**, **1e22** & **1e24**).

Ellenberger, T.E., Brandl, C.J., Struhl, K. and Harrison, S.C. (1992). The GCN4 basic region leucine zipper binds DNA as a dimer of uninterrupted alpha helices: crystal structure of the protein-DNA complex. *Cell* 71: 1223–1237 (pdb: **1ysa**).

Finzel, B.C., Clancy, L.L., Holland, D.R., Muchmore, S.W., Watenpaugh, K.D. and Einspahr, H.M. (1989). Crystal structure of recombinant human interleukin-1β at 2.0 Å resolution. *J. Mol. Biol.* 209: 779–791 (pdb: **1i1b**).

Harel, M., Quinn, D.M., Nair, H.K., Silman, I. and Sussman, J.L. (1996). The X-ray structure of a transition state analog complex reveals the molecular origins of the catalytic power and substrate specificity of acetylcholinesterase. *J. Am. Chem. Soc.* 118: 2340–2346 (pdb: **1amn**).

Heal, J.R., Bino, S., Ray, K.P., Christie, G., Miller, A.D. and Raynes, J.G. (1999). A search within the IL-1 type 1 receptor reveals a peptide with hydropathic complementarity to the IL-1β trigger loop which binds to IL-1 and inhibits *in vitro* responses. *Mol. Immunol.* 36: 1141–1148.

Heal, J.R., Bino, S., Roberts, G.W., Raynes, J.G. and Miller, A.D. (2002a). Mechanistic investigation into complementary (antisense) peptide mini-receptor inhibitors of cytokine interleukin-1. *ChemBioChem* 3: 76–85.

Heal, J.R., Roberts, G.W., Christie, G. and Miller, A.D. (2002b). Inhibition of β-amyloid aggregation and neurotoxicity by complementary (antisense) peptides. *ChemBioChem* 3: 86–92.

Heal, J.R., Roberts, G.W., Raynes, J.G., Bhakoo, A. and Miller, A.D. (2002c). Specific interactions between sense and complementary peptides; the basis for the proteomic code. *ChemBioChem* 3 (136–151): 271.

Hughes, S.J., Tanner, J.A., Hindley, A.D., Miller, A.D. and Gould, I.R. (2003). Functional asymmetry in the lysyl-tRNA synthetase explored by molecular dynamics, free energy calculations and experiment. *BMC Struct. Biol.* 3: 5.

Hughes, S.J., Tanner, J.A., Miller, A.D. and Gould, I.R. (2006). Molecular dynamics simulations of LysRS: an asymmetric state. *Proteins* 62: 649–662.

Keller, M., Tagawa, T., Preuss, M. and Miller, A.D. (2002). Biophysical characterization of the DNA binding and condensing properties of adenoviral core peptide μ (mu). *Biochemistry* 41: 652–659.

Kleinjung, J., Petit, M.C., Orlewski, P., Mamalaki, A., Tzartos, S.J., Tsikaris, V. *et al.* (2000). The third-dimensional structure of the complex between an Fv antibody fragment and an analogue of the main immunogenic region of the acetylcholine receptor: a combined two-dimensional NMR, homology and molecular modeling approach. *Biopolymers* 53: 113–128 (pdb: **1f3r**).

Miller, A.D. (2015). Sense-antisense (complementary) peptide interactions and the proteomic code; potential opportunities in biology and pharmaceutical science. *Exp. Opin. Biol. Ther.* 15: 245–267.

Onesti, S., Miller, A.D. and Brick, P. (1995). The crystal structure of the lysyl-tRNA synthetase (LysU) from *Escherichia coli*. *Structure* 3: 163–176 (pdb: **1lyl**).

Preuss, M., Hutchinson, J.P. and Miller, A.D. (1999). Secondary structure forming propensity coupled with amphiphilicity is an optimal motif in a peptide or protein for association with Chaperonin 60 (GroEL). *Biochemistry* 38: 10272–10286.

Preuss, M. and Miller, A.D. (1999). Interaction with GroEL destabilises non-amphiphilic secondary structure in a peptide. *FEBS Lett.* 461: 131–135.

Pullen, J.R., Dalmaris, J., Serapian, S.A. and Miller, A.D. (2013). Assessing the preferred solution conformation of an interacting sense-antisense (complementary) peptide pair. *Bioorg. Med. Chem. Lett.* 23: 496–502.

Vigers, G.P., Anderson, L.J., Caffes, P. and Brandhuber, B.J. (1997). Crystal structure of the type-I interleukin-1 receptor complexed with interleukin-1β. *Nature* 386: 190–194 (pdb: **1itb**).

Vigers, G.P., Caffes, P., Evans, R.J., Thompson, R.C., Eisenberg, S.P. and Brandhuber, B.J. (1994). X-ray structure of interleukin-1 receptor antagonist at 2.0 Å resolution. *J. Biol. Chem.* 269: 12874–12879 (pdb: **1ilt**).

Wright, M., Boonyalai, N., Tanner, J.A., Hindley, A.D. and Miller, A.D. (2006). The duality of LysU, a catalyst for both Ap_4A and Ap_3A formation. *FEBS J.* 273: 3534–3544.

Chapter 8

General Reading

Atkins, P.W. (1995) *Physical Chemistry*, 5e. Oxford, UK: Oxford University Press.

Atkins, P.W., De Paula, J. and Keller, J. (2019). *Physical Chemistry*, 11e. Oxford, UK: Oxford University Press.

Eisenthal, R. and Danson, M.J. (eds.) (2002). *Enzyme Assays: A Practical Approach*, 2e. Oxford, UK: Oxford University Press.

Fersht, A.R. (1985). *Enzyme Structure and Mechanism*, 2e. New York, NY, USA: W.H. Freeman & Co.

Fersht, A.R. (2017). *Structure and Mechanism in Protein Science: A Guide to Enzyme Catalysis and Protein Folding*, 4e. Hackensack, NJ, USA: World Scientific Publishing.

Goodfellow, J.M. and Moss, D.S. (eds.) (1992). *Computer Modelling of Biomolecular Processes*. Chichester, UK: Ellis Horwood Ltd.

Gul, S., Sreedharan, S.K. and Brocklehurst, K. (1998). *Enzyme Assays: Essential Data*. Chichester, UK: John Wiley & Sons.

Segel, I.H. (1975). *Enzyme Kinetics*. New York, NY, USA: John Wiley & Sons.

Extra Reading

Abbate, F., Casini, A., Owa, T., Scozzafava, A. and Supuran, C.T. (2004). Carbonic anhydrase inhibitors: E7070, a sulfonamide anticancer agent, potently inhibits cytosolic isozymes I and II and transmembrane, tumor-associated isozyme IX. *Bioorg. Med. Chem.* 14: 217–223.

Banner, D.W., Bloomer, A., Petsko, G.A., Phillips, D.C. and Wilson, I.A. (1976). Atomic coordinates for triose phosphate isomerase from chicken muscle. *Biochem. Biophys. Res. Commun.* 72: 146–155 (pdb: **1tim**).

Birktoft, J.J., Rhodes, G. and Banaszak, L.J. (1989). Refined crystal structure of cytoplasmic malate dehydrogenase at 2.5-Å resolution. *Biochemistry* 28: 6065–6081 (pdb: **4mdh**).

Borgstahl, G.E., Parge, H.E., Hickey, M.J., Beyer, W.F., Jr, Hallewell, R.A. and Tainer, J.A. (1992). The structure of human mitochondrial Mn^{3+} superoxide dismutase reveals a novel tetrameric interface of two 4-helix bundles. *Cell* 71: 107–118 (pdb: **1n0j**).

Burbaum, J.J., Raines, R.T., Albery, W.J. and Knowles, J.R. (1989). Evolutionary optimization of the catalytic effectiveness of an enzyme. *Biochemistry* 28: 9293–9305.

Capitani, G., Biase, D.D., Aurizi, C., Gut, H., Bossa, F. and Grutter, M.G. (2003). Crystal structure and functional analysis of *Escherichia coli* glutamate decarboxylase. *EMBO J.* 22: 4027–4037 (pdb: **1pmo** & **1pmm**).

Chen, X., Boonyalai, N., Lau, C., Thipayang, S., Xu, Y., Wright, M. *et al.* (2013). Multiple catalytic activities of *Escherichia coli* lysyl-tRNA synthetase (LysU) are dissected by site directed mutagenesis. *FEBS J.* 280: 102–114.

Day, P.J. and Shaw, W.V. (1992). Acetyl coenzyme A binding by chloramphenicol acetyltransferase. *J. Bio. Chem.* 267: 5122–5127.

Desogus, G., Todone, F., Brick, P. and Onesti, S. (2000). Active site of lysyl-tRNA synthetase: structural studies of the adenylation reaction. *Biochemistry* 39: 8418–8425 (pdb: **1e1t, 1e22 & 1e24**).

Diamond, R. (1974). Real-space refinement of the structure of hen egg-white lysozyme. *J. Mol. Biol.* 82: 371–391 (pdb: **6lyz**).

Fields, P.A., Rudomin, E.L. and Somero, G.N. (2006). Temperature sensitivities of cytosolic malate dehydrogenases from native and invasive species of marine mussels (genus *Mytilus*): sequence-function linkages and correlations with biogeographic distribution. *J. Exp. Biol.* 209: 656–667.

Harel, H., Quinn, D.M., Nair, H.K., Silman, I. and Sussman, J.L. (1996). The X-ray structure of a transition state analog complex reveals the molecular origins of the catalytic power and substrate specificity of acetylcholinesterase. *J. Am. Chem. Soc.* 118: 2340–2346 (pdb: **1amn**).

Kannan, K.K., Ramanadham, M. and Jones, T.A. (1984). Structure, refinement and function of carbonic anhydrase isozymes: refinement of human carbonic anhydrase I. *Ann. N. Y. Acad. Sci.* 429: 49–60 (pdb: **2cab**).

Kirk, S.R., Luedtke, N.W. and Tor, Y. (2001). 2-Aminopurine as a real time probe of enzymatic cleavage and inhibition of hammerhead ribozymes. *Bioorg. Med. Chem.* 9: 2295–2301.

Knowles, J.R. (1991). Enzyme catalysis – not different, just better. *Nature* 350: 121–124.

Lewendon, A., Murray, I.A., Shaw, W.V., Gibbs, M.R. and Leslie, A.G. (1990). Evidence for transition-state stabilization by serine-148 in the catalytic mechanism of chloramphenicol acetyltransferase. *Biochemistry* 29: 2075–2080 (pdb: **1cla**).

Masaki, K., Aizawa, T., Koganesawa, N., Nimori, T., Bando, H., Kawano, K. *et al.* (2001). Thermal stability and enzymatic activity of a smaller lysozyme from silk moth (*Bombyx mori*). *J. Prot. Chem.* 20: 107–113.

Morollo, A.A., Petsko, G.A. and Ringe, D. (1999). Structure of a Michaelis complex analogue: propionate binds in the substrate carboxylate site of alanine racemase. *Biochemistry* 38: 3293–3301 (pdb: **2sfp**).

Onesti, S., Miller, A.D. and Brick, P. (1995). The crystal structure of the lysyl-tRNA synthetase (LysU) from *Escherichia coli*. *Structure* 3: 163–176 (pdb: **1lyl**).

Smith, D.L., Almo, S.C., Toney, M.D. and Ringe, D. (1989). 2.8-Å-resolution crystal structure of an active-site mutant of aspartate aminotransferase from *Escherichia coli*. *Biochemistry* 28: 8161–8167 (pdb: **2aat**).

Stroupe, M.E., DiDonato, M. and Tainer, J.A. (2001). Manganese superoxide dismutase. In: *Handbook of Metalloproteins* (eds. A. Messerschmidt, R. Huber, T. Poulos and K. Wieghardt). Chichester, UK: John Wiley & Sons.

Theoclitou, M.-E., Wittung, E.P.L., Hindley, A.D., El-Thaher, T.S.H. and Miller, A.D. (1996). Characterisation of stress protein LysU. Enzymic synthesis of diadenosine 5′, 5‴-P¹, P⁴-tetraphosphate (Ap₄A) analogues by LysU. *J. Chem. Soc., Perkin Trans.* 1: 2009–2019.

Tsukada, H. and Blow, D.M. (1985). Structure of α-chymotrypsin refined at 1.68Å resolution. *J. Mol. Biol.* 184: 703–711 (pdb: **4cha**).

Wlodawer, A., Svensson, L.A., Sjolin, L. and Gilliland, G.L. (1988). Structure of phosphate-free ribonuclease A refined at 1.26Å. *Biochemistry* 27: 2705–2717 (pdb: **7rsa**).

Wright, M., Boonyalai, N., Tanner, J.A., Hindley, A.D. and Miller, A.D. (2006). The duality of LysU, a catalyst for both Ap₄A and Ap₃A formation. *FEBS J.* 273: 3534–3544.

Chapter 9

General Reading

Dass, C. (2001). *Principles and Practice of Biological Mass Spectrometry*, New York, NY, USA: Wiley-Interscience.

Dass, C. (2012). *Principles and Practice of Biological Mass Spectrometry*, 2e. Chichester, UK: Wiley-Blackwell.

De Hoffmann, E. and Stroobant, V. (2009). *Mass Spectrometry: Principles and Applications*, 3e. Chichester, UK: John Wiley & Sons.

Hamdan, M.H. and Righetti, P.G. (2005). *Proteomics Today: Protein Assessment and Biomarkers Using Mass Spectrometry, 2D Electrophoresis and Microarray Technology*. Hoboken, NJ, USA: John Wiley & Sons.

Kinter, M. and Sherman, N.E. (2000). *Protein Sequencing and Identification Using Tandem Mass Spectrometry*. New York, NY, USA: Wiley-Interscience.

Laskin, J. and Lifshitz, C. (eds.) (2006). *Principles of Mass Spectrometry Applied to Biomolecules*. Hoboken, NJ, USA: JohnWiley & Sons.

Marte, B. (ed.) (2003). Nature insight proteomics. *Nature* 422: 193–237.

Pennington, S.R. and Dunn, M.J. (2001). *Proteomics: From Protein Sequence to Function*. Oxford, UK: BIOS Scientific Publishers.

Twyman, R.M. (2014). *Principles of Proteomics*, 2e. Oxford, UK: BIOS Scientific Publishing.

Yates, J.R., 3rd. (2019). Recent technical advances in proteomics. *F1000Res.* 8. F1000 Faculty Rev–351.

Extra Reading

Aebersold, R. and Mann, M. (2003). Mass spectrometry-based proteomics. *Nature* 422: 198–207.

Azhar, M.A., Wright, M., Kamal, A., Nagy, J. and Miller, A.D. (2014). Biotin-c10-AppCH$_2$ppA is an effective new chemical proteomics probe for diadenosine polyphosphate binding proteins. *Bioorg. Med. Chem. Lett.* 24: 2928–2933.

Biemann, K. (1988). Contributions of mass-spectrometry to peptide and protein-structure. *Biomed. Environ. Mass Spectrom.* 16: 99–111.

Glish, G.L. and Vachet, R.W. (2003). The basics of mass spectrometry in the twenty-first century. *Nat. Rev. Drug Discov.* 2: 140–150.

Grotjahn, L. (1986). *In Mass Spectrometry in Biomedical Research* (ed. S.J. Gaskell), 215–234. New York: John Wiley & Sons.

Guo, W., Azhar, M.A., Xu, Y., Wright, M., Kamal, A. and Miller, A.D. (2011). Isolation and identification of diadenosine 5′, 5‴-P^1, P^4-tetraphosphate binding proteins using magnetic bio-panning. *Bioorg. Med. Chem. Lett.* 21: 7175–7179.

Jensen, N.J., Tomer, K.B. and Gross, M.L. (1985). Gas-phase ion decompositions occurring remote to a charge site. *J. Am. Chem. Soc.* 107: 1863–1868.

Jensen, N.J., Tomer, K.B. and Gross, M.L. (1987). FAB MS/MS for phosphatidylinositol,-glycerol, phosphatidylethanolamine and other complex phospholipids. *Lipids* 22: 480–489.

MacBeath, G. (2002). Protein microarrays and proteomics. *Nat. Genet.* 32: 526–532.

Palmer, E.L., Miller, A.D. and Freeman, T.C. (2006). Identification and characterization of human apoptosis inducing proteins using cell-based transfection microarrays and expression analysis. *BMC Genomics* 7: 145.

Pandey, A. and Mann, M. (2000). Proteomics to study genes and genomes. *Nature* 405: 837–846.

Patterson, S.D. and Aebersold, R.H. (2003). Proteomics: the first decade and beyond. *Nat. Genet.* 33: 311–323.

Reinhold, V.N., Reinhold, B.B. and Costello, C.E. (1995). Carbohydrate molecular-weight profiling, sequence, linkage and branching data: ES-MS and CID. *Anal. Chem.* 67: 1772–1784.

Rogawski, R. and Sharon, M. (2022). Characterizing endogenous protein complexes with biological mass spectrometry. *Chem. Rev.* 122: 7386–7414.

Tyers, M. and Mann, M. (2003). From genomics to proteomics. *Nature* 422: 193–197.

Wu, C.C. and Yates, J.R. (2003). The application of mass spectrometry to membrane proteomics. *Nat. Biotechnol.* 21: 262–267.

Zubarev, R.A. and Makarov, A. (2013). Orbitrap mass spectrometry. *Anal. Chem.* 85: 5288–5296.

Chapter 10

General Reading

Cevc, G. (ed.) (2007). *Phospholipids Handbook*, 2e. New York, NY, USA: Taylor & Francis.

Janda, K.D., Shevlin, C.G. and Lo, C.H.L. (1996). Catalytic antibodies chemical and biological approaches, in comprehensive supramolecular chemistry. In: *Supramolecular Reactivity and Transport: Bioorganic Systems*, 4 (eds. Y. Murakami, J.L. Atwood and J.-M. Lehn), 43–72. New York, NY, USA: Pergamon Press.

Kauffman, S.A. (1996). *At Home in the Universe; the Search for Laws of Self-organization and Complexity*. London, UK: Penguin Books.

Small, D.M. (1986). *Handbook of Lipid Research-4: The Physical Chemistry of Lipids*. New York, NY, USA: Plenum Press.

Extra Reading

Breaker, R.R. (1997). DNA aptamers and DNA enzymes. *Curr. Opin. Chem. Biol.* 1: 26–31.

Eschenmoser, A. (1999). Chemical etiology of nucleic acid structure. *Science* 284: 2118–2124.

Eschenmoser, A. and Kisakurek, M.V. (1996). Chemistry and the origin of life. *Helv. Chimica Acta* 79: 1249–1259.

Eschenmoser, A. and Krishnamurthy, R. (2000). Chemical etiology of nucleic acid structure. *Pure Appl. Chem.* 72: 343–345.

Eschenmoser, A. and Loewenthal, E. (1992). Chemistry of potentially prebiological natural products. *Chem. Soc. Rev.* 21: 1–16.

Flynn-Charlebois, A., Wang, Y.M., Prior, T.K., Rashid, I., Hoadley, K.A., Coppins, R.L. *et al.* (2003). Deoxyribozymes with 2′-5′ RNA ligase activity. *J. Am. Chem. Soc.* 125: 2444–2454.

Hermann, T. and Patel, D.J. (2000). Biochemistry – adaptive recognition by nucleic acid aptamers. *Science* 287: 820–825.

Liebeton, K., Zonta, A., Schimossek, K., Nardini, M., Lang, D., Dijkstra, B.W. *et al.* (2000). Directed evolution of an enantioselective lipase. *Chem. Biol.* 7: 709–718.

Miller, A.D. (2002). Order for free: molecular diversity and complexity promote self-organisation. *ChemBioChem* 3: 45–46.

Williamson, J.R. (1994). G-Quartet structures in telomeric DNA. *Annu. Rev. Biophys. Biomol. Struct.* 23: 703–730.

Xiang, Y.-B., Drenkard, S., Baumann, K., Hickey, D. and Eschenmoser, A. (1994). Chemistry of α-amino-nitriles. 12. Exploratory experiments on thermal-reactions of α-amino-nitriles. *Helv. Chimica Acta* 77: 2209–2250.

Chapter 11

General Reading

Ashton, R.S., Keung, A.J., Peltier, J. and Schaffer, D.V. (2011). Progress and prospects for stem cell engineering. *Annu. Rev. Chem. Biomol. Eng.* 2: 479–502.

Cacace, E., Kritikos, G. and Typas, A. (2017). Chemical genetics in drug discovery, Curr. *Opin. Systems Biol.* 4: 35–42.

Spring, D.R. (2005). Chemical genetics to chemical genomics: small molecules offer big insights. *Chem. Soc. Rev.* 34: 472–482.

Stockwell, B.R. (2000). Chemical genetics: ligand-based discovery of gene function. *Nat. Rev. Genet.* 1: 116–125.

Weissman, T.A. and Pan, Y.A. (2015). Brainbow: new resources and emerging biological applications for multicolor genetic labeling and analysis. *Genetics* 199: 293–306.

Extra Reading

Azhar, M.A., Wright, M., Kamal, A., Nagy, J. and Miller, A.D. (2014). Biotin-c10-AppCH$_2$ppA is an effective new chemical proteomics probe for diadenosine polyphosphate binding proteins. *Bioorg. Med. Chem. Lett.* 24: 2928–2933.

Gadek, T.R., Burdick, D.J., McDowell, R.S., Stanley, M.S., Marsters Jr., J.C., Paris, K.J. *et al.* (2002). Generation of an LFA-1 antagonist by the transfer of the ICAM-1 immunoregulatory epitope to a small molecule. *Science* 295: 1086–1089.

Guo, W., Azhar, M.A., Xu, Y., Wright, M., Kamal, A. and Miller, A.D. (2011). Isolation and identification of diadenosine 5′, 5′′′-P^1, P^4-tetraphosphate binding proteins using magnetic bio-panning. *Bioorg. Med. Chem. Lett.* 21: 7175–7179.

Kuczenski, B., Ruder, W.C., Messner, W.C. and LeDuc, P.R. (2009). Probing cellular dynamics with a chemical signal generator. *PLoS One* 4: e4847; doi: 10.1371/journal.pone.0004847.

Kuruvilla, F.G., Shamji, A.F., Sternson, S.M., Hergenrother, P.J. and Schreiber, S.L. (2002). Dissecting glucose signaling with diversity-oriented synthesis and small-molecule microarrays. *Nature* 416: 653–657.

Lepourcelet, M., Chen, Y.-N.P., France, D.S., Wang, H., Crews, P., Peterson, F. *et al.* (2004). Small-molecule antagonists of the oncogenic Tcf/β-catenin protein complex. *Cancer Cell* 5: 91–102.

MacRae, C.A. and Peterson, R.T. (2003). Zebrafish-based small molecule discovery. *Chemistry & Biology* 10: 901–908.

Nelson, C.M. and Chen, C.S. (2003). VE-cadherin simultaneously stimulates and inhibits cell proliferation by altering cytoskeletal structure and tension. *J. Cell Sci.* 116: 3571–3581.

Unzue, A., Cribiú, R., Hoffman, M.M., Knehans T., Lafleur K., Caflisch, A. *et al.* (2018). Iriomoteolides: novel chemical tools to study actin dynamics. *Chem. Sci.* 9: 3793–3802.

Ward, G.E., Carey, K.L. and Westwood, N.J. (2002). Using small molecules to study big questions in cellular microbiology. *Cellular Microbiol.* 4: 471–482.

Winn, M., Reilly, E.B., Liu, G., Huth, J.R., Jae, H.-S., Freeman, J. *et al.* (2001). Discovery of novel p-arylthio cinnamides as antagonists of leukocyte function-associated antigen-1/intercellular adhesion molecule-1 interaction. 4. Structure-activity relationship of substituents on the benzene ring of the cinnamide. *J. Med. Chem.* 44: 4393–4403.

Chapter 12

General Reading

Ashton, R.S., Keung, A.J., Peltier, J. and Schaffer, D.V. (2011). Progress and prospects for stem cell engineering. *Annu. Rev. Chem. Biomol. Eng.* 2: 479–502.

Extra Reading

Aiba, K., Sharov, A.A., Carter, M.G., Foroni, C., Vescovi, A.L. and Ko, M.S.H. (2006). Defining a developmental path to neural fate by global expression profiling of mouse embryonic stem cells and adult neural stem/progenitor cells. *Stem Cells* 24: 889–895.

Ashton, R.S., Peltier, J., Fasano, C.A., O'Neill, A., Leonard, J., Temple, S. *et al.* (2007). High-throughput screening of gene function in stem cells using clonal microarrays. *Stem Cells* 25: 2928–2935.

Doudna, J.A. and Charpentier, E. (2014). The new frontier of genome engineering with CRISPR-Cas9. *Science* 346: 1258096-1-9.

Engler, A.J., Sen, S., Sweeney, H.L. and Discher, D.E. (2006). Matrix elasticity directs stem cell lineage specification. *Cell* 126: 677–689.

Gao, L., McBeath, R. and Chen, C.S. (2010). Stem cell shape regulates a chondrogenic versus myogenic fate through Rac1 and N-cadherin. *Stem Cells* 28: 564–572.

Gómez-Sjöberg, R., Leyrat, A.A., Pirone, D.M., Dean, C.S. and Quake, S.R. (2007). Versatile, fully automated, microfluidic cell culture system. *Anal. Chem.* 79: 8557–8563.

Khademhosseini, A., Langer, R., Borenstein, J. and Vacanti, J.P. (2005). Microscale technologies for tissue engineering and biology. *Proc. Natl. Acad. Sci. USA* 103: 2480–2487.

Li, X.-L., Li, G.-H., Fu, J., Fu, Y.-W., Zhang, L., Chen, W. *et al.* (2018). Highly efficient genome editing via CRISPR–Cas9 in human pluripotent stem cells is achieved by transient BCL-XL overexpression. *Nucleic Acids Res.* 46: 10195–10215.

Liang, J. and Qian, H. (2010). Computational cellular dynamics based on the chemical master equation: a challenge for understanding complexity. *J. Comput. Sci. Technol.* 25: 154–168.

Mašek, J., Lubasová, D., Lukáč, R., Turánek-Knotigová, P., Kulich, P., Plocková, J. *et al.* (2017). Multi-layered nanofibrous mucoadhesive films for buccal and sublingual administration of drug-delivery and vaccination nanoparticles – important step towards effective mucosal vaccines. *J. Control. Rel.* 249: 183–195.

McBeath, R., Pirone, D.M., Nelson, C.M., Bhadriraju, K. and Chen, C.S. (2004). Cell shape, cytoskeletal tension and RhoA regulate stem cell lineage commitment. *Develop. Cell* 6: 483–495.

Mei, Y., Saha, K., Bogatyrev, S.R., Yang, J., Hook, A.L., Kalcioglu, Z.I. *et al.* (2010). Combinatorial development of biomaterials for clonal growth of human pluripotent stem cells. *Nat. Mater.* 9: 768–778.

Prudhomme, W., Daley, G.Q., Zandstra, P. and Lauffenburger, D.A. (2004). Multivariate proteomic analysis of murine embryonic stem cell self-renewal versus differentiation signaling. *Proc. Natl. Acad. Sci. USA* 101: 2900–2905.

Sharov, A.A., Piao, Y., Matoba, R., Dudekula, D.B., Qian, Y., VanBuren, V. *et al.* (2003). Transcriptome analysis of mouse stem cells and early embryos. *PLoS Biol.* 1: 410–419.

Woolf, P.J., Prudhomme, W., Daheron, L., Daley, G.Q. and Lauffenburger, D.A. (2005). Bayesian analysis of signaling networks governing embryonic stem cell fate decisions. *Bioinformatics* 21: 741–753.

Yang, J., Mei, Y., Hook, A.L., Taylor, M., Urquhart, A.J., Bogatyrev, S.R. *et al.* (2010). Polymer surface functionalities that control human embryoid body cell adhesion revealed by high throughput surface characterization of combinatorial material microarrays. *Biomaterials* 31: 8827–8838.

Chapter 13

General Reading

Escriou, V., Mignet, N. and Miller, A.D. (2014). Auto-associative lipid-based systems for nonviral nucleic acid delivery. In: *Advanced Textbook on Gene Transfer, Gene Therapy and Genetic Pharmacology* (ed. D. Scherman), 221–254. London, UK: Imperial College Press.

Kurreck, J. (2009). RNA interference: from basic research to therapeutic applications. *Angew. Chem. Int. Ed. Engl.* 48: 1378–1398.

Miller, A.D. (1999). Nonviral delivery systems for gene therapy. In: *Understanding Gene Therapy* (ed. N.R. Lemoine), 43–69. Oxford, UK: Bios Scientific Publishers.

Miller, A.D. (2004). Nonviral liposomes. In: *Methods in Molecular Medicine*, 90 (ed. C.J. Springer), 107–137. Totowa, NJ, USA: Humana Press.

Miller, A.D. (2017). Synthetic nucleic acid delivery systems in gene therapy. In: *Encyclopedia of Life Sciences*, Chichester, UK: John Wiley & Sons; doi: 10.1002/9780470015902.a0005745.pub3.

Thanou, M., Waddington, S. and Miller, A.D. (2007). Gene therapy. In: *Comprehensive Medicinal Chemistry II*, 1 (eds. J.B. Taylor and D.J. Triggle), 297–319. Oxford, UK: Elsevier.

Extra Reading

Aissaoui, A., Chami, M., Hussein, M. and Miller, A.D. (2011). Efficient topical delivery of plasmid DNA to lung *in vivo* mediated by putative triggered, PEGylated pDNA nanoparticles. *J. Control. Rel.* 154: 275–284.

Allen, T.M. and Cullis, P.R. (2004). Drug delivery systems: entering the mainstream. *Science* 303: 1818–1822.

Allen, T.M. and Martin, F.J. (2004). Advantages of liposomal delivery systems for anthracyclines. *Semin. Oncol.* 31: 5–15.

Alton, E.W.F.W., Armstrong, D.K., Ashby, D., Bayfield, K.J., Bilton, D., Bloomfield, E.V. *et al.* (2015). Repeated nebulisation of non-viral CFTR gene therapy in patients with cystic fibrosis: a randomised, double-blind, placebo-controlled, phase 2b trial. *Lancet Respir. Med.* 3: 684–691.

Alton, E.W.F.W., Middleton, P.G., Caplen, N.J., Smith, S.N., Steel, D.M., Munkonge, F.M. *et al.* (1993). Non-invasive liposome-mediated gene delivery can correct the ion transport defect in cystic fibrosis mutant mice. *Nat. Genet.* 5: 135–142.

Alton, E.W.F.W., Stern, M., Farley, R., Jaffe, A., Chadwick, S.L., Phillips, J. *et al.* (1999). Cationic lipid-mediated CFTR gene transfer to the lungs and nose of patients with cystic fibrosis: a double-blind placebo-controlled trial. *Lancet* 353: 947–954.

Andreu, A., Fairweather, N. and Miller, A.D. (2008). Clostridium neurotoxin fragments as potential targeting moieties for liposomal gene delivery to the CNS. *Chembiochem* 9: 219–231.

Argyros, O., Wong, S.P., Niceta, M., Waddington, S.N., Howe, S.J., Coutelle, C. *et al.* (2008). Persistent episomal transgene expression in liver following delivery of a scaffold/matrix attachment region containing non-viral vector. *Gene Ther.* 15: 1593–1605.

Bai, J., Liu, Y., Sun, W., Chen, J., Miller, A.D. and Xu, Y. (2013). Down-regulated lysosomal processing improved PEGylated lipopolyplex-mediated gene transfection. *J. Gene Med.* 15: 182–192.

Carmona, S., Jorgensen, M.R., Kolli, S., Crowther, C., Salazar, F.H., Marion, P.L. *et al.* (2009). Controlling HBV replication *in vivo* by intravenous administration of triggered PEGylated siRNA-nanoparticles. *Mol. Pharmaceut.* 6: 706–717.

Centelles, M.N., Wright, M., So, P.-W., Amrahli, M., Xu, X.-Y., Stebbing, J. *et al.* (2018). Image-guided thermosensitive liposomes for focused ultrasound drug delivery: using NIRF-labelled lipids and topotecan to visualize the effects of hyperthermia in tumours. *J. Control. Rel.* 280: 87–98.

Chen, J., Jorgensen, M.R., Thanou, M. and Miller, A.D. (2011). Post-coupling strategy enables true receptor-targeted nanoparticles. *J. RNAi Gene Silencing* 7: 449–455.

Coates, C.J., Kaminski, J.M., Summers, J.B., Segal, D.J., Miller, A.D. and Kolb, A.F. (2005). Site-directed genome modification: derivatives of DNA-modifying enzymes as targeting tools. *Trends Biotechnol.* 23: 407–419.

Drake, C.R., Aissaoui, A., Argyros, O., Serginson, J.M., Monnery, B.D., Thanou, M. *et al.* (2010). Bioresponsive small molecule polyamines as non-cytotoxic alternative to polyethylenimine. *Mol. Pharmaceut.* 7: 2040–2055.

Drake, C.R., Aissaoui, A., Argyros, O., Thanou, M., Steinke, J.H.G. and Miller, A.D. (2013). Examination of the effect of increasing the number of intra-disulfide amino functional groups on the performance of small molecule cyclic polyamine disulfide vectors. *J. Control. Rel.* 171: 81–90.

Duraisamy, G. S., Bhosale, D., Lipenská, I., Huvarová, I., Růžek, D., Windisch, M.P. *et al.* (2020). Advanced therapeutics, vaccinations and precision medicine in the treatment and management of hepatitis B viral infections; where are we and where are we going? *Viruses* 12: 998; doi:10.3390/v12090998.

Fellowes, R., Etheridge, C.J., Coade, S., Cooper, R.G., Stewart, L., Miller, A.D. *et al.* (2000). Amelioration of established collagen induced arthritis by systemic IL-10 gene delivery. *Gene Ther.* 7: 967–977.

Fletcher, S., Ahmad, A., Perouzel, E., Heron, A., Miller, A.D. and Jorgensen, M.R. (2006a). *In vivo* studies of dialkynoyl analogues of DOTAP demonstrate improved gene transfer efficiency of cationic liposomes in mouse lung. *J. Med. Chem.* 49: 349–357.

Fletcher, S., Ahmad, A., Perouzel, E., Jorgensen, M.R. and Miller, A.D. (2006b). A dialkanoyl analogue of DOPE improves gene transfer of lower-charged, cationic lipoplexes. *Org. Biomol. Chem.* 4: 196–199.

Fletcher, S., Ahmad, A., Price, W.S., Jorgensen, M.R. and Miller, A.D. (2008). Biophysical properties of CDAN/DOPE-analogue lipoplexes account for enhanced gene delivery. *ChemBioChem* 9: 455–463.

Gabizon, A., Shmeeda, H. and Barenholz, Y. (2003). Pharmacokinetics of PEGylated liposomal doxorubicin: review of animal and human studies. *Clin. Pharmacokinetics* 42: 419–436.

Harbottle, R.P., Cooper, R.G., Hart, S.L., Ladhoff, A., McKay, T., Knight, A.M. *et al.* (1998). An RGD-oligolysine peptide: a prototype construct for integrin-mediated gene delivery. *Hum. Gene Ther.* 9: 1037–1047.

Jayaraman, M., Ansell, S.M., Mui, B.L., Tam, Y.-K., Chen, J., Du, X. *et al.* (2012). Maximizing the potency of siRNA lipid nanoparticles for hepatic gene silencing *in vivo*. *Angew. Chem. Int. Ed. Engl.* 51: 8529–8533.

Jenke, A.C.W., Eisenberger, T., Baiker, A., Stehle, I.M., Wirth, S. and Lipps, H.J. (2005). The nonviral episomal replicating vector pEPI-1 allows long-term inhibition of bcr-abl expression by shRNA. *Hum. Gene Ther.* 16: 533–539.

Kamaly, N., Kalber, T., Ahmad, A., Oliver, M.H., So, P.-W., Herlihy, A.H. *et al.* (2008). Bimodal paramagnetic and fluorescent liposomes for cellular and tumor magnetic resonance imaging. *Bioconjugate Chem* 19: 118–129.

Kamaly, N., Kalber, T., Thanou, M., Bell, J.D. and Miller, A.D. (2009). Folate receptor targeted bimodal liposomes for tumor magnetic resonance imaging. *Bioconj. Chem.* 20: 648–655.

Kamaly, N., Kalber, T.L., Kenny, G.D., Bell, J.D., Jorgensen, M.R. and Miller, A.D. (2010a). A novel bimodal lipidic contrast agent for cellular labelling and tumour MRI. *Org. Biomol. Chem.* 8: 201–211.

Kamaly, N., Miller, A.D. and Bell, J.D. (2010b). Chemistry of tumour targeted T_1 based MRI contrast agents. *Curr Topics Med. Chem.* 10: 1158–1183.

Kamaly, N. and Miller, A.D. (2010). Paramagnetic liposome nanoparticles for cellular and tumour imaging. *Int. J. Mol. Sci.* 11: 1759–1776.

Kenny, G.D., Kamaly, N., Kalber, T.L., Brody, L.P., Sahuri, M., Shamsaei, E. *et al.* (2011). Novel multifunctional nanoparticle mediates siRNA tumour delivery, visualisation and therapeutic tumour reduction *in vivo*. *J. Control. Rel.* 149: 111–116.

Kolb, A.F., Coates, C.J., Kaminski, J.M., Summers, J.B., Miller, A.D. and Segal, D.J. (2005). Site-directed genome modification: nucleic acid and protein modules for targeted integration and gene correction. *Trends Biotechnol.* 23: 399–406.

Kolli, S., Wong, S.-P., Harbottle, R., Johnston, B., Thanou, M. and Miller, A.D. (2013). pH-Triggered nanoparticle mediated delivery of siRNA to liver cells *in vitro* and *in vivo*. *Bioconj. Chem.* 24: 314–332.

Kostarelos, K. and Miller, A.D. (2005). Synthetic, self-assembly ABCD nanoparticles; a structural paradigm for viable synthetic non-viral vectors. *Chem. Soc. Rev.* 34: 970–994.

Kranz, L.M., Diken, M., Haas, H., Kreiter, S., Loquai, C., Reuter, K.C. *et al.* (2016). Systemic RNA delivery to dendritic cells exploits antiviral defense for cancer immunotherapy. *Nature* 534: 396–401.

Maeda, H., Wu, J., Sawa, T., Matsumura, Y. and Hori, K. (2000). Tumor vascular permeability and the EPR effect in macromolecular therapeutics: a review. *J. Control. Rel.* 65: 271–284.

Magini, D., Giovani, C., Mangiavacchi, S., Maccari, S., Cecchi, R., Ulmer, J.B. *et al.* (2016). Self-amplifying mRNA vaccines expressing multiple conserved influenza antigens confer protection against homologous and heterosubtypic viral challenge. *PloS One* 11: e0161193; doi: 10.1371/journal.pone.0161193.

Martin-Herranz, A., Ahmad, A., Evans, H.M., Ewert, K., Schulze, U. and Safinya, C.R. (2004). Surface functionalized cationic lipid–DNA complexes for gene delivery: PEGylated lamellar complexes exhibit distinct DNA–DNA interaction regimes. *Biophys. J.* 86: 1160–1168.

Miller, A.D. (1998). Cationic liposomes for gene therapy. *Angew. Chem. Int. Ed. Engl.* 37: 1768–1785.

Miller, A.D. (2003). The problem with cationic liposome/micelle-based non-viral vector systems for gene therapy. *Current Med. Chem.* 10: 1195–1211.

Miller, A.D. (2008). Towards safe nanoparticle technologies for nucleic acid therapeutics. *Tumori* 94: 234–245.

Miller, A.D. (2013). Delivery of RNAi therapeutics: work in progress. *Exp. Rev. Med. Dev.* 10: 781–811.

Miller, A.D. (2014). Delivering the promise of small ncRNA therapeutics. *Ther. Delivery* 5: 569–589.

Miller, A.D. (2016). Nanomedicine therapeutics and diagnostics are the goal. *Ther. Delivery* 7: 431–456.

Morrissey, D.V., Lockridge, J.A., Shaw, L., Blanchard, K., Jensen, K., Breen, W. *et al.* (2005). Potent and persistent *in vivo* anti-HBV activity of chemically modified siRNAs. *Nat. Biotechnol.* 23: 1002–1007.

Naito, Y., Yamada, T., Ui-Tei, K., Morishita, S. and Saigo, K. (2004). siDirect: highly effective, target-specific siRNA design software for mammalian RNA interference. *Nucl. Acids Res.* 32: W124–W129.

Papahadjopoulos, D., Allen, T.M., Gabizon, A., Mayhew, E., Matthay, K., Huang, S.K. *et al.* (1991). Sterically stabilized liposomes: improvements in pharmacokinetics and antitumor therapeutic efficacy. *Proc. Natl. Acad. Sci. USA* 88: 11460–11464.

Pardi, N., Hogan, M.J., Naradikian, M.S., Parkhouse, K., Cain, D.W., Jones, L. *et al.* (2018). Nucleoside-modified mRNA vaccines induce potent T follicular helper and germinal center B cell responses. *J. Exp. Med.* 215: 1571–1588; doi.org/10.1084/jem.20171450.

Pardi, N., Hogan, M.J., Pelc, R.S., Muramatsu, H., Andersen, H., DeMaso, C.R. *et al.* (2017). Zika virus protection by a single low-dose nucleoside-modified mRNA vaccination. *Nature* 543: 248–251.

Robert, N.J., Vogel, C.L., Henderson, I.C., Sparano, J.A., Moore, M.R., Silverman, P. *et al.* (2004). The role of the liposomal anthracyclines and other systemic therapies in the management of advanced breast cancer. *Semin. Oncol.* 31: 106–146.

Rosca, E.V., Wright, M., Gonitel, R., Gedroyc, W., Miller, A.D. and Thanou, M. (2015). Thermosensitive, near-infrared-labeled nanoparticles for topotecan delivery to tumors. *Mol. Pharmaceut.* 12: 1335–1346.

Semple, S.C., Akinc, A., Chen, J., Sandhu, A.P., Mui, B.L., Cho, C.K. *et al.* (2010). Rational design of cationic lipids for siRNA delivery. *Nat. Biotechnol.* 28: 172–176.

Song, S., Liu, D., Peng, J., Deng, H., Guo, Y., Xu, L.X. *et al.* (2009). Novel peptide ligand directs liposomes toward EGF-R high-expressing cancer cells *in vitro* and *in vivo*. *FASEB J.* 23: 1396–1404.

Spagnou, S., Miller, A.D. and Keller, M. (2004). Lipidic carriers of siRNA: differences in the formulation, cellular uptake and delivery with plasmid DNA. *Biochemistry* 43: 13348–13356.

Stadler, C.R., Bähr-Mahmud, H., Celik, L., Hebich B., Roth, A.S., Roth, R.P. *et al.* (2017). Elimination of large tumors in mice by mRNA-encoded bispecific antibodies. *Nat. Med.* 23: 815–817.

Stewart, L., Manvell, M., Hillery, E., Etheridge, C.J., Cooper, R.G., Stark, H. *et al.* (2001). Physico-chemical analysis of cationic liposome-DNA complexes (lipoplexes) with respect to *in vitro* and *in vivo* gene delivery efficiency. *J. Chem. Soc., Perkin Trans.* 2: 624–632.

Straubinger, R.M., Lopez, N.G., Debs, R.J., Hong, K. and Papahadjopoulos, D. (1988). Liposome-based therapy of human ovarian cancer: parameters determining potency of negatively charged and antibody-targeted liposomes. *Cancer Res.* 48: 5237–5245.

Tagawa, T., Manvell, M., Brown, N., Keller, M., Perouzel, E., Murray, K.D. *et al.* (2002). Characterisation of LMD virus-like nanoparticles self-assembled from cationic liposomes, adenovirus core peptide μ (mu) and plasmid DNA. *Gene Ther.* 9: 564–576.

Torchilin, V.P. (2005). Recent advances with liposomes as pharmaceutical carriers. *Nat. Rev. Drug Discovery* 4: 145–160.

Treat, J., Greenspan, A., Forst, D., Sanchez, J.A., Ferrans, V.J. Potkul, L.A. *et al.* (1990). Antitumor activity of liposome-encapsulated doxorubicin in advanced breast cancer: phase II study. *J. Natl. Cancer Inst.* 82: 1706–1710.

Ui-Tei, K., Naito, Y., Takahashi, F., Haraguchi, T., Ohki-Hamazaki, H., Juni, A. *et al.* (2004). Guidelines for the selection of highly effective siRNA sequences for mammalian and chick RNA interference. *Nucl. Acids Res.* 32: 936–948.

Valdés, J.J. and Miller, A.D. (2019). New opportunities for designing effective small interfering RNAs. *Sci. Rep.* 9: 16146; doi.org/10.1038/s41598-019-52303-5.

Verwey, E.J.W. and Overbeek, J.T.G. (1948). *Theory of the Stability of Lyophobic Colloids*. Amsterdam: Elsevier.

Wang, M., Löwik, D.W., Miller, A.D. and Thanou, M. (2009a). Targeting the urokinase plasminogen activator receptor with synthetic self-assembly nanoparticles. *Bioconj. Chem.* 20: 32–40.

Wang, M., Miller, A.D. and Thanou, M. (2008). Organising peptide ligands on the surface of gene vectors. In: *NSTI Nanotech 2008 Technical Proceedings*, 2, 394–397.

Wang, M., Takeshita, F., Ochiya, T., Miller, A.D. and Thanou, M. (2009b). Engineering and optimization of peptide-targeted nanoparticles for DNA and RNA delivery to cancer cells. In: *13th International Conference on Biomedical Engineering*, 1–3, 1503–1507.

Wang, M., Miller, A.D. and Thanou, M. (2013). Effect of surface charge and ligand organization on the specific cell-uptake of uPAR-targeted nanoparticles. *J. Drug Targeting* 21: 684–692.

Waterhouse, J.E., Harbottle, R.P., Keller, M., Kostarelos, K., Coutelle, C., Jorgensen, M.R. *et al.* (2005). Synthesis and application of integrin targeting lipopeptides in targeted gene delivery. *ChemBioChem* 6: 1212–1223.

Wheeler, J.J., Palmer, L., Ossanlou, M., MacLachlan, I., Graham, R.W., Zhang, Y.P. *et al.* (1999). Stabilized plasmid-lipid particles: construction and characterization. *Gene Ther.* 6: 271–281.

Yingyuad, P., Mevel, M., Prata, C., Furigati, S., Kontogiorgis, C., Thanou, M. *et al.*, (2013). Enzyme-triggered PEGylated pDNA-nanoparticles for controlled release of pDNA in tumours. *Bioconj. Chem.* 24: 343–362.

Yingyuad, P., Mevel, M., Prata, C., Kontogiorgis, C., Thanou, M. and Miller, A.D. (2014). Enzyme-triggered PEGylated siRNA-nanoparticles for controlled release of siRNA. *J. RNAi Gene Silencing* 10: 490–499.

Zhang, N.-N., Li, X.-F., Deng, Y.-Q., Zhao, H., Huang, Y.-J., Yang, G. *et al.* (2020). A thermostable mRNA vaccine against CO-VID-19. *Cell* 182: 1271–1283.

Zhang, Y.P., Sekirov, L., Saravolac, E.G., Wheeler, J.J., Tardi, P., Clow, K. *et al.* (1999). Stabilized plasmid-lipid particles for regional gene therapy: formulation and transfection properties. *Gene Ther.* 6: 1438–1447.

Zimmermann, T.S., Lee, A.C.H., Akinc, A., Bramlage, B., Bumcrot, D., Fedoruk, M.N. *et al.* (2006). RNAi-mediated gene silencing in non-human primates. *Nature* 441: 111–114.

Chapter 14

General Reading

Peer, D., Karp, J.M., Hong, S., Farokhzad, O.C., Margalit, R. and Langer, R. (2020). Nanocarriers as an emerging platform for cancer therapy. In: *Nano-enabled Medical Applications*, 1 (ed. L.P. Balogh), Chap. 2. New York, NY, USA: Jenny Stanford Publishing.

Thanou, M. and Duncan, R. (2003). Polymer-protein and polymer-drug conjugates in cancer therapy. *Curr. Opin. Investig. Drugs* 4: 701–709.

Thanou, M. and Miller, A.D. (2014). Nanomedicine in cancer diagnosis and therapy: converging medical technologies impacting healthcare. In: *Nanomedicine: Perspectives of Nanomedicine*, 2 (eds. Y. Ge, S. Li, S. Wang and R. Moore), 365–384. New York, NY, USA: Springer.

Extra Reading

Averitt, R.D., Westcott, S.L. and Halas, N.J. (1999). Linear optical properties of gold nanoshells. *J. Opt. Soc. Am. B* 16: 1824–1832.

Bai, J., Liu, Y., Sun, W., Chen, J., Miller, A.D. and Xu, Y. (2013). Down-regulated lysosomal processing improved PEGylated lipopolyplex-mediated gene transfection. *J. Gene Med.* 15: 182–192.

Bardhan, R., Chen, W., Bartels, M., Peres-Torres, C., Botero, M.F., McAninch, R.W. *et al.* (2010). Tracking of multimodal therapeutic nanocomplexes targeting breast cancer *in vivo*. *Nano Lett.* 10: 4920–4928; doi:10.1021/nl102889y.

Bardhan, R., Lal, S., Joshi, A. and Halas, N.J. (2011). Theranostic nanoshells: from probe design to imaging and treatment of cancer. *Acc. Chem. Res.* 44: 936–946.

Centelles, M.N., Wright, M., So, P.-W., Amrahli, M., Xu, X.-Y., Stebbing, J. *et al.* (2018). Image-guided thermosensitive liposomes for focused ultrasound drug delivery: using NIRF-labelled lipids and topotecan to visualize the effects of hyperthermia in tumours. *J. Control. Rel.* 280: 87–98.

de Smet, M., Langereis, M., van den Bosch, S. and Grull, H. (2010). Temperature-sensitive liposomes for doxorubicin delivery under MRI guidance. *J. Control. Rel.* 143: 120–127.

Drummond, D.C., Noble, C.O., Guo, Z., Keeling, H., Park, J.W. and Kirpotin, D.B. (2006). Development of a highly active nano-liposomal irinotecan using a novel intraliposomal stabilization strategy. *Cancer Res.* 66: 3271–3277.

Flexman, J.A., Yung, A., Yapp, D.T.T., Ng, S.S.W. and Kozlowski, P. (2008). Assessment of vessel size by MRI in an orthotopic model of human pancreatic cancer. *Conf. Proc. IEEE Eng. Med. Biol. Soc.* 851–854.

Harrington, K.J., Mohammadtaghi, S., Uster, P.S., Glass, D., Peters, A.M., Vile, R.G. *et al.* (2001). Effective targeting of solid tumors in patients with locally advanced cancers by radiolabeled PEGylated liposomes. *Clin. Cancer Res.* 7: 243–254.

Hirsch, L.R., Stafford, R.J., Bankson, J.A, Sershen, S.R., Rivera, B., Price, R.E. *et al.* (2003). Nanoshell-mediated near-infrared thermal therapy of tumors under magnetic resonance guidance. *Proc. Natl. Acad. Sci. USA* 100: 13549–13554.

Kamaly, N., Kalber, T., Ahmad, A., Oliver, M.H., So, P.-W., Herlihy, A.H. *et al.* (2008). Bimodal paramagnetic and fluorescent liposomes for cellular and tumor magnetic resonance imaging. *Bioconjugate Chem* 19: 118–129.

Kamaly, N., Kalber, T.L., Kenny, G.D., Bell, J.D., Jorgensen, M.R. and Miller, A.D. (2010a). A novel bimodal lipidic contrast agent for cellular labelling and tumour MRI. *Org. Biomol. Chem.* 8: 201–211.

Kamaly, N., Kalber, T., Thanou, M., Bell, J.D. and Miller, A.D. (2009). Folate receptor targeted bimodal liposomes for tumor magnetic resonance imaging. *Bioconj. Chem.* 20: 648–655.

Kamaly, N. and Miller, A.D. (2010). Paramagnetic liposome nanoparticles for cellular and tumour imaging. *Int. J. Mol. Sci.* 11: 1759–1776.

Kamaly, N., Miller, A.D. and Bell, J.D. (2010b). Chemistry of tumour targeted T_1 based MRI contrast agents. *Curr Topics Med. Chem.* 10: 1158–1183.

Kenny, G.D., Kamaly, N., Kalber, T.L., Brody, L.P., Sahuri, M., Shamsaei, E. *et al.* (2011). Novel multifunctional nanoparticle mediates siRNA tumour delivery, visualisation and therapeutic tumour reduction *in vivo*. *J. Control. Rel.* 149: 111–116.

Kong, G., Anyarambhatla, G., Petros, W.P., Braun, R.D., Colvin, O.M., Needham, D. *et al.* (2000). Efficacy of liposomes and hyperthermia in a human tumor xenograft model: importance of triggered drug release. *Cancer Res.* 60: 6950–6957.

Matsumura, Y. and Maeda, H. (1986). A new concept for macromolecular therapeutics in cancer chemotherapy: mechanism of tumoritropic accumulation of proteins and the antitumor agent smancs. *Cancer Res.* 46: 6387–6392.

Medarova, Z., Pham, W., Farrar, C., Petkova, V. and Moore, A. (2007). *In vivo* imaging of siRNA delivery and silencing in tumors. *Nat. Med.* 13: 372–377.

Miller, A.D. (2013). Delivery of RNAi therapeutics: work in progress. *Exp. Rev. Med. Dev.* 10: 781–811.

Miller, A.D. (2014). Delivering the promise of small ncRNA therapeutics. *Ther. Delivery* 5: 569–589.

Miller, A.D. (2016). Nanomedicine therapeutics and diagnostics are the goal. *Ther. Delivery* 7: 431–456.

Moore, A. and Medarova, Z. (2009). Imaging of siRNA delivery and silencing. *Methods Mol. Biol.* 487: 93–110.

Needham, D., Anyarambhatla, G., Kong, G. and Dewhirst, M.W. (2000). A new temperature-sensitive liposome for use with mild hyperthermia: characterization and testing in a human tumor xenograft model. *Cancer Res.* 60: 1197–1201.

Needham, D. and Dewhirst, M.W. (2001). The development and testing of a new temperature-sensitive drug delivery system for the treatment of solid tumors. *Adv. Drug Deliv. Rev.* 53: 285–305.

Negussie, A.H., Yarmolenko, P.S., Partanen, A., Ranjan, A., Jacobs, G., Woods, D. *et al.* (2011). Formulation and characterisation of magnetic resonance imageable thermally sensitive liposomes for use with magnetic resonance-guided high intensity focused ultrasound. *Int. J. Hyperthermia* 27: 140–155.

Oliver, M., Ahmad, A., Kamaly, N., Perouzel, E., Caussin, A., Keller, M. *et al.* (2006). MAGfect: a novel liposome formulation for MRI labelling and visualization of cells. *Org. Biomol. Chem.* 4: 3489–3497.

Partanen, A., Yarmolenko, P.S., Viitala, A., Appanaboyina, S., Haemmerich, D., Ranjan, A. *et al.* (2012). Mild hyperthermia with magnetic resonance-guided high-intensity focused ultrasound for applications in drug delivery. *Int. J. Hyperthermia* 28: 320–336.

Peer, D., Karp, J.M., Hong, S., Farokhzad, O.C., Margalit, R. and Langer, R. (2007). Nanocarriers as an emerging platform for cancer therapy. *Nat. Nanotechnol.* 2: 751–760.

Poon, R.T. and Borys, N. (2009). Lyso-thermosensitive liposomal doxorubicin: a novel approach to enhance efficacy of thermal ablation of liver cancer. *Exp. Opin. Pharmacother.* 10: 333–343.

Ranjan, A., Jacobs, G.C., Woods, D.L., Negussie, A.H., Partanen, A., Yarmolenko, P.S. *et al.* (2012). Image-guided drug delivery with magnetic resonance guided high intensity focused ultrasound and temperature sensitive liposomes in a rabbit Vx2 tumor model. *J. Control. Rel.* 158: 487–494.

Rosca, E.V., Wright, M., Gonitel, R., Gedroyc, W., Miller, A.D. and Thanou, M. (2015). Thermosensitive, near-infrared-labeled nanoparticles for topotecan delivery to tumors. *Mol. Pharmaceut.* 12: 1335–1346.

Ting, G., Chang, C.H. and Wang, H.E. (2009). Cancer nanotargeted radiopharmaceuticals for tumor imaging and therapy. *Anticancer Res.* 29: 4107–4118.

van Tilborg, G.A.F., Strijkers, G.J., Pouget, E.M., Reutelingsperger, C.P.M., Sommerdijk, N.A.J.M., Nicolay, K. *et al.* (2008). Kinetics of avidin-induced clearance of biotinylated bimodal liposomes for improved MR molecular imaging. *Magn. Reson. Med.* 60: 1444–1456.

Wood, B.J., Poon, R.T., Locklin, J.K., Dreher, M.R., Ng, K.K., Eugeni, M. *et al.* (2012). Phase I study of heat-deployed liposomal doxorubicin during radiofrequency ablation for hepatic malignancies. *J. Vasc. Interv. Radiol.* 23: 248–255..e7.

Yu, M.K., Jeong, Y.Y., Park, J., Park, S., Kim, J.W., Min, J.J. *et al.* (2008). Drug-loaded superparamagnetic iron oxide nanoparticles for combined cancer imaging and therapy *in vivo*. *Angew. Chem. Int. Ed. Engl.* 47: 5362–5365.

Zhou, Y., Chakraborty, S. and Liu, S. (2011). Radiolabeled cyclic RGD peptides as radiotracers for imaging tumors and thrombosis by SPECT. *Theranostics* 1: 58–82.

Chapter 15

General Reading

Chen, Y.J., Groves, B., Muscat, R.A. and Seelig, G. (2015). DNA nanotechnology from the test tube to the cell. *Nat. Nanotech.* 10: 748–760.

Hu, Q., Li, H., Wang, L., Gu, H. and Fan, C. (2019). DNA nanotechnology-enabled drug delivery systems. *Chem. Rev.* 119: 6459–6506.

Rothemund, P.W.K. (2006). Folding DNA to create nanoscale shapes and patterns. *Nature* 440: 297–302.

Seeman, N.C. and Sleiman, H.F. (2017). DNA nanotechnology. *Nat. Rev. Mater.* 3: 17068.

Extra Reading

Benson, E., Mohammed, A., Gardell, J., Masich, S., Czeizler, E., Orponen, P. *et al.* (2015). DNA rendering of polyhedral meshes at the nanoscale. *Nature* 523: 441–444.

Chen, J. and Seeman, N.C. (1991). Synthesis from DNA of a molecule with the connectivity of a cube. *Nature* 350: 631–633.

Douglas, S.M., Bachelet, I. and Church, G.M. (2012). A logic-gated nanorobot for targeted transport of molecular payloads. *Science* 335: 831–834.

Douglas, S.M., Dietz, H., Liedl, T., Hogberg, B., Graf, F. and Shih, W.M. (2009). Self-assembly of DNA into nanoscale three-dimensional shapes. *Nature* 459: 414–418.

Han, D., Pal, S., Nangreave, J., Deng, Z., Liu, Y. and Yan, H. (2011). DNA origami with complex curvatures in three-dimensional space. *Science* 332: 342–346.

He, Y. and Liu, D.R. (2010). Autonomous multistep organic synthesis in a single isothermal solution mediated by a DNA walker. *Nat. Nanotech.* 5: 778–782.

He, Y., Ye, T., Su, M., Zhang, C., Ribbe, A.E., Jiang, W. *et al.* (2008). Hierarchical self-assembly of DNA into symmetric supramolecular polyhedral. *Nature* 452: 198–202.

Jones, H.R., Seeman, N.C. and Mirkin, C.A. (2015). Programmable materials and the nature of the DNA bond. *Science* 347: 1260901.

Ke, Y., Ong, L.L., Shih, W.M. and Yin, P. (2012). Three-dimensional structures self-assembled from DNA bricks. *Science* 338: 1177–1183.

Seeman, N.C. (2003). DNA in a material world. *Nature* 421: 427–431.

Shin, J.S. and Pierce, N.A. (2004). A synthetic DNA walker for molecular transport. *J. Am. Chem. Soc.* 126: 10834–10835.

Sun, W., Ji, W., Hall, J.M., Hu, Q., Wang, C., Beisel, C.L. *et al.* (2015). Efficient delivery of CRISPR-Cas9 for genome editing via self-assembled DNA nanoclews. *Angew. Chem. Intl. Ed.* 54: 12029–12033.

Wang, X., Chandrasekaran, A.R., Shen, Z., Ohayon, Y.P., Wang, T., Kizer, M.E. *et al.* (2019). Paranemic crossover DNA: There and back again. *Chem. Rev* 119: 6273–6289.

Wei, B., Dai, M. and Yin, P. (2012). Complex shapes self-assembled from single-stranded DNA tiles. *Nature* 485: 623–627.

Yurke, B., Turberfield, A.J., Mills, A.P., Simmel, F.C. and Neumann, J.F. (2000). A DNA-fuelled molecular machine made of DNA. *Nature* 406: 605–608.

Zhang, F., Jiang, S., Wu, S., Li, Y., Mao, C., Liu, Y. *et al.* (2015). Complex wireframe. DNA origami nanostructures with multi-arm junction vertices. *Nat. Nanotech.* 10: 779–784.

Index

Note: Page numbers for figures are in *italics*, tables are in **bold.**

Essentials of Chemical Biology: Structures and Dynamics of Biological Macromolecules In Vitro *and* In Vivo, Second Edition. Andrew D. Miller and Julian A. Tanner.
© 2024 John Wiley & Sons, Inc. Published 2024 by John Wiley & Sons, Inc.
Companion Website: www.wiley.com/go/miller/essentialschembiol2